5

AF698576

Developmental Biology Protocols

METHODS IN MOLECULAR BIOLOGY™

John M. Walker, SERIES EDITOR

METHODS IN MOLECULAR BIOLOGY™

Developmental Biology Protocols

Volume II

Edited by

Rocky S. Tuan

Thomas Jefferson University, Philadelphia, PA

and

Cecilia W. Lo

University of Pennsylvania, Philadelphia, PA

Humana Press **Totowa, New Jersey**

© 2000 Humana Press Inc.
999 Riverview Drive, Suite 208
Totowa, New Jersey 07512

All rights reserved. No part of this book may be reproduced, stored in a retrieval system, or transmitted in any form or by any means, electronic, mechanical, photocopying, microfilming, recording, or otherwise without written permission from the Publisher. Methods in Molecular Biology™ is a trademark of The Humana Press Inc.

All authored papers, comments, opinions, conclusions, or recommendations are those of the author(s), and do not necessarily reflect the views of the publisher.

This publication is printed on acid-free paper. ∞
ANSI Z39.48-1984 (American Standards Institute)
Permanence of Paper for Printed Library Materials.

Cover design by Patricia F. Cleary

Cover illustration taken from Fig. 1 in Chapter 20 of Vol. I in this series. It is a composite three-dimensional view of epifluorescence micrographs of representative stages of the developing rat embryo coated with acridine orange and imaged in toto by confocal laser-scanning microscopy.

For additional copies, pricing for bulk purchases, and/or information about other Humana titles, contact Humana at the above address or at any of the following numbers: Tel.: 973-256-1699; Fax: 973-256-8341; E-mail: humana@humanapr.com; or visit our Website: http://humanapress.com

Photocopy Authorization Policy:
Authorization to photocopy items for internal or personal use, or the internal or personal use of specific clients, is granted by Humana Press Inc., provided that the base fee of US $10.00 per copy, plus US $00.25 per page, is paid directly to the Copyright Clearance Center at 222 Rosewood Drive, Danvers, MA 01923. For those organizations that have been granted a photocopy license from the CCC, a separate system of payment has been arranged and is acceptable to Humana Press Inc. The fee code for users of the Transactional Reporting Service is: [0-89603-853-X/00 (hardcover) or 0-89603-575-1/00 (paper) $10.00 + $00.25].

Printed in the United States of America. 10 9 8 7 6 5 4 3 2 1

Library of Congress Cataloging in Publication Data

Main entry under title:

Methods in molecular biology™.

Developmental biology protocols/edited by Rocky S. Tuan and Cecilia W. Lo.
p. cm.—(Methods in molecular biology; v. 135, 136, 137).
Includes bibliographical references and index.
ISBN 0-89603-852-1 (v. 1: alk. paper) (hardcover), ISBN 0-89603-574-3 (paper); ISBN 0-89603-853-X (v. 2: alk. paper) (hardcover), ISBN 0-89603-575-1 (paper); ISBN 0-89603-854-8 (v. 3: alk. paper) (hardcover), ISBN 0-89603-576-X (paper); ISBN 0-89603-855-6 (hc: alk paper); ISBN 0-89603-578-6 (comb: alk paper)
1. Developmental biology—Laboratory manuals. 2. Molecular biology—Laboratory manuals. I. Tuan, Rocky S. II. Lo, Cecilia W. III. Series.
QH491.D4625 2000
571.8—dc21

99-23532
CIP

Preface

Developmental biology is one of the most exciting and fast-growing fields today. In part, this is so because the subject matter deals with the innately fascinating biological events—changes in form, structure, and function of the organism. The other reason for much of the excitement in developmental biology is that the field has truly become the unifying melting pot of biology, and provides a framework that integrates anatomy, physiology, genetics, biochemistry, and cellular and molecular biology, as well as evolutionary biology. No longer is the study of embryonic development merely "embryology." In fact, development biology has produced important paradigms for both basic and clinical biomedical sciences alike.

Although modern developmental biology has its roots in "experimental embryology" and the even more classical "chemical embryology," the recent explosive and remarkable advances in developmental biology are critically linked to the advent of the "cellular and molecular biology revolution." The impressive arsenal of experimental and analytical tools derived from cell and molecular biology, which promise to continue to expand, together with the exponentially developing sophistication in functional imaging and information technologies, guarantee that the study of the developing embryo will contribute one of the most captivating areas of biological research in the next millennium.

There is a demonstrated need for students of developmental biology to be knowledgeable of the breadth and depth of the available experimental methodologies, by necessity derived from multiple disciplines, which are applicable to the study of the developing embryo. In particular, because developmental biology deals with multiple model systems, from organismal to tissue and cell levels, as well as a wide range of "change"-related biological activities, the investigator is often frustrated as to how his/her findings relate to those obtained in another model system and/or by using different reagents or functional markers. Compared to other more strictly defined fields of biological research, the number of "reference" publications that deal specifically with the practical aspects of experimental developmental biology are, however, relatively scarce.

Developmental Biology Protocols grows out of the need for a comprehensive laboratory manual that provides the readers the principles, background, rationale, as well as the practical protocols, for studying and analyzing the events of embryonic development. This three-volume set, consisting of 142 chapters, is intentionally broadin scope, because of the nature of modern developmental biology. Information is grouped into the following topics: (1) systems—production, culture, and storage; (2) developmental pattern and morphogenesis; (3) embryo structure and function; (4) cell lineage analysis; (5) chimeras; (6) experimental manipulation of embryos; (7) application of viral vectors; (8) organogenesis; (9) abnormal development and teratology; (10) screening and mapping of novel genes and mutations; (11) transgenesis production and gene knockout; (12) manipulation of developmental gene expression and function; (13)

analysis of gene expression; (14) models of morphogenesis and development; and (15) in vitro models and analysis of differentiation and development.

Throughout *Developmental Biology Protocols*, the authors have consistently striven for a balanced presentation of both background information and actual laboratory details. It is believed that this highly practical format will permit readers to bring the concepts and principles we present into their personal research practices in a most efficient manner. Specifically, the wide range of model systems and multidisciplinary experimental techniques presented here should lower the "activation energy" for the student of developmental biology to become a contributing member of this exciting scientific discipline. In addition, teachers of developmental biology at all levels should also readily find relevant and useful information to enrich the experience of their students.

The practice of developmental biology is currently in a state of constant change, reflecting the close relationship of the field to other rapidly developing fields of biological research, particularly cell and molecular biology, and imaging and information technology. The materials presented in this three-volume set are therefore the beginning of a project that will involve continuous update and upgrade to reach and enhance the scientific endeavors of developmental biologists at large.

The production of *Developmental Biology Protocols* would not have been possible without the outstanding work of the contributing authors who share here with the readers the hands-on wisdom they have earned in the laboratory. We are grateful for their intellectual contributions as well as their remarkable tolerance to our constant reminders. Tom Lanigan and his staff at Humana Press worked diligently on the project to ensure a final product of the highest quality. Chuck, our young son, persevered throughout the gestation period of the project, and constantly demonstrated to us the meaning of "developmental biology."

Our final, heartfelt thanks go to Lynn Stierle, who expertly and singlehandedly maintained the massive organization of the manuscripts and the correspondence (snail-mail and e-mail), as well as the sanity of the editors! Michelle Levinski also provided valuable assistance in proofreading the final production.

Finally, we hope that these volumes will find their place on the laboratory shelves, with their pages well soiled and their contents tried and tested, and prove their utility as an everyday resource for the students of developmental biology, the most exciting discipline of biology for many decades to come!

Rocky S. Tuan, PhD
Cecilia W. Lo, PhD

List of Color Plates

Color plates 1–11 appear as an insert following p. 262.

Contents

TABLES OF CONTENTS FROM VOLUMES I AND III

VOLUME I

Part V Cell Lineage Analysis

Part VI Chimeras

VOLUME III

Contributors

BARBARA D. ABBOTT • *Reproductive Toxicology Division, National Health Effects and Environmental Research Laboratory, US Environmental Protection Agency, Research Triangle Park, NC*

TAKASHI ARIIZUMI • *CREST, Japan Science and Technology Corp., Tokyo, Japan*

MAKOTO ASASHIMA • *Department of Life Science (Biology), Graduate School of Arts and Sciences, The University of Tokyo, and CREST, Japan Science and Technology Corp., Tokyo, Japan*

RICHARD R. BEHRINGER • *Department of Molecular Genetics, The University of Texas M. D. Anderson Cancer Center, Houston, TX*

AMY BEJSOVEC • *Department of Biochemistry, Northwestern University, Evanston, IL*

SHEILA M. BELL • *Department of Pediatrics, Children's Hospital Research Foundation, Cincinnati, OH*

MICHAEL R. BLACKBURN • *Department of Biochemistry and Molecular Biology, University of Texas Health Science Center–Houston Medical School, Houston, TX*

NANCY M. BONINI • *Department of Biology, University of Pennsylvania, Philadelphia, PA*

JACQUELINE A. BONSELAAR • *Department of Animal and Poultry Science, University of Guelph, Guelph, Ontario, Canada*

SIEW-SIM CHEAH • *Department of Molecular Genetics, The University of Texas M. D. Anderson Cancer Center, Houston, TX*

ALVIN J. CHIN • *Department of Pathology, University of Southern California School of Medicine, Los Angeles, CA*

CHENG-MING CHUONG • *Department of Pathology, University of Southern California School of Medicine, Los Angeles, CA*

MARY ELLEN CLARK • *Department of Animal and Poultry Science, University of Guelph, Guelph, Ontario, Canada*

PATRICIA COGRAM • *Neural Development Unit, Institute of Child Health, University College London, London, UK*

ANDREW COPP • *Neural Development Unit, Institute of Child Health, University College London, London, UK*

JOSHUA DANIELS • *University of Wisconsin School of Veterinary Medicine, Madison, WI*

ELENA DE LA CASA • *The Fels Institute for Cancer Research and Molecular Biology Temple University School of Medicine, Philadelphia, PA*

MARY E. DESMOND • *Department of Biology, Villanova University, Villanova, PA*

ROBERT J. ETCHES • *Origin Therapeutics, Burlingame, CA*

MARCIA C. FIELD • *Department of Biology, Villanova University, Villanova, PA*

ANGELEEN FLEMING • *Neural Development Unit, Institute of Child Health, University College London, London, UK*

MASAHIKO FUJINAGA • *Department of Anesthesia, Stanford University School of Medicine, Palo Alto, CA*

MIKI FUJIOKA • *Kimmel Cancer Institute, Thomas Jefferson University, Philadelphia, PA*

DIANNE GERRELLI • *Neural Development Unit, Institute of Child Health, University College London, London, UK Environmental Protection Agency, Research*

PATRICIA A. (SHARPLES) HAAS • *Department of Biology, Villanova University, Villanova, PA*

MELISSA A. HAENDEL • *Neuroscience Training Program, University of Wisconsin Medical School, Madison, WI*

THEODORUS B. M. HAKVOORT • *Department of Anatomy and Embryology, University of Amsterdam, Amsterdam, The Netherlands*

DANIEL L. HARTL • *Department of Organismic and Evolutionary Biology, Harvard University, Cambridge, MA*

GARY A. HELD • *Reproductive Toxicology Division, National Health Effects and Environmental Research Laboratory, US Environmental Protection Agency, Research Triangle Park, NC*

DEBORAH HENDERSON • *Neural Development Unit, Institute of Child Health, University College London, London, UK*

ANDREW HYNES • *Neural Development Unit, Institute of Child Health, University College London, London, UK*

TAKAYUKI INAGAKI • *Department of Neurosurgery, Kansai Medical University, Osaka, Japan*

AYA JAKOBOVITS • *Abgenix Inc., Fremont, CA*

JAMES B. JAYNES • *Kimmel Cancer Institute, Thomas Jefferson University, Philadelphia, PA*

JASPAL KHILLAN • *Mouse Genetics & Genome Manipulation Laboratory, The Wistar Institute, Philadelphia, PA*

MARY LOU KING • *Department of Cell Biology and Anatomy, University of Miami School of Medicine, Miami, FL*

THOMAS B. KNUDSEN • *Department of Pathology, Anatomy and Cell Biology, Jefferson Medical College, Philadelphia, PA*

MARIA KOLATSI-JOANNOU • *Neural Development Unit, Institute of Child Health, University College London, London, UK*

BRUCE T. LAMB • *Department of Genetics, School of Medicine, Case Western Reserve University, Cleveland, OH*

WOUTER H. LAMERS • *Department of Anatomy and Embryology, University of Amsterdam, Amsterdam, The Netherlands*

KEITH E. LATHAM • *The Fels Institute for Cancer Research and Molecular Biology and Department of Biochemistry, Temple University School of Medicine, Philadelphia, PA*

CHRISTOPHER LAU • *Reproductive Toxicology Division and Research and Administrative Support Division, NHEERL, US Environmental Protection Agency, Research Triangle Park, NC*

LISA L. LAU • *The Wistar Institute, Philadelphia, PA*

YUNZHENG LE • *Developmental Biology Program, Oklahoma Medical Research Foundation, Oklahoma City, OK*

YEFU LI • *Department of Growth and Development, Harvard School of Dental Medicine, Bostln, MA.*

LIN LIN • *Department of Biology, Queens College, Flushing, NY*

SHUO LIN • *Institute of Molecular Medicine and Genetics, Medical College of Georgia, Augusta, GA*

CECILIA W. LO • *Department of Biology, University of Pennsylvania, Philadelphia, PA*

GARY E. LYONS • *Department of Anatomy, University of Wisconsin Medical School, Madison, WI*

GEORGE M. MALACINSKI • *Department of Biology, Indiana University, Bloomington, IN*

ROGER MARKWALD • *Department of Cell Biology and Anatomy, Medical University of South Carolina, Charleston, SC*

MANSOOR MOHAMMED • *Department of Animal and Poultry Science, University of Guelph, Guelph, Ontario, Canada*

STEFAN MUNDLOS • *Children's Clinic, Mainz, Germany*

JENNIFER MURDOCH • *Neural Development Unit, Institute of Child Health, University College London, London, UK*

GLEN NUCKOLLS • *Craniofacial Development Section, National Institute of Arthritis, Musculoskeletal and Skin Diseases, National Institutes of Health, Bethesda, MD*

ROY C. OGLE • *Department of Neurosurgery, University of Virginia School of Medicine, Charlottesville, VA*

KENNETH R. PETERSON • *Department of Biochemistry and Molecular Biology, University of Kansas Medical Center, School of Medicine, Kansas City, KS*

JONATHAN P. RAST • *Division of Biology, California Institute of Technology, Pasadena, CA*

JEAN RICHA • *Department of Genetics, University of Pennsylvania, Philadelphia, PA*

CALVIN D. ROSKELLEY • *Department of Anatomy, University of British Columbia, Vancouver, BC, Canada*

BRIAN SAUER • *Developmental Biology Program, Oklahoma Medical Research Foundation, Oklahoma City, OK*

SCOTT SCHAU • *Department of Animal and Poultry Science, University of Guelph, Guelph, Ontario, Canada*

GARY C. SCHOENWOLF • *Department of Neurobiology and Anatomy, University of Utah School of Medicine, Salt Lake City, UT*

Claire M. Schreiner • *Department of Pediatrics, Children's Hospital Medical Center, Cincinnati, OH*

Richard M. Schultz • *Department of Biology, University of Pennsylvania, Philadelphia, PA*

William J. Scott • *Department of Pediatrics, Children's Hospital Research Foundation, Cincinnati, OH*

R. Woodrow Setzer • *Research and Administrative Support Division, National Health and Environmental Effects Research Laboratory, US Environmental Protection Agency, Research Triangle Park, NC*

Akihiko Shimono • *Department of Molecular Genetics, The University of Texas M. D. Anderson Cancer Center, Houston, TX*

Lillian Shum • *Craniofacial Development Section, National Institute of Arthritis, Musculoskeletal and Skin Diseases, National Institutes of Health, Bethesda, MD*

Mark L. Siegal • *Department of Biological Sciences, Stanford University, Palo Alto, CA*

Harold Slavkin • *Craniofacial Development Section, National Institute of Arthritis, Musculoskeletal and Skin Diseases and National Institute of Dental Research, National Institutes of Health, Bethesda, MD*

Bradley R. Smith • *Department of Radiology, Duke University Medical Center, Durham, NC*

Jodi L. Smith • *Department of Neurosurgery, University of Utah Medical Center, Salt Lake City, UT*

Aruna M. Somasiri • *Department of Anatomy, University of British Columbia, Vancouver, BC, Canada*

Lisa M. Spain • *The Wistar Institute, Philadelphia, PA*

Johannes A. A. Spijkers • *Department of Anatomy and Embryology, University of Amsterdam, Amsterdam, The Netherlands*

David L. Stocum • *Departament of Biology, Indiana University–Purdue University School of Science, Indianapolis, IN*

Bradley J. Swanson • *Program in Cellular and Molecular Biology, University of Wisconsin Medical School, Madison, WI*

Yousuke Takahama • *Department of Immunology and PRESTO Research Project, Institute of Basic Medical Sciences, University of Tsukuba, Tsukuba, Japan*

Shuji Takahashi • *CREST, Japan Science and Technology Corp., Tokyo, Japan*

Rocky S. Tuan • *Department of Orthopaedic Surgery, Thomas Jefferson University, Philadelphia, PA*

Marion L. Walker • *Department of Neurosurgery, University of Utah Medical Center, Salt Lake City, UT*

Qingxia Wei • *Department of Animal and Poultry Science, University of Guelph, Guelph, Ontario, Canada*

Michael P. Weir • *Department of Biology, Hall-Atwater Laboratories, Wesleyan University, Middletown, CT*

ANDY WESSELS • *Department of Cell Biology and Anatomy, Medical University of South Carolina, Charleston, SC*

MARK E. WESTHUSIN • *Department of Veterinary Physiology and Pharmacology, Texas A & M University, College Station, TX*

KRISTIN L. WOODS • *Department of Animal and Poultry Science, University of Guelph, Guelph, Ontario, Canada*

COLLEEN WU • *Department of Anatomy, University of British Columbia, Vancouver, BC, Canada*

LIN XU • *Department of Drug Development, Harvard School of Dental Medicine, Cambridge, MA.*

PATRICIA YBOT-GONZALEZ • *Neural Development Unit, Institute of Child Health, University College London, London, UK*

H. JOSEPH YOST • *Department of Oncological Sciences, Huntsman Cancer Institute, University of Utah, Salt Lake City, UT*

LAURA D. ZAJCHOWSKI • *Department of Animal and Poultry Science, University of Guelph, Guelph, Ontario, Canada*

ZAHRA F. ZAKERI • *Department of Biology, Queens College, Flushing, NY*

JIAN ZHANG • *Department of Cell Biology and Anatomy, University of Miami School of Medicine, Miami, FL*

I

INTRODUCTION

1

Developmental Biology Protocols

Overview II

Rocky S. Tuan and Cecilia W. Lo

1. Introduction

The discipline of developmental biology covers scientific investigations aimed at deciphering the underlying mechanisms responsible for diversity and order within tissues and organs. It is the goal of Developmental Biology Protocols to present to the readers a set of contemporary, practical experimental tools dealing with a wide-ranging spectrum of topics in developmental biology research. This second volume of the three-volume set begins by presenting the tissue and organ models currently being studied, and the characteristics of abnormal development. The volume concludes with detailed description of the technologies used to identify developmentally important genes and the methods for transgenesis, including gene knockout.

2. Organogenesis

Studies on *Drosophila* (Chapter 2) and *Xenopus* (Chapter 3) are first presented to illustrate the power of these systems for deciphering the experimental principles of developmental biology. Practical details on a number of organ/tissue systems in vertebrates are presented, including mammary gland (Chapter 4), heart (Chapter 5), skeleton (craniofacial: Chapters 6 and 7; axial and appendicular, Chapter 8), limb (Chapter 9), thymus (Chapter 10), liver (Chapter 11), and skin (Chapter 12). These systems exemplify the diversity of developmental mechanisms involved in organogenesis, as well as the experimental techniques applicable to their analysis. In addition, because programmed cell death has emerged as a common mechanistic step in many aspects of morphogenesis, two chapters are devoted to the methods of the analysis of apoptosis (Chapters 13 and 14).

3. Abnormal Development and Teratology

Birth defects are the leading cause of infant mortality and are responsible for substantial morbidity and disability. In the United States alone, over 120,000 babies are born each year with a structural birth defect or malformation. At present, the causes remain unknown for most of these cases. One of the long-term goals of the science of developmental biology is, indeed, the discovery of the mechanisms responsible for

From: *Methods in Molecular Biology, Vol. 136: Developmental Biology Protocols, Vol. II*
Edited by: R. S. Tuan and C. W. Lo © Humana Press Inc., Totowa, NJ

abnormal development. This section contains chapters representing a cross-section of many such examples and the experimental approaches currently used in analyzing the underlying mechanisms. Experimental studies of neurulation and neural tube defects are first described (Chapters 16–19). Abnormalities in placentation, which is critical for a range of physiological interactions required for fetal growth and development, are often associated with early embryonic mortality as well as serious pregnancy disorders such as pre-eclampsia; methods for examining placentation is presented in Chapter 20. Craniofacial malformations, such as cleft palate, are among the most frequent birth defects in live-born human infants, and two chapters (Chapters 21 and 22) are devoted to the analysis of palatal dysmorphogenesis. The method of interspecies tissue grafting, also detailed in Volume I of this series, is described here in the context of understanding the basis of developmental limb anomalies (Chapter 23). Techniques used to study cardiac morphogenesis and dysmorphogenesis, particularly related to laterality defects, are detailed in Chapters 24–26. Finally, in assessing the developmental toxicity of potentially harmful substances, it is crucial to utilize valid dose-response models; Chapter 27 summarizes the biologically based risk assessment models for developmental toxicity.

4. Screening and Mapping of Novel Genes and Mutations

The utility of applying molecular biology techniques to the study of development is well documented in this Part. Specifically, developmental biologists have successfully adopted differential gene screening and cloning approaches to relate specific gene expression events to spatiotemporally defined stages of development and embryogenesis, such as cellular commitment, differentiation, and morphogenesis. Such correlations then provide a rational basis for further analysis of the putative functions of specific genes and their products. Examples of differential screening of developmental gene expression are presented in Chapters 31–33. Gene cloning may be carried out by positional cloning (Chapter 28) and gene trapping in embryonic stem cells (Chapter 29). In most cases, cloning is accomplished using polymerase chain reaction (PCR)-based methods, as described in Chapter 30.

5. Transgenesis: Production and Gene Knockout

The development of techniques that permit the introduction of gene sequences into the zygote or embryonic stem cells to produce transgenic animals ranks as one of the seminal achievements of the molecular biology revolution. The chapters in this Part provide a panoramic portfolio of the many diverse and powerful approaches that have been developed to produce transgenic animals and the manipulations that permit targeted gene knockouts. Protocols used in producing transgenic *Drosophila* (Chapter 34), sea urchin (Chapter 35), and zebrafish (Chapter 36) are first described. Recent technical advances in the production of avian transgenics are also summarized (Chapters 37–39). Methods for the production of mammalian, particularly mouse, transgenic animals are described in Chapters 40–42, the last dealing specifically with the use of yeast artificial chromosomes (YACs). A novel method for high-efficiency formation of chimeric animals with germline gene transmission is presented in Chapter 44. Finally, exciting recent developments in gene-targeting strategies (Chapter 43), and the application of the *Cre/LoxP* site-specific recombination for conditional gene knockout (Chapter 45) and transgene coplacement for *Drosophila* (Chapter 46) are also presented.

II

ORGANOGENESIS

2

Drosophila as a Genetic Tool to Define Vertebrate Pathway Players

Nancy M. Bonini

I. Introduction

In many instances, the strength of *Drosophila melanogaster* genetics can be used to enhance our understanding of complex vertebrate signaling systems. The general success of this approach is underscored by the large number of vertebrate signaling components whose very names derive in part from the names of *Drosophila* mutants. Examples include the vertebrate pathway components *Sonic Hedgehog*, *Son of Sevenless*, *Lunatic Fringe*, *Notch*, the SMAD family of transforming growth factor-β (TGF-β) signaling, and many others. Given the powerful genetics of *Drosophila melanogaster* (*see* **ref.** ***1***), it can be of interest to test functional equivalence of vertebrate homologs with fly genetic pathway components, or to re-create in *Drosophila* transgenic models for vertebrate or human gene function. If such complementation can be established, then the strength of *Drosophila* genetics can be brought to bear on defining additional components of the particular pathway of interest; for example, through enhancer and suppressor screens. Subsequently, one can then clone such modifier genes from *Drosophila*, as a springboard from which to identify their vertebrate counterparts. To establish a genetic model for a vertebrate gene function in *Drosophila*, there are a number of considerations with respect to expressing foreign genes in the fly, establishing whether and how the foreign proteins function, and using the transgenic lines in genetic screens.

Examples of functional complementation in flies with vertebrate genes include the ability of domains of human bone morphogenetic protein to substitute in the related fly protein Dpp ***(2)***, effects of vertebrate *fringe* homologs to establish boundaries like the fly gene ***(3)***, functional complementation of *orthodenticle* homeobox gene homologs ***(4–6)***, and functional complementation of mammalian counterparts of eye determination genes *eyeless* and *eyes absent* ***(7,8)***. In addition, dominant effects can be generated in flies with vertebrate genes, such as phenocopying fly homeotic mutants with the appropriate vertebrate *Hox* homologs ***(9,10)***, and generating genetically tractable human disease models by expressing mutant human disease proteins in flies ***(11)***.

From: *Methods in Molecular Biology, Vol. 136: Developmental Biology Protocols, Vol. II*
Edited by: R. S. Tuan and C. W. Lo © Humana Press Inc., Totowa, NJ

2. Experimental Approaches

2.1. Design of Constructs for Transformation

As noted, there are a number of examples of expressing vertebrate cDNA counterparts of a particular gene of interest in flies. In general, it appears that simply taking a human, mouse, or other vertebrate cDNA and expressing it in flies will usually generate a functional protein. Thus, it has not proven necessary to be overly concerned about possible differential codon usage from flies to vertebrates. However, because the insect body plan is quite different from the typical mammalian body plan, gene regulatory sequences cannot be transferred between species so easily (but *see* **refs. *10*** and ***12***). Consequently, care must be taking to select an expression system compatible with *Drosophila* to achieve adequate protein levels in the relevant tissues, as discussed below.

An important consideration is the ability to detect the foreign protein when expressed in flies, especially when one considers that the *Drosophila* genome displays position effects such that some transgenic insertions will express at higher levels than others as a result of the location in which the transgene has inserted in the genome ***(13)***. Transformation vectors can include insulators, such that the transgene will be much less sensitive to genomic position effects ***(14)***. Otherwise, it is typically necessary to generate a number of different transgenic lines in order to obtain a sufficient number with strong expression; a minimum estimate is about four lines. Having different lines that express at weak, moderate, or strong levels to give a weak, moderate, or strong phenotype, respectively, can be of benefit, however, especially when performing genetic screens for modification of the phenotype (e.g., *see* **refs. *15–17***). If the experiments require the construction of a large number of genetic stocks, having the ability to detect expression of the foreign protein can be extremely valuable to allow selection of strongly expressing transgenic lines. In addition, if a phenotype is not observed, unless transgene expression can be monitored directly, it might be difficult to distinguish whether the foreign protein does not function in flies or if there is simply a technical problem with expression.

It is of course possible to use *in situ* hybridization to detect expression of the transcript for the transgene. A disadvantage of this approach is that it is, in general, more laborious than detecting protein expression and, moreover, does not indicate whether the protein is being translated appropriately in the fly. In some cases, an antibody to the foreign protein may already be available; one can then test for crossreactivity to potential fly counterparts to determine the utility and limitations of the antibody. When testing for antibody crossreactivity, it is frequently necessary to preadsorb an antibody against fixed fly tissue (e.g., a 1:10 antibody dilution preadsorbed with 50 μL of 4% paraformaldehyde-fixed, dechorionated, devitellinized embryos) to lower potential background crossreactivity. This is particularly necessary for rabbit antisera, which are notorious for giving a high background on fly tissue. It is also necessary to determine whether the antibody to the vertebrate counterpart crossreacts to the fly counterpart; if so, one must be able to distinguish expression of the vertebrate counterpart from the fly gene by some other means, such as expression in a novel tissue where the fly gene is not normally expressed or by tagging the vertebrate protein with a peptide domain to which antibodies are available.

An alternative approach is to tag the foreign protein with a small peptide sequence for which antibodies are available. Examples include FLAG, c-Myc, and hemaglutinin

(HA), for which antibodies can be purchased commercially; alternatively, fusion to a protein with endogenous fluorescence, such as green fluorescent protein (GFP) or one of its derivatives, can be used. If electron microscopy is ultimately of interest, then a glutaraldehyde-resistant epitope tag is particularly useful, as many antibodies lose reactivity to tissue treated with glutaraldehyde [although we have successfully performed immunoelectron microscopy with HA-tagged protein ***(11)***]. Such epitope tags can be added by polymerase chain reaction (PCR) or by subcloning into various commercially available vectors which have these epitopes upstream of a number of convenient restriction enzyme sites. We have added HA and GFP to the N- or C-terminus of a number of proteins successfully ***(18)***. Frequently, we add HA to the C-terminus using PCR. To do this, we design a C-terminal primer which deletes the stop codon and adds a linker of a few small amino acids (glycine and alanine, plus a convenient restriction site), followed by the HA sequence and a stop codon. One can also multimerize the exogenous tag (3–5×) to boost sensitivity of detection ***(19)***. If the tag is at the C-terminus, then one can be assured that the entire protein is being produced if the introduced protein can be detected with the relevant antibody. Alternatively, Western immunoblotting can be used to confirm the synthesis of a protein of the appropriate size.

2.2. Expression Systems

There are a number of different expression systems available, the simplest of which couples a standard transformation vector with an appropriate promoter. Such a promoter may be conditional, such as a heat-shock promoter which is inducible by heat pulsing the animal at 37°C for a short time ***(20)***. Alternatively, it may be a constitutive promoter expressed in a tissue of interest, such as actin or ubiquitin which will be expressed in most cells of the animal ***(21)***, or a promoter that targets gene expression to a particular tissue, such as the *gmr (glass multiple reporter)* or *sevenless* promoter elements which target gene expression to developing eye cells ***(22–24)***. Such constructs have the advantages of simplicity and, depending on the promoter used, yield a transgenic line with a constant and consistent phenotype. Conditional promoters allow one to express the protein at any desired time; however, in general, expression will vary over time (although the heat-shock promoter can give a constant basal level of expression at normal growth temperatures, depending on insertion site, which can be sufficient for a phenotype at normal growth temperatures [e.g., **ref.** ***25***]). If one suspects that ubiquitous or early expression of the protein may be lethal to the animal, then conditional or tissue-restricted expression is essential.

Another approach is a two-component system, the GAL4-UAS system ***(26)***. In this system, the gene of interest is cloned downstream of the yeast UAS–GAL4 DNA-binding regulatory sequences in a fly transformation vector pUAST, and transgenic lines are generated. Then, upon crossing the transgenic line to any of a large collection of fly lines that express GAL4 in tissue-specific patterns, one can express the gene of interest in different tissues at different times of development. One advantage of this system is versatility, as there are many GAL4 lines with different expression patterns available from *Drosophila* stock centers or research laboratories. In this system, a *UAS–lacZ* tester strain can be used to monitor promoter strength and tissue-specific expression of the GAL4 lines being used. In addition, an advantage is the ease of determining the viability or other features of the phenotype—even if expressing the protein widely is lethal, one may be able to obtain transgenic lines because expression is only induced when the transgenic line is crossed to a GAL4 expression line. Conversely, the fact that crosses must be made in order to express the transgene represents a disadvantage of the GAL4–UAS system. Furthermore, the double-insert line of interest is not of itself stable unless one takes the trouble to generate an appropriate stable recombinant fly line. This requirement can become particularly unwieldy when testing the phenotype of a foreign gene in a fly mutant background—performing a single experiment can require many crosses to assemble a complex combination of mutant alleles and transgenic constructs. Again, one must consider the different potential uses of the transgenic line in the long run to determine which approach or approaches will be best suited for the experiments.

2.3. Testing for Function

There are a number of ways to test for function of a foreign protein in transgenic *Drosophila*. If testing homologs of a known fly gene for which mutants exist, then one test for function is ability of the foreign gene to rescue the fly mutant phenotype. If the fly counterpart has dominant effects or if one might expect dominant effects as a result of the function of the protein in vertebrates (such as for a dominant oncogene or disease gene), then another test is to determine whether the vertebrate homolog can induce similar dominant phenotypes in flies. There are examples of dominant oncogenic mutations leading to a form of the protein that also functions dominantly in the fly ***(27–33)***.

In some cases, expression of vertebrate genes in flies has demonstrated that a conserved function of the vertebrate and fly genes is autoregulation; thus, the vertebrate protein (frequently a transcription factor) turns on expression of the endogenous fly counterpart ***(9,34)***. If one has mutants in the fly gene involved, then it is possible to test for functional conservation in the genetic background of a protein null of the fly gene and, hence, address broader aspects of functional conservation (e.g., **ref. *5***).

When expressing a foreign gene in the fly in a tissue that normally does not express any such gene, one must consider if screens to identify interacting proteins will be useful for understanding the function of the gene in its normal cellular context. It is important to assess whether any phenotypic effects observed in the fly accurately reflect conserved functions of the vertebrate protein under scrutiny. For example, will vertebrate anti-apoptotic genes block *Drosophila* programmed cell death? Will the vertebrate homolog, like its fly counterpart, direct ectopic tissue formation in the fly? If the vertebrate cDNA induces a dominant effect, is that effect the result of elevated levels of a normal activity of the protein (a hypermorphic effect) or of a new activity of the protein that may have little to do with its normal function (a neomorphic effect). Neomorphic effects, for example, might be the result of subcellular mislocalization of the vertebrate protein in the fly. To what degree does the pathology of a human disease gene reflect biological effects known to occur in humans or vertebrate models, and can these effects be faithfully replicated in the fly model? These are, of course, specific issues that vary for any one gene of interest, and they are critical to consider.

2.4. Genetic Screens for Modifier Mutations

A major goal of expressing a foreign protein in flies is to be able to apply *Drosophila* genetics to further understand the biological problem. The basic idea is to find mutations in fly genes that enhance or suppress the phenotype, and use these mutations to identify vertebrate genes that function in the same pathways or biological process. By this means, one can, therefore, define additional genes that elucidate or indirectly influence the biological pathway of interest.

There are two general approaches for identifying modifier mutations: (1) to screen collections of existing mutations or deficiencies to define interacting genes and (2) do a *de novo* mutagenesis in flies to define interacting genes. Usually, both approaches are performed as screens for dominantly modifying mutations on the autosomes and recessive or dominant mutations on the X chromosome. These screens allow direct analysis of modifying effects in the progeny of mass fly matings, enabling a large number of potential mutants to be rapidly screened relative to other methods.

One approach is to look for enhancers or suppressors by crossing the flies bearing the foreign gene of interest to a collection of *Drosophila* deficiency chromosomes. This collection, available from the Bloomington *Drosophila* Stock Center, consists of about 190 fly lines, which uncover, in total, approximately 70–80% of the *Drosophila* euchromatin. By this approach, one searches for regions of the chromosomes that harbor genes that, when reduced in dosage by 50%, will modify the phenotype of interest. Thus, to test all regions of the genome uncovered by available deficiencies, one simply performs fly crosses and examines the resulting progeny flies. Once a deficiency region of interest is found, then the genetic interaction can be confirmed and the cytological region of the chromosome narrowed down as much as possible using smaller available deficiencies. Eventually, one can test for interactions with all available known mutations in the region and/or perform a mutagenesis to define genes in the region. Hay et al. ***(15)*** have successfully used this strategy to identify a conserved gene that is involved in programmed cell death pathways.

A disadvantage of this technique is that the deficiency lines tend to show variable genetic background effects; that is, it is difficult to determine whether any observed effect on the phenotype of interest is the result of the deficiency itself or to the fact that the cross is made between nonisogenic fly lines. Thus, the success of the approach can depend on the strength and variability of the phenotype being modified. If the modifier effect is very strong, then this approach can be quite successful; however, if the modifier effect is subtle, then it can be difficult to distinguish modification of the phenotype in the widely variable backgrounds of the deficiency lines. Another disadvantage is that eventually after narrowing down a region to the smallest possible extent, it may still be necessary to perform many molecular biological manipulations before having a defined gene in hand.

A variation on this approach is to look for modifier mutations among the large collection of P-element-induced mutation lines ***(36,37)***. Should an interaction be found by using P-element-induced mutations to look for dosage-sensitive modifier interactions, then the gene can easily be cloned if the P element has inserted into it or nearby. In addition, some of the P lethals have been generated using reporter gene constructs, such that one can stain the line for the reporter gene expression (β-galactosidase), which may reveal interesting aspects of the expression pattern of the potentially interacting gene. However, the P-element-induced mutations have a similar background problem as the deficiency lines, which, again, can often be too variable in practice to make such a screen successful (e.g., *see* **ref. *17***).

In general, both deficiencies and the P lethals test for the same type of interaction: an interaction resulting from reduction of a gene dosage by 50%. An alternative approach is to use a point mutagen (ethyl methanesulfonate [EMS] is commonly used in *Drosophila*) to identify dominantly interacting mutations. With a point mutagen, one can obtain both loss-of-function mutations that reduce gene function and gain-of-function mutations resulting from single amino acid substitution in a critical region of the interacting protein. Thus, using EMS as a mutagen may select for different types of interacting mutations than a deficiency or P-element lethal screen. X-ray mutagenesis can also be of interest, because X-rays will, in general, produce chromosomal rearrangements that can affect very large genes or gene complexes as well as result in gain-of-function mutations, depending on the particular rearrangement (*see* **ref.** ***17***). Most crucially, by doing such a mutagenesis, one has greater control over the genetic background: one can select an isogenic background that, when crossed to the line of interest, gives a uniform phenotype such that the effect of any modifier interaction will be readily seen. Such an approach has proven successful for a number of different types of modifier screens ***(16,17)***.

By any of these approaches, the real challenge comes in the analysis of the modifiers obtained to identify those that are most interesting with respect to the question of interest. In all of these approaches, it is essential to have good controls to eliminate modifiers that interact with the expression system or the promoter expressing the gene rather than with the protein being expressed, and so forth. Thus, secondary screens are critical to classify mutants to distinguish those modifiers more directly involved in the question of interest, from those that are only peripherally involved. An excellent example of this is **ref.** ***38***, where 30,000 mutagenized chromosomes were screened for modification of a *sevenless* receptor tyrosine kinase mutant phenotype. Of seven complementation groups identified, four of seven also modified the mutant phenotype of a second tyrosine kinase receptor (the EGF receptor), thus defining those genes that were common signaling components of receptor tyrosine kinase pathways. Some argue that the best approach is to do different types of modifier screens and then focus on those subsets of new genes that are repeatedly identified in multiple screens, indicating that they are likely to be centrally important in the biological pathway of interest.

By these means, one can apply the ease and rapidity of genetics in a simple model system like *Drosophila* to questions of fundamental interest and importance in vertebrates. With the advent of genomic sequencing, the importance of model systems like *Drosophila* to reveal protein function and define biological pathways becomes ever more important.

Acknowledgments

Thanks to A. Cashmore and M. Fortini for critical comments. N.B. is supported by grants from the John Merck Fund, the National Eye Institute, the Alzheimer's Association, and the David and Lucile Packard Foundation.

References

1. Rubin, G. M. (1988) *Drosophila melanogaster* as an experimental organism. *Science* **240,** 1453–1459.
2. Padgett, R. W., Wozney, J. M., and Gelbart, W. M. (1993) Human BMP sequences can confer normal dorsal-ventral patterning in the *Drosophila* embryo. *Proc. Natl. Acad. Sci. USA* **90,** 2905–2909.

3. Johnston, S. H., Rauskolb, C., Wilson, R., Prabhakaran, B., Irvine, K. D., and Vogt, T. F. (1997) A family of mammalian *Fringe* genes implicated in boundary determination and the *Notch* pathway. *Development* **124,** 2245–2254.
4. Acampora, D., Avantaggiato, V., Tuorto, F., Barone, P., Reichert, H., Finkelstein, R., and Simeone, A. (1998) Murine *Otx1* and *Drosophila otd* genes share conserved genetic functions required in invertebrate and vertebrate brain development. *Development* **125,** 1691–1702.
5. Leuzinger, S., Hirth, F., Gerlich, D., Acampora, D., Simeone, A., Gehring, W. J., Finkelstein, R., Furukubo-Tokunaga, K., and Reichert, H. (1998) Equivalence of the fly *orthodenticle* gene and the human *OTX* genes in embryonic brain development of *Drosophila. Development* **125,** 1703–1710.
6. Nagao, T., Leuzinger, S., Acampora, D., Simeone, A., Finkelstein, R., Reichert, H., and Furukubo-Tokunaga, K. (1998) Developmental rescue of *Drosophila* cephalic defects by the human *Otx* genes. *Proc. Natl. Acad. Sci. USA* **95,** 3737–3742.
7. Halder, G., Callaerts, P., and Gehring, W. (1995) Induction of ectopic eyes by targeted expression of the *eyeless* gene of *Drosophila. Science* **267,** 1788–1792.
8. Bonini, N. M., Bui, Q. T., Gray-Board, G. L., and Warrick, J. M. (1997) The *Drosophila eyes absent* gene directs ectopic eye formation in a pathway conserved between flies and vertebrates. *Development* **124,** 4819–4826.
9. McGinnis, N., Kuziora, M. A., and McGinnis, W. (1990) Human *Hox-4.2* and *Drosophila* deformed encode similar regulatory specificities in *Drosophila* embryos and larvae. *Cell* **63,** 969–976.
10. Malicki, J., Cianetti, L. C., Peschle, C., and McGinnis, W. (1992) A human *HOX4B* regulatory element provides head-specific expression in *Drosophila* embryos. *Nature* **358,** 345–347.
11. Warrick, J. M., Paulson, H., Gray-Board, G. L., Bui, Q. T., Fischbeck, K., Pittman, R. N., and Bonini, N. M. (1998) Expanded polyglutamine protein forms nuclear inclusions and causes neural degeneration in Drosophila. *Cell* **93,** 939–949.
12. Haerry, T. E. and Gehring, W. J. (1996) Intron of mouse *Hoxa-7* gene contains conserved homeodomain binding sites that can function as an enhancer in *Drosophila. Proc. Natl. Acad. Sci. USA* **93,** 13,884–13,889.
13. Spradling, A. C. and Rubin, G. M. (1983) The effect of chromosomal position on the expression of the Drosophila xanthine dehydrogenase gene. *Cell* **34,** 47–57.
14. Kellum, R. and Schedl, P. (1991) A position-effect assay for boundaries of higher order chromosomal domains. *Cell* **64,** 941–950.
15. Hay, B. A., Wassarman, D. A., and Rubin, G. M. (1995) *Drosophila* homologs of baculovirus inhibitor of apoptosis proteins function to block cell death. *Cell* **83,** 1253–1262.
16. Dickson, B. J., van der Straten, A., Dominguez, M., and Hafen, E. (1996) Mutations modulating Raf signaling in *Drosophila* eye development. *Genetics* **142,** 163–171.
17. Karim, F. D., Chang, H. C., Therrien, M., Wassarman, D. A., Laverty, T., and Rubin, G. M. (1996) A screen for genes that function downstream of Ras1 during Drosophila eye development. *Genetics* **143,** 315–329.
18. Bui, Q. and Bonini, N., unpublished results.
19. Panin, V., Papayannopoulos, V., Wilson, R., and Irvine, K. (1997) Fringe modulates Notch-ligand interactions. *Nature* **387,** 908–912.
20. Schneuwly, S., Klemenz, R., and Gehring, W. (1987) Redesigning the body plan of *Drosophila* by ectopic expression of the homeotic gene *Antennapedia. Nature* **325,** 816–818.
21. Sekelsky, J. J., Newfeld, S. J., Raftery, L. A., Chartoff, E. H., and Gelbart, W. M. (1995) Genetic characterization and cloning of *mothers against dpp*, a gene required for *decapentaplegic* function in Drosophila melanogaster. *Genetics* **139,** 1347–1358.

22. Basler, K., Siegrist, P., and Hafen, E. (1989) The spatial and temporal expression pattern of *sevenless* is exclusively controlled by gene-internal elements. *EMBO J.* **8,** 2381–2386.
23. Bowtell, D. D., Lila, T., Michael, W. M., Hackett, D., and Rubin, G. M. (1991) Analysis of the enhancer element that controls expression of *sevenless* in the developing *Drosophila* eye. *Proc. Natl. Acad. Sci. USA* **88,** 6853–6857.
24. Ellis, M., O'Neill, E., and Rubin, G. (1993) Expression of *Drosophila* glass protein and evidence for negative regulation of its activity in non-neuronal cells by another DNA-binding protein. *Development* **119,** 855–865.
25. Fehon, R. G., Oren, T., LaJeunesse, D. R., Melby, T. E., and McCartney, B. M. (1997) Isolation of mutations in the Drosophila homologues of the human *Neurofibromatosis 2* and yeast *CDC42* genes using a simple and efficient reverse-genetic method. *Genetics* **146,** 245–252.
26. Brand, A. H. and Perrimon, N. (1993) Targeted gene expression as a means of altering cell fates and generating dominant phenotypes. *Development* **118,** 401–415.
27. Bishop, J. G., III and Corces, V. (1988) Expression of an activated ras gene causes developmental abnormalities in transgenic *Drosophila melanogaster. Genes Dev.* **2,** 567–577.
28. Fortini, M., Simon, M., and Rubin, G. (1992) Signalling by the *sevenless* protein tyrosine kinase is mimicked by Ras1 activation. *Nature* **355,** 559–561.
29. Fortini, M. E., Rebay, I., Caron, L. A., and Artavanis-Tsakonas, S. (1993) An activated Notch receptor blocks cell-fate commitment in the developing *Drosophila* eye. *Nature* **365,** 555–557.
30. Lieber, T., Kidd, S., Alcamo, E., Corbin, V., and Young, M. W. (1993) Antineurogenic phenotypes induced by truncated Notch proteins indicate a role in signal transduction and may point to a novel function for Notch in nuclei. *Genes Dev.* **7,** 1949–1965.
31. Rebay, I., Fehon, R. G., and Artavanis-Tsakonas, S. (1993) Specific truncations of Drosophila Notch define dominant activated and dominant negative forms of the receptor. *Cell* **74,** 319–329.
32. Struhl, G., Fitzgerald, K., and Greenwald, I. (1993) Intrinsic activity of the Lin-12 and Notch intracellular domains in vivo. *Cell* **74,** 331–345.
33. Casanova, J., Llimargas, M., Greenwood, S., and Struhl, G. (1994) An oncogenic form of human *raf* can specify terminal body pattern in *Drosophila. Mech. Dev.* **48,** 59–64.
34. Glardon, S., Callaerts, P., Halder, G., and Gehring, W. (1997) Conservation of Pax-6 in a lower chordate, the ascidian *Phallusia mammillata. Development* **124,** 817–825.
35. Hay, B. A., Wolff, T., and Rubin, G. M. (1994) Expression of baculovirus P35 prevents cell death in *Drosophila. Development* **120,** 2121–2129.
36. Cooley, L., Kelley, R., and Spradling, A. (1988) Insertional mutagenesis of the *Drosophila* genome with single P elements. *Science* **239,** 1121–1128.
37. Török, T., Tick, G., Alvarado, M., and Kiss, I. (1993) P-lacW insertional mutagenesis on the second chromosome of Drosophila melanogaster: isolation of lethals with different overgrowth phenotypes. *Genetics* **135,** 71–80.
38. Simon, M. A., Bowtell, D. D. L., Dodson, G. S., Laverty, T. R., and Rubin, G. M. (1991) Ras1 and a putative guanine nucleotide exchange factor perform crucial steps in signaling by the Sevenless protein tyrosine kinase. *Cell* **67,** 701–716.

3

Bioassays for Studying the Role of the Peptide Growth Factor Activin in Early Amphibian Embryogenesis

Makoto Asashima, Takashi Ariizumi, Shuji Takahashi, and George M. Malacinski

1. Introduction

Activin, a peptide growth factor, is a member of the transforming growth factor-β (TGF-β) superfamily. It was originally isolated from follicle fluid as a gonadal hormone that stimulates follicle-stimulating hormone (FSH) secretion and is identical to EDF, the erythroid differentiation factor (which stimulates erythroleukemia cells to differentiate into hemoglobin-producing cells) *(1)*. It also appears to be related, if not identical, to the so-called "vegetalizing factor" described originally by Tiedemann and colleagues, which can induce amphibian embryonic tissue rudiments to display various differentiation patterns (reviewed in **ref.** *2*). In addition, it appears to be identical to the XTC factor isolated by Smith et al. *(3)* from transformed *Xenopus* fibroblasts. Thus, the identification of activin as a potential morphogen in amphibian embryos *(4)* solved several mysteries surrounding the puzzle regarding the molecular nature of various hitherto ill-characterized "inducing substances."

Activin is a dimeric protein and has been described in mammals as sharing subunits with a related dimeric protein—inhibin (**Fig. 1**). Inhibin is another peptide, which, in contrast to activin, *inhibits* the secretion of FSH by pituitary cells. Although inhibins have been identified in mammals, they have not yet been described in amphibia.

Activin warrants nomination as a candidate endogenous inducer, for it is found in embryos at the expected place and at the appropriate time to play a role in mesoderm formation. For example, activin proteins have been detected in unfertilized eggs *(5,6)* and have been demonstrated to be bound to the vitellogenin component of yolk platelets *(7)*. We know that the maternal supply of activin protein is most likely synthesized outside of the oocyte (e.g., in follicle cells) and transported into the oocyte, because activin mRNA has been localized to follicle cells *(8,9)*. It may actually be transported "piggyback" fashion into the oocyte on yolk proteins which are synthesized in the liver during the vitellogenic stage of oogenesis. Later, during the blastula/gastrula stage, activin genes are expressed from the zygotic genome *(10)*.

From: *Methods in Molecular Biology, Vol. 136: Developmental Biology Protocols, Vol. II*
Edited by: R. S. Tuan and C. W. Lo © Humana Press Inc., Totowa, NJ

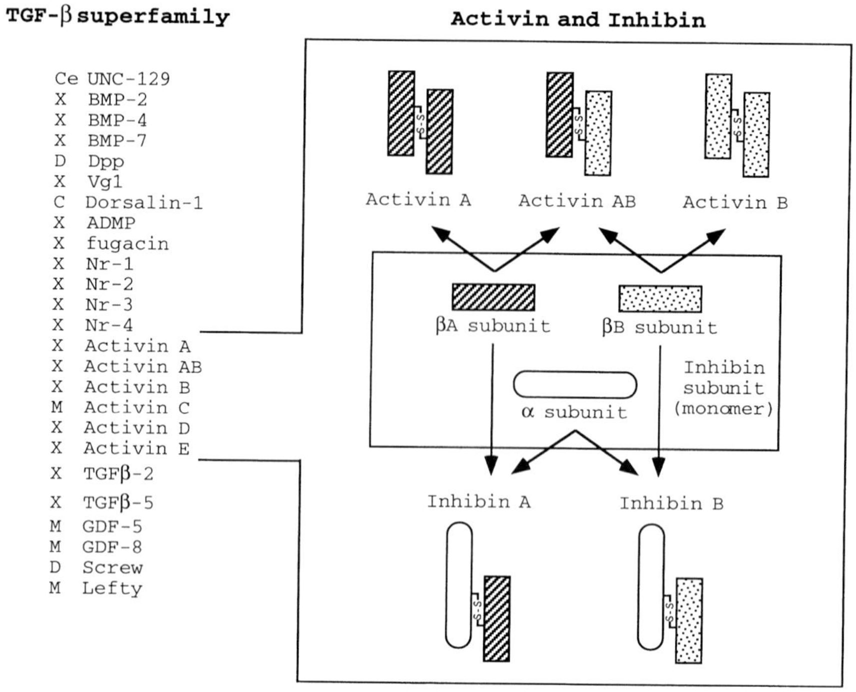

Fig. 1. Molecular structure of activin and inhibin proteins. Activin A is a homodimer consisting of two inhibin β_A chains; activin AB is a heterodimer of inhibin β_A and β_B chains; and activin B is a homodimer of two inhibin β_B subunits. All three forms have been identified in amphibian embryos (*see* **ref. 5**). The native molecular weight of activin is 25,000 D.

The potential importance of activin studies is substantial: Activin induces mesoderm development, and mesoderm induction appears to be a common early feature of the patterning of virtually all vertebrate embryos. Thus, to be able to study the mechanism of action of a purified protein that can generate simulated mesoderm induction in vitro is a major experimental advantage. Indeed, a wide range of mesodermal tissues can be induced with activin and profound effects on embryonic patterning can be generated, depending upon the stage of the responding tissue, concentration of activin ***(11,12)***, and type of bioassay. For example, in the tissue-culture assay to be described below, with relatively low concentrations of activin (e.g., 0.3 ng/mL), only ventral mesoderm such as blood cells and coelomic epithelium is induced, whereas at 5–10 ng/mL, muscle is induced, and at higher concentrations (e.g., 50 ng/mL), notochord (a distinctly dorsal structure) is induced. A range of dramatic patterning effects can also be observed using one of the injection assays described below.

Consideration of those bioassay results makes it easy to conceive of a model for pattern formation that uses varying concentrations of activin as a key molecule in concentration gradient-dependent patterning scenarios. As well, models for patterning that involve *serial* inductions initiated by activin action can be readily conceptualized. When activin-treated newt animal cap is sandwiched between nontreated animal caps, an induction cascade reminiscent of classical Spemann-embryonic organizer action is observed ***(13)***. The type of axial tissue induced (i.e., head vs tail) depends on the duration of the preculture period of activin-treated animal cap. This observation is also reminiscent of the original primary embryonic organizer grafting experiments, which demonstrated that early-stage organizer tissue induces head structures, whereas a later-stage organizer induces mostly tail tissue. Verification of the nature of the induced tissues can be achieved with the use of either histological or molecular indices of gene expression *(see below)*.

In addition to using activin's cell differentiation effects as a model system for understanding the molecular basis of patterning in the early embryo, activin can be employed either alone or in combination with other prospective regulatory molecules for two other purposes as well: (1) studying differentiation of specific tissues/organs in culture [e.g., renal tissue ***(14)*** and heart ***(15)***] and (2) understanding theoretical aspects of embryonic development. Concerning this latter point, we are presently initiating an ambitious undertaking, the so-called A-FEAT project: an attempt to *assemble a functioning embryo from activin-treated components*. It is briefly described in **Subheading 4.**

2. Types of Bioassay Protocols

Several different bioassay schemes for activin studies have been developed. Each measures a different aspect of the action of activin. The *tissue-culture assays*, for example, focus on the effects of activin on cell differentiation. The *injection assays* are designed to recognize activin effects on embryonic patterning. Finally, the *interference–injection assays* target endogenous activin mechanisms. Each will be described in turn.

2.1. Tissue-Culture Assay

This assay protocol is perhaps the simplest, for it does not require expertise in microinjection techniques. However, as will be mentioned later, it is important that the preparation of the explanted tissue be carefully monitored. Essentially, animal cap tissue is bathed in a simple salt (culture) medium containing an appropriate concentration of activin. Nontreated blastomeres differentiate into "atypical epidermis," whereas treated animal cap differentiates into ventral mesodermal cell types, such as blood, mesenchyme, and coelomic epithelium, or into dorsal mesoderm, such as muscle and notochord.

Figure 2 illustrates the general experimental protocol for using activin to generate that range of cell differentiations in cultured animal cap tissue.

Scoring the results can be accomplished by either observing histological indices of differentiation (**Fig. 3**) or by detecting the appearance of specific molecular markers (**Table 1**).

Thus, this bioassay can be employed to determine the extent to which activin promotes the development of specific differentiation profiles in cultured tissues. Although it has been used primarily with *Xenopus* test systems, we have also used it often for analyzing organogenesis in urodele animal cap tissue. For example, urodele animal caps, when treated with high concentrations of activin, developed beating hearts ***(15)***.

This *tissue-culture assay* offers several advantages to the experimenter. Primarily, of course, it is relatively easy to perform. The culturing of amphibian tissues is remarkably simple because early-stage embryonic tissue is preloaded with an abundant supply

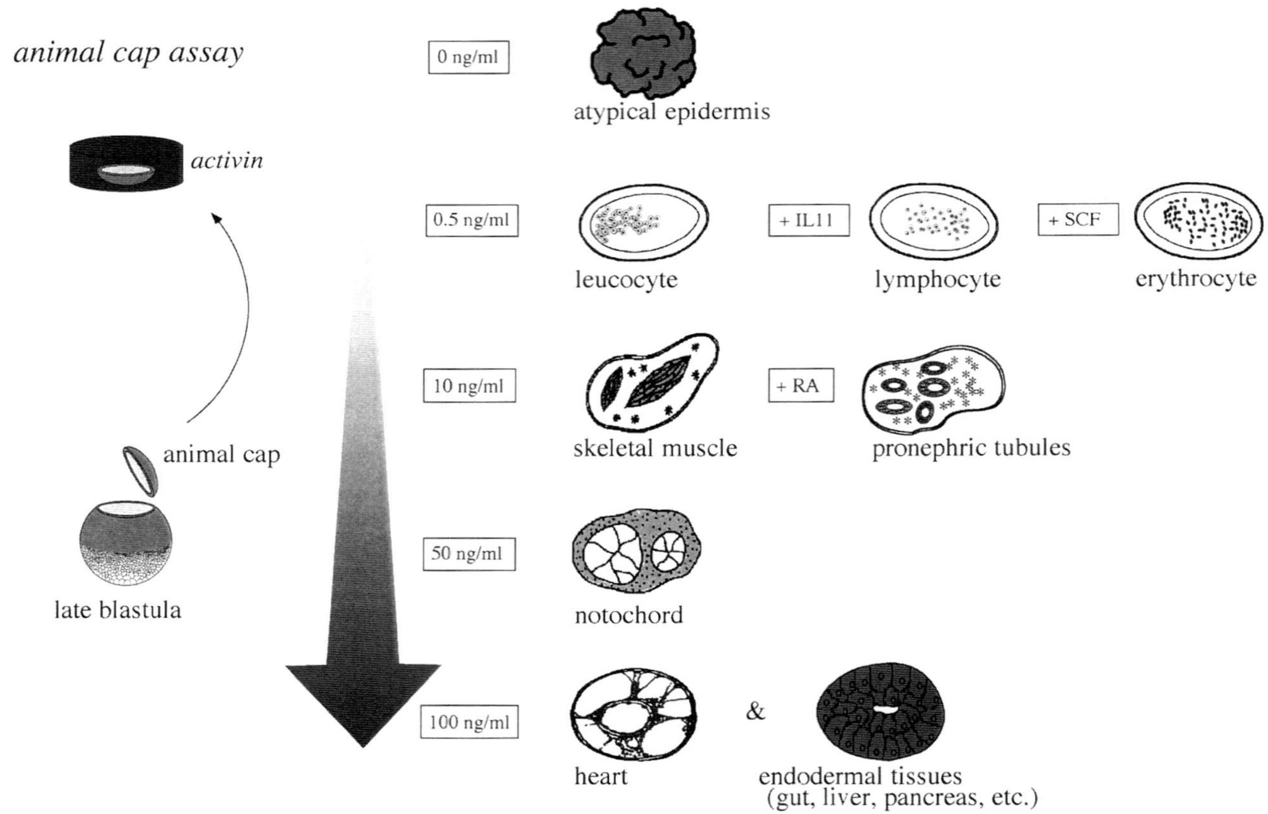

Fig. 2. The animal cap assay generates a broad range of cell differentiations, depending on the concentration of activin in the culture medium. Typically, a *Xenopus* animal cap is cultured in the presence of activin for 3 d (20°C). Low concentrations (e.g., 0.5 ng/mL) induce ventral mesodermal tissue. Mid-range concentrations (e.g., 10 ng/mL) induce muscle, whereas higher concentrations (e.g., 50 ng/mL) induce dorsal mesodermal structures (e.g., notochord). IL11: interleukin 11; SCF: stem cell factor; RA: retinoic acid.

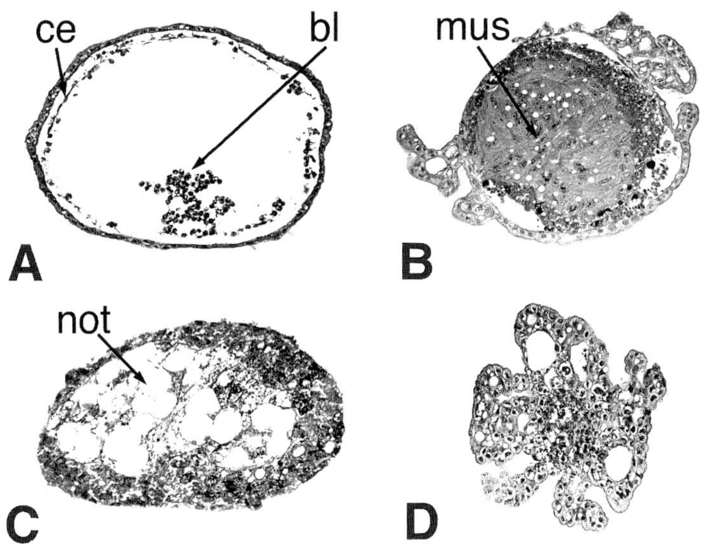

Fig. 3. Histological examination of *Xenopus* animal caps. treated with activin for 3 d. Using the fixation and staining methods described in **Subheading 3.** differentiated tissues were easily recognized. These included blood cells (bl) and coelomic epithelium (ce) **(A)**, muscle (mus) **(B)**, and notochord (not) **(C)**. A section through an animal cap that received no activin treatment (control) is shown in **(D)**.

of nutrients, so the culture medium need only consist of a simple buffered salt solution. Purified activin is available and its concentration in a culture medium is easily adjusted. It is also easy to add additional factors to the culture medium and thereby search for synergistic effects on tissue differentiation. For example, we have added retinoic acid or other morphogens to some cultures (**Fig. 2**) and observed renal tubule differentiation. We have also tested the combination of activin and bFGF on animal caps. Those factors functioned synergistically for blood cell differentiation (unpublished data).

The foregoing procedure describes "direct, single-culture" assays in which animal caps are treated with activin and, after a suitable culturing period, are assayed for type of differentiation. It is also possible, however, to harvest that "direct, single-cultured" tissue and use it as an inducing tissue by combining it with naive (uninduced) tissue in so-called "animal cap combination" assays. **Figure 4** illustrates the general procedure. Such combination assays are especially valuable for studying *serial* inductive interactions, as described in Chapter 11 in Volume I.

2.1.1. Cautionary Notes

When preparing to perform a tissue-culture assay or to compare the data an assay has generated with information collected by other laboratories, the following checklist can serve as a guide. Often, data collected in different laboratories has been compared without careful reference to possible differences in the experimental conditions used to collect it, thus these cautions should be carefully considered.

_____ (1) Has the activin preparation been biologically calibrated so that doses used by different laboratories can be compared?

Table 1
Candidate Molecular Indices of Gene Expression that Are Likely Associated with Activin Treatment of Animal Caps

Gene	Onset of expression	Defining feature of gene
Early-response genes		
Mix 1	30 min	Homeobox gene, relative of *Drosophila paired*
goosecoid	30 min	Homeobox gene, relative of *Drosophila goosebery* and *bicoid*
Xbra	1.5 h	Homolog of mouse *Brachyury (T)*
Xlim-1	2 h	Homeobox gene with LIM domain
XFKHI		Homeobox gene, relative of *Drosophila forkhead*
Middle-response genes		
Xwnt-8	3 h[a]	Relative of *Drosophila*/mouse oncogene int-1
Xnot	3 h[a]	Homeobox gene, relative of *Drosophila empty spiracles*
noggin	3 h[a]	BMP-binding protein
chordin	3 h[a]	BMP-binding protein
follistatin	3 h[a]	Activin and BMP-binding protein
cerverus	3 h[a]	Head inducer. BMP, Wnt, nodal-binding protein
Xhox3	6h*	Homeobox gene, relative of *Drosophila even skipped*
Xprx-1	6 h	*Drosophila paired* related homeobox gene
XMyoD	6 h[a]	Homolog of mouse MyoD
Late-response genes		
X1Hbox1	11 h[a]	Homeobox gene, relative of *Drosophila antennapedia*, homologue of mouse *Hox C6*
X1Hbox6	13 h[a]	Homeobox gene, relative of *Drosophila abdominal-B*, homologue of mouse *Hox B9*
a-actin	19 h	Cardiac and early axial muscle actin
N-CAM	19 h	Neural cell-adhesion molecule gene
X1Hbox8	36 h[a]	Homeobox gene, expressed in pancreas

[a]Approximate time after initiation of activin treatment.

_____ (2) Are similar species of amphibia employed?
_____ (3) Are animal caps isolated from similar stage embryos?
_____ (4) Are similarly sized animal caps used by different laboratories?
_____ (5) Are the animal caps uniform in age* and size?
_____ (6) Are the histological criteria or molecular markers used to score differentiation comparable?

2.2. Injection Assays

Microinjection of either activin mRNA or activin protein into blastomeres or blastocoel of the early embryo can be carried out without much difficulty if microinjection equipment is available. By injecting into the uncleaved egg, the advantage

*Because *Xenopus* embryos develop so quickly, unless temperature-regulated conditions are employed it is likely that the last batch of animal caps prepared for assay will have developed a significant extent beyond the stage of the first-collected batch.

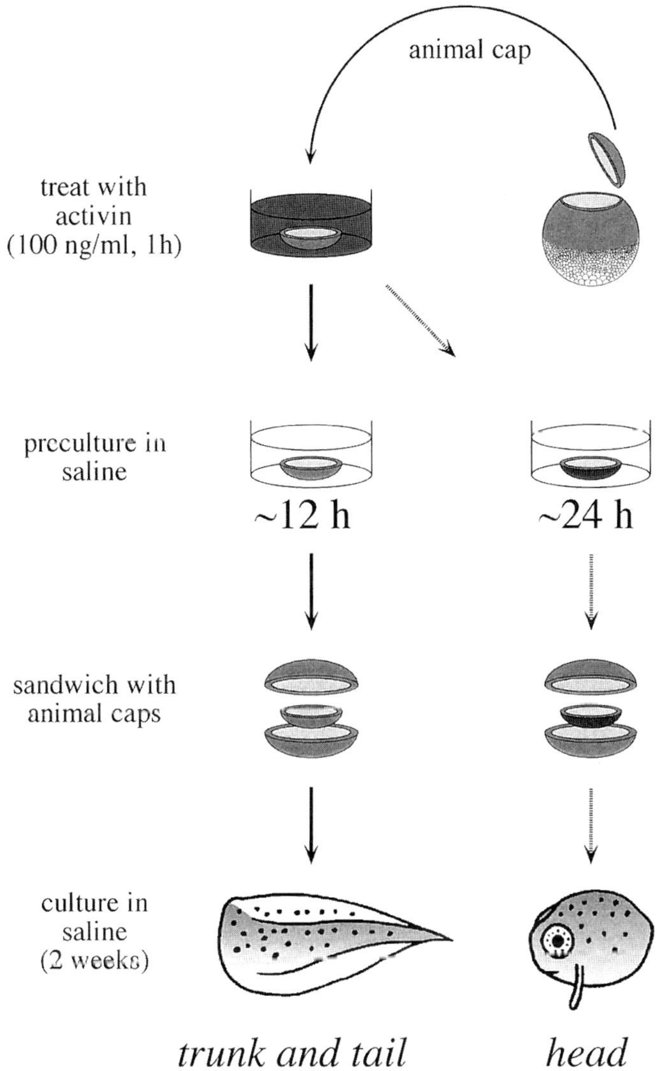

Fig. 4. "Animal cap combination" assay in which an ectodermal cap (in this example, from the urodele [newt] *Cynops pyrrhogaster*) is treated with activin in the tissue culture assay, and precultured for either a short or long period of time, then inserted as a component of a sandwich culture with naive ectoderm and further cultured. In this instance, the activin-treated animal cap mimics the properties of early- and late-stage Spemann primary embryonic organizers.

accrues—in principle—that all cells of the early embryo are exposed to the injected component.

Thus, this assay can be employed to measure the effects of activin on patterning processes (e.g., axis formation) in the whole, intact embryo because virtually all cells in the early embryo are exposed to activin. **Figure 5** illustrates a typical protocol for a simple injection assay. As well, by dissection of the animal cap from an embryo that previously had been injected with either activin protein (*see* **ref. *16***) or activin mRNA and using that animal cap as prospective inducing tissue in the manner illustrated in **Fig. 4**, the utility of injection assays can be expanded.

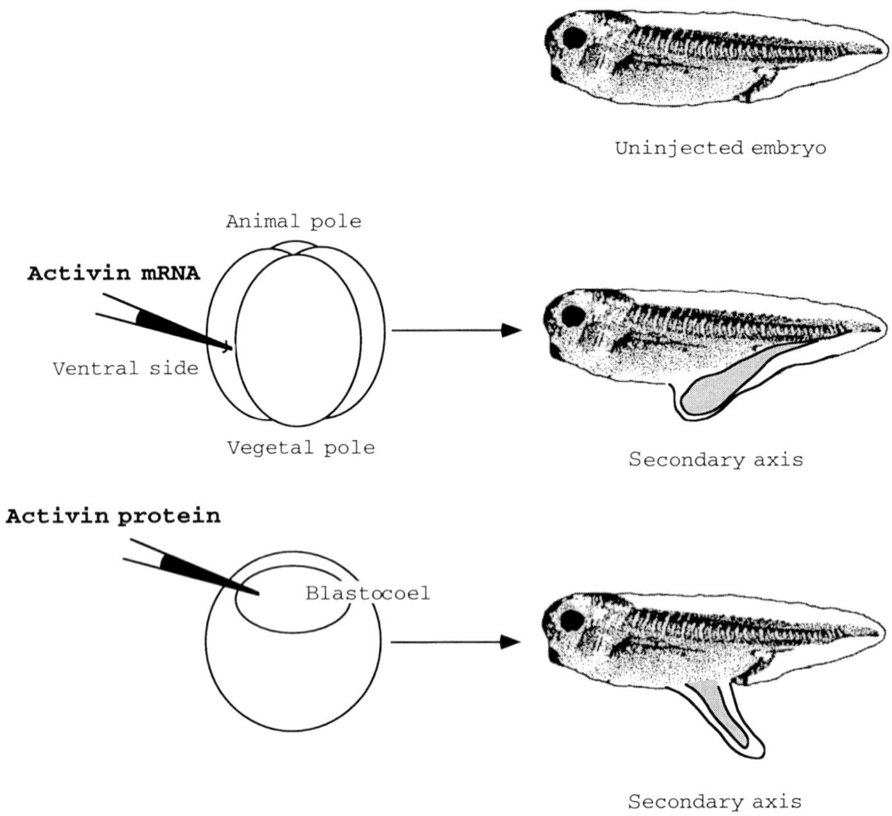

Fig. 5. Microinjection of either activin mRNA or protein into the four-cell stage or the blastocoel of an early embryo has effects on subsequent pattern formation.

2.2.1. Cautionary Notes

For the injection of activin protein, many of the same above-mentioned *cautions* for the tissue-culture assay should be considered. For mRNA injection, yet another set of *cautions* should be carefully considered:

_____ (1) Is the mRNA preparation fully intact, or partially degraded?
_____ (2) Is the mRNA actually translated into a functional protein once inside the egg?*
_____ (3) Is the mRNA preparation distributed uniformly throughout the egg, or does it remain as a bolus (near the injection site)?
_____ (4) Is the mRNA sufficiently long-lived, once injected into an egg, to give a satisfactory result?
_____ (5) Is a "no effect" result meaningful?

2.3. Interference–Injection Assays

Because it is well known that peptide growth factors such as activin must interact with specific protein receptors in order to exert their ultimate effects on differentiation and morphogenesis, microinjection of either excess amounts of the normal

*This criterion is, unfortunately, seldom addressed.

activin receptor (mRNA) or defective variants of that receptor (mRNA) provides a way to test the role endogenous activin plays in differentiation and morphogenesis. A typical assay protocol would be the same as shown in **Fig. 5**. For example, Kondo et al. ***(17)*** and Hemmati-Brivanlou et al. ***(18)*** overexpressed the activin receptor in *Xenopus* embryos and observed dramatic effects on patterning. Their strategy was to inject copious amounts of the mRNA into fertilized, uncleaved eggs. More recently, Armes and Smith ***(19)*** have observed the development of a secondary axis when an activin receptor mRNA was injected into the ventral side of 32-cell stage embryos. Thus, patterning effects can be obtained by the appropriate administration of receptor mRNA.

In addition, microinjection of defective variants of the activin receptor mRNA has been performed. Such *dominant negative activin receptor mRNA* injections have been accomplished by several laboratories. For example, Hemmati-Brivanlou and Melton ***(20)*** observed that embryonic cells normally fated to become epidermis developed neural traits when they were injected with a truncated activin receptor mRNA. New et al. ***(21)*** discovered that defective receptor mRNA caused axial defects in some circumstances and secondary axes in others.

Thus, the use of these bioassay formats designed to interfere with endogenous functions of activin circuitry provide fascinating chances for further experimentation and a plethora of opportunities for speculation regarding the mechanism(s) of action of activin!

2.3.1. Cautionary Notes

Similar considerations to those expressed for the *injection assays* should, of course, be monitored in these *interference–injection assays*. In addition, the following points are pertinent:

_____ (1) Is the presumptive activin receptor mRNA translated into a protein which binds activin in vivo?

_____ (2) Does the pattern of accumulation of the activin receptor protein coincide with tissue locations where activin is expected to act?

_____ (3) What is the meaning of a "no effect" result?

3. Methods

General conditions for operating on amphibian embryos are described in Chapter 11, Volume I. Many of the same methods can be used at the "front end" of the activin bioassays described herein. For example, many of the *animal cap assay* methods are described in that chapter, as are details and photographs of the microinjection techniques which would be used in the injection bioassays described herein. The details of additional methods are described below.

Histological methods vary of course, depending on the size and source of the tissue, as well as the types of observations intended. Our companion chapter in Volume I can be consulted for some protocols.

Activin/antiactivin antibody suppliers include Innogenetics, S.A., Ghent, Belgium and Austral Biologicals, San Raman, CA, USA.

For marking cells to trace their fate in these bioassays 10–50 nL of 1% Texas Red–dextran–amine (TRDA, D-1863; Molecular Probes, Eugene, OR) or 10–50 nL of 1%

fluorescein–dextran–amine (FDA, D-1820; Molecular Probes) are injected at the early cleavage stage (e.g., two or four cells) into single blastomeres.

4. Concluding Remarks

Activins have been reported to display a variety of effects on many different tissues and cell types at numerous stages in the growth and development of vertebrates. A remarkable array of effects has been reported in a broad range of test systems (reviewed in **ref.** *22*). In no instances, however, is the molecular basis of action fully understood. Activin likely functions in the context of an assemblage of regulatory molecules, including activin receptors and inhibitors (e.g., follistatin), and in combination with other regulatory molecules (e.g., other peptide growth factors). Unraveling its mechanism(s) of action will probably require the use of several different experimental approaches, many of which might include the use of the types of bioassays described in this chapter.

The inherent simplicity of the bioassays described herein should be attractive for *reductionist* approaches to discovering the details associated with activin action. It will certainly be interesting to—ultimately—establish the similarities and differences between activin action in, say, the amphibian egg compared to smooth muscle. These bioassays have the advantage that they can serve as a model for the characterization of not only the effects on differentiation when activin is studied alone, but also the action of combinations of growth factors.

In addition to employing these assays for the reductionist-type analyses for which they are especially well suited, it is possible to use them to generate *holistic* approaches to understanding embryonic patterning. For example, we have embarked on the A-FEAT (Assembling a Functioning Embryo from Activin-Treated components) project, an attempt to answer the following question:

> What are the minimal morphogen and cell-mass requirements for complete patterning of the tissues and organs in an early embryo?

Restated as a hypothesis:

> Activin and animal cap cells are sufficient, when manipulated in an appropriate fashion, to construct a whole, functioning embryo!

The observation of the remarkable ability of activin to induce in animal caps a wide range of tissues and organs (e.g., **Fig. 2**) is providing a foundation for the A-FEAT project, an attempt to determine *what components are sufficient for patterning a whole embryo*. Activin alone at various concentrations, as well as in combination with other prospective patterning molecules (e.g., other peptide growth factors), will provide a research platform, based largely on the bioassays described in this chapter.

Acknowledgments

We are grateful to members of our laboratories and our colleagues for useful discussions (e.g., Prof. Anton Neff) and editing (Susan Duhon) of this manuscript. Most of this work was supported by Grants-in-Aid for Scientific Research from the Ministry of Education, Science, Sports and Culture of Japan, and by CREST (Core Research for Evolutional Science and Technology) of the Japan Science and Technology Corporation. G.M.M.'s research is funded by the National Science Foundation (USA).

References

1. Murata, M., Eto, Y., Shibai, H., Sakai, M., and Muramatsu, M. (1988) Erythroid differentiation factor is encoded by the same mRNA as that of inhibin βA chain. *Proc. Natl. Acad. Sci. USA* **85,** 2434–2438.
2. Tiedemann, H., Asashima, M., Born, J., Grunz, H., Knochel, W., and Tiedemann, H. (1996) Determination, inductin and pattern formation in early amphibian embryos. *Dev. Growth Differ.* **36,** 233–246.
3. Smith, J. C., Price, B. M. J., van Nimmen, K., and Huylebroek, D. (1990) Identification of a potent *Xenopus* mesoderm inducing factor as a homologue of activin A. *Nature* **345,** 729–731.
4. Asashima, M., Nakano, H., Shimada, K., Kinoshita, K., Ishi, K., Shibai, H., and Ueno, N. (1990) Mesodermal induction in early amphibian embryos by activin A. *Roux's Arch. Dev. Biol.* **198,** 330–335.
5. Fukui, A., Nakamura, T., Uchiyama, H., Sugino, K., and Asashima, M. (1994) Identification of activins A, AB, and B and follistatin proteins in *Xenopus* embryos. *Dev. Biol.* **163,** 279–281.
6. Asashima, M., Nakano, H., Uchiyama, H., Sugino, H., Nakamura, T., Eto, Y, Ejima, D., Nishimatsu, S., Ueno, N., and Kinoshita, K. (1991) Presence of activin (erythroid differentiation factor) in unfertilized eggs and blastulae of *Xenopus laevis*. *Proc. Natl. Acad. Sci. USA* **88,** 6511–6514.
7. Uchiyama, H., Nakamura, T., Komazaki, S., Takio, K., Asashima, M., and Sugino, H. (1994) Localization of activin and follistatin proteins in the *Xenopus* oocyte. *Biochem. Biophys. Res. Commun.* **202,** 484–489.
8. Dohrmann, C. E., Hemmati-Brivanlou, A., Thomsen, G. H., Fields, A., Woolf, T. M., and Melton, D. A. (1993) Expression of activin mRNA during early development in *Xenopus laevis*. *Dev. Biol.* **157,** 474–483.
9. Okabayashi, K., Shoji, H., Nakamura, O., Nakamura, T., Hashimoto, O., Asashima, M., and Sugino, H. (1996) cDNA cloning and expression of the *Xenopus laevis* vitellogenin receptor. *Biochem. Biophys. Res. Commun.* **224,** 406–413.
10. Dohrmann, C. A., Hemmati-Brivanlou, A., Thomsen, G. H., Fields, A., Woolf, T. M., and Melton, D. A. (1993) Expression of activin mRNA during early development of *Xenopus laevis*. *Dev. Biol.* **157,** 474–483.
11. Ariizumi, T., Moriya, N., Uchiyama, H., and Asashima, M. (1991) Concentration dependent inducing activity of activin A. *Roux's Arch. Dev. Biol.* **200,** 230–233.
12. Green, J. B. and Smith, J. C. (1990) Graded changes in dose of a *Xenopus* activin A homologue elicit stepwise transitions in embryonic cell fate. *Nature* **347,** 391–394.
13. Ariizumi, T. and Asashima, M. (1995) Head and trunk-tail organizing effects of the gastrula ectoderm of *Cynops pyrrogaster* after treatment with activin A. *Roux's Arch. Dev. Biol.* **204,** 427–435.
14. Uochi, T., and Asashima, M. (1996) Sequential gene expression during pronephric tubule formation *in vitro* in *Xenopus* ectoderm. *Dev. Growth Differ.* **38,** 625–634.
15. Ariizumi, T., Komazaki, S., Asashima, M., and Malacinski, G. M. (1996) Activin treated urodele ectoderm: a model experimental system for cardiogenesis. *Int. J. Dev. Biol.* **40,** 715–718.
16. Ariizumi, T., Sawamura, K., Uchiyama, H., and Asashima, M. (1991) Dose and time-dependent mesoderm induction and outgrowth formation by activin A in *Xenopus laevis*. *Int. J. Dev. Biol.* **35,** 407–414.
17. Kondo, M., Tashiro, K., Fujii, G., Asano, M., Miyoshi, R., Yamada, R., Muramatsu, M., and Shiokawa, K. (1991) Activin receptor mRNA is expressed early in *Xenopus* embryo-

genesis and the level of the expression affects the body axis formation. *Biochem. Biophys. Res. Commun.* **181,** 684–690.

18. Hemmati-Brivanlou, A., Wright, D. A., and Melton, D. A. (1992) Embryonic expression and functional analysis of a *Xenopus* activin receptor. *Dev. Dyn.* **194,** 1–11.
19. Armes, N. A. and Smith, J. C. (1997) The ALK-2 and ALK-4 activin receptors transduce distinct mesoderm-inducing signals during early *Xenopus* development but do not cooperate to establish thresholds. *Development* **124,** 3797–3804.
20. Hemmati-Brivanlou, A., and Melton, D. A. (1994) Inhibition of activin receptor signaling promotes neuralization in *Xenopus. Cell* **77,** 273–281.
21. New, H. V., Kavka, A. I., Smith, J. C., and Green, J. B. (1997) Differential effects on *Xenopus* development of interference with type IIA and type IIB activin receptors. *Mech. Dev.* **61,** 175–186.
22. Ying, S.-Y., Zhang, Z., Furst, B., Batres, Y., Huang, G., and Li, G. (1997) Activins and activin receptors in cell growth. *Proc. Soc. Exp. Biol. Med.* **214,** 114–122.

4

Analysis of Mammary Gland Morphogenesis

Calvin D. Roskelley, Colleen Wu, and Aruna M. Somasiri

1. Introduction

Mammary gland development is marked by numerous cellular changes that begin in the embryo, continue postnatally, are modified at puberty, and come to full fruition during the adult cycles of pregnancy, lactation, and involution. These changes are initiated by hormones produced at distant sites and by the microenvironment within the gland itself. The latter, which includes soluble factors and the insoluble extracellular matrix (ECM), will be the focus of this chapter.

In the embryo, mammary gland morphogenesis begins when epidermal protrusions branch within an underlying adipocytic mesenchyme to form the preliminary ductal tree *(1)*. At birth, the epithelial and mesenchymal compartments of the developing gland are separated by a continuous layer of specialized ECM, the basement membrane *(2)*. The deposition of laminin and heparan sulfate proteoglycan into the basement membrane by the adipocytic mesenchyme is crucial for further epithelial development. In males, androgen-mediated conversion to a fibroelastic mesenchyme results in fibronectin and tenascin deposition that sparks ductal regression *(2–4)*. Thus, the specific molecular composition of the basement membrane is an important mediator of early ductal outgrowth.

Postnatally, a second phase of morphogenesis occurs as growth and branching of the ductal tree accelerates. Epithelial proliferation at puberty is directed by ovarian hormones that trigger mesenchymal production of locally acting growth factors such as transforming growth factor-α (TGF-α) *(5)*. Other factors produced by the mesenchyme alter ECM-dependent branching decisions. For example, localized production of TGF-β causes the deposition of a fibroelastic ECM that prevents branching along the walls of the ducts *(6)*. In contrast, site-specific side branching is induced by hepatocyte growth factor (HGF; *see* **ref. 7**). Despite the fact that HGF is produced by the mesenchyme, its phenotypic actions are mediated by receptors on the surface of epithelial cells *(8)*. Although it is not yet known whether HGF interacts directly with the ECM, it is clear that the growth factor alters interactions between mammary epithelial cell interactions and collagen matrices *(9,10)*.

Lobulo-alveolar development takes place during pregnancy. Estrogens, progestins, and prolactin are critical regulators of the massive epithelial proliferation that takes

From: *Methods in Molecular Biology, Vol. 136: Developmental Biology Protocols, Vol. II*
Edited by: R. S. Tuan and C. W. Lo © Humana Press Inc., Totowa, NJ

place early in this third morphogenic phase. This short-lived hyperplasia is kinetically coupled to the formation of small alveolar outpocketings that bud from the lateral margins of the terminal portions of the ductal tree. The latter process is regulated, at least in part, by another mesenchymally produced growth factor, neuregulin *(7)*. As alveolar budding proceeds, an intimate association between the epithelial cells and the underlying basement membrane is maintained. This association is absolutely required for further alveolar morphogenesis, milk protein production, and lactational secretion ***(11)***. Later, after weaning, the targeted destruction of the basement membrane induces epithelial apoptosis, which is the major force driving glandular involution ***(12–14)***.

We have long been interested in those interactions between the mammary epithelium and the surrounding basement membrane that regulate alveolar morphogenesis ***(15,16)***. Mammary epithelial cells isolated from mid-pregnant mice rapidly dedifferentiate in monolayer culture. However, when they are placed on a reconstituted basement membrane gel, these cells round up, aggregate, form cell–cell junctions, and become polarized. This produces small alveolar "mammospheres" that cavitate to form a central lumen. In an appropriate hormonal milieu, cells within these mammospheres fully differentiate. Specifically, the cells express milk protein genes, package the products in secretory vesicles, transport the vesicles to the apical cell surface, and release the vesicular products vectorially into the central lumen.

In an effort to identify the mechanisms responsible for regulating basement-membrane-dependent alveolar morphogenesis, we developed a battery of specialized culture models. First, we isolated a functional mouse mammary epithelial cell line that is not capable of depositing its own endogenous basement membrane. We have used this cell line for the following:

1. Suppress or enhance epithelial-to-mesenchymal transitions (EMT).
2. Manipulate cell shape.
3. Identify the basement membrane ligand and the integrin receptor responsible for milk protein gene expression.
4. Examine integrin-mediated signal transduction.
5. Determine how each of these events contributes to the complete alveolar morphogenesis that occurs during mammosphere formation.

These models are now described in detail (*see* **Fig. 1**).

2. Experimental Models and Approaches

2.1. Model 1: Epithelial-to-Mesenchymal Transition

2.1.1. Rationale

To examine ECM-dependent differentiation, we isolated a basement-membrane-responsive mouse mammary epithelial cell line by limited dilution cloning ***(17)***. This line, designated scp2, is made up of cells that contain keratin, but not vimentin, intermediate filaments. Approximately 90% of these cells respond to a combined treatment with lactogenic hormones and basement membrane by expressing the milk protein β-casein ***(18)***.

When they are routinely maintained and passaged in a two-dimensional (2-D) monolayer culture, a small proportion of scp2 cells "drift" from an epithelial to a fibroblastic morphology. The latter exhibit decreased keratin expression and increased vimentin

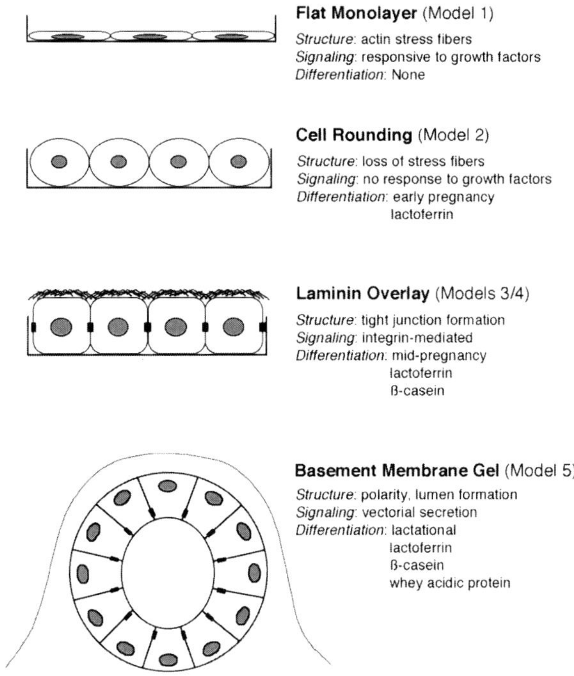

Fig. 1. A schematic representation of the five tissue culture models used to study structural and functional events that initiate alveolar morphogenesis by mammary epithelial cells.

expression and cannot be induced to express milk protein genes. In addition, cell–cell interactions are disrupted, and the cells invade basement membrane gels. It is not yet clear whether this phenotypic drift, which resembles an epithelial-to-mesenchymal transition (EMT) ***(19)***, indicates that the original scp2 clone had stem-cell characteristics. Regardless, this potential can be used to test the efficacy of suspected EMT-inducing agents or to examine epithelial—mesenchymal interactions in the emerging mixed cultures. However, during routine culture, it is best to minimize this conversion, because it leads to a decreased stringency in the response to exogenously added basement membrane.

We have identified three factors that contribute to the phenotypic drift of mammary epithelial cells during routine 2-D culture: high serum concentrations, growth past confluence, and overtrypsinization. Thus, we have developed a protocol to minimize these factors. Additionally, to ensure that our stocks remain epithelial, we use differential adhesion to remove fibroblastic cells at the time of passaging.

2.1.2. Methods

1. Rinse a subconfluent (approx 80%) monolayer culture of mammary epithelial cells (i.e., mouse epithelial clone scp2; *see* **ref. *17***) in a 100-mm dish with serum-free Dulbecco's modified Eagle's medium (DMEM)/F12 medium, maintain at 37°C for 10 min, and repeat 3×. This removes any adhesive serum factors from the cultures.
2. Rinse with Ca^{2+}/Mg^{2+}-free DMEM/F12 and maintain at 37°C for 10 min. This reduces cell–cell adhesion and dramatically decreases the time required for trypsinization.
3. Add 2 mL of 0.05% trypsin/0.02% EDTA in Hank's balance salt solution (HBSS) at room temperature and leave until the cells begin to round on the plate. This takes approx 2–5 min.

4. Remove trypsin, add 2 mL fresh trypsin, maintain at 37°C until all cells detach (5–15 min). Using this low concentration of trypsin (0.05%), the cells often detach as clusters, and, unless a specific experiment requires single cells, this is optimal, as it slows fibroblastic conversion.
5. Add 8 mL of 5% FBS containing DMEM/F12 to quench trypsin; centrifuge at 200*g* for 5 min.
6. Resuspend in 12-mL of attachment medium. This is 5% fetal bovine serum (FBS) DMEM/F12 containing insulin (5 μg/mL) and gentamycin (50 μg/mL). Insulin at this concentration acts as a growth factor supplement and prevents apoptosis in routine culture.
7. Plate in 1 × 100-mm dish, incubate 30 min at 37°C. Gently remove unattached cells and plate in 4 × 100-mm dishes (3 mL per dish, 1:4 split), add 7 mL of medium to each. The four dishes containing the cells that did not rapidly adhere to the original dish are epithelial. The rapidly adherent cells left behind in the original culture dish have converted to a fibroblastic phenotype.
8. After overnight attachment of the epithelial cell population, change the medium to routine growth medium. This is DMEM/F12 supplemented with 3% newborn calf serum, insulin, and gentamycin. This medium is changed every second day until the cells are once again subconfluent, which takes approx 4–6 d. The maintenance of the epithelial cells in medium containing a low percentage of calf serum is sufficient to ensure robust cell growth (with insulin), but it decreases the fibroblastic conversion observed when fetal serum is used. It is also important not to overgrow the monolayers past confluence to prevent fibroblastic conversion.
9. Fibroblastic cells in the original adherent dish can also be maintained with growth medium and passaged as above except that trypsinization times will be increased.

2.1.3. Experimental Uses of the Model

1. Phenotypic drift occurs in other mammary epithelial lines. Despite the fact that the HC-11 mouse mammary cell line was isolated by limited dilution cloning, it is heterogeneous, and only approx 10–20% of the cells are capable of expressing the milk protein genes ***(20)***. As a result of this heterogeneity, when HC-11 cultures are maintained at very high cell densities, they deposit an endogenous basement membrane. Thus, β-casein expression can be induced in the absence of exogenous basement membrane addition ***(21)***. This has proven very useful in experiments designed to elucidate functional lactogenic hormone signaling ***(22)***.
2. Mixing of epithelial and fibroblastic cultures. Epithelial and fibroblastic mouse mammary cell lines have been intentionally mixed to induce the deposition of endogenously produced extracellular matrix proteins. This was used to demonstrate that the basement membrane glycoprotein laminin is deposited at sites of direct contact between the two cell types ***(23)***.
3. Epithelial-to-mesenchymal transition and invasion. EMT plays an important role in tumorigenesis ***(24)***. Therefore, we have assayed the ability of various agents to facilitate this transition in scp2 mammary epithelial cells. Two efficient inducers of EMT are an activated form of the metalloprotease stomelysin-1 ***(25)*** and the integrin-linked kinase ***(26)***. The latter is a signaling molecule that disrupts integrin-mediated cell adhesion and accelerates the mesenchymal conversion of scp2 cells by decreasing E-cadherin expression and translocating the resulting free β-catenin to the nucleus.

2.2. Model 2: Cell-Shape Manipulation

2.2.1. Rationale

Mammary epithelial cells form flat, nondescript monolayers that do not express milk proteins after attachment to chemically crosslinked basement membrane matrices ***(27)***.

A critical differentiative component that is lost after attachment to such rigid matrices is cell rounding. This has been demonstrated by forcing the cells to round in the absence of ECM. Under the latter conditions, proliferation ceases and expression of the iron-binding milk protein lactoferrin is initiated ***(28)***. It is important to point out that this mechanical (i.e., mediated by cell shape) induction of differentiation is only partial. Expression of other milk protein genes does not occur. Cell rounding also acts to initiate the differentiation of hepatocytes ***(29)***, retinal pigment epithelial cells ***(30)***, keratinocytes ***(31)***, and steroidogenic cells ***(32)***.

Cell shape can be manipulated in a number of ways. Adherent cells can be maintained in suspension, which causes them to become spheroidal, although this often initiates apoptotic cell death within hours ***(33,34)***. Cells can also be plated on flexible filters. Alternatively, rounded cells can be produced by restricting cell spreading after attachment to a rigid substratum. This has been accomplished using "elastomeric stamps" that produce microscopic adhesive islands of precise area. Such methodology has been used to demonstrate definitively that cell shape alone regulates a shift between proliferative and differentiative phenotypes in hepatocytes ***(35)***. Generalized decreases in adhesivity, which allow for cell attachment but not for spreading, have also been used to produce rounded, differentiated keratinocytes ***(31)***. We have used the latter approach to culture rounded mammary epithelial cells. To do this, the cells are plated on the inert, nonadhesive substatum poly-hydroxyethyl methacrylate (polyHEMA; *see* **ref.** ***18***). The cells attach to low concentrations of this substratum and form rounded-cell clusters. Alternatively, single rounded cells can be produced by lowering the calcium concentration during cell plating to prevent cell–cell interactions ***(28)***.

2.2.2. Methods

1. Dissolve the antiadhesive polymer polyHEMA (Sigma Chemical Co., St. Louis, MO, USA) in 95% ethanol at 50 mg/mL. Maintain this stock in a water bath at 37°C overnight to ensure that all polyHEMA is in solution.
2. Dilute the stock polyHEMA solution 1:100 with 95% ethanol (0.5 mg/mL final concentration) and coat regular tissue culture dishes with 125 µL/cm^2 surface area (i.e., 1.0 mL/35-mm dish, 2.5 mL/60-mm dish, 5.0 mL/100-mm dish). Evaporate the ethanol overnight in a dry, sterile incubator at 37°C. This will leave a clear polyHEMA coating on the surface of the dish.
3. Plate scp2 (or other mammary epithelial) cells on polyHEMA-coated plates at 5×10^4 cells/cm^2 surface area (i.e., 4×10^5/35-mm dish, 1.0×10^6/60-mm dish, 2.0×10^6/100-mm dish) in DMEM/F12 medium containing 1% FBS and gentamycin (50 µg/mL) supplemented with the lactogenic hormones insulin (5 µg/mL), hydrocortisone (1.0 µg/mL), and prolactin (3 µg/mL).
4. Allow cells to attach overnight and then change to serum-free DMEM/F12 containing lactogenic hormones.
5. Lactoferrin expression is initiated within 48 h, and this can be assessed by Northern blotting, immunofluorescence, Western blotting, or immunoprecipitation of metabolically labeled cell proteins (*see* **refs.** ***18*** and ***28*** for specific methods and reagents).

2.2.3. Experimental Uses of the Model

1. The functional importance of changes to the cytoskeleton can be assessed in response to cell rounding. The loss of actin stress fibers induced by cell rounding or by treatment with the f-actin-disrupting agent cytochalasin D contributes to lactoferrin induction ***(28)***.

2. Cell rounding inhibits proliferative responses to growth factors. Insulin-mediated signal transduction is downregulated in rounded scp2 cells. Specifically, MAP kinase pathways are curtailed, which results in a downregulation of *ets* and AP1 transcription factor activities. Artificial stimulation of this pathway using an activated form of the *ras* oncogene prevents lactoferrin expression (C. Roskelley, unpublished observations).
3. Cell rounding transcriptionally activates the lactoferrin promoter. Stable transfection of a reporter gene construct has been used to demonstrate that cell rounding transcriptionally activates a 2.6-kb fragment of the mouse lactoferrin promoter ***(28)***.

2.3. Model 3: Integrin Interactions with the Basement Membrane

2.3.1. Rationale

When flat, nonfunctional mammary epithelial monolayers on tissue culture plastic are overlaid with a solution of basement membrane matrix, the cells round and cluster in a manner similar to that observed when they are plated on antiadhesive polyHEMA ***(17)***. Cells under both conditions also express lactoferrin. However, a crucial differentiative difference in the basement-membrane-overlaid condition is the expression of a second milk protein, β-casein ***(18)***. Induction of the latter is specifically initiated by laminin, a glycoprotein component of the basement membrane overlay ***(36)***.

2.3.2. Method

1. Plate scp2 (or other mammary epithelial cells) on tissue culture dishes at 2.5 × 10^4/cm^2 surface area (i.e., 2 × 10^5/35-mm dish, 5 × 10^6/60-mm dish, 1 × 10^6/100-mm dish) in DMEM/F12 medium containing 1% FBS supplemented with the lactogenic hormones insulin (5 µg/mL), hydrocortisone (1.0 µg/mL), and prolactin (3 µg/mL).
2. Allow cells to attach overnight and then change to serum-free DMEM/F12 containing lactogenic hormones for 24 h to remove growth factors. This ensures that the cells are not proliferating when the ECM overlay is added.
3. Dilute ECM to the desired concentration in cold DMEM/F12 containing lactogenic hormones. Basement membrane ECM (*see* **ref. *37***; Matrigel; Collaborative Research Inc., New Bedford, MA, USA) is thawed on ice and is diluted in the medium to a final total protein concentration of 150 µg/mL (approx 1:100 dilution). For purified laminin overlays, we routinely use a concentration of 50 µg/mL. Concentrations for laminin fragments and other ECM proteins vary (*see* **ref. *36*** for specific methods and reagents).
4. Warm the medium containing the ECM proteins to 37°C in a water bath, vortex briefly, mix by pipeting (always use plastic pipets when working with ECM-containing medium), and add the medium to the cells immediately. This constitutes the ECM overlay. Medium change the overlay every 2 d.
5. The cells will round and cluster after 2–3 d in response to the overlay. Lactoferrin and β-casein induction occurs soon thereafter, usually between 3 and 5 d. Expression of both milk proteins can be assessed by Northern blotting, immunofluorescence, Western blotting, or immunoprecipitation of metabolically labeled cell proteins (*see* **refs. *18*** and ***28*** for specific methods and reagents).

2.3.3. Experimental Uses of the Model

1. A specific domain of laminin is involved in β-casein induction. Overlays with purified ECM proteins were used to determine that laminin is the specific basement ligand that induces cell rounding and β-casein expression ***(36)***. A specific fragment of laminin, derived from the E3 domain, is able to inhibit overlay-mediated β-casein induction.

2. β-Casein promoter activation. A basement-membrane-responsive 160-bp element derived from the bovine β-casein promoter *(38)* is activated by the laminin overlay in this assay, as is a 300-bp fragment from the rat β-casein promoter *(39)*. These elements must be stably integrated into the mammary epithelial cell genome to be ECM responsive. In transient transfection assays, they are only lactogenic hormone responsive *(40)*.
3. More than one cell-surface receptor mediates the differentiative effects of a laminin overlay. Cell-rounding-mediated lactoferrin expression does not require integrin signaling *(28)*, whereas β-casein expression is integrin dependent *(41)*. Therefore, the same basement membrane glycoprotein induces the differentiative expression of two tissue-specific genes using at least two different receptors and/or signaling pathways.

2.4. Model 4: Integrin-Mediated Signaling

2.4.1. Rationale

Cell clustering and scp2 mammary epithelial cell rounding are necessary—but on their own, these morphological changes are not sufficient—to induce the β-casein gene expression *(18)*. A second integrin-mediated signal is also required. This signal can be assayed in isolation by overlaying preclustered cells with laminin. Under the latter conditions, β-casein expression is induced very quickly, within hours *(16)*. This short time frame makes it possible to identify specific transduction events that contribute to β-casein induction. This is an important and novel model system for assessing differentiative integrin signaling, as the great majority of other models focus on ECM-dependent cell attachment, spreading, migration, and/or proliferation.

2.4.2. Method

1. Plate scp2 (or other mammary epithelial) cells on polyHEMA-coated plates (*see* **Subheading 2.2.2.**) at 5×10^4/cm^2 surface area (i.e., 4×10^5/35-mm dish, 1.0×10^6/60-mm dish, 2.0×10^6/100-mm dish) in DMEM/F12 medium containing 1% FBS supplemented with lactogenic hormones insulin (5 µg/mL), hydrocortisone (1.0 µg/mL), and prolactin (3 µg/mL).
2. Allow cells to attach overnight and then change to serum-free DMEM/F12 containing lactogenic hormones; leave for 48 h. These "naked" cell clusters express lactoferrin but not β-casein.
3. Dilute laminin in cold, serum-free DMEM/F12 medium containing lactogenic hormones (*see* **step 1**). Warm medium to 37°C and add this overlay to the preclustered cells immediately. There is no further overt morphological change, but β-casein expression is rapidly initiated.

2.4.3. Experimental Uses of the Model

1. Cell rounding "primes" the cells to rapidly respond to laminin: Once they are rounded and clustered (i.e., on polyHEMA), a laminin overlay initiates rapid induction of β-casein expression within 4–8 h *(16,18)*. This rapid induction is regulated, at least in part, transcriptionally, when a 300-bp fragment of the rat β-casein promoter is activated using this protocol (C. Roskelley, unpublished observations).
2. Integrin-mediated tyrosine phosphorylation contributes to β-casein expression. The laminin overlay initiates an increase in tyrosine phosphorylation in at least seven protein species *(18)*, a signaling event that is required for efficient β-casein induction *(42)*.
3. Activation of the focal adhesion kinase. The laminin overlay causes increased phosphorylation of the focal adhesion kinase, which is an indication of functional activation *(16)*.

4. Tight-junction formation. Tight junctions are found in laminin-overlaid cell clusters. If these junctions are disrupted, β-casein induction does not occur (A. Somasiri and C. Roskelley, unpublished observations).

2.5. Model 5: Complete Alveolar Morphogenesis

2.5.1. Rationale

In vivo, lactoferrin and β-casein expression are initiated during early and mid-pregnancy, respectively. At these developmental stages, alveolar morphogenesis is not complete and the mammary epithelium is not lactational. In culture, laminin-overlaid cell clusters expressing lactoferrin and β-casein are also not lactational. There is no central lumen and no vectorial secretion. The latter only occurs when the cells are cultured as alveolus-like mammospheres on reconstituted basement membrane gels. Alveolar morphogenesis in culture also induces whey acidic protein (WAP) production, a third milk protein gene that is first expressed just prior to lactation in vivo.

Prior to mammosphere formation, the basement-membrane gel causes mammary epithelial cells to round up, aggregate, and form cell–cell junctions. Each of these morphogenic events has been examined in isolation using the models in **Subheadings 2.1.–2.4.** However, additional morphogenic changes must occur prior to lactational WAP expression. These include the acquisition of apical basal polarity, endogenous basement-membrane deposition, and cavitation of a central lumen.

2.5.2. Method

1. Thaw basement-membrane ECM (Matrigel, Collaborative Research) on ice. Place tissue culture dishes on ice in a laminar flow hood. Spread cold, undiluted liquid Matrigel on the bottom of the dish (200 μL/35-mm dish, 500 μL/60-mm dish, 1 mL/100-mm dish). Place dishes in a 37°C humidified incubator for 1 h prior to adding cells. This produces a malleable basement-membrane gel on the bottom of the dish.
2. Plate scp2 (or other mammary epithelial cells) on Matrigel-coated plates at $5 \times 10^4/cm^2$ surface area (i.e., 4×10^5/35-mm dish, 1.0×10^6/60-mm dish, 2.0×10^6/100-mm dish) in DMEM/F12 medium containing 1% FBS supplemented with the lactogenic hormones insulin (5 μg/mL), hydrocortisone (1 μg/mL), and prolactin (3 μg/mL). The cells attach to the gel and begin to form clusters almost immediately.
3. Allow cells to attach overnight and then change to serum-free DMEM/F12 containing lactogenic hormones. This medium should be changed every second day. The cells begin to express milk proteins after 2–3 d. Polarized "mammospheres," which secrete milk proteins vectorially into a central lumen, form after 5–7 d.

2.5.3. Experimental Uses of the Model

1. Downregulation of growth factor production. As mammospheres form, the expression of autocrine growth factors by the mammary epithelial cells is downregulated. The growth factors affected include TGF-α *(43)*, which causes alveolar hyperplasia when force-expressed in transgenic mice *(44)*, and TGF-β *(45)*, which inhibits ductal branching during postpubertal development *(6)*.
2. Disruption of integrin signaling initiates apoptosis. Weaning triggers a massive apoptosis of the lactational mammary epithelium that drives alveolar involution in vivo *(14)*. This process is initiated by a metalloprotease-mediated destruction of the basement membrane *(12,13,46)*. Forced overexpression of an activated metalloprotease or function-blocking integrin antibodies both initiate apoptosis in the cells of cultured mammospheres *(47,48)*.

3. Distinguishing between normal and tumor-derived mammary epithelial cells. In (2-D) monolayer culture, it is very difficult to distinguish normal from tumor-derived human mammary epithelial cells. However, when they are cultured within basement membrane gels, normal cells form polarized 3-D mammospheres, whereas tumor-derived cells form loose, disorganized proliferating masses that do not exhibit apical–basal polarity ***(49)***. This differential response to the basement membrane is mediated by altered integrin expression and signaling ***(50)***. Similarly, forced-expression of oncogenes causes morphogenic changes in 3-D culture that are integrin dependent ***(51)***.

3. Conclusion

Mammary gland development is directed by long-range-acting hormones and microenvironmental factors within the gland itself. One component of the latter, the basement membrane, plays a key role in alveolar morphogenesis throughout developmental cycles of pregnancy, lactation, and involution. In culture, cellular interactions with the basement membrane induce the formation of "mammospheres" which resemble lactational alveoli in vivo. We have developed five specialized culture models to examine individual aspects of this ECM-dependent process, and experimental manipulation of each model has been used to identify many of the mechanisms responsible for regulating alveolar morphogenesis. Not surprisingly, the co-optation of these regulators plays a significant role in the emergence of a number of functionally and structurally important tumor-associated phenotypes. These include uncontrolled proliferation, disruption of cell–cell junctions, loss of epithelial polarity, and increased invasiveness.

Acknowledgments

Work in the authors' laboratory is supported by the Canadian Breast Cancer Research Initiative, the National Cancer Institute of Canada, and the British Columbia Health Research Foundation.

References

1. Sakakura, T. (1983) Epithelial–mesenchymal interactions in mammary gland development and its perturbation in relation to tumorigenesis, in *Understanding Breast Cancer* (Rich, M. A., Hager, J. C., and Furmanski, P., eds.), Marcel Dekker, New York, pp. 261–284.
2. Kimata, K., Sakakura, T., Inaguma, K., Kato, M., and Nishizuka, Y. (1985) Participation of two different mesenchymes in the developing mouse mammary gland: synthesis of basement membrane components by fat pad precursor cells. *J. Embryol. Exp. Morphol.* **89,** 243–257.
3. Hueberger, B., Fitzka, I., Wasner, G., and Kratochwil, K. (1982) Induction of androgen receptor formation by epithelium–mesenchyme interaction in embryonic mouse mammary gland. *Proc. Natl. Acad. Sci. USA* **79,** 2957–2961.
4. Chiquet-Ehrismann, R., Mackie, E. J., Pearson, C. A., and Sakakura, T. (1986) Tenascin: an extracellular matrix protein involved in tissue interactions during fetal development and carcinogenesis. *Cell* **47,** 131–139.
5. Snedeker, S. M., Brown, C. F., and DiAugustine, R. P. (1991) Expression and functional properties of transforming growth factor α and epidermal growth factor during mouse mammary gland ductal morphogenesis. *Proc. Natl. Acad. Sci. USA* **88,** 276–280.
6. Silberstein, G. B., Flanders, K. C., Roberts, A. B., and Daniel, C. W. (1992) Regulation of mammary morphogenesis: evidence for extracellular matrix-mediated inhibition of ductal budding by transforming growth factor-beta 1. *Dev. Biol.* **152,** 354–362.

7. Yang, Y., Spitzer, E., Meyer, D., Sachs, M., Niemann, C., Hartmann, G., Weidner, K. M., Birchmeier, C., and Birchmeier, W. (1995) Sequential requirement of hepatocyte growth factor and neuregulin in morphogenesis and differentiation of the mammary gland. *J. Cell Biol.* **131,** 215–226.
8. Brinkmann, V., Foroutan, H., Sachs, M., Weidner, K. M., and Birchmeier, W. (1995) Hepatocyte growth factor/scatter factor induces a variety of tissue-specific morphogenic programs in epithelial cells. *J. Cell Biol.* **131,** 1573–1586.
9. Niranjan, B., Buluwela, L., Yant, J., Perusinghe, N., Atherton, A., Phippard, D., Dala, T., Gusterson, B., and Kamalati, T. (1995) HGF/SF: a potent cytokine for mammary growth, morphogenesis and development. *Development* **121,** 2897–2908.
10. Soriano, J. V., Pepper, M. S., Nakamura, T., Orci, L., and Montesano, R. (1995) Hepatocyte growth factor stimulates extensive development of branching duct-like structures by cloned mammary gland epithelial cells. *J. Cell Sci.* **108,** 413–431.
11. Howlett, A. R. and Bissell, M. J. (1993) The influence of tissue microenvironment on the development and function of mammary epithelium. *Epithelial Cell Biol.* **2,** 79–89.
12. Talhouk, R. S., Chin, J. R., Unemori, E. N., Werb, Z., and Bissell, M. J. (1991) Proteinases of the mammary gland: developmental regulation in vivo and vectorial secretion in culture. *Development* **112,** 439–449.
13. Talhouk, R. S., Bissell, M. J., and Werb, Z. (1992) Coordinated expression of extracellular matrix-degrading proteinases and their inhibitors regulates mammary epithelial function during involution. *J. Cell Biol.* **118,** 1271–1282.
14. Strange, R., Li, F., Saurer, S., Burkhardt, A., and Friis, A. A. (1992) Apoptotic cell death and tissue remodelling during mouse mammary gland involution. *Development* **115,** 49–58.
15. Bissell, M. J., Hall, G. J., and Parry, G. (1982) How does the extracellular matrix direct gene expression? *J. Theoret. Biol.* **99,** 31–68.
16. Roskelley, C. D. and Bissell, M. J. (1995) Dynamic reciprocity revisited: a continuous, bidirectional flow of information between cells and the extracellular matrix regulates mammary epithelial cell function. *Biochem. Cell Biol.* **73,** 391–397.
17. Desprez, P. Y., Roskelley, C. D., and Bissell, M. J. (1993) Isolation of functional cell lines from a mouse mammary epithelial cell strain: the importance of basement membrane and cell–cell interactions. *Mol. Cell. Differ.* **1,** 99–110.
18. Roskelley, C. D., Desprez, P. Y., and Bissell, M. J. (1994) Extracellular matrix-dependent tissue-specific gene expression in mammary epithelial cells requires both physical and biochemical signal transduction. *Proc. Natl. Acad. Sci. USA* **91,** 12,378–12,382.
19. Hay, E. D. (1995) An overview of epithelio-mesenchymal transformation. *Acta Anatomica* **154,** 8–20.
20. Ball, R. K., Friis, R. R., Schoenenberger, C. A., Doppler, W., and Groner, B. (1988) Prolactin regulation of B-casein gene expression and of a cytosolic 120-kd protein in a cloned mouse mammary epithelial cell line. *EMBO J.* **7,** 2089–2095.
21. Chammas, R., Taverna, D., Cell, N., Santos, C., and Hynes, N. E. (1994) Laminin and tenascin assembly regulate HC-11 mouse mammary cell differentiation. *J. Cell. Sci.* **107,** 1031–1040.
22. Wortmann, M., Cella, N., Hofer, P., Groner, B., Lieu, X., Heninghausen, L., and Hynes, N. E. (1996) Lactogenic hormone activation of Stat5 and transcription of the β-casein gene in mammary epithelial cells is independent of p42 ERK2 mitogen-activated protein kinase activity. *J. Biol. Chem.* **271,** 31,863.
23. Reichmann, E., Ball, R., Groner, B., and Friss, R. R. (1989) New mammary epithelial and fibroblastic cell clones in co-culture form structures competent to differentiate functionally. *J. Cell Biol.* **108,** 1127–1138.

24. Birchmeier, C., Birchmeier, W., and Brand-Sabri, B. (1996) Epithelial-mesenchymal transitions in cancer progression. *Acta Anatomica* **156,** 217–226.
25. Lochter, A., Galsosy, S., Muschler, J., Freedman, N., Werb, Z., and Bissell, M. J. (1997) Matrix metalloproteinase stromelysin-1 triggers a cascade of molecular alterations that leads to stable epithelial-mesenchymal conversion and pre-malignant phenotype in mammary epithelial cells. *J. Cell Biol.* **139,** 1861–1872.
26. Novak, A., Hsu, S., Leung-Hagesteijn, C., Radeva, G., Papkoff, J., Montesano, R., Roskelley, C., Grosschedl, R., and Dedhar, S. (1998) Cell adhesion and the integrin linked kinase (ILK) regulate the LEF-1 and B-catenin signalling pathways. *Proc. Natl. Acad. Sci. USA* **95,** 4374–4379.
27. Li, M., Aggeler, J., Farson, D. A., Hatier, C., Hassell, J., and Bissell, M. J. (1987) Influence of reconstituted basement membrane and its components on casein gene expression and secretion. *Proc. Natl. Acad. Sci. USA* **82,** 1419–1493.
28. Close, M. J., Howlett, A. R., Roskelley, C. D., Desprez, P. Y., Bailey, N., Rowning, B., Teng, C. T., Stampfer, M. R., and Yaswen, P. (1997) Lactoferrin expression in mammary epithelial cells is mediated by changes in cell shape and actin cytoskeleton. *J. Cell Sci.* **110,** 2861–2871.
29. Mooney, D., Hansen, L., Vacanti, J., Langer, R., Farmer, S., and Ingber, D. (1992) Switching from differentiation to growth in hepatocytes: control by extracellular matrix. *J. Cell. Physiol.* **151,** 497–505.
30. Opas, M. (1989) Expression of the differentiated phenotype by epithelial cells is regulated by both biochemistry and mechanics of substratum. *Dev. Biol.* **131,** 281–293.
31. Watt, F., Jordan, P. W., and O'Neill, C. H. (1988) Cell shape controls terminal differentiation of human epidermal keratinocytes. *Proc. Natl. Acad. Sci. USA* **85,** 5576–5588.
32. Roskelley, C. D. and Auersperg, N. (1993) Mixed parenchymal-stromal populations of rat adrenocortical cells support the proliferation and differentiation of steroidogenic cells. *Differentiation* **55,** 37–45.
33. Meredith, J. E., Fazeli, B., and Schwartz, M. A. (1993) The extracellular matrix as a cell survival factor. *Mol. Biol. Cell* **4,** 953–961
34. Frisch, S. M. and Francis, H. (1994) Disruption of epithelial cell–cell interactions induces apoptosis. *J. Cell Biol.* **124,** 619–626.
35. Wang, N, Butler, J. P., and Ingber, D. E. (1993) Mechanotransduction across the cell surface and through the cytoskeleton. *Science* **260,** 1124.
36. Streuli, C. H., Schmidhauser, C., Bailey, N., Yurchenco, P., Skubitz, P. N., Roskelley, C. D., and Bissell, M. J. (1995) Laminin mediates tissue-specific gene expression in mammary epithelia. *J. Cell Biol.* **129,** 591–603.
37. Kleinman, H. K., McGarvey, M. L., Hassell, J. R., Star, V. L., Cannon, F. B., Laurie, G. W., and Martin, G. R. (1986) Basement membrane complexes with biological activity. *Biochemistry* **25,** 312–318.
38. Schmidhauser, C., Bissell, M. J., Myers, C. A., and Casperson, G. F. (1990) Extracellular matrix and hormones transcriptionally regulate bovine β-casein 5' sequences in stably transfected mouse mammary cells. *Proc. Natl. Acad. Sci. USA* **87,** 9118–9122.
39. Doppler, W., Welte, T., and Phillipp, S. (1995) CCAAT/enhancer-binding protein isoforms are expressed in mammary epithelial cells and bind to multiple sites in the β-casein gene promoter. *J. Biol. Chem.* **270,** 17,962–17,969.
40. Myers, C. A., Schmidhauser, C., Mellintin-Michelotti, J., Fragoso, G., Roskelley, C. D., Casperson, G., Mossi, R., Pujuguet, P., Hager, G., and Bissell, M. J. (1998) Characterization of BCE-1: a transcriptional enhancer regulated by prolactin and extracellu-

lar matrix and modulated by the state of histone acetylation. *Mol. Cell Biol.* **18,** 2184–2195.

41. Streuli, C. H., Bailey, N., and Bissell, M. J. (1991) Control of mammary epithelial differentiation: the separate roles of cell–substratum and cell–cell interaction. *J. Cell Biol.* **115,** 1383–1385.
42. Roskelley, C. D., Srebrow, A., and Bissell, M. J. (1995) A hierarchy of ECM-mediated signalling regulates tissue-specific gene expression. *Curr. Opin. Cell Biol.* **7,** 736–747.
43. Lin, C. Q., Dempsey, P. J., Coffey, R. J., and Bissell, M. J. (1995) Extracellular matrix regulates whey acidic protein gene expression by suppression of TGF-alpha in mouse mammary epithelial cells—Studies in culture and in transgenic mice. *J. Cell Biol.* **129,** 1115–1125.
44. Matsui, Y., Halter, S. A., Holt, J. T., Hogan, B. L., and Coffey, R. J. (1990) Development of mammary hyperplasia and neoplasia in MMTV-TGF alpha transgenic mice. *Cell* **61,** 1147.
45. Streuli, C. H., Schmidhauser, C., Kobrin, M., Bissell, M. J., and Derynck, R. (1993) Extracellular matrix regulates expression of the TGFβ1 gene. *J. Cell Biol.* **120,** 253–260.
46. Sympson, C. J., Talhouk, R. S., Alexander, C. M., Chin, S. K., Clift, S. M., Bissell, M. J., and Werb, Z. (1994) Targeted expression of stromelysin-1 in mammary gland provides evidence for a role of proteinases in branching morphogenesis and the requirement for an intact basement membrane for tissue-specific gene expression. *J. Cell Biol.* **125,** 681–692.
47. Boudreau, N., Sympson, C. J., Werb, Z., and Bissell, M. J. (1995) Suppression of ICE and apoptosis in mammary epithelial cells by extracellular matrix. *Science* **267,** 891–895.
48. Boudreau, N., Werb, Z., and Bissell, M. J. (1996) Suppression of apoptosis by basement membrane requires three-dimensional tissue organization and withdrawal from the cell cycle. *Proc. Natl. Acad. Sci. USA* **93,** 3509–3513.
49. Peterson, O. W., Ronnov-Jessen, L., Howlett, A. R., and Bissell, M. J. (1992) Interaction with basement membrane serves to rapidly distinguish growth and differentiation pattern of normal and malignant human breast epithelial cells. *Proc. Natl. Acad. Sci. USA* **89,** 9064–9068.
50. Weaver, V. M., Peterson, O. W., Wang, F., Larabell, C. A., Briand, P., Damsky, C., and Bissell, M. J. (1997) Reversion of the malignant phenotype of human breast cells in three-dimensional culture and in vivo by integrin blocking antibodies. *J. Cell Biol.* **137,** 231–245.
51. Berdichevsky, F. Wetzels, R. Shearer, M., Maringnone, S., Ramaekers, F. C. S., and Taylor-Papdimitriou, J. (1994) Integrin expression in relation to cell phenotype and malignant change in the human breast. *Mol. Cell Differ.* **2,** 255–274.

5

Specification of Cardiac Mesenchyme and Heart Morphogenesis In Vitro

H. Joseph Yost

1. Introduction

One of the key issues in developmental biology is the question of specification. Specification is the commitment of a group of cells to proceed down a developmental pathway without further extraneous signals. In order identify the development period in which cells have acquired all the necessary information to become committed to a developmental pathway, such as the formation of an organ or tissue, cells are explanted from embryos at various stages and cultured in vitro in a neutral medium. Analysis of the tissues that are formed by the explants allows the investigator to define the developmental period during which specification of a tissue is completed. From these results, one can infer that if cell-to-cell signals are involved in specification of the tissue, they must have occurred earlier than the defined developmental period.

It is important to clarify the distinctions between analysis of specification by explantation and analysis of cell fate by fate-mapping. Fate-mapping identifies, at a specific stage in development, the location of cells that will contribute to an organ or tissue at a later stage in development. Fate-mapping does not identify the developmental period during which cells acquire a specific developmental identity nor does it identify the mechanisms by which cell identities are conferred. In contrast, explant analysis identifies the extent to which cells are committed to a developmental pathway *at the time of their explantation*. Because cells might be exposed to additional signals at subsequent stages of development, explant analysis does not necessarily define the developmental fate of the cells if they were left in an unperturbed embryo.

The explantation of cells from *Xenopus* embryos in order to assess their ability to form cardiac structures in vitro has a long history and has allowed investigators to identify early inductive interactions that lead to cardiac mesoderm formation ***(1,2)***, developmental restriction of the cardiogenic field ***(3)***, regulation of looping morphogenesis ***(4)***, and specification of cardiac left–right orientation ***(5)***. Early *Xenopus* embryonic cells contain internal yolk platelets that provide nutrition, which allows explants to be cultured in minimal culture, avoiding the confounding effects of external supplements. In addition, the robust formation of a looped, beating cardiac tube from a

From: *Methods in Molecular Biology, Vol. 136: Developmental Biology Protocols, Vol. II*
Edited by: R. S. Tuan and C. W. Lo © Humana Press Inc., Totowa, NJ

sheet of mesoderm cultured in vitro serves as a valuable teaching tool and a dramatic example of morphogenesis for students of developmental biology.

2. Materials

1. Culture media: Embryos and explants are grown in 1/3 × Marc's Modified Ringers solution (MMR), referred to hereafter as R/3. 1× MMR is 0.1*M* NaCl, 1.8 m*M* KCl, 2 m*M* $CaCl_2$, 1 m*M* $MgCl_2$, and 5 m*M* HEPES, pH 7.8. A 40 × stock salt solution is prepared without HEPES and pH adjustment, filtered, and stored. The 40 × stock is diluted to 1 × strength, HEPES is added (1.2 g/L) and the 1 × stock is brought to pH 7.8. This 1 × MMR can be stored at 4°C and is diluted to 1/3 × to make R/3. For explant culture, 50 µg/mL gentamicin sulfate is added to R/3.
2. Fixative: MEMFA is 0.1*M* MOPS, pH 7.4, 2 m*M* EGTA, 1 m*M* $MgSO_4$, and 3.7% formaldehyde. A 5 × stock without formaldehyde is filtered and stored for extended periods at 4°C, protected from light. Stock is diluted and formaldehyde is added on the day of use.
3. Forceps: The leading tip of dissecting forceps (Dumont #5, Roboz Surgical, Rockville MD) is beveled at 45° by gently grinding against a flat diamond-coated knife stone. The angle presents a broad surface at the tip to the forceps, which facilitates removal of the vitelline envelope without puncturing the embryo.
4. Tungsten wire knives: Handles are made from 9-in Pasteur pipets. The glass pipets are heated in a flame near the middle of the narrow half of the pipet, drawn out, and snapped off to form a narrow tip at the end of the handle. A 2-cm length of tungsten wire (size 0.001 Ω/ft; California Fine Wire Company, Grover City, CA) is inserted approximately 1 cm into the tip and sealed by heating briefly over a flame. Heating should be brief, sufficient to melt the glass tip around the tungsten wire but not to sublimate the wire. Several knives should be made in advance. Immediately before use, the wire is trimmed with fine scissors so that approximately 0.3 cm extends from the glass handle.
5. Agarose dishes: Dissections are performed in 60 × 15-mm Petri dishes coated on the bottom with agarose. Agarose (1 g, medium grade) in 100 mL water is melted and poured to the level of one-third of the dish; this amount is sufficient to coat the bottoms of approximately 40 dishes. After solidifying, the agarose dishes can be returned to the plastic sleeve (or wrapped in parafilm or plastic wrap) to prevent dehydration and stored at 4°C for up to several months. Before use, dishes are returned to room temperature and filled with R/3.
6. Dissecting microscopes: A good dissecting binocular microscope with a wide range of magnifications (from 6 × to 50×) and fiber light illumination from the sides greatly facilitates embryo and explant manipulation. A stage that can be cooled to 16°C by recirculating water serves to slow embryo development, allowing a longer period for dissections within the desired embryo stages.
7. Protease. Protease (Qiagen, Valencia, CA) is resuspended to 20 mg/mL in R/3 and stored at 4°C.

3. Methods

1. Preparation of embryos: *Xenopus laevis* females are injected in the dorsal lymph sac with 50 units of pregnant mare's serum gonadotropin (Sigma) 2 d before egg laying and 800 units of human chorionic gonadotropin (Sigma) 12–16 h before egg laying. Injected females are kept at 16°C overnight. In the morning, eggs are squeezed from the ovulating female into a Petri dish and immediately fertilized with a minced testis suspension in R/3.
2. Removal of jelly coat: The jelly coat must be completely removed to make the embryos accessible to dissection. Embryos are dejellied by pouring them into an Erlenmeyer flask, decanting the R/3, adding 2% cysteine (pH 8.0, in water), and swirling until embryos

rapidly settle to the bottom. Then, the cysteine solution is decanted, embryos are rinsed three times by swirling in and decanting R/3. Embryos in R/3 are then poured into a clean Petri dish. The jelly coat can be removed at any time after 10 min of fertilization. Embryos tend to be healthier if the jelly coat is left intact for the first few cell division cycles, and then removed so that healthy embryos can be separated from unfertilized eggs or abnormal embryos. Embryos are raised at temperatures ranging from 16°C to 22°C, which allows targeted stages to be obtained at convenient times. Embryos are staged according to Nieuwkoop and Faber *(6)*.

3. Removal of vitelline membrane (*see* **Note 1**): Place embryos at appropriate stage in agar dishes filled with R/3. Remove the vitelline by pinching the top of the embryo at the surface, allowing the ground angle of the forceps to be tangential to the sphere of the embryo. With the second forceps, pinch the vitelline close to the first forceps, and pull the two forceps away from each other. This should release the embryo from the vitelline. For some investigators and some batches of embryos, it is easier to remove embryos without damage at some stages (early blastula and tailbud) than others (gastrula and early neurula). If this is the case, vitellines can be removed early and embryos can be incubated in R/3 in agarose-coated dishes until the appropriate developmental stage is obtained.
4. Microdissection of precardiac mesoderm: As indicated in **Fig. 1A**, a series of cuts through the embryo with tungsten wire knives and/or the tips of forceps quickly releases the region of the embryo containing precardiac mesoderm, which is subsequently separated from surrounding tissues. First, place the embryo on its side. Cuts 1 and 2 can be done with the wire knife or simultaneously with the tips of the forceps. With practice, the tips of the forceps can be spaced (and measured through a reticule in the microscope eyepiece) so that the size of the explants are consistent from embryo to embryo. Cut 3 releases a block of tissue, the anterior ventral region of the embryo, which contains an outer ectoderm layer, a middle layer that consists of precardiac mesoderm, and a deep layer of endoderm. The positioning of cut 3 will determine the extent of lateral plate mesoderm included in the explant. Examination of Nkx2.5 expression patterns by *in situ* hybridization in whole-mount embryos *(7)* is useful to align the boundaries of the cuts with the region of cardiogenic mesoderm for the stages under consideration. In open neural plate stages (e.g., N&F stage 15), the precardiac mesoderm is a single layer of translucent mesoderm cells which tightly adheres to the lightly pigmented ectoderm layer. The deep, opaque, yolk-filled endoderm layer can be peeled away from the mesoderm without disturbing in vitro cardiac morphogenesis *(5)*. To remove endoderm from the explant, begin by peeling the deeper layer at the edge of cut 2, the posterior end of the explant. The layers will readily separate toward the anterior edge (near cut 1), at which the endoderm is more tightly adhered to the ectoderm and mesoderm. This junction can be cut away with a tungsten knife.
5. Culturing cardiac explants (*see* **Note 2**): The edges of the ectoderm begin to curl inward immediately after explantation. The ventral midline of the explant appears to flex (and can be assisted with forceps) so that the lateral edges begin to join; that is, the edge of cut 3 on the left joins with the edge of cut 3 on the right side. After approximately 30 min, the edges have sealed together, forming an oblong explant in which general anterior–posterior relationships are maintained. Loose cells and debris are removed from the vicinity with a Pasteur pipet. Explants are maintained in R/3 in agarose dishes until the edges of the explants have sealed. After this, the explants can be transferred with a Pasteur pipet to R/3 in dishes without an agarose base. It is convenient to maintain up to four explants in each well of a covered 24-well culture dish. Explants can be cultured at temperatures between 16°C and 22°C. Control embryos, siblings of the embryos from which explants

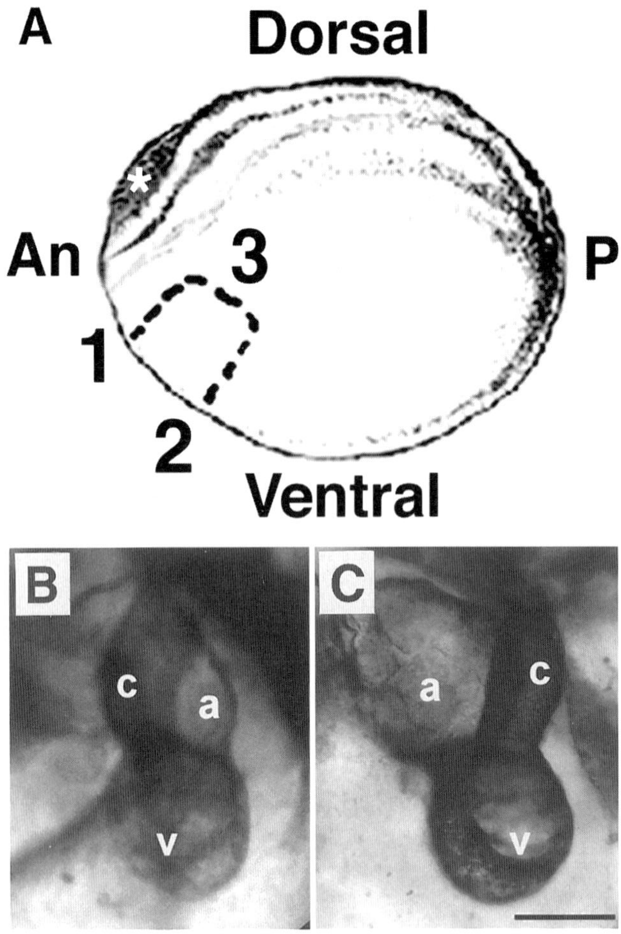

Fig. 1. Explantation of cardiogenic region from stage 18 (open neural plate) *Xenopus* embryos. **(A)** Diagram depicts left lateral view of an embryo at stage 18, with dorsal at the top, anterior (An) to the left, and posterior (P) to the right. The anterior neural plate is indicated by an asterisk. Cardiogenic region is removed by cuts described in the text and depicted as numbers 1 to 3. **(B,C)** Photomicrographs of explants cultured to sibling embryo stage 45, fixed and stained with MF20 antibody, in which the left–right orientation of cardiac looping is normal (B) or inverted (C). The atrium (a), conus or outflow (c), and ventricle (v) are indicated. Scale bar represents 50 μm. Adapted with permission from (**ref. 5**).

were derived, are maintained in parallel to assess normal cardiac development in vivo. Usually, explants are cultured until control embryos are between stages 42 and 45. By these stages, the outer ectoderm layer of the explant has become transparent, allowing direct observation of cardiac morphogenesis in the explant.

6. Analysis of in vitro cardiac development: Once the outer layer of the explant (ectoderm) has cleared, cardiac morphogenesis can be directly observed. The direction of rhythmic pulsing can be traced by direct observation and by videomicroscopy *(4)*. The left–right orientation of cardiac tube looping can be determined by comparison of the orientation of the three chambers (**Fig. 1B**,**C**) in living explants *(5)*. Explants are fixed in MEMFA for 2 h and stepped through an series of rinses of increasing ethanol concentrations, and they can

be stored in 100% methanol at –20°C for several months. Explants can be analyzed by immunohistochemistry or by RNA *in situ* hybridization by standard methods *(8,9)*. Stained explants are cleared in benzyl benzoate : benzyl alcohol (2:1) for photography.

4. Notes

1. Many find that the most difficult part of the explant protocol is removal of the vitelline membrane. An alternative to removal of the vitelline membrane by forceps is to loosen the vitelline membrane by light treatment with protease. Place 1.9 mL R/3 and 0.1 mL of 20 mg/mL protease in a 2-mL Eppendorf flat-bottom tube. Add approximately 30 embryos and cap the tube so that air bubbles are not trapped. Put the tube on end-over-end rotator for 2–4 min, checking embryos intermittently. Remove embryos to an agarose-coated dish and remove loosened vitelline membranes with forceps. Protease solution can be stored at 4°C for subsequent use. If protease is used to remove the vitelline membrane, it would be best do so well before the stage of explantation, so that protease can be rinsed away.
2. During transfers and culturing, explants and embryos without vitelline membranes must not contact the medium meniscus, as surface tension will lyse cells. In transfer pipets, draw up some medium before drawing up embryos or explants, and insert the tip of the pipet into medium before expelling embryos or explants from pipet.

Acknowledgment

I thank Dr. Maria Danos for participation in development and refinement of this protocol and Dr. Jeff Essner for discussion of protease treatment and assistance with the figure. This work is supported in part by a grant from the National Institutes of Health (R01 HL57840) and an American Heart Association Established Investigator award.

References

1. Sater, A. K. and Jacobson, A. G. (1990) The role of the dorsal lip in the induction of heart mesoderm in Xenopus laevis. *Development* **108,** 461–470.
2. Nascone, N. and Mercola, M. (1995) An inductive role for the endoderm in *Xenopus* cardiogenesis. *Development* **121,** 515–523.
3. Sater, A. K. and Jacobson, A. G. (1990) The restriction of the heart morphogenetic field in Xenopus laevis. *Dev. Biol.* **140,** 328–336.
4. Yost, H. J. (1990) Inhibition of proteoglycan synthesis eliminates left–right asymmetry in *Xenopus laevis* cardiac looping. *Development* **110,** 865–874.
5. Danos, M. C. and Yost, H. J. (1996) Role of notochord in specification of cardiac left–right orientation in zebrafish and Xenopus. *Dev. Biol.* **177,** 96–103.
6. Nieuwkoop, P. D. and Faber, J. (1967) *Normal Table of* Xenopus laevis *(Daudin)*. North-Holland, Amsterdam.
7. Tonissen, K. F., Drysdale, T. A., Lints, T. J., et al. (1994) XNkx-2.5, a Xenopus gene related to Nkx-2.5 and tinman: evidence for a conserved role in cardiac development. *Dev. Biol.* **162,** 325–328.
8. Klymkowsky, M. W. and Hanken, J. (1991) Whole-mount staining of *Xenopus* and other vertebrates. *Methods Cell Biol.* **36,** 419–441.
9. Harland, R. M. (1991) In situ hybridization: an improved whole-mount method for Xenopus embryos. *Methods Cell Biol.* **36,** 685–695.

6

Craniofacial Development and Patterning

Harold Slavkin, Glen Nuckolls, and Lillian Shum

1. Introduction

During early craniofacial development, cranial neural crest cells emigrate from segmentally distinct divisions of the hindbrain (rhombomeres) to populate the various branchial arches and subsequently differentiate into multiple neuronal and non-neuronal cell lineages of the head and neck region. Organ culture of the mandibular portion of the developing first branchial arch *(1,2)* facilitates investigations of molecular signaling events of crest cell differentiation during chondrogenesis, osteogenesis, odontogenesis, and myogenesis *(3–5)*. In addition, organs explanted in culture are assessable physically by microdissections, implantations, transplantations, and microinjections, and can be readily manipulated by antisense oligonucleotides strategies, viral delivery, and immunoperturbation methodologies.

Cell growth and differentiation are often examined at the animal model level using transgenic overexpression and targeted disruption approaches or at cell culture level using transfection and infection techniques. However, organ culture methods provide unique opportunities for the examination of time- and position-dependent differential gene expression related to pattern formations and instructive tissue interactions. Explant organ culture circumvents problems with whole animal modeling, such as lengthy and labor-intensive preparations and early embryonic lethality. It also avoids many of the cell culture artifacts, such as adherence to plasticware and loss of normal cell–cell and/or cell–extracellular matrix interactions.

This protocol describes a methodology that employs a serumless, chemically defined medium permissive for normal differentiation programs; both morphogenesis and cytodifferentiation are comparable to in vivo development. With the absence of confounding variations introduced by serum factors, this protocol is particularly attractive for studies of endogenous growth factors and their cognate receptors. Organ culture coupled with implantation of growth-factor-soaked beads allow for focal delivery of exogenous ligands, which offers a number of approaches. Therefore, this protocol enables a venue for developmental biologists interested in studies of intrinsic morphoregulatory programs during critical stages of early vertebrate embryogenesis.

From: *Methods in Molecular Biology, Vol. 136: Developmental Biology Protocols, Vol. II*
Edited by: R. S. Tuan and C. W. Lo © Humana Press Inc., Totowa, NJ

2. Materials

2.1. Organ Culture

1. Mouse embryos: Timed pregnant female mice can be purchased from laboratory animal suppliers or bred at in-house facilities. Males and females are caged together overnight; the presence of the vaginal sperm plug on the next morning indicates mating and the day designated as gestation day 0 (*see* **Notes 1–3**).
2. Microdissecting instruments: One pair of surgical scissors, one pair of blunt forceps, one pair of spring scissors, two pairs of fine forceps #55, and disposable microscalpels.
3. BGJb medium: Fitton-Jackson modification (cat. no. 12591-038; Life Technologies, Gaithersburg, MD). Basic BGJb medium can be purchased, but the culture medium should be made just prior to culture. Aliquot basic BGJb medium in 10-mL aliquots and store at 4°C. Just prior to use, supplement with 0.1 µg/mL ascorbic acid, 100 U penicillin, and 100 U streptomycin. Work in the laminar flow hood to maintain sterility.
4. Hanks' balanced salt solution (HBSS): Without phenol red (cat. no. 14025-092; Life Technologies).
5. Supporting filter: Type AA, 0.8-µm pore size (cat. no. AABP-047-00; Millipore, Bedford, MA). Filters are supplied as 47-mm diameter discs. First, stack approx 5 filters together, use a hole punch to punch out 6-mm diameter discs and collect them in a 500-mL glass beaker. Wet and rinse the filters with 5 changes of 200-mL distilled, deionized water (ddH_2O). Boil the filters in 200 mL of ddH_2O for 5 min and rinse in another two changes of 200 mL ddH_2O. Rinse once with 70% ethanol. Work in a laminar flow hood to ensure sterility. Pour the filters and ethanol in several 10-cm diameter petri dishes. Aspirate ethanol from the dishes. Dry the filters by removing the dish cover and allowing the remaining ethanol to evaporate. These filters can be stored at room temperature (RT) indefinitely if maintained sterile.
6. Supporting stainless steel grid: 30-mesh, openings 0.015 in. (cat. no. 5335-00-B55-0105; Cambridge Wirecloth Company, Cambridge, MD). Steel mesh is supplied as sheets. Use metal-cutting scissors to cut square grids of 2-cm sides. Cut a small notch on one side. Clean the grids thoroughly with water and detergent after each use and store in 70% ethanol.
7. Gauze: 2 × 2 in., 4-ply (cat. no. 7632; Johnson and Johnson, Arlington, TX).
8. Organ culture dish: 60 × 15-mm style with center well (cat. no. Falcon 3037; Becton Dickinson, Franklin Lakes, NJ).
9. Tissue culture dish, 10-cm diameter: 100 × 20-mm style (cat. no. Falcon 3003; Becton Dickinson).
10. Dissection dish: Tissue culture dish coated with a 5-mm thick 2% agar on the bottom of the dish. Melt 2 g of agar in 100 mL ddH_2O by gentle boiling in a 500-mL glass beaker; avoid foaming and overflowing. Swirl to melt the agar completely, and let sit at RT to cool. In the laminar flow hood, pour approx 10 mL of melted agar into each dish to coat the dish bottom, and allow to cool completely. Seal the dish with Parafilm, store at 4°C, and use within 1 mo.
11. Sterile transfer pipets: (cat. no. 202-1S; Samco Scientific, San Fernando, CA).
12. Water-jacketed carbon dioxide incubator set to 37°C and 5% carbon dioxide.

2.2. Bead Implantation

1. Beads: Affi-Gel Blue Gel at 100–200 mesh (cat. no. 153-7302; BioRad, Hercules, CA). Store at 4°C.
2. Siliconized microfuge tubes: 1.5 mL (Fisher Scientific Co., Pittsburgh, PA).
3. Bovine serum albumin (BSA): Fraction V (cat. no. A1933; Sigma, St. Louis, MO), cell-culture tested. Store powder at 2–8°C.

4. BSA-coated pipet tips: Prepare 1 mg/mL BSA solution in distilled, deionized water. Coat the inner barrel of pipet tips with BSA by pipetting the BSA solution up and down 2–3 times. Expel the solution completely from the pipet tips and let dry at RT in a laminar flow hood.
5. Phosphate buffered saline (PBS) at pH 7.4.
6. Manipulation of glass capillary for micropipet: Hard glass capillary, 75 mm in length (Drummond Scientific; Broomall, PA). Using a small flame from a bunsen burner, rotate the glass capillary and heat to soften the middle portion. Remove from flame and pull the ends apart, thus drawing the capillary out to a finer diameter to be used as micropipet. Break off the tapered end gently and smoothen the jagged edges by gently flaming the ends. Monitor the inner diameter under a stereomicroscope. The inner diameter of the capillary should be approx 50% of the diameter of the beads used for optimum holding and ease of manipulation. Sterilize by standard autoclaving.
7. Micropipet: Mouth-controlled micropipet can be assembled by connecting the mouthpiece (cat. no. 258616; Curtin Matheson Scientific, Jessup, MD), tubing, glass Pasteur pipet, micropipet holder, and micropipet.

3. Methods

3.1. Organ Culture

1. Assembly of the organ culture setup (**Fig. 1A,B**). This should be performed in the laminar flow hood and completed prior to the beginning of dissection. Place one piece of gauze and 6 mL of ddH_2O in a 10-cm diameter tissue culture plate so that the piece of gauze is thoroughly soaked. Place one organ culture dish on top of the piece of gauze and 3 mL of ddH_2O in the outer compartment of the organ culture dish. Use forceps to pick up one supporting grid, drain off excess ethanol, flame to dry and sterilize, and allow to cool briefly. Place this supporting grid over the inner compartment of the organ culture dish. Using a sterile transfer pipet, place enough culture medium, approx 1.5 mL, into the inner compartment so that the meniscus of the medium is even with the top of the grid. Replace the covers and allow the setup to equilibrate to 37°C and 5% CO_2 in the incubator.
2. Sacrifice gestation day 10 timed-pregnant female mice by cervical dislocation or carbon dioxide inhalation. Wet the abdominal area with 70% ethanol for sterilization and open the abdomen with bilateral incisions through the skin and body wall. Displace the bowels to reveal the uterus. Excise the bicornate uterus and place immediately in ice-cold HBSS. Rinse the uterus several times with ice cold HBSS to remove as much blood as possible to create a clearer field of microdissection.
3. Isolate each embryo from the implantation site by opening the site with forceps and spring scissors, extracting the embryo, and cleaning away from extraembryonic membranes. Stage the embryos according to Theiler stages (1972). The optimum stage for mandibular process explant is stage 18, with 40–44 somite pairs (**Fig. 1C,D**) (*see* **Notes 4–6**).
4. Microdissect the mandibular portion of the first branchial arch in the dissection dish with HBSS by making the following four incisions:
   a. Transverse cut immediately superior to the heart to separate the cranial region from the trunk region (**Fig. 1E**).
   b. Transverse cut between the first and second branchial arch (**Fig. 1F**).
   c. Transverse cut between the maxillary and mandibular portion of the first branchial arch (**Fig. 1G**).
   d. Frontal cut to separate the mandibular process from the neural tube (**Fig. 1H**) (*see* **Notes 7–9**).

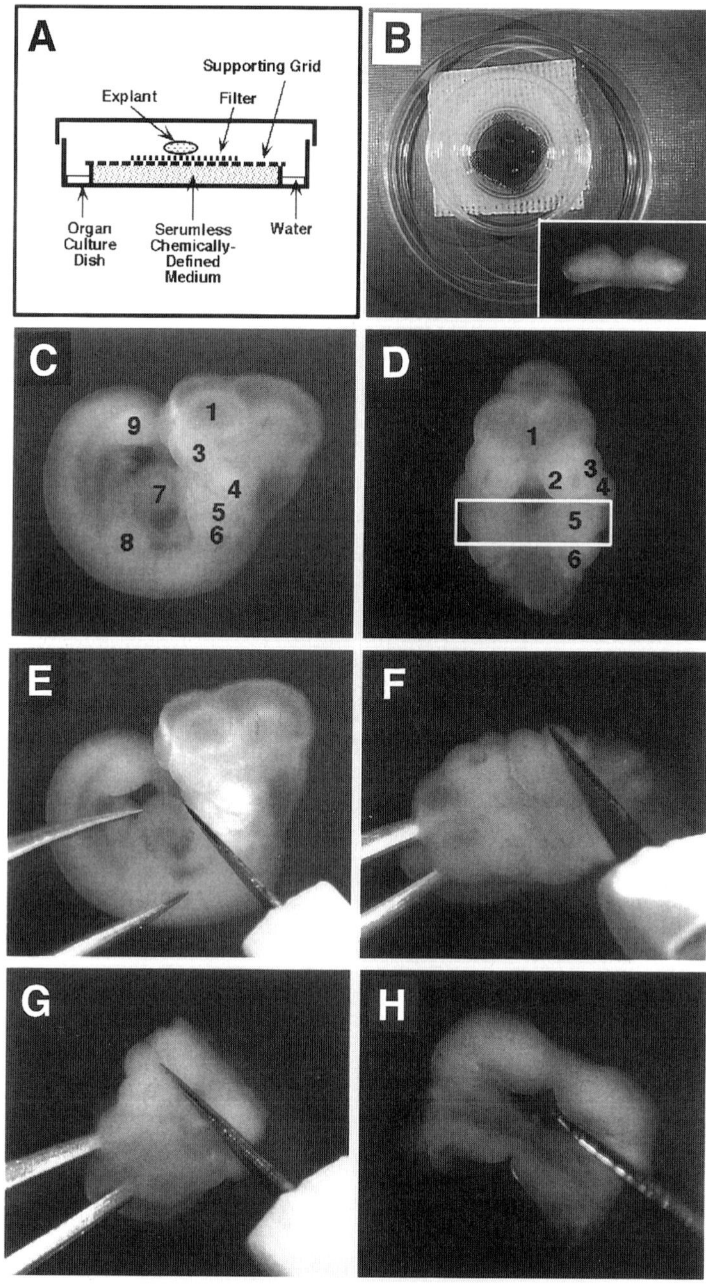

Fig. 1. Microdissection and organ culture assembly. Schematic representation **(A)**, and actual image **(B)** of the organ culture system illustrate the explant being supported by filter placed on grid and cultured at the atmosphere-medium interface in the culture dish. Water in the outer compartment of the culture dish and water-soaked gauze placed in the outer tissue culture dish serve to maintain optimum humidity for growth and development of the explant. The explant is oriented with the oral epithelium facing upward (B, inset). The optimum stage for collecting mandibular process explants from mouse embryos is Theiler stage 18; 40–44 somite pairs shown at the **(C)** lateral and **(D)** frontal views. The mandibular process (D, boxed) is isolated by a series of four incisions (**E–H**, *see text*) using fine forceps and

5. Select a piece of filter, wet and submerge it into HBSS in the dissection dish. Gently place the mandibular process onto the filter with the oral epithelial side facing up. Use two pairs of fine forceps to pick up the filter with the mandible and raise out of the HBSS, taking care to maintain the filter at a horizontal plane and to not dislodge the orientation of the mandibular process. Place the mandibular process explant on the supporting grid of the organ culture assembly (**Fig. 1B**, inset) (*see* **Notes 10–15**).
6. The assembly can be returned to the incubator. Change to fresh medium every 2 d. Aspirate and replace medium by using a Pasteur pipet accessing the inner compartment of the organ culture dish through the notched side of the supporting grid. Adjust the amount of medium so that the mandibular explants are placed at the atmosphere-medium interface to achieve optimum growth and development (*see* **Note 16**).

3.2. Bead Implantation (See Notes 17–19)

1. Preparation of beads: Growth-factor-soaked beads are prepared fresh on the day of use. Swirl the bottle of Affi-Gel Blue beads to resuspend. Take 10 µL of bead suspension and place it in a siliconized microfuge tube. Add 100 µL of PBS into the tube, mix, and allow to sit at RT for 5 min. Collect the beads by centrifugation at 14K for 5 s. Aspirate the supernatant. Repeat washing with PBS once more. Work in the laminar flow hood to maintain sterility from here on. Using a BSA-coated pipet tip, add 5 µL of a 100 ng/µL solution of BMP4 to the beads, mix, and allow to incubate at RT for at least 1 h.
2. With fine forceps, make a small well in the agar of the dissection dish. Cover the agar with a thin layer of BGJb medium, approx 1 mL. Transfer the beads into the well.
3. From the incubator, retrieve one specimen of mandibular process with the filter and place it in the dissection dish.
4. Under the dissecting microscope, using gentle suction on the mouth-controlled micropipet, pick up and hold one bead.
5. Position the tip of the micropipet with the attached bead to the desired point of implantation. Release suction and gently push the bead through the epithelium with the micropipet. Return the specimen to the incubator.

4. Notes

1. Handling of the animals and disposition of carcasses and tissues should be performed in accordance with Animal and Care Guidelines and the institution's Animal Study Protocols.
2. If timed pregnant mice are purchased from a vendor, then the investigator should ascertain the protocol of counting gestation days as determined by the vendor. Most vendors designate the day of finding a vaginal sperm plug in the female as day 0 of gestation. However, day 0.5 and day 1 are designated by others. The method of assignment is neither universal nor standardized.
3. Frequently, even when particular gestation stage embryos are requested, there is a wide range of variation in actual stage of the embryos among litters and within a litter. The variation can be as much as 12 h of development. Some investigators breed animals for a limited time in their laboratory vivarium (e.g., caged together for only 2–3 h). However, the time of fertilization is often dependent on the time of ovulation (which, in turn, is

microscalpels. 1, frontal process; 2, medial nasal process; 3, lateral nasal process; 4, maxillary process; 5, mandibular process; 6, second branchial (hyoid) arch; 7, heart; 8, forelimb bud; 9, hindlimb bud.

dependent on the light–dark cycle of the housing facility) and not the time of mating. Therefore, the development variation is an unavoidable and inherent factor in the experiment. The investigator is encouraged to determine the exact stage of the embryos by external morphology, such as employing the number of somite pairs according to the Theiler staging system *(7)*.

4. In mandible explant cultures, the ideal initial stage is between 40 and 44 somite pairs, corresponding to Theiler stage 18. The mandibular process develops and produces multiple phenotypes, including cartilage, osteoid material, incisor and molar tooth buds, and muscular tongue, all within 9 d in serumless medium (**Fig. 2A–H**). These phenotypes are detected by in 3–6 d of culture. After 9 d, the explant can be kept alive but without further differentiation. For example, the tooth organs will reach the bell stage (but not beyond that) regardless of the length of culture. Therefore, most experiments are terminated after a maximum of 9 d.
5. In addition to the mandibular processes, many other organs from the craniofacial region have been cultured successfully in the system; rhombomeres *(8)*, inner ear *(9)*, palatal shelves *(10)*, and tooth explants *(11)*. Organs from the trunk region are also used; limb bud *(12)*, lung *(13)*, and metatarsal explants *(14)*. Several variations in methodology include the stage of embryo the explant is isolated from, the method of microdissection, and duration of culture. Rhombomeres are the early segmentations of the hindbrain, isolated from E8 embryos. Because of the early stage of these tissues, explant cultures require serum-containing medium for optimal growth and differentiation.
6. The explant culture system not only supports normal growth and development but also allows for the study of dysmorphology when coupled with perturbation strategies such as the use of antisense oligonucleotides (**Fig. 3**).
7. Microdissection is usually done under a stereomicroscope in a laminar flow hood. However, if the working surface is sufficiently sterilized by thorough wiping with 70% ethanol, the procedure can be performed on a regular laboratory bench.
8. Instruments are thoroughly sterilized by dipping in 70% ethanol and then flaming to dryness and allowed to cool. Often, the flaming may cause discoloration of the instrument without adverse effects to the microdissection. Instruments can also be sterilized by standard autoclaving.
9. Inherently, there is a small but noticeable variation in the amount of tissue isolated for cultures; that is, where exactly to make the cut. Therefore, we suggest that for one complete set of experiments, either only one person should perform the microdissection, for consistency and reproducibility, or efforts should be made to calibrate several scientists in microdissection or both.
10. The mandibular process should be explanted into culture with the oral epithelial side up. This orientation may be challenging to identify for the beginner. One clue is to use the cut surfaces between the maxillary and mandibular process as a guide, because these should be facing up as well, aligned with the oral side.
11. During the step when a cut is made to separate the first and second branchial arches, we suggest that the dissection include a small amount of tissue from the second arch with the explant (**Fig. 2A**). Empirical data revealed that the inclusion of second arch tissue enhances the symmetrical development of the mandible and Meckel's cartilage.
12. In one culture assembly, six filters are usually placed on the supporting grid. With some careful arrangement, up to eight filters can be accommodated. On each filter, a maximum of three mandible explants can be supported. If one were to design an experiment that would require a large number of explants, then arranging more mandible explant

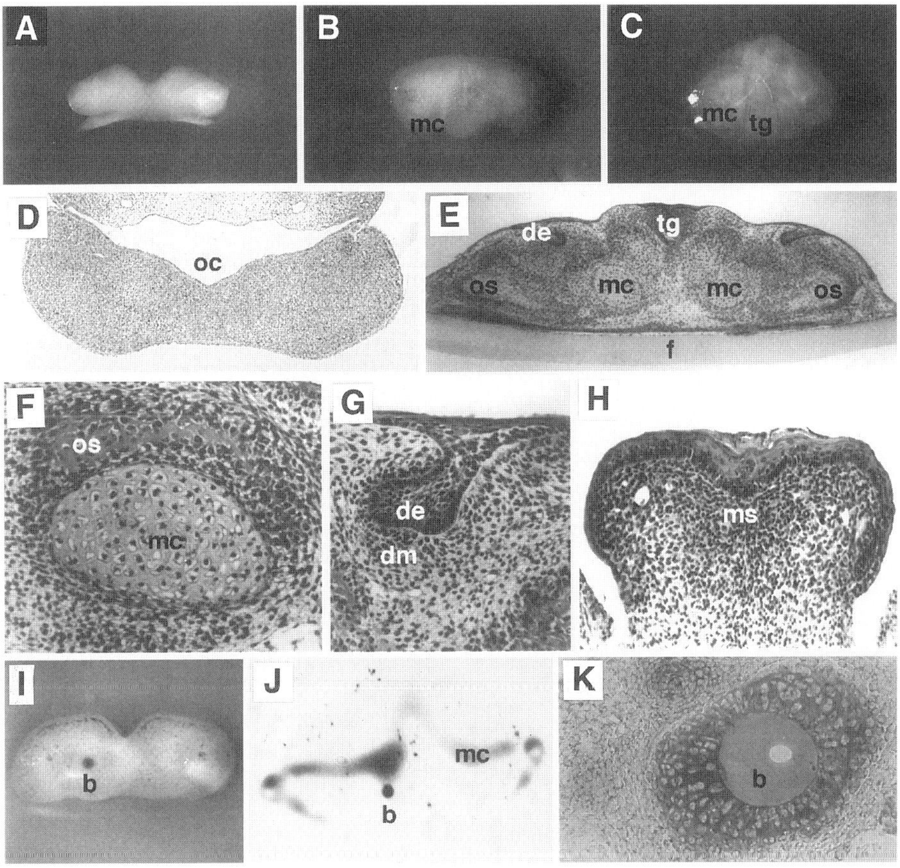

Fig. 2. Phenotypic differentiation of the mandibular process cultured in serumless, chemically defined medium is comparable to in vivo development. The isolated mandibular process explant at the beginning of culture **(A)** consists of a simple cuboidal epithelium enclosing a population of ectomesenchymal cells derived from cranial neural crest **(D)**. After 3 d of culture **(B)**, a rudiment of Meckel's cartilage (mc) at the posterior lateral aspect of the explant is readily observable at the macroscopic level. This unique cartilage further grows into symmetrically positioned rod-shaped structures by 7 d of culture **(C)**. The tongue (tg) is centrally placed. Histological analyses **(E–H)** illustrate additional structural differentiation, such as the developing tooth bud, which is composed of the dental epithelium (de), surrounding condensing dental mesenchyme (dm), and osteoid materials (os) subsequently to be the mandibular bone deposited adjacent to Meckel's cartilage. Myoblasts and myotubes can be seen in the tongue. oc, oral cavity; f, filter; ms, median sulcus. **(I)** Exogenous BMP4 delivered by Affi-Gel Blue beads induces ectopic cartilage formation in vitro. BMP4-soaked bead (b) is implanted in the mandibular process at the beginning of culture. **(J)** After 6 d of culture, ectopic cartilage surrounding the bead (b) can be detected by whole-mount alcian blue staining. **(K)** Toluidine blue staining of histological sections demonstrate for cartilage-specific glycosaminoglycans surrounding BMP4-soaked beads (b).

per culture assembly would be more convenient during subsequent handling. However, if more than 10 mandibles are placed in one assembly, then daily medium change would be required.

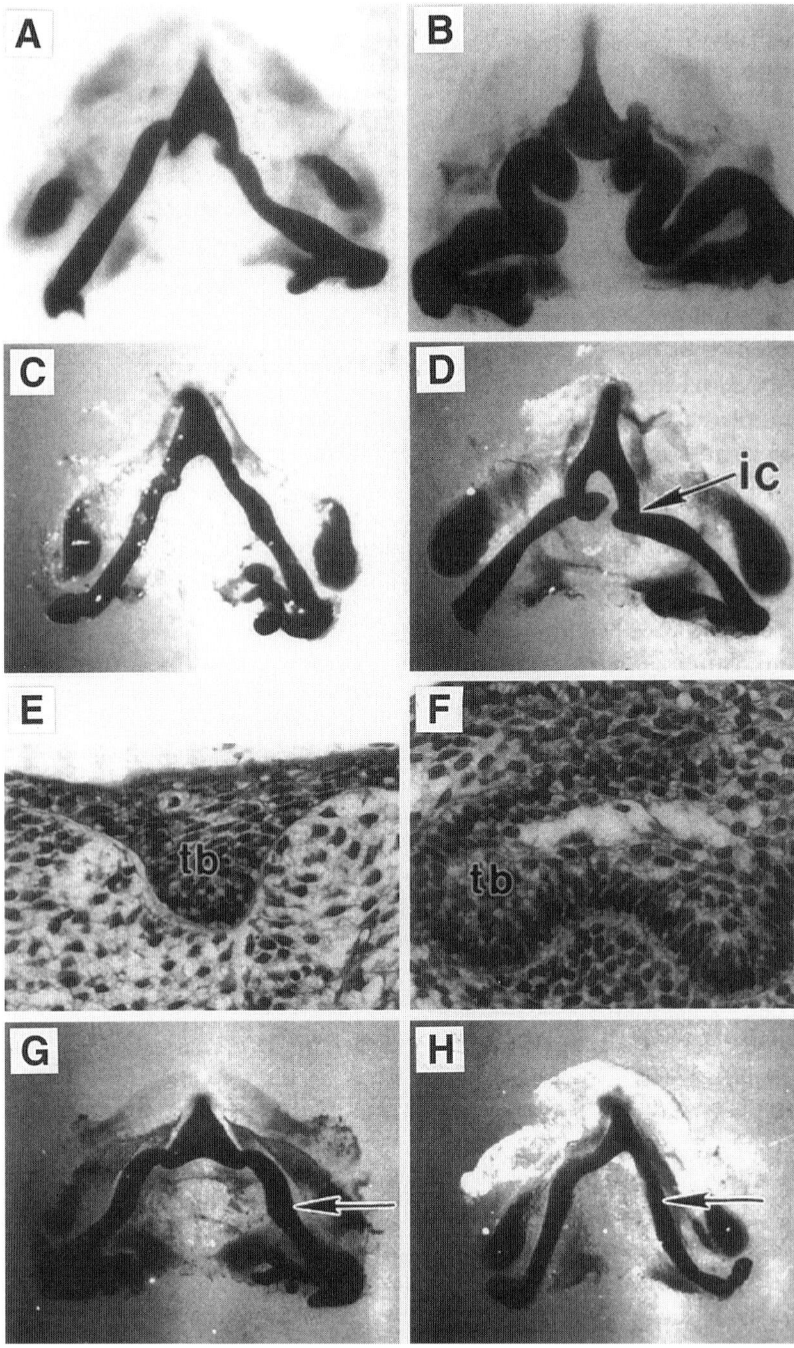

Fig. 3. Study of mandibular process dysmorphology using explant culture system coupled with antisense deoxynucleotide strategies (**refs.** ***4,5***). Mouse E10-stage mandibular processes were explanted into culture in the presence of 30 μ*M* antisense oligonucleotides directed against epidermal growth factor (EGF), (B), transforming growth factor-beta 1 (TGFβ1) (D), TGFβ2 (F), or TGFβ3 (H). After 9 d in culture, growth and development of Meckel's cartilage or tooth buds were compared with the respective sense oligonucleotide-treated controls (A, C, E, G). Antisense inhibition of EGF induced fusilli-form dysmorphogenesis of Meckel's cartilage (B)

13. HBSS can be purchased in a variety of modified components. We found that HBSS without phenol red was particularly suitable for use as a dissection buffer, because the solution is clear and allows for better visibility.
14. Dissection is performed using a dissection dish coated with a thin layer of agar to preserve the sharp cutting edge of the microscalpel during the duration of each microdissection session. It also provides a clean cut through of the tissues compared with dissecting against a hard plastic surface of the plate. To provide maximum visual contrast for the identification of embryonic parts during microdissection, it is possible to add 0.5 mL of black India ink into the agar during preparation of the dissection dish.
15. During the preparation of supporting filter, always wear gloves, as grease from the scientist's fingers causes contamination and uneven wetting of the surface. The particular type of filter is selected because the dark color surface provides visual contrast to the otherwise opaque explants.
16. After several days of culture, the mandible explants will have enough fibroblastic outgrowth so that the explants become adherent to the filters. It is not necessary to separate the explant from the filter for fixation for subsequent histological evaluation, because the filter is porous and that the tissue is thin. In fact, the filter provides a useful substrate for handling the mandibles and for orientation to define the plane of section for histology.
17. The mesh size of Affi-Gel Blue beads corresponds to effective diameter of beads of 50–75 μm. The gel is stored in PBS buffer with sodium azide as preservative. Therefore, do not breathe vapor from the solution. The gel has a shelf life of 1 yr when stored at 2–8°C. The matrix is cross-linked agarose coupled to Cibacron blue dye F3GA. The dye provides active binding sites to proteins and peptides with a capacity of more than 11 mg/mL. Affi-Gel Blue beads have a high affinity for proteins and therefore are suitable for soaking and delivering many other different peptide growth factors in addition to BMP4 used in this protocol's investigations (**Fig. 2I–K**).
18. Growth factors are small polypeptide molecules that have a high affinity to plastic. Therefore, to minimize the loss of exogenous proteins through "stickiness" to handling plasticware, we use siliconized microfuge tubes and BSA-coated pipet tips. Wherever possible, also add carrier BSA to a concentration of 0.1 mg/mL to growth-factor solutions and store growth factors in small aliquots at a high stock concentration. Dilute to working concentration just before use. Avoid freeze–thaw cycles.
19. During microdissection and bead implantation, it is essential to perform the procedures as swiftly as possible. Dissection should be done using ice cold buffer. Viability of the explants decreases with exposure to suboptimal temperature, atmosphere, and medium.

Acknowledgment

We thank Pablo Bringas, Jr. for his participation in the development of the organ culture protocol and Kazuaki Nonaka and Ichiro Semba for their contributions to the development of the bead implantation procedures. Human recombinant BMP4 was supplied by Genetics Institutes. This work is supported by NIH funding Z01-AR41114.

---

and reduction in tooth bud size (data not shown). Abrogation of TGFβ1 resulted in malformation of the rostral region of Meckel's cartilage and caused the appearance of inward curvature (ic) between the rostral and mid-segment (D). TGFβ2 antisense treatment produced a threefold enlargement of tooth buds (tb) and accelerated the development to cap stage (F). TGFβ3 abrogation resulted in reduction in length and thickness of Meckel's cartilage (H, *arrow*).

References

1. Slavkin, H. C., Bringas, P., Jr., Sasano, Y., and Mayo, M. (1989) Early embryonic mouse mandibular morphogenesis and cytodifferentiation in serumless, chemically-defined medium: a model for studies of autocrine and/or paracrine regulatory factors. *J. Craniofac. Genet. Dev. Biol.* **9,** 185–205.
2. Slavkin, H. C., Sasano, Y., Kikunaga, S., Bessem, C., Bringas, P., Jr., Mayo, M., Luo, W., Mak, G., Rall, L., and Snead, M. L. (1990) Cartilage, bone and tooth induction during early embryonic mouse mandibular morphogenesis using serumless, chemically-defined medium. *Connect. Tissue Res.* **24,** 41–51.
3. Mayo, M. L., Bringas, P., Jr., Santos, V., Shum, L., and Slavkin, H. C. (1992) Desmin expression during mouse tongue morphogenesis. *Int. J. Dev. Biol.* **36,** 255–263.
4. Shum, L., Sakakura, Y., Bringas, P., Jr., Luo, W., Snead, M. L., Mayo, M., Crohin, C., Millar, S., Werb, Z., Buckley, S., Hall, F. L., Warburton, D., and Slavkin, H. C. (1993) EGF abrogation induced fusilli-form dysmorphogenesis of Meckel's cartilage during embryonic mouse mandibular morphogenesis in vitro. *Development* **118,** 903–917.
5. Chai, Y., Mah, A., Crohin, C., Groff, S., Bringas, P., Jr., Le, T., Santos, V., and Slavkin, H. C. (1994) Specific transforming growth factor beta subtypes regulate embryonic mouse Meckel's cartilage and tooth development. *Dev. Biol.* **162,** 85–103.
6. Tabata, M. J., Kim, K., Liu, J. G., Yamashita, K., Matsumura, T., Kato, J., Iwamoto, M., Wakisaka, S., Matsumoto, K., Nakamura, T., Kumegawa, M., and Kurisu, K. (1996) Hepatocyte growth factor is involved in the morphogenesis of tooth germ in murine molars. *Development* **122,** 1243–1251.
7. Theiler, K. (1989) *The House Mouse: Atlas of Embryonic Development.* Springer-Verlag, New York.
8. Chai, Y., Bringas, P., Jr., Shuler, C., Devaney, E., Grosschedl, R., and Slavkin, H. C. (1997) An early embryonic mouse mandibular culture model permits the study of cranial neural crest cell migration and transcription factor in regulating mouse tooth development during craniofacial morphogenesis. *Int. J. Dev. Biol.* **42,** 87–94.
9. Hoffman, D. S., Bringas, P., Jr., and Slavkin, H. C. (1996) Co-culture of contiguous developmental fields in a serumless, chemically-defined medium: an in vitro model permissive for coordinate development of the mouse ear. *Int. J. Dev. Biol.* **40,** 953–964.
10. Shuler, C. F., Guo, Y., Majumder, A., and Luo, R. Y. (1991) Molecular and morphologic changes during the epithelial-mesenchymal transformation of palatal shelf medial edge epithelium in vitro. *Int. J. Dev. Biol.* **35,** 463–472.
11. Evans, J., Bringas, P., Jr., Nakamura, M., Nakamura, E., Santos, V., and Slavkin, H. C. (1988) Metabolic expression of intrinsic developmental programs for dentine and enamel biomineralization in serumless, chemically-defined, organotypic culture. *Calcif. Tissue Int.* **42,** 220–230.
12. Canoun, C., Ma, C., Halpern, D., Shum, L., Bringas, P., Jr., Sank, A., and Slavkin, H. C. (1993) Endogenous epidermal growth factor regulates limb development. *J. Surg. Res.* **54,** 638–647.
13. Jaskoll, T. F., Johnson, R., Don, G., and Slavkin, H. C. (1986) Embryonic mouse lung morphogenesis in serumless, chemically-defined medium in vitro. *Prog. Dev. Biol.*, part A, 381–384.
14. Minkin, C., St. James, S., Tao, H., Yu, X., Pockwinse, S., MacKay, C., and Marks, S. C., Jr. (1991) Skeletal development and formation of osteoclast-like cells from in situ progenitors in fetal mouse metatarsals cultured in chemically defined medium. *Bone Mineral* **12,** 141–155.

7

Craniofacial Skeletal Morphogenesis In Vitro

Roy C. Ogle

1. Introduction

All the bones of the craniofacial skeleton are united by the fibrous sutures, and those of the neurocranium are lined additionally by the dura mater. The sutures and dura mater are not only integral structural elements of the formed skeleton, but during morphogenesis they also are sites of appositional growth of the membranous bones of the vault and inductive interactions that regulate the process of suture obliteration, respectively *(1–5)*. When sutures fail to form or are prematurely fused prior to cessation of rapid brain growth, morphogenesis of the entire head is severely altered *(6)*. Although a variety of approaches employing transplantation of embryonic rudiments and surgical perturbation of elements of the developing craniofacial skeleton have demonstrated that cellular interactions among the perisutural tissues are critical to morphogenesis *(1)*, recent development of media and methods for growing fetal rat and mouse calvaria in vitro from non-mineralized rudiments may allow identification of the molecules and mechanisms underlying normal craniofacial development *(3,7–11)*.

The cranial vault sutures of rats and mice form during a 72-h period immediately prior to birth. At fetal day 18 (F18) in rats and a day earlier in mice, calcification of the bones is restricted to a small ossification center in each bone separated widely by embryonic mesenchymal tissues. Between each of the paired frontal and parietal bones the coronal sutures develop, with overlapping bones and fibrous sutures that remain open throughout the lives of rodents. Between the frontal bones the intrafrontal suture is formed, which is a butt or nonoverlapping suture that fuses in the third week of life.

As described in this protocol, culture of the calvaria under defined conditions, allowing both mineralization and formation of tissues as seen in vivo, can be used to identify the factors originating in the dura mater and developing bones that serve as signals guiding the component processes of morphogenesis—proliferation, differentiation, and apoptosis (*see* **Fig. 1**). This method has proven amenable to approaches with antisense oligonucleotide, antibody and pharmacological inhibition of growth factors and their receptors to perturb elements of the system, as well as attempts to reconstitute the system by supplementing defined media with purified factors. An example of the use of a neutralizing antibody to one of the fibroblast growth factor receptors

From: *Methods in Molecular Biology, Vol. 136: Developmental Biology Protocols, Vol. II*
Edited by: R. S. Tuan and C. W. Lo © Humana Press Inc., Totowa, NJ

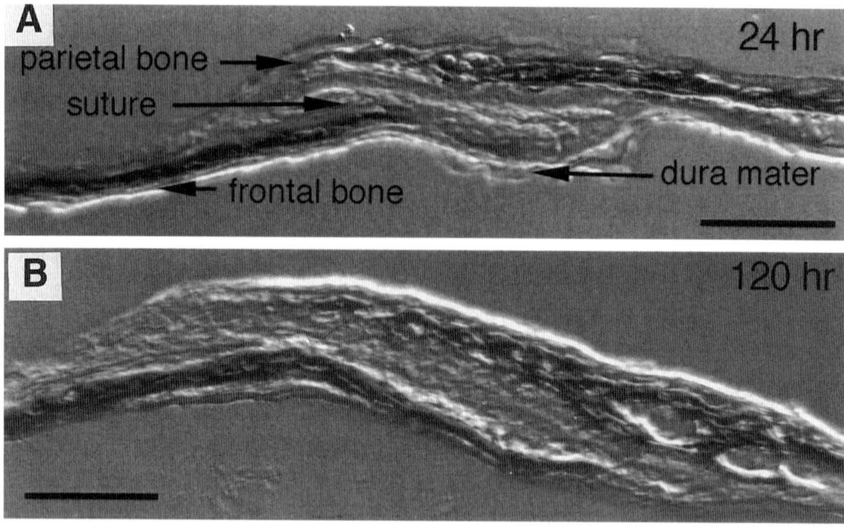

Fig. 1. Development of the coronal suture region of the rat calvaria in vitro. Photomicrographs (Hoffman modulation contrast optics) of frozen sections (7 μ*M*) stained with Alizarin red to visualize mineralized bone. **(A)** By 24 h in culture, the frontal and parietal bones have begun to overlap and are partially calcified. **(B)** At 120 h, the bones and suture have continued growing. Bone mineralization and thickness have increased and the suture remains nonossified. If the dura mater is removed prior to culture, the bones fuse (not shown, similar to Fig. 2B) unless appropriate factors are supplied in the defined medium. Bars = 100 μ*M*.

(FGFR) is included to illustrate this approach. This system may also be employed to investigate morphogenesis of the axial and appendicular skeletons and other events of appropriate or abnormal bony fusion, such as the formation of the digits.

2. Materials

1. Serum-free culture medium (*see* **Notes 1** and **2**): Dulbecco's high-glucose minimum essential medium with pyruvate (DMEM; Gibco, BRL, Gaithersburg, MD) containing 1 μg/mL gentamicin, 5 m*M* glutamine, 1X nonessential amino acids (100X stock; Gibco-BRL), 1 m*M* insulin, transferrin, and selenium (ITS+; Collaborative Research, Inc., New Bedford, MA), and 3 m*M* inorganic phosphate (stock solution = 0.1*M* NaH_2PO_4, pH 7.0). Cultures were supplemented daily with 100 μg/mL ascorbic acid (stock solution = 0.1*M* sodium ascorbate in H_2O; Sigma Chemical Co., St. Louis, MO).
2. F19 Sprague-Dawley rat fetuses, 1-day-old (N1) or 17-day-old (N17) male rat pups from laboratory breeding colony.
3. Hank's balanced salt solution (HBSS): 0.14 g $CaCl_2$, 0.4 g KCl, 0.06 g KH_2PO_4. 0.1 g $MgCL_2(6H_2O)$, 0.1 g $MgSO_4$ $(7H_2O)$, 8 g NaCl, 0.35 g $NaHCO_3$, 0.06 g Na_2HPO_4 $(2H_2O)$. and 1.25 g glucose in 1 L dH_2O.
4. Cultureware: 24-well culture dishes and 0.45-μm polycarbonate membrane inserts (Corning-Costar, Corning, NY).
5. Carnoy's fixative: 60 mL absolute ethanol, 30 mL chloroform, and 10 mL acetic acid.
6. 0.2% Alizarin Red-S (Sigma) 0.2 g in 100 mL dH_2O plus 1 drop concentrated ammonium hydroxide (Fisher Scientific Co., Pittsburgh, PA).
7. Decalcification solution: ready-to-use EDTA/HCl solution (Stephens Scientific, Riverdale, NJ).

8. Statistical analysis software: SAS (SAS Institute, Cary, NC).
9. Calcium analysis: 5% trichloroacetic acid (Fisher Scientific), Sigma kit #587-A and Sigma calcium standard #360-5.

3. Methods

1. Preparation of explants for culture (*see* **Note 3**): F19 Sprague-Dawley rat fetuses or (N17) male neonates are used. Male and female rats are caged as mating pairs in the late afternoon and females are checked for plugs the following morning. On discovery of a plug, males and females are immediately separated. Day of plug is denoted as day 0. At the appropriate developmental stage, pregnant females are killed by overadministration of Halothane and pups aseptically removed and placed on ice prior to removal of calvaria. The scalp is opened and whole calvaria removed from pups and placed in HBSS on ice. The frontal and parietal bones, including presumptive coronal suture, sagittal, and intrafrontal sutures are carefully dissected away from the surrounding tissues. Using a dissecting microscope, the dura mater is removed from those calvaria to be cultured denuded of dura mater, with care being taken to eliminate all traces of dura mater without damaging the underlying suture. In experiments focusing only on the coronal suture, F19 calvaria are divided in half by cutting through the sagittal and intrafrontal sutures. In cases in which the intrafrontal is of sole interest, the N17 calvaria are cut through the coronal suture.
2. Culture and harvesting of calvaria: Although calvaria were typically cultured with dura mater intact, in reconstitution experiments, calvaria with dura mater removed are used, occasionally separated from (but in liquid contact with) the dissected dura mater by a 0.45-μm polycarbonate membrane insert. Calvaria are placed dura side down into 24-well culture dishes and covered with 400 μL medium (*see* **Note 4**). If antibodies or other test substances are to be added, they and vehicle controls, if needed, are added at this point. In the example shown in **Fig. 2B**, 4 μg per 400 μL of medium of neutralizing IgG against FGFR2 (Santa Cruz Labs) was employed. All groups receive serum-free medium, which is replaced by one half with fresh medium and supplemented daily with 100 μg/mL ascorbic acid. Individuals from each group are harvested at 24 h or more frequent intervals and prepared either for histology or calcium analysis.
3. Preparation of tissues for histology: Explants are fixed in Carnoy's fixative overnight and transferred to 70% ethanol, stained with 0.02% Alizarin red overnight (*see* **Note 5**), embedded in OCT compound (Tissue-tek), and frozen sectioned (7 μ*M*). In situations requiring optimal morphology, specimens are decalcified (20 min) with EDTA/HCl, and processed for paraffin sectioning. Typically, thirty-six 5-μ*M* sagittal sections are cut through the central portion of each suture and stained with hematoxylin and eosin (H & E). Assessment of the patent state of the suture serves as a bioassay for establishing the ability of various conditioned media, their fractions, and known growth factors to promote fusion in the coronal or prevent sutural obliteration of the intrafrontals.
4. Histomorphometry and statistical analysis (*see* **Note 6**): Sections are randomized according to random numbers generated by SAS (SAS Institute, Cary, NC) and examined by two independent observers such that observers are blinded to experimental manipulation. Sections exhibiting poor histology or incorrect plane of section are excluded from analysis. Each section is scored for degree of osseous obliteration of sutures (intact sutures = 3; bone fronts touching, but not fused = 2; bone fronts fused in region of suture = 1) and thickness of bones. Because osseous obliteration of sutures occurs in several ways, intact sutures are scored as 100% and degree of osseous obliteration calculated as a percentage thereof. Results are tested by analysis of variance as previously described ***(1,3)***. Data are presented as mean + SEM.

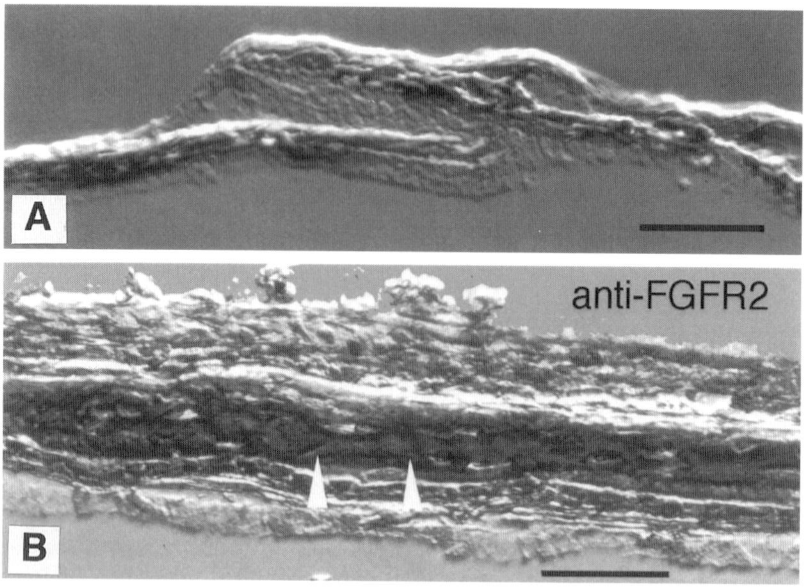

Fig. 2. Development of rat calvaria in vitro—effects of neutralizing antibody to FGFR. Photomicrographs (Hoffman modulation contrast optics) of frozen sections (7 μ*M*) stained with Alizarin red to visualize mineralized bone. **(A)** Control calvaria cultured for 48 h as in Fig. 1. **(B)** Calvaria cultured in presence of purified rabbit immunoglobulin G against FGFR 2 (4 μg in 400 μL) for 48 h. Note the fusion of the parietal and frontal bones in the region indicated by *arrowheads* and the apparent increased growth of bone and fibrous tissues.

5. Calcium analysis: Calcium is extracted from calvaria with two 1-h incubations in 5% trichloroacetic acid (1 mL each), which are pooled for analysis using Sigma kit #587-A and Sigma calcium standard #360-5. The tissues are then washed 3X in acetone and allowed to air dry overnight prior to weighing on a Mettler 6300 balance.

4. Notes

1. A wide variety of methods and additives for culture were tested in developing this methodology. Whereas other media are more commonly used for bone cell culture, the high-glucose DMEM used as described worked best on intact calvaria. Inclusion of 1–10% fetal calf serum actually inhibited growth and should be avoided. Use of beta-glycerol phosphate instead of inorganic phosphate and concentrations of phosphate lower than 3 m*M* in addition to that present in DMEM resulted in ectopic calcification within the periosteum.
2. The source of transferrin in the insulin, transferrin and selenium (ITS+) used in the media is human, and precautions taken when working with human blood products should be observed. This laboratory has used transferrin that was later determined to have been prepared from pooled blood that contained viruses (Creutzfeld-Jacob), although the transferrin preparation was reportedly free of virus.
3. It is essential to take care not to damage dura or suture during dissection. A small amount of dura remaining after removal can be enough to block fusion of the suture, while damage to the suture mesenchyme can prevent its normal development. If litter size is sufficient, several dissected rudiments should be analyzed by histology to assure the quality of dissection.
4. Although 200 μL of medium is sufficient to cover the explants, 400 μL is the minimal amount that gave good reproducible growth. As the explants grow, by 72 h in culture

sufficiently acidic byproducts are secreted into the medium to lower pH as indicated by yellow color of the phenol red indicator dye. In these cases, sterile sodium bicarbonate was added in a dropwise fashion to titrate the medium back to neutrality (cherry red).

5. Alizarin staining of whole skulls did not work well unless the 70% ethanol soak between fixation and staining was used.
6. The type of histomorphometric analysis employed to evaluate suture fusion is very laborious, and others have attempted to evaluate outcomes more rapidly by whole-mount immunostaining or *in situ* hybridization. That method does not work with the coronal or other overlapping sutures. The geometry of the tissues is such that only cross-sectional analysis indicates the degree of suture fusion.

Acknowledgments

The author thanks Lynne Opperman, Ralph Passarelli, Ellen Morgan, Amber Nolen, Chandan Chopra, Anikar Chhabra, Richard Cho, and Mark Reintjes for their roles in developing this protocol. This work was supported by NIH grant DEI0369.

References

1. Opperman, L. A., Sweeney, T. M., Redmon, J., Persing, J. A., and Ogle, R. C. (1993) Tissue interactions with underlying dura mater inhibit osseous obliteration of developing cranial sutures. *Dev. Dyn.* **198,** 312–322.
2. Opperman, L. A., Persing, J. C., Sheen, R., and Ogle, R. C. (1994) In the absence of periosteum, transplanted fetal and neonatal rat coronal sutures resist osseous obliteration. *J. Craniofac. Surg.* **5,** 327–332.
3. Opperman, L. A., Passarelli, R. W., Morgan, E. P., Reintjes, M., and Ogle, R. C. (1995) Cranial sutures require tissue interactions with dura mater to resist osseous obliteration *in vitro*. *J. Bone Min. Res.* **10,** 1978–1987.
4. Opperman, L. A., Passarelli, R. W, Nolen, A. A., Lin, K. Y., Gampper, T. G., and Ogle, R. C. (1996) Dura mater secretes soluble, heparin-binding factors required for cranial suture morphogenesis. *In Vitro Cell. Dev. Biol.* **32,** 627–632.
5. Opperman, L. A., Nolen, A. A., and Ogle, R. C. (1997) TGF-β1, TGF-β2 and TGF-β3 exhibit distinct patterns of expression during cranial suture formation and obliteration in vivo and in vitro. *J. Bone Min. Res.* **3,** 301–310.
6. Enlow, D. H. (1989) Normal and abnormal patterns of craniofacial growth, in *Scientific Foundations and Surgical Treatment of Craniosynostosis* (Persing, J. A., Edgerton, M. T., and Jane, J. A., eds.), Williams & Wilkins, Baltimore, MD, pp. 83–86.
7. Gronowicz, G., Woodiel, F. N., McCarthy, M.-B., and Raisz, L. G. (1989) In vitro mineralization of fetal rat parietal bones in defined serum-free medium: effect of β-glycerol phosphate. *J. Bone Min. Res.* **4,** 313–324.
8. Chen, T. L. and Bates, R. L. (1993) Recombinant human transforming growth factor β1 modulates bone remodeling in a mineralizing bone organ culture. *J. Bone Min. Res.* **8,** 423–427.
9. Dieudonne, S. C., Foo, P., van Zoelen, E. J., and Burger, E. H. (1991) Inhibiting and stimulating effects of TGF-β1 on osteoclastic bone resorption in fetal mouse bone organ cultures. *J. Bone Min. Res.* **6,** 479–487.
10. Hock, J. M., Canalis, E., and Centrella, M. (1990) Transforming growth factor-β stimulates bone matrix apposition and bone cell replication in cultured fetal rat calvaria. *Endocrinology* **126,** 421–426.
11. Wilkie, A. O. M. (1997) Craniosynostosis: genes and mechanisms. *Hum. Mol. Gen.* **6,** 1647–1656.

8

Skeletal Morphogenesis

Stefan Mundlos

1. Introduction

The formation of a skeleton with its numerous bones of various shapes and sizes and the growth of these bones from embryonic to adult size is a complex process involving a multitude of genes. The complexity of the process is reflected by the large number of inherited diseases with skeletal phenotypes (in man and mouse) as well as by the ever-increasing number of genes shown to be involved in skeletal morphogenesis (for review, *see* **refs. *1,2***). The inactivation of genes through transgenic mouse technology has become a popular method of analyzing the function of newly discovered genes. Such mice may have expected or unexpected phenotypes and, considering the large number of genes involved, skeletal alterations are commonly found. Naturally, the question arises, What is the role of this gene in skeletal morphogenesis and how can the phenotype be explained. This chapter will address these questions and will supply the investigator with a conceptual framework for how to investigate skeletal defects in the mouse.

Three distinct lineages contribute to the early skeleton. Most cells of the craniofacial bones are of neural crest origin. Neural crest cells migrate from the dorsal side of the neural tube through the sclerotome into the head region. They contribute to the calvarium, midface, mandible and the teeth. Vertebrae and ribs originate from the sclerotome, a structure that forms through the differentiation of somites. The appendicular skeleton is derived from lateral plate mesoderm. Skeletal morphogenesis from these three lineages involves, in principle, four different steps: patterning, organogenesis, growth, and homeostasis (**Fig. 1**).

Patterning determines the future size, shape (gestalt), and number of individual skeletal elements. This is achieved through the expression of patterning genes in progenitor cells long before overt skeletogenesis.

Organogenesis involves the generation of bone/cartilage as an organ (skeleton). During this process, undifferentiated cells condense (condensation) according to a preset pattern, differentiate into chondrocytes or osteoblasts (differentiation), and start producing their specific extracellular matrix (histogenesis). These aspects of early skeletogenesis, the "anlage," represent the outlines of the future skeletal elements. Bones develop either by direct formation of osteoblastic progenitor cells or within a

From: *Methods in Molecular Biology, Vol. 136: Developmental Biology Protocols, Vol. II*
Edited by: R. S. Tuan and C. W. Lo © Humana Press Inc., Totowa, NJ

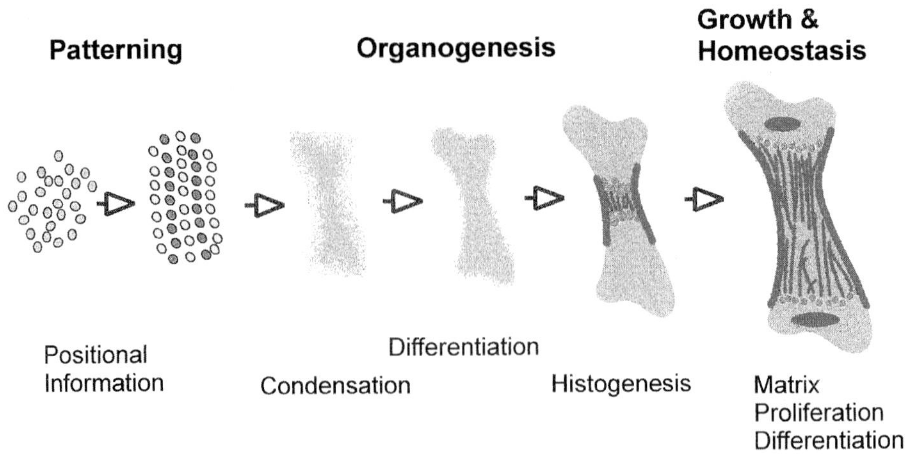

Fig. 1. (*See* color plate 1 appearing after p. 262.) Schematic showing the different steps of skeletal morphogenesis.

temporary framework of hyaline cartilage. The direct formation of bone is called "desmal" or "intramembranous" and produces the flat bones of the skull, part of the clavicle, and the mandible. In contrast, all other bones develop through endochondral bone formation, a process that involves the proliferation and differentiation of chondrocytes to form a temporary cartilaginous template that is subsequently replaced by bone.

All longitudinal *growth* takes place by proliferation and differentiation of chondrocytes. Chondrocytes in the middle of the cartilaginous anlage hypertrophy and their matrix calcifies, while newly differentiated osteoblasts deposit a thin layer of bone around the central part of the anlage. Blood vessels invade this center of primary ossification via the periosteal bone shaft. Some of the invading cells differentiate into hematopoietic stem cells, others into osteoclasts or osteoblasts. In what will become long bones, the primary center of ossification expands toward the two ends. In addition, secondary centers will form at one or both ends. At each side of the primary ossification center, a growth plate forms, highly specialized regions of cartilage where all longitudinal growth takes place via endochondral ossification. Histologically and functionally, growth plate chondrocytes can be subdivided into several zones with specific functions (**Fig. 2**).

Bone mass, shape, and strength are maintained throughout development and adult life through an equilibrium between the forces of bone destruction and those of bone formation. *Homeostasis* is the process that controls the continuous remodeling of bone.

Molecular defects in any of these steps can lead to defects in skeletal morphogenesis. Defects of patterning are characterized by changes in gestalt or number of bones, whereas the remaining skeleton is structurally and functionally normal. Such defects are frequently caused by early patterning genes such as Hox or Pax genes. Defects of growth are characterized by stunted, disproportionate size with the majority of skeletal elements affected. Bone, and/or growth plates show abnormal histology and/or abnormal expression of marker genes. Molecular defects that lead to growth defects are frequently caused by mutations in genes that encode for extracellular matrix (ECM) products or for regulatory signal peptides. Remodeling of bone requires resorption.

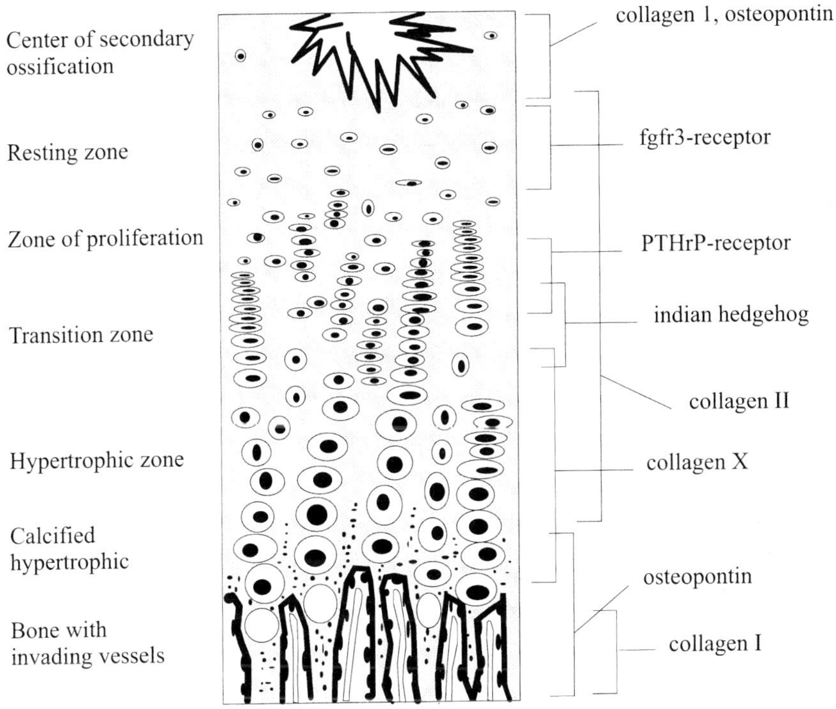

Fig. 2. Growth plate with different zones of chondrocyte differentiation. Expression of marker genes is shown on the right side.

Bone resorption is carried out by osteoclasts, highly specialized cell type derived from monocyte precursors. Defects in osteoclast function lead to increased bone density (osteopetrosis). Decreased bones density resulting from increased absorption or decreased production results in osteoporosis.

2. Materials

1. Staining solution: Prepare stock solutions of Alcian Blue (Sigma A3157; Sigma Chemical Co., St. Louis, MO), (0.14% in 60% ethanol) and Alizarin Red (Sigma A5533), (0.06% in 60% ethanol). For 1 L of staining solution, add components in following order and stir: 33 mL stock solution Alcian Blue, 33 mL stock solution Alizarin Red, 230 mL H_2O, 130 mL glacial acetic acid, 573 mL ethanol 100%.
2. Fixative: 4% formalin in 1X phosphate buffered saline (PBS), with pH adjusted to 7.2.
3. Van Kossa stain: Prepare solutions of: silver nitrate (Sigma S0139) 1%, pyrogallol (Sigma P0381) 0.5%, and, as fixative, sodiumthiosulfate (Sigma S8503) 5%. Staining is Neutral Red (Sigma N6634).
4. Decalcification: Dissolve EDTA tetrasodium 20% in formalin 4%, and adjust pH to 7.2 with powdered citric acid.

3. Methods

In many instances, the first question will be, Is there a skeletal phenotype and how can it be visualized and documented? What the X-ray is in the human, the Alizarin Red stain is in the mouse. The combination of Alizarin Red and Alcian blue allows a differential staining of cartilage (blue) and bone (red). The treatment with KOH and glycerol dissolves muscle and connective tissue but spares bone and cartilage. We use two methods. Method A gives best results for embryos and young mice up to 3 wk of age. Method B stains only bone and is best for adult mice. For best results, mutants should be stained, along with their normal litter mates, at several time points (e.g., E15.5, newborn, 2 wk, adults).

3.1. Preparation and Staining of Skeletons

3.1.1. Method A

Embryos of E13.5–E14.5 are isolated and adjacent membranes removed and fixed in ethanol 95%. Older stages are killed and eviscerated. Using a delicate forceps and a dissecting microscope, remove as much of the skin and underlying adipose tissue as possible. The quality of preparations is very dependent on the care taken here. Bone will not stain if covered by skin. In older stages (>1 wk), the skin can be peeled off from tail to head. Carefully remove as much adipose tissue as possible, because it will not be removed by the KOH digestion. Keep specimen in ethanol 95% overnight (3 h minimum). Remove ethanol, keep in staining solution for 2 d. Remove staining solution and clear with KOH. The intensity and duration of KOH digestion has to be adapted to the developmental stage of the embryo. For 3-wk-old mice, we use 2% KOH for 1 d, followed by 1.8% KOH for 3 d, and 0.3% KOH for 4 d. A 1-wk-old mouse will need 2% KOH for 6 h, 1.8% KOH for 2 d, and 0.3% KOH for 1 d. For earlier stages, the times have to be considerably shortened and only 1.8% KOH is used.

3.1.2. Method B

This method is used for adult mice only. Skin and eviscerate specimen. Cover specimen with 1% KOH for 3–4 d, followed by 2% KOH for 1 d. Pour out solution, rinse carcass with water, clean tail, cover with 1.8% KOH containing 0.004% Alizarin Red, stain for 1–2 d. Pour out solution, rinse specimen, and clear in (1:1) glycerol: 70% ethanol (several days), transfer into 80% ethanol, and store in 100% glycerol.

Once a phenotype has been described and adequately documented, the question How can the phenotype be explained? can be asked. Before starting a set of complicated experiments, it is important to realize what sort of defect the investigator is dealing with. As discussed in the introduction, skeletal defects can be subdivided into defects of patterning, organogenesis, growth, and homeostasis (**Fig. 1**) Does the mutant show abnormalities in gestalt or number of single bones of embryologically related elements (i.e., vertebrae/ribs, limbs, or cranial skeleton), or are the changes present in the majority of bones regardless of their origin? Is there stunted growth or is the growth potential normal? Is the density of bone normal, increased, or decreased? The answer to these questions leads the way to the experimental design.

3.2. Patterning of the Skeletal Elements

If the phenotype turns out to be a defect of patterning, there is little use in looking at histological sections of affected bones. They will be normal. In contrast, whole-mount *in situ* experiments with genes that are known to pattern this part of the skeleton will be useful to determine where and when alterations occur. The methods used for whole-mount *in situ* hybridization have been described in detail elsewhere ***(3a)***. We generally amplify our probes from embryonic RNA with gene-specific primers with a T7 or T3 promoter added to the 3' primer. The PCR product is cleaned on spin columns (Qiagen, Inc., Chatsworth, CA) and used as a template to transcribe Dig-labeled cRNA according to standard protocols (e.g., Genius Kit, Boehringer Mannheim, Mannheim, Germany).

Patterning of individual skeletal elements is an extremely complex process regulated by many, mostly unknown genes. It is beyond the scope of this chapter to discuss all the factors, pathways, and mechanisms. It is, however, possible to give a framework that allows the investigator to perform a series of basic experiments that may lead the way to a more detailed analysis.

Ribs and vertebrae originate from the sclerotome, a structure derived from the somites. During gastrulation, somites form from the paraxial mesoderm as balls of undifferentiated cells on both sides of the neural tube. Signals from the notochord lead to differentiation of the somite into a dorsal dermomyotome and a ventral sclerotome. Sclerotome cells migrate toward the notochord where they condense to form the vertebrae. The notochord regresses and eventually contributes to the intervertebral discs.

Mox1 is a good marker of the early, undifferentiated somite. Mox1 is expressed in all cell types of somites, exhibiting a domain of weak expression in the anterior portion and a domain of strong expression in the posterior portion ***(3)***. Sclerotome markers are Pax1, and Pax9 ***(4)***, and later in the precondensation and condensation stage, Col2a1. Paraxis can be used as a marker for dermomyotome and sclerotome ***(5)***. The myotome can be identified through expression of myogenin and myf5 ***(6)***. Notochord tissue can be detected using brachyury (T) ***(7)***, or Hnf-3β ***(8)*** as markers. Shh is the signal from the notochord that induces the sclerotome and is thus also an excellent marker ***(9)***. Pax3 is expressed along the dorsal neural tube and neural crest cells and later, in the lateral half of the dermomyotome ***(10)***.

The appendicular skeleton originates from a dual contribution of the lateral plate and the somitic mesoderm. After the initial outgrowth of the limb, bud cells from the lateral edges of nearby somites migrate into the limb and contribute to limb muscles, nerves, and vasculature. All other limb tissues, including skeletogenic mesenchyme, derive from lateral plate mesoderm. The skeletal elements of the limbs are laid down sequentially as cartilaginous templates in a proximal-to-distal fashion so that the humeral anlage forms first, followed by the radius and ulna, and finally the digits.

Cells of the limb bud are embedded in a three-dimensional structure of signaling molecules that determine their fate (for a detailed description of limb development, *see* **ref. *11***). Many of these molecules may be used as markers to detect alterations in expression patterns that may explain the observed phenotype. Fgf8 (early and throughout the ridge) ***(12)*** and Fgf4 (later and more posterior) ***(13)*** are good markers for the apical ectodermal ridge (AER), a structure of specialized cells at the outer most edge of the limb bud that control outgrowth and proliferation in the proximal-distal dimension. Shh mediates the activity of the polarizing region (ZPA), a group of cells at the posterior margin of the limb bud that establish anteroposterior pattern ***(13)***. Hox genes of the A and D clusters show characteristic stage-dependent expression patterns ***(14)*** that determine the skeletal pattern of zeugopod (A11, D11) and autopod (A13, D12, D13). The segmentation of the digital anlage into individual phalanges and the separation of digits from one another is achieved through controlled cell death, a process that is likely to be Bmp mediated ***(15,16)***.

3.3. Organogenesis

Signaling centers provide positional information that is recorded on a cell-by-cell basis. With this information, cells are now able to differentiate into specific tissues, a process that results in organogenesis. Genes that are involved in the process of organogenesis are likely to play a role in growth as well. The expected phenotype may therefore involve defects in patterning as well as growth. As a first step in organogenesis, skeletal precursors condense and will then differentiate into chondrocytes or osteoblasts to form the anlage of the future bone. The molecular mechanisms that control condensation and differentiation are largely unknown. Initial adhesion is likely to be mediated by adhesive interactions between cell-surface and extracellular matrix molecules. Transmembrane proteoglycans, such as syndecan-3, have been implicated in this process ***(17)***.

Recent experiments have shown that the differentiation of skeletal precursor cells into osteoblasts is controlled by Cbfa1, a transcription factor that is expressed in skeletal precursor cells as well as in osteoblasts. Mice with inactive Cbfa1 alleles develop cartilage but no bone. Cbfa1 regulates the expression of many of the bone-related genes such as osteocalcin, collagen type I, and osteopontin ***(18,19)***. Whether Sox9, a transcription factor expressed in prechondrocytes and chondrocytes ***(20)***, has a similar function in controlling chondrocyte differentiation remains to be determined. Both genes are good markers for skeletal organogenesis.

3.4. Histology of Bones

Defects of growth generally affect the growth plate as an organ and result in stunted, disproportionate growth. In this case, histology of growth plates and entire bones at various stages is extremely useful to determine the nature of the defect. Sections of long bones (i.e., humerus or femur) at stages 14.5 and after birth stained with standard H&E will give a first idea whether this is a defect of matrix or a defect of proliferation and/or differentiation. In addition, the material can be used for *in situ* hybridizations to be discussed. All tissues are fixed in phosphate buffered solution (PBS) 4% formalin. Frozen sections or sections from paraffin-embedded tissue can be used. Paraffin has the advantage that histology is well maintained, paraffin blocks keep forever, and sections can be stored for many months before use (regular stain or ISH). Embryos of less than E17.5 do not need to be decalcified and can be processed as for regular tissue specimens. Older specimens should be decalcified either with 0.2*N* HCl or in formic acid-sodium citrate until they become soft (depending on the size of the specimen, usually several days). Standard staining procedures include H&E, toluidine blue for staining of cartilage, and van Kossa staining for visualization of calcified tissue (nondecalcified tissue only!). For van Kossa stain, use deparaffinized or frozen sections, stain in 1% silver nitrate for 1–3 min, followed by pyrogallol 30 s. Fix in 5% sodium thiosulfate for 3–5 min, wash in tap water for 10 min, aqua dest. for 30 s, counterstain with neutral red or nuclear fast red.

With these sections in hand, the following questions can be asked. What is the overall aspect of the growth plate? Is the cellular architecture maintained, or is there loss of cartilage differentiation? Is the size of the different zones maintained, are zones expanded, or have they disappeared? Is the staining of cartilage or bone matrix altered? The general aspect of histology can be substantiated by *in situ* hybridization with stage-specific marker genes. The methods used for *in situ* hybridization have been described elsewhere in detail *(22a)*.

Figure 2 shows a schematic diagram of the different zones of cartilage and bone differentiation as they are generally observed in the growth plate. Chondrocytes of the resting zone are kept in a quiescent, undifferentiated state by expressing fibroblast growth factor receptor 3 (Fgfr3) *(21)*. Growth hormone and IGFs lead to a recruitment of resting cells to the zone of proliferation, where cells rapidly divide. In the transition zone, indian hedgehog and PTHrP regulate, through a feedback loop involving ptc and PTHrP-receptor, the differentiation of proliferating cells into hypertrophic cells *(22)*. Once determined to hypertrophy, chondrocytes start expressing collagen type X. In the lower hypertrophic zone, the extracellular matrix begins to calcify and cells express osteopontin. Collagen type II and aggrecan are expressed throughout the cartilage except in the lower calcified hypertrophic zone. Collagen type I is found in the perichondrium and osteoblasts but not in cartilage. Osteocalcin is a marker for mature osteocytes. Sparc, or osteonectin, is expressed in perichondrium, osteoblasts, and to a lesser extent, in nonhypertrophic cartilage (for a review of the expression of matrix genes, *see* **ref.** *23*).

The testing with these and possibly other genes allows a detailed analysis of growth plate function. If there is evidence for altered extracellular matrix, EM histology of different parts of the growth plate should be performed. Excellent cellular morphology and good fixation of cartilaginous matrix is achieved by ruthenium red fixation, as described by Hunsicker and Herrmann *(24)*.

The rate of proliferation can be tested by BrdU incorporation in dividing cells. Inject pregnant females intraperitoneally with 50 µg BrdU (cat, no. 280 879; Boehringer Mannheim) per gram of body weight in PBS. Animals are killed after 2 h, and embryos isolated, fixed, and embedded in paraffin. Sections of growth plates are performed and BrdU-positive cells can be detected by an anti-BrdU-AP monoclonal antibody (cat. no. 1 758 756; Boehringer Mannheim). Labeled cells should be found within the zone of proliferation. Quantification of labeled cells and comparison to controls allows the detection of altered rates of proliferation.

Bone homeostasis is dependent on bone production by osteoblasts on one side and bone resorption by osteoclasts on the other. The regulation of this intricate process is largely unknown. Regular histology of entire, decalcified bones will give first evidence of altered bone density and/or structure. In earlier stages, undecalcified sections should be processed for van Kossa stains to document ossification. In osteoporotic bones, the trabecular and, later, the diaphyseal bone will be reduced and thinned. In contrast, in osteopetrosis, much of the marrow will be filled with dense bone, leaving little place for marrow cells. Such alterations can be quantified by counting the number and measuring the width of trabecles in a given area or by measuring diaphyseal bone strength. The number of osteoclasts can be estimated by detection of tartrate-resistant acid phosphatase (TRAP). The most effective and specific method is detection through TRAP-RNA *in situ* hybridization.

3.5. Primer Sequences

The following sequences give 5' and 3' primers to amplify specific probes from Mouse cDNA for *in situ* hybridization. PCR conditions are: 94°C 4' followed by

35X (94°C 1', 60°C 45", 72°C 45"), followed by 72°C 10'. The T7 promoter sequence (5'> CCT ATA GTG AGT CGT ATT AGG <3') can be added to the 3' primer to generate templates that can be used directly for RNA transcription. All reactions will generate products between 400 and 600 bp.

Aggrecan 5': CAC CAC TGA CCT CAG CTA C
Aggrecan 3': CTC TTT GGC AGT GGA CAG TC
Brachyury (T) 5': TGT TCT ACA GCC TCT TGT TTG A
Brachyury (T) 3': TTT CTG CAG ATT GTC TTT GGC
Collagen type I (Col1a2) 5': CGA GAT CGA GCT CAG AGG C
Collagen type I (Col1a2) 3': CTG GGG CAC CAA TGT CCA A
Collagen type II (Col2a1) 5': CCT GTC TGC TTC TTG TAA AAC
Collagen type II (Col2a1) 3': CAG AGG TGT TTG ACA CAG A
Collagen type X (Col10a1) 5': CAC TTT AGA AAG TGT AAG TTG TTC
Hoxa13 5': CCA GGA GTT TGC TTT CTA CCA
Hoxa13 3': TAG CGT ATT CCC GTT CGA G
Ihh 5': GAG CGC TTC AAA GAG CTC AC
Ihh 3': GAT CCT GAG TCT CGA TGA C
Mox1 5': AGT TGC CCA GTA TGT GGG AG
Mox1 3': TTG GAG AAC ACA AGA CGC TG
MyoD 5': CTG TGT CGA GCC TCG ACT G
MyoD 3': GAG GGA GGG CTC CAG AAA GTG AC
Pax1 5': AGC AGC CAC AGT CCC AAG
Pax1 3': AAC CTC ACC ACC CTG AAG C
Pax3 5': GGA GCC AAT CAA CTG ATG G
Pax3 3': CAT ACT GCC CAT ACT GGT AG
Osp 5': GAT GAA TCT GAC GAA TCT CAC
Osp 3': CTG CTT AAT CCT CAC TAA CAC
Sox9 5': GAC GTG CAA GCT GGC AAA G
Sox9 3': TGG TCA GCG TAG TCG TAT TC
Sparc 5': GCA TTC CTG CAG CCC TTC A
Sparc 3': GCA CTT GGT GGC AAA GAA GT
TRAP 3': CAT GGA TAC GTG GAT GGT GC
TRAP 5': CTC GGC GAT GGA CCA GAT G
(rat cDNA for TRAP, GenBank accession no. M76110)

References

1. Mundlos, S. and Olsen, B. R. (1997) Heritable diseases of the skeleton. Part I: Molecular insights into skeletal development-transcription factors and signaling pathways. *FASEB J.* **11,** 125–132.
2. Mundlos, S. and Olsen, B. R. (1997) Heritable diseases of the skeleton. Part II: Molecular insights into skeletal development-matrix components and their homeostasis. *FASEB J.* **11,** 227–233.
3. Candia, A. F., Hu, J., Crosby J., Lalley, P. A., Noden, D., Nadeau, J. H., and Wright, C. V. (1992) Mox-1 and Mox-2 define a novel homeobox gene subfamily and are differentially expressed during early mesodermal patterning in mouse embryos. *Development* **116,** 1123–1136.

3a. Henrique, D., Adam, J., Myat, A., Chitnis, A., Lewis, J., and Ish-Horowicz, D. (1995) Expression of a Delta homologue in prospective neurons in the chick. *Nature* **375,** 787–790.

4. Neubuser A., Koseki, H., and Balling, R. (1995) Characterization and developmental expression of Pax9, a paired-box-containing gene related to Pax1. *Dev. Biol.* **170,** 701–716.
5. Burgess, R., Cserjesi, P., Ligon, K. L., Olson, E. N. (1995) Paraxis: a basic helix-loop-helix protein expressed in paraxial mesoderm and developing somites. *Dev. Biol.* **168,** 296–306.
6. Montarras, D., Chelly, J., Bober, E., Arnold, H., Ott, M. O., Gros, F., and Pinset, C. (1991) Developmental patterns in the expression of Myf5, MyoD, myogenin, and MRF4 during myogenesis. *New Biol.* **3,** 592–600.
7. Wilkinson, D. G., Bhatt, S., and Herrmann, B. G. (1990) Expression pattern of the mouse T gene and its role in mesoderm formation. *Nature* **343,** 657–659.
8. Ang, S. L. and Rossant, J. (1994) HNF-3 beta is essential for node and notochord formation in mouse development. *Cell* **78,** 561–574.
9. Echelard, Y., Epstein, D. J., St.-Jacques, B., Shen, L., Mohler, J., McMahon, J. A., and McMahon, A. P. (1993) Sonic hedgehog, a member of a family of putative signaling molecules, is implicated in the regulation of CNS polarity. *Cell* **75,** 1417–1430.
10. Gruss, P. and Walther, C. (1992) Pax in development. *Cell* **69,** 719–722.
11. Johnson, R. L. and Tabin, C. J. (1997) Molecular models for vertebrate limb development. *Cell* **90,** 979–990.
12. Vogel, A., Rodriguez, C., and Izpisua-Belmonte, J. C. (1996) Involvement of FGF-8 in initiation, outgrowth and patterning of the vertebrate limb. *Development* **122,** 1737–1750.
13. Niswander, L., Jeffrey, S., Martin, G. R., and Tickle, C. (1994) A positive feedback loop coordinates growth and patterning in the vertebrate limb. *Nature* **371,** 609–612.
14. Nelson, C. E., Morgan, B. A., Burke, A. C., Laufer, E., DiMambro, E., Murtaugh, L. C., Gonzales, E., Tessarollo, L., Parada, L. F., and Tabin, C. (1996) Analysis of Hox gene expression in the chick limb bud. *Development* **122,** 1449–1466.
15. Zou, H. and Niswander, L. (1996) Requirement for BMP signaling in interdigital apoptosis and scale formation. *Science* **272,** 738–741.
16. Storm, E. E. and Kingsley, D. M. (1996) Joint patterning defects caused by single and double mutations in members of the bone morphogenetic protein (BMP) family. *Development* **122,** 3969–3979.
17. Gould, S. E., Upholt, W. B., and Kosher, R. A. (1992) Syndecan 3: a member of the syndecan family of membrane-intercalated proteoglycans that is expressed in high amounts at the onset of chicken limb cartilage differentiation. *Proc. Natl. Acad. Sci. USA* **89,** 3271–3275.
18. Otto, F., Thornell, A. P., Crompton, T., Denzel, A., Gilmour, K. C., Rosewell, I. R., Stamp, G. W., Beddington, R. S., Mundlos, S., Olsen, B. R., Selby, P. B., and Owen, M. J. (1997) Cbfa1, a candidate gene for cleidocranial dysplasia syndrome, is essential for osteoblast differentiation and bone development. *Cell* **89,** 765–771.
19. Ducy, P., Zhang, R., Geoffroy, V., Ridall, A. L., and Karsenty, G. (1997) Osf2/Cbfa1: a transcriptional activator of osteoblast differentiation. *Cell* **89,** 747–754.
20. Wright, E., Hargrave, M. R., Christiansen, J., Cooper, L., Kun, J., Evans, T., Gangadharan, U., Greenfield, A., and Koopman, P. (1995) The Sry-related gene Sox9 is expressed during chondrogenesis in mouse embryos. *Nat. Genet.* **9,** 15–20.
21. Peters, K., Ornitz, D., Werner, S., and Williams, L. (1993) Unique expression pattern of the FGF receptor 3 gene during mouse organogenesis. *Dev. Biol.* **155,** 423–430.

22. Wallis, G. A. (1996) Bone growth: coordinating chondrocyte differentiation. *Curr. Biol.* **6,** 1577–1580.

22a. Huang, L.-F., Fukai, N., Selby, P. B., Olsen, B. R., and Mundlos, S. (1997) Mouse clavicular development: analysis of wild-type and cleidocranial dysplasia mutant mice. *Dev. Dyn.* **210,** 33–40.

23. Mundlos, S. (1994) Expression patterns of matrix genes during human skeletal development. *Prog. Histochem. Cytochem.* **28,** 1–47.

24. Hunziker, E. B. and Herrmann, W. (1987) In situ localization of cartilage extracellular matrix components by immunoelectron microscopy after cryotechnical tissue processing. *J. Histochem. Cytochem.* **35,** 647–655.

9

Transplantation and Culture Techniques for the Analysis of Urodele Limb Regeneration

David L. Stocum

1. Introduction

Altering the spatial relationships of embryonic or postembryonic tissues is one of the oldest and most valuable techniques for investigating mechanisms of development. *Urodele amphibians* (salamanders and newts) are unique among vertebrates in their ability to regenerate several complex structures, including limbs, via the dedifferentiation of mature cells to embryonic-like progenitor cells, which proliferate to form a *blastema* that self-organizes into the missing structures. The *spatial relationships of tissues* in the regenerating limb can be altered surgically either prior to amputation or after blastema formation or chemically by the administration of *retinoic acid* (RA). These alterations evoke regenerative abnormalities that can be correlated with cell interactions, leading to the expression of specific genes in organized spatiotemporal patterns ***(1–3)***. Such experiments are thus valuable as a first level of functional genomic analysis, as well as providing a wealth of information about *cell interactions* during regeneration.

This chapter describes *transplantation*, *implantation*, and in vitro *culture* techniques for analyzing the mechanisms of salamander limb regeneration. In addition, the use of RA combined with transplantation to study limb regeneration is described.

2. Materials

1. Larvae: *Ambystoma mexicanum* (Indiana University Axolotl Colony, Bloomington, IN), 50–70-mm snout-tail length, or *maculatum* (Mike Tolley, Nashville, TN), 25–50-mm snout-tail length.
2. Solutions
   a. Nile blue sulfate (NBS) or neutral red (NR) dye in 2% agar: Mix 20 mg agar (Sigma, Chemical Co., St. Louis, MO) in 1 mL distilled water and heat until completely dissolved. While the solution is still warm (liquid), add 10 mg NBS or NR powder (Sigma) and dissolve the dye completely.
   b. Holtfreter solution: Dissolve 3.5 g NaCl, 100 mg $CaCl_2$, and 20 mg $NaHCO_3$ in 1 L of distilled water. The pH should be approx 7.5 and can be adjusted up or down with 1*N* NaOH or 1*N* HCl.

From: *Methods in Molecular Biology, Vol. 136: Developmental Biology Protocols, Vol. II*
Edited by: R. S. Tuan and C. W. Lo © Humana Press Inc., Totowa, NJ

c. Anesthetic solution: Dissolve 200 mg ethyl p-aminobenzoate (benzocaine; Sigma) in 10 mL of 100% ethyl alcohol (ETOH) and dilute to 1 L with 10% Holtfreter solution to make a 0.02% solution (*see* **Note 1**).
d. Operating solution: Dissolve 20 mg benzocaine in 1 mL ETOH and dilute to 1 L with 100% Holtfreter solution to make a 0.002% solution.
e. Stock all-trans RA solution: Add 50 mg all-trans RA (type XX, Sigma) to 1 mL spectrophotometric-grade dimethyl sulfoxide (DMSO, Mallinckrodt-Baker, Chesterfield, MO) in an Eppendorf tube and vortex until dissolved (solution will be a clear yellow color). The solution should be made under yellow light or subdued daylight, because RA is photosensitive. Aliquots of 100 µL can be stored frozen in the dark until use. Any RA remaining in an aliquot after use should be discarded.
f. Gregg's formalin-alcohol fixative: Dilute 50 mL of stock 40% formalin to 1 L with 950 mL of 70% ETOH.
g. Methylene blue stain for whole mounts: Dissolve 250 mg methylene blue powder (Sigma) in 100 mL of 70% ETOH; then add 0.5 mL of 1*N* HCl.
h. Chelating solution: Dissolve 200 mg ethylenediaminetetraacetic acid, disodium salt (EDTA); (Sigma) in 100 mL calcium-free 100% Holtfreter solution to make a 0.2% solution.

3. Blastema culture medium ***(4)***: Mix 22 mL 80% Leibovitz L-15 medium ***(5)*** (Grand Island Biological), 2.5 mL denatured fetal bovine serum (Microbiological Associates) and 0.5 mL beef embryo extract (Microbiological Associates). pH is approx 7.4, osmolality, approx 260 mOsm (*see* **Note 2**).
4. Iridectomy scissors and #5 watchmakers forceps (Fine Science Tools).
5. Vital dye pipets: Touch the tip of a disposable borosilicate glass pipet to the surface of a NBS or NR/agar solution. Dyed agar solution will be taken up into the tip by capillary action and will solidify as it cools. Place the pipet tip up in a small beaker and allow to dry for several days.

3. Methods

1. Anesthesia: Immerse larvae in anesthetic solution until they become inactive. Larvae are completely anesthetized when they no longer respond to pinching the tail tip with watchmaker's forceps.
2. Limb amputation: Sever the limb at the desired level with iridectomy scissors, using watchmaker's forceps to hold the limb steady. When amputation is through the long bones of the limb, the soft tissue will retract within 1 or 2 min, leaving the ends of the bones protruding. The protruding ends may be left that way, which may delay regeneration somewhat, or they can be trimmed with iridectomy scissors flush with the soft tissue. After amputation, maintain the animal in 10% Holtfreter solution.
3. Blastema transplantation: After anesthetizing the animal, transfer it to a Petri dish containing 0.002% benzocaine/Holtfreter operating solution. This solution allows prolonged low-level anesthesia. Sever the blastema from the limb with iridectomy scissors. If the axial orientation of the graft is important, the dorsal anterior or dorsal posterior edge of the blastema can be stained with NBS or NR dye by gently pressing a dyed agar pipet tip against the blastema for 30 s to 1 min, prior to severing the blastema. With watchmaker's forceps, maneuver the blastema close to the graft site. For best results, the host site should bleed. If bleeding does not occur during preparation of the host site, nick one or two blood vessels with the watchmaker's forceps. Gently press the graft onto the host wound surface; the clotted blood plasma will serve as a glue to hold the graft in place (*see* **Note 3**). Allow the transplant to heal in place for 3–4 h at room temperature (RT). Transfer the animal to a finger bowl containing 10% Holtfreter solution at 4°C by gently submerging

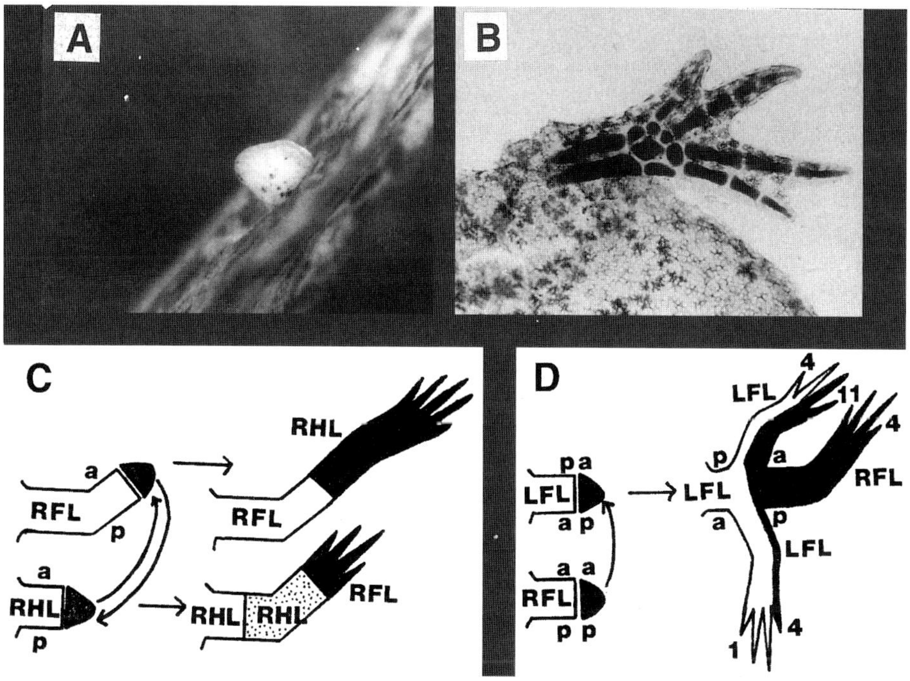

Fig. 1. **(A)** Dorsal view of a cone-stage *Ambystoma maculatum* forelimb blastema (derived from the distal upper arm) autografted to a wound bed made by removing a patch of skin from the dorsal fin, 4 d after grafting. The blastema is growing and is beginning to undergo morphogenesis. **(B)** A similar graft that has developed a nearly normal set of skeletal elements (methylene blue stain for cartilage). **(C)** Reciprocal autografts of a wrist-level right forelimb blastema (RFL) and a mid-thigh level right hindlimb blastema. Both transplants develop according to their limb type and proximodistal (PD) level of origin. In the case of the wrist blastema grafted to the mid-thigh of the hindlimb, the missing intermediate parts (stippled) are filled in by intercalary regeneration from the host hindlimb. a, anterior; p, posterior. **(D)** Right forelimb blastema autografted to the same PD level of a left forelimb stump with reversal of the anteroposterior (AP) axis of the blastema. One or two supernumerary regenerates arise on the anterior and/or posterior side of the limb at the graft/host junction because of intercalary regeneration between cells with anterior (a) and posterior (p) positional identities. Numbers indicate digits. The transplanted blastema develops with the AP polarity of its origin. The supernumeraries have host chirality and are formed by varying contributions from host and graft tissue (black and white color).

the Petri dish in the finger bowl and pushing the animal out of the Petri dish. Leave overnight at 4°C, then maintain at RT for development of the graft. Revascularization of the graft will be evident within 1–2 d. Three different types of blastema transplant experiments and their results are illustrated in **Fig. 1** (*see* **refs.** ***8–11***).

4. Constructing symmetrical limbs (**Fig. 2**) (*see* **refs.** ***12,13***): Anesthetize the animal and transfer it onto 0.02% benzocaine-soaked filter paper lining the bottom of a Petri dish. Place anesthetic-soaked strips of Kimwipes (Kimberly-Clark, Roswell, GA) over the head and gills of the animal to keep it moist and to maintain anesthesia. For double anterior and posterior zeugopodia (*see* **Note 4**) make a midline cut with iridectomy scissors through the carpals or tarsals between digits 2 and 3 and between the radius and ulna or the tibia and fibula almost to the elbow or knee joint. With a transverse cut, remove the posterior half of

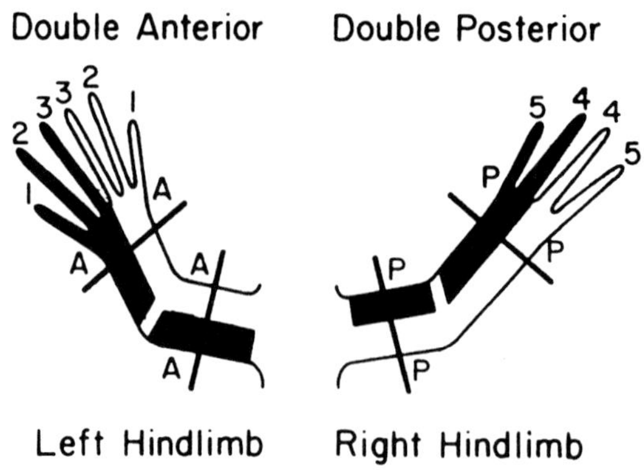

Fig. 2. Diagram of symmetrical hindlimb constructs made by exchange of anterior (a) and posterior (p) halves. Because the cut separating the halves of the lower limb (zeugopodium) passes through digits 2 and 3, and the hindlimb has five digits, the double anterior construct will have six digits and the double-posterior construct four digits.

one limb and the anterior half of the other and exchange the halves. A small piece of the zeugopodial bone of the removed half should be left proximally to impart mechanical stability to the construct. Double dorsal and ventral zeugopodia are made by removing the skin and flexor muscles from one limb and the skin and extensor muscles from the other and exchanging. Double anterior, posterior, dorsal, and ventral stylopodia (*see* **Note 4**) are constructed by making exchanges between anterior and posterior or between dorsal and ventral skin and muscles. The host wound surfaces should bleed so that grafts will adhere when pressed into place (*see* **Note 5**). Allow the grafts to heal for 4–6 h at RT. Transfer the animal to a finger bowl containing 10% Holtfreter solution at 4°C and leave overnight at 4°C. Allow the grafts to heal for 7–10 d before conducting further procedures. Revascularization of the grafts will be evident within 2–3 d. An extra limb may arise proximally where graft and host tissues are confronted (*see* **Note 6**).

5. In vitro blastema culture ***(4)***: Sever the blastema and an adjacent segment of the limb with iridectomy scissors and soak it for 30–45 min in a tissue-culture well containing chelating solution (*see* **Note 7**). Transfer the blastema to another well containing blastema culture medium. Holding the stump segment with one pair of watchmaker's forceps, remove the loosened epidermis with another pair of watchmaker's forceps. Transfer the tissue into a third well containing culture medium and sever the blastema from the stump. Wash the blastema in four successive changes of culture medium. For short-term cultures (a few days), make a hanging drop culture by placing the blastema in a drop of culture medium on an acid-washed, sterile glass cover slip. Invert the cover slip over the well of a depression slide ringed with silicone grease and seal the edges with grease. For long-term cultures, explant the blastema onto an acid-washed sterile glass cover slip in a small culture dish (internal diameter, 30 mm, height 12 mm; Bellco Glass Co.) and cover the bottom of the dish with just enough medium to gently press the blastema onto the cover slip. Seal the lid of the culture dish with stopcock grease. More medium can be added after 1 or 2 d when the blastema cells have attached to the glass. The explants differentiate cartilage and muscle, but cells also migrate as a monolayer from the edges of the explant. **Figure 3** illustrates living migrated cells as viewed with an inverted microscope equipped with phase contrast optics.
6. Culturing blastemas as fin tunnel implants ***(14)***: Loosen the epidermis in chelating solution as previously described (*see* **Note 8**). Remove the epidermis from the blastema and

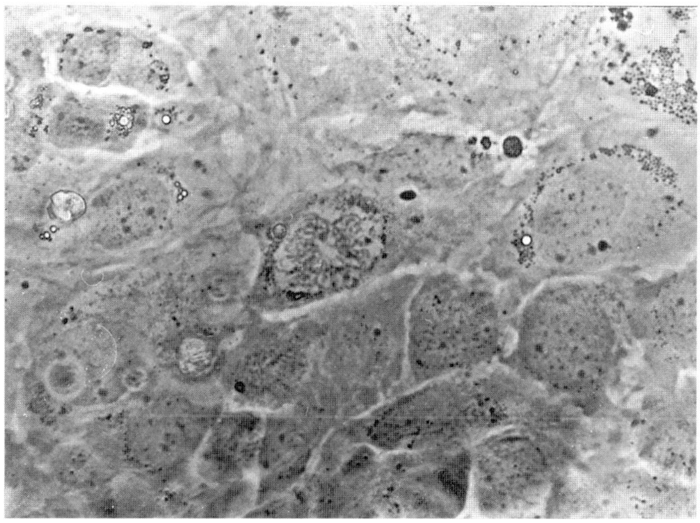

Fig. 3. Living blastema cells migrated from a cone-stage blastema, 2 wk after explantation, phase contrast optics. In the center of the field is a cell in metaphase.

sever it from the adjacent limb tissue as previously described. Use one tine of the watchmaker's forceps to pierce the skin of the dorsal fin and make a tunnel in the connective tissue of the fin. Run the tine through the skin again at the end of the tunnel to relieve air and fluid pressure, and enlarge the tunnel using a glass ball tip of appropriate size. Maneuver the blastema to the mouth of the tunnel with watchmaker's forceps and push it into the interior with the ball tip (*see* **Note 9**). **Figure 4** illustrates an experiment using this procedure and the results.

7. Injection of RA *(15)*: Weigh the animal on a top-loading balance. In yellow light or subdued daylight, place the animal on its back on a wet paper towel under the dissecting microscope. With a 27-gage needle, make a prepuncture through the skin and abdominal muscles into the peritoneal cavity, a few millimeters anterior to one of the hindlimbs and lateral to the midline. Draw into a microsyringe the amount of stock RA solution that will deliver the desired dose of RA/g body wt. (*see* **Note 10**). Insert the microsyringe needle through the prepuncture into the peritoneal cavity and inject the RA solution. RA, like other retinoids, is insoluble in an aqueous environment and will precipitate at the injection site, being visible through the skin as a compact or diffuse yellow mass that will slowly be solubilized by retinoid binding proteins in the plasma. Inject control animals with an equivalent amount of DMSO. Return the animal to 10% Holtfreter. When the blastema has developed, it can be used in transplantation or implantation experiments (*see* **refs.** ***17,18***). **Figure 5** illustrates an experiment involving transplantation of RA-treated blastemas and the result.
8. Whole-mount preparation for cartilage: Fix each specimen for 3–24 h in a small vial of Gregg's fixative. Pipet off the fixative and replace with a small amount of methylene blue stain. Stain for 30 min. Pipet off the stain and rinse with 95% ETOH. Destain in fresh 95% ETOH, changing the alcohol as it becomes colored. Periodically agitate the specimen until it no longer gives off stain. Dehydrate in three changes of 100% ETOH, 10 min each. Clear the specimen in methyl salicylate for 1 h. The tissues can be stored indefinitely in the methyl salicylate (keep the vials tightly capped) or, after replacing the methyl salicylate with three changes of toluene, 10 min each, they can be mounted in mounting medium

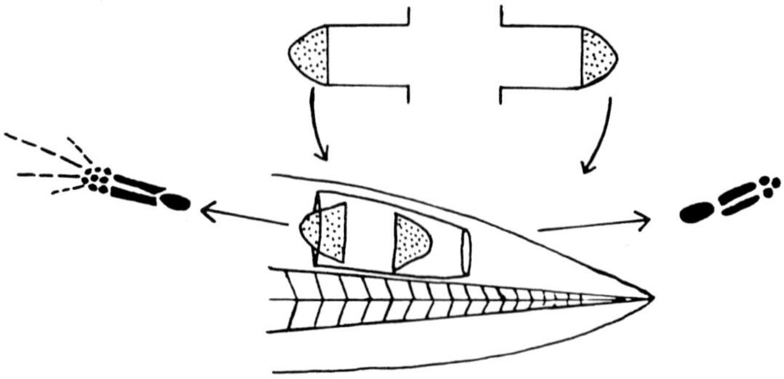

Fig. 4. Diagram showing right and left forelimb *A. maculatum* epidermis-free blastema mesenchymes autoimplanted into a dorsal fin tunnel. The tip of the left mesenchyme protrudes from the mouth of the tunnel and is recovered with fin epidermis, which promotes completion of the PD pattern of skeletal elements. The right mesenchyme resides completely within the tunnel and develops with a distally truncated pattern of skeletal elements.

under a cover slip on a glass slide. Cartilage appears blue and other tissues are colorless (*see* **Note 11**).

4. Notes

1. Two other useful anesthetics are: **(A)** ethyl m-aminobenzoate, methane sulfonate salt (MS:222, Finquel; Sigma) at a concentration of 1:1000 in 10% Holtfreter; **(B)** 1,1,1-trichloro-2-methyl-2-propanol (chlorobutanol, Chlorotone; Sigma) at a concentration of 0.075% in 10% Holtfreter. Operating solutions (*see* **Subheading 2.2.**) would be one tenth these concentrations in 100% Holtfreter. It is important that dose-response data be obtained for any of the anesthetics used, as the speed and depth of anesthesia and the length of time the animals can be kept under prolonged anesthesia may vary with species, size, and other factors.
2. This medium is designed to approximate the osmolality of salamander body fluid, 258 mOsm as calculated from the freezing point depression of adult *Salamandra maculata* blood. L-15 medium was chosen because it is buffered by its free base amino acids and thus requires no CO_2 gassing. Beef embryo extract, which provides a crude source of growth factors, is no longer advertised by companies that sell sera. A suitable substitute may be bovine embryonic fluid (Sigma), which originates from the chorion of the embryo, but this has not been verified experimentally. Two other media for blastema cell culture that do not use embryo extract are described in ***(6,7)***.
3. If the blastema is transplanted to a body surface, remove a patch of skin to make a wound bed for the graft. If the blastema is transplanted to the stump of an amputated limb, the protruding skeletal elements of the host site must be trimmed flush with the soft tissue for the graft to adhere to the wound surface.
4. Zeugopodium refers to the second segment of the vertebrate limb containing the radius and ulna or tibia and fibula; stylopodium refers to the first segment of the limb containing the humerus or femur.
5. Good adherence is facilitated by propping the limbs up with small balls of modeling clay so they are above the surface of the filter paper. Gravity and surface tension then press the grafts onto the host wound bed.
6. A supernumerary regenerate often arises at this site as a result of the juxtaposition of anterior and posterior tissue, which triggers formation of an accessory blastema.

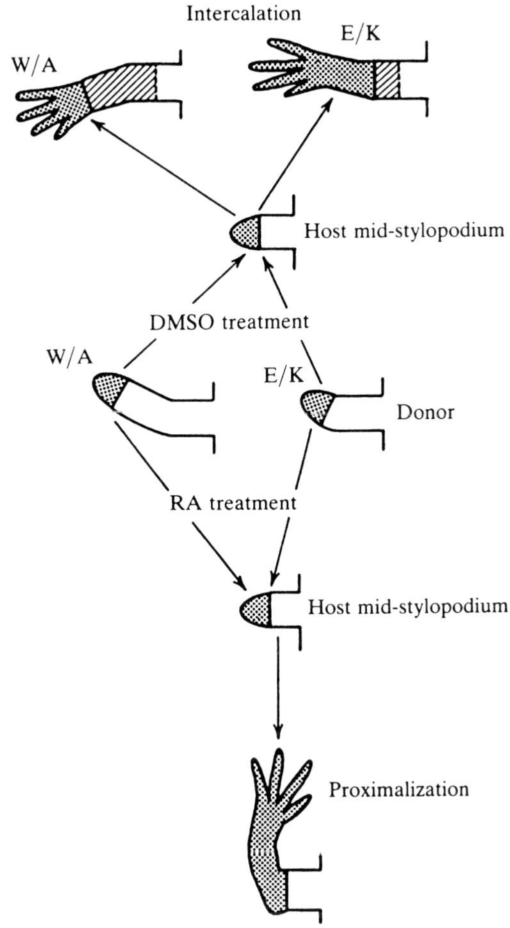

Fig. 5. Experiment in which a RA or DMSO-treated blastema (stipple) from the wrist or ankle (W/A), or the elbow or knee (E/K) is heterografted to the mid-stylopodium (upper arm or leg) of an untreated animal. The positional identity of the RA-treated blastema is proximalized to the mid-stylopodial level as evidenced by the fact that intercalary regeneration does not occur. The DMSO-treated blastema, however, evokes intercalary regeneration from the host stump (crosshatch).

7. All instruments, solutions, and media must be sterile. The forceps and iridectomy scissors can be resterilized periodically by wiping them with cotton soaked in 95% ETOH and flaming in a Bunsen burner. The washes in culture medium are effective in removing microbial contamination; however, antibiotics can be added to the medium if desired, and the epidermis can be sterilized by immersion in 1% sodium hypochlorite for 90 s.
8. None of the instruments or solutions used in this method need be sterilized, as the animal's immune system will prevent infection.
9. Fin tunnel implants have been done only in *Ambystoma maculatum* and *Ambystoma mexicanum* larvae. The implants develop successfully only in *maculatum*. For unknown reasons, blastemas implanted into the dorsal fin of *Ambystoma mexicanum* larvae resorb (personal unpublished research).
10. RA causes proximalization, posteriorization, and ventralization of blastema cell positional identity ***(12,13,15,16)***. The most effective dose is 100–150 μg RA/g body wt administered

during the period of dedifferentiation-initial blastema formation, which is approx 4 d for the size of larvae described here. At these doses, RA delays blastema formation by 7-10 days. In controls, a cone-stage blastema is present by 6–7 days after amputation, whereas in an RA-treated animal, a cone-stage blastema is not attained until 14–17 days after amputation. Higher doses result in inhibition of regeneration.

11. This relatively rapid staining technique gives a global view of cartilage patterns. For preparation and staining of sections, refer to histological technique manuals such as ***(19)***.

References

1. Gardiner, D. M., Blumberg, B., Komine, Y., and Bryant, S. V. (1995) Regulation of *Hox A* expression in developing and regenerating limbs. *Development* **121,** 1731–1741.
2. Pecorino, L. T., Entwistle, A., and Brockes, J. P. (1996) Activation of a single retinoic acid receptor isoform mediates proximodistal respecification. *Curr. Biol.* **6,** 563–569.
3. Imokawa, Y. and Yoshizato, K. (1997) Expression of *Sonic hedgehog* gene in regenerating newt limb blastemas recapitulates that in developing limb buds. *Proc. Natl. Acad. Sci. USA* **94,** 9159–9164.
4. Stocum, D. L. (1968) The urodele limb regeneration blastema: a self-organizing system. I. Differentiation *in vitro*. *Develop. Biol.* **18,** 441–456.
5. Leibovitz, A. (1963) The growth and maintenance of tissue/cell cultures in free gas exchange with the atmosphere. *Am. J. Hyg.* **78,** 173–180.
6. Hinterberger, T. and Cameron, J. A. (1983) Muscle and cartilage differentiation in axolotl limb regeneration blastema cultures. *J. Exp. Zool.* **226,** 399–407.
7. Ferretti, P. and Brockes, J. P. (1988) Culture of newt cells from different tissues and their expression of a regeneration associated antigen. *J. Exp. Zool.* **147,** 77–91.
8. Stocum, D. L. (1968) The urodele limb regeneration blastema: a self-organizing system. II. Morphogenesis and differentiation of autografted whole and fractional blastemas. *Develop. Biol.* **18,** 457–480.
9. Stocum, D. L. and Melton, D. A. (1977) Self-organizational capacity of distally transplanted limb regeneration blastemas in larval salamanders. *J. Exp. Zool.* **201,** 451–462.
10. Pescitelli, M. J. and Stocum, D. L. (1980) The origin of skeletal structures during intercalary regeneration in larval *Ambystoma* limbs. *Develop. Biol.* **79,** 255–275.
11. Stocum, D. L. (1982) Determination of axial polarity in the urodele regeneration blastema. *J. Embryol. Exp. Morphol.* **71,** 193–214.
12. Kim, W.-S. and Stocum, D. L. (1986) Retinoic acid modifies positional memory in the anteroposterior axis of regenerating axolotl limbs. *Develop. Biol.* **114,** 170–179.
13. Ludolph, D. C., Cameron, J. A., and Stocum, D. L. (1990) The effect of retinoic acid on positional memory in the dorsoventral axis of regenerating axolotl limbs. *Develop. Biol.* **140,** 41–52.
14. Stocum, D. L. and Dearlove, G. E. (1972) Epidermal-mesodermal interaction during morphogenesis of the limb regeneration blastema in larval salamanders. *J. Exp. Zool.* **181,** 49–62.
15. Thoms, S. D. and Stocum, D. L. (1984) Retinoic acid-induced pattern duplication in regenerating urodele limbs. *Develop. Biol.* **103,** 319–328.
16. Maden, M. (1982) Vitamin A and pattern formation in the regenerating limb. *Nature (London)* **195,** 672–675.
17. Crawford, K. and Stocum, D. L. (1988a) Retinoic acid coordinately proximalizes regenerate pattern and blastema differential affinity in axolotl limbs. *Development* **102,** 687–698.
18. Crawford, K. and Stocum, D. L. (1988b) Retinoic acid proximalizes level-specific properties responsible for intercalary regeneration in axolotl limbs. *Development* **104,** 703–712.
19. Humason, G. L. (1972) *Animal Tissue Techniques.* W. H. Freeman, San Francisco, CA.

10

Retroviral Infection of T-Cell Precursors in Thymic Organ Culture

Lisa M. Spain, Lisa L. Lau, and Yousuke Takahama

1. Introduction

The ex vivo culture of intact fetal thymus lobes, fetal thymic organ culture (FTOC) has proven to be a useful technique for the elucidation of the molecular and cellular bases of T-cell development and T-cell receptor (TCR) repertoire selection (reviewed in **ref.** ***1***. Most dramatically, FTOC has been used to determine the role of peptides in positive and negative selection, which has helped to shed light on the complex mechanism whereby the T-cell repertoire is tolerized to self (reviewed in **ref.** ***2***).

Gene transfer into developing T-cell precursors in FTOC provides a fast and economical method (relative to germ line transgenics) to explore gene function during T-cell development. Gene transfer in FTOC can also serve as a replacement for the more expensive and technically difficult retroviral infection of hematopoietic stem cells followed by bone marrow transplantation. Alternatively, retrovirus-mediated gene transduction in vitro can complement these other in vivo technologies.

Direct retrovirus-mediated gene transfer to thymus in FTOC has been reported ***(3)*** and has provided information on the role of MAP kinases in T-cell development. However, we have developed modified techniques for gene transfer that are more efficient and reproducible in our hands than the direct infection method. In this chapter, we will describe retrovirus-mediated gene transfer of fetal liver and thymic T-cell progenitors and reconstitution of depleted thymic lobes.

2. Materials

2.1. Medium

1. Staining medium: Hank's buffered saline (HBS; Life Technologies, Bethesda, MD) plus 5% fetal bovine serum (FBS). For analysis, use 0.05% sodium azide; for collecting tissues and sterile sorts, omit the azide.
2. DME+: Dulbecco's modified Eagles medium (DMEM, 4.5 g glucose/L; Life Technologies), 10% FBS, 50 µg/mL gentamycin, $5 \times 10^{-5} M$ βME, 1 m*M* HEPES, pH 7.3, nonessential amino acids, L-glutamine.
3. RPMI-15: RPMI (Life Technologies), 15% prescreened FBS (*see* **Note 1**), L-glutamine, 50 µg/mL gentamycin, $5 \times 10^{-5} M$ βME, 10 m*M* HEPES, pH 7.3.

From: *Methods in Molecular Biology, Vol. 136: Developmental Biology Protocols, Vol. II*
Edited by: R. S. Tuan and C. W. Lo © Humana Press Inc., Totowa, NJ

4. RPMI-10: Same as above except just 10% FBS.
5. Infection medium, per 100 mL: 65 mL RPMI-10 (as above except 10% FBS), 30 mL DME/10% calf serum, 5 mL FBS, 0.5 mL 3X supernatant (*see* **Subheading 2.2.**). OPTIONAL: 2 µg/mL polybrene (cat. no. 10,768-9; Aldrich, Milwaukee, WI), make up an 8 mg/mL stock solution).
6. Ficoll plaque (#17-0840-02; Pharmacia, Uppsala, Sweden) or lympholyte M (#ACL-5030, more expensive; Accurate (Chemical and Scientific Corp., Westbury, NY).

2.2. Growth Factors

SCF (a.k.a. stem cell factor, *Steel* factor, c-kit ligand) from Peprotech (Rocky Hill, NJ) and IL-7 (Genzyme, Boston, MA) at 5 ng/mL each final concentration. IL-3 supplied as 3X cell-conditioned medium *(4)*.

2.3. Mice

Strains congenic for Ly 5 alleles are used to discriminate between donor and recipient thymocytes. These congenic mice are available from The Jackson Laboratories (Bar Harbor, ME) on the C57BL/6 background. Set up cages containing one male with two females for timed pregnancies. Check for plugs early the next 3 d, using a flame-polished capillary tube as a probe. The day of plug observation is counted as day 0. To make an ongoing timed pregnancy colony, be sure to rest males (house without females) for at least 3 d out of every week. Unplugged females can be reused if they are not visibly pregnant by 12 d. For B10.BRSgSn/J (C57BL/10 congenics; Jackson Laboratories) mice in our colony, we observe that on average, 75% of females are plugged by 4 d, and of these, on average 50–75% become pregnant. Embryos are harvested from gestational days 14–17 as required.

2.4. Cell Lines

Ψ2 cells *(5)* or other appropriate retroviral producer cell line such as GP+E-86 cells *(6)*. Ψcre *(7)*, or BOSC23 *(8)*. Titers of virus produced should be above 1×10^6 (*see* **Note 6**).

2.5. Miscellaneous Reagents

1. Antibodies: KJ25-biotin (anti-Vβ3, Pharmingen, San Diego, CA), Streptavidin-Red 670™ (Gibco-BRL/Life Technologies, Gaithersburg, MD), CD4-PE (Gibco-BRL/Life Technologies), CD8-FITC (Gibco-BRL/Life Technologies or Caltag Laboratories, Burlingame, CA), biotinylated anti-human c-fms (a gift from C. Sherr).
2. Sterile gelatin sponges: Gelfoam (NDC0009-0342-01 size 100; Upjohn, Kalamazoo, MI). Cut into small pieces (1 × 1 cm) using sterile scissors, store at room temperature.
3. Nucleopore filters: 0.8-mm pore, 13-mm discs, (#11409 Costar, Cambridge, MA). Autoclave dry wrapped in foil.
4. Terasaki dishes (from Nunc, Wiesbaden-Biebrich, Germany) (#R53359064 Fisher Scientific Co., Pittsburgh, PA).
5. Fine forceps (#5 Roboz, #11 Dumont), scalpels (#11 Fisher), cone homogenizer, (#05-559-26; Fisher).
6. 2-deoxyguanosine from SIGMA, 50X in water, heat to 50°C to dissolve. Final concentration for depletions = 1.35 µ*M* in medium.

2.6. Specialized Equipment

Access to a flow cytometer for sterile sorting and 3-color analysis and a ^{137}Cs source for irradiations.

3. Methods

3.1. Retroviral Infection of T-Cell precursors

3.1.1. Fetal Liver Stem Cells

3.1.1.1. Day 0

Split the producer cell lines into 100-mm dishes so that they will be confluent by the end of the cocultivation (3 d). This is approx 4–5 × 10^5 cells/dish, although it will probably be necessary to empirically determine the best density for seeding each individual producer subline and for different packaging lines.

3.1.1.2. Day 1

Euthanize the pregnant female (gestational day 14; *see* **Note 2**) using CO_2 inhalation and remove the uterus. Remove each embryo and place into sterile (no azide) staining medium on ice. Pin each embryo on a styrofoam board ventral side up. The dark red liver will be visible just below the skin at the midsection. Using fine tweezers, pierce the abdomen covering the liver and rip a small hole through skin and peritoneum. Press down gently on the abdomen on either side of the hole to force the four lobes of the liver out, then lift the liver and place into a dish containing dissecting medium. Disrupt the livers by pipetting up and down in medium (*see* **Note 3**).

Wash the liver cells 1X with RPMI-15. The number of nucleated cells recovered per liver varies from 10^5 to 10^6 cells. Plate cells from 1–2 livers (5 × 10^5 cells) per 10-cm dish in Infection Medium (*see* **Note 4**).

3.1.1.3. Day 3

After 48 h, harvest the fetal liver cells by gently pipetting the medium over the surface of the monolayer, but not so hard as to remove the producer cells. If many producer cells contaminate the fetal liver preparation, flow cytometry can be used to remove them based on size (*see* **Subheading 3.2.**). Purify cells on a ficoll paque or Lympholyte M gradient, and wash 3X. Recovery can be variable, but is usually approx 5 × 10^5 cells per liver.

Kill a pregnant female (gestational day 15–17) for recipient thymus lobes. Harvest embryos as above and pin down ventral side up, so that there is some tension through the neck area. Using a dissecting microscope, insert the point of a No. 11 scalpel just behind the top of sternum and cut an anterior-posterior incision using outward motions. The tension should help to draw open the upper thorax. The thymic lobes will be visible on either side of the trachea on top of or just above the major heart vessels; pull back and cut away surrounding tissues until clear access to thymus is obtained. Using the tip of the scalpel, lightly tear the membranes around the thymic lobes so they can be easily lifted out separately using fine tweezers. Place the lobes in a well of a 24-well dish containing 0.2 mL of sterile (no azide) staining medium. Irradiate the lobes with 2250 rad, and rinse in RPMI-15.

Resuspend the fetal liver cells at 3 × 10^5 cells per 30–40-μL medium (7.5–10 × 10^6 cells per millileter). Pipet 35 μL into wells (well-spaced) of a Terasaki dish. Be sure to include some wells with no cells for radiation controls. Place one irradiated thymic lobe per well, invert dish, and place in a plastic box containing wet paper towels within the incubator (*see* **Note 5**). The box is to increase local humidity to prevent evaporation of the medium in the hanging drops.

3.1.1.4. Day 4

After 15–24 h, prepare the air-medium interface thymic organ cultures. Add 2.5 mL of RPMI-15 to each well of a 6-well culture dish. Add one cube of gelfoam to each well and soak it with medium by pressing with a sterile pipet or forceps. Make sure the gelfoam absorbs the culture medium well. Rinse a nucleopore filter in medium and place it, shiny-side up, on top of the wet sponge. Avoid trapping an air bubble between the gelfoam and the nucleopore filter. Revert the Terasaki dishes and transfer each lobe carefully using fine forceps (*see* **Note 6**). Arrange the lobes on the filter so that the lobes are not touching each other. Approximately 4–6 lobes can be cultured together on the filter. Incubate lobes for the required time (10–18 d) replacing 2 mL of the medium in each well every 2–3 d.

3.1.2. Thymic Stem Cells

3.1.2.1. Day 4

Prepare recipient lobes for reconstitution. Harvest gestational day 15–16 thymus lobes and place in organ culture with 1.35 μ*M* 2-deoxyguanosine (*see* **Note 7**).

3.1.2.2. Day 0

One day before obtaining fetal thymocytes, trypsin treat and harvest producer cells, and plate in a flat-bottom 96-well dish at 2–4 × 10^3 per well (*see* **Note 8**). As above, the exact number of producers cells plated might need to be empirically determined for different producer sublines.

3.1.2.3. Day 1

Dissect out day 14 fetal thymocytes (for day 15 thymocytes, *see* **Note 8**), pool, and make single-cell suspensions using a small cone homogenizer. Place 200 μL of medium into a sterile Eppendorf tube and gently press the lobe between the homogenizer and the side of the tube and twist. Add cells to cultures at 0.5–2 × 10^4/well. Add SCF and IL-7 at 5 ng/mL each final concentration. Coculture the thymocytes and producer cells for 2 d. Viral titer in wells should reach about 1–3 × 10^6 cfu/mL by day 2 (*see* **Note 9**). The producer cell monolayer should be approx one-half confluent at day 1 and densely confluent by day 2. If 2-deoxyguanosine was used to deplete the recipient thymic lobes, they should be rinsed free of 2-deoxyguanosine and placed in a fresh well overnight to allow all residual drug to diffuse away.

3.1.2.4. Day 2

Harvest the cocultured thymocytes—it is okay if the producer monolayer dislodges a bit when collecting thymocytes. Cells will be sorted by size and density using forward and side scatter; thymocytes are smaller and less dense than are producer cells. Count cells. (Significant cell expansion is usually not observed.) Using a fluorescence-activated cell sorter, sort viable thymocytes; they should be easily separated from the producers based on forward and side scatter (**Fig. 1**). Alternatively, staining with antibody against the hematopoietic cell marker CD45 can serve to separate thymocytes from producer fibroblasts. At this step, if the retrovirally transduced gene product is a

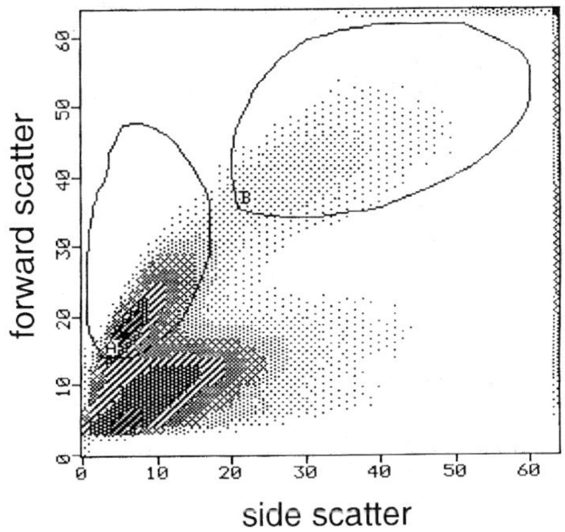

Fig. 1. Presort analysis of thymocytes cultured with ψ2 producer cells for 2 d in the presence of SCF and IL-7, then cell-sorted by size based on forward and side scatter parameters. Viable thymocytes were collected in region A, and residual producer cells (in region B) were excluded.

surface marker expressed on immature thymocytes, it is possible to also sort the thymocytes for the presence of the marker by fluorescent antibody staining. Another marker that has worked well for one of us (Y.T.) is the green fluorescent protein (GFP) expressed via an IRES site in the retroviral vector. This allows direct sorting of transduced thymocytes based on green fluorescence. After the sort, wash thymocytes and combine equal numbers (from 1000–2500 if day 14 thymocytes are used, 1×10^3–1×10^4 cells if day 15 thymocytes are used) with each depleted thymic lobe in hanging drops (30 μL DME$^+$ in Terasaki wells). Each experimental condition should be done in triplicate or quadruplicate and analyzed separately.

3.1.2.5. Day 3

Transfer to nucleopore filters in standard FTOC as above and culture ≥ 10 d before staining. Feed/replenish DME+ culture medium every 3 d.

3.2. Analysis of T-Cell Development and Retroviral Infection

3.1.3.1. Days 10–18

Remove each thymic lobe to a separate eppendorf tube in 200 μL of staining medium on ice. Disrupt the thymus using a small cone homogenizer as above. Usually up to 10^5 thymocytes are recovered, enough for up to three separate antibody stainings (*see* **Note 10**). Stain the cells using fluorescently labeled antibodies against CD4, CD8, and the transduced gene product, in the present examples, TCR β chain. In addition, the congenic marker Ly 5.1 can be used to discriminate the donor from host thymocytes, although normally, if the depletions have worked adequately, there should be no remaining host thymocytes. Other combinations of antibodies and T-cell markers can of course be used based on the capabilities of the available flow cytometer. Analyze the staining pattern by flow cytometry using standard procedures.

Thymic lobes reconstituted using thymic precursors will develop faster than those reconstituted by fetal liver producers. At day 10, we saw that thymic lobes reconsti-

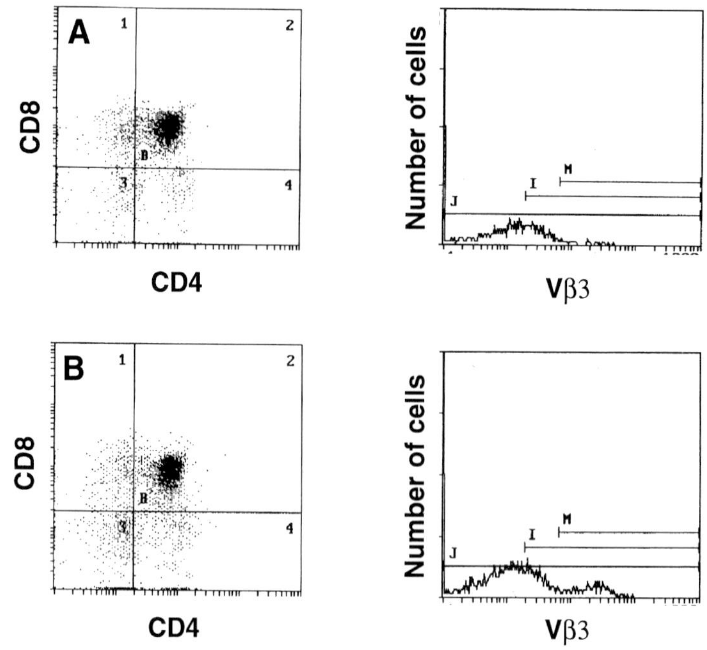

Fig. 2. Vβ3 expression by fetal thymocytes following retrovirus-mediated gene transfer to fetal thymus progenitors. Gestational day 15 fetal thymocytes were cocultured for 2 d with **(A)** control or **(B)** MFG-Vβ3 retrovirus-producing cells in the presence of SCF and IL-7. Viable thymocytes were then selected by cell sorting and allowed to differentiate in FTOC for 10 d. Flow cytometry was performed to analyze expression of CD4 versus CD8 *(left)* and Vβ3 *(right)* by the thymocytes. Approximately 10% of total thymocytes in (B) were Vβ3+, compared with 2% in (A).

tuted by thymic precursors had developed significant numbers of mature single-positive as well as double-positive thymocytes. A significant fraction (50% of $CD4^+8^-$ thymocytes) of these expressed our transduced TCR β chain gene, and 10% of the total thymocytes expressed the gene (**Fig. 2**). In contrast, at day 10 of the fetal liver reconstitutions, the majority of thymocytes were still $CD4^-8^-$. By day 16, the fetal liver precursors had given rise to more differentiated thymocytes, expressing the transduced TCR β gene appropriately (**Fig. 3**). Although, in the experiments shown, the efficiency of transduction was higher in fetal liver compared with thymocytes, fetal liver transduction may be less reproducible from lobe to lobe. This may be because infection of fetal liver targets a more primitive progenitor. The decision to infect fetal liver or thymocyte progenitors will depend on the gene to be expressed. In some cases, expression in an earlier progenitor is desired, whereas in others, early expression might be detrimental.

4. Notes

1. Different lots of FBS to be used for FTOC should be screened. Set up quadruplicate FTOC in different lots of serum in RPMI-15 media. Harvest gestational day 15 thymuses, set up in FTOC directly as described below, changing medium every 2 d. Harvest at 5–6 d and stain for T-cell markers. Select only those lots of serum that support efficient development of especially the $CD4^+8^+$ and single-positive subsets (≥65% should be $CD4^+8^+$, ≤20% should be $CD4^-8^-$, 5% $CD4^+8^-$, and 10% $CD4^-8^+$).

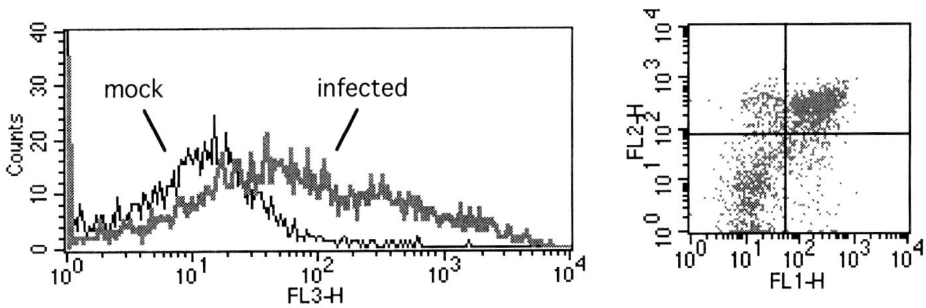

Fig. 3. Vβ3 expression by fetal thymocytes following retrovirus-mediated gene transfer to fetal liver stem cells. Gestational day 14 fetal liver cells were cocultured with mock transfected or MFG-Vβ3 retrovirus-producing cells as indicated *(left)*. Gating on the Vβ3+ cells and displaying CD4 and CD8 expression *(right)*, it is clear that Vβ3-expressing cells develop appropriately. Shown is the single lobe with the highest transfection efficiency obtained, results ranged from 5% to 60% transduction efficiency in 10 samples.

2. It is useful to keep a list of the range of crown-rump lengths of the embryos at various gestational ages. This can come in handy for estimating the gestational age of embryos resulting from "unplanned" pregnancies.
3. Some very small liver tissue fragments may remain after pipetting. There is no need to remove these at this stage because they will be removed later, and anecdotal evidence suggests their presence in the ensuing cocultivation may be beneficial.
4. As an alternative to Infection Medium and DME+, the producer lines can be adapted for growth in RPMI-15.
5. Set up the box in the incubator before beginning so it will equilibrate with respect to temperature and CO_2. One-half hour after transferring the dishes to the box, seal the lid.
6. Successful manipulation of the small thymic lobes requires carefully maintained forceps. In particular, if the forceps tips become bent, the required transfers of the lobes from dish to dish can be made difficult.
7. It is also acceptable to use irradiation to deplete the recipient lobes when thymic stem cells are used. Follow the procedures in the previous section.
8. Alternatively, day 15 fetal thymus can be used; in this case, it is necessary to scale up the experiment. For producers, plate 5×10^4/well in 24-well plates in 1 mL DME^+. Usually 6×10^4 thymocytes are recovered from a day 15 thymic lobe. Add cells to cultures at 7×10^5/well in 1.5 mL or 2.0 mL final volume.
9. High titers are important for the success of this technique. Retrovirus titers are determined by infection of NIH 3T3 fibroblasts. For viruses containing selectable markers, this is done by diluting a 12–24 h culture supernatant into wells of a 6-well dish containing 6×10^4 3T3 cells. Allow infection in the presence of polybrene overnight, then change medium. After 48 h, replace with fresh medium containing selective drug and incubate for 10–14 d. Stain the resulting colonies with crystal violet and count. For viruses with no selectable markers, titers can be determined by Southern blotting. Infect 3T3 fibroblasts with 1–2 mL of test virus and control (viral stock of known titer). After infection, make genomic DNA from the infected populations and cut using a restriction enzyme with sites within the provirus. Transfer the digested DNA from gel to blot and hybridize. Using a phosphorimager, compare the intensity of the signal to control to estimate relative viral titer. If a marker is available such as green fluorescent protein, transfected producer cells can be selected based on high expression of the marker and cloned. This step increases the efficiency of obtaining a high titer producer.

10. Reserve approx 10% of the cells from several well-repopulated lobes—when pooled, these cells can be used for staining controls to set up the cytometer.

Acknowledgments

This work was supported in part by American Cancer Society (grant no. JFRA-551, NIH grant no. AI36453), and NASA (grant no. NAG9-832) (L.M.S.), the Immunobiology Training Grant CA 09140 (L.L.L), and University of Tsukuba Research Projects, the PRESTO Research Projects, and the Ministry of Education, Science, Sports and Culture (Y.T.).

References

1. Jenkinson, E. J. and Owen, J. J. T. (1990) T-cell differentiation in thymus organ cultures. *Semin. Immunol.* **2,** 51–58.
2. Jameson, S. C., Hogquist, K. A., and Bevan, M. J. (1995) Positive selection of thymocytes. *Annu. Rev. Immunol.* **13,** 93–126.
3. Crompton, T., Gilmour, K. C., and Owen, M. J. (1996) The MAP kinase pathway controls differentiation from double-negative to double-positive thymocytes. *Cell* **86,** 243–251.
4. Karasuyama, H. and Melchers, F. (1988) Establishment of mouse cell lines which constitutively secrete large quantities of interleukin 2, 3, 4 or 5, using modified cDNA expression vectors. *Eur. J. Immunol.* **18,** 97–104.
5. Mann, R., Mulligan, R. C., and Baltimore, D. (1983) Construction of a retroviral packaging mutant and its use to produce helper-free defective retroviruses. *Cell* **33,** 153–159.
6. Markowitz, D., Goff, S., and Bank, A. (1988) A safe packaging line for gene transfer: separating viral genes on two different plasmids. *J. Virol.* **62,** 1120–1124.
7. Danos, O. and Mulligan, R. C. (1988) Safe and efficient generation of recombinant retroviruses with amphotropic and ecotropic host ranges. *Proc. Natl. Acad. Sci. USA* **85,** 6460–6466.
8. Pear, W. S., Nolan, G. P., Scott, M. L., and Baltimore, D. (1993) Production of high-titer helper-free retroviruses by transient transfection. *Proc. Natl. Acad. Sci. USA* **90,** 8392–8396.

11

Assay for the Isolation of Hepatogenic Factors

Key Molecules in Hepatocyte Formation and Liver Morphogenesis

Johannes A. A. Spijkers, Theodorus B. M. Hakvoort, and Wouter H. Lamers

1. Introduction

During embryonic development, pluripotent cells from the endoderm layer are directed toward specific cell lineages, leading to the formation of highly specialized organs, such as liver, lung, and pancreas. In case of liver, hepatocyte specification and maturation is the result of a tissue interaction between prehepatic endoderm and adjacent pre-cardiac mesoderm and, subsequently, the mesenchyme of the *septum transversum*. During these interactions, molecules regulating hepatogenesis ("hepatogenic factors") are presented to the endoderm by the mesoderm *(1,2)*. In rat, the inductive potential of the mesoderm (hepatogenic activity) is highest until embryonic day (ED) 12 and subsequently fades to background level, reached at ED 15 *(3)*.

Our laboratory is interested in isolating the molecules involved in hepatocyte formation and maturation. Their identification, in combination with an analysis of their spatio-temporal expression pattern during embryonic development, will provide insight in the molecular pathways underlying liver formation and regeneration and possibly organ formation in general. The technique we developed for isolating the hepatogenic factors involves the following components: a representative cDNA expression library prepared from the inductive tissue, a mammalian expression system for in vitro production of the proteins encoded by the library and a responder tissue for identification of cDNA clones encoding the inducing molecules. Our method requires the preparation of only a single cDNA library, provides a comprehensive way to study all cDNA clones present in this library for their biological activity, and yields full-length cDNA clones suited for immediate further study. In more detail, four methodological key steps can be distinguished (**Fig. 1**).

1. Construction of a full-length cDNA library *(4)*, unidirectionally cloned into an eukaryotic expression vector and representative for the inducing tissue, in this case hearts from ED 12 rats (*see* **Notes 1** and **2**). Selection of an expression vector containing the SV40 origin of replication, such as pcDNA$_3$ (InVitrogen, Leek, The Netherlands), provides the possibility to induce replication of the plasmid and, hence, to enhance the level of expression of the inserted cDNA (*see* **step 3**).

From: *Methods in Molecular Biology, Vol. 136: Developmental Biology Protocols, Vol. II*
Edited by: R. S. Tuan and C. W. Lo © Humana Press Inc., Totowa, NJ

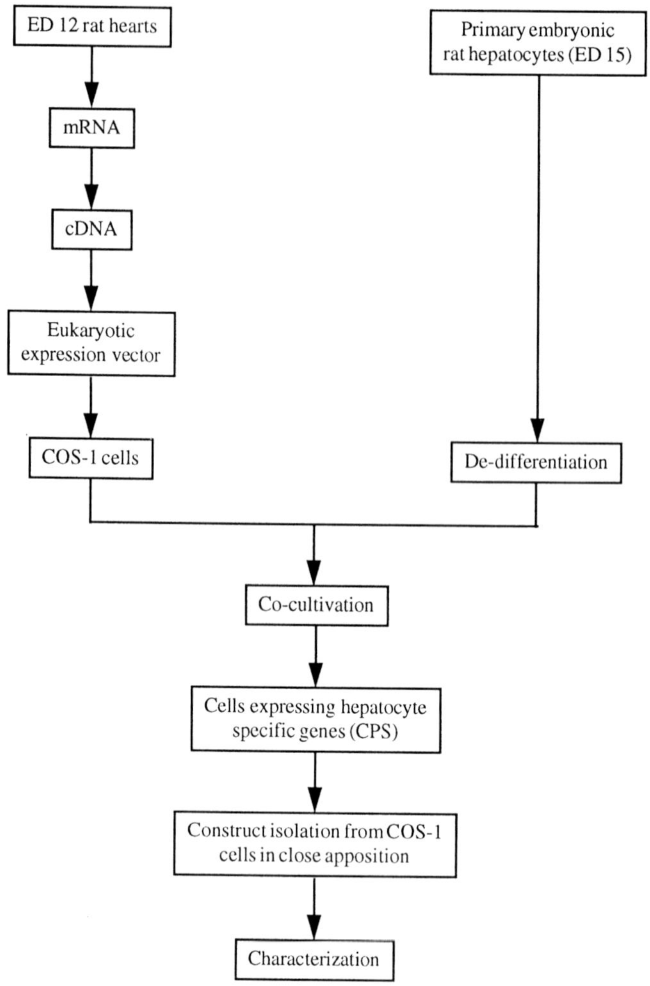

Fig. 1. Schematic representation of the strategy followed to isolate hepatogenic factors: molecules involved in the formation of hepatocytes from pluripotent endoderm. A cDNA expression library, derived from ED 12 rat hearts, is transfected into COS-1 cells. Some of the cDNA-encoded proteins are presented to primary dedifferentiated ED 15 rat hepatocytes following cocultivation. In case of a hepatogenic factor, the dedifferentiated cell will redifferentiate and start (re-)expressing hepatocyte differentiation markers. CPS-positive cells, identified via immunohistochemistry, and surrounding COS-1 cells are harvested and cDNA constructs are recovered.

2. Setup of a system for in vitro culture of the responder tissue or an analog, such as a pluripotent cell line. Because primary undifferentiated endoderm could not be obtained in sufficient quantities, and a pluripotent endodermal cell line that can differentiate into hepatocytes is not available, dedifferentiated ED 15 rat hepatocytes were used. These cells no longer express hepatocyte-specific genes such as glutamate dehydrogenase (GDH) and have a fibroblast-like appearance ***(5)***. Therefore, dedifferentiated hepatocytes may be regarded as an intermediate cell-type between undifferentiated endoderm and embryonic hepatocytes, primed for differentiation.
3. Identification of cDNA clones encoding the inductive biological property via a functional assay employing the cocultivation of two different cell types. One is a mammalian cell

line transfected with the cDNA expression library *(6)*, thus serving as producer cells. If a large T-antigen-expressing cell line like COS-1 is selected, then this will result, in combination with a transfected plasmid carrying the SV40 origin of replication, in high replication of the plasmid and therefore in a high level of protein expression *(7)*. The other cells are the natural responder tissue or an analog, in our case dedifferentiated hepatocytes. During cocultivation, the transfected cDNA constructs are expressed by the COS-1 cells, and some of the encoded proteins will be presented to the dedifferentiated hepatocytes. After 36–48 h of culturing, the mixed cell population is screened, by means of immunohistochemistry, for the (re)expression of a hepatocyte differentiation marker, in our case carbamoylphosphate synthetase (CPS), the first enzyme of the ornithine cycle. The reason CPS is used is twofold: First, CPS is not detectable before ED 15 in rat hepatocytes and only starts accumulating thereafter *(8)*. This provides an excellent possibility for monitoring hepatocyte differentiation. Second, CPS can be induced in primary cultures of ED 14 rat hepatocytes by hormonal stimulation *(8)*. CPS therefore represents a composite and, hence, sensitive parameter for complete hepatocyte dedifferentiation. In all experiments (>20), control cultures of dedifferentiated hepatocytes under hormonal stimulation failed to express CPS, showing the loss of this characteristic behavior (*see* **Note 2**). A positive hepatocyte in the coculture indicates that the adjacent transfected COS-1 cell produces the inductive molecule of interest *(5)*.

4. Isolation of the cDNA clones of interest. Positive responder cells and surrounding COS-1 cells are harvested and cDNA constructs are isolated *(9)*. This initially yields an enriched cDNA library and eventually leads, after several rounds, to purification of the construct of interest.

This chapter provides a quick method for identifying and isolating cDNA clones encoding a protein with known biological transacting function from a eukaryotic cDNA expression library. In this particular case, the identification of molecules involved in the development of hepatocytes from pluripotent endodermal cells is described, but the method should also be applicable for the isolation of other molecules involved in cell lineage determination and cell maturation, either via direct tissue interactions or paracrine pathways.

2. Materials

2.1. Hepatocyte Isolation and Dedifferentiation

1. Dulbecco's Modified Eagle's Medium: Ham's Nutrient Mixture F12 (1:1) (DMEM/F12, with L-glutamine and 15 m*M* HEPES, without sodium bicarbonate; Gibco-BRL, Gaithersburg, MD) with antibiotics. Dissolve powder medium in 90% of final required volume deionized double-distilled water (NPBI; Emmer-Compascuum, The Netherlands) at room temperature (RT). Dissolve per liter 1.125 g $NaHCO_3$ (Merck; Darmstadt, Germany). Adjust pH with 1*M* NaOH or 1*M* HCl to 0.2 below desired final working pH 7.4, as the pH rises with filtration. Add per liter medium 10 mL penicillin/streptomycin solution (10,000 U penicillin and 10 mg streptomycin/mL) (Gibco-BRL) and 0.1 g kanamycin sulfate (Gibco-BRL). Adjust to final volume with deionized double-distilled water (NPBI). Sterilize by filtration through a 0.2-µm acrocapfilter (Gelman Science Inc., Ann Arbor, MI). Dispense into sterile containers and store in the dark at 4°C (2–4 wk) or at –20°C (long term storage).
2. Digestion solution: DMEM/F12 (Gibco-BRL) containing 0.25% trypsin (Gibco-BRL) and 1 m*M* ethylene glycol-bis(β-aminoethyl ether)N,N,N',N'-tetraacetic acid (EGTA; Merck). Store in 1 mL aliquots at –20°C.
3. Fetal bovine serum (FCS; Gibco-BRL).

4. Deoxyribonuclease I (DNase I, from bovine pancreas Grade II; Boehringer-Mannheim, Mannheim, Germany).
5. Rat-tail collagen (Becton Dickinson, Erembodegem, Belgium).
6. Phosphate-buffered saline (PBS), pH 7.4: Add 0.5*M* NaH_2PO_4 (Merck) to 0.5*M* Na_2HPO_4 (Merck) until pH reaches 7.4. Dilute 5X with double-distilled water and add 87.5 g NaCl per liter. This gives a 10X PBS stock solution. Dilute 10X with double-distilled water before use. Store at 4°C.
7. Trypsinization solution: DMEM/F12 (Gibco-BRL) containing 0.25% trypsin (Gibco-BRL) and 1 m*M* EGTA) (Merck). Prepare fresh by diluting a 1-mL aliquot of 2.5% trypsin 10X with DMEM/F12 containing 1.1 mM EGTA.

2.2. COS-1 Cell Transfection and Cocultivation

1. DMEM/F12 (*see* **Subheading 2.1.**).
2. FCS (*see* **Subheading 2.1.**).
3. PBS (*see* **Subheading 2.1.**).
4. Trypsinization solution (*see* **Subheading 2.1.**).
5. "Cytomix" ***(10)***: Prepare the following stock solutions in deionized double-distilled water (NPBI): *A*, 2.4*M* KCl (Merck); *B*, 30 m*M* $CaCl_2$ (Merck); *C*, a mixture of 200 m*M* K_2HPO_4 (Merck) and 200 m*M* KH_2PO_4 (Merck) at a ratio yielding pH 7.6; *D*, 0.5*M* N-2-hydroxyethylpiperazine-N'-2-ethanesulfonic acid (HEPES; Sigma, St. Louis, MO) set to pH 7.6 with 5*M* KOH; *E*, 80 m*M* EGTA (Merck) set to pH 7.6 with 5*M* KOH; *F*, 100 m*M* $MgCl_2$ (Merck). Mix to prepare 2X cytomix: 10 mL *A*, 1 mL *B*, 10 mL *C*, 10 mL *D*, 5 mL *E*, 10 mL *F* and 54 mL deionized double-distilled water. Store at 4°C for up to 4 wk. Shortly before use add 3 mg glutathione (reduced form, Merck) per mL 2X cytomix.
6. Glutathione (reduced form, Merck).
7. Deionized double-distilled water (NPBI).
8. Dibutyryl-A-3,5-MP, cyclic (cyclic AMP; Boehringer-Mannheim).
9. Dexamethasone (cell culture tested, Sigma). Prepare a 20,000X stock solution in ethanol (*see* **Note 2**) and store at 4°C in the dark.
10. 3-isobutyl-1-methylxanthine (IBMX; Sigma). Prepare a 100X stock solution in DMEM/F12 (*see* **Note 2**) and store at 4°C in the dark.

2.3. Detection of Specific Gene Products

2.3.1. Immunohistochemistry on Cultured Cells

1. PBS (*see* **Subheading 2.1.**).
2. Fixation solution: 40% methanol (Merck), 40% acetone (Merck), 20% double-distilled water. Chill on ice before use.
3. Dehydration solutions: double-distilled water:ethanol (Merck) series of composition 70:30, 50:50, and 30:70, respectively.
4. Hydrogen peroxide (30%; Merck). Prepare fresh 0.3% and 2% hydrogen peroxide in 70% ethanol.
5. TENG-T: 100 m*M* Tris-HCl (Gibco-BRL), pH 8.0, 50 m*M* ethylenediaminetetraacetic acid (EDTA; Merck), 1.5*M* NaCl (Merck), 2.5% gelatin (Merck), 0.5% polyoxyethylene sorbitan monolaurate (Tween-20; Merck). This gives a 10X TENG-T stock solution. To prepare a 1X solution, melt the mixture in a water bath at 60°C, homogenize, dilute 10X with double-distilled water and set pH to 8.0 with HCl or NaOH. Store at 4°C.
6. Primary antibody directed against gene product of interest.
7. Rabbit-peroxidase-anti-peroxidase conjugated antibody directed against the final antibody on the specimen (R-PAP; Nordic, Tilburg, The Netherlands) (*see* **Note 3**).

8. Peroxidase reaction buffer: add 17 mL of a 0.2*M* solution of acetic acid (HAc; Merck) in double-distilled water to 33 mL of a 0.2*M* solution of NaAc (Merck) in double-distilled water. This yields a 2X HAc buffer pH 4.9. Dilute 25 mL of 2X HAc buffer with 25 mL double-distilled water and add 25 mg 3-amino-9-ethyl-carbazole (AEC; Sigma) dissolved in 2.5 mL N,N'-dimethyl formamide (Merck). Mix, filter immediately, and add H_2O_2 (Merck) to a final concentration of 0.1%. Prepare fresh. Buffer should be slightly orange after filtration. Use fumehood when working with dimethylformamide (volatile, toxic) and avoid direct contact with AEC (carcinogenic) and H_2O_2.

2.3.2. In Situ *Hybridization on Cultured Cells*

1. Rat-tail collagen (*see* **Subheading 2.1.**).
2. PBS (*see* **Subheading 2.1.**).
3. Formaldehyde fixative: 4% formaldehyde (Merck) in 1X PBS, pH 7.5. Prepare fresh. Keep at 4°C until use.
4. Dehydration solutions: double-distilled water:ethanol (Merck) series of composition 65:35, 50:50, 30:70, 4:96, and 0:100, respectively.
5. 0.2*M* HCl (Lamers en Pleuger, Den Bosch, The Netherlands).
6. Permeabilization solution: 2X SSC, pH 7.0. (1X SSC: 150 m*M* NaCl [Merck], 15 m*M* sodium citrate [Merck]).
7. Pepsin solution: 0.1% pepsin (800–2500 U/mg; Sigma) in 0.01*M* HCl. Prepare a 10% pepsin stock solution in sterile double-distilled water and incubate for 2 h at 37°C. Aliquots stored at –20°C are stable for at least one year. Working solutions, prepared from aliquots, should be incubated for an additional 45 min at 37°C and used within 3 h.
8. 0.2% glycine (Brunschweig, Amsterdam, The Netherlands) in PBS.
9. 10 m*M* EDTA (Merck), pH 8.0. Set pH to 8.0 with NaOH.
10. 10 m*M* Dithiotreitol (DTT; Merck). Store in aliquots at –20°C and keep on ice prior to use. Aliquots may be freeze-thawed for 5X at maximum.
11. Herring testes DNA (Sigma). Dissolve in double-distilled water at 10 mg/mL by stirring overnight at 4°C. Adjust NaCl concentration to 0.1*M*, extract 1X with phenol (Biosolve, Amsterdam, The Netherlands) and 1X with phenol:chloroform (Merck):isoamylalcohol (Merck) of composition 25:24:1, respectively. Recover aqueous phase and shear DNA by passing 12X rapidly through a 17-gage hypodermic needle. Precipitate with 2 vol 96% ethanol, collect by centrifugation, dry, dissolve in double-distilled water at 10 mg/mL, boil for 10 min, chill on ice, aliquot, and store at –20°C.
12. Hybridization solution: 50% formamide (deionized; Merck), 10% dextran sulfate (Pharmacia, Roosendaal, The Netherlands), 2X SSC, 2X Denhardt's solution (1X Denhardt's: 0.02% bovine serum albumin [BSA, fraction V; Sigma], 0.02% polyvinylpyrrolidone [Sigma], 0.02% Ficoll 400 [Pharmacia]), 0.1% Triton X-100 (Merck), 10 m*M* DTT (Merck), 200 ng/μL herring testes DNA (Sigma) and labeled cRNA probe at a concentration of 2–4 × 10^4 cpm/μL. Herring testes DNA and cRNA probe must be heated for 5 min at 100°C and 3 min at 80°C, respectively, and quickly chilled on ice before adding to the hybridization solution.
13. ^{35}S-labeled ([α-^{35}S]-UTP, 1000Ci/mmol; Amersham, Buckinghamshire, England) cRNA probe for gene product of interest, prepared according to standard in vitro transcription protocols (e.g., *see* **ref. *11***). The probe should be of high specific activity (indication for a single-labeled probe: 1.67 × 10^9 cpm/μg).
14. Wash solution 1: 50% formamide (deionized; Merck), 1X SSC.
15. Wash solution 2: 1X SSC.
16. Wash solution 3: 0.1X SSC.
17. RNase buffer: 10 m*M* Tris-HCl (Gibco-BRL), pH 8.0, 5 m*M* EDTA (Merck), 0.5*M* NaCl (Merck).

18. Ribonuclease A (RNase A; Boehringer-Mannheim): To prepare DNase free RNase A, dissolve RNase A at 10 mg/mL in 10 m*M* Tris-HCl, pH 7.5, 15 m*M* NaCl, and heat for 15 min at 100°C. Equilibrate slowly to RT, aliquot, and store at –20°C.
19. 0.3*M* ammonium acetate (Merck).
20. Ilford Nuclear Research Emulsion G-5 (Ilford, Leiden, The Netherlands). Store at 4°C, protected from radiation and light.
21. 2% glycerol (Brunschwig) in double-distilled water.
22. Developing solution: Dissolve 1.13 g of 4-hydroxy 1,3-phenylenediammonium-dichloride (Amidol; Fluka, Zwijndrecht, The Netherlands) and 4.5 g Na_2SO_3 (Merck) in 250 mL double-distilled water, filter, and add 2 mL 10% KBr (Merck). Prepare fresh just prior to use.
23. Fixation solution: 30% $Na_2S_2O_3 \cdot 5H_2O$ (Merck) in double-distilled water.

2.4. Plasmid Recovery from Immunohistochemically Stained Cells

1. Lysis buffer: 100 m*M* NaCl (Merck), 10 m*M* Tris-HCl (Gibco-BRL), pH 8.0, 25 m*M* EDTA (Merck), 0.5% sodium dodecylsulfate (SDS; USB, Cleveland OH), 100 µg/mL proteinase K (fungal; Gibco-BRL). Add proteinase K shortly before use.
2. Herring testes DNA (Sigma).
3. DNA purification: phenol (Biosolve):chloroform (Merck):isoamylalcohol (Merck) of composition 25:24:1, respectively. Avoid direct contact with phenol and work in fumehood.
4. TE: 10 m*M* Tris-HCl (Gibco-BRL), pH 8.0, 1 m*M* EDTA (Merck), pH 8.0.
5. 3*M* NaAc (Merck). Set pH to 5.2 with glacial acetic acid (HAc; Merck).
6. Glycogen (from muscle; Boehringer-Mannheim).

3. Methods

3.1. Hepatocyte Isolation and Dedifferentiation

Rat embryos of desired age are harvested from pregnant rats as follows. Sacrifice rats by decapitation under CO_2 anesthesia. Isolate embryos, with surrounding membranes, by cutting the antimesometrical wall of the uterus with blunt scissors and transfer to DMEM/F12 in a Petri dish at RT. Release the embryos from the membranes and decapitate immediately with a scalpel. Extract the livers under the dissecting microscope, cut into small pieces (2 mm^3) and transfer to a 2-mL Eppendorf tube on ice containing 1 mL digestion solution (*see* **Note 4**). Incubate at 37°C in a waterbath for 15–20 min. Triturate 10–15X with a Pasteur pipet and incubate for an additional 10 min. Swirl every 5 min during incubation (*see* **Note 5**). Add 100 µL fetal bovine serum (FCS) and 100 µL 0.1% DNase I in DMEM/F12 and triturate as above (*see* **Note 6**). Transfer the suspension to a tube containing 9 mL DMEM/F12 with 10% FCS and 0.002% DNase I at RT. Collect cells by centrifugation for 7 min, 100*g* at RT (100*g* corresponds to 480 rpm in a Jouan GR 4.11 centrifuge). Resuspend cells in 10 mL DMEM/F12 with 10% FCS at RT and filter the suspension through a double-layered nylon mesh (pore size 50–100 µm) to remove large debris. Collect cells as above. Resuspend cells in DMEM/F12 with 10% FCS at RT and seed the cells in rat-tail collagen-coated 6-well culture plates (Nunc, Wiesbaden-Biebrich, Germany) at appropriate density (day 1) (*see* **Notes 7** and **8**). Incubate overnight at 37°C, 5% CO_2–95% air, 90% relative humidity, and replace the culture medium by fresh, prewarmed (37°C) DMEM/F12 with 10% FCS. Passage cells at day 3 as follows: wash 1X with PBS (37°C), 1X with trypsinization solution (RT), and incubate at 37°C, 5% CO_2–95% air,

90% relative humidity. Monitor cell detachment under the microscope at regular intervals, (3–5 min). Harvest cells in 10 mL DMEM/F12 with 10% FCS at RT and collect by centrifugation for 7 min, 100*g* at RT. Seed and culture cells as above. Wash cells at day 5, passage at day 6, wash at day 7, and harvest at day 8 for cocultivation (*see* **Note 9**).

3.2. COS-1 Cell Transfection and Cocultivation

Culture COS-1 cells in DMEM/F12 with 10% FCS at 37°C, 5% CO_2–95% air, 90% relative humidity, until subconfluent (approx 3–4 × 10^6 cells in a 80 cm^2 tissue-culture flask [Nunc]); (*see* **Note 10**). Wash the cells 1X with 4.5 mL PBS (37°C), 1X with 1.5 mL trypsinization solution at RT, and incubate at 37°C, 5% CO_2–95% air, 90% relative humidity until cell detachment. Collect cells in 10 mL DMEM/F12 with 10% FCS at RT. Swirl and load a sample into a cell-counting chamber. Collect cells by centrifugation for 7 min, 100*g* at RT; meanwhile, count cells. Resuspend cell pellet in 10 mL PBS at RT and collect cells as above. Resuspend cells in 2X "cytomix," containing 3 mg per mL glutathione, at a density of 3.5 × 10^6 cells per 250 µL (*see* **Notes 11** and **12**). Add cDNA (plasmids) to be transfected (dissolved in 250 µL deionized double-distilled water), mix gently, transfer to a sterile cuvette with a 2-mm electrode gap and incubate for 10 min at RT (*see* **Notes 13** and **14**). Electro-shock at 1200 µF and 260 V (*see* **Note 12**). Incubate for 10 min at RT and transfer the sample to a tube containing 12 mL of a 1:1 mixture of DMEM/F12 with 10% FCS and NIH3T3 mouse fibroblast-conditioned medium at RT containing appropriate hormones (*see* **Notes 2** and **15**). Wash cuvette once with 500 µL DMEM/F12 with 10% FCS at RT to collect remaining cells. Distribute the cell suspension evenly over 10 wells of two 6-well culture plates (Nunc) and incubate at 37°C, 5% CO_2–95% air, 90% relative humidity. Simultaneously harvest the responder cells in the cocultivation system (dedifferentiated hepatocytes), via trypsinization as described above, and resuspend the cell pellet in 13 mL of the medium-mixture containing the hormones. Add equal portions of the cell suspension to the transfected cells, mix by gentle swirling, and incubate overnight at 37°C, 5% CO_2–95% air, 90% relative humidity (*see* **Note 16**). Next morning, replace half the culture medium by fresh prewarmed (37°C) medium-mixture containing hormones, culture for 36–48 h in total and analyze for expression of marker enzymes ([re-]differentiation; *see* **Note 17**).

3.3. Detection of Specific Gene Products

3.3.1. Immunohistochemistry on Cultured Cells

The following protocol does not require that cells be grown on cover slips; they can be applied directly on cells grown in culture dishes.

Wash the cells 3X with PBS at RT and fix by incubation in precooled fixation solution for 10 min at 4°C (*see* **Note 18**). Wash 2X with 70% ethanol at 4°C. Treat the cells with 0.3% H_2O_2 in 70% ethanol and subsequently with 2% H_2O_2 in 70% ethanol for 10 min each at 4°C (*see* **Note 19**). Rehydrate cells by passing through 50% ethanol, 30% ethanol and PBS, respectively, for 2 min each. Wash 3X with PBS and incubate in TENG-T for 30 min at RT with gentle agitation (*see* **Note 20**). Wash 3X with PBS. Incubate with the primary, intermediate, and final (labeled) antibody in PBS for required time span and at appropriate dilutions. Wash 3X with PBS at RT following each antibody incubation. Detect the conjugated label with peroxidase reaction buffer and analyze using a light microscope (*see* **Note 3**).

3.3.2. *In Situ Hybridization on Cultured Cells*

If *in situ* hybridization is the method of choice to detect specific gene products, cells should be cultured on rat-tail collagen-coated microscope slides with flexiperm 8-well mounts (Heraeus, Hanau, Germany). One should keep in mind that it is not possible to recover DNA from cells subjected to *in situ* hybridization.

Wash the cells 3X for 5 min with PBS at RT and fix by incubation in formaldehyde fixative for 1 h at RT with gentle agitation. Wash 3X for 5 min with PBS at RT, dehydrate the cells by passing through a graded series of ethanol (35, 50, 70, 96, and 100%, respectively), for 10 min each, air dry for 30 min in a filtered air stream (0.2-µm filter), and store at –20°C with desiccant until use. Equilibrate to RT and incubate for 5–10 min in 0.2*M* HCl to remove basic proteins, 5 min in double-distilled water, 5 min at 70°C in permeabilization solution to improve penetration of the probe, 5 min in double-distilled water, and finally 10 min at 37°C in pepsin solution to enhance accessibility of mRNAs. Then in sequence, 30 s in 0.2% glycine in PBS to terminate proteolytic activity, 2X for 30 s in PBS, 20 min in formaldehyde fixative, 5 min in PBS, 10 min in 10 m*M* EDTA to dissociate the ribosomes from the RNA, and finally 5 min in 10 m*M* DTT. Air dry in a filtered air stream as short a time as possible and proceed within 3 h. Hybridize overnight at 54°C in hybridization solution containing ^{35}S-labeled cRNA probe (approx 4×10^4 cpm/µL). Next morning, rinse the cells with 2 mL 1X SSC to remove excess probe. Wash, while gently shaking, 2X for 15 min at 54°C with wash solution 1, and once for 10 min at RT with wash solution 2. Rinse with 1X SSC. Incubate for 30 min at RT in RNase buffer containing 10 µg/mL RNase A. Rinse with 1X SSC, wash for 10 min with wash solution 2 and for 10 min with wash solution 3, both at RT. Dehydrate by passing through a graded series of ethanol (50, 70, and 96%, respectively), containing 0.3*M* ammonium acetate for 3 min each. Air dry in a filtered air stream for 1 h. Visualize probe as follows: Dilute Ilford Nuclear Research Emulsion G-5 2.5X with 2% glycerol in water and warm to 39°C. Apply a thin film of emulsion onto the cells, place the slide for 10 min on an ice-cooled glass plate, dry for 60 min at RT, and expose for 10–12 d at 4°C in the dark. Equilibrate specimens to RT and develop the emulsion, in the dark, at 18°C as follows: Incubate for 4 min at 18°C with developing solution with gentle agitation, for 1 min at 18°C with double-distilled water to terminate reaction, for 10 min at 18°C with fixation solution with gentle agitation, and finally wash for 1 h in the dark with running tap water at 18°C. Counterstain the cells (e.g., with toluidine-blue or nuclear-fast-red) and mount under a cover slip in embedding medium (e.g., Malinol) (**Fig. 2**).

3.4. Plasmid Recovery from Immunohistochemically Stained Cells

Harvest area of positive cells (**Fig. 3**) by scraping with a yellow tip, transfer to a 2-mL Eppendorf tube containing 300 µL lysis buffer, and incubate for at least 1 h at 50°C in a waterbath (*see* **Notes 21** and **22**). Add 15 µg herring testes carrier DNA to the lysate, extract 1X with phenol:chloroform:isoamylalcohol (25:24:1, respectively), and re-extract the organic phase 2X with 0.5 vol TE. Precipitate the recovered DNA overnight at –20°C with 0.1 vol 3*M* NaAc, pH 5.2, and 2 vol of 96% ethanol, with 10 µg glycogen as a carrier (*see* **Notes 23** and **24**). Collect the DNA by centrifugation for 30 min, 16,000*g* in a Eppendorf centrifuge at 4°C. Dissolve the pellet (enriched cDNA library) in 20 µL of TE, pH 8.0, and store at 4°C (*see* **Notes 25** and **26**).

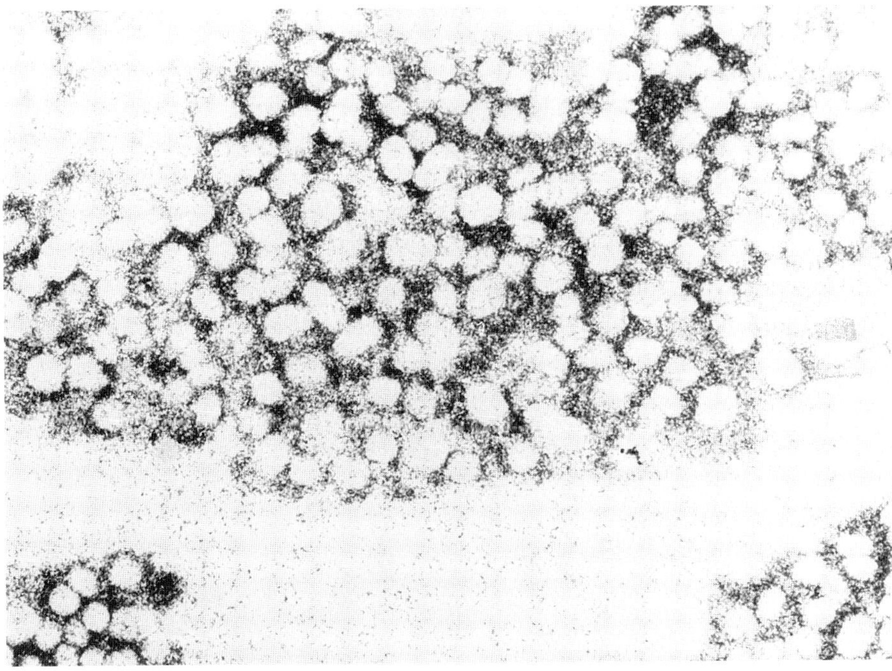

Fig. 2. Localization of albumin mRNA in ED 14 rat hepatocytes, cultured for 24 h in the presence of dexamethasone, cAMP, and IBMX. Albumin transcripts are detected by means of *in situ* hybridization with a ^{35}S-labeled cRNA probe (bright field illumination). The hybridization signal is restricted to the cytoplasm of the hepatocytes, showing the nuclei as punched-out holes.

4. Notes

1. In total, 363 ED 12 rat hearts were used for RNA isolation, yielding 405 µg of total RNA. The amount of mRNA isolated from total RNA, utilizing Promega's polyATtract mRNA isolation system (Promega, Leiden, The Netherlands), represented 0.7–1% the amount of total RNA. To save mRNA, its integrity was monitored indirect via analysis of the extracted total RNA fraction (polyA$^-$), in two ways: First, via denaturing agarose gel electrophoresis, to determine the ratio between ribosomal 28S and 18S RNA (the ratio should be approx 2:1, respectively). Second, via Northern blotting, applying the sarcoplasmic reticulum Ca^{2+}ATPase-probe (SERCA2) for visualization of residual cardiomyocyte mRNA. Information on mRNA isolation efficiency can be obtained by loading an equal amount of total RNA on the Northern blot. To enhance both uniform and full-length cDNA synthesis, a primer was used containing one additional nucleotide, either cytidine, guanosine, or adenosine, at the 3' end of a stretch of 25 thymidine residues, hence providing site-directed hybridization at the 5' end of the polyA tail of the mRNA. For unidirectional cloning purposes, a *Not*I recognition site was introduced in the primer at the 5' end of the stretch of 25 thymidine residues. The unidirectional cDNA expression library contained 2.2×10^7 independent clones. Insert analysis of 12 randomly picked clones yielded cDNAs varying from 500 to 2300 bp (median: 1200–1500 bp).
2. Hormones, in combination with the inductive protein presented by the transfected cell, may be required to obtain expression of the marker enzyme for (re-)differentiation. As we have chosen CPS as a marker, the culture medium contains cAMP ($10^{-3}M$), dexamethasone ($10^{-6}M$), and IBMX ($10^{-4}M$) (final concentrations; *see* **ref. *13***). (GDH, another early

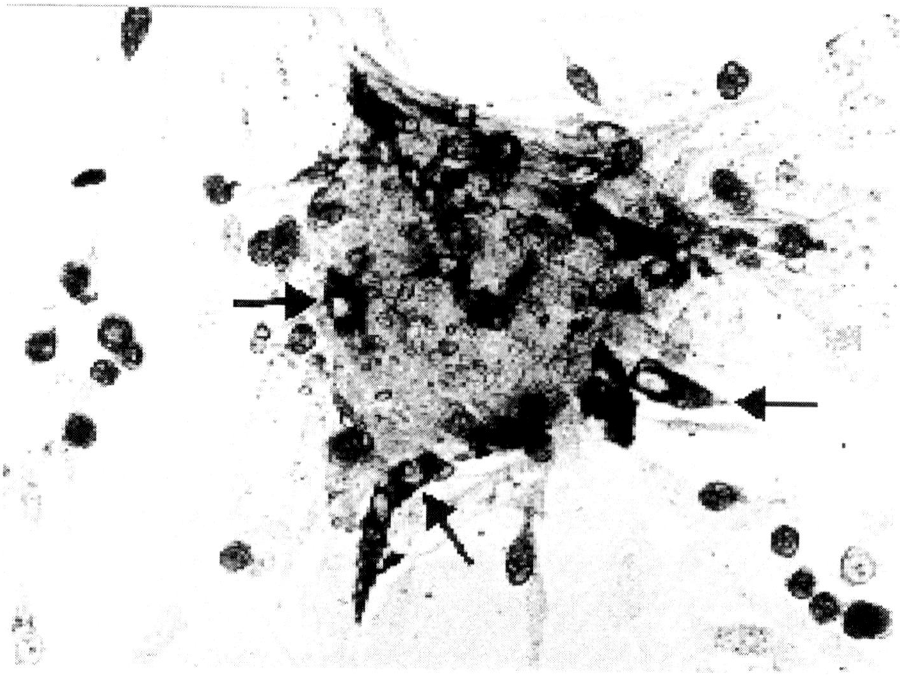

Fig. 3. Expression of the hepatocyte differentiation marker carbamoylphosphate synthetase (CPS) in a coculture of dedifferentiated hepatocytes and COS-1 cells transfected with the ED 12 rat heart cDNA expression library. Marker enzyme expression is detected via immunohistochemistry. The negatively stained cells in the center are most likely COS-1 cells producing the inductive protein, whereas the CPS-positive cells *(arrows)* are redifferentiated hepatocytes. Note that the nucleus of CPS-positive cells is negative, whereas dead cells are homogeneously stained.

hepatocyte marker, is not suited for our system, because it is endogenously expressed by COS-1 cells.) It is essential to investigate whether hormonal stimulation by itself is sufficient for induction of marker enzyme expression. To test this, primary ED 15 rat hepatocytes were cultured to dedifferentiation (*see* **Subheading 3** and **Note 9**) and changed to hormone containing medium. After 3 d of culturing, in the presence of hormones, cells remained fibroblast-like, and both CPS and GDH could not be detected via immunohistochemistry. In addition, the same result was obtained with dedifferentiated cells cocultivated under hormonal stimulation with either untransfected or mock transfected COS-1 cells. We also tried to induce (re-)differentiation in dedifferentiated hepatocytes by coculturing with ED 12 rat hearts. However, this experiment could not be properly evaluated, as it proved technically impossible to isolate the embryonic heart without hepatocyte contamination.

3. In our experiment, the final antibody is labeled with peroxidase, and enzyme activity can be detected with the 3-amino-9-ethyl-carbazole (AEC) substrate ***(19)***. The progression of the staining is monitored by using a light microscope and terminated by washing with bidest when specific staining is profound and background staining is still low. Store the immunostained cells under bidest at 4°C. Plasmid DNA should be isolated within 3 d following immunohistochemistry; otherwise, it will be degraded. Another commonly used peroxidase substrate, 3,3'-diamino-benzidine (DAB), should not be used, because no DNA can be recovered from DAB-treated cells. Both AEC and DAB are carcinogenic.

4. All nonhepatic tissues should be removed carefully but quickly from the livers to diminish the number of fibroblasts in the primary culture and to prevent overgrowth of the hepatocytes during dedifferentiation. In an effort to remove nonparenchymal liver cells, hepatocytes were also cultured in a selective medium deficient in arginine ***(12)*** but supplemented with ornithine. In theory, nonparenchymal cells die under these conditions as a result of a lack of essential arginine, whereas hepatocytes survive by metabolizing ornithine to arginine via the ornithine cycle. For up to 12 d we observed clusters of cells with characteristic hepatocyte morphology, positive for GDH, while very few fibroblasts were present. During this period, the hepatocytes did not dedifferentiate and displayed hardly any mitotic activity. A medium change to DMEM/F12 with 10% FCS, resulted in growth and dedifferentiation of the hepatocytes but also in a normal percentage of fibroblast contamination, indicating survival. Therefore, this approach is not suited to eliminate nonparenchymal cells from the culture. Another alternative to suppress growth of nonparenchymal cells, addition of hormones to the culture medium (*see* **Note 2**), is also not applicable, as this will retain the hepatocytes in the differentiated state ***(13)***.
5. These incubation times are optimized for ED 15 rat livers. Liver tissue of earlier stages can be suspended with shorter incubation times and less trypsin (0.05%). Livers of later stages can be suspended as mentioned but with inclusion of 0.025% collagenase.
6. FCS is added to inactivate the trypsin. DNase I prevents the solution from becoming viscous as a result of DNA release from disrupted cells.
7. Rat-tail collagen is used to enhance hepatocyte attachment.
8. Plating density greatly affects cell behavior. A high cell density promotes cell attachment but inhibits hepatocyte dedifferentiation. As a rule of thumb, seeding of 2.0–2.5 ED 15 rat livers in 2.5-mL culture medium per well of a 6-well tissue culture plate (Nunc) yields cultures that are 70% confluent. Following overnight incubation, islands of hepatocytes should be present, comprising 40–70 cells. Some nonparenchymal cell contamination is allowed inbetween islands but should not exceed 10–15% of the initial cell population.
9. The dedifferentiation of ED 15 rat hepatocytes in primary culture follows a highly reproducible time scale if initial plating density is not changed. At day 3, the hepatocytes still display characteristic morphology and are positive for GDH. Islands (*see* **Note 8**) have expanded to 100–150 closely packed cells. The space between the islands is confluent with nonparenchymal cells. At day 4, after first passage, islands of GDH-positive hepatocytes with characteristic morphology are present, comprising 70–100 cells with nonparenchymal cells in between. At day 6, the culture has grown confluent. Hepatocytes are still weakly expressing GDH but are clearly losing their characteristic morphology and start to look like fibroblasts. At day 7, after second passage, all cells have a fibroblast-like appearance. Occasionally a single cell shows weak GDH expression, slightly above background. At day 8, all cells in the culture have a fibroblast-like appearance and are no longer expressing the hepatocyte-specific marker-molecule GDH.
10. Cells from subconfluent cultures give a higher transfection efficiency compared with cells from confluent cultures, probably because of differences in protein content of the cell membrane and the composition of the extracellular matrix ***(6)***. In addition, exponentially growing cells give the highest transient expression of a transfected construct ***(14)***.
11. During electrotransfection, large pores are created in the cell membrane, making the cells very sensitive to the composition of the medium. "Cytomix" resembles the intracellular concentration of most important ions ***(10)***. Glutathione is added (shortly before transfection) to prevent oxidation of cytoplasmic components and to enhance cell membrane restoration and preservation ***(15)***.

12. In case of COS-1 cells, 3.5×10^6 cells per 250 µL gave best results. However, optimal cell density (and electroporation conditions) must be empirically determined for every cell type separately *(14)*.
13. The amount of DNA that gives best results has to be established empirically. In general 3–10 µg per kilobase of DNA is used *(6)*. In case of the cDNA expression library, we use 6 µg plasmid DNA supplemented with 20 µg linearized pBluescript as carrier. The volume of the DNA solution (dissolved in deionized double-distilled water) is also of great importance, because this influences the osmolarity of the electroporation medium and, as a consequence, the expression of the transfected construct and cell viability. The osmolarity of the electroporation medium should be 300 mOsm *(16)*.
14. It is very important to use cuvettes with the prescribed interelectrode gap. Modifying the distance between the electrodes will affect the electric field. This is one of the two important parameters in electroporation, because the electric field strength is what determines the size and number of pores created per cell. (The other parameter is the sample's conductivity and thus the pulse time; *see* **ref.** ***6***).
15. The NIH3T3 mouse fibroblast-conditioned medium enhances attachment of (dedifferentiated) embryonic hepatocytes during passaging. To obtain it, culture NIH3T3 fibroblasts *(17)* under standard conditions, collect the medium at confluency, centrifuge for 7 min, 100*g* at RT to spin down any cells, and take the supernatant and store at 4°C for up to 2 wk or at –20°C for long-term storage.
16. Optimal cell mixing, resulting in optimal producer/responder cell interaction, is obtained with suspended cells. This was investigated by mixing COS-1 cells with primary hepatocytes in the differentiated state. When cells were mixed in suspension, clusters of 10–15 cells of a single cell type were formed. If either cell type was allowed to attach before adding the second cell type, clusters of a single cell type consisted of up to 100 cells. Following attachment, the mixed cell culture should be close to confluent if cell-to-cell contact is needed for the inductive interaction.
17. The culture medium may contain the inducing molecule in a soluble form. To avoid removing the active component from the culture and to reduce the chance of dropping below biologically active thresholds, only one half of the culture medium is refreshed. In case the inductive protein is secreted into the medium, it may become too diluted to exert its effect on the responder cell. In such case, a hierarchical coculture can be tried. This implies culturing the responder cell on top of the producing cell with a 0.025% collagen layer sandwiched in between. With this setup, the inductive protein will become trapped in the collagen matrix, leading to locally higher concentrations.
18. Cell fixation at low temperature diminishes shrinkage of the cells as a result of dehydration *(18)*.
19. This step is performed to reduce endogenous peroxidase activity *(18)*.
20. TENG-T acts as a blocking agent, reducing nonspecific binding of antibody *(18)*.
21. When harvesting the positive cells, the culture dish should be as dry as possible, with only a thin film of fluid on top of the cells. This prevents the scraped cells from floating away and makes them stick to the yellow tip. To suspend the cells in lysis buffer, carefully attach the tip with the cells onto a 200-µL pipet, set at 80 µL, and pipet up and down in the buffer 25X.
22. In our procedure, the total DNA content (plasmids and genomic DNA) is isolated from the harvested cells. Per harvested area (transfected cell) approx 5 plasmids can be recovered. A different method for the isolation of the DNA of interest from the immunohistochemically stained cells is that of Hirt *(20)*. It utilizes differential precipitation of genomic DNA and plasmid DNA and, possibly because of this, gives approximately a twofold lower yield than our method (two to three plasmids recovered per transfected cell). In addition, time-consuming centrifugation steps are involved, which is not the case with our method.

23. If precipitation is performed over time periods exceeding overnight, it will result in the formation of insoluble complexes.
24. To avoid losing the plasmid DNA during isolation, the use of carrier material is of critical importance.
25. To further purify the cDNA clones of interest, the enriched cDNA library is amplified in bacteria (e.g., XL1-Blue MRF') and used for another round of screening.
26. The fold-enrichment that can be obtained depends on the initial concentration of the plasmid of interest in the library. If the initial concentration is high, the enrichment factor will be low. To validate our procedure, we performed an experiment employing a ED 13 rat heart cDNA library, from which we isolated two abundant cDNA clones: α-smooth muscle actin (α-SMA) (1300 bp) and β-myosin heavy chain (β-MHC) (6000 bp). After one round of enrichment, we found a twofold enrichment for β-MHC and a threefold enrichment for α-SMA. If further enrichment of the plasmid of interest can no longer be obtained via repeated transfections and isolations, one can divide the plasmids in the enriched cDNA library into subpopulations (e.g., according to cDNA insert size) and screen the resulting subdivisions individually or turn to screening of separate cDNA clones.

References

1. Le Douarin, N. M. (1975) An experimental analysis of liver development. *Med. Biol.* **53,** 427–455.
2. Zaret, K. S. (1996) Molecular genetics of early liver development. *Annu. Rev. Physiol.* **58,** 231–251.
3. Fukuda-Taira, S. (1981) Hepatic induction in the avian embryo: Specificity of reactive endoderm and inductive mesoderm. *J. Embryol. Exp. Morphol.* **63,** 111–125.
4. Chenchik, A., Diatchenko, L., Chang, C., and Kuchibhatla, S. (1994) Great lengths cDNA synthesis kit for high yields of full-length cDNA. *Clontechniques* **IX,** 9–12.
5. Houssaint, E. (1976) Réversibilité de la dédifférenciation observée dans les cultures cellulaires de foie embryonnaire d'oiseau. *J. Embryol. Exp. Morphol.* **35,** 227–240.
6. Chang, D. C., Chassy, B. M., Saunders, J. A., and Sowers, A. E. (1992) *Guide to Electroporation and Electrofusion.* Academic, San Diego, CA.
7. Fiers, W., Contreras, R., Haegeman, G., Rogiers, R., Voorde, A. van de, Heuverswyn, H. van, Herreweghe, J. van, Volckaert, G., and Ysebaert, M. (1978) Complete nucleotide sequence of SV40 DNA. *Nature* **273,** 113–120.
8. Dingemanse, M. A., Boer, P. A. J. de, Moorman, A. F. M., Charles, R., and Lamers, W. H. (1994) The expression of liver-specific genes within rat embryonic hepatocytes is a discontinuous process. *Differentiation* **56,** 153–162.
9. Horst, E., Wijngaard, P. L. J., Metzelaar, M., Bast, E. J. E. G., and Clevers, H. C. (1991) A method for cDNA cloning in COS cells irrespective of subcellular site of expression. *Nucleic Acids Res.* **19,** 4556.
10. Hoff, M. J. B. van den, Moorman, A. F. M., and Lamers, W. H. (1992) Electroporation in "intracellular" buffer increases cell survival. *Nucleic Acids Res.* **20,** 2902.
11. Melton, D. A., Krieg, P. A., Rebagliati, M. R., Maniatis, T., Zinn, K., and Green, M. R. (1984) Efficient *in vitro* synthesis of biologically active RNA and RNA hybridization probes from plasmids containing a bacteriophage SP6 promotor. *Nucleic Acids Res.* **12,** 7035–7056.
12. Leffert, H. L. and Paul, D. (1972) Studies on primary cultures of differentiated fetal liver cells. *J. Cell Biol.* **52,** 559–568.
13. Lamers, W. H., Zonneveld, D., and Charles, R. (1984) Inducibility of carbamoylphosphate synthetase (ammonia) in cultures of embryonic hepatocytes: Ontogenesis of the responsiveness to hormones. *Dev. Biol.* **105,** 500–508.

14. Hoff, M. J. B. van den, Christoffels, V. M., Labruyere, W. T., Moorman, A. F. M., and Lamers, W. H. (1995) Electrotransfection with "intracellular" buffer, in *Methods in Molecular Biology 48: Animal Cell Electroporation and Electrofusion Protocols.* Humana, Totowa, NJ.
15. Gankema, H. S., Laanen, E., Groen, A. K., and Tager, J. M. (1981) Characterization of isolated rat-liver cells made permeable with filipin. *Eur. J. Biochem.* **119,** 409–414.
16. Hoff, M. J. B. van den, Labruyere, W. T., Moorman, A. F. M., and Lamers, W. H. (1990) The osmolarity of the electroporation medium affects the transient expression of genes. *Nucleic Acids Res.* **18,** 6464.
17. Jainchill, J. L., Aaronson, S. A., and Todaro, G. J. (1969) Murine sarcoma and leukemia viruses: assay using clonal lines of contact-inhibited mouse cells. *J. Virol.* **4,** 549–553.
18. Sternberger, L. A. (1986) *Immunocytochemistry*, 3rd ed. Wiley, New York.
19. Graham, R. C., Jr., Lundholm, U., and Karnovsky, M. J. (1965) Cytochemical demonstration of peroxidase activity with 3-amino-9-ethylcarbazole. *J. Histochem. Cytochem.* **13,** 150–152.
20. Hirt, B. (1967) Selective extraction of polyoma DNA from infected mouse cell cultures. *J. Mol. Biol.* **26,** 365–369.

12

Skin Morphogenesis

Embryonic Chicken Skin Explant Cultures

Cheng-Ming Chuong

1. Introduction

Chicken skin development is an excellent model to study the mechanisms of morphogenesis. It has a long experimental history and has been well characterized phenotypically. Chicken skin offers distinct patterns, large numbers of different cutaneous appendages, availability of in vitro culture systems, accessibility to microsurgery *in ovo*, and the existence of skin mutants. All these factors make feather an ideal model for the investigation of epithelial-mesenchymal interactions and the regeneration of epithelial appendages ***(1–3)***.

In this chapter, we describe the model of feather explant cultures (*see* **Fig. 1**). Skin explant cultures were originally started by Wolff and Haffen ***(4)***. Sengel ***(3)*** has continued to develop the explant culture procedure and has stated that explants from E6.5 embryos still need fetal calf serum (FCS), but after E6.75 serum-free culture is possible. When epithelia and mesenchyma are separated, the organization of both components are lost. However, if the epithelium and mesenchyme are recombined shortly after separation, feather germs regenerate ***(5,6)***. The location of new buds are in accord to that of the mesenchyme. If the epithelium is rotated relative to the mesenchyme in the recombinants, the anterior-posterior orientation of the feather buds will follow that of the epithelium. This provides a unique opportunity to study the dependence of molecular expression on E-M contact and the sequence of molecular expression required for skin appendage formation. The following is the culture procedure we use routinely in our laboratory.

2. Materials

1. Eggs: Pathogen-free fertilized chicken eggs are from SPAFAS Connecticut Hatchery (Preston, CT). Non-pathogen-free eggs can be acquired from a local farm.
2. Media.
   a. Hanks's buffered saline solution (HBSS) (Gibco-BRL, Gaithersburg, MD).
   b. Dulbecco's modified Eagle's medium (DMEM) (Gibco-BRL supplemented with 2% FCS and gentamicin (1:1000).

From: *Methods in Molecular Biology, Vol. 136: Developmental Biology Protocols, Vol. II*
Edited by: R. S. Tuan and C. W. Lo © Humana Press Inc., Totowa, NJ

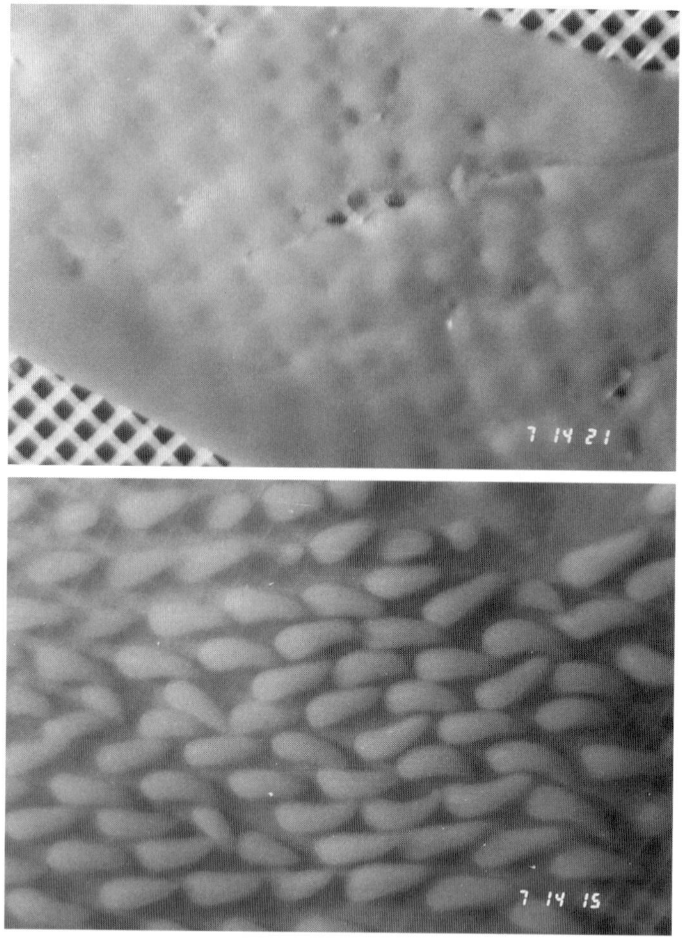

Fig. 1. An example of feather buds formation from cultured explants. A piece of dorsal skin from stage 31 chicken embryo was cultured for 4 d. Note the hexagonally arranged feather buds emerging from the homogeneous skin explant. Also note that most of the buds elongate toward the posterior end (caudal end of the dorsal skin, toward right edge of the panel).

c. Calcium-magnesium free saline (CMFS 10X)
  $NaCl$ (1.37 *M*) 80 g
  KCl (0.04 *M*) 3 g
  NaH_2PO_4 (0.004 *M*) 0.5 g
  KH_2PO_4 (2 *M*) 0.25 g
  $NaHCO_3$ (0.12 *M*) 10 g
  Glucose (0.1 *M*) 20 g in 1000 mL distilled water, pH 7.3

3. Tissue culture supply: Dishes and insert are from Falcon (Los Angeles) (#3090, pore size 0.4 μm). Filters are from Millipore (Bedford, MA) (#110608, pore size 0.6 μm).
4. Tools
   Spatula
   Regular forceps
   Fine-point forceps
   Watchmaker's forceps (#5, titanium)

3. Methods

3.1. Skin Explant Culture

1. Eggs are incubated in a humidified incubator at 38°C and staged according to Hamburger and Hamilton *(7)*. For feather development, stage 28–34 chicken embryo dorsal skins are most frequently used (*see* **Note 1**).
2. The outside of the egg is cleaned with alcohol. The shell is cracked with regular forceps. Gently remove the shell with sterile regular forceps. Avoid dropping shell debris into the egg.
3. Use another set of sterile forceps and spatula to get into the egg and remove the chicken embryo. Transfer the embryo to a 60-mm dish with HBSS. Rinse the embryo and transfer the embryo to different dishes until HBSS is clear.
4. Place embryo under the dissection microscope with the head toward the left (assuming you are right handed). With the left hand, use sterile fine-point forceps to hold the embryo at the neck. With the right hand, use scalpels to make two longitudinal cuts along the flank region parallel to the midline.
5. With the right hand switched to a watchmaker's forceps, grab the skin and gently rip it off toward the back. When the skin is near the tail, make a transverse cut anterior to the tail. In case there are muscles attached to the bottom, trim them off. Using transillumination, dermal condensations are visible. Record how many rows of dermal condensations have already formed (*see* **Note 2–4**).
6. Place tissue culture insert in the 60-mm dish. Add 2 mL/well of DMEM supplemented with 2% FCS, gentamicin (1:1000).

7a. Use a smoothed-edge spatula to transfer the skin to the insert. Spread the skin explant flat gently. Make sure that the bottom side is down. Mark the cephalic end of the dorsal skin so that you know the orientation later.

7b. If the filter method is used *(8)*, spread the skin flat on the filter. Then lift the filter and transfer it to a dish with the same culture medium. Let the explant/filter float on top of the air-liquid surface. Alternatively, the explant filter can be placed on top of a metal meshwork that is supported on an organ culture dish (Costar, Cambridge, MA, #32360 (*see* **Note 5**).

8. For insert cultures, media should be placed in both the outside well and the inner chamber. In the inner chamber, a thin layer of media is left just to cover the explant so that the explant is moist and close to the air-liquid interface.
9. The skin explant cultures are incubated at 37°C in 5% carbon dioxide and 95% air. The medium is changed every 2 d.

3.2. Epithelial-Mesenchymal Recombination

1. Following the dissection of the skin, place the skin in twofold concentrated calcium-magnesium free saline with 0.25% EDTA. The dish is left on ice for 10 min. Some regions should have shown signs of separation of epithelium and mesenchyme. If not, leave it longer, or transfer to room temperature. Then from the separated region, use two watchmaker's forceps to grab the epithelium and mesenchyme and peel them off gently.
2. The epithelium is transparent. Leave it in the medium but remember which side is up and which end is cephalic. The mesenchyme is thicker and can be transferred to the insert as described. Also remember which end is cephalic.
3. Use a spatula to pick up the epithelium with the upper side facing up. Slide the epithelium, in desired orientation, on top of the denuded mesenchyme on the culture insert.
4. Press the transplanted epithelium down gently while the media is being withdrawn until a thin layer covers the explant. We use surface tension to hold them together. Leave the recombined explant semidry for 2 h in the incubator (*see* **Note 6**).

5. Gently add more media from the side. For recombinants, 10% FCS is usually required for better growth.

3.3. Perturbing Functions with Beads Coated with Growth Factors or Drugs

Localized delivery of growth factors or drugs ***(9–12)*** can be delivered by Affi-Gel Blue beads (Bio-Rad, Hercules, CA, 100–250 µm in diameter) or heparin-acrylic beads (200–250 µm diameter, Sigma Chemical, Co., St. Louis, MO).

1. Beads are washed three times in sterile phosphate-buffered saline (PBS) by short centrifugation.
2. Approximately 20–100 beads are added to 5 µL aliquots of growth factor solution.
3a. For the Affi-Gel Blue beads, incubate at 37°C for 1 h.
3b. For the heparin-acrylic beads, incubate at room temperature for 2 h.
4. Treated beads are picked up with the watchmaker's forceps and placed on top of the skin explants or sandwiched between the epithelium and mesenchyme.
5. If treated beads are not immediately used, they can be stored at 4°C for up to 1 wk.

3.4. Reconstitution of Skin Appendages with Dissociated Mesenchymal Cells and Intact Epithelium

The epithelial-mesenchymal recombination reset the development of epithelium, but not the mesenchyme. Previous placode epithelium disappears and the state, but the locations of new feather buds are in accord to that of the previous mesenchyma ***(5,6)***. To reset mesenchyme to the initial state, we have developed a new procedure to dissociate mesenchymal cells, reaggregate them, and allow them to reform feather buds with a new periodic pattern ***(13)***. This model allows us to study the most initial events that occur in the dermal mesenchyme before periodic patterning.

1. Stage 29–35 dorsal skins containing the spinal tract were incubated at 4°C in 2% trypsin for 15–20 min and then washed in medium containing 10% fetal calf serum.
2. Under a dissection microscope, the epithelium and the mesenchyme were separated using watchmaker's forceps.
3. The epithelia were trimmed to be of equal same size using transparent graph paper beneath the disk as a guide. The epithelia were left in the media in ice.
4. The mesenchyme was pooled (usually from five pieces of skin) and gently triturated into single cells by drawing them through fine pore pipets with decreasing diameters. Cells were filtered through Nitex netting when necessary. The viability of cells and completion of dissociation were checked microscopically with trypan blue inclusion.
5. The dissociated cells were counted, repelleted by mild centrifugation (6500 rpm for 4 min) and allowed to reaggregate at specific cell densities for 1 h at 37°C on culture insert dishes (Falcon). The number of explant cultures to set up is equivalent to the original number of pieces of skin used.
6. The epithelium was then placed on top of the mesenchyme and the reconstituted explants were cultured on tissue culture inserts (Falcon, LA).
7. At designated times, explants were observed, photographed or fixed. Feather buds usually re-form in about 24 h.

3.5. Morphological and Molecular Analysis

1. At day 2, many dermal condensations have formed. At day 4, many feather buds have formed (see **Fig. 1**). Photographs can be taken daily from live insert cultures.
2. The contoured buds can be traced and analyzed by image analysis software, where the average size of the buds can be determined. If the variation is big, expressing them as a

histogram can be more informative. The density of the buds can be calculated by dividing the number of buds by the total area of the explant. The total interbud space can be calculated by subtracting the total bud area from the total explant area, which can be expressed as percentage of the explant. The regularity of the feather pattern can be judged by connecting the center of each bud to form a grid pattern. Orientation can be determined by measuring the angle between the anterior-posterior axis of a certain bud and the original midline of the dorsal skin.

3. Whole-mount *in situ* hybridization can be carried out on the explants. Wash the explant three times with DEPC (diethyl/pyrocarbonate)/PBS (1:1000), then process the explants for *in situ* hybridization as usual ***(14)***.
4. For sectioning, wash explants with DEPC/PBS, after which the explants are fixed. After fixation, the explant can be removed gently from the insert and put in a histology cassette. During processing, keep the explant as flat as possible. The sections can be stained following sectioning for *in situ* hybridization or immunochemistry ***(14)***.

4. Notes

1. Shipped eggs are equivalent to day 1 incubation. When these eggs are stored at 4–10°C, they do not develop. To have embryos of certain ages at desired dates, these eggs can be stored in lower temperature and transferred to the incubator within a 1-wk period without too much drop in viability. Once development starts at 38°C, trying to stop it by lowing the temperature will be detrimental.
2. There are many tracts on chicken skin ***(2,3)***. The major tracts we have studied are spinal, femoral, humeral, and caudal. Different tracts develop at different times. Within the tract, there is also a temporal sequence of development. For the spinal tract, the feather buds start from the primary row over the midline of the body and then spread bilaterally toward the flank region. For the femoral tract, the feather buds start as a primary row at the junction between the leg and trunk and then spread medially and upward toward the body midline. It should be kept in mind that there are feather buds of different stages over the same piece of skin explant. This is important, as the tested reagents may have different effects on buds of different stages or from different tracts.
3. Although the Hamburger-Hamilton staging ***(7)*** indicated in both embryonic days and stages, stages are more accurate because the days can be ±1 d. For feather study, this is not even sufficient, as the development progresses faster than stages between stages 28–35. Therefore it is useful to indicate the number of rows of formed dermal condensations. The explant can then be named "stage X dorsal skin explants with Y rows of dermal condensations."
4. There are two ways to visualize the feather buds in the whole-mount view ***(1)***. Transillumination with a dissection microscope is useful for visualizing dermal condensations and distinguishing bud and interbud spaces ***(2)***. Oblique epi-illumination using fiberoptic light is useful in observing the surface morphology and three-dimensional view of the skin appendages.
5. One useful point for using the insert culture over the filter culture is that the live cultures can be photographed daily.
6. Recombination can be carried out between different stages of skin explants, different types of skin appendage, or even appendages from different classes of animals ***(15,16)***.

Acknowledgment

The author thanks Dr. Ting-Xin Jiang and Dr. Han-Sung Jung for help in setting up these cultures. This work was supported by grants from the National Institutes of Health AR42177 and the National Science Foundation IBN 9808874.

References

1. Jiang, T.-X., Stott, S., Widelitz, R. B., and Chuong, C.-M. (1998) Current methods in the study of avian skin appendages, in *Molecular Basis of Epithelial Appendage Morphogenesis* (Chuong, C.-M., ed.), Landes Bioscience, Austin, TX, pp. 359–370.
2. Chuong, C.-M. and Widelitz, R. B. (1998) Feather morphogenesis: a model of the formation of epithelial appendages, in *Molecular Basis of Epithelial Appendage Morphogenesis* (Chuong, C.-M., ed.), Landes Bioscience, Austin, TX, pp. 57–74.
3. Sengel, P. (1976) *Morphogenesis of Skin.* Cambridge University Press, Cambridge, UK.
4. Wolff, E. and Haffen, K. (1952) Sur une méthode de culture d'organes embryonnaires in vitro. *Texas Rep. Biol. Med.* **10,** 463–472.
5. Novel, G. (1973) Feather pattern stability and reorganization in cultured skin. *J. Embryol. Exp. Morphol.* **30,** 605–633.
6. Chuong, C.-M., Widelitz, R. W., Ting-Berreth, S. A., and Jiang, T.-X. (1996). Early events during the regeneration of skin appendages: dependence of epithelial-mesenchymal interaction and order of molecular reappearance. *J. Invest. Dermatol.* **107,** 639–646.
7. Hamburger, V. and Hamilton, H. L. (1951) A series of normal stages in development of the chick embryo. *J. Morphol.* **88,** 49–91.
8. Song, H., Wang, Y., and Goetinck, P. F. (1996) Fibroblast growth factor 2 can replace ectodermal signaling for feather development. *Proc. Natl. Acad. Sci. USA* **93,** 10,246–10,249.
9. Hayamizu, T. F., Sessions, S. K., Wanek, N., et al. (1991) Effects of localized application of transforming growth factor beta 1 on developing chick limbs. *Dev. Biol.* **145,** 164–173.
10. Vogel, A. and Tickle, C. (1993) FGF-4 maintains polarizing activity of posterior limb bud cells in vivo and in vitro. *Development* **119,** 199–206.
11. Chuong, C-M, Ting, S. A., Widelitz, R. B., et al. (1992) Mechanism of skin morphogenesis: II. Retinoic acid modulates axis orientation and phenotypes of skin appendages. *Development* **115,** 839–852.
12. Noveen, A., Jiang, T.-X., and Chuong, C.-M. (1995) Protein kinase A and protein kinase C modulators have reciprocal effects on mesenchymal condensation during skin appendage morphogenesis. *Dev. Biol.* **171,** 677–683.
13. Widelitz, R. B., Jiang, T.-X., Chen, C.-W. J., Stott, N. S., Jung, H. S., and Chuong, C.-M. (1999) Wnt-7a in feather morphogenesis: involvement of anterior-posterior asymmetry and proximal-distal elongation demonstrated with an in vitro reconstitution model. *Development* **126,** 2577–2587.
14. Ting-Berreth, S. A. and Chuong, C.-M. (1996) Sonic hedgehog in feather morphogenesis: induction of mesenchymal condensation and association with cell death. *Dev. Dynamics.* **207,** 157–170.
15. Dhouaily, D. (1975) Formation of cutaneous appendages in dermo-epidermal interactions between reptiles, birds and mammals. *Roux Arch. Dev. Biol.* **177,** 323–340.
16. Rawles, M.E. (1963) Tissue interactions in scale and feather development as studied in dermal-epidermal recombination. *J. Embryol. Exp. Morphol.* **11,** 765–789.

13

Apoptosis in Development

Lin Lin and Zahra F. Zakeri

1. Introduction

Development involves a delicate orchestration of cell division, movement, differentiation, and death. Cell death in mammalian embryogenesis occurs as early as inner cell-mass differentiation, and it continues to be part of the developing embryo during the formation and functional completion of the different organs *(1–3)*. The importance of cell death was first recognized by developmental biologists and teratologists *(4,5)*. More recently, with the new recognition of its possible role in many other areas such as aging, infectious diseases, and cancer, the field has received a burst of energy and, therefore, many investigators are in search of the signals that regulate cell death. The goal of this quest is to identify the dead or dying cells. The ability to identify and characterize the type of cell death has increased interest in cell death under different circumstances. Cell death has been divided into different classes based on the morphology and biochemical function of the cell during its demise. In necrotic cell death, which is not programmed, the cell typically loses energy resources or membrane integrity, swells, and osmotically lyses, losing or destroying contents in a chaotic manner *(6)*. In programmed cell death, the cell appears to participate in its own demise. This type of cell death has been further divided into type I and type II cell death. In type I (apoptosis), a characteristic coalescence and margination of chromatin, related to degradation of the DNA ultimately to a nucleosomal ladder, is an early and prominent feature *(7)*. In type II, although the nuclear collapse eventually occurs, it typically is late and modest, well preceded by many cytoplasmic changes. In cells with large amounts of cytoplasm, these changes likely include lysosomal (autophagic) degradation of massive amounts of cytoplasmic constituents *(8,9)*. Although there are specific markers unique to each type of cell death, there are many gray areas and points of overlap, reflecting the different cell types that might undergo cell death.

Although in the developing embryo all three types of cell death are found, it is simpler to identify the more common apoptotic cell death. For this reason we will focus our discussion on the identification of apoptotic cell death in the developing embryo. The approaches described can also be used to identify cell death in adult tissues. The different approaches that are presented can be used as tools to monitor the events that

From: *Methods in Molecular Biology, Vol. 136: Developmental Biology Protocols, Vol. II*
Edited by: R. S. Tuan and C. W. Lo © Humana Press Inc., Totowa, NJ

take place. These events include DNA fragmentation, engulfment of cells by phagocytic cells (indicating activation of cell-surface markers), activation of lysosomal enzymes (for recognition and destruction of the cells), and alteration in the pattern of gene expression. In the following section, we describe methodology that can be used in serial tissue sections to identify cell death by apoptosis combined with identification of specific gene products. These methods examine DNA fragmentation *in situ*, activation of the lysosomal enzyme acid phosphatase, and detection of phagocytic cells by the use of F4/80 surface antibody, which recognizes mature macrophages and monocytes. Several gene products are upregulated during cell death. We have shown that the cell-cycle-dependent protein kinase 5 (CDK5) is strongly upregulated at the level of protein in the dying cells ***(10)***. The expression of CDK5 can be colocalized in the cells undergoing apoptosis.

2. Materials

2.1. Buffers

1. 1X Phosphate-buffered saline (PBS), 0.85% NaCl, 0.02% KCl; 10 m*M* Na_2HPO_4, 1.8 m*M* KH_2PO_4, pH 7.4).
2. 1X PBST: 0.1% Tween-20 (polyoxyethylenesorbitan monolaurate, Sigma Chemical Co., St. Louis, MO) in 1X PBS.

2.2. Embryo Fixation and Slide Preparation

1. 4% Paraformaldehyde: For 50 mL of fixer, add 2 g of paraformaldehyde (Fisher Scientific, Pittsburgh, PA) to 25 mL of DEP–H_2O (1 mL of diethyl pyrocarbonate [Sigma] is applied to 1 L of H_2O, mixed, allowed to stand for 1 h, and autoclaved for 1 h). Add 5 mL of 10 *M* NaOH and 5 mL of 10X PBS. Stir the mixture at 55–65°C (no more than 65°C) for 2 h. Filter the solution through a 0.45-µm filter (Gelman Sciences, Ann Arbor, MI, Cat. # 4154). Then adjust to pH 8.0 and use DEP–H_2O to adjust the final volume to 50 mL. This solution can be stored at 4°C no more than 1 wk.
2. Poly-L-lysine-coated slides: Place clean slides in autoclaved metal racks. Dip them in 0.2 *M* HCl, DEP–H_2O, and acetone for 30 s each. Leave slides under a hood to dry for 2 h. Dip in a solution of 50 mg poly-L-lysine (Sigma) in 1 L of 0.01 *M* Tris, pH 8.0 for 5 min, and dry overnight in a hood. Coated slides are good to use within 4 wk. An alternative is Vectabond-coated slides: Place clean slides in autoclaved metal racks. Dip them in 7 mL Vectabond reagent in 350 mL acetone for 5 min. Wash by dipping them several times over 30 s in dH_2O or DEP–H_2O and air-dry overnight. Slides are good for at least 4 wk.

2.3. Lysosomal Activity

1. Fast garnet stain: 0.6 mL $NaNO_3$ and 0.6 mL Fast Garnet stain are mixed and allowed to stand for 5 min. Then, 22.8 mL of prewarmed dH_2O, 3 mL of acetate solution and 3 mL of naphthol are added.
2. Citrate–acetone–formaldehyde solution: Make the mixture solution of citrate (pH 3.6), acetone, and 37% formaldehyde at volume ratio of 13:33:4. (The citrate solution is provided as part of the kit.)
3. 0.3% Hydrogen peroxide solution: Dilute 30% v/v hydrogen peroxide (Sigma) with methanol. Make fresh!

2.4. Detection of Phagocytic Cells

1. 1X PBST–gelatin solution: 0.1% gelatin (Fisher Scientific) in 1X PBST.
2. 5% Milk blocker: 5% dry milk (Carnation) in 1X PBST–gelatin solution.

3. F4 Primary antibody: Dilute F4/80 (Serotec, Oxford, UK) at 1:10 in 1X PBST–gelatin solution.
4. Secondary antibody solution: A 1:50 dilution of peroxidase labeled $F(ab)_2$ fragment of goat anti-rat IgG (H+L) (Jackson ImmunoResearch, West Grove, PA) in 1X PBST–gelatin.

2.5. Nonfluorescent DNA Fragmentation

(ApopTag, Oncor Inc., Gaithersburg, MD)

1. Working strength TdT solution: Mix well 2 drops (76 µL) of reaction buffer (S7100-2) with 1 drop (32 µL) TdT enzyme (S7100-3) by vortexing, and keep on ice for no more than 6 h.
2. Working strength stop/wash buffer: Add 1 mL of stop/wash buffer (S7100-4) into 34 mL distilled water. This reagent may be stored at 4°C for up to 1 yr.
3. Methyl Green: 0.5% methyl green in 0.1*M* sodium acetate, pH 4.0.

2.6. Fluorescent DNA Fragmentation

(ApopTag, Oncor)

1. Working strength TdT enzyme: Mix 38 µL of reaction buffer (S7111-2) with 16 µL of TdT enzyme (terminal deoxynucleotidyl transferase) (S7111-3) by vortexing and keep on ice for no more than 6 h.
2. Working strength stop/wash buffer: Add 1 mL of stop/wash buffer (S7111-4) into 34 mL distilled water. This reagent may be stored at 4°C for up to 1 yr.
3. Working strength anti-digoxygenin–fluorescein: Mix well 56 µL blocking solution (S7111-5) with 49 µL anti-digoxigenin–fluorescein (S7111-6) and keep on ice for no more than 3 h.

2.7. Double Labeling

1. Blocker: 1 drop goat serum (Vectastain ABC Kit, Vector Lab Inc., Burlington, CA) in 10 mL 1X PBST.
2. Diluted primary antibody: For Cdk5, it is 1 µg/mL. This is prepared as 40 µg BSA (bovine serum albumin) +10 µL Cdk5 antibody stock (Santa Cruz Biotechnology, Inc., Santa Cruz, CA, 100 µg/mL) +1 mL 1X PBST.
3. Biotinylated secondary antibody: 1 drop of biotinylated antibody (Vectastain ABC Kit, Vector Lab Inc.) in 10 mL 1X PBST.
4. Cy3-conjugated IgG mouse anti-biotin (Jackson ImmunoResearch): 1:250 dilution in 1X PBS.

3. Methods

1. Embryo fixation and slide preparation (*see* **Note 1**): Mate male and female mice overnight and check the females for the presence of a vaginal plug. The time of plug detection is designated gestational day 0.5. Embryos are removed from pregnant females at different days of gestation depending on the specific period of interest. Embryos or embryonic tissues are fixed in 4% paraformaldehyde overnight. After overnight fixation, embryos or tissues are placed in 20% sucrose in 1X PBS overnight (filtered and stored at 4°C), blotted on a filter paper quickly to eliminate excess liquid around the tissue sample, and embedded in aluminum cups half-filled with OCT (TissueTek; Miles Inc. [Bayer Corp.], Tarrytown, NY) embedding compound. The samples are then gently frozen by immersing the cup in isopentane placed in liquid N_2. Frozen sections (5 µm) are cut in a cryostat at –20°C, placed on poly-L-lysine- or Vectabond-coated slides and stored at –70°C prior to use. For all the steps outlined below, the slides are allowed to come to room temperature before use.

2. Detection of phagocytic cells: For detection of phagocytic cells, frozen sections are brought to room temperature, rehydrated in a graded series of 20%, 10%, and 5% sucrose in 1X PBS for 5 min each and washed in dH2O for 5 min. Endogenous peroxidase is inactivated by immersing the sections in 0.3% hydrogen peroxide solution and washing in dH_2O and 1X PBST for 5 min each, followed by blocking the nonspecific binding of the primary antibody with a mixture of 5% milk in 1X PBS for 30 min at 37°C. After washing the slides two times with 1X PBST–gelatin for 5 min, the primary antibody F4/80 is applied and the slides are incubated overnight at 4°C. To detect the primary antibody the slides, are washed twice with 1X PBST–gelatin for 5 min each, the secondary antibody solution is applied, and slides are incubated for 2 h at room temperature. The secondary antibody is removed by washing the slides three times in 1X PBST–gelatin for 5 min each. Immunoreactivity is visualized by DAB (diaminobenzidine, provided in predissolved solution by Research Genetics, Inc., Huntsville, AL) staining for 2 min. Slides are then counterstained with hematoxylin (Sigma) for 30 s, washed in dH_2O three times for 1 min each, and mounted with CrystalMount (Fisher). The phagocytic cells are apparent by a dark brown staining of their cell membranes (**Fig. 1**, **panels 1** and **2**).
3. Measurement of lysosomal activity: To measure lysosomal activity, one can analyze acid phosphatase activity by the use of an acid phosphatase kit (Sigma, 181-A). Sections are passed through serial-graded solutions of 20%, 10%, and 5% sucrose in 1X PBS for 5 min each, postfixed with citrate–acetone–formaldehyde solution for 30 s, and washed in dH_2O for 1 min. Sections are treated with naphthol AS-BI phosphate and fast garnet stain for 1 h at 37°C. Slides are washed in running tap H_2O for 2 min, air-dried for 15 min and counterstained with methylene blue (1:100 dilution for 1 to 2 minutes depending on the tissue), rinsed in dH_2O two times for 1 min each, and mounted with CrystalMount (Fisher). The acid phosphatase activity is detected as a distinct red focal precipitate (**Fig. 1**, **panels 3** and **4**).
4. DNA fragmentation assay: Fragmented DNA can be detected using both fluorescence and nonfluorescence methodology. We present the methodology for both.
   a. Nonfluorescence method (Colorimetric TUNEL—Terminal transferase deoxyUridine Nick End Labeling): ApopTag Plus *In Situ* Apoptosis Detection Kit (Peroxidase, Oncor, S7101-KIT) is used. Tissue sections are treated and the endogenous peroxidase is quenched as above. After rinsing with 1X PBS for 5 min, the excess liquid is gently tapped off and equilibration buffer (13 $\mu L/cm^2$) is applied. The section is covered with a plastic cover slip and incubated for 10 min (can go up to 30 min) in a humidity chamber at room temperature. The equilibration buffer is removed by gently tapping it off and the plastic cover slip is blotted dry to be reused. The working strength of TdT solution is applied (10 $\mu L/cm^2$) and the preparation is covered with the cover slip and incubated for at least 60 min in a humid chamber at 37°C. Slides are placed in prewarmed stop/wash buffer (prewarm the stop/wash buffer to 37°C for 30 min) and incubated for 30 min at 37°C. The slides should be agitated by dipping in and out of buffer once every 10 min. The slides are rinsed three times more in 1X PBS for 5 min/rinse. To detect the attached digoxygenin-labeled UTP, apply anti-digoxygenin–peroxidase to the section (13 $\mu L/cm^2$), re-cover with plastic cover slips and incubate in a humid chamber for 30 min at room temperature. Wash four times in 1X PBS for 5 min/rinse. To detect the peroxidase, place the slides in a Coplin jar containing DAB for up to 20 min at room temperature. Wash the DAB three times with tap water, 1 min/wash, and with dH_2O for 5 min. Counterstain with Methyl Green for 5 min. Rinse slides with dH_2O twice, agitating each slide 10 times up and down per rinse. Rinse slides with dH_2O for 30 s without agitation. Dehydrate by rinsing slides in *n*-butanol twice, agitating each slide 10 times up and down per rinse, in fresh *n*-butanol for 30 s without agitation and in toluene three times for 2 min/rinse. Do not let slides dry.

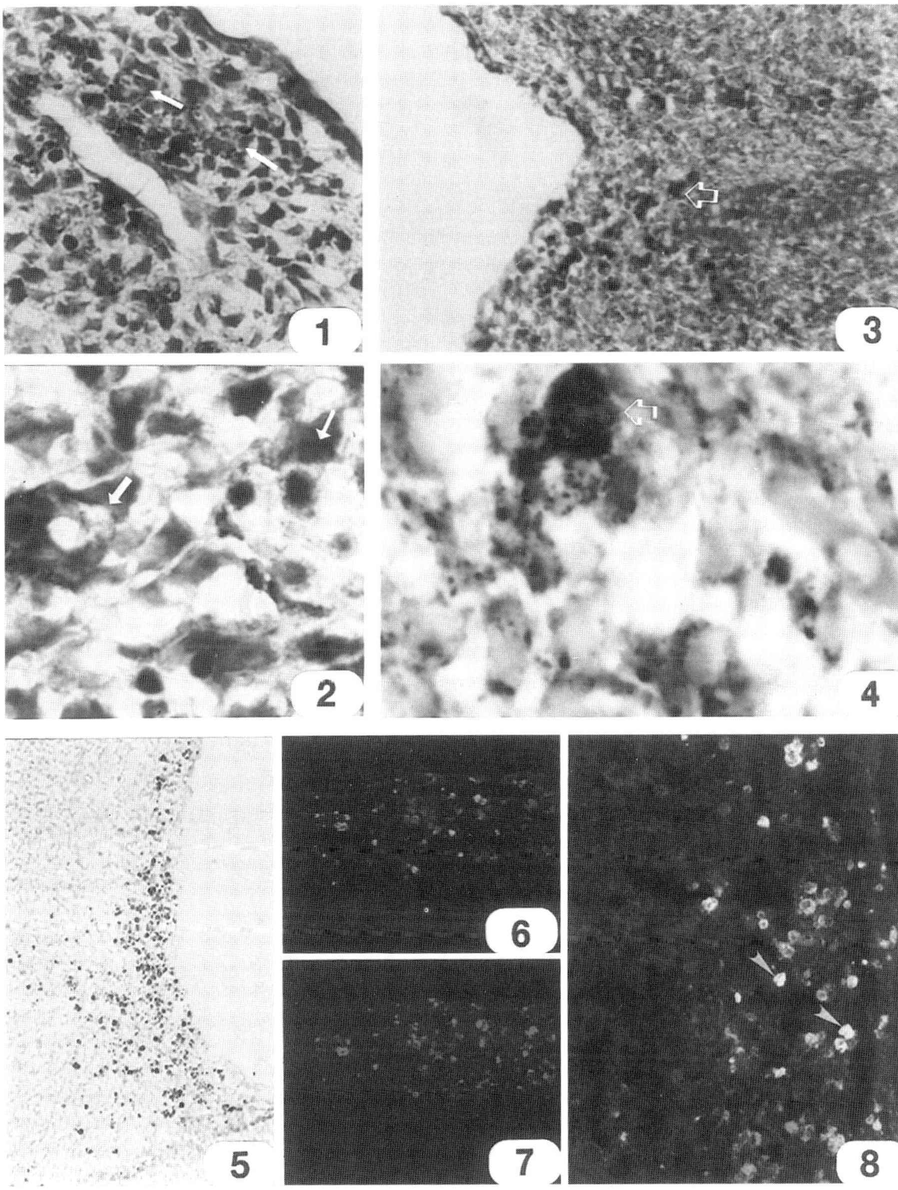

Fig. 1. (*See* color plate 3 appearing after p. 262.) Identification of cell death in d 14.5 mouse limb. Mouse embryonic limbs of gestation d 14.5 were used for the identification of cell death using markers indicated in text. In this study, the embryonic limbs were fixed and processed for frozen sectioning; 5-µm sections were used. (**Panels 1** and **2**) detection of phagocytic cells by the antibody F4/80, (1) = 40X, (2) = 100X. The brown color indicated by solid arrows is peroxidase labeling of the sites of phagocytic cells. (**Panels 3** and **4**) lysosomal activity (acid phosphatase; [3]=40X, [4] = 100X). The distinct red focal precipitates (empty arrows) indicate the activity of acid phosphatase in the lysosomes. (**Panel 5**) DNA fragmentation using nonfluorescent end labeling, 40X. (**Panel 6**) DNA fragmentation using fluorescence end labeling. (**Panel 7**) detection of Cdk5 by Cy3. (Panel 8) Double labeling and confocal overlay of DNA fragmentation and Cdk5, of panels 6 and 7 at 40X. The yellow color indicated by arrowheads indicates the overlays of DNA fragmentation and Cdk5 expression.

Immediately mount using Permount (Fisher Scientific) and a glass cover slip; be careful to avoid bubbles. The labeled cells have a brown staining that covers the nuclei and sometimes parts of the cytoplasm (**Fig. 1**, **panel 5**).

b. Fluorescence DNA fragmentation detection (fluorescence TUNEL): ApopTag Plus *In Situ* Apoptosis Detection Kit (fluorescein) (Oncor, S7111-KIT) is employed here. Rinse slides twice in 1X PBS at 5 min/rinse. Gently tap off excess liquid and carefully blot around sections. Apply equilibration buffer (13 μL/cm^2) directly to specimen, cover with plastic cover slips, and incubate for 5 min in a humidity chamber. Tap off equilibration buffer. Rinse and dry plastic cover slips for later reuse. Blot around sections and apply working strength TdT enzyme (11 μL/cm^2). Cover sections again with plastic cover slips and incubate for at least 60 min in a humid chamber at 37°C. Prewarm Stop/Wash Buffer to 37°C for 30 min before next step. Remove plastic cover slips. Place slides in a Coplin jar containing prewarmed stop/wash buffer and incubate for 30 min at 37°C. Agitate slides by dipping in and out of buffer once every 10 minutes. Tap off liquid. Rinse three times in 1X PBS at 3 min/rinse. Apply Oncor propidium iodide/Antifade (in kit). Mount under glass coverslip. If storage is required, apply clear nail polish to edges of cover slip. Store at –20°C in the dark. At this point, you can stop and examine the slides for DNA fragmentation, using a fluorescence microscope for fluorescein, or continue as in **step 5** for double labeling for gene expression. The dead cells with fragmented DNA appear as green fluorescent dots as seen in **Fig. 1**, **panel 6**.

5. Detection of DNA fragmentation and gene expression (*see* **Notes 2** and **3**)—Double-labeling fluorescence assay: Follow the same procedure as in fluorescence DNA fragmentation up to the mounting step. Without applying propidium iodide and Permount, put slides in a Coplin jar containing 0.3% hydrogen peroxide for 20 min. Wash slides two times in 1X PBST at 10 min/wash. Apply blocker to cover slides and incubate for 1 h. Tap off excess blocker. Cover section with 30 μL/cm^2 diluted primary antibody. (The final concentration of primary antibody depends on different target proteins.) Incubate at 4°C overnight in humidity chamber. Wash three times in 1X PBST at 10 min/wash. Agitate occasionally. Apply biotinylated secondary antibody to cover. Incubate at 4°C overnight in a humid chamber. Wash three times in 1X PBST at 10 min/wash. Incubate with Cy3-conjugated IgG mouse anti-biotin for 30 min. Wash three times in 1X PBST at 10 min/wash. Mount slides with 90% glycerol and cover slips. The specific primary and secondary antibodies to be used for the detection of CDK5 gene are described in the previous section. The steps outlined here are the general steps and can be applied to other genes of interest. The DNA fragmentation is fluorescein isothiocyanate labeled and will fluoresce green. In the example, we have used the expression of CDK5 labeled with Cy3 and fluorescing red (**Fig. 1**, **panel 7**). One can use confocal microscopy or other overlapping imaging systems to correlate the expression of the given gene and fragmented DNA. As seen in **Fig. 1**, **panel 8**, the overlapping signals come out yellow.

4. Notes

1. Tissue on slides may be fixed or unfixed, frozen or paraffin. For unfixed tissue cryosections, the following steps should be followed: Fix the sections on slides in a Coplin jar containing 10% neutral buffered formalin for 10 min at room temperature; blot off excess liquid and wash in 1X PBS twice for 5 min each change; postfix in ethanol: acetic acid (2:1) for 5 min at –20°C; drain, but do not let slides dry; rinse in 1X PBS twice for 5 min/rinse. For paraffin-embedded tissue, the samples must be deparaffinized as follows. Prewarm xylene to 60°C for at least 30 min. The remaining steps are performed at room temperature. Incubate slides in prewarmed xylene twice for 10 min each change; in 100% ethanol twice for 5 min each; in 50% ethanol, twice for 5 min each; wash slides in 1X PBS

once for 5 min. Circle the area of tissue sections or cells with a peroxidase-antiperoxidase pen (PAP) (Research Products International, Mt. Prospect, IL).

2. The method of double labeling uses a two-step approach to confirm that the cell undergoing DNA fragmentation is also expressing specific messages. In the first step, the free 3'-OH ends of fragmented DNA are labeled by the indirect fluorescence TUNEL technique described earlier. In the second step, specific gene expression is localized by means of a specific antibody. The technique described here combines the use of separate fluorescent markers to identify both DNA fragmentation and gene expression. Alternatively, DNA fragmentation may be labeled and gene expression may be studied separately using this methodology. To do the combined staining it is necessary to perform each procedure separately as well. This controls that each step has worked on its own and provides information regarding the level of background noise that has to subtracted from the combined procedure.
3. It is very important to have both positive and negative control slides for each part of this assay. Negative controls should be run as duplicates of each section. For this purpose, sham staining can be performed by substituting distilled water for the TdT enzyme in the DNA end-labeling study, and no primary antibody (PBS) for the anti-Cdk5 antibody. As a positive control for the method, one can use short exposure of the control slide to DNase I, which causes fragmentation of DNA, and then look for DNA end labeling.

Acknowledgments

We thank Dr. Richard Lockshin for critical reading of this manuscript. We also thank Dr. R. Halaby, Dr. H. Singh-Ahuja, and Mr. Y. Zhu for their roles in developing the different methods described. This work has been supported by the National Institute of Health (National Institute on Aging K04-AG 0031) to ZZ, and a Professional Staff Congress City University of New York Award (to ZZ).

References

1. Walker, N. I., Harmon, B. V., Gove, G. X., and Kerr, J. F. R. (1988) Patterns of cell death, in *Methods and Achievements in Experimental Pathology* (Jasmin, G., ed.), Karger, Basel, pp. 18–54.
2. Raff, M. C. (1992) Social control on cell survival and cell death. *Nature* **356,** 397–400.
3. Coucouvanis, E. C., Sherwood, S. W., Carswell-Crumpton, C., Spack, E. G., and Jones, P. P. (1993) Evidence that the mechanism of prenatal germ cell death in the mouse is apoptosis. *Methods Cell Biol.* **46,** 387–440.
4. Glücksmann, A. (1951) Cell death in normal development. *Biol. Rev. Camb. Phil. Soc.* **26,** 59–86.
5. Saunders, J. W. (1966) Death in embryonic systems. *Science* **154,** 604–612.
6. Kerr, J. F. R. (1969) An electron-microscope study of liver cell necrosis due to heliotrine. *J. Pathol.* **97,** 557–562.
7. Zakeri, Z., Bursch, W., Tenniswood, M., and Lockshin, R. A. (1995) Cell death: programmed, apoptosis, necrosis, or other. *Cell Death Differ.* **2,** 87–96.
8. Halaby, R., Zakeri, Z., and Lockshin, R. A. (1994) Metabolic events during programmed cell death in insect labial glands. *Biochem. Cell Biol.* **72,** 597–601.
9. Zakeri, Z. (1993) *In vitro* limb differentiation as an experimental model, in *Limb Development and Regeneration* (Fallon, J. F., Kelley, R. O., and Stocum, D. L., eds.), Wiley–Liss, New York, pp. 361–370.
10. Zhang, Q., Ahuja, H. S., Zakeri, Z., and Wolgemuth, D. (1997) Cyclin-dependent kinase 5 is associated with apoptotic cell death during development and tissue remodeling. *Dev. Biol.* **183,** 222–233.

14

Methods to Detect Patterns of Cell Death in *Drosophila*

Nancy M. Bonini

1. Introduction

There are many times during *Drosophila* development or in the adult when cell death contributes to a phenotype of interest (e.g., **refs. *1–5***, and many others). The cell death may be primary to gene function or a secondary consequence of failure of proper differentiation or cell maintenance (e.g., **ref. *6***). To detect whether cells are dying and how, there are a number of assays that can be used to determine cell viability. In general, there are two fundamental ways in which cells die: necrosis or apoptosis *(7,8)*. Necrosis is pathological death characterized morphologically by swelling of the organelles, including mitochondria, with ultimate rupture of the nuclear and plasma membranes. In contrast, apoptosis—also called programmed cell death—is characterized by condensation of the chromatin, nucleus, and cytosol, with cell fragmentation into distinct condensed bodies (apoptotic bodies), which ultimately become engulfed and removed by surrounding cells. The mitochondria in cells undergoing apoptosis remain morphologically healthy, and the DNA is degraded into nucleosomal fragments *(7,8)*.

In a model organism like *Caenorhabditis elegans*, it is possible to readily distinguish death as a result of programmed cell death based on gene dependence. The invariant lineage pattern allows identification of all cells fated to die by apoptosis: during normal *C. elegans* development, 131 of 1090 somatic cells are eliminated by apoptosis, resulting in an adult hermaphrodite consisting of precisely 959 cells *(9)*. Programmed cell death in *C. elegans* uses common gene activities of *ced-3* and *ced-4* *(9,10)*. Therefore, by establishing dependence of any specific cell death on these gene activities, it is possible to definitively assign the death to programmed cell death pathways. *Drosophila*, too, has a number of genes that are involved in controlling programmed cell death in the animal *(11–13)*. However, *Drosophila* has added complications in that the genes known to date function, at least in part, redundantly, because all can induce death in the absence of the other genes, and that removal of all of these gene activities (with a deficiency referred to as *H99*) is embryonic lethal. This makes performing gene dependence experiments more complicated, because one is essentially limited to examining dominant gene dosage interactions when addressing functional relevance with programmed cell death genes, or making genetic mosaics. By

From: *Methods in Molecular Biology, Vol. 136: Developmental Biology Protocols, Vol. II*
Edited by: R. S. Tuan and C. W. Lo © Humana Press Inc., Totowa, NJ

contrast, in *C. elegans*, *ced-3*, and *ced-4* mutants are viable. Also, in *Drosophila*, genes not currently identified appear involved in inducing apoptosis, in addition to those defined by the *H99* deficiency region ***(14,15)***. In organisms like zebrafish, chick, and mouse, as well as *Drosophila*, tissue-staining methods and mutant analysis, coupled with studies of expression of known apoptotic genes, can be used to address the nature of cell loss and the involvement of specific gene functions.

2. Methods and Materials

2.1. Acridine Orange Staining

One of the simplest and most rapid techniques to examine tissue for an increase in programmed cell death is to stain the tissue with the vital dye Acridine Orange (AO). AO specifically labels cells dying by programmed cell death in *Drosophila* ***(16)*** and is now also used for vertebrate studies ***(17–19)***. AO shows selective affinity for chromatin of apoptotic cells, resulting in bright fluorescence. AO has been shown in *Drosophila* to be specific for cells dying by programmed cell death and will not label chromatin of cells dying by necrosis due to oxygen starvation ***(16)***.

The advantages of AO staining are that it is simple and quick. The disadvantages are that the tissue must be living and the preparations are not permanent. It is necessary to perform the staining and view the tissue by fluorescent optics as rapidly as possible. Because the tissue is unfixed, it is fragile and manipulations should be kept to a minimum. By this technique, one can readily visualize patterns of cell death when staining whole mounts (*see* **refs.** ***16–20***). Note that it is very important to carefully stage wild-type and mutant tissue to the same developmental time, as cell death patterns can vary greatly during development.

2.1.2. AO Staining of Dissected Tissue (Adapted from **ref. 2**)

1. Dissect tissue in a physiological solution such as Ringer's solution. PBS (phosphate-buffered saline) also works. *Drosophila* Ringer's solution: 128 m*M* NaCl, 2 m*M* KCl, 35.5 m*M* sucrose, 5 m*M* HEPES, 1.8 m*M* $CaCl_2$, 4 m*M* $MgCl_2$, pH to 7.1 with HCl, sterilize.
2. Incubate 5 min in AO solution (about 0.5 µg/mL for thin tissue, 5 µg/mL for thick whole mounts). Concentration should be determined empirically. Adjust up or down depending on the staining (too high a concentration will give a high general background fluorescence; too low a concentration will quench too rapidly when viewed by fluorescence optics).
3. Mount in Ringer's solution and view by fluorescence optics. Because the tissue is alive, it is very fragile, put cover slips on carefully or view without a cover slip. Fluorescein filters generally give a stronger signal, but a higher background. Rhodamine filters give a better signal to noise, but, overall, the signal appears dimmer. I have seen examples of ultraviolet (UV) excitation, as well. Because the preparations are not permanent, be prepared to take pictures immediately. Apoptotic staining is apparent as brightly fluorescent bodies.

2.1.3. AO Protocol for staining Embryos (Adapted from **ref. 16**)

1. Dechorionate embryos in 50% bleach.
2. Rinse in water.
3. Incubate 5 min with shaking in a solution of equal volumes heptane and acridine orange (5 µg/mL in 0.1 *M* phosphate buffer, pH 7.2).
4. Place embryos on a slide under series 700 Halocarbon oil (Halocarbon Products Corp., River Edge, NJ, USA) and view by fluorescent optics, as before.

Variations on the AO protocol include the use of other vital dyes, such as Nile Blue A (*see* **ref. 16**). With Nile Blue, the tissue can be viewed with normal light optics.

2.2. TUNEL

TUNEL (terminal deoxynucleotidyl transferase [TdT]-mediated dUTP–biotin nick end labeling) is widely used to detect apoptotic cells. Based on the principle of fragmentation of the DNA during apoptotic death, the enzyme TdT will label the ends of the DNA fragments with a deoxynucleotide, which is labeled fluorescently or with biotin, for example *(21)*. The label is then directly viewed or detected with a secondary reagent, as appropriate. The advantages include the option of a permanent preparation, as well as a three-dimensional view if done as a whole mount. In addition, TUNEL is widely accepted as being distinct for apoptosis because it relies on the characteristic DNA fragmentation that occurs with apoptosis. The disadvantages include that the protocol is longer and more tedious than vital dye staining. Numerous companies now make reagents or kits for TUNEL labeling, including Roche Diagnostics Corp. (Indianapolis, IN), Gibco/BRL (Gaithersburg, MD), and Oncor (Gaithersburg, MD), among others. A standard protocol is as follows.

2.2.1. TUNEL Protocol for Dissected tissues (After **ref. 11**, with Modifications by S. Robinow, University of Hawaii, Honolulu)

1. Dissect tissue in PBS or phosphate-buffer solution. Fix in 2–4% paraformaldehyde in PBS or 0.1 *M* phosphate buffer, pH 7.4, 30 min to 2 h, depending on the tissue (for dissected imaginal disks, use 2% paraformaldehyde, 30 min; for large tissue whole mounts such as the brain or thorax, use 4% paraformaldehyde, incubating several hours to overnight).
2. Rinse 3 × 5 min in PBS-Tx (1× PBS + 0.3% Triton X-100). For a positive control, pretreat a tissue sample with DNase I (1 µg/mL to 1 mg/mL, 10–30 min at 37°C), prior to **step 3**.
3. Rinse 10 min in 1× TdT buffer.
   TdT buffer (final concentrations)
   - 200 m*M* potassium cacodylate
   - 1 m*M* $CoCl_2$
   - 0.25 mg/mL bovine serum albumin
   - 0.1% Triton X-100
   - 25 m*M* Tris–HCl, pH 6.6

   Note: This buffer contains cacodylate and cobalt and is hazardous.
4. Incubate in TdT reaction mix, 37°C, 2–3 h to overnight.
   TdT reaction mix:
   - 200 m*M* potassium cacodylate
   - 1 m*M* $CoCl_2$
   - 0.25 mg/mL bovine serum albumin
   - 0.1% Triton X-100
   - 25 m*M* Tris–HCl, pH 6.6
   - 0.1 m*M* biotin–dUTP or fluorescein–dUTP
   - 0.6 units/µL TdT enzyme
5. Wash extensively (6× 10 min) in PBS-Tx.
6. Process as appropriate for labeled nucleotide. For fluorescent labeling, either use fluorescein–dUTP or incubate tissue in streptavidin conjugated with the appropriate fluorophore (Texas Red, fluorescein, or other; Jackson ImmunoResearch Laboratories, Inc., West Grove, PA) 1:200, 1 h, wash in PBS-Tx, mount in 80% glycerol/2% propyl-gallate.

If using immunoperoxidase, develop tissue according to standard peroxidase methods: 1 h in A + B reagents of Vectastain ABC kit (Vector Labs, Burlingame, CA, USA), followed by development of DAB reaction product (0.5 mL of 0.5 mg/mL DAB in PBS, plus 1 µL 3% H_2O_2).

Note that sometimes cells that are polyploid label; these cells appear distinct from apoptotic cells because the labeling is observed to be uniform throughout the nucleus.

Modifications to protocol for embryo staining ***(11)***:

1. Dechorionate embryos with 50% bleach.
2. Rinse well in water.
3. Fix 30 min in 4% paraformaldehyde in 0.1 *M* phosphate buffer, pH 7.4.
4. Remove vitelline membrane with methanol.
5. Rehydrate through a graded methanol series (75%, 50%, 25% methanol in PBT [PBS + 0.1% Tween-20], 5 min each step).
6. Proteinase K treat 5 min with 10 µg/mL in PBS.
7. Wash in PBT.
8. Postfix 20 min in 4% paraformaldehyde in PBS.
9. Go to **step 2** in preceding TUNEL protocol.

2.3. Histological Staining Methods

A classical method for detecting cells dying by programmed cell death is to stain thick plastic sections with Toluidine Blue or a related compound. To do this, tissue is fixed in a buffered paraformaldehyde–glutaraldehyde combination, embedded in plastic, thick sectioned, and collected on a slide. The sectioned material is then stained with Toluidine Blue, Methylene Blue, or a Toluidine Blue–Methylene Blue mixture (Methylene Blue yields better tissue detail). Tissue sections are then mounted with a permanent mounting medium under a cover slip, and the morphology viewed using conventional light microscopy. This technique is particularly good for finding appropriate regions to then do thin sections to view by transmission electron microscopy. Some nice examples of tissue stained to highlight dead cells in this manner, as well as the comparison with AO staining, can be found in **refs. *16,19,20***, and ***22***.

2.3.1. Staining of Tissues with Toluidine and/or Methylene Blue

1. Fix and embed tissue in plastic (Epon or Durcupan, Electron Microscopy Sciences, Fort Washington, PA), according to standard protocols for the tissue of interest (for *Drosophila* embryos, *see* **ref. *16***; for eye and brain, *see* **refs. *22–24***). Cut standard thick sections (1–10 µm).
2. Stain sections as follows:

   Cover the slide with stain solution.

   Put on a warm hot plate (about 60°C or so—too hot to touch, but not hot enough to boil water) for 5–15 s.

   Rinse the solution off with 95% ethanol, then water, either using squirt bottles or dipping into Coplin jars.

   Let slide air-dry.
3. Tissue can now be viewed under light optics.

If tissue is appropriately stained, permanently mount slide under a cover slip with Permount (Fisher Scientific, Pittsburgh, PA), DPX (Electron Microscopy Sciences, Fort Washington, PA), or other permanent mounting medium. If the material is too lightly stained, the tissue can be stained again until it is the desired contrast is obtained. Apoptotic bodies stain very darkly. We have found that Methylene Blue gives better detail of tissue morphology, whereas Toluidine Blue will generally stain darker.

Solutions
- Toluidine Blue:
  - 1 g toluidine blue (Sigma)
  - 6 g sodium borate
  - 1 g boric acid
  - Distilled water to 100 mL
  - Stir until dissolved (overnight in dark or foil-wrapped glass bottle)
  - Gravity filter to remove residual, undissolved dye particles
  - To use: mix 3 parts stain solution with 1 part 70% ethanol.
- Methylene Blue:
  - 1 g sodium borate
  - 1 g Methylene Blue (Sigma)
  - 1 g Azur II (Sigma)
  - Distilled water to 100 mL
  - Stir until dissolved (overnight in dark or foil-wrapped glass bottle)
  - Filter to remove residual undissolved dye particles
  - Use straight

Some people prefer a mix of Toluidine Blue and Methylene Blue. Mix Methylene Blue solution 1:1 with 0.2% Toluidine Blue solution, dilute 1:10 with water and use as above.

Ultimately, electron microscopy should be performed to examine dying cells for ultrastructural features characteristic of apoptotic cells and bodies, such as intact membranes, condensed nuclei, and healthy mitochondria. *See* **refs.** *7* and *8* for electromicrographs of apoptotic cells in general, and **refs.** ***16*** and ***25*** for electromicrographs of apoptotic cells in *Drosophila* in particular.

2.4. Interactions with Programmed Cell Death Genes

When "killer" genes are ectopically expressed in *Drosophila*, they induce cell death. A particularly tractable situation is when gene expression is targeted to the eye, using the *gmr* promoter element [*glass multiple reporter* ***(26)***], although other promoters have also been used. When targeted to the eye, the killer genes known in *Drosophila* to date (*grim*, *reaper*, and *hid*) ablate the eye in a dosage-sensitive manner ***(12,13,27,28)***. There are also known genes that can block programmed cell death in *Drosophila*, including the baculoviral gene *P35*, as well as endogenous genes *DIAP1* and *DIAP2* ***(27,29)***. These genes will not only block normally occurring cell death in the fly but also ablation of the eye induced by killer gene activity. In addition, there are mutants or deficiencies available for the endogenous genes (the *H99* deficiency deletes *grim*, *reaper* and *hid*; *thread* mutants are loss-of-function alleles of the *DIAP1* gene). Thus, with these available tools, it is possible to address whether a gene or mutant of interest interacts with these components of programmed cell death pathways. This can be done through genetic mosaic analysis (e.g., **refs.** ***14*** and ***15***), or through directed gene expression to determine whether killer gene activity is modified ***(12,13,27,28)***. Note that not all genes involved in programmed cell death have been found yet in *Drosophila*; some examples of programmed cell death do not require gene function within the *H99* deficiency ***(14,15)***, as noted originally given that apoptosis can be induced in embryos homozygous for the *H99* deficiency region ***(11)***.

3. Conclusions

With these various techniques, it is possible to examine tissue for increased or decreased patterns of programmed cell death and potentially define which genes of programmed cell death pathways are involved in the effect by simple genetic tests. Because cell death patterns are dynamic during development, it is critical in any study to use controls that are developmentally matched with the experimental tissue. Many signals can lead to programmed cell death ***(6,30)***; thus, the relationship of any one gene function to programmed cell death can be very close to or many steps removed from the activation of killer or survival gene activity. The relationships among proliferation signals, developmental signals, and cell death continue to be a fruitful and interesting direction. As one example, some recent studies indicate nonautonomous regulation of cell death during development, an area that promises to be interesting to define in detail at the molecular and biological levels (e.g., **refs.** ***31*** and ***32***).

Acknowledgments

Thanks go to A. Cashmore, M. Fortini, and S. Robinow for critical comments. N.B. is supported by grants from the John Merck Fund, the National Eye Institute, the Alzheimer's Association, and the David and Lucile Packard Foundation.

References

1. Fristrom, D. (1969) Cellular degeneration in the production of some mutant phenotypes in *Drosophila melanogaster*. *Mol. Gen. Genet.* **103,** 363–379.
2. Spreij, T. E. (1971) Cell death during the development of the imaginal discs of *Calliphora erythrocephala*. *Nether. J. Zool.* **21,** 221–264.
3. Martinez-Arias, A. (1985) The development of *fused*-embryos of *Drosophila melanogaster*. *J. Embryol. Exp. Morph.* **87,** 99–114.
4. Magrassi, L. and Lawrence, P. A. (1988) The pattern of cell death in *fushi tarazu*, a segmentation gene of *Drosophila*. *Development* **104,** 447–451.
5. Bryant, P. J. (1988) Localized cell death caused by mutations in a *Drosophila* gene coding for a transforming growth factor-beta homolog. *Dev. Biol.* **128,** 386–395.
6. Bonini, N. M. (1997) Surviving *Drosophila* eye development. *Cell Death Differ.* **4,** 4–11.
7. Kerr, J. F. R., Wyllie, A. H., and Currie, A. R. (1972) Apoptosis: a basic biological phenomenon with wide ranging implications in tissue kinetics. *Br. J. Cancer* **26,** 239–257.
8. Wyllie, A. H., Kerr, J. F. R., and Currie, A. R. (1980) Cell death: the significance of apoptosis. *Int. Rev. Cytol.* **68,** 251–306.
9. Ellis, R. E., Yuan, J., and Horvitz, H. R. (1991) Mechanisms and functions of cell death. *Annu. Rev. Cell Biol.* **7,** 663–698.
10. Driscoll, M. and Chalfie, M. (1992) Developmental and abnormal cell death in *C. elegans*. *Trends Neurosci.* **15,** 15–19.
11. White, K., Grether, M. E., Abrams, J. M., Young, L., Farrell, K., and Steller, H. (1994) Genetic control of programmed cell death in *Drosophila*. *Science* **264,** 677–683.
12. Grether, M. E., Abrams, J. M., Agapite, J., White, K., and Steller, H. (1995) The *head involution defective* gene of *Drosophila melanogaster* functions in programmed cell death. *Genes Dev.* **9,** 1694–1708.

13. Chen, P., Nordstrom, W., Gish, B., and Abrams, J. (1996) *grim*, a novel cell death gene in *Drosophila. Genes Dev.* **10,** 1773–1782.
14. Foley, K. and Cooley, L. (1998) Apoptosis in late stage *Drosophila* nurse cells does not require genes within the *H99* deficiency. *Development* **125,** 1075–1082.
15. Staveley, B. E., Rue, L., Jin, J., Stambolic, V., Mastronardi, F. G., Heitzler, P., Woodgett, J. R., and Manoukian, A. S. (1998) Genetic analysis of protein kinase B (AKT) in *Drosophila. Curr. Biol.* **8,** 599–602.
16. Abrams, J. M., White, K., Fessler, L. I., and Steller, H. (1993) Programmed cell death during *Drosophila* embryogenesis. *Development* **117,** 29–43.
17. Graham, A., Heyman, I., and Lumsden, A. (1993) Even-numbered rhombdomeres control the apoptotic elimination of neural crest cells from odd-numbered rhombdomeres in the chick hindbrain. *Development* **119,** 233–245.
18. Chen, Z.-F. and Behringer, R. R. (1995) *twist* is required in head mesenchyme for cranial neural tube morphogenesis. *Genes Dev.* **9,** 686–699.
19. Furutani-Seiki, M., Jiang, Y. J., Brand, M., Heisenberg, C. P., Houart, C., Beuchle, D., van Eeden, F. J., Granato, M., Haffter, P., Hammerschmidt, M., Kane, D. A., Kelsh, R. N., Mullins, M. C., Odenthal, J., and Nusslein-Volhard, C. (1996) Neural degeneration mutants in the zebrafish, *Danio rerio. Development* **123,** 229–239.
20. Bonini, N. M., Leiserson, W. M., and Benzer, S. (1993. The *eyes absent* gene: genetic control of cell survival and differentiation in the developing Drosophila eye. *Cell* **72,** 379–395.
21. Gavrieli, Y., Sherman, Y., and Ben-Sasson, S. A. (1992) Identification of programmed cell death in situ via specific labeling of nuclear DNA fragmentation. *J. Cell Biol.* **119,** 493–501.
22. Fischbach, K. F. and Technau, G. (1984) Cell degeneration in the developing optic lobes of the sine oculis and small-optic-lobes mutants of *Drosophila melanogaster. Dev. Biol.* **104,** 219–239.
23. Tomlinson, A. and Ready, D. F. (1987) Cell fate in the *Drosophila* ommatidium. *Dev. Biol.* **123,** 264–275.
24. Renfranz, P. J. and Benzer, S. (1989) Monoclonal antibody probes discriminate early and late mutant defects in development of the *Drosophila* retina. *Dev. Biol.* **136,** 411–429.
25. Wolff, T. and Ready, D. F. (1991) Cell death in normal and rough eye mutants of *Drosophila. Development* **113,** 825–839.
26. Ellis, M., O'Neill, E., and Rubin, G. (1993) Expression of *Drosophila* glass protein and evidence for negative regulation of its activity in non-neuronal cells by another DNA-binding protein. *Development* **119,** 855–865.
27. Hay, B. A., Wassarman, D. A., and Rubin, G. M. (1995) Drosophila homologs of baculovirus inhibitor of apoptosis proteins function to block cell death. *Cell* **83,** 1253–1262.
28. White, K., Tahaoglu, E., and Steller, H. (1996) Cell killing by the *Drosophila* gene *reaper. Science* **271,** 805–807.
29. Hay, B. A., Wolff, T., and Rubin, G. M. (1994) Expression of baculovirus P35 prevents cell death in *Drosophila. Development* **120,** 2121–2129.
30. Steller, H. (1995) Mechanisms and genes of cellular suicide. *Science* **267,** 1445–1449.
31. Milan, M., Campuzano, S., and Garcia-Bellido, A. (1997) Developmental parameters of cell death in the wing disc of Drosophila. *Proc. Natl. Acad. Sci. USA* **94,** 5691–5696.
32. Karim, F. D. and Rubin, G. M. (1998) Ectopic expression of activated *Ras1* induces hyperplastic growth and increased cell death in *Drosophila* imaginal tissues. *Development* **125,** 1–9.

III

Abnormal Development and Teratology

15

Mechanisms of Neurulation

Gary C. Schoenwolf and Jodi L. Smith

1. Introduction

The process of neurulation results in the formation of the neural tube, the rudiment of the adult central nervous system. Neurulation occurs in two phases in vertebrate embryos, called primary and secondary neurulation. Primary neurulation, the formation of the neural plate and subsequent morphogenetic movements that transform it into a neural tube, forms the entire neural tube in amphibians and reptiles. Secondary neurulation, the formation of an epithelial cord and its subsequent cavitation to form a neural tube, forms the entire neural tube in fishes. Both primary and secondary neurulations occur in birds and mammals. The brain and trunk level of the spinal cord form by primary neurulation, whereas the tail spinal cord forms by secondary neurulation. In humans, primary neurulation is the most significant phase of neurulation from a clinical perspective because this phase, when it occurs abnormally, results in open neural tube defects—serious malformations of the central nervous system (*see* Chapters 15 and 17, in this volume). In this chapter, we will emphasize primary neurulation because of its clinical importance and because far more is known about primary neurulation that about secondary neurulation.

2. Stages of Primary Neurulation

Primary neurulation occurs in four stages (**Fig. 1**): formation of the neural plate; shaping of the neural plate; bending of the neural plate and fusion of the neural folds, resulting in closure of the neural groove ***(1,2)***. Because primary neurulation begins in the future brain region and progresses rostrocaudally—eventually establishing and closing the caudal neuropore—and caudorostrally—closing the cranial neuropore—the four stages occur simultaneously, with different stages occurring along different rostral–caudal levels of the neurulating embryo. For example, in the chick at mid-neurula stages (**Fig. 1D**), bending of the neural plate and fusion of the neural folds occur at the future midbrain and hindbrain levels of the neural tube, whereas shaping of the neural plate occurs more caudally, and formation of the neural plate occurs more caudally still. The cranial neuropore is broadly patent, but during subsequent development, bending of the neural plate and fusion of the neural folds occur in this area, obliterating the cranial neuropore and separating the cavity of the rostral neural tube (i.e., the neurocoel) from the amniotic cavity (**Fig. 1E**).

From: *Methods in Molecular Biology, Vol. 136: Developmental Biology Protocols, Vol. II*
Edited by: R. S. Tuan and C. W. Lo © Humana Press Inc., Totowa, NJ

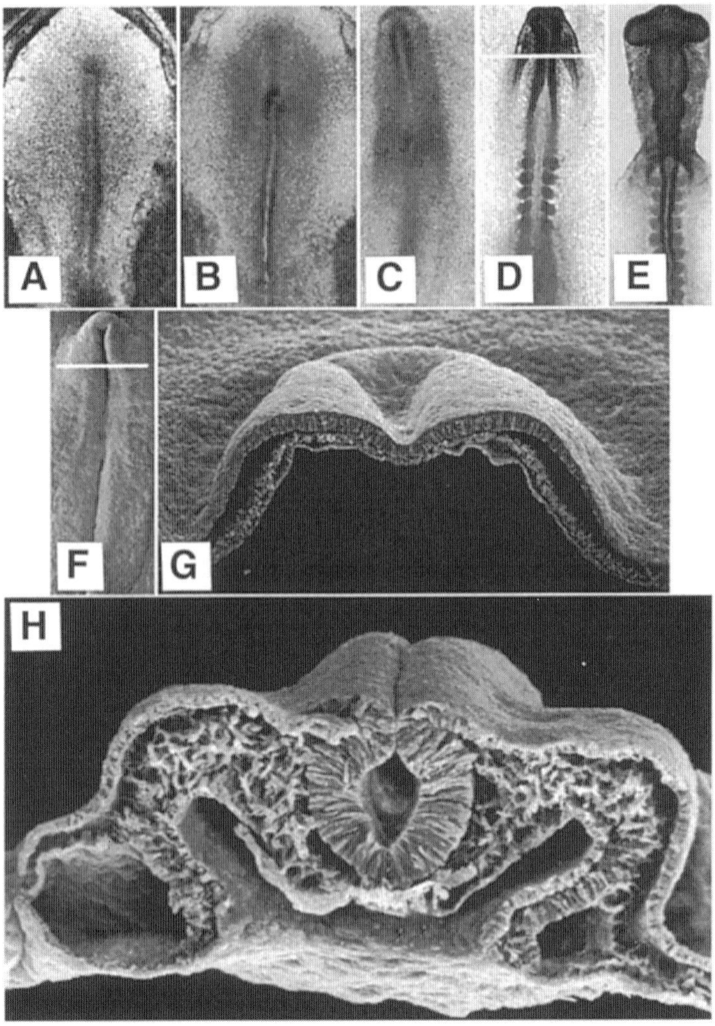

Fig. 1. Stages of primary neurulation in the chick embryo. **(A)** Whole mount shortly after formation of the neural plate. **(B)** Whole mount showing early shaping of the neural plate. **(C)** Whole mount showing late shaping of the neural plate; a scanning electron micrograph (SEM) of the same stage is shown in **(F)**. **(D)** Whole mount showing bending of the neural plate with fusion of the neural folds at the future midbrain/hindbrain level; the line indicates the transverse level of the section (SEM) shown in **(H)**. **(E)** Whole mount showing the incipient neural tube after fusion of the neural folds throughout the brain and rostral trunk. (F) SEM at the stage shown in (C); the line indicates the transverse level of the section (SEM) shown in (G). **(G)**, SEM of a transverse section through the level indicated by the line in (F). (H) SEM of a transverse section through the level indicated by the line in (D).

2.1. Formation of the Neural Plate

The neural plate consists of thickened ectoderm arranged in a pseudostratified columnar epithelium. Formation of the neural plate requires two types of inductive interactions: vertical interaction between the ingressed endoderm and mesoderm underlying the prospective neural plate ectoderm, and horizontal or planar interaction between the organizer and the surrounding neural plate. The organizer is a localized area of the blastoderm located at the rostral end of the primitive streak—the site of ingression of prospective endodermal and mesodermal cells into the interior of the embryo (*see* Chapter 12 in Vol. I in this series). The organizer has many unique properties ***(3)***, including the prospective fate of its cells (e.g., rostral endoderm, notochord, and floor plate of the neural tube), the morphogenetic movements of its cells (i.e., convergent extension driven in large part by cell–cell intercalation), the state of commitment of its cells (i.e., prospective notochordal cells are committed to their fate at the time the organizer becomes well established—significantly earlier than other cells of the gastrula), its patterns of gene expression (e.g., *Lim-1*, *HNF-3β*, *Sonic hedgehog*, *cNot-1*, and *cNot-2*) and its most unique ability—the ability to induce and organize an ectopic embryo containing a rostrocaudally and mediolaterally patterned neural tube. In addition, the organizer can serve as an inductive signaler for limb outgrowth ***(4)***, a polarizer for limb rostrocaudal patterning ***(5)***, and a suppressive signaler preventing reconstitution of the organizer and notochord in lateral blastoderm isolates lacking organizer and primitive streak ***(6)***.

The molecular basis of organizer inductive, polarizing, and suppressive signaling is still poorly understood, but, recently, some candidate molecules have emerged ***(7)***. A general model is that antagonism occurs between the dorsal (organizer) and ventral molecules that results in both ectodermal and mesodermal patterning. For example, the organizer secretes at least two neural inducers, chordin and noggin, which bind to bone morphogenic proteins (BMPs) (and/or their receptors) secreted by the epidermal ectoderm, competitively blocking BMP signaling. Thus, neural induction occurs because BMP signaling, which promotes epidermal differentiation, is blocked. Similarly, the organizer produces Frz-B, a soluble (i.e., not membrane bound) Wnt receptor (i.e., a frizzled family member), whereas ventral mesoderm secretes Wnt8. Thus, Wnt8 activity, which promotes ventral mesoderm differentiation, is blocked dorsally, allowing dorsal mesoderm (i.e., notochord and somites) to differentiate. This emerging model for ectodermal and mesodermal patterning is likely too simplistic; other factors probably directly promote differentiation of the neural plate and formation of the dorsal mesoderm (e.g., activin).

2.2. Shaping of the Neural Plate

Formation of the neural plate is quickly followed by its shaping. During shaping, the neural plate continues to thicken apicobasally; concomitantly, it narrows transversely and extends rostrocaudally ***(8)***. These three morphogenetic events of shaping occur autonomously to the neural plate (i.e., they are intrinsic to the neural plate) and are generated by the collective changes that occur in the behaviors of the constituent neuroepithelial cells ***(9,10)***; Namely neuroepithelial cells elongate, increasing the thickness or apicobasal dimension of the neural plate; decrease their diameters as they elongate, narrowing the neural plate; divide mitotically, with about half of their divisions oriented to increase the length of the neural plate as cell division and subsequent cell growth increase the volume of the neural plate exponentially with each cell division; and intercalate mediolaterally, thereby decreasing the width of the neural plate and increasing its length ***(8,11,12)***. How these cell behaviors are coordinated in both time and space, and how cell behaviors are oriented to yield a neural plate with the correct dimensions are unknown, except to say that the organizer *(13,14)* is likely involved.

2.3. Bending of the Neural Plate

Bending of the neural plate involves at least three key morphogenetic events: formation of the hinge points, formation of the neural folds, and folding of the neural plate *(15)*. Two types of hinge points form during bending: median and dorsolateral. The median hinge point is a single hinge point extending the entire rostrocaudal length of the neuraxis. It consists of the midline neural plate and underlying axial tissue (i.e., the prechordal plate beneath the forebrain and notochord beneath the rest of the neural tube). The dorsolateral hinge points are paired hinge points present only at certain rostrocaudal levels of the neuraxis: throughout the future brain level of the neuraxis and at the caudal part of the neural groove (primary spinal cord) within an area called the sinus rhomboidalis. The dorsolateral hinge points consist of the dorsolateral neural plate and the adjacent contacting epidermal ectoderm of the neural fold (*see* next paragraph). Two additional features characterize the hinge points. During their formation, neuroepithelial cells undergo wedging, changing their shapes from spindlelike to wedgelike, and folding of the neural plate occurs concomitantly at the apices of the wedging cells.

As the hinge points are forming, a bilaminar fold is established on each side of the neural plate. These are the neural folds, and each consists of an inner layer of neural plate and an outer layer of epithelial ectoderm. Subsequent bending involves folding of the neural plate around the hinge points, establishing the neural groove and bringing the neural folds into apposition in the dorsal midline where fusion occurs. Folding involves one step (i.e., elevation) or two steps (i.e., elevation and convergence) depending on the level of the neuraxis at which it occurs. Elevation of the neural plate and incipient neural folds occurs around the median hinge point, forming a V-shaped neural groove, and convergence of the definitive neural folds occurs around the dorsolateral hinge points, forming a diamond-shaped neural groove. The former process occurs at all levels of the neuraxis. The latter process modifies the shape of the neural groove at the future brain and sinus rhomboidalis levels. Between these two levels of the neuraxis, only the first step of bending occurs (i.e., elevation). As a consequence, the apical sides of the lateral walls of the neural groove are brought into apposition in the midline along most of the rostrocaudal extent of the spinal cord, temporarily occluding the lumen of the neural tube. Because of the presence of the dorsolateral hinge points and convergence of the neural folds at the future brain and sinus rhomboidalis levels, the lumen of the incipient neural tube is broadened transversely and occlusion does not occur at these levels of the neuraxis (**Fig. 1H**).

Bending involves the cooperation of both intrinsic and extrinsic neurulation forces *(9,10,16,17)*. Formation of the hinge points presumably involves extracellular matrix-mediated changes in cell adhesion between the neural plate and adjacent tissues. Subsequent wedging of neuroepithelial cells within at least the median hinge point requires an inductive interaction to occur *(18)*. Cell wedging involves at least two main events: apical constriction, mediated by circumferentially oriented apical bands of microfilaments *(19)*, and basal expansion, mediated by the translocation of cell nuclei to the bases of wedging cells during cell-cycle-regulated interkinetic nuclear migration *(20)*.

Changes in cell–cell adhesion of neuroepithelial cells within the hinge points likely also acts in cell wedging. Cell wedging, at least within the median hinge point, is an active event that occurs independently of lateral forces generated by the forming and elevating neural folds *(9)*. Such cell wedging within the median hinge point provides the motive force for neural plate furrowing within the hinge point.

In addition to intrinsic forces, extrinsic forces also function in bending. Expansion of the epidermal ectoderm plays a major role in bending, being both sufficient *(14,16)* and necessary *(9,13,17)* for bending. Expansion of the epidermal ectoderm is driven intrinsically within this layer *(10)* and involves multiple cell behaviors, including changes in cell shape, position, and number *(21)* as well as oriented mitotic divisions *(12)*.

2.4. Fusion of the Neural Folds

The final stage of primary neurulation is fusion of the neural folds. As a result of bending, the neural folds are brought into apposition in the dorsal midline, where they adhere due to the presence of cell-surface coats *(22)*, functionally closing the neural groove and separating the neurocoel from the amniotic cavity. Subsequent fusion of the neural folds closes the neural groove anatomically and establishes the roof of the neural tube (future roof plate), overlying epidermal ectoderm (future dorsal or back skin) and the neural crest, cells that contribute to the peripheral nervous system as well as to a variety of other structures (e.g., melanocytes and skeletal elements of the head and neck). The details of the fusion process remain unclear. For example, the precise composition of the extracellular surface coats mediating adhesion and fusion are unknown, as are the molecular cues that cause the epithelial neural folds to break down, reorganize, and become reestablished as epithelial sheets. In both birds and mammals, rather than having closure occurring in a smooth zipperlike progression from rostral to caudal, multiple closure sites occur along the neuraxis *(23,24)*. The presence of multiple closure sites likely correlates with the temporal sensitivity (i.e., critical periods) in humans for developing various kinds of neural tube defects at different levels of the neuraxis.

3. Stages of Secondary Neurulation

Secondary neurulation occurs in three stages (**Fig. 2**): formation of the medullary cord from the tail bud; cavitation of the medullary cord; and coalescence of the multiple lumina formed by cavitation into a single lumen *(25)*. Secondary neurulation begins at the future lumbosacral level in chick embryos and at the sacral or tail levels of mammals *(26,27)*. In chick embryos (but not in mouse embryos), an overlap zone exists in the future lumbosacral region between the caudal end of the neural plate, undergoing primary neurulation dorsally within the overlap zone, and the tail bud, undergoing second neurulation ventrally within the overlap zone *(28)*. The overlap zone is oblique rostrocaudally: within this overlap zone, progressively less of the neural tube is formed by primary neurulation dorsally and progressively more of the neural tube is formed by secondary neurulation ventrally as it differentiates rostrocaudally. Within the rostral end of the overlap zone, virtually the entire neural tube (except its most ventral part) is formed from the neural plate during primary neurulation. Within the caudal end of the overlap zone, virtually the entire neural tube (except its most dorsal part) is formed from the tail bud during secondary neurulation. Caudal to the overlap zone, the entire neural tube forms from the tail bud during secondary neurulation.

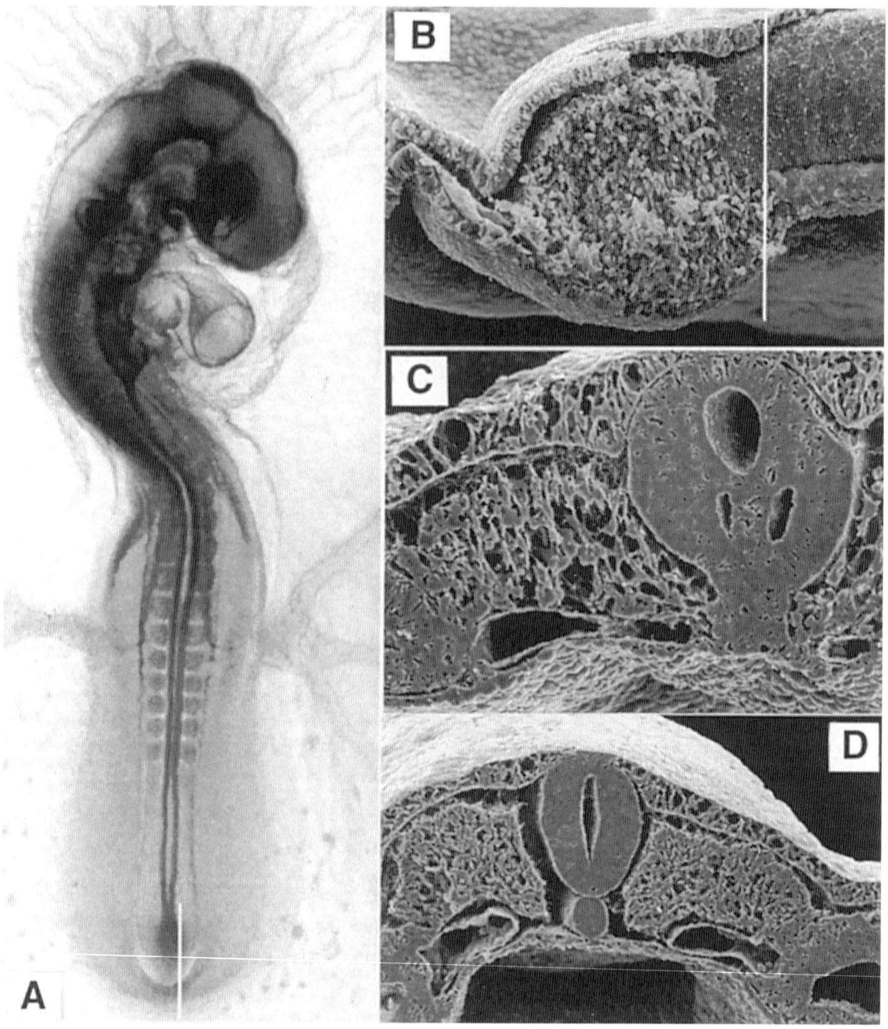

Fig. 2. Stages of secondary neurulation in the chick embryo. **(A)** Whole mount shortly after closure of the caudal neuropore and formation of the tail bud; the line indicates the sagittal level of the section (SEM) shown in **(B)**. (B) SEM of a sagittal section through the level indicated by the line in a; rostral is toward the right; the line indicates the transverse level of the section (SEM) shown in **(C)**. (C) SEM of a transverse section through the level indicated by the line in (B). **(D)** SEM of a transverse section of the caudal neural tube after the completion of neurulation at this level. Note the presence of a single lumen.

3.1. Formation of the Medullary Cord

As primary neurulation culminates, gastrulation is also culminating. During gastrulation, cells ingress toward and through the primitive streak (*see* Chapter 12 in Vol. I in this series). As they do so, the rostral end of the primitive streak regresses caudally and more caudal levels of the primitive streak are depleted as cells move into the interior of the blastoderm to form the ingressed endoderm and mesoderm. At about the time that the caudal neuropore is closing, Hensen's node and persisting remnants of the primitive streak become consolidated into a spherical mesenchymal mass at the caudal end of the embryo called the tail bud *(29)*. The tail bud is continuous dorsally with the overlying epidermal ectoderm and ventrally with the underlying endoderm. Laterally, it is continuous with the paraxial mesoderm on each side of the midline (i.e., the segmental plate mesoderm, the future caudal somites). Rostrally, it is continuous with the caudal part of the incipient primary neural tube as well as an underlying area containing precursors of the caudal end of the notochord. Its caudal boundary becomes delineated with the formation of the tail fold of the body. The tail bud gives rise to three major derivatives: the caudal neural tube (i.e., the most caudal spinal cord), the associated spinal ganglia (via neural crest derived from the roof of the caudal neural tube), and the caudal somites *(26,30,31)*.

The medullary cord forms from the mesenchymal tail bud through a mesenchymal-to-epithelial transformation. During this process, mesenchymal cells become interconnected apically via gap junctions, elongate apicobasally, and elaborate a basal surface with a subjacent basement membrane *(25,32,33)*.

The medullary cord forms in dorsal-to-ventral sequence. As the medullary cord forms, neural crest cells differentiate from its dorsal side, and shortly thereafter they undergo migration to form the caudal spinal ganglia *(31)*.

3.2. Cavitation of the Medullary Cord

As tail bud cells which are destined to form medullary cord become apicobasally polarized and interconnected apically, lumina develop centrally with respect to their apices. Multiple lumina (four to five) frequently form in chick (**Fig. 2C**) and human embryos, whereas only a single lumen forms in mouse embryos.

3.3. Coalescence of Lumina

The multiple lumina formed during secondary neurulation progressively coalesce into a single, centrally located lumen. Within the overlap zone, the multiple lumina formed during secondary neurulation also coalesce with the single dorsal lumen formed during primary neurulation, again creating a single central lumen (**Fig. 2D**).

4. Neural Patterning

The neural tube formed during primary and secondary neurulation becomes patterned during its formation as a result of interactions with the adjacent mesodermal and endodermal regions. Patterning occurs in two general planes: dorsoventral and rostrocaudal. Neural patterning is a complex and poorly understood process. The topic of neural patterning is beyond the scope of this brief review except to mention that studies on dorsoventral patterning have provided the most insight, and progress in this area has been substantial *(34,35)*. These studies have shown that both dorsal and ventral signals are involved. Ventrally, the notochord, through the action of a secreted signaling molecule, *Sonic hedgehog*, induces the floor plate of the neural tube. The floor plate, in turn, also becomes a signaling center, inducing motoneurons at the spinal cord level (as well as other types of neurons at various other rostrocaudal levels) through the action of *Sonic hedgehog* (and in cooperation with the notochord and its secreted *Sonic hedgehog*). Dorsal signaling occurs from the epidermal ectoderm overlying the neural tube and consists of secreted growth factors of the transforming growth factor-β family (e.g., dorsalin-1 and BMP 4). Collectively, dorsalizing and ventralizing signals act to induce/suppress neuronal development within the neural tube, thereby establishing unique clusters of neurons along the dorsoventral extent of the neural tube.

5. Conclusions

In this brief review, we have discussed the process of neurulation. Although much remains to be learned about this complex process, what is now clear is that neurulation is a multifactorial process driven by the coordination of forces generated both intrinsic to and extrinsic to the neural plate. These forces are driven by changes in cell behaviors, including changes in cell shape, size, position, number, and adhesivity. Along the rostrocaudal extent of the developing neural tube, different combinations of forces are generated by the multiple cell behaviors accounting for regional differences and presumably differing susceptibilities for neural tube defects. Moreover, as exemplified by the major differences in primary and secondary neurulation, totally different morphogenetic events can be used at different levels of the neuraxis and in different organisms, but the outcome is the same: a neural tube composed of a pseudostratified columnar epithelium. How multiple cell behaviors are coordinated in both space and time and how regional differences are established within the incipient neural tube will remain as questions for well into the 21st century.

Acknowledgments

Original work described herein from the Schoenwolf laboratory was supported by Grants NS 18112 and HD 28845 from the National Institutes of Health.

References

1. Schoenwolf, G. C. and Smith, J. L. (1990) Mechanisms of neurulation: traditional viewpoint and recent advances. *Development* **109,** 243–270.
2. Smith, J. L. and Schoenwolf, G. C. (1997) Neurulation: coming to closure. *Trends Neurosci.* **20,** 510–516.
3. Smith, J. L. and Schoenwolf, G. C. (1998) Getting organized: new insights into the organizer of higher vertebrates. *Curr. Topics Dev. Biol.* **40,** 79–110.
4. Hornbruch, A. and Wolpert, L. (1986) Positional signaling by Hensen's node when grafted to the chick limb bud. *J. Embryol. Exp. Morphol.* **94,** 257–265.
5. Dealy, C. N. (1997) Hensen's node provides an endogenous limb-forming signal. *Dev. Biol.* **188,** 216–223.
6. Yuan, S., Darnell, D. K., and Schoenwolf, G. C. (1995) Identification of inducing, responding, and suppressing regions in an experimental model of notochord formation in avian embryos. *Dev. Biol.* **172,** 567–584.
7. Zorn, A. M. (1997) Cell–cell signaling: frog frizbees. *Curr. Biol.* **7,** R501–R504.
8. Schoenwolf, G. C. (1985) Shaping and bending of the avian neuroepithelium: Morphometric analyses. *Dev. Biol.* **109,** 127–139.
9. Schoenwolf, G. C. (1988) Microsurgical analyses of avian neurulation: separation of medial and lateral tissues. *J. Comp. Neurol.* **276,** 498–507.
10. Moury, J. D. and Schoenwolf, G. C. (1995) Cooperative model of epithelial shaping and bending during avian neurulation: autonomous movements of the neural plate, autonomous movements of the epidermis, and interactions in the neural plate/epidermis transition zone. *Dev. Dyn.* **204,** 323–337.

11. Schoenwolf, G. C. and Alvarez, I. S. (1989) Roles of neuroepithelial cell rearrangement and division in shaping of the avian neural plate. *Development* **106,** 427–439.
12. Sausedo, R. A., Smith, J. L., and Schoenwolf, G. C. (1997) Role of nonrandomly oriented cell division in shaping and bending of the neural plate. *J. Comp. Neurol.* **381,** 473–488.
13. Keller, R., Shih, J., Sater, A. K., and Moreno, C. (1992) Planar induction of convergence and extension of the neural plate by the organizer of *Xenopus*. *Dev. Dyn.* **193,** 218–234.
14. Poznanski, A., Minsuk, S., Stathopoulos, D., and Keller, R. (1997) Epithelial cell wedging and neural trough formation are induced planarly in *Xenopus* without persistent vertical interactions with mesoderm. *Dev. Biol.* **189,** 256–269.
15. Schoenwolf, G. C. and Smith, J. L. (1990) Epithelial cell wedging: a fundamental cell behavior contributing to hinge point formation during epithelial morphogenesis, in *Control of Morphogenesis by Specific Cell Behaviors.* Seminars in Developmental Biology, Vol. 1 (Keller, R. E. and Fristrom, D., eds.), W. B. Saunders, London, pp. 325–334.
16. Alvarez, I. S. and Schoenwolf, G. C. (1992) Expansion of surface epithelium provides the major extrinsic force for bending of the neural plate. *J. Exp. Zool.* **261,** 340–348.
17. Hackett, D. A., Smith, J. L., and Schoenwolf, G. C. (1997) Epidermal ectoderm is required for full elevation and for convergence during bending of the avian neural plate. *Dev. Dyn.* **210,** 397–406.
18. Smith, J. L. and Schoenwolf, G. C. (1989) Notochordal induction of cell wedging in the chick neural plate and its role in neural tube formation. *J. Exp. Zool.* **250,** 49–62.
19. Nagele, R. G., Bush, K. T., Kosciuk, M. C., Hunter, E. T., Steinberg, A. B., and Lee, H.-Y. (1989) Intrinsic and extrinsic factors collaborate to generate driving forces for neural tube formation in the chick: A study using morphometry and computerized three-dimensional reconstruction. *Dev. Brain Res.* **50,** 101–111.
20. Smith, J. L. and Schoenwolf, G. C. (1988) Role of cell-cycle in regulating neuroepithelial cell shape during bending of the chick neural plate. *Cell Tissue Res.* **252,** 491–500.
21. Schoenwolf, G. C. and Alvarez, I. S. (1991) Specification of neurepithelium and surface epithelium in avian transplantation chimeras. *Development* **112,** 713–722.
22. Sadler, T. W. (1978) Distribution of surface coat material on fusing neural folds of mouse embryos during neurulation. *Anat. Rec.* **191,** 345–350.
23. Van Allen, M. I., Kalousek, D. K., Chernoff, G. F., Juriloff, D., Harris, M., McGillivray, B. C., Yong, S.-L., Langlois, S., MacLeod, P. M., Chitayat, D., Friedman, J. M., Wilson, R. D., McFadden, D., Pantzar, J., Ritchie, S., and Hall, J. G. (1993) Evidence for multisite closure of the neural tube in humans. *Am. J. Med. Genet.* **47,** 723–743.
24. Straaten van, H. W. M., Jaskoll, T., Rousseau, A. M. J., Terwindt-Rouwenhorst, E. A. W., Greenberg, G., Shankar, K., and Melnick, M. (1993) Raphe of the posterior neural tube in the chick embryo: Its closure and reopening as studied in living embryos with a high definition light microscope. *Dev. Dyn.* **198,** 65–76.
25. Schoenwolf, G. C. and DeLongo, J. (1980) Ultrastructure of secondary neurulation in the chick embryo. *Am. J. Anat.* **158,** 43–63.
26. Schoenwolf, G. C. (1977) Tail (end) bud contributions to the posterior region of the chick embryo. *J. Exp. Zool.* **201,** 227–246.
27. Schoenwolf, G. C. (1984) Histological and ultrastructural studies of secondary neurulation in mouse embryos. *Am. J. Anat.* **169,** 361–376.
28. Criley, B. B. (1969) Analysis of the embryonic sources and mechanisms of development of posterior levels of chick neural tubes. *J. Morphol.* **128,** 465–501.
29. Schoenwolf, G. C. (1979) Histological and ultrastructural observations of tail bud formation in the chick embryo. *Anat. Rec.* **193,** 131–148.

30. Schoenwolf, G. C. (1978) Effects of complete tail bud extirpation on early development of the posterior region of the chick embryo. *Anat. Rec.* **192,** 289–296.
31. Schoenwolf, G. C., Chandler, N. B., and Smith, J. (1985) Analysis of the origins and early fates of neural crest cells in caudal regions of avian embryos. *Dev. Biol.* **110,** 467–479.
32. Griffith, C. M. and Wiley, M. J. (1989) Distribution of cell surface glycoconjugates during mouse secondary neurulation. *Anat. Embryol.* **180,** 567–575.
33. Griffith, C. M. and Wiley, M. J. (1990) Distribution of glycoconjugates during chick secondary neurulation. *Anat. Rec.* **226,** 81–90.
34. Placzek, M. and Furley, A. (1996). Neural development: patterning cascades in the neural tube. *Curr. Biol.* **6,** 526–529.
35. Sasai, Y. and Robertis, E. M. D. (1997) Ectodermal patterning in vertebrate embryos. *Dev. Biol.* **182,** 5–20.

16

Neurulation and Neural Tube Closure Defects

Andrew Copp, Patricia Cogram, Angeleen Fleming, Dianne Gerrelli, Deborah Henderson, Andrew Hynes, Maria Kolatsi-Joannou, Jennifer Murdoch, and Patricia Ybot-Gonzalez

1. Introduction

The purpose of this chapter is to summarize some technical approaches and aspects of interpretation that are important when studying the process of neurulation and the origin of neural tube defects in mouse embryos. The techniques are not exclusive to studies of neurulation, so few protocols are described in practical detail. Instead, we emphasize the overall type of approach that appears most effective and give tips on the best application of standard techniques such as those described elsewhere in this and other volumes.

2. Dissection and Handling of Embryos

Studies of mouse neurulation begin with explantation of embryos from the pregnant uterus. This may be a preliminary to immediate analysis by a variety of methods including morphology, histology, whole-mount *in situ* hybridization or biochemistry, or it may precede whole-embryo culture or maintenance of embryonic parts in organ culture. In any event, embryos must be explanted in good condition in order to permit further analysis and/or continued development. The following aspects of the dissection process are important.

2.1. Forceps

Precisely sharpened forceps are essential for accurate explantation. Operations such as removal of Reichert's membrane from the outside of the yolk sac are extremely difficult without the use of delicate forceps whose tips meet exactly. The risk with blunt and poorly matched forceps is a clumsy dissection in which mechanical stresses are applied to the embryo, causing the rupture of blood vessels in the yolk sac and damage to the embryo itself. Watchmakers' forceps (Dumont 5ST Inox, Cat. No. 11254-20, InterFocus, Haverhill, Suffolk, UK) are essential and should be sharpened as described by Cockroft *(1)*.

2.2. Dissection Approach

Removal of the neurulation-stage embryo from the uterus is most efficiently and safely performed by an approach from the mesometrial surface. The placenta lies toward this side of the uterus, so the inexperienced dissector is least likely to damage

From: *Methods in Molecular Biology, Vol. 136: Developmental Biology Protocols, Vol. II*
Edited by: R. S. Tuan and C. W. Lo © Humana Press Inc., Totowa, NJ

the embryo by this approach. Moreover, the rapidly accumulating fluid inside the yolk sac of the neurulation-stage embryo (especially at E9.5 and E10.5) imparts a surprisingly high pressure with the result that if the thin yolk sac is exteriorized through a small hole in the uterine wall during dissection, it will burst and the embryo will be ejected through the opening. This will usually damage the embryo beyond any further use.

2.3. Removing the Decidual Swelling from the Uterus

After removing the uterus from the pregnant female, place it in a Petri dish of culture medium (*see* **Subheading 2.7.**) on the stage of a stereomicroscope and trim fat and blood vessels from the mesometrial surface of the uterus (**Fig. 1A**). Separate the uterine horns and then, while keeping each horn intact, open the uterus over each implantation site in turn, beginning at the mesometrial surface (**Fig. 1B**). Start at one end of the horn and work along, dissecting each embryo in turn. Keep both pairs of forceps close together during dissection, to minimize mechanical stresses on the embryo. As you create a hole in the elastic uterine wall, the decidual swelling will bulge out. The hole should be gently enlarged until it is large enough for the decidua to be fully exteriorized (**Fig. 1B**). Then, clamp a pair of forceps around the uterus, just to one side of the decidual swelling, and press gently against the decidua to "milk" it out of the uterus.

2.4. Removing the Embryos from the Decidua

All decidual swellings should be dissected to this point before proceeding. Transfer them to a fresh Petri dish of medium and then open each in turn, beginning at the broad, fluffy end of the decidual swelling (i.e., the former mesometrial end; **Fig. 1C**). Once the ectoplacental cone trophoblast is encountered (**Fig. 1D**), separate the decidua from the conceptus as if peeling an orange. Keep the two pairs of forceps close together at all times. Remove decidual debris from the dish immediately, as damaging enzymes may be released from ruptured decidual cells that may adversely affect embryonic viability.

2.5. Removing Trophoblast and Reichert's Membrane from the Embryo

After removal of the decidua, the mural trophoblast (i.e., that surrounding the yolk sac) will usually be intact; if it is ruptured in places, this is not important provided that the underlying yolk sac is not damaged (**Fig. 1E**). Using your finest forceps, grasp the trophoblast and tear it open (or enlarge an existing hole). The aim is also to rupture the elastic Reichert's membrane (which is practically invisible) lying beneath the trophoblast without damaging the yolk sac that underlies both layers. If the trophoblast comes off in pieces, this indicates that Reichert's membrane is still intact, whereas if it comes off in a sheet (**Fig. 1F**), you can be sure that Reichert's membrane has ruptured. Trim both layers around the edge of the ectoplacental cone, where Reichert's membrane is continuous with the yolk-sac endoderm (**Fig. 1G**). The embryo is now ready for culture, but should be dissected further, with removal of the yolk sac and amnion, if it is to be analyzed immediately.

2.6. Removal of Ectoplacental Cone Trophoblast

This can be attempted from E9.5 onward, but it is worth the additional effort only if some further embryonic manipulation is to be carried out prior to culture (*see* **Subheading 8.2.**). With one pair of forceps, grasp the torn edge of the ectoplacental cone trophoblast and, with the other forceps, grasp the adjacent torn edge of Reichert's mem-

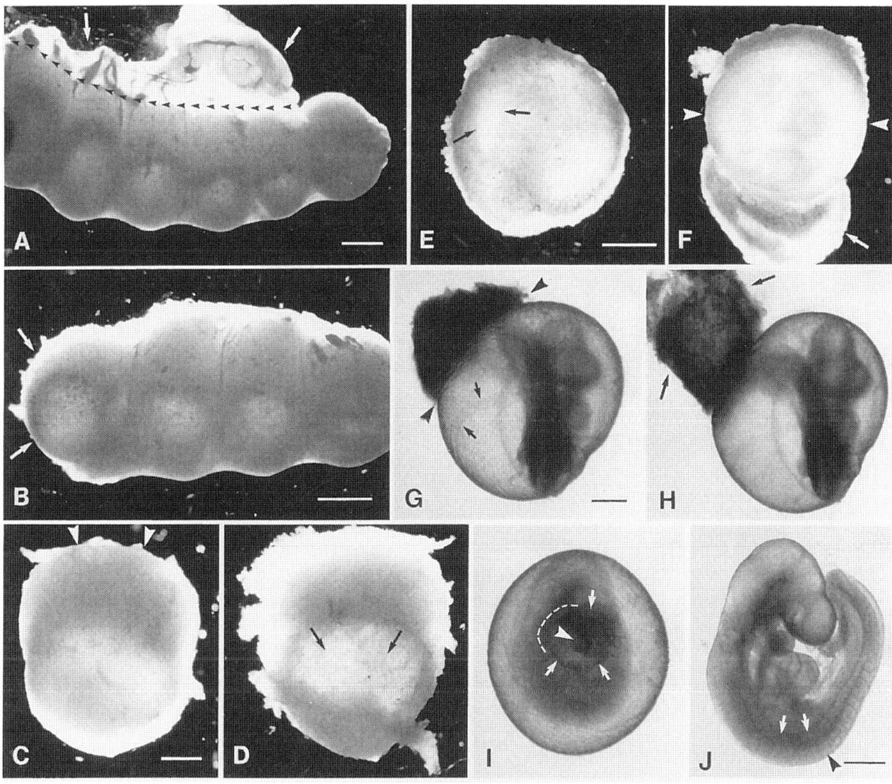

Fig. 1. Dissection of neurulation-stage mouse embryos at E9.5 as illustrated by dark-field (**A** to **F**) and bright field (**G** to **J**) photomicrographs taken on a Zeiss SV11 stereomicroscope. (A) Uterus after removal from the pregnant female. Implantation sites are closely spaced and the mesometrial surface is marked by entry of blood vessels (arrows). Mesometrium should be trimmed off using scissors (along the line indicated by arrowheads). (B) An implantation site in the early stages of being dissected open from the mesometrial surface. The decidual swelling bulges from the opened uterine wall (arrows) indicating high internal pressure. (C) Decidual swelling after removal from the uterus. Arrowheads indicate the surface from which the next stage of the dissection should begin. (D) Decidual swelling partially opened to reveal the trophoblast layer beneath (arrows). (E) Dissected conceptus with trophoblast intact apart from one torn region (arrows). The yolk sac is intact beneath, making this embryo suitable for subsequent culture. (F) Trophoblast layer (arrow) has been opened and partially removed, revealing the embryo enclosed in its yolk sac (arrowheads). (G) Embryo completely explanted for culture. Trophoblast has been trimmed up to the edge of the ectoplacental cone (arrowheads). Vessels can be seen within the yolk sac (arrows). (H) Ectoplacental cone (arrows) partially removed by peeling apart from Reichert's membrane. (I) Embryo after removal of ectoplacental cone. Outline of the chorion edge appears as a dark ring (arrows) with the attachment of the allantois at its center (arrowhead). A window in the yolk sac can be opened in the avascular region at the edge of the chorion (e.g., along dashed line), permitting access to the embryo for surgical manipulations. (J) E9.5 embryo after removal from yolk sac and amnion. Somite 11 (arrowhead) is located at the caudal edge of the forelimb bud (between the arrows). This embryo has 21 somites. Scale bars represent 2 mm in (A) and (B); 1 mm in (C) (also [D]); 1 mm in (E) (also [F]); 0.5 mm in (G) (also [H and I]); 0.5 mm in (J).

brane. Then, peel the two layers apart, working up over the ectoplacental cone region (**Fig. 1H**). The cone trophoblast will lift off, exposing a space (the ectoplacental cavity) between it and the chorion, which borders the superior surface of the exocoelom. The ectoplacental cone trophoblast can be completely peeled away without blood loss, exposing the edge of the chorion, which appears as a dark ring (**Fig. 1I**). Embryos prepared in this way undergo excellent development in culture, and a window in the nonvascular part of the yolk sac (adjacent to the chorion) can be opened, providing access to the caudal part of the embryo (*see* **Subheading 8.2.**).

2.7. Medium for Handling Embryos During Dissection and Manipulation

Many suitable variations exist; we use Dulbecco's modified Eagle's medium containing 25 m*M* HEPES (DMEM, Gibco-BRL, Gaithersburg, MD) and 10% fetal calf serum (FCS). This maintains the embryo in good condition (inclusion of 10% serum) and can be used in an ambient air environment without quickly becoming alkaline (because of the HEPES pH 7.0 buffering system). FCS cannot be used to culture mouse embryos, but they do not suffer if they are dissected in FCS-containing medium. If embryos are to be cultured, they should be rinsed in fresh DMEM+FCS and then transferred in a minimal volume into prewarmed rat serum (*see* **Subheading 3.2.**). If they are to be immediately fixed or frozen for further analysis, they are first rinsed in at least two changes of ice-cold phosphate-buffered saline (PBS). If RNA or protein is to be extracted or analyzed, the entire dissection is performed in ice-cold medium, whereas if embryos are to be cultured, they are dissected at room temperature.

3. Culturing Intact Embryos During Neurulation

The development of a technique for culturing rodent embryos throughout the period of neural tube closure has made feasible a variety of experimental studies in mammalian embryos that were possible previously only with lower vertebrates. We use the technique as it was first described for the rat and mouse *(2,3)*. The following issues are essential in order to achieve optimal development *in vitro*.

3.1. Preparation of Rat Serum

Whole or diluted rat serum serves as the standard medium, although various authors have described the use of serum from other species *(4,5)*. Serum should be prepared by puncture of the abdominal aorta of terminally ether-anesthetized rats, as described by Cockroft *(1)*. We have evaluated inhalational anesthesia using halothane and intraperitoneal injection of a mixture of 3.4 mg/kg body weight midazolam, 0.22 mg/kg fentanyl citrate, 6.75 mg/kg fluanisone (Hypnorm/Hypnovel, Janssen Cilag Ltd., Highwycombe, UK) as alternatives to ether. However, embryonic development is suboptimal in serum produced using either of these agents. Great care must be taken when withdrawing the blood from the aorta to avoid hemolysis: Withdraw the blood smoothly and steadily into the syringe and decant it gently down the inside of the centrifuge tube. Immediately centrifuge the blood at 1000*g* and then allow the plasma to clot at room temperature before expressing the serum with sterile forceps. We discard any serum that is not clear and pale straw colored. After heat inactivation (56°C for 30 min), store the serum in aliquots at –20°C. Extensive dialysis of serum against saline prior to culture removes small molecules and, after replacement of glucose and certain vitamins, dialyzed serum can support relatively normal development. This technique can facilitate molecular studies of cultured embryos ***(1,6–8)***.

3.2. Culture Technique

Good quality serum and well-prepared embryos are the key to successful culture. Thaw the serum immediately prior to use, pass it through a 0.45-μm filter (Whatman, Clifton, NJ) and pipet it into 30-mL Universal tubes (Nalgene Nunc, Rochester, NY), with the inside rim of the cap smeared lightly with silicone grease to create a gas-tight seal. Serum from different batches should be pooled at this stage so that all embryos in the experiment are exposed to identical serum conditions. Gas with the appropriate mixture *(see below)* and prewarm to 38°C in a roller incubator (30–40 rpm). Dissect the embryos as described in **Subheading 2**, minimizing the length of time between explantation from the uterus and the start of rat serum culture. Embryos can be dissected on the bench, without the use of a laminar flow hood or antibiotics in the serum, provided that instruments are flame-sterilized regularly and dissection medium, pipets, and dishes are sterile. As a general rule, we culture one E9.5 embryo per milliliter of serum, although this can vary depending on the stage (more embryos per milliliter at earlier stages) and length of the culture period (more embryos per milliliter for cultures of less than 12 h). Embryos cultured between E8.5 and E9.5 should be gassed with a mixture comprising 5% O_2, 5% CO_2, and 90% N_2. Between E9.5 and E10.5, use 20% O_2, 5% CO_2, and 75% N_2 (i.e., 5% CO_2 in air), and for embryos older than E10.5, use 40% O_2, 5% CO_2, and 55% N_2. Gently blow the gas mixture into the culture tube from a Pasteur pipet attached to the gas cylinder with the gas jet directed at the wall of the tube, not at the serum. Replenish the gas atmosphere at least every 12 h and change the serum at least every 24 h throughout the culture.

3.3. Assessing the Success of Cultures

The most sensitive index of healthy cultures is development and maintenance of a functional blood circulation in the yolk sac. E8.5 embryos should develop a circulation during the first 12 h of culture and E9.5 embryos should maintain it for at least 24 h in culture. At the end of culture, score the yolk-sac circulation immediately, before the embryos cool down. We use a simple scoring system where ++ represents a vigorous circulation, + represents a slow or patchy circulation, and – represents no circulation. Embryos without a circulation should be excluded from analysis, as they are effectively dead and, depending on when the circulation ceased, may exhibit signs of degeneration. These include expansion of the fluid-filled cavities, particularly the pericardial cavity and yolk-sac cavity (exocoelom), stunting of the forebrain and caudal extremity, irregularity of the most caudal somites, waviness of the neural tube, and subepidermal fluid-filled blebs, particularly on the cranial region. A heart beat *per se* is not a good sign of a healthy embryo, as the heart's intrinsically contractile nature means that it will go on beating after the embryo is dead. Healthy embryos should be further assessed as described in **Subheadings 4** and **5**. If development is normal, embryos should gain one somite every 2 h in culture, as *in utero*.

4. Determining Embryonic Stage

Staging of embryos must be the most vexed subject in mammalian embryology. No standard convention exists and, all too often, authors cite numerical ages of embryos without defining their staging system, thereby complicating comparisons between studies. We stage embryos as follows.

4.1. Chronological Staging

Mice are conventionally mated overnight, and the finding of a copulation plug is taken as evidence of mating midway through the preceding dark period of the light cycle. Some workers designate the day of finding a copulation plug as d 0 postcoitum and others as d 1, leading to considerable confusion. An unambiguous alternative is to designate the day of finding a plug as 0.5 d postcoitum or embryonic day 0.5 (E0.5 for short). This is logical, as noon on the day of finding a plug is usually 12 h (0.5 d) after mating the previous night. We use this convention in all our studies, referring to embryos dissected on the morning of E9.5 as "early E9.5," not "E9.25," which indicates a precision of staging that is not usually matched in reality. We stage more precisely using somite number, as follows.

4.2. Somite Number as a Staging System

Neural tube closure (primary neurulation) begins at the 6–7 somite stage (E8.5) and finishes at the 29–30 somite stage (early E10.5), therefore taking just less than 2 d to be completed. Somite number relates precisely to the events of neurulation, making it a much more useful measure than chronological age. Mouse strains vary considerably in the postcoital age at which they reach a particular stage of the neurulation process (e.g., outbred embryos are developmentally more advanced than those from inbred strains) and, within a strain, there is variability between pregnant females, for instance, depending on the precise time that mating takes place. Even littermates can vary considerably. Ideally, therefore, all embryos should have their somite number determined and this should constitute the basis of staging for comparative purposes. Downs and Davies *(9)* have described a useful staging system for presomite, gastrulation-stage mouse embryos, which can be cultured to neurulation stages.

4.3. Determining Somite Number After Removal of the Yolk Sac

Somites can readily be counted in embryos after removal of the yolk sac and amnion. The lighting on the stereomicroscope should be adjusted to give the best definition. With living embryos, this usually means use of transmitted light in bright-field mode, whereas in fixed (i.e., more opaque) embryos, dark-field or incident illumination may give best results. When staging embryos with fewer than 20 somites (i.e., broadly, embryos at E9.5), it is necessary to count from the first somite, which is located in the occipital region, ventro-caudal to the otic pit. Only complete somites are counted. In more advanced embryos, the occipital somites become ill-defined and may be missed, giving a falsely low somite count. This problem can be circumvented by beginning the count at somite 11, which is situated level with the caudal edge of the forelimb bud (**Fig. 1J**), and proceeding caudally. When studying embryos with 30 somites or more, many of the rostrally positioned somites may be indistinct and it is advisable to begin with somite 29, which is opposite the caudal edge of the hindlimb bud, and proceed caudally. Early-somite-stage embryos that are unturned (i.e., have not yet undergone axial rotation) are often analyzed without removal from the yolk sac. Somites can be readily counted through the ventral surface of the embryo, except at stages just prior to

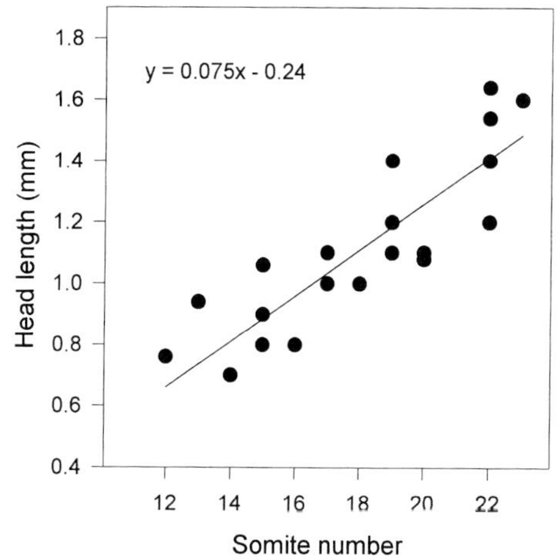

Fig. 2. Relationship between antero-posterior head length, measured by trans-illumination through the intact yolk sac, and somite number (determined after removal of the yolk sac) in E9.5 mouse embryos. The regression line allows somite number to be accurately estimated in embryos prior to culture.

axial rotation (i.e., 8–12 somites) when the caudal-most somites are not clearly visible from the ventral surface. Somite number at this stage needs to be established following removal of the extraembryonic membranes, or by indirect estimation from the general stage of embryonic development reached.

4.4. Determining Somite Number Prior to Embryo Culture

The yolk sac and amnion must be kept intact for embryo culture, precluding a direct count of the somites in embryos that are enclosed in their membranes (i.e., beyond the 8 somite stage). Although the somite number can be found at the end of the culture period, it is advantageous to know the starting somite number so that embryos of a precise somite range can be used for experiments and parameters such as rate of increase of somite number in culture can be calculated. We have described a method based on measurement of antero-posterior head length, by observing through the yolk sac in transilluminated embryos *(10)*. Head length is linearly correlated with somite number in embryos that have undergone axial rotation (**Fig. 2**). It is advisable for each laboratory to construct its own standard graph for this purpose, in order to allow for differences in the precise method of measuring head length.

4.5. Measuring Embryonic growth

Although an increase in embryonic mass correlates precisely with embryonic age and somite stage *(11)*, it is possible that growth and developmental advancement may become dissociated in embryos under experimental or pathological conditions. Thus, some measure of growth is a useful component of embryonic analysis. Total protein content (*see* **Subheading 7.4.**) or total DNA content *(12)* are traditional measures of embryonic growth, but these techniques are of limited usefulness, as they generally preclude other types of analysis on the same embryo. Alternatively, the relatively crude measure of crown–rump length *(11)* in fully turned embryos is easily performed and compatible with other modes of analysis. We routinely use crown–rump length to normalize measurements of neural tube closure such as length of the posterior neuropore, in order to permit comparisons between embryos of different sizes.

4.6. Brown–Fabro Morphological Score

This scoring system *(13)* provides an overall morphological score for individual embryos based on a system-by-system assessment of developmental advancement. It is based on the whole embryo, not just on one body part, but this very fact means that the score will often not highlight specific major malformations. For instance, isolated midbrain exencephaly will reduce the morphological score of an embryo at E10.5 by less than 5%, which may not be a statistically significant difference. The technique is very time-consuming, but it is unparalleled in its usefulness for teaching embryonic anatomy.

5. Assessing the Neurulation Status of Embryos

Having explanted embryos and performed a general developmental analysis, the next step is to assess neurulation status. This involves determining whether the neural folds are open, which closure events have occurred successfully, which have failed to be accomplished, whether any abnormalities are the result of developmental defect or dissection artifact, and the extent to which abnormalities at a particular embryonic stage are predictive of future developmental defects.

5.1. The Progression of Neurulation Events

Mouse neurulation is now firmly established to be a discontinuous, piece-meal process involving considerable heterogeneity along the body axis. Closure is initiated at three distinct sites (closures 1–3) within the brain and upper spinal cord, and neurulation progresses (by "zippering") from these sites in both rostral and caudal directions. The open regions of neural plate between the closure points are termed neuropores. These regions are the last to close in a particular body region. **Table 1** presents a classification of the main events and gives suggestions on how best to observe them. It is important to remember that the neurulation process is continuous, that several events occur simultaneously at certain somites stages, and that failure in an early event may preclude accomplishment of later events. It can be appreciated, therefore, that a wide variety of defects of neural tube closure are possible depending on the precise event(s) affected. This variety is reflected in the heterogeneity observed among neural tube defects in the mouse (**Subheading 9**) and human *(14,15)*.

5.2. Recognizing Artifactual "Defects" of Neural Tube Closure

The recently closed neural tube is delicate, so that rough treatment can cause it to reopen, giving the appearance of a neural tube defect. Trauma during dissection usually produces artifacts of the brain, as the embryonic head is easily squashed laterally. Depending on the embryonic stage, this can yield an open neural tube or a closed neural tube with an apparently defective structure (e.g., loss of the pontine flexure). Chemical treatment during processing for histology and whole-mount *in situ* hybridization can readily cause the spinal neural tube to reopen, even after fixation. These are not problems provided they are recognized as artifacts. Paradoxically, growth retardation,

Table 1
Assessing Neurulation Status in Mouse Embryos

Closure event	Description of event	Somite stage	Method of assessment
Closure 1	Initiation of closure at occipital/cervical boundary	6–7	Requires removal of embryo from yolk sac and amnion, usually with transection of body axis to enable inspection of the neural tube cross-section.
Closure 2	Initiation of closure at forebrain/midbrain boundary	12–15	Gross observation from anterior view; with experience, can be assessed in cultured embryos by observation through yolk sac.
Closure 3	Initiation of closure at the rostral extremity	12–15	Gross observation from anterior view, usually in embryos removed from yolk sac and amnion.
Anterior neuropore closure	Completes closure in forebrain, by progression from closures 2 and 3	14–17	Gross observation from anterior view; with experience, can be assessed in cultured embryos by observation through yolk sac.
Hindbrain neuropore[a]	Completes closure in hindbrain, by progression from closures 1 and 2	17–19	Gross observation from posterior view in embryos removed from yolk sac and amnion; owing to the thin roof of the hindbrain, it is often necessary to confirm closure by gentle probing with a blunt forcep.
Posterior neuropore	Completes closure in the spinal region, by progression from closure 1	29–30	Gross observation by a dorso-lateral oblique view of the caudal region in embryos removed from yolk sac and amnion; patency of very small neuropores can be ascertained by injection of dye into the neural tube higher in the neuraxis ***(93)***.
Secondary neurulation	Formation of the neural tube in the low spinal region, caudal to the posterior neuropore	31–60	Requires histological analysis.

[a]Called closure 4 by some authors, although this event is clearly a neuropore-type closure, not an initiation process, as in closures 1–3.

which is common when embryo culture conditions are substandard, can produce *premature* closure of the posterior neuropore by virtue of the decreased angle of curvature of the body axis that occurs in stunted embryos ***(16)***. This emphasizes the general rule that control (e.g., unmanipulated) embryos *must* be demonstrated to develop normally in all studies, otherwise, interpretation of experimental manipulations is invalidated.

5.3. Neural Tube Closure in the Context of the Body Axis

The caudal region at E9.5 is frequently referred to in the literature as the "tail" of the embryo, even though the posterior neuropore at the 15 somite stage, in fact, corresponds to the future mid-thoracic level of the fetus and postnatal individual. Therefore, abnormalities of neuropore closure at this level are expected to affect the thorax, not the extreme caudal end. **Figure 3** summarizes the fate of the various levels of the body axis and shows how somite stage, level of the posterior neuropore (in terms of future somite level), and axial level correspond in terms of position along the vertebral column. It can be seen that primary neurulation is completed with final closure of the posterior neuropore at the 29 somite stage. The site of the closing posterior neuropore corresponds to the level of the future 34th somite, which is within the upper sacral region. Thus, open neural tube defects above this level correspond to faulty closure of the posterior neuropore, whereas defects below this level must originate in abnormalities of secondary neurulation.

6. Structural Examination of Neurulation-Stage Embryos

Many types of structural analysis are possible with neurulation-stage embryos, frequently requiring efficient production of sections. For instance, routine histological examination requires embedding in paraffin wax and sectioning, transmission electron microscopy requires embedding in resin, immunocytochemistry may require cryostat sections, and analysis of whole-mount *in situ* hybridization preparations is facilitated by vibratome sectioning.

6.1. Optimal Orientation of Embryos for Sectioning

Despite the popularity of sagittal embryonic sections, they are of very limited value when studying neurulation, as they provide almost no information on the closure status of the neural tube. We recommend careful use of transverse sections. The curvature of the embryonic body axis precludes obtaining transverse sections through the entire body axis in a single run of serial sections. This is usually not a major problem, as a particular level of the body axis is normally of overriding interest for analysis and the embryo can be orientated during embedding in order to achieve transverse sections through this region. We often dissect the embryo prior to embedding so that the critical fragments can be embedded more accurately. An advantage is that fewer sections need to be collected, although the smaller tissue fragments are more likely to be lost during histological processing.

6.2. Embedding for Paraffin Sections

A typical requirement is to produce serial transverse paraffin sections through the posterior neuropore of E9.5 embryos. After fixation of the whole embryo, cut off the caudal region with fine forceps and process it through dehydrating steps in a single tube, by pouring or carefully pipeting off each alcohol and replacing it with the next, taking care not to allow fragments to dry out. Inspect the tube under a stereomicroscope after each alcohol change to ensure that the fragments are still present. After clearing in Histoclear (National Diagnostics, Atlanta, GA), pipet the fragments in a minimal volume into solid watchglasses containing molten paraffin wax and incubate to allow wax penetration. It is important that the temperature of the embedding oven is

I	II	III	IV	V	VI
3					
4					
5	C1				
6	C2				
7	C3		Closure 1		
8	C4	13			
9	C5	14			
10	C6	15			
11	C7	16			
12	T1	17			Mode 1
13	T2	18	Closures 2 and 3		
14	T3	19			
15	T4	20			
16	T5	21		Primary neurulation	
17	T6	22			
18	T7	23			
19	T8	24	Closure of HNP		
20	T9	25			Mode 2
21	T10	26			
22	T11	27			
23	T12	28			
24	T13	29			
25	L1	30			
26	L2	31			Mode 3
27	L3	32			
28	L4	33			
29	L5	34	Closure of the PNP		
30	L6				
31	S1				
32	S2			Secondary neurulation	
33	S3				
34	S4				
35	Ca1				

Fig. 3. Correspondence between the events of axial development and neurulation in the mouse. Unsegmented mesoderm (i.e., the segmental plate) intervenes between the somite rows and the posterior neuropore at most stages of spinal neurulation. Organ culture experiments demonstrate that this mesoderm will give rise to approximately five to six future somites *(17)*. Thus, at any particular stage, the level of the posterior neuropore (column III) can be estimated as the sum of the somite number already present (column I) plus the approximately five future somites in the segmental plate. Column II indicates the level of the vertebral column corresponding to the somite level given in Column I. Columns IV, V, and VI show the events of neurulation timed according to the somite stages in Column I. Modes 1, 2, and 3 comprise a morphological classification of spinal neural tube closure *(18)*.

kept stable at not more than 60°C and that the period of incubation of the embryos in molten paraffin wax is kept to a minimum, or vacuum embedding is used, otherwise reopening artifacts of the neural tube and antigenic denaturation may occur. Minimal times to allow adequate wax penetration need to be determined in preliminary trials. Old wax is carefully poured off and fresh wax is added. To embed, heat two pairs of old forceps in a flame and use them to keep the wax molten while observing under a stereomicroscope at room temperature. Keep the tissue fragments afloat in the molten wax until a layer of wax has solidified at the base of the watchglass, then allow them to settle and, one at a time, orient the fragments by gently positioning them on the soft wax base using the heated forceps. To obtain transverse sections, position caudal embryonic fragments with their long axis vertical in the watchglass, with the posterior neuropore uppermost, so that this is the last part of the tissue to be sectioned. This means that the most critical sections are reached late in the process of serial sectioning, when the block is usually cutting optimally.

6.3. "Embedding" for Cryosections

Fresh, unfixed embryo fragments are processed through increasing concentrations of sucrose (e.g., 7.5%, 15%, 30%, and 50% in PBS) and then placed in a plastic mold containing liquid OCT compound (Raymond Lamb, London, UK). The high viscosity of this embedding medium allows tissue fragments to retain their position for several seconds and the aim is to place the fragments in the desired orientation and then quickly and gently to place the mold onto a slurry of dry ice in isopentane in order to quickly freeze the embedding medium. Blocks are stored in their molds at –70°C for up to 3 mo prior to sectioning. We routinely prepare cryosections of 7 μm thickness, which can be stored at –20°C for 1 mo prior to staining.

6.4. Embedding for Vibratome Sections

Embryos are embedded in a mixture of 0.45% gelatin, 27% albumen, and 18% glucose, which is solidified by the addition of 1/10th volume of 25% glutaraldehyde. Orientation is more problematic than for paraffin or cryosections, as the solidification process is rapid. If embryo orientation is not optimal, pare down the block and re-embed it in fresh embedding medium, allowing reorientation. We routinely produce 30–50-μm vibratome sections of embryos following whole-mount *in situ* hybridization.

6.5. Embedding for Transmission Electron Microscopy

An excellent review of this subject has appeared previously ***(19)***.

6.6. Preparation for Scanning Electron Microscopy

Scanning electron microscopy can be used to illustrate the morphological events of neurulation. Embryos are fixed for not longer than 24 h at 4°C in half-strength Karnovsky's fixative (freshly prepared 2% paraformaldehyde, 2.5% glutaraldehyde, 0.07 *M* cacodylate buffer, pH 7.4), then rinsed twice in cacodylate buffer and postfixed in 4% osmium tetroxide for 1 h. After dehydration, the embryos are critical point dried by the replacement of 100% ethanol with liquid CO_2 ***(20)*** and mounted on metal stubs coated with carbon adhesive tape. Orientation of embryos cannot be altered after mounting, and embryos are extremely fragile at this stage, so great care is needed during mounting. Sputter coating is carried out immediately: a gold coat of 100 nm is sufficient to allow embryos to be viewed without obscuring surface details. After coating, specimens should be stored with silica gel under vacuum to prevent rehydration.

7. Examining Gene and Protein Expression in Neurulation-Stage Embryos

The analysis of any developmental event, not least neurulation, requires that spatial and temporal patterns of gene expression are accurately assessed in embryos. Currently, the majority of studies use techniques designed to localize the mRNA for a given gene, but it must be remembered that the distribution of the transcript may not correspond to the actual distribution of the protein product, for instance, in cases where the protein fails to be synthesized, is rapidly degraded or is exported from the cell. It is important, therefore, to combine mRNA and protein analyses on embryonic material, wherever possible.

7.1. Preparing RNA

We extract embryonic RNA using a simple protocol based on the Trizol Reagent (Gibco-BRL) performed according to the manufacturers instructions. Embryos are suspended in Trizol and homogenized by repeatedly drawing them in and out of a syringe needle fitted onto a 1-mL disposable syringe, using needles of decreasing diameter (19 gage to 25 gage) to achieve a fine suspension. A rough guide for the volume of Trizol is 200 µL for a single E8.5 embryo, 500 µL for E9.5, 1 mL for E10.5, and 2 mL for E11.5. As with any RNA work, precautions should be taken to minimize the effect of RNase. Solutions (e.g., PBS) should be pretreated with diethylpyrocarbonate (DEPC) to inactivate RNase, all dissecting dishes or storage tubes should be sterile, and dissecting instruments should be alcohol-sterilized and briefly flamed at regular intervals. The worker should wear gloves at all times. Embryos should be kept on ice throughout the dissection procedure. We find that embryos prepared in this way can be stored "dry," or in Trizol Reagent, for several weeks at –70°C prior to RNA isolation without significant RNA degradation. Subsequent analysis of RNA (for instance, by reverse-transcriptase polymerase chain reaction [RT-PCR]) is carried out using standard protocols.

7.2. In Situ *Hybridization—Localizing Gene Transcripts in Embryos*

Both whole-mount *in situ* hybridization using digoxygenin-labeled riboprobes and hybridization using ^{35}S-labeled riboprobes on histological sections have been used extensively for the analysis of neural tube closure in mouse embryos.

7.2.1. Whole Mount In-Situ *Hybridization*

Advantages include the ease and speed of analysis, the numbers of embryos that can be processed in one experiment, and the elimination of radioactivity from the protocol. Whole-mount studies give an excellent overall impression of gene expression in the embryo, although histological examination of hybridized embryos is not easy and artifacts, such as opening of the neural tube, may occur during processing. Whole-mount *in situ* hybridization is usually the method of choice with embryos of E10.5 or younger, whereas the protocol does not work so well at E11.5 and older, because of the poor penetration of probe into the increased bulk of larger embryos. The protocol in most common use is that of Wilkinson ***(21)***. Particular care should be taken with the following aspects of this protocol.

The concentration and timing of proteinase K treatment are critical: If proteinase K is inadequate, the staining will be faint, as the probe will not penetrate the tissue adequately, whereas if the treatment is too aggressive then the integrity of the embryo can be affected (for instance, causing opening of the closed neural tube). For E10.5 mouse embryos, 10 µg/mL proteinase K for 4–5 min at room temperature is appropriate. For smaller embryos, dilute the proteinase K and reduce the time of incubation (e.g., to 5 µg/mL for 3 min for E9.5 mouse embryos). Proteinase K treatment should be optimized for the age and size of embryo under analysis and the degree of staining and integrity of the embryo achieved.

Trapping of probe in the ventricles of the brain and to a lesser extent in the neural tube is a frequent artifact encountered during whole-mount *in situ* hybridization. This can be avoided by dissecting open any cavities prior to processing, in order to allow free passage of reagents. However, when closure of the neural tube is the subject of the study, it may not be possible to do this without losing valuable morphological information. In these cases, it may help to increase the length of the washing steps. Vibratome sectioning of the embryos is then essential to distinguish between real staining and trapped probe within the lumen.

7.2.2. *Radioactive* In Situ *Hybridization on Sections*

This is generally thought to be a more sensitive technique than whole mount and permits a more detailed analysis of gene expression in internal embryonic tissues. We use the method of Angerer and Angerer *(22)* using paraffin sections (*see* **Subheading 6.2.**). Although this procedure is more labor intensive than whole-mount *in situ* hybridization, it may be more applicable to cases where the orientation of the tissue is critical or when the embryos are very small, as, once fixed onto the slide, they are unlikely to be lost during processing.

7.3. Preparing Embryonic Proteins for Analysis

Many methods exist for the isolation of proteins from embryonic tissues. If proteins are to be separated and further analyzed (for instance, by Western blotting), then attention must be paid to methods of homogenization and solubilization. Embryos are often disrupted simply by boiling for 5 min in sodium dodecyl sulfate–polyacrylamide gel electrophoresis (SDS-PAGE) sample buffer, but this procedure will denature the proteins and prevent subsequent assay for activity. A more flexible approach is to disrupt the embryos in a buffer that preserves enzyme activity and then mix aliquots with SDS-PAGE sample buffer for Western blotting, while retaining aliquots for activity studies. The method of disruption must reflect the lability of the protein of interest, but, in general, we homogenize neurulation-stage embryos either by drawing them through 19–25-gage syringe needles (gentle disruption) or by ultrasonication using the microprobe of a sonicator (harsher but more complete disruption). All procedures are conducted at 4°C in the presence of protease inhibitors to minimize proteolytic activity and prevent denaturation. A suitable homogenization buffer is 60 m*M* Tris-HCl (pH 7.5), 0.25*M* sucrose, 10 m*M* EGTA, 5 m*M* EDTA, 500 µ*M* leupeptin, 10 m*M* benzamidine, 1 m*M* phenylmethylsulfonyl fluoride, 0.1 U/mL aprotinin, 1 m*M* dithiothreitol, 100 µ*M* sodium vanadate, and 50 m*M* sodium fluoride.

7.4. Quantifying Embryonic Proteins

If protein content is to be used solely as a measure of embryonic growth (*see* **Subheading 4.5.**), embryos can simply be dissolved in sodium hydroxide and assayed direct [for instance, using a microscale Lowry method *(12)*]. However, if quantitation is to be combined with subsequent protein analysis, proteins must be extracted using a method as described in **Subheading 7.3.** and this raises the problem that the commonly used protein assays (e.g., Biuret, Lowry, bicinchoninic acid, Bradford, silver binding) all suffer from interference by commonly used reagents (e.g., dithiothreitol). To overcome this problem, we use the bicinchoninic acid (BCA, Pierce Chemical, Rockford, IL) assay combined with a double precipitation step using tricholoroacetic acid (TCA) to remove interfering substances from the samples being assayed. Prepare duplicate dilutions of protein samples and duplicate buffer blanks in a final volume of 50 µL in 1.5-mL microfuge tubes. Add 950 µL of ice-cold TCA (25% w/v) to each tube and stand on ice for 10 min. Spin the tubes at 3000*g* for 15 min (4°C). Pour off supernatant and drain tubes by inverting them over tissue paper. Repeat the TCA precipitation, then resuspend the pellet in 50 µL SDS (5% w/v) containing 0.1*N* NaOH and proceed to the BCA assay as described by the manufacturer (Pierce).

7.5. Immunocytochemistry on Embryonic Sections

Localizing antigenic determinants in embryonic tissues is a powerful technique for studying gene expression, and the increasing availability of specific antibodies makes this a viable approach for studies of neurulation. We use standard techniques *(23)* for application of primary and secondary antibodies and the biotin–streptavidin–alkaline phosphatase system to reveal sites of antibody binding. The most critical issues to consider when applying these techniques to mouse embryos are the type of embedding/sectioning to be used, the method of tissue fixation and the problem of using monoclonal primary antibodies that are generally prepared in mice.

Some antibodies require antigenic determinants to be presented in their native form, in which case it is often necessary to prepare cryosections, although the relatively poor preservation of embryonic structure makes it desirable to avoid cryosections if possible. Paraffin histology yields the best morphological results and facilitates the production of serial sections. Use of low-melting-point wax minimizes the denaturation of antigens for subsequent immunocytochemistry.

In conjunction with paraffin embedding, we find that fixation in Sainte Marie's solution (95% ethanol, 1% glacial acetic acid, 4% water) at 4°C overnight is compatible with the use of many antibodies, although specific antibodies may require other conditions for optimal usage. Monoclonal antibodies, raised in mice, present a particular challenge when studying mouse embryonic tissues, as the secondary antibody (e.g., biotin-conjugated anti-mouse IgG) may bind nonspecifically to immunoglobulins in the embryonic tissues. There are two main ways to overcome this potential problem. The sections can be blocked with serum of the species in which the secondary antibody was raised, in order to saturate these potential nonspecific binding sites before applying the secondary antibody. Alternatively, the primary antibody can be conjugated (e.g., to biotin), thereby making it unnecessary to use a secondary antibody *(23)*.

8. Experimental Manipulation of Neurulation-Stage Embryos

So far, we have dealt with methods for carrying out descriptive studies of neurulation in the mouse embryo. However, experimental analysis is essential if developmental mechanisms are to be established. Although methods are available for manipulation

Table 2
Surgical Interventions Using Neurulation Stage Embryos in Culture

Operation	Stage[a]	Ref.
Reopening of the posterior neuropore	E10.5	***27***
Insertion of a splint into the caudal region	E9.5	***28***
Tying together the neural folds at site of closure 1	E8.5	***29***
Neural crest grafting	E9.5 (rat)	***30***
Grafting of the node	E7.5	***31***
Allantoic removal and grafting	E8.5, E9.5 (rat)	***32,33***
Limb bud amputation	E9.5	***34***
Removal of the surface ectoderm	E9.5	(AJC unpublished)

[a]Refers to mouse days of gestation unless specified.

of development *in utero*, and using the *exo utero* technique ***(24,25)***, they are principally relevant to later-stage embryos and fetuses and have not been used extensively for studies of neurulation. The embryo culture method, in contrast, is optimally effective at the time of neurulation and, therefore, provides an opportunity for a variety of experimental interventions with embryos during neural tube closure.

8.1. Identifying Individual Cultured Embryos

A prerequisite for analysis of experimentally manipulated embryos is that they can be distinguished individually after the culture period. Although experimental embryos can be cultured separately from controls, this is not ideal, as culture conditions may vary subtly between embryo groups. Culture of embryos in individual tubes is expensive in terms of serum usage. We use an alternative approach: A short length (e.g., 0.5 cm) of braided silk suture material is teased into individual strands and a single strand is inserted through the yolk sac and trimmed with fine scissors. Loops can be tied in the strand, before insertion, to identify different embryos ***(26)***. In this way, control and experimental embryos can be marked uniquely and cultured in the same tube, ensuring precise comparisons after culture.

8.2. Surgical Operations on Cultured Embryos

A variety of microsurgical manipulations have been described with cultured mouse and rat embryos (**Table 2**). In some cases, these were specifically designed to answer questions about neurulation, whereas, in other cases, they were aimed at understanding development in other systems. The range of these techniques demonstrates that the surgical manipulation of the rodent embryo in culture is limited only by the ingenuity of the investigator. In general, for embryos at E9.5 and E10.5, it is necessary to open the yolk sac in order to gain access to the embryonic region that is to be manipulated. We prepare the embryo by completely removing the ectoplacental cone (*see* **Subheading 2.6.**), then opening the yolk sac along approximately 25% of the circumference of its circular junction with the chorion, where vascularization is minimal. The yolk-sac (exocoelomic) fluid is expressed and the embryo comes close to the window. This approach is especially advantageous for surgical access to the right side of the caudal region. The amnion is then torn open locally over the embryonic region, which can be exteriorized, allowing manipulation. After surgery, it is gently pushed back into the exocoelom and the incised edges of the yolk sac are pulled together over it. A silk suture thread (*see* **Subheading 8.1.**) can be used to loosely connect the yolk-sac edges. Operated embryos maintain a good yolk-sac circulation and do not appear to suffer from the loss of exocoelomic fluid.

8.3. Adding Substances to the Culture Medium

The embryo culture method permits a wide range of molecules to be administered to the embryo in isolation from maternal metabolism, which can be a complicating factor *in utero*. For instance, radioisotopes are cleared from the maternal circulation in less than 1 h, making it difficult to label embryos continuously *in utero*. In contrast, embryos become reproducibly labeled when radioisotopes are added to the culture serum, as described previously *(26)*. In principle, any agent can be administered to the embryo *in vitro*. Water-soluble agents are dissolved in PBS or sterile water and water-insoluble substances can be dissolved in either ethanol or dimethyl sulfoxide (DMSO). Adding ethanol or DMSO to a final concentration of 1–2% in rat serum is not usually toxic to cultured embryos, although a vehicle-only control should always be included in the experimental design.

Studies in which proteins are administered to cultured embryos can be more problematic. The yolk sac has extensive proteolytic capacity and digestion of ^{3}H-labeled proteins can yield labeled amino acids for protein synthesis in the embryo *(35)*. Thus, it can be misleading to localize radioactivity within the embryo as an indication of the uptake of, say, a ^{3}H-labeled growth factor. Amino acids labeled with ^{125}I are not reutilized in protein synthesis *(36)* and, thus, administration of iodinated proteins is the method of choice.

8.4. Bypassing the Yolk Sac

In cases where extensive yolk-sac metabolism is likely, molecules can be injected into the amniotic cavity, thus bypassing the yolk sac and gaining direct access to the embryo *(37–39)*. A micropipet is pulled using an electromagnetic puller and broken so that its tip is approximately 10–20 µm in diameter. The micropipet can be controlled by a micromanipulator and electronic injection controller (e.g., picoinjector), although we have found that a hand-held micropipet is quite satisfactory. It remains to be demonstrated, convincingly, that intra-amniotic injection achieves superior access of most molecules to the embryo compared with addition to the culture serum. Pratten and colleagues *(40)* have pioneered the technique of intravitelline injection, in which a major vessel in the yolk sac is cannulated and substances are administered directly into the circulation. This is a highly specialized technique that probably offers superior access to the embryo compared with other approaches.

8.5. Use of Antisense Oligonucleotides In Vitro

These are just beginning to be used with mouse embryos in the culture system *(41,42)* and can potentially offer short-term inhibition of gene function at any stage of neurulation. This is especially useful in cases where a knockout may disturb an early event of embryogenesis so that the mutants die before reaching neurulation. As with all antisense studies, careful attention must be paid to controls *(43)*. When using antisense oligo-

Table 3
Genes That Cause Neural Tube Defects When Mutated in the Mouse

Mouse gene (mutation)	Gene symbol (mutation symbol)	Chromosome	Type of mutant alleles	Neural tube defect phenotype	Refs.
1. Transcription factors					
Cartilage homeoprotein 1	*Cart1*	10	Knockout	Exencephaly	***45***
Gli3 (extra-toes)	*Gli3 (Xt)*	13	Spontaneous and radiation induced	Exencephaly	***46,47***
Hairy and Enhancer of Split homolog-1	*HES-1*	16	Knockout	Exencephaly	***48***
Hox-a1	*Hox-a1*	6	Knockout	Exencephaly	***49***
Mf3	*Mf3*	9	Knockout	Exencephaly	***50***
Pax3 (Splotch)	*Pax3 (Sp)*	1	Spontaneous and radiation induced	Exencephaly and spina bifida	***51,52***
RBP-jκ	*RBP-jκ*	ND	Knockout	Persistently open "anterior neuropore"	***53***
Transcription factor AP-2 alpha	*Tcfap2a*	13	Knockout	Exencephaly	***54,55***
Twist	*Twist*	12	Knockout	Exencephaly	***56***
2. Receptors					
Fibroblast growth factor receptor 1	*Fgfr-1*	8	Knockout chimera	Spina bifida	***57***
Notch3	*Notch3*	17	Overexpression	Exencephaly	***58***
Platelet derived growth factor receptor alpha (Patch)	*PDGFrα (Ph)*	5	Spontaneous	Exencephaly	***59,60***
Retinoic acid receptors, α/γ double mutants	*RARα* *RARγ*	11 15	Knockout	Exencephaly	***61***
3. Cancer related					
jumonji	*jmj*	13	Gene trap insertion	Exencephaly	***62***
patched	*Ptch*	13	Knockout	Complete failure of closure at time of death at E8.75	***63***

p53	*Trp53*	11	Knockout	Exencephaly	***64,65***
ski	*ski*	4	Knockout	Exencephaly	***66***
XPC (double mutants with p53)	*XPC*	6	Knockout	Exencephaly	***67***
4. Protein kinase related					
csk	*csk*	9	Knockout	Persistently open "anterior neuropore"	***68***
MARCKS	*Macs*	10	Knockout	Exencephaly	***69***
MARCKS-related protein	*Mrp, MacMARCKS, F52*	4	Knockout	Exencephaly and spina bifida	***70,71***
5. Miscellaneous					
Apolipoprotein B	*ApoB*	12	Knockout	Exencephaly	***72,73***
Connexin 43	*Cx43, Gja1*	10	Overexpression	Exencephaly	***74***

nucleotides in embryo culture, it is important to establish whether embryonic abnormalities result from specific effects of the treatment or from nonspecific influences (for instance, mediated via compromise of yolk-sac function). In this context, one should look for the hallmarks of yolk-sac dysfunction (*see* **Subheading 3.3.**).

8.6. Other Culture Methods

Culture of isolated embryonic fragments (for instance, on filters or within collagen gels) is a useful technique for studying tissue interactions *(44)*. This approach has so far been little used in neurulation studies, although we have demonstrated that the neuroepithelium from the mouse posterior neuropore can continue closure in vitro after it is cultured in isolation from adjacent tissues *(16)*. Organ culture approaches may be useful, but neurulation, as a global morphogenetic event affecting the entire body axis, may be most appropriately studied in intact embryos.

9. Genetic Analysis of Normal and Abnormal Neural Tube Closure

The most direct way to evaluate the role of specific genes in neural tube closure is to study embryos lacking functional gene product. Neurulation is well endowed with genetic mutants in the mouse. These are of two types: targeted or insertional mutants (**Table 3**) and spontaneous or radiation-induced mutants (**Table 4**). In the former, the identity of the mutated gene is usually known, whereas the causative gene in many cases of spontaneous mutation is not yet established. The majority of the targeted mutants have defects in cranial neural tube closure, with few exhibiting spina bifida, whereas more instances of spina bifida are available among the spontaneous and radiation-induced group. The earliest closure event, closure 1 (**Table 1**), appears relatively resistant to disturbance by genetic mutation, with only one mutant so far described *(loop–tail)* causing disruption of this event.

Relatively few of the mutants have been studied in detail to determine the developmental abnormality that results in failure of neural tube closure. We have demonstrated that the posterior neuropore is delayed in its closure in the *curly tail* mutant as a result of an imbalance of cell proliferation within the caudal region that causes ventral curvature of the body axis *(28)*. This defect can be prevented by administration of myoinositol to the neurulation-stage embryo both in vitro and in utero *(8)*. We recently demonstrated that failure of closure 1 in the *loop–tail* mutant is related to the presence of an enlarged floor-plate region in the neuroepithelium at early somite stages *(75)*.

9.1. Genotyping Embryos for Genetic Studies

It is vital to determine the genotype of embryos in litters containing homozygous, heterozygous, and wild-type embryos independently of the morphological defect. Where the identity of the mutant gene is known, it is usually possible to identify the mutation directly, preferably by PCR, because it can be carried out using the small amount of DNA obtainable from the yolk sac (after careful removal of trophoblast, which may be contaminated with maternal blood), leaving the embryo for other types of analysis. Where the causative gene is not known, we use a closely linked, polymorphic microsatellite marker (e.g., *Crp* in the *loop–tail* mutant) *(76)*, again demonstrated by PCR. As these procedures do not require high-quality DNA, a suitable protocol for DNA extraction involves proteinase K digestion of the embryo or yolk sac, potassium-

Table 4
Mouse Mutations with Neural Tube Defects for Which the Mutated Gene Is Unknown

Mutant name	Gene symbol	Chromosome	Type of mutant allele	Neural tube defect phenotype	Refs.
Axial defects	*Axd*	ND	Spontaneous	Spina bifida	*78*
Bent-tail	*Bn*	X	Spontaneous	Exencephaly and spina bifida	*79*
Cranioschisis	*crn*	ND	Spontaneous	Exencephaly	*80*
Crooked tail	*Cd*	6	Spontaneous	Exencephaly	*81*
Curtailed	T^{ct}	17	Radiation-induced	Spina bifida	*82*
Curly tail	*ct*	4	Spontaneous	Exencephaly and spina bifida	*83,84*
Exencephaly	*xn*	ND	Spontaneous	Exencephaly	*85*
Forebrain overgrowth	*Fog*	10	Spontaneous	Exencephaly and spina bifida	*86*
Loop-tail	*Lp*	1	Spontaneous	Craniorachischisis	*87,88*
Open brain	*opb*	1	Spontaneous	Exencephaly	*89*
Rib fusions	*Rf*	ND	Spontaneous	Exencephaly	*90*
SELH*	—	ND	Polymorphism	Exencephaly	*91*
Vacuolated lens	*vl*	1	Spontaneous	Spina bifida	*92*

*SELH/Bc is an inbred strain in which several interacting genes predispose to neural tube defect.

acetate-mediated precipitation of the proteins, then ethanol precipitation of the DNA from the supernatant.

9.2. Interpreting Abnormal Neural Tube Closure

The finding of an open neural tube in late fetal stages should always prompt a study of development at earlier stages to determine whether this defect indeed represents failure of neural tube closure, as opposed to reopening of the closed neural tube, as has been described in rat embryos treated with cyclophosphamide *(77)*. Moreover, it is important to categorize mutants with cranial defects according to which closure site or neuropore (*see* **Table 1**) is compromised. This type of analysis has rarely been carried out to date, making cranial closure much less well understood than spinal closure. It is crucial to ask whether mutant embryos with an open neural tube have progressed far enough in development to have achieved neural tube closure. For instance, although E9.5 embryos homozygous for the *csk* knockout ***(68)*** have open cranial neural tubes, they are developmentally retarded and may die before reaching the stage at which the neural tube would have closed in normal development. Ultimately, the analysis of mutants will enable us to establish the genetic requirement for neural tube closure at each level of the body axis, providing valuable information on the integration of genetic pathways in this complex morphogenetic event.

References

1. Cockroft, D. L. (1990) Dissection and culture of postimplantation embryos, in *Postimplantation Mammalian Embryos: A Practical Approach* (Copp, A. J. and Cockroft, D. L., eds.), IRL Press, Oxford, pp. 15–40.

2. New, D. A. T., Coppola, P. T., and Terry, S. (1973) Culture of explanted rat embryos in rotating tubes. *J. Reprod. Fertil.* **35,** 135–138.
3. Sadler, T. W. and New, D. A. T. (1981) Culture of mouse embryos during neurulation. *J. Embryol. Exp. Morphol.* **66,** 109–116.
4. Flynn, T. J., Friedman, L., Black, T. N., and Klein, N. W. (1987) Methionine and iron as growth factors for rat embryos cultured in canine serum. *J. Exp. Zool.* **244,** 319–324.
5. Chatot, C. L., Klein, N. W., Piatek, J., and Pierro, L. J. (1980) Successful culture of rat embryos on human serum: use in the detection of teratogens. *Science* **207,** 1471–1473.
6. Cockroft, D. L. (1979) Nutrient requirements of rat embryos undergoing organogenesis in vitro. *J. Reprod. Fertil.* **57,** 505–510.
7. Cockroft, D. L. (1988) Changes with gestational age in the nutritional requirements of postimplantation rat embryos in culture. *Teratology* **38,** 281–290.
8. Greene, N. D. E. and Copp, A. J. (1997) Inositol prevents folate-resistant neural tube defects in the mouse. *Nat. Med.* **3,** 60–66.
9. Downs, K. M. and Davies, T. (1993) Staging of gastrulating mouse embryos by morphological landmarks in the dissecting microscope. *Development* **118,** 1255–1266.
10. Peeters, M. C. E., Shum, A. S. W., Hekking, J. W. M., Copp, A. J., and Van Straaten, H. W. M. (1996) Relationship between altered axial curvature and neural tube closure in normal and mutant *(curly tail)* mouse embryos. *Anat. Embryol.* **193,** 123–130.
11. Brown, N. A. (1990) Routine assessment of morphology and growth: scoring systems and measurements of size, in *Postimplantation Mammalian Embryos: A Practical Approach* (Copp, A. J. and Cockroft, D. L., eds.), IRL Press, Oxford, pp. 93–108.
12. Estibeiro, J. P., Copp, A. J., Cockroft, D. L., Brown, N. A., and Clarke, D. O. (1990) Extraction of macromolecules from embryonic materials, in *Postimplantation Mammalian Embryos: a Practical Approach* (Copp, A. J. and Cockroft, D. L. eds), IRL Press, Oxford, pp. 173–204.
13. Brown, N. A. and Fabro, S. (1981) Quantitation of rat embryonic development in vitro: a morphological scoring system. *Teratology* **24,** 65–78.
14. Copp, A. J., Brook, F. A., Estibeiro, J. P., Shum, A. S. W., and Cockroft, D. L. (1990) The embryonic development of mammalian neural tube defects. *Prog. Neurobiol.* **35,** 363–403.
15. Van Allen, M. I., Kalousek, D. K., Chernoff, G. F., Juriloff, D., Harris, M., McGillivray, B. C., et al. (1993) Evidence for multi-site closure of the neural tube in humans. *Am. J. Med. Genet.* **47,** 723–743.
16. Van Straaten, H. W. M., Hekking, J. W. M., Consten, C., and Copp, A. J. (1993) Intrinsic and extrinsic factors in the mechanism of neurulation: effect of curvature of the body axis on closure of the posterior neuropore. *Development* **117,** 1163–1172.
17. Tam, P. P. L. (1986) A study on the pattern of prospective somites in the presomitic mesoderm of mouse embryos. *J. Embryol. Exp. Morphol.* **92,** 269–285.
18. Shum, A. S. W. and Copp, A. J. (1996) Regional differences in morphogenesis of the neuroepithelium suggest multiple mechanisms of spinal neurulation in the mouse. *Anat. Embryol.* **194,** 65–73.
19. Martins-Green, M. M. (1990) Transmission electron microscopy and immunolabelling for light and electron microscopy, in *Postimplantation Mammalian Embryos: A Practical Approach* (Copp, A. J. and Cockroft, D. L., eds.), IRL Press, Oxford, pp. 127–154.
20. Stanisstreet, M. (1990) Scanning electron microscopy and morphometry, in *Postimplantation Mammalian Embryos: A Practical Approach* (Copp, A. J. and Cockroft, D. L., eds.), IRL Press, Oxford.
21. Wilkinson, D. G. (1992) *In Situ Hybridisation: A Practical Approach.* IRL Press, Oxford.
22. Angerer, L. M. and Angerer, R. C. (1992) *In situ* hybridisation to cellular RNA with radiolabelled RNA probes, in *In Situ Hybridisation. A Practical Approach* (Wilkinson, D. G., ed.), IRL Press, Oxford, pp. 15–32.

23. Harlow, E. and Lane, D. (1988) *Antibodies. A Laboratory Manual.* Cold Spring Harbor Laboratory Press, Cold Spring Harbor, NY.
24. Muneoka, K., Wanek, N., Trevino, C., and Bryant, S. V. (1990) Exo utero surgery, in *Postimplantation Mammalian Embryos: A Practical Approach* (Copp, A. J. and Cockroft, D. L., eds.), IRL Press, Oxford, pp. 41–59.
25. Papaioannou, V. E. (1990) In utero manipulations, in *Postimplantation Mammalian Embryos: A Practical Approach* (Copp, A. J. and Cockroft, D. L., eds.), IRL Press, Oxford, pp. 61–80.
26. Copp, A. J. (1990) Studying developmental mechanisms in intact embryos, in *Postimplantation Mammalian Embryos: A Practical Approach* (Copp, A. J. and Cockroft, D. L., eds.), IRL Press, Oxford, pp. 293–316.
27. Copp, A. J. (1985) Relationship between timing of posterior neuropore closure and development of spinal neural tube defects in mutant (curly tail) and normal mouse embryos in culture. *J. Embryol. Exp. Morphol.* **88,** 39–54.
28. Brook, F. A., Shum, A. S. W., Van Straaten, H. W. M., and Copp, A. J. (1991) Curvature of the caudal region is responsible for failure of neural tube closure in the curly tail (ct) mouse embryo. *Development* **113,** 671–678.
29. Gerrelli, D. and Copp, A. J. (1997) Failure of neural tube closure in the *loop–tail (Lp)* mutant mouse: analysis of the embryonic mechanism. *Dev. Brain Res.* **102,** 217–224.
30. Tan, S. S. and Morriss-Kay, G. M. (1986) Analysis of cranial neural crest cell migration and early fates in postimplantation rat chimaeras. *J. Embryol. Exp. Morphol.* **98,** 21–58.
31. Beddington, R. S. P. (1994) Induction of a second neural axis by the mouse node. *Development* **120,** 613–620.
32. Downs, K. M. and Gardner, R. L. (1995) An investigation into early placental ontogeny: allantoic attachment to the chorion is selective and developmentally regulated. *Development* **121,** 407–416.
33. Fujinaga, M., Hoffman, B. B., and Baden, J. M. (1995) Axial rotation in rat embryos: morphological analysis and microsurgical study on the role of the allantois. *Teratology* **51,** 94–106.
34. Lee, K. K. H. and Chan, W. Y. (1991) A study on the regenerative potential of partially excised mouse embryonic fore-limb bud. *Anat. Embryol.* **184,** 153–157.
35. Freeman, S. J. and Lloyd, J. B. (1983) Evidence that protein ingested by the rat visceral yolk sac yields amino acids for synthesis of embryonic protein. *J. Embryol. Exp. Morphol.* **73,** 307–315.
36. McArdle, H. J. and Priscott, P. K. (1984) Uptake and metabolism of transferrin and albumin by rat yolk sac placenta. *Am. J. Physiol.* **247,** C409–C414.
37. Chan, W. Y. and Tam, P. P. L. (1988) A morphological and experimental study of the mesencephalic neural crest cells in the mouse embryo using wheat germ agglutinin–gold conjugate as the cell marker. *Development* **102,** 427–442.
38. Stark, K. L. and Juchau, M. R. (1989) Microinjection of cultured rat embryos. A new technique for studies in chemical dysmorphogenesis. *Toxicol. Appl. Pharmacol.* **100,** 411–416.
39. Chan, W. Y. and Lee, K. K. H. (1992) The incorporation and dispersion of cells and latex beads on microinjection into the amniotic cavity of the mouse embryo at the early-somite stage. *Anat. Embryol.* **185,** 225–238.
40. Tebbs, C. A., Cumberland, P. F. T., and Pratten, M. K. (1997) The role of maternally derived epidermal growth factor and the epidermal growth factor receptor during organogenesis in the rat embryo. *J. Anat.* **190,** 491–503.
41. Chen, B. Y. and Hales, B. F. (1995) Antisense oligonucleotide down-regulation of E-cadherin in the yolk sac and cranial neural tube malformations. *Biol. Reprod.* **53,** 1229–1238.

42. Augustine, K., Liu, E. T., and Sadler, T. W. (1993) Antisense attenuation of *Wnt-1* and *Wnt-3a* expression in whole embryo culture reveals roles for these genes in craniofacial, spinal cord, and cardiac morphogenesis. *Dev. Genet.* **14,** 500–520.
43. Wagner, R. W. (1994) Gene inhibition using antisense oligodeoxynucleotides. *Nature* **372,** 333–335.
44. Greenburg, G. and Hay, E. D. (1982) Epithelia suspended in collagen gels can lose polarity and express characteristics of migrating mesenchyme cells. *J. Cell Biol.* **95,** 333–339.
45. Zhao, Q., Behringer, R. R., and De Crombrugghe, B. (1996) Prenatal folic acid treatment suppresses acrania and meroanencephaly in mice mutant for the *Cart1* homeobox gene. *Nature Genet.* **13,** 275–283.
46. Schimmang, T., Lemaistre, M., Vortkamp, A., and Rüther, U. (1992) Expression of the zinc finger gene *Gli*3 is affected in the morphogenetic mouse mutant *extra-toes (Xt). Development* **116,** 799–804.
47. Hui, C. and Joyner, A. L. (1993) A mouse model of Greig cephalopolysyndactyly syndrome: the *extra-toesJ* mutation contains an intragenic deletion of the *Gli3* gene. *Nat. Genet.* **3,** 241–246.
48. Ishibashi, M., Ang, S.-L., Shiota, K., Nakanishi, S., Kageyama, R., and Guillemot, F. (1995) Targeted disruption of mammalian *hairy* and *Enhancer of split* homolog-1 *(HES-1)* leads to up-regulation of neural helix-loop-helix factors, premature neurogenesis, and severe neural tube defects. *Genes Dev.* **9,** 3136–3148.
49. Lufkin, T., Dierich, A., LeMeur, M., Mark, M., and Chambon, P. (1991) Disruption of the *Hox-1.6* homeobox gene results in defects in a region corresponding to its rostral domain of expression. *Cell* **66,** 1105–1119.
50. Labosky, P. A., Winnier, G. E., Jetton, T. L., et al. (1997) The winged helix gene, *Mf3*, is required for normal development of the diencephalon and midbrain, postnatal growth and the milk-ejection reflex. *Development* **124,** 1263–1274.
51. Goulding, M. D., Chalepakis, G., Deutsch, U., Erselius, J. R., and Gruss, P. (1991) Pax-3, a novel murine DNA binding protein expressed during early neurogenesis. *EMBO J.* **10,** 1135–1147.
52. Epstein, D. J., Vekemans, M., and Gros, P. (1991) splotch (Sp^{2H}), a mutation affecting development of the mouse neural tube, shows a deletion within the paired homeodomain of Pax-3. *Cell* **67,** 767–774.
53. Oka, C., Nakano, T., Wakeham, A., et al. (1995) Disruption of the mouse *RBP-J*kappa gene results in early embryonic death. *Development* **121,** 3291–3301.
54. Schorle, H., Meier, P., Buchert, M., Jaenisch, R., and Mitchell, P. J. (1996) Transcription factor AP-2 essential for cranial closure and craniofacial development. *Nature* **381,** 235–238.
55. Zhang, J., Hagopian-Donaldson, S., Serbedzija, G., et al. (1996) Neural tube, skeletal and body wall defects in mice lacking transcription factor AP-2. *Nature* **381,** 238–241.
56. Chen, Z.-F. and Behringer, R. R. (1995) *twist* is required in head mesenchyme for cranial neural tube morphogenesis. *Genes Dev.* **9,** 686–699.
57. Deng, C. X., Bedford, M., Li, C. L., et al. (1997) Fibroblast growth factor receptor-1 (FGFR-1) is essential for normal neural tube and limb development. *Dev. Biol.* **185,** 42–54.
58. Lardelli, M., Williams, R., Mitsiadis, T., and Lendahl, U. (1996) Expression of the Notch 3 intracellular domain in mouse central nervous system progenitor cells is lethal and leads to disturbed neural tube development. *Mech. Dev.* **59,** 177–190.
59. Morrison-Graham, K., Schatteman, G. C., Bork, T., Bowen-Pope, D. F., and Weston, J. A. (1992) A PDGF receptor mutation in the mouse *(Patch)* perturbs the development of a non-neuronal subset of neural crest-derived cells. *Development* **115,** 133–142.

60. Orr-Urtreger, A., Bedford, M. T., Do, M.-S., Eisenbach, L., and Lonai, P. (1992) Developmental expression of the a receptor for platelet-derived growth factor, which is deleted in the embryonic lethal *Patch* mutation. *Development* **115,** 289–303.
61. Lohnes, D., Mark, M., Mendelsohn, C., et al. (1994) Function of the retinoic acid receptors (RARs) during development. (I) Craniofacial and skeletal abnormalities in RAR double mutants. *Development* **120,** 2723–2748.
62. Takeuchi, T., Yamazaki, Y., Katoh-Fukui, Y., *et al.* (1995) Gene trap capture of a novel mouse gene, *jumonji*, required for neural tube formation. *Genes Dev.* **9,** 1211–1222.
63. Goodrich, L. V., Milenkovic, L., Higgins, K. M., and Scott, M. P. (1997) Altered neural cell fates and medulloblastoma in mouse *patched* mutants. *Science* **277,** 1109–1113.
64. Schmid, P., Lorenz, A., Hameister, H., and Montenarh, M. (1991) Expression of p53 during mouse embryogenesis. *Development* **113,** 857–865.
65. Sah, V. P., Attardi, L. D., Mulligan, G. J., Williams, B. O., Bronson, R. T., and Jacks, T. (1995) A subset of *p53*-deficient embryos exhibit exencephaly. *Nature Genet.* **10,** 175 180.
66. Berk, M., Desai, S. Y., Heyman, H. C., and Colmenares, C. (1997) Mice lacking the ski proto-oncogene have defects in neurulation, craniofacial, patterning, and skeletal muscle development. *Genes Dev.* **11,** 2029–2039.
67. Cheo, D. L., Meira, L. B., Hammer, R. E., Burns, D. K., Doughty, A. T. B., and Friedberg, E. C. (1996) Synergistic interactions between *XPC* and *p53* mutations in double-mutant mice: neural tube abnormalities and accelerated UV radiation-induced skin cancer. *Curr. Biol.* **6,** 1691–1694.
68. Imamoto, A. and Soriano, P. (1993) Disruption of the *csk* gene, encoding a negative regulator of Src family tyrosine kinases, leads to neural tube defects and embryonic lethality in mice. *Cell* **73,** 1117–1124.
69. Stumpo, D. J., Bock, C. B., Tuttle, J. S., and Blackshear, P. J. (1995) MARCKS deficiency in mice leads to abnormal brain development and perinatal death. *Proc. Natl. Acad. Sci. USA* **92,** 944–948.
70. Wu, M., Chen, D. F., Sasaoka, T., and Tonegawa, S. (1996) Neural tube defects and abnormal brain development in F52-deficient mice. *Proc. Natl. Acad. Sci. USA* **93,** 2110–2115.
71. Chen, J., Chang, S., Duncan, S. A., Okano, H. J., Fishell, G., and Aderem, A. (1996) Disruption of the MacMARCKS gene prevents cranial neural tube closure and results in anencephaly. *Proc. Natl. Acad. Sci. USA* **93,** 6275–6279.
72. Meehan, R. R., Barlow, D. P., Hill, R. E., Hogan, B. L. M., and Hastie, N. D. (1984) Pattern of serum protein gene expression in mouse visceral yolk sac and foetal liver. *EMBO J.* **3,** 1881–1885.
73. Homanics, G. E., Smith, T. J., Zhang, S. H., Lee, D., Young, S. G., and Maeda, N. (1993) Targeted modification of the apolipoprotein B gene results in hypobetalipoproteinemia and developmental abnormalities in mice. *Proc. Natl. Acad. Sci. USA* **90,** 2389–2393.
74. Ewart, J. L., Cohen, M. F., Meyer, R. A., et al. (1997) Heart and neural tube defects in transgenic mice overexpressing the Cx43 gap junction gene. *Development* **124,** 1281–1292.
75. Greene, N. D., Gerrelli, D., Van Straaten, H. W., and Copp, A. J. (1998) Abnormalities of floor plate, notochord and somite differentiation in the *loop-tail (Lp)* mouse: a model of severe neural tube defects. *Mech. Dev.* **73,** 59–72
76. Copp, A. J., Checiu, I., and Henson, J. N. (1994) Developmental basis of severe neural tube defects in the *loop-tail (Lp)* mutant mouse: Use of microsatellite DNA markers to identify embryonic genotype. *Dev. Biol.* **165,** 20–29.
77. Padmanabhan, R. and Hameed, M. S. (1985) Malformations of the axial skeleton in cranioschisis aperta and exencephaly in rat fetuses induced after neural tube closure. *Virchows Arch. A Path. Anat. Hist.* **408,** 61–73.

78. Essien, F. B., Haviland, M. B., and Naidoff, A. E. (1990) Expression of a new mutation (*Axd*) causing axial defects in mice correlates with maternal phenotype and age. *Teratology* **42,** 183–194.
79. Johnson, D. R. (1976) The interfrontal bone and mutant genes in the mouse. *J. Anat.* **121,** 507–513.
80. Kalter, H. (1985) Experimental teratological studies with the mouse CNS mutations cranioschisis and delayed splotch. *J. Craniofac. Genet. Dev. Biol.* **Suppl. 1, 1,** 339–342.
81. Morgan, W. (1954) A new crooked-tail mutation involving distinct pleiotropism. *J. Genet.* **52,** 354–373.
82. Park, C. H., Pruitt, J. H., and Bennett, D. (1989) A mouse model for neural tube defects: the curtailed (Tc) mutation produces spina bifida occulta in Tc/+ animals and spina bifida with meningomyelocele in Tc/t. *Teratology* **39,** 303–312.
83. Gruneberg, H. (1954) Genetical studies on the skeleton of the mouse. VIII. Curly tail. *J. Genet.* **52,** 52–67.
84. Neumann, P. E., Frankel, W. N., Letts, V. A., Coffin, J. M., Copp, A. J., and Bernfield, M. (1994) Multifactorial inheritance of neural tube defects: localization of the major gene and recognition of modifiers in *ct* mutant mice. *Nat. Genet.* **6,** 357–362.
85. Wallace, M. E., Knights, P. J., and Anderson, J. R. (1978) Inheritance and morphology of exencephaly, a neonatal recessive with partial penetrance, in the house mouse. *Genet. Res.* **32,** 135–149.
86. Harris, B. S., Franz, T., Ullrich, S., Cook, S., Bronson, R. T., and Davisson, M. T. (1997) Forebrain overgrowth (*fog*): a new mutation in the mouse affecting neural tube development. *Teratology* **55,** 231–240.
87. Strong, L. C. and Hollander, W. F. (1949) Hereditary loop-tail in the house mouse. *J. Hered.* **40,** 329–334.
88. Stanier, P., Henson, J. N., Eddleston, J., Moore, G. E., and Copp, A. J. (1995) Genetic basis of neural tube defects: the mouse gene *loop–tail* maps to a region of Chromosome 1 syntenic with human 1q21-22. *Genomics* **26,** 473–478.
89. Günther, T., Struwe, M., Aguzzi, A., and Schughart, K. (1994) *open brain*, a new mouse mutant with severe neural tube defects, shows altered gene expression patterns in the developing spinal cord. *Development* **120,** 3119–3130.
90. Theiler, K. and Stevens, L. C. (1960) The development of rib fusions, a mutation in the house mouse. *Am. J. Anat.* **106,** 171–183.
91. Juriloff, D. M., MacDonald, K. B., and Harris, M. J. (1989) Genetic analysis of the cause of exencephaly in the SELH/Bc mouse stock. *Teratology* **40,** 395–405.
92. Wilson, D. B. and Wyatt, D. P. (1986) Pathogenesis of neural dysraphism in the mouse mutant vacuolated lens (vl). *J. Neuropath. Exp. Neurol.* **45,** 43–55.
93. Copp, A. J., Seller, M. J., and Polani, P. E. (1982) Neural tube development in mutant (curly tail) and normal mouse embryos: the timing of posterior neuropore closure in vivo and in vitro. *J. Embryol. Exp. Morphol.* **69,** 151–167.

17

Neural Tube Defects

Takayuki Inagaki, Jodi L. Smith, Marion L. Walker, and Gary C. Schoenwolf

1. Introduction

We start with a quote by Rupert Willis *(1)*, who said "Malformations of every conceivable kind, degree, and combination occur, and no two of them are ever alike," to emphasize the difficulty of what we attempt to do in this chapter: provide a simplified, useful scheme for classifying neural tube defects. The difficulty results from the fact, as stated above, that individual birth defects are part of a continuum of anomalies rather than unique members of distinct classes of defects; terminology is not always logical nor is it universally accepted among embryologists, pathologists, neurosurgeons, and so forth; and the etiology of birth defects, including neural tube defects, is virtually unknown. Nevertheless, in this chapter, we provide a classification scheme, albeit an imperfect one, to help embryologists begin to understand the clinical consequences of development when it goes awry. In addition, we believe this classification scheme will be useful to clinicians for categorizing neural tube defects and will help to facilitate communication among basic scientists and clinicians as they work together in the important task of understanding neural tube defects.

In this chapter, we briefly discuss open and closed neural tube defects. A neural tube defect is an anomaly of the central nervous system and/or its membranes, resulting either from faulty neurulation or from abnormal development of the neural tube during the early postneurulation period. Simply speaking, an *open* neural tube defect is an abnormal neural tube with abnormal membranous coverings, which lacks the overlying skin. In contrast, a *closed* neural tube defect is defined as for the open one, except that it is skin covered. In addition, in this chapter, we discuss anomalies occurring secondarily to open or closed neural tube defects, such as tethered cords, the Chiari II malformation, and hydrocephalus, as well as other major anomalies of the neural tube, such as holoprosencephaly and lissencephaly. Ideally, the classification of congenital malformations of the brain and spinal cord would be based on the cause of abnormality. However, as stated earlier, the causes of such defects are not known in many cases. Therefore, in this chapter, we classify neural tube defects based on their anatomical features. More exhaustive coverage of neural tube defects appears elsewhere *(2,3)*.

From: *Methods in Molecular Biology, Vol. 136: Developmental Biology Protocols, Vol. II*
Edited by: R. S. Tuan and C. W. Lo © Humana Press Inc., Totowa, NJ

Our classification scheme is as follows:

A. Open neural tube defects
   1. Anencephaly
   2. Spinal rachischisis
      a. Myeloschisis
      b. Myelomeningoceles (meningoceles)
   3. Iniencephaly
   4. Encephaloceles
B. Closed neural tube defects
   1. Spina bifida occulta
   2. Lipomas and lipomeningoceles
   3. Split cords (diastematomyelia, diplomyelia)
   4. Neurenteric cysts
   5. Dermal sinuses
   6. Sacral agenesis (caudal regression)

Each of these neural tube defects will now be defined briefly.

2. Classification of Neural Tube Defects

2.1. Open Neural Tube Defects

2.1.1. Anencephaly

Failure of the cephalic part of the neural tube to close results in a condition called exencephaly. Because the malformed brain tissue is exposed on the surface of the body and is not covered by skin, the vault (i.e., roof or calvarium) of the skull does not form. During subsequent gestation, the exposed brain tissue degenerates, resulting in anencephaly (literally, absence of the brain).

2.1.2. Spinal Rachischisis (Spina Bifida Operta)

Failure of the neural tube to close in the spinal cord region results in spinal rachischisis (literally, cleft spine). The term spina bifida aperta refers to the fact that the spines of the vertebrae are bifid. Thus, spinal rachischisis involves defects both of soft tissues (i.e., neural, meninges, muscle, and skin) and hard tissues (bone). There are two main types of spinal rachischisis:

1. Myeloschisis: The neural tube at the site of the defect is broadly open and the central canal is absent. This is the most deleterious type of neural tube defect. Myeloschisis commonly occurs at the thoracolumbar junction.
2. Myelomeningoceles: A central canal forms, although it may be open on its dorsal surface. The membranous coverings (i.e., meninges), as well as the malformed spinal cord (i.e., myelo), protrude through a defect in the skin. Myelomeningoceles tend to occur in the lumbar and sacral regions. Rarely, another malformation occurs at the spinal cord level in which the membranous coverings of the spinal cord protrude through the defect in the skin, but the spinal cord does not; this defect is called a *meningocele*. In meningoceles, the spinal cord can still be abnormal, but the abnormality is usually less severe than in myelomeningoceles.

2.1.3. Iniencephaly

Iniencephaly is a defect in which the occipital bone is deficient and the brain is exposed. It is usually combined with cervicothoracic spinal retroflexion and may be accompanied by rachischisis. Iniencephaly differs from anencephaly in that it is associated with the formation of the cranial vault and spinal retroflexion.

2.1.4. Encephaloceles

Encephaloceles are characterized by herniation of the brain and/or meninges through an opening in the skull. If the hernia contains both brain tissue and meninges, the defect would best be described by the term encephalomeningocele. However, the term encephalocele has been used traditionally by clinicians to describe this defect, and this tradition continues today. If the hernia contains only meninges, it is called a cranial meningocele.

2.2. Closed Neural Tube Defects

2.2.1. Spina Bifida Occulta

The term spina bifida occulta refers to the fact that the spines of the vertebrae are bifid. In contrast to spina bifida aperta (spinal rachischisis), the spinal cord and its meninges are normal in spina bifida occulta. Thus, spina bifida occulta is usually only an asymptomatic defect of the bony components of the spine. As such, it is not a true neural tube defect.

2.2.2. Lipomas and Lipomeningoceles

A lipoma is a benign fatty tumor. It can form along any level of the spinal cord. Lipomas associated with the cauda equina (the spinal nerve roots that descend from the inferior part of the spinal cord) and with the presence of an caudal cyst (i.e., an abnormal expansion of the dural sac that contains cerebrospinal fluid) are called lipomeningoceles. Despite the fact that the name for this defect includes the term meningocele, lipomeningoceles are *closed* neural tube defects not open neural tube defects; that is, they are covered with skin.

2.2.3. Split Cords

The use of the term split cord is growing in popularity and is equivalent to the more cumbersome term diastematomyelia. *Diastematomyelia* is a localized abnormality of the spinal canal in which the dura mater is perforated by a bony spike or fibrous band, creating two chambers. Each chamber contains a half-spinal cord, split by the central partition, but joined together both superiorly and inferiorly to the partition. *Diplomyelia* is a type of split cord that is considered to be a "true" duplication. Thus, each cord has two dorsal and two ventral horns and paired nerve roots, but the two cords are contained within a single dural sac.

2.2.4. Neurenteric Cysts

Neurenteric cysts are rare cysts occurring within the spinal canal or cranial cavity. These cysts are lined with epithelium resembling that of the gastrointestinal tract and are of endodermal origin. Neurenteric cysts most frequently occur at low cervical and upper thoracic levels. Their origin is not well understood.

2.2.5. Dermal Sinuses

A dermal sinus is a fistula extending from the skin to some deeper structure. The inside of the fistula is lined with skin. The dermal sinus can be a blind pouch or can attach to the dura mater or to the spinal cord. Dermal sinuses may occur any level of the neuraxis, although they tend to occur in the sacrococcygeal and occipital levels.

2.2.6. Sacral Agenesis

Sacral agenesis or *caudal regression* is a variable defect of the lumbar vertebrae, sacrum, and coccyx. The severest form is *sirenomelia* (mermaid deformity; not to be confused with syringomyelia described below) in which there are abnormalities of the anorectal region, urogenital system, and lower limbs—the latter consisting of a fused thigh and leg with partial or complete fusion of the feet. The caudal spinal cord is also abnormal, and the severity of its abnormality is proportional to the degree of abnormality of the vertebral column. Sacral agenesis may be combined with a presacral mass, which can include a teratoma (a tumor of embryonic origin containing cells derived from all three germ layers) or an anterior sacral meningocele

3. Defects Occurring Secondarily to Neural Tube Defects

Three defects that occur secondarily to neural tube defects will now be defined: tethered cords, hydrocephalus, and the Chiari II malformation.

3.1. Tethered Cords

Tethered cords may occur secondarily to an open neural tube defect or to a closed neural tube defect such as spinal lipoma, lipomyelomeningocele, or diastematomyelia. As the vertebral column grows during early childhood, the inferior end of the spinal cord (the conus medullaris) normally “ascends” within the vertebral canal. Tethered cord results when the abnormal spinal cord adheres to its meninges and prevents this ascension. Thus, the growth of the vertebral column exerts traction on the lower spinal cord and spinal nerves, causing neurological deficits.

3.2. Chiari II (Arnold Chiari) Malformation

Four types of Chiari malformations are recognized. Type I malformations are characterized by minor herniation of the cerebellar tonsils. Patients with type I malformations often have hydromyelia (dilation of the central canal of the spinal cord) and syringomyelia (presence of ectopic, fluid-filled cavities within the spinal cord tissue; not to be confused with sirenomyelia described earlier). Type II malformations are more extensive, involving descent of both cerebellar tonsils and the vermis below the level of the foramen magnum. The medulla and at times the pons are also displaced below the foramen magnum, and the cervical cord is displaced caudally by the herniated tissue. Type II malformations are associated with open neural tube defects of the spinal cord, namely myeloschisis, myelomeningoceles, and meningoceles. This association suggests that Chiari II malformations occur secondarily to the open neural tube defect. Type III malformations also involve herniation of the cerebellum (and possibly medulla and pons) into the spinal canal and are associated with lower occipital/high cervical encephaloceles. Type IV malformations are rather different. In type IV malformations, most of the cerebellum is hypoplastic.

3.3. Hydrocephalus

In patients with hydrocephalus, the cerebral ventricles are dilated with an excess of cerebrospinal fluid. Flow of cerebrospinal fluid from the ventricles to the subarachnoid space is blocked, often at the foramen of Magendie. Like the Chiari II malformation, hydrocephalus is associated with open neural tube defects, suggesting that it occurs secondarily to such defects.

4. Other Major Anomalies of the Neural Tube

4.1. Holoprosencephaly

The term holoprosencephaly indicates that only a single brain ventricle, rather than paired ventricles, forms in the area that arises from the prosencephalon. Holoprosencephaly has three important clinical subclassifications: (1) alobar holoprosencephaly, the subclassification used when only one hemisphere is present; (2) semilobar holoprosencephaly, the subclassification used when there is a slight midline depression in the single hemisphere; and (3) lobar holoprosencephaly, the subclassification used when two distinct hemispheres (but with only one ventricle) are present. Holoprosencephaly, especially the alobar type, is frequently associated with cyclopia (a single midline eye).

4.2. Lissencephaly

Lissencephaly—literally, smooth brain—is considered to be a defect of cell migration within the wall of the cerebral hemispheres. As a consequence, the normal histological layers of the brain do not form and gyri and sulci are absent.

5. Conclusions

During the process of neurulation, the neural plate rolls up, forming a neural groove flanked laterally by neural folds, which, in turn, fuse to establish the closed neural tube. The cause of open neural tube defects is generally considered to be nonclosure of the neural tube during neurulation. However, in rare cases reopening of the closed neural tube may occur shortly after neurulation. The cause of closed neural tube defects is understood even less. Although a neural tube may form normally during neurulation, many other developmental events must occur properly to convert this simple epithelial tube into a regionally organized, complex central nervous system. Thus, many opportunities exist for development to go awry and generate neural tube defects.

Overall, neural tube defects occur in about 1–8 per 1000 live births *(4)*, but this incidence is dropping with improvements in prenatal detection of neural tube defects (e.g., with ultrasound) and with periconception supplementation of vitamins containing folic acid *(5)*. Nevertheless, neural tube defects are still a major cause of postnatal morbidity and mortality and are responsible for significant health care costs. Continued research on mechanisms underlying neural tube defects offers hope that some day these serious developmental anomalies may be fully prevented.

Acknowledgments

Original work described herein from the Schoenwolf laboratory was supported by Grant Nos. NS 18112 and HD 28845 from the National Institutes of Health.

References

1. Willis, R. A. (1962) The Borderland of Embryology and Pathology, 2nd ed., Butterworth, London.

2. Norman, M. G., McGillivray, B. C., Kalousek, D. K., Hill, A., and Poskitt, K. J. (1995) Congenital Malformation of the Brain. Pathologic, Embryologic, Clinical, Radiologic and Genetic Aspects. Oxford University Press, Oxford.
3. Warkany, J. (1971) Congenital Malformations. Notes and Comments. Year Book Medical, Chicago.
4. Campbell, L. R., Dayton, D. H., and Sohal, G. S. (1986) Neural tube defects: a review of human and animal studies on the etiology of neural tube defects. *Teratology* **34,** 171–187.
5. Czeizel, A. E. (1996) Prevention of neural tube defects. *CNS Drugs* **6,** 399–412.

18

Experimental Manipulation and Morphometric Analysis of Neural Tube Development

Mary E. Desmond and Patricia A. Haas

1. Introduction

The morphology of the early embryonic vertebrate neural tube differs significantly in the brain and spinal cord anlages. Although both are hollow, the early embryonic brain has a larger cavity volume than tissue volume, whereas the spinal cord consists mostly of tissue and a reduced cavity. Growth of these two central nervous system (CNS) anlages therefore requires attention to both fluid and tissue dynamics as well as to tissues outside the CNS, namely the mesenchyme and skin ectoderm. By measuring the volumes of the tissue and cavity compartments of the brain as well as the cell number, rates of DNA synthesis, and mitosis of the neuroepithelium, it is possible to gain a better understanding of the cellular mechanisms involved in growth of the CNS. Volume assessment over time gives us an idea of change in size, whereas the details of cell behavior tell us how this changes occurs. Collectively, these changes in size and shape of the CNS are analyzed by morphometric techniques that, to date, have required performing measurements on serial sections of properly fixed and prepared embryos. Most likely in the near future, imaging techniques utilizing magnetic resonance proper will obviate the need to fix and section embryos. However, because these sophisticated imaging techniques are not yet available to mainstream research laboratories, this chapter focuses on performing morphometrics on sectioned and whole embryos.

Morphometrics refers to the measurement of anatomical structure and usually involves determining the volume of a tissue, counting the total number of cells in a particular area, and counting some type of uniquely labeled cells in the same area. These labeled cells may be those in mitosis, in DNA synthesis, in protein synthesis, or with some type of receptor. Because the embryo is a three-dimensional (3-D) structure, it is essential to assess the entire structure for morphometrics, not just a selected few sections to obtain an accurate assessment of growth. The embryonic brain offers a unique challenge to morphometric analysis because it changes shape while it increases in size. This necessitates measuring no fewer than every fourth section for volumes ***(1)***. This is true whether using a mechanical polarizing compensating planimeter or a computer software program, because they both calculate areas by integrating the areas of

From: *Methods in Molecular Biology, Vol. 136: Developmental Biology Protocols, Vol. II*
Edited by: R. S. Tuan and C. W. Lo © Humana Press Inc., Totowa, NJ

spheres or ellipses *(2)*. In fact, all of the morphometrics that are commonly used are based on principles of stereology, which is a body of mathematical methods that relates 3-D parameters defining a structure to 2-D measurements obtainable on sections of the structure *(2)*. One needs to know this at the onset and make the necessary corrections for the irregular shapes that the brain vesicles assume as development proceeds *(1)*. It is also important to fix the embryos with the same fixatives, as different fixatives are known to shrink tissues to different degrees *(3)*. On the other hand, it has also been established that shrinkage affects different stages and different embryos to the same extent *(4)*. Even though computer-assisted image analysis greatly speeds up the acquisition of morphometric data, it still is tedious; thus, we have adopted a method to calculate changes in brain volume without sectioning. This method is particularly valuable if the changes are extreme, as is often the case in pharmacological studies *(5,6)*. The method involves measuring diameters from photomicrographs of living embryos *(5)* and has the advantage of allowing other measurements of the embryo such as whole-mount *in situ* (WISH) of some particular mRNA. Finally, we have also adopted a method to assess physiological mechanisms of fluid transport across the neuroepithelium. This method produces neuroepithelial spheres that can be grown in miniwell plates in defined media *(7)*. Although the preparation of the spheres requires some skill in microsurgery, it has the advantage of using pure neuroepithelium and requiring only a diameter measurement to determine the volume. This chapter discusses basic embryo preparation and morphometric treatments from sections and photomicrographs of chick embryos in the first part of this chapter, followed by the more delicate preparation of neuroepithelial spheres (Chapter 19). The same techniques used on the chick embryo can be used for other vertebrate embryos and for organs other than the CNS.

2. Materials

1. Physiological saline (p. saline): 0.9% NaCl, or 0.9 g NaCl per 100 mL distilled H_2O. Use NaCl that is biologically pure and is noted as such on the bottle. Young embryos are particularly sensitive to impurities in the nonbiological grade NaCl.
2. Carnoy's fluid: 75 mL absolute alcohol plus 25 mL glacial acetic acid.
3. Various percentage ethanols: Use 95% ethanol to make lower percentage ethanols, treating it as if it were absolute ethanol. Aliquot absolute ethanol into smaller containers and seal with parafilm to assure it does not become hydrated. Ideally, purchase 500-mL bottles.
4. Tissue prep from Fisher Scientific, Pittsburgh, PA, catalog #T565: A purified paraffin plus synthetic polymer with a melting point of 56–57°C.
5. Slides and cover slips: Gold Seal #3050, from Becton Dickinson Labware (Rutherford, NJ); Corning No. 1.5 thickness cover slips (Corning Glassworks, Corning, NY).
6. Stains: Harris' Hematoxylin modified with acetic acid (mercury-free) (Fisher) and 2% Eosin Y.
7. Colchicine: FW 399.4, Sigma C-9754 (Sigma Chemical, St. Louis, MO).
8. 5-Bromo-21-deoxyuridine (BrdUr): Calbiochem 203806 (La Jolla, CA); analog of thymidine; MW = 07.1. BrdUr staining kit from Calbiochem, catalog #HCS24, utilizing a biotinylated monoclonal anti-BrdUr, thus eliminating the need for a species-specific antibody. The staining is visualized using a streptavidin–peroxidase and diaminobenzadine (DAB), which stains BrdUr incorporated into the DNA, thus staining the nuclei a dark brown.
9. ^{3}H-Thymidine: Specific activity needs to be very high to assure the most concentrated signal possible by the radioactive marked molecule. A specific activity > 40 C/mmol is common. Tritium must be the isotope used because of its short tracking distance. We use

methyl 1', 2',-^{3}H from Amersham (Arlington, Heights, IL), catalog #TRK.565, which is 100–130 Ci/mmol or 37 MBq/mCi.

10. Kodak NTB II emulsion: Catalog # 1654433 from International Biotechnologies, Inc. (New Haven, CT).
11. Siliconizing solution: Prosil 28, organosilane concentrate, product No. 11975-0, Thomas Scientific (Swedesboro, NJ) catalog #7805525.
12. D.P.X. a clear mountant available from Aldrich Chemicals (Milwaukee, WI) catalog #31, 761-6.
13. Embryo instruments: iris scissors; embryo transfer spatula (section lifter) from Carolina Biological Supply Co., Burlington, SC, catalog #K3-62-7510; stainless-steel dissecting probes; watchmaker's forceps (Dumont #55); medium-point, serrated forceps (high-quality instruments can be purchased from Fine Science Tools, Inc., Foster City, CA), disposable pipets with 7 in, tip; latex bulbs for disposable pipets in 1- and 2-mL sizes.
14. Lab-made instruments: Small glass rods (0.5 mm in diameter) made by pulling out the end of 1-mm glass rods and rounding off the end in a hot flame; hair loops made by attaching a small loop of baby hair to the end of a Pasteur pipet using paraffin. Be sure to use baby hair only because it is extremely fine. Even children's hair is too coarse. Surgical sponges: Coil up a 1-cm piece of Kimwipe (Kimberly-Clark Corp., Roswell, GA) into a tight wick and hold in place with a hemostat.
15. Microburner for pulling microglass rods and tubes: Attach a 16-gage needle to rubber tubing that is attached to a gas outlet for a standard-sized Bunsen burner. Seal the attachment site with electrical tape. *Cautionary note:* Be sure that the free end of the needle is pointing into free space, as the flame can be very long (10–12 in.) before adjusting it after you light the needle.
16. Assortment of sterile Petri plates: 90-, 60-, and 30-mm in diameter; standard 3- and 4-in. diameter glass culture dishes. To keep glass dishes sterile, cover the tops with aluminum foil (shiny side out) before autoclaving.

3. Methods

3.1. Embryo Preparation and Morphometrics from Sections

1. Precisely time the age of the embryo by staging it according to the criteria established by Hamburger and Hamilton ***(8)***. Counting the number of somites is the most accurate way to stage young embryos. In general, they add a discrete pair every hour between stages 8 and 26 ***(8)***.
2. Position the egg on its horizontal axis for at least 15 min to assure rotation of the embryo to the top of the yolk. Crack the shell by hitting the bottom of the egg on a sharp edge (e.g., the edge of a Petri dish). Dump the contents into a culture dish half-filled with physiological saline (p. saline) by placing both thumbs next to the crack and rotating your hands outward from the crack (**Note 1**).
3. Excise the embryo off the yolk by pinching the vitelline membrane outside of the embryonic disk on the left with a pair of medium forceps while cutting counterclockwise around the outside of the blastodisk, starting on the right side.
4. Transfer the blastodisk to a small Petri dish (60 mm) containing sterile p. saline by using a transfer pipet constructed from a disposable Pasteur pipet. To prepare the pipet, break off the narrow tip and put a 2-mL latex bulb on this end. Use the other end for transferring the embryo. Once transferred, swirl the dish gently to remove excess yolk and the vitelline membrane, which will appear as a translucent film. Then, transfer the embryo to another Petri dish containing p. saline.
5. Flatten the embryo by aspirating off the p. saline. Be sure that the embryo is alive when adding the fixative, Carnoy's fluid, to the dish. Be careful not to place the fixative directly

on top of the embryo, as it may cause the embryo to wrinkle. For chick embryos younger than 4 d, be sure that the blastodisk is flat after the fixative is added. Flattening is best achieved by aspirating around the entire circumference of the blastodisk using a disposable pipet. It is important to add ample fixative, dehydrants, and clearing agents to assure adequate processing. This is achieved by adding a volume of fluid that is at least 40X the volume of the embryo (e.g., immerse a stage 10 embryo in 15–20 mL of fluid and a stage 25 embryo in 25–30 mL).

6. Keep the embryos in fixative 10–12 h, being careful to not exceed 24 h. To prevent evaporation of the fixative, place the small Petri dishes inside a larger covered container such as a plastic food-storage dish. Do not cover containers containing fixatives with aluminum foil because the fixatives will cause the foil to disintegrate and the products will fall into the container with your specimens.
7. Remove the Carnoy's fluid using a disposable pipet. Add 70% ethanol. Swirl the embryo, then aspirate off the ethanol and add fresh 70% ethanol. The embryo may be stored in this solution for months as long as care is taken to avoid desiccation. This is a good time to trim the blastodisk into a rectangular shape. If planning to continue the dehydration series, leave the embryo in this first ethanol solution for 1 h. Ideally, it is best to place the samples on a shaker. You may want to add some eosin stain to the 70% ethanol so that the embryo will be easier to see when processing. The eosin will be removed once the sections are processed through a graded series of ethanols.
8. Continue the dehydration of the embryo by making the following changes of solutions at 1-h intervals: 70% ethanol; two changes of 95% ethanol; two changes of absolute ethanol. Now, transfer the embryo to a glass vial such as a scintillation vial and make the following changes: two changes of histosol/hemode (*see* **Note 2**); two changes of 50:50 hemode:paraffin; three changes of pure paraffin (*see* **Note 3**).
9. Embed the embryos for sagittal sections, as they are the easiest from which to discern parts of the embryonic brain accurately. Use a dissecting microscope to assure proper orientation of the embryo in the embedding mold (**Note 4**). Partially fill the mold and place the embryo to the bottom using a fine pair of forceps and a dissecting probe. Then, fill up the mold to its top and allow it to cool and solidify by putting it on top of a cooling platform.
10. Section the embryos at uniform thickness from 8 to 15 μm. Prior to sectioning, preclean slides to remove the oily film which is even present on commercial slides that are identified as precleaned (*see* **Note 5**). Place a uniform number on all slides. We put five sections in a row and two rows on a slide. This facilitates ease in accounting for all of the sections. If a section is lost, leave a space where that section should be. Once the embryo is completely sectioned, process the slides for staining (*see* **Note 6**) and mounting.
11. Screen the slides for the beginning and end of the sections containing brain. Circle every other section with a fine-point permanent marker. It helps to use a different colored marker to indicate the beginning and end sections from the color used to mark the interim sections.
12. To measure using a computerized image analysis system, you will need a high-resolution camera connected to a high-quality compound microscope and computer with at least 16 MB RAM and abundant ROM. We use a MT1-CCD72, B & W camera connected to a Nikon Optiphot-2 (Nikon/Image Systems Inc., Columbia, MD) via a trinocular head and transmitting a split signal to an Ikegami monitor (Pike Technologies, Madison, WI) and a Power Mac 7500/100 (Apple Computer, Cupertino, CA). The monitor is an extra output, allowing us to view large images while the computer is being used. It is wise to setup the computer with a backup system for both the images and spreadsheet of data as soon as

you begin. A Zip drive or Jazz drive are good ways to save the data, or a local network hard drive if you have access. A convenient program, NIH Image, can be downloaded for free from the Internet. At the moment, this program is only available for a Macintosh. However, Sigma Plot (Jandel Scientific, San Rafael, CA), a commercial enterprise, has one similar that is available for both Macintosh and PC.

13. Calibrate the computer for the objective and camera relay lens used for your measurements. Make sure the numbers make sense. A good way to check the output of your measurements is to measure a square of graph paper. You should also calibrate a mechanical polarizing planimeter the same way. Measure each section three times and only keep those measurements that are within 2–3% of each other. If you have several people measuring, which is most likely on a large project, make sure that their measurements agree with each other within 2–3% and check their results frequently. Weekly checks will save a lot of aggravation when you are trying to end a project. If your project requires many morphometric measurements, you will need to dedicate a computer to that purpose.
14. Set up the spreadsheet to calculate the volume from the sections. Remember, you need to account for the total number of sections as well as the thickness of the sections. Be sure that all of the units are consistent; that is, if your volume is in cubic micrometers, be absolutely certain that the thickness is in micrometers. The formula for determining the volume is (Sum of the area) × (Section factor) × (Thickness of the sections). If you are using a mechanical planimeter or computerized planimeter that requires you to measure images projected by the microscope, you will need to divide the value obtained above by the magnification factor cubed (*see* **Notes 7** and **8**).
15. Scan in the images to be measured into the computer, then identify regions that need measuring by using morphological markers. This is easy to do in an early chick brain because there are defined morphological boundaries between the first three vesicles and the next five vesicles. The boundaries are the telencephalomesencephalic fissure, the mesencephalorhombencephalic fissure, and the rostral edge of the otocyst. The program allows you to draw a straight line across from the indentation on the dorsal surface to that on the ventral surface. Several studies can be used as resources for morphological boundaries ***(3,9)***. Trace the outside perimeter of the brain and the inside. Then, subtract the inside from the outside. The outside tracing will give the total area of that brain section; the inside tracing will yield the total area of the cavity for that section. The difference between the two will be the total tissue area for that section (*see* **Fig. 1A**).
16. Determine the cell number of a particular portion of the brain (e.g., the midbrain) by counting nonrandomized quadrats of the midbrain throughout its entire rostral to caudal extent. It is important to include a cell count from the rostral, mid, and caudal portions of the embryonic brain because it has been well documented for the chick embryo that rates of cell proliferation differ for these areas and for different regions of the brain; that is, the rates are not identical for the forebrain and midbrain ***(10–12)***. In other tissues, a nonrandomized sampling might be required because the density differs in various regions. To assure that you are not counting the same cells twice you need to skip enough sections to account for the thickness of a cell (*see* **Note 9**). Place an ocular reticle into one eyepiece of a compound microscope and count cells within each small unit of the reticle. The reticle will come with its dimensions (e.g., a $1 \times 1\text{-mm}^2$ square divided into 10 units such that the area of each small unit is $0.1\ \text{mm}^2$). Count the number of cells that lie in each square using a mechanical cell counter to keep track. (Actually, you will be calculating the number of nuclei, as it is nearly impossible to discern cell membranes of embryonic neuroepithelial cells.) Compare the cell numbers obtained for the different regions. If

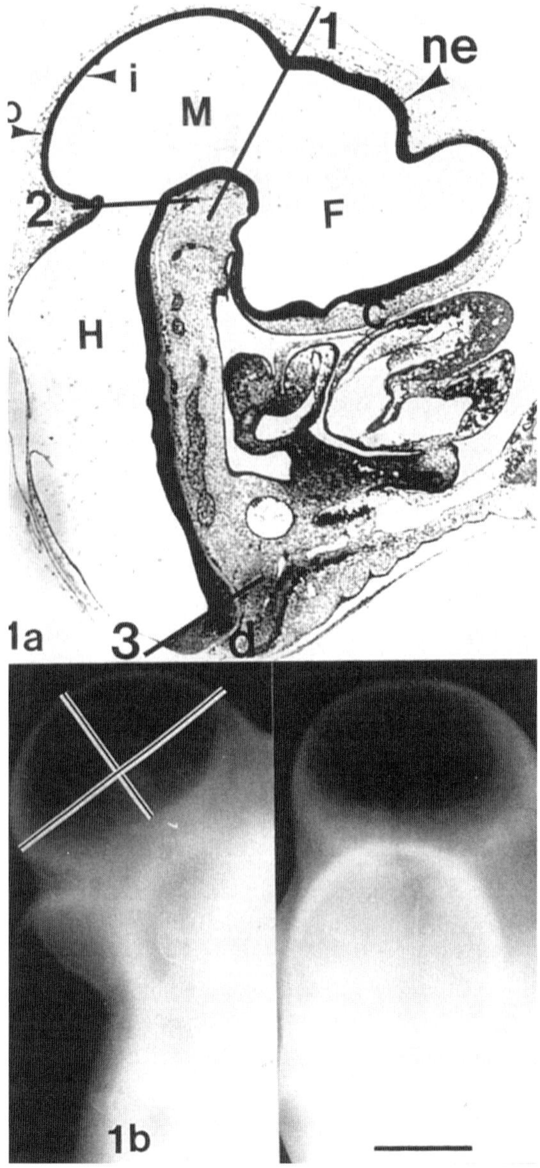

Fig. 1. The midsagittal section of a stage 24 chick embryo head in **(A)** illustrates the distinct borders separating the forebrain from the midbrain **(1)**, the midbrain from the hindbrain **(2)**, and the caudal end of the hindbrain **(3)**. The large spaces surrounded by the neuroepithelium (ne) are the presumptive ventricles. The inner lining or ependyma of the neuroepithelium is labeled "i" and the outer border or apical surface is labeled "o." The left and right photomicrographs in **(B)** feature lateral and dorsal views of the mesencephalon from a living, stage 24 chick embryo. In the lateral view (left view), the horizontal line = the anterior-posterior axis and the vertical line = the dorsal-ventral axis whereas a horizontal line drawn between the midpoints of the sides in the right view is the bilateral axis. These are the axes a, b, and c of an oblate spherioid used to calculate the volume.

they are not significantly different (i.e., if they do not vary by 5%), use them all as an average. Extrapolate the total number of cells for the entire midbrain based on the relationship (Number of cells)/1 mm^3 = Number of cells/(Midbrain volume in mm^3). You need to have calculated the volume of the midbrain as previously described. Finally, a cautionary note is important here: Cell proliferation does not always lead to immediate growth in embryos; therefore, the measurement of cell proliferation rates is an inadequate indicator of growth. This is why cell numbers reported in the context of tissue volume is so important.

16a. Another way to determine the cell number of the embryonic brain is to determine the total DNA content of the neuroepithelium using the standard diphenylamine reaction ***(13–15)*** and converting this value to diploid equivalents by dividing this value by 2.5 pg, which is the amount of DNA reported in chick diploid cells ***(16)***. Unfortunately, this technique has the major disadvantage of destroying the embryonic brains because they are homogenized in the process, and the minor inconveniences of requiring total isolation of the neuroepithelium from surrounding mesenchyme and skin ectoderm plus estimating the proportion of cells that are not diploid. Such is the case because the cell cycle is asynchronous in the embryonic neuroepithelium so that a certain proportion of neuroepithelial cells will be haploid and tetraploid. If you assume that cells have the 4 *N* amount of DNA during all of mitosis (M), G2, and half of synthesis (S), the actual number of cell can be obtained with the formula: $C = d/(1 + p)$ where d is the number of diploid equivalents and p is the proportion of cells expected to be 4 *N* (obtained by dividing the duration in hours of ½ S + G2 + M by the total generation time of the brain cells in hours at that stage ***(17)***.

16b. To determine the total DNA of a preweighed pooled sample of whole chick brains, (1) homogenize a known weight of tissue in p. saline with a hand-held Potter–Elvehjem homogenizer; (2) precipitate the macromolecules with cold 0.5*N* perchloric acid (PCA) followed by centrifugation (1290*g* for 15 min); (3) hydrolyze the nucleic acid (NA) with 0.5*N* PCA at 80°C for 40 min; (4) cool the tubes on ice for 5 min; (5) centrifuge and collect the supernatants for reaction with fresh diphenylamine reagent at 30°C overnight; (6) read the colored product formed at 595 and 700 nm visible light in a spectrophotometer; (7) determine the amount of DNA on a standard curve.

16c. To determine the number of diploid equivalents from the total DNA, calculate the amount of DNA per head by dividing the total amount of DNA by the number of heads in the pool. Then, divide this value by 2.5 pg, which is the mean of the highest and lowest values reported for chick diploid nuclei by Davidson et al. ***(18)***.

16d. If the tissues investing the brain are impossible to remove, then the above DNA concentration can be determined for a pool of whole heads from which the proportion of cells of nonbrain tissue is subtracted from total cell number. The proportion of brain and nonbrain cells can be determined by multiplying the cell density times the volume of that particular tissue. Cell density is determine by counting the number of nuclei in 100 squares of an ocular reticule in the dorsal and ventral regions of each of the brain vesicles included in the original brain sample excised and of regions of mesenchyme and epidermis adjacent to these brain regions. Volumes can be determined as described in **step 15**.

17. Determine the mitotic index or density by counting colchicine-induced metaphases per 100 cells or volume, respectively. Prior to sacrificing embryos, treat embryos with 0.2 mL of colchicine (3.25 mg/mL); that is, the embryo receives 0.65 mg of the colchicine. This concentration has been worked out experimentally in our lab. Do not inject the colchicine into the yolk, but, rather, inject it right on top of the embryo as detailed in **Note 10**. Treat the embryos for 2–24 h. If more than 24 h, be sure to account for the added age of the embryo in your analysis. After sectioning the embryos, count the number of metaphase

figures per total cells, as outlined in **step 16**. Then multiply this figure by 100. This is the mitotic index. Calculate the total number of metaphase cells per unit volume. This is the mitotic density.

18. Determine the DNA synthetic index and density by prelabeling the cells with either BrdUr or tritiated thymidine (^{3}H-thymidine). For many reasons, the most notable of which is the cost of getting rid of radioactive waste and long exposure times required, we prefer the BrdUr method. However, it is possible to use both to get a precise measure of the actual generation time of the cells in the population *(19)*. Inject the living embryo, as detailed in **Note 10**, with BrdUr *(19)* 2–6 h before sacrifice. Prepare sections as described and treat these sections with the primary antibody, anti-BrdUr, followed by a secondary antibody conjugated with horseradish peroxidase that will degrade the substrate diaminobenzidine, forming a brown–black product in the nucleus. Several kits are available for detection of BrDU. Although we have used the alkaline phosphatase–NBT, BCIP regime as well as the true blue substrate (3,3',5,5'-tetramethylbenzidine) by Kirkegaard and Perry Laboratories (Gaithersburg, MD), we find the kit by Calbiochem particularly reliable. We have also found the brown–black precipitate easier to detect against other staining. We use a very light hematoxylin counterstain so that nonlabeled cells can be detected. Once the cells are labeled, calculate the percentage of labeled cells and the volume to determine the DNA synthetic index and DNA synthetic density respectively. Be careful not to infer that DNA synthesis is an indication of mitosis. Synthesis and mitosis are two discretely separate phases of the cell cycle.
19. Label DNA using ^{3}H-thymidine by injecting the embryos 1–6 h prior to sacrifice with 0.2 mL of ^{3}H-thymidine having a high specific activity (i.e., 40 C/mmol). The high specific activity is to assure that an adequate signal will be generated by the nucleus. Thymidine is required because it has a very short tracking distance and, thus, it can be inferred that it is located within 0.2 μm from the place it is detected. Process the embryos for sectioning as already described. Once the sections are put on the slides, emulse them in liquid photographic emulsion (Kodak, Rochester, NY, NTBII). This must be done in a darkroom with only a **red safelight** on. Do not use the regular yellow darkroom safelights. Two to four hours prior to using the emulsion, place it in an incubator at 37°C or oven at 4°C. **Gently** shake the emulsion to see if it has liquefied. Then, take the following into the darkroom: a tissue flotation water bath preset to 48°C; preheated water for the bath; a heavy clamp to hold a plastic slide mailer tube upright in the water bath; test tube racks that can hold the slides once emulsed; a small container of distiller water; a red safelight; the premelted emulsion and slides to be emulsed. Add 7 mL of distilled water to the mailer, followed by 7 mL of emulsion. Close the cap of the mailer and gently invert it back and forth to mix. Be careful not to make bubbles that will interfere with the coating process. Place the mailer into the clamp and then put the apparatus into the water in the water bath. Be sure that the water covers at least two-thirds of the mailer height. Place two slides back to back and dip once into the diluted emulsion. Immediately pull apart and place vertically in separate spaces in the test tube rack for drying. Emulsing two slides together has the advantage of saving on emulsion by not coating the backside of the slides and cutting the time for emulsing in half. (A mailer full of emulsion will coat approximately 40 slides.) Once all of the slides have been emulsed and placed in a rack(s) to dry at room temperature (RT), place the rack(s) in a light-tight drawer or box to dry for 2–4 h. You may simply cover the racks with aluminum foil. Do not leave at RT for more than 4 h, as you risk getting excessive background. Now, transfer the dried, emulsed slides to slide boxes, label the box, and cover it with aluminum foil to prevent light leakage. Store the covered slide box at 4°C for 4–6 wk. After 3 wk, develop a test slide to see how much the emulsion has been exposed by the radioactivity.

20. Develop the slide(s) as follows. Have four containers for photographic chemicals. Use Coplin staining jars if only planning to develop a few slides; otherwise, use standard staining dishes with appropriate holders. The four containers will hold developer, distilled water (DW), and photographic fixer. In general, 200 mL of solution will process 100–200 slides before becoming exhausted and needing replacement. We prefer Kodak D-19 developer (Eastman Kodak, Rochester, NY) but many other labs use Dektol (Kodak) with excellent results. Whichever developer is used, be sure to make the concentration used for film. Set up the solutions with only a red safelight on and, once ready, open the slide box and place the emulsed slides into slide holders. Develop 5 min, rinse 5 min in DW, fix 5 min, and rinse for 5 min in DW. After all of the slides have been fixed, the slides may be exposed to regular room light. All of the emulsion should be removed from the slides because of the fixing process. Now the slides are ready to be stained and permanently mounted for viewing.

3.2. Morphometrics from Whole Embryos

1. Repeat **steps 1–7** of the above procedure. Thus, the embryos will be fixed and stored in 70% EtOH. Begin with embryos that are between stages 20 and 28 of development.
2. Prepare an embryo holder for the head as follows. Place a circle of black construction paper in the bottom of a 60-mm Petri dish and fill the dish with paraffin. Once the paraffin in solidified, use a small metal spatula or rounded glass rods to make the appropriate size and shape depression to hold the head in its various orientations: lateral, dorsal, ventral, and frontal planes. Small glass rods made from small capillaries are helpful to provide weights in various locations on the embryo. To keep the rod from sticking to the embryo, siliconize it (**Note 11**). The objective is to hold the head in its perfect orientation so that a photograph can be taken. All of the measurements will be taken from the photographs.
3. Three axes of the midbrain can be measured: median anteroposterior, dorsoventral, and bilateral from identically magnified photographs (**Fig. 1B**). Using these measurements, the volume of the midbrain can be calculated using the mathematical formula for an oblate spheroid, $(4/3)\ \pi\ abc$, where $a = c > b$ *(5)*. We have shown that the upper and lower hemispheres are noncongruent; however, volume differences based on the midbrain having congruent hemispheres vs. non-congruent hemispheres are negligible. Differences in volumes based on $a = c$ compared to those where $a = c$ are also negligible.
4. Setting up the embryo for photographing in the four planes is time-consuming and requires tremendous patience. Not only is positioning the embryo tedious, but setting up reflected lighting without glares is also. Each position takes about 20–30 min to be set up. Thus, to get all of the planes photographed for one embryo will take 80–120 min. However, the payoff is tremendous when you think that it takes a minimum of 4–5 d to determine the same information from sections. A major use of this technique would be for pharmacological studies in which one wants to compare brain volumes of treated embryos with untreated embryos. Another use might be that of measuring volumes of the midbrain from embryos, which had been used for WISH.

4. Notes

1. Avoid dumping the egg residue, yolks, and whites excluding shells down the sink without abundant rinsing of the pipes with excess cold water. It is very easy to clog long segments of piping by running hot water into a line that still has egg residue. The best way to discard of the egg residue is to dump it down the toilet or freeze it in a plastic bag and discard with animal waste. The shells can be disposed of by placing in a plastic bag and dumping in the trash. Check with the animal waste committee of the institution about egg disposal.

2. Hemode/histosol are both synthetic xylenes available from commercial vendors. They have the advantage of smelling less offensively but still do the same liver damage; therefore, it should be used in a chemical hood.
3. Be sure that the timing of the solutions containing either pure paraffin or partial paraffin are based on when the paraffin is melted. Be very careful to have the oven or hot block just 3–5°C above the melting temperature of the paraffin so as not to "cook" the embryos and lose all morphology. Finally, only leave the embryos in the melted paraffin for 1 h. However, you may store them in solidified paraffin at 4°C for as long as necessary.
4. Embedding molds and rings: The easiest system for embedding is to use stainless-steel base molds on top of which sits a plastic ring. Another commonly used system is the peel-away plastic molds. Use of these molds necessitates placing a paper identification tag into the top of the liquid paraffin before it solidifies. Both are readily available at either Fisher or VWR Scientific (Atlanta, GA).
5. Cleaning of slides: Place slides in slide carriers and transfer for 5 min each through the following solutions: three changes of sodium dichromate–sulfuric acid cleaning solution; three changes of distilled water; three changes of absolute alcohol. Air-dry by placing a beaker over the slide carrier with an inch gap at the bottom to allow airflow. Be sure to keep them covered or the slides will get dusty. Once dry, place the slides in a slide box for storage.
6. Preliminary to staining, deparaffinize the sections by placing them in a slide box, then placing the box into an oven at the melting temperature of the paraffin for no longer than 30 min. Remove to room temperature and allow to cool. Transfer the slides to a slide carrier and begin the chemical hydration process: (a) pass the slides through the following chemicals for 5-min intervals: three changes of histosol/hemode; two changes of absolute ethanol; two changes of 95% ethanol; two changes of 70% ethanol; one change of DW; (b) stain the slides 5–8 min in Harris' Hematoxylin that has been prefiltered; (c) rinse in running tap water until no color leeches out into the water bath; (d) dip 10–20 s in 1% acid alcohol; in running water bath for 15 s; in saturated lithium carbonate for 15 s; in running tap water for 15 s; (e) stain in alcoholic Eosin for 5–10 min; (f) rinse through three changes of 95% ethanol, 15–20 dips each; (g) transfer through two changes of absolute ethanol, 5 min each; finally, three changes of hemode for 5 min each. Mount with permount or DPX. Allow to dry thoroughly.
7. The magnification factor is determined by projecting the focused image of a stage micrometer onto a piece of paper and tracing the lines. It can also be photographed. Be sure that it is projected at the same magnification as the sections. This is the magnified image. Measure several (six to eight) of some known increment of the traced micrometer with a good ruler. Be sure to measure the same side of the lines, not the middle. Take the average of these measurements and you have the average magnified distance. Divide this value by the value given for the same increment on the stage micrometer. Often the smallest increment is 0.01 mm. If the magnified image of the smallest increment is 4.5 mm, then the magnification factor would be 4.5 divided by 0.01 mm, or 450.
8. The details may vary for your particular computer setup, but a stepwise procedure for capturing and measuring images with a Mac can be obtained from the Internet together with the NIH image.
9. Neuroepithelial cells are elliptical in shape in early embryos. Thus, we measured the short and long axes of several cells. We took the average of each and then calculated the volume of the cell. A best approximation of the thickness of a cell is 25 μm based on the longitudinal axis. Thus, if the sections are cut at 10 μm, at least 2.5 sections need to be skipped in between the ones used for cell counts. A common comparator for cell size is to remember that a human red blood cell is 8 μm in diameter.

10. Candle the fertilized egg and, with a pencil, draw a circle on the shell that indicates the edge of the blastodisk. Swab this inscribed area with an alcohol sponge, then make a small hole (0.5 mm) in the shell using a sanding disk attached to a Dremel drill (Dremel, Racine, WI). Insert the end of a 18-gage needle that is attached to a 1.0-mL sterile syringe containing the colchicine solution. Slowly inject the colchicine. Remove the needle and seal the hole with 3M transparent tape.
11. Dip glass probes for 2 s into a 1:100 diluted Prosil solution (Thomas Scientific) followed by 2-s dips in three changes of DDW. Place the wet probes onto the right half of an opened Kimwipe that has been placed on the bottom of a plastic container. Cover the probes with the left half and place more siliconized probes on the top of the left half. Add another opened Kimwipe for more tips. Cover the plastic container when finished to keep tips clean.

Acknowledgment

The authors thank Antone G. Jacobson for introducing one of us (MED) to the art of embryo microsurgery and the intricacies of morphometric analyses. Both authors thank the many graduate and undergraduate students, especially Bernard Martin, who have developed techniques for working with embryos. This work is supported by grants from the NIH HD18143 and NS24136.

References

1. Desmond, M. E. and Jacobson, A. G. (1977) Embryonic brain enlargement requires cerebrospinal fluid pressure. *Dev. Biol.* **57,** 188–198.
2. Weibel, E. R. (1979) *Stereological Methods. Vol. 1. Practical Methods for Biological Morphometry.* Academic, New York.
3. Desmond, M. E. and O'Rahilly, R. (1981) The growth of the human brain during the embryonic period proper. 1. Linear axes. *Anat. Embryol.* **162,** 137–151.
4. Summerbell, D. (1976) A descriptive study of the rate of elongation and differentiation of the skeleton of the developing chick wing. *J. Embryol. Exp. Morphol.* **35,** 241–260.
5. Desmond, M. E., New, M. S., Martin, B. G., and Fleischman, W. M. (1990) A rapid reliable calculation of brain expansion in living chick embryos to reflect fluid transport across the neuroepithelium. *Anat. Rec.* 226, 34A.
6. Gibson, K. D., Segen, B. J., and Doller, H. J. (1979) B-D-xylosides cause abnormalities of growth and development in chick embryos. *Nature* **273,** 151–157.
7. Desmond, M. E. and Field, M. C. (1990) Bubble-like minispheres of avian embryonic neuroepithelium: a model for studying fluid transport. *Soc. Neurosci. Abstr.* **16,** 1150A.
8. Hamburger, V. and Hamilton, H. (1951) A series of normal stages in the development of the chick embryo. *J. Morphol.* **88,** 49–92.
9. Pacheco, M. A., Marks, R. W., Schoenwolf, G. C., and Desmond, M. E. (1986) Quantification of the initial phases of rapid brain enlargement in the chick embryo. *Am. J. Anat.* **175,** 403–411.
10. Kallen, B. (1961) Studies on cell proliferation in the brain of chick embryos with special reference to the mesenchephalon. *Z. Anat. Entwicklungsgesch.* **122,** 388–401.
11. Cowan, W., Martin, A., and Wenger, E. (1968) Mitotic patterns in the optic tectum of the chick during normal development and after early removal of the optic vesicle. *J. Exp. Zool.* **169,** 71–92.
12. Wilson, D. B. (1973) Chronological changes in the cell cycle of chick neuroepithelial cells. *J. Embryol. Exp. Morphol.* **29,** 745–751.
13. Burton, K. (1956) A study of the conditions and mechanism of the diphenylamine reaction for the colorimetric estimation of deoxyribonucleic acid. *Biochem. J.* **62,** 315–323.

14. Giles, K. and Myers, A. (1965) An improved diphenylamine method for the estimation of deoxyribonucleic acid. *Nature* **206,** 93.
15. Dische, Z. (1930) Ueber einige neue characteristische farbreaktionen der thymonukleinsaure und eine mikromethode zur bestimmung derselben in tierischen organen mit hilfe dieser reaktionen. *Mikrochemie* **8,** 4–32.
16. Desmond, M. E. (1985) Reduced number of brain cells in so-called neural overgrowth. *Anat. Rec.* **212,** 195–198.
17. Wilson, D. B. (1974) The cell cycle of ventricular cells in the overgrown optic tectum. *Brain Res.* **69,** 41–48.
18. Davidson, J., Leslie, I., Smellie, R., and Thomson, R. (1950) Chemical changes in the developing chick embryo related to the desoxyribonucleic acid content of the nucleus. *Biochem. J.* **47,** XL.
19. Hyatt, G. A. and Beebe, D. C. (1992) Use of a double-label method to detect rapid change in the rate of cell proliferation. *J. Histochem. Cytochem.* **40,** 619–627.

19

Isolation of Neuroepithelium and Formation of Minispheres

Mary E. Desmond and Marcia C. Field

1. Introduction

The brain of the early chick embryo initially grows by expansion of its fluid-filled cavity much as a balloon expands *(1–3)*. Such expansion occurs rapidly over 3–4 d, beginning at around 43 h of development *(1,3)*, and is under the direct control of internal fluid pressure, as shown from brain intubation experiments *(1,4–8)*. Intraluminal pressure is generated by the accumulation of fluid within a closed system and promotes expansion of the brain. The source of the fluid within the early neural tube still remains unknown, although Weiss *(9)* reported that the neuroepithelium is secretory as soon as it forms a tube. Weiss's experiments are only suggestive, however, and need to be repeated using contemporary methods in order to establish for certain whether the neuroepithelium is secretory, and if it is, at what stage it first becomes so. In addition, no information exists about how fluid crosses the neuroepithelium during this early expansion period. Although several reports exist summarizing the secretory nature of the choroid plexus, the choroid plexus does not develop until much later in embryogenesis.

Our laboratory has adopted a technique devised by Stern et al. *(10)*, in which he cultured ministrips of embryonic epiblast that became fluid-filled spherical vesicles. We cultured explants from whole-head segments and strips of either neuroepithelium or skin ectoderm *(11)*. Although we found that the whole-head segments sorted out into two primary tissues and configurations (a mesenchymal glob and ectodermal fluid-filled sphere [minispheres], we were unable to distinguish those spheres that were purely epiblast from those that were purely neuroepithelium. To address this question, we developed a microsurgery technique that produces pure neuroepithelium and epiblast strips. Although such a technique requires some skill in microsurgery, it provides pure neuroepithelial minispheres that can be cultured long term and used for various physiological and pharmacological studies involving assessment of fluid transport across the neuroepithelium.

2. Materials

1. Crude trypsin: Type III from Sigma (St. Louis, MO), catalog #T8253. Prepare immediately or within the hour of use.

From: *Methods in Molecular Biology, Vol. 136: Developmental Biology Protocols, Vol. II*
Edited by: R. S. Tuan and C. W. Lo © Humana Press Inc., Totowa, NJ

2. Tungsten needles: Prepare these needles from the finest tungsten wire available (i.e., that used in electric light filaments). The outside diameter of this wire should be no greater than 0.2 mm. Cut a 3-in. segment from the wire supply and thread it through a 23-gage needle, allowing 5–10 mm to extend from the sharp beveled end of the needle. Taper one end of a 6- to 8-in. length of wood dowel to fit into the plastic end of the needle used to attach the needle to a syringe. Be sure that the dowel is at least 6 in long so that it is comfortable as a handle for doing fine manipulations. Insert the tapered end into the dowel, wrap the excess tungsten wire around the wood dowel near the junction with the plastic end of the needle, and then wrap the junction site with electrical tape. Electrical tape is preferable to other tapes because it has more stretch and makes a tight wrapping around the junction. We like to use enough tape to make a comfortable holding position along the dowel. This means that the thickness where you grasp the dowel is like that of a comfortable pen. Once the wire is secured into the dowel, sharpen the free end of the wire by inserting it in and out of a solution of saturated sodium nitrite. In order for ionic displacement to occur, either the wire or the sodium nitrite needs to be very hot. It is easier and faster to sharpen the wire by heating it as it is inserted. This is done by applying an electrode to the metal portion of the needle while attaching the other part of the electrode to a reference carborundum rod from a D-cell battery. The easiest source for the electrical reference and sample electrodes is to use a power supply for an external microscope light (e.g., Bausch and Lomb, Rochester, NY, 125 V). Make the reference electrode by attaching copper wire around the corborundum rod and bending the free end of the wire into a hook that can be hung at the top of a 100-mL straight-sided glass beaker holding the saturated sodium nitrite. Once the needle is attached to the sample electrode and the reference carborundum rod is inserted into the solution, turn on the power of the power supply to 12 V and insert the last 2 mm of the tungsten wire in and out of the sodium nitrite solution. Bubbles will appear at the end of the wire. Continue doing this until you get the thinness of point that you desire. *Caution:* Do not simply leave the end segment of the wire in the solution while current passes through it, because the wire will be uniformly displaced and become a thin filament too flexible for using as a cutting device. By inserting the end of the wire in and out the solution, a tapered end like a pencil point is formed. The saturated solution of sodium nitrite can be used over and over for years. Although the preparation of these tungsten needles takes some time, the results are cutting needles that keep their sharpness for several operations and never break, in contrast to glass needles that dull more quickly and easily break. To store these needles, insert the base of the wood handle into a 4 × 5 × 2 in. piece of styrofoam and cover the entire construct with an inverted plastic beaker. Another storage method involves putting the wood base into a 15-mL plastic conical tube, covering the rest of the needle with another conical tube, and taping the two tubes at the junction. This forms a protector similar to a cigar storage cylinder. To keep the tungsten needle from moving back and forth, place some styrofoam in the bottom of the protective tube that will hold the base securely. Finally, the tungsten can be sharpened by boiling the solution of saturated sodium nitrite. Although both methods can be dangerous, this one emits noxious fumes and takes at least 10 time longer than the electrical method. The solution needs to be in a porcelain crucible and the heat needs to be at a constantly high temperature. It is essential to use in a chemical hood and heat-protective gloves.
3. Hanks balanced salt solution, modified (10X) Catalog #19-101 (without $NaHCO_3$ from Flow Laboratories, Inc. (Rockville, MD, USA).
4. Enriched balanced salt solution: 94 mL Medium 199 plus 2.2 g $NaHCO_3$ plus 5 mL fetal bovine serum or 5 mL Chang C medium plus 1 mL penicillin (25 units/mL), streptomycin

(0.25 mg/mL). Penicillin–streptomycin is from Gibco-BRL, Gaithersburg, MD, catalog # 600-5140-AG.

5. Medium 199 (10X) is from Gibco-BRL, catalog #330-1180AG (100 mL).
6. Chang C medium: Catalog #T104 from Irvine Scientific (Santa Ana, CA). Developed for the primary culture of human amniotic fluid, this fluid contains salts, glucose, amino acids, polypeptides, vitamins, ribonucleotides, deoxyribonucleotides, sodium pyruvate, steroid hormones, and bovine serum proteins.
7. Miniwell plates with 24 wells.

3. Methods

3.1. Minisphere Formation from Isolated Head Segments

1. Repeat **steps 1–4** inclusive of the methods in Chapter 18. This technique can be applied to chick embryos of HH stages 7–15. Use sterile techniques throughout (*see* **Note 1**). Aspirate off the p. saline using a Pasteur pipet and add 2–3 mL of 1% *crude* trypsin in p. saline. Place the Petri dish on an ice bath, being careful that the water of the bath does not get into the Petri dish. A standard glass embryo culture dish makes a nice container for the ice bath. Incubate at 4°C for 20 min, followed by five rinses in Hanks' balanced salt solution (BSS) maintained at room temperature (17–18°C).
2. Flatten the embryo by aspirating around the perimeter of the blastodisk with a Pasteur pipet. Hold the head of the embryo in position by placing the tines of a pair of straight tipped forceps on either side of the head. Draw the point of a sharp tungsten needle along the boundary between the forebrain and midbrain, midbrain and hindbrain and at the caudal border of the hindbrain (cf. **Fig. 1** in Chapter 18).
3. Transfer each of the head isolates into separate miniwells of a 24-well culture plate containing 1 mL of BSS plus 10% calf serum plus 1% penicillin and streptomycin. Several (four to five) head isolates may be cultured in one well.
4. Place the miniwell cultures in a 37°C humidified incubator, biological oxygen demand. Check daily for the presence of minispheres. The minispheres take a minimum of 24 h to form and will continue living for as long as 2 wk if the medium is changed daily. The viable minispheres will appear transparent and under tension. The volume can be monitored on the living spheres by measuring the diameter of the sphere using an ocular micrometer and applying the formula, $V = 4°\pi = 3.1416$ and r = radius = ½ d (*see* **Note 2**).
5. For a permanent record, the minispheres can be photographed through a dissecting microscope using transmitted light, then printing the photographs. The diameter can be easily measured from the photomicrographs. However, now the volume must be corrected for the magnification factor in the same manner as described in **Note 7** of Chapter 18.

3.2. Preparation of Pure Neuroepithelial and Skin Ectoderm Explants

1. Repeat **steps 1–4** of the method described in Chapter 18 and **step 1** above using stage 5 to stage 15 embryos.
2. Prepare the embryo explant by cutting away all of the tissue lateral, anterior and posterior to the embryonic axis. First, using a sharp tungsten needle, transect the top edge of the embryo axis by cutting an arc at the border of the top of the head, extending laterally along the top edge of the skin ectoderm. Then, on both sides of the embryo axis proper, cut away the lateral tissues by making a longitudinal cut parallel to the embryo axis at the border of the skin ectoderm with a less differentiated epiblast (lateral cut). Next, make a longitudinal cut parallel to the embryo axis at the border of the neuroepithelium with the skin ectoderm (medial cut). Finally, make a posterior cut across the longitudinal segment of the embryo axis at the level of Hensen's node. There are now three distinct segments of

tissue present in the Petri dish: two fairly identical rectangles of skin ectoderm atop endomesoderm plus the embryo axis proper, also consisting of neuroectoderm (topmost layer), mesoderm (middle layer), and endoderm (innermost layer).

3. Separate the topmost layer of neuroectoderm of the embryonic axis from its underlying endomesoderm layers by holding the endomesoderm down with a dissecting probe and gently pulling away the neuroectoderm with another set of forceps. Once the neuroectoderm of the entire length of the neural plate has been separated from its underlying tissues, cut away the underlying tissues using a tungsten needle. Then, place the neural plate down on the Petri dish surface and transect it from the rest of the embryo axis at the superior level of Hensen's node. Transfer this pure neuroectoderm to a miniwell containing 1 mL of enriched BSS by using a Pasteur pipet (*see* **Fig. 1A–F**).
4. Similarly, separate the top layer of skin ectoderm from each of the previously prepared rectangles of embryonic tissue lateral to the neural tube axis by holding the endomesoderm down onto the surface of the Petri dish with a dissecting probe while pulling away the skin ectoderm. Then, place these two rectangles of skin ectoderm into a miniwell containing enriched medium.
5. Monitor the cultures of pure neuroepithelial and skin ectoderm explants daily for lack of contamination, viability of the tissues, and formation of minispheres. The skin explants will most likely form minispheres after 1 d in culture, whereas the neuroepithelial explants will take a minimum of 2 d. Minispheres produced from each of the tissues can be cultured long term (i.e., at least 2 wk and no doubt longer if the medium is changed every other day). Volumes can be determined from these minispheres exactly as described in **steps 4** and **5** of **Subheading 3.1.**

4. Notes

1. Although embryo cultures are more forgiving than mammalian cell culture, sterile technique needs to be quite rigidly followed. The following procedures should be followed as a minimum. Ideally, work in a laminar-flow hood. Whereever you work, swab the working surface with 70% ethanol before working. Have the microscope positioned close to the incubator. Use sterile glassware or plasticware throughout. Sterilize the tungsten needles in between touching the embryo by dipping them in soapy water, followed by distilled water (DW), followed by ethanol, followed by flaming in a low Bunsen burner flame or alcohol flame. We use 100-mL glass beakers for our sterilizing solutions. Sterilize all instruments using this method. This is very important. The reason we use soapy water is because chick embryo tissue is very sticky and does not easily come off with just water. Keep your nose away from the operating site (i.e., use the power of your eyes to see and do not put your face down into the culture). Wear a mask and gloves, as the procedure takes 20–30 min.
2. Because the micrometer is inside the ocular, it will not change size when different objectives are used. However, the diameter of the minisphere will increase in length as the sphere is magnified by a higher-power objective. Therefore, to determine the amount of magnification, multiply the power of the objective by the power of the ocular, then divide the diameter length by this number. Then, calculate the actual volume by the formula noted in **item 4** of **Subheading 3.1.**

Acknowledgment

Both authors thank the many graduate and undergraduate students who have developed techniques for working with living chick embryos. This work is supported by grants from the NIH HD18143 and NS 24136.

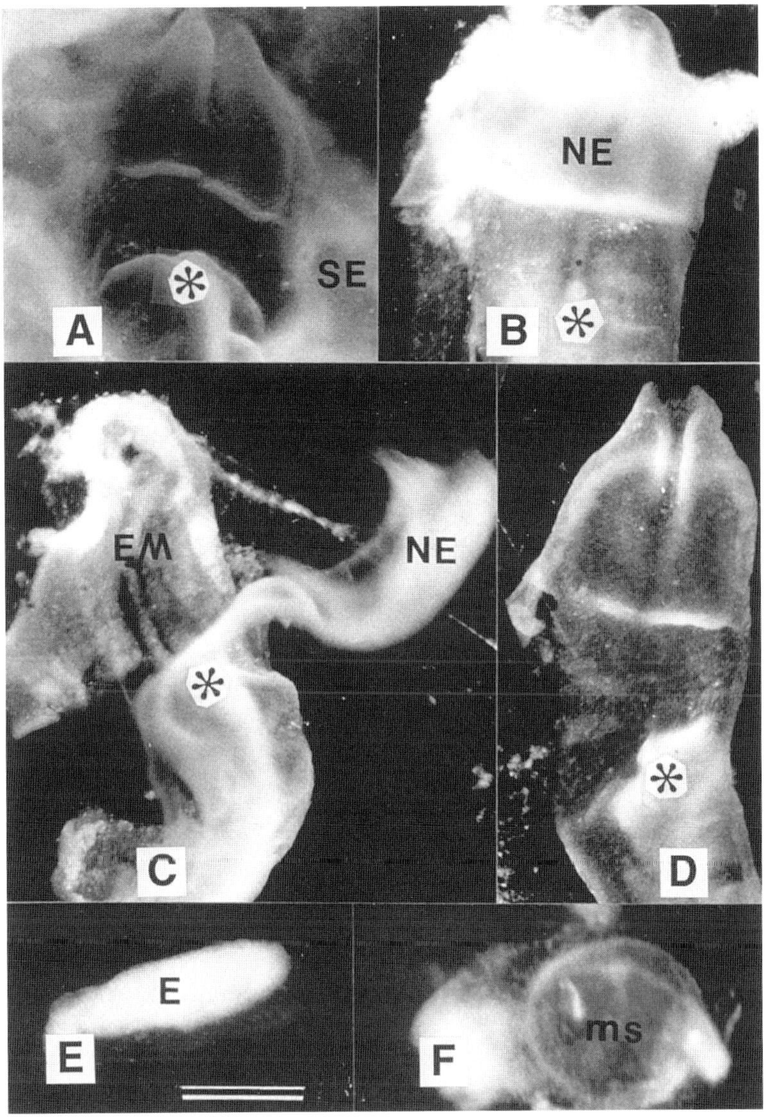

Fig. 1. This series of photomicrographs illustrate the formation of pure neuroepithelial minispheres **(F)**. The embryo in **(A)** is a stage 5 embryo, dorsal side up, that has been excised off the yolk, ready for isolation of the neuroectoderm from the underlying endomesoderm. Hensen's node is marked by the asterisk. View **(B)** shows the absence of the skin ectoderm lateral to the neural plate as well as the neuroectoderm pulled away from the underlying endomesoderm. In **(C)**, the neuroectoderm (right side of the forked tissue) is clearly identified, apart form the underlying endomesoderm on the left. View **(D)** shows the isolated neural ectoderm before it has been cut away from the embryonic axis at the superior level of Hensen's node. This is the same piece of tissue shown on the rights in view (C). Once the isolated neuroectoderm is put into liquid culture, it curls up as shown in **(E)** and after 2 d, it sorts out into a hollow neuroepithelial minisphere (ms in F) with associated extracellular material (glob). The bar represents 2.50 mm for (A)—(D) and 2.54 mm for (E) and (F). SE = skin ectoderm; NE = neuroepithelium (neural ectoderm); E = explant; ms = minisphere; asterisk = Hensen's node.

References

1. Desmond, M. E. and Jacobson, A. J. (1977) Embryonic brain enlargement requires cerebrospinal fluid pressure. *Dev. Biol.* **57,** 188–198.
2. Desmond, M. E. (1985) Reduced number of brain cells in so-called neural overgrowth. *Anat. Rec.* **212,** 195–198.
3. Pacheco, M. A., Marks, R. W., Schoenwolf, G. C., and Desmond, M. E. (1985) Quantification of the initial phases of rapid brain enlargement in the chick embryo. *Am. J. Anat.* **175,** 403–411.
4. Coulombre, A. J. and Coulombre, J. L. (1956) The role of intraocular pressure in the development of the chick eye. 1. Control of eye size. *J. Exp. Zool.* **133,** 211–225.
5. Coulombre, A. J. and Coulombre, J. L. (1958) The role of intraocular pressure in the development of the chick eye. IV. Corneal curvature. A.M.A. *Arch. Ophthalmon.* **59,** 502–506.
6. Coulombre, A. J. and Coulombre, J. L. (1958) The role of mechanical factors in the brain morphogenesis. *Anat. Rec.* **130,** 289–290.
7. Jelinek, R. and Pexieder, T. (1968) The pressure of encephalic fluid in chick embryos between the 2nd and 6th day of incubation. *Physiol. Bohemoslav.* **17,** 297–305.
8. Jelinek, R. and Pexieder, T. (1970) Pressure of the CSF and the morphogenesis of the CNS. *Folia Morphol.* **18,** 102–110.
9. Weiss, P. (1934) Secretory activity of the inner layer of the embryonic mid-brain of the chick as revealed by tissue culture. *Anat. Rec.* **58,** 299–302.
10. Stern, C., Manning, S., and Gillespie, J. (1985) Fluid transport across the epiblast of the early chick embryo. *J. Embryol. Exp. Morphol.* **88,** 365–384.
11. Desmond, M. E. and Field, M. C. (1990) Bubble-like minispheres of avian embryonic neuroepithelium: A model for studying fluid transport. *Soc. Neurosci. Abstr.* **16,** 1150A.

20

Examination of Normal and Abnormal Placentation in the Mouse

Michael R. Blackburn

1. Introduction

Placental development, also known as placentation, is orchestrated by precise molecular and cellular interactions between extraembryonic cells and cells of the pregnant uterus. The major extraembryonic cell type of the placenta is the trophoblast cell, which arises from the trophectoderm cell lineage *(1)*. Trophoblast cells make the physical connection between the embryo and the maternal environment and play important roles in the implantation process and placental function *(2)*. In mice, trophoblast cells lead the embryonic invasion into the uterine stroma to achieve implantation, and contribute to the formation of a temporary yolk-sac placenta (choriovetelline placenta) during early postimplantation stages. Trophoblast cells eventually provide the major extraembryonic cell lineage that contribute to the formation of the mature chorioallantoic placenta *(3)*. In the chorioallantoic placenta, the fetal and maternal circulatory systems are brought into close apposition to allow for a range of physiological interactions that are critical for fetal growth and development. Abnormal placentation is often associated with early embryonic mortality *(4–6)* and can lead to serious pregnancy disorders such as preeclampsia *(7)*.

The importance of placental development in the mouse has recently been highlighted by the targeted disruption of genes that lead to placental phenotypes that, in turn, cause embryolethality *(4–6)*. An example of this is the targeted disruption of the *mash-2* gene, which encodes a basic helix–loop–helix transcription factor that is expressed in developing trophoblast cells *(4)*. Absence of this transcription factor results in the absence of a population of trophoblast cells known as spongiotrophoblast cells, which in turn, results in abnormal placentation and the death of the embryo by mid-gestation. In addition, genes that have not previously been known to serve a role during placentation are found to be important when their disruption leads to a placental phenotype *(6)*. It is therefore critical that researchers closely investigate the placenta when assessing the phenotypic outcome of gain of function or loss of function mutations in transgenic mice. Typically, extraembryonic tissues are discarded in the process of dissecting the embryo or fetus from the uterus for analysis. In this chapter, a procedure will be

From: *Methods in Molecular Biology, Vol. 136: Developmental Biology Protocols, Vol. II*
Edited by: R. S. Tuan and C. W. Lo © Humana Press Inc., Totowa, NJ

described for the processing and analysis of gestation sites with a particular emphasis on analyzing intact extraembryonic tissues for the analysis of maternal–embryonal cellular interactions. In addition to describing the techniques for processing gestation sites for histopathological analysis, a technique for studying giant trophoblast cell populations will be presented.

2. Materials

2.1. Fixation and Embedding of Gestation Sites

1. Phosphate buffered saline (PBS): Make 10X stock solution: Start with 800 mL H_2O, add 80 g NaCl, 2 g KCl, 14.4 g $NaHPO_4 \cdot 7H_2O$, and 2.4 g KH_2PO_4. pH to 7.4 using HCl. Bring volume to 1 L with H_2O, and autoclave. The week of use, make 1X working buffer and autoclave.
2. 2*M* NaOH. 8 g NaOH into 100 mL H_2O.
3. 2*M* HCl. 16.5 mL 12.1*N* HCl into 83.5 mL H_2O.
4. 4% Paraformaldehyde (PFA) in PBS (*see* **Note 1**). Heat 90 mL H_2O and 100 µL 2*M* sodium hydroxide to 50°C in a microwave oven. Add 4 g PFA (Fisher Scientific, Pittsburgh, PA) avoiding inhalation (wear mask) and stir in a hood until the PFA is dissolved. Add 10 mL of 10X PBS, mix well, and then filter solution through Whatman filter paper. Let the solution cool to room temperature and pH to 7.4 by adding 100 µL 2*M* HCl. Store tightly sealed at 4°C and use for up to 1 wk.
5. 20-mL glass scintillation vials (Kimball [Kimble/Kontes, Vineland, NJ]).
6. Graded ethanols.
7. Histoclear (National Diagnostics, Atlanta, GA) or Xylenes (Fisher).
8. Paraffin (Parablast, Fisher).
9. Embedding molds (Fisher).
10. Positively charged microscope slides, cover glasses (Fisher).
11. Slide-staining racks and dishes (Baxter, Muskegon, MI).
12. Fine scissors, blunt forceps, and a small hemostat (Thomas Scientific, Swedesboro, NJ).
13. 35-mm Plastic Petri dishes (Fisher).

2.2. Monitoring Placentation

1. Hematoxylin and Eosin Staining Kit (Shandon, Inc., Pittsburgh, PA).
2. Paramount mounting media (Fisher).
3. Hoechst, bisbenzimide (Sigma, St. Louis, MO). Prepare a 10-mg/mL stock solution by dissolving 100 mg bisbenzimide in 10 mL dimethylsulfoxide (DMSO, Sigma). Aliquot into 500 µL samples in 1.5-mL microfuge tubes, wrap tubes in foil, and store at –70°C until needed.
4. Methyl salicylate (Sigma).
5. Canada balsam (Sigma). Prepare Canada balsam mounting solution by mixing 5 g Canada balsam with 10 mL methyl salicylate. Store at room temperature in a light-protected container such as a bottle wrapped in foil.

3. Methods

3.1. Timed Pregnancies and Dissection of Gestation Sites

1. Mate two or three females (at least 7 wk of age), with one male (at least 7 wk of age) (*see* **Note 2**). Place females into the male's cage at 16:00 and check for the presence of a vaginal plug the following morning (*see* **Note 3**). Detection of vaginal plug designates 0.5 d post coitum (dpc) (*see* **Note 4**). Keep plugged females caged apart from males.

2. At the desired stage of pregnancy, sacrifice mothers by cervical dislocation or carbon monoxide asphyxiation. Place sacrificed animal on its back and drench animal with 80% ethanol.
3. Lift the skin at the base of the abdomen (between hind legs) with blunt forceps and cut through the skin using dissecting scissors, revealing the abdominal muscle wall. Next, grasp the muscle wall, lift, and cut a hole in the muscle wall into the abdominal cavity. Continue to lift the muscle wall and cut through both the muscle and skin up both sides of the abdomen and resect away the muscle and skin to fully expose the abdominal cavity. Take care not to cut any internal organs or the uterus. The uterus can be located by lifting the intestines and viscera up and forward, out of the way. Locate the cervix at the base of the abdomen between the hind legs, grasp it with forceps, and cut between the cervix and vagina. Lift the cervix and cut along the underside of both uterine horns, cutting through the mesentery and uterine artery on the dorsal side of each uterine horn without cutting through the myometrium of the uterus itself. Remove the uterus by cutting between the ovary and the uterus. Place the intact uterus containing gestation sites into a 100-mm Petri dish containing ice-cold PBS.
4. Using forceps and scissors, trim away excess fat and vessels from the mesometrial side of the uterus (*see* **Note 5**). Attach a small hemostat to the cervix to help stabilize the uterus. Next, grasp the end of a uterine horn, stretch it slightly, and cut between each gestation site (*see* **Fig. 1A**). This will generate individual gestation sites contained in the myometrium (*see* **Note 6**). Rinse separated gestation sites in fresh ice-cold PBS before transferring to fixative.

3.2. Tissue Fixation, Embedding, and Viewing

1. Place gestation sites into glass scintillation vials containing ice-cold 4% PFA in PBS (approx 20 mL). The number of gestation sites in each vial will depend on the stage of gestation (*see* **Note 7**). Fix gestation sites overnight at 4°C, with gentle rocking.
2. Wash gestation sites 2 × 30 min (*see* **Note 8**) in PBS, rocking at 4°C.
3. Dehydrate samples by incubating for 2 × 30 min, rocking at 4°C in a mixture of 30% ethanol in PBS, 50% ethanol in PBS, 70% ethanol in H_2O, 85% ethanol in H_2O, a mixture of 95% ethanol in H_2O, and 100% ethanol. Gestation sites can be stored at this stage at –20°C or cleared and embedded.
4. Clear gestation sites by incubating 20 min, rocking at room temperature in a 1:1 mixture of ethanol and histoclear (or xylenes; *see* **Note 9**), followed by 2 × 20 min in 100% histoclear, rocking at room temperature.
5. Infiltrate gestation sites with paraffin by incubating 20 min (*see* **Note 9**) at 56°C in a 1:1 mixture of histoclear and paraffin, followed by 3 × 30 min at 57°C in 100% paraffin.
6. Embedding sections: Warm paraffin wax ahead of time by placing in a beaker overnight at 58°C. Do not overheat wax and discard heated wax after 2–3 d.
7. Fill embedding mold with melted paraffin. Using forceps, heated slightly over an ethanol flame, immediately remove a single gestation site and place in the mold. Using heated forceps, orient the gestation site such that the lumen of the uterus is facing the bottom of the mold (like a football standing on end, *see* **Fig. 1B** and **Note 10**). Do not allow the formation of air bubbles. If the gestation site is not oriented properly, remelt the paraffin block at 58°C and re-embed.
8. Allow blocks to cool to room temperature. Gestation sites in the blocks can be stored at room temperature indefinitely.

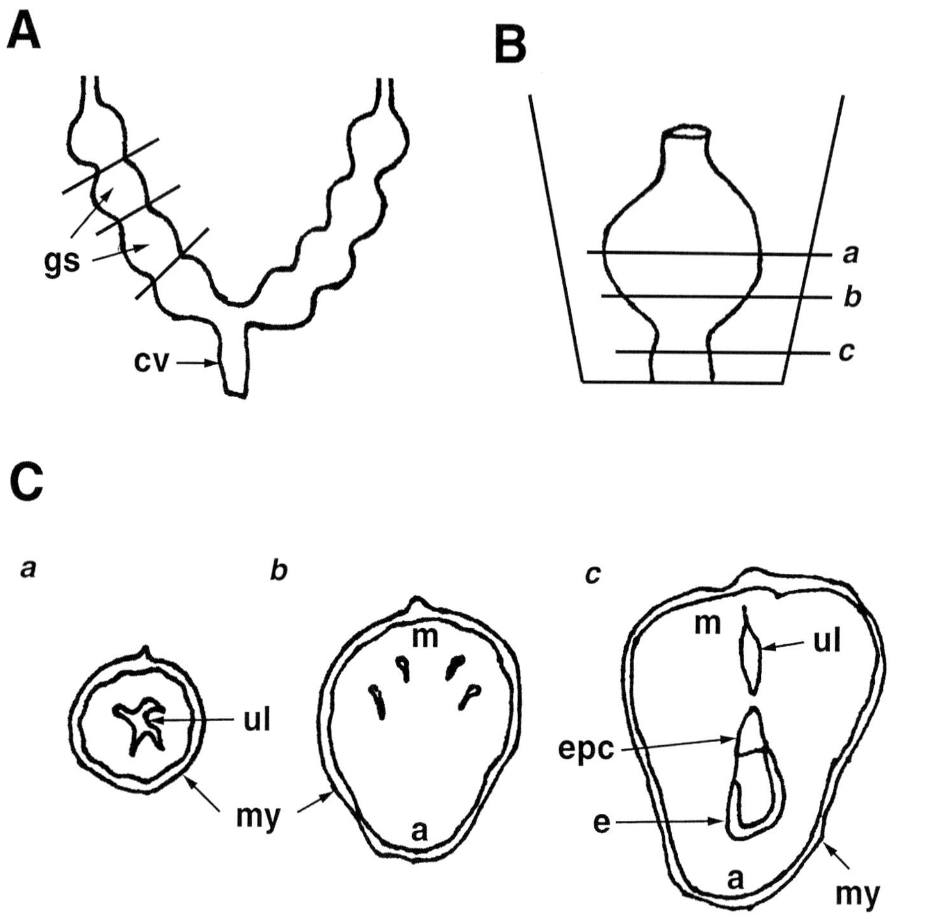

Fig. 1. Dissection and orientation of murine gestation sites for optimal visualization of extraembryonic membranes and placenta. **(A)** Diagram of uterine horns on approx 8.5 dpc. Gestation sites (gs) are defined as the developing embryo and its extraembryonic membranes contained within the uterus. To obtain an individual gs cut between decidualized areas of the uterus as depicted with lines. cv, cervix. **(B)** When embedding, orient the gs standing on end, with the uterine lumen facing down. **(C)** When sectioning, initial cuts are through the undecidualized uterus *(a)*, then through decidualized uterine tissue devoid of embryonic of extraembryonic tissue *(b)*. Begin collecting sections once the implantation chamber becomes evident *(c)*: e, embryo; epc, ectoplacental cone; ul, uterine lumen; my, myometrium; m, mesometrial deciduum; a, antimesometrial deciduum.

9. Sectioning. Mount the embedded gestation site on a wooden block for sectioning (or comparable mounting apparatus). The bottom of the embedded block should be facing out (*see* **Fig. 1B,C**).
10. Using a rotary microtome, section through the gestation site and start collecting sections 5 μm in thickness when you enter the implantation chamber (*see* **Fig. 1B,C**). Collect three to five serial sections onto positively charged microscope slides using a floaton water bath set at 42°C. Dry slides overnight at 37°C on a slide warmer. Store slides desiccated at 4°C until use.

11. Place slides in rack and deparaffinize sections by immersing slides in a slide-staining dish containing histoclear for 2 × 10 min.
12. Rehydrate sections by immersing slides for 2 min in each of the following: 100% ethanol; 100% ethanol; 95% ethanol in H_2O; 85% ethanol in H_2O; 70% ethanol in H_2O; 50% ethanol in H_2O; 30% ethanol in H_2O and 100% H_2O.
13. Rehydrated sections should be used immediately for the analysis of histopathology by Hematoxylin and Eosin staining (*see below* and **Fig. 2**), or Hoechst staining of nuclei (*see below* and **Fig. 3B**). In addition, sections processed in this manner can be used for analysis of gene expression by *in situ* hybridization ***(8)*** or immunohistochemistry ***(9)***.
14. Hematoxylin and Eosin staining: We use a kit from Shandon Inc. to rapidly stain sections following manufactures instructions. However, Hematoxylin and Eosin can be purchased and solutions made according to standard protocols ***(10)***.
15. After staining with hematoxylin and eosin, dehydrate sections by immersing in the following for 2 min each: 95% ethanol; 100% ethanol; 100% ethanol.
16. Immerse slides in histoclear or xylenes for 2 × 5 min and then cover slip using paramount, taking care not to allow the formation of air bubbles under the cover glass.
17. Allow the slides to dry in a flat horizontal position until the paramount dries (approx 2 d).
18. View and photograph stained gestation sites using bright-field microscopy. This will allow for the assessment of general placental structure and cell morphology (*see* **Fig. 2**).

3.3. Using Hoechst Staining to Monitor Giant Trophoblast Cells

1. Place slides in rack and deparaffinize sections by immersing slides in a slide-staining dish containing histoclear for 2 × 10 min.
2. Rehydrate sections by immersing slides for 2 min in each of the following: 100% ethanol; 100% ethanol; 95% ethanol in H_2O; 85% ethanol in H_2O; 70% ethanol in H_2O; 50% ethanol in H_2O; 30% ethanol in H_2O and 100% H_2O.
3. Immerse slides in 2 μg/mL Hoechst dye in H_2O for 2 min.
4. Immerse slides for 2 min in H_2O and then an additional 2 min in running H_2O (*see* **Note 11**)
5. Blot slides dry using tissue paper and lay slides flat and allow to dry thoroughly by placing at 37°C for 1 h.
6. Once slides are completely dry, overlay sections with Canada balsam (approx 90 μL per slide), and cover slip, taking care not to allow the formation of air bubbles under the cover glass.
7. Blot away excess Canada balsam and keep slides in a flat position until completely dry (approx 2 wk).
8. View and photograph stained nuclei using ultraviolet epifluorescence. Nuclei of giant cells in particular, but diploid nuclei as well, will appear brilliant blue, allowing for easy identification and analysis of these cells (*see* **Fig. 3B**).

4. Notes

1. Four percent PFA is a commonly used fixative for the analysis of embryonic and extraembryonic tissue. Proper fixation of gestation sites with 4% PFA will preserve histological detail important for histopathological examination of the embryo and placenta. In addition, this fixation is suitable for the analysis of gene expression by *in situ* hybridization ***(8,11)*** and immunohistochemistry ***(9,12)***.
2. To ensure that males and females are of adequately mature reproductive status, use mice at least 7 wk of age. It is preferable to use stud males of proven reproductive competency. Reproductive productivity of males is optimized by mating each male only once or twice

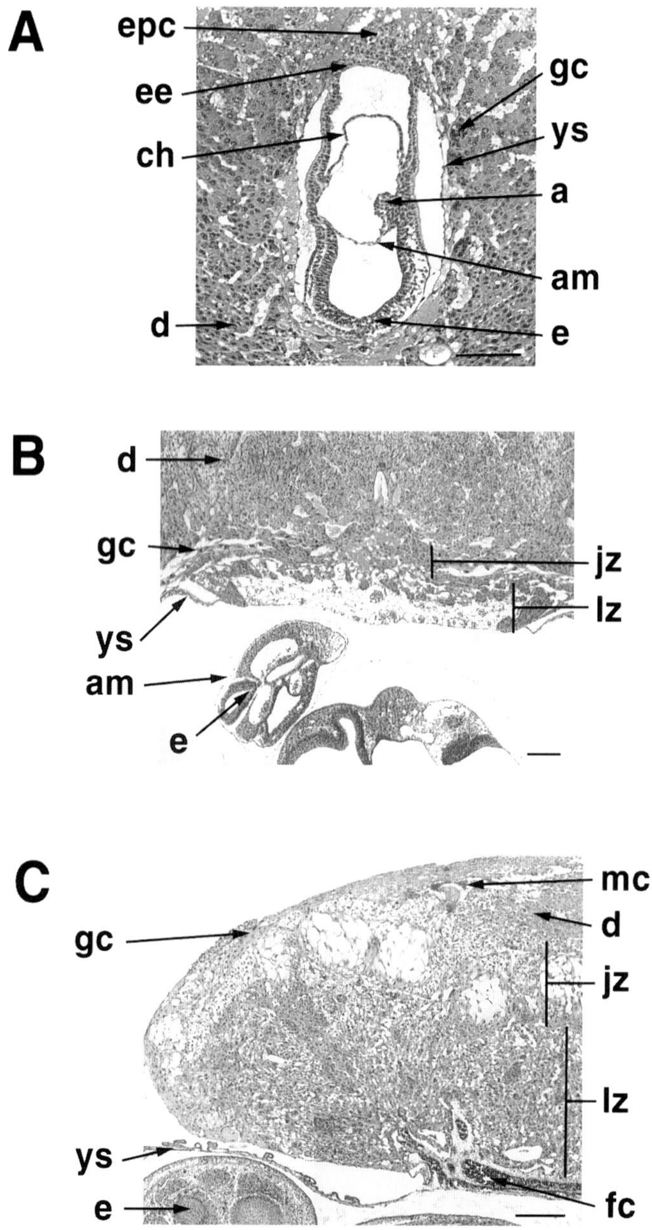

Fig. 2. (*See* color plate 5 appearing after p. 262.) Developing placenta in the mouse. Gestation sites were collected and processed for hematoxylin and eosin staining as described in **Subheading 3. (A)** 7.5 dpc gestation site. At this stage, the embryo (e) is undergoing gastrulation. A mass of trophoblast stem cells derived from the trophectoderm form the ectoplacental cone (epc), which is located at the mesometrial pole of the implantation chamber. Giant trophoblast cells (gc) derived from the trophectoderm and epc surround the implantation chamber, as do the yolk sacs (ys). A bud of extraembryonic mesoderm known as the allatoic bud (a) will grow toward the chorion (ch) which will fuse with the extraembryonic ectoderm (ee). am, amnion; d, deciduum. Scale bar = 75 µm. **(B)** A 9.5-dpc gestation site. The fusion of the ch and ee with the a will form the labyrinthine zone (lz) of the mature chorioallantoic placenta. It is in this region where fetal and maternal circulatory systems are

each week. Female reproductivity can be increased and more adequately timed using hormone superovulation protocols ***(13)***.

3. A vaginal plug appears as a white or yellow hard and waxy deposit in the vaginal opening. Females should be examined for the presence of vaginal plugs in the morning, because the plug will dislodge with time, causing a pregnancy to be missed.
4. The detection of a vaginal plug the morning following mating is designated 0.5 dpc on the assumption that coitus occurred around the middle of the dark cycle or 24:00. It is not uncommon for this estimate to be off by as much as 0.5 d between litters. If precise staging is required, gestation sites can be staged using established staging features of the embryo and uterus ***(14,15)***.
5. The pregnant uterus can be separated into two halves. The half of the uterus in proximity to the mesentery that attaches the uterus to the dorsal wall of the abdomen is referred to as the mesometrium and contains the mesometrial deciduum. The half of the uterus opposing the mesometrium contains the developing embryo and is referred to as the antimesometrium and contains the antimesometrial deciduum (*see* **Fig. 1C**).
6. A gestation site is defined as the developing embryo and its extraembryonic membranes contained within the uterus.
7. The number of gestation sites per vial for fixation and subsequent dehydration, clearing, and infiltration will vary depending on the developmental stage or size of the gestation site. On 4.5–8.5 dpc, up to 10 gestation sites can be placed per vial, which holds 20 mL fixative or subsequent solutions. On 9.5–11.5 dpc, six gestation sites per vial should be used, on 12.5–15.5 dpc, four gestation sites per vial should be used, and on 16.5–19.5 dpc, two to three gestation sites per vials should be used. This will allow for proper fixation with 4% PFA as well as proper dehydration, clearing, and infiltration using the times described in the methods.
8. Make 1X PBS and keep at 4°C for use in washing and mixing with ethanol for dehydration.
9. Clearing of gestation sites is a critical and empirical step in the procedure. Overclearing will make the tissue brittle and difficult to section, whereas underclearing will cause improper infiltration of paraffin and, subsequently, pour sectioning and morphology of tissue. The times given in the protocol are optimized for gestation sites on 4.5–9.5 dpc. When analyzing older gestation sites, it may be necessary to increase the incubation times. A good rule of thumb is to increase times by 5 min for each additional dpc.
10. Orienting the gestation site in this manner is important for obtaining sections through the implantation chamber that show the development of all the extraembryonic membranes, specialized structures of the developing placenta, and the embryo itself (**Fig. 2**).
11. Washing or destaining of nuclei is critical for adequate visualization. The times described work best on gestation sites cut at 5 μm. Thicker sections may require more extensive washing. Inadequate washing results in high-background and nonspecific cytoplasmic staining, whereas overwashing will make visualization of nuclei difficult. The intensity of staining can be checked by viewing the freshly rinsed slide under the microscope at low

---

brought into close apposition. Distal to the lz is the junctional zone (jz) that consists of hormone-producing diploid trophoblast cells derived from the epc. A thin layer of gc surrounds the developing placenta and implantation chamber. Scale bar = 150 μm. **(C)** A 13.5-dpc placenta. After 9.5 dpc, there is a large growth and expansion of the lz and jz. By 13.5 dpc, the jz may appear acellular due to large amounts of glycogen accumulation in certain trophoblast cells. A line of gc is found surrounding the placenta. mc, maternal circulation; fc, fetal circulation. Scale bar = 300 μm.

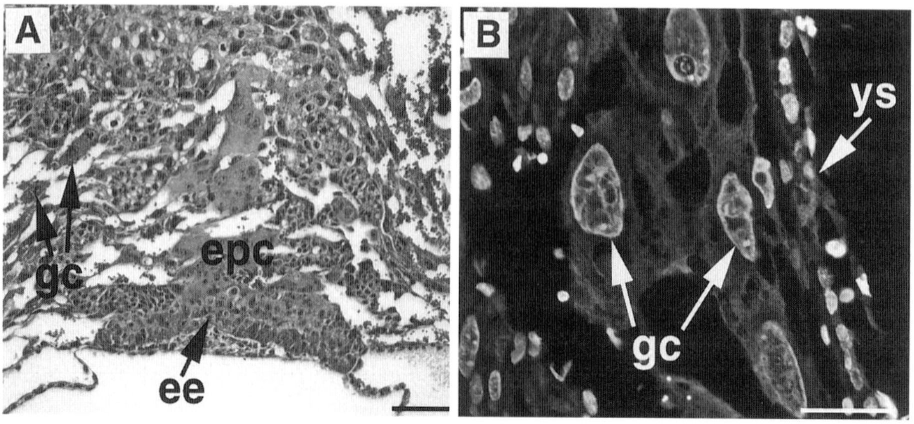

Fig. 3. (*See* color plate 6 appearing after p. 262.) Visualization of abnormal placentation and giant trophoblast cells. **(A)** Hematoxylin and Eosin stained 9.5-dpc gestation site that has been genetically engineered to lack the purine catabolic enzyme adenosine deaminase *(12)*. The absence of this enzyme causes the buildup of substrates that are embryotoxic. Therefore, the allantois does not make contact with the extraembryonic ectoderm (ee) and the placenta does not develop (compare with **Fig. 2B**). epc, ectoplacental cone; gc, giant trophoblast cells. Scale bar = 100 μm. **(B)** Wild-type 9.5-dpc gestation site stained with Hoechst according to methods. The enlarged nuclei of gc are easily detected surrounding the implantation chamber. Nuclei of diploid cells appear much smaller. ys, yolk sacs. Scale bar = 50 μm.

power using ultraviolet epifluorescence. Sections can then be washed again or restained accordingly, before drying and mounting.

References

1. Soraes, M. J. Faria, T. N., Hamlin, G. P., Lu, X. J., and Deb, S. (1993) Trophoblast cell differentiation: expression of the placental prolactin family, in *Trophoblast Cells: Pathways For Maternal–Embryonic Communication* (Soares, M. J., Handwerger, S., and Talamantes, F., eds.), Springer-Verlag, New York, pp. 45–67.
2. Cross, J. C., Werb, Z., and Fisher, S. J. (1994) Implantation and the placenta: Key pieces of the development puzzle. *Science* **266,** 1508–1518.
3. Peel, S. and Bulmer, D. (1977) Proliferation and differentiation of trophoblast in the establishment of the rat chorioallantoic placenta. *J. Anat.* **124,** 675–687.
4. Guillemot, F., Nagy, A., Auerbach, A., Rossant, J., and Joyner, A. L. (1994) Essential role of *Mash-2* in extraembryonic development. *Nature* **371,** 333–336.
5. Gurtner, G. C., Davis, V., Li, H., McCoy, M. J., Sharpe, A., and Cybulsky, M. I. (1995) Targeted disruption of the murine VCAM1 gene: essential role of VCAM-1 in chorioallantoic fusion and placentation. *Genes Dev.* **9,** 1–14.
6. Uehara, Y., Minowa, O., Morl, C., Shiota, K., Kuno, J., Noda, T., and Kitamura, N. (1995) Placenta defect and embryonic lethality in mice lacking hepatocyte growth factor/scatter factor. *Nature* **373,** 702–705.
7. Naeye, R. L. (1992) Disorders of the placenta and decidua, in *Disorders of the Placenta, Fetus, and Neonate: Diagnosis and Clinical Significance* (Gay, S., ed.), Mosby Year Book, St. Louis, MD.
8. Albrecht U., Eichele, G., Helms, J. A., and Lu, H.-C. (1997) Visualization of gene expression patterns by in situ hybridization, in *Molecular and Cellular Methods in Developmental Toxicology* (Daston, G. P., ed.), CRC, Boca Raton, FL, pp. 22–47.

9. Johnstone, A. and Thorpe, R. (1996) *Immunochemistry in Practice*, 3rd ed., Blackwell Science, Oxford, UK.
10. Humason, G. L. (1979) *Animal Tissue Techniques*, 4th ed., W. H. Freeman, San Francisco, CA.
11. Shi, D., Winston, J. H., Blackburn, M. R., Datta, S. K., Hanten, G., and Kellems, R. E. (1997) Diverse genetic regulatory motifs required for murine adenosine deaminase gene expression in the placenta. *J. Biol. Chem.* **272,** 2334–2341.
12. Blackburn, M. R., Knudsen, T. B., and Kellems, R. E. (1997) Genetically engineered mice suggest that adenosine deaminase is essential for early postimplantation development in the mouse. *Development* **124,** 3089–3097.
13. Hogan, B., Beddington, R., Costantini, F., and Lacy, E. (1994) *Manipulating the Mouse Embryo: A Laboratory Manual*, 2nd ed. Cold Spring Harbor Laboratory Press, Cold Spring Harbor, NY.
14. Theiler, K. (1972) *The House Mouse.* Springer-Verlag, New York.
15. Kaufman, M. H. (1992) *The Atlas of Mouse Development.* Academic, London.

21

Palatal Dysmorphogenesis

Palate Organ Culture

Barbara D. Abbott

1. Introduction

Clefts of the secondary palate are among the most frequent birth defects in live-born human infants. Across the United States between 1981 and 1995, a compilation of data from states with birth defects monitoring programs shows the incidence of cleft palate without cleft lip to range from 2.01 to 14.2 per 10,000 live births *(1)*. Thus, the formation of the secondary palate and the mechanism(s) for induction of cleft palate have been the focus of extensive research. Palatogenesis also presents an interesting model for many of the processes involved in morphogenesis in the embryo. The formation of the secondary palate requires neural crest cell migration, interaction of the palatal cells with the surrounding extracellular matrix, interaction and signaling between epithelial and mesenchymal cells, adhesion and fusion of morphological structures, which also involves cell death and transformation of cells from epithelial to mesenchymal phenotypes, and, finally, differentiation into bone and stratified oral and ciliated nasal epithelia *(2)*. In order to study these processes and the potential of exogenous agents to disrupt them, in vitro models have been developed, including mesenchymal cell culture, epithelial cell culture, and palatal organ culture. Palatal organ culture can be a system in which the palatal shelves are supported on a membrane above the medium; our laboratory has used this method in the past *(3–5)*. However, the model described in this chapter is a submerged culture of the entire midfacial region. This model offers several advantages over the earlier system, as it permits the palatal shelves to grow, elevate, and fuse in the culture medium. This model has been tested in our laboratory for several strains of rat and mouse, and it has supported development of the palate in human embryonic midfacial tissue *(6)*. Examples of the application of this culture model include studies of the effects of methanol and 5-fluorouracil on palatogenesis *(7–9)*.

The information in this document has been funded wholly (or in part) by the U.S. Environmental Protection Agency. It has been reviewed by the National Health and Environmental Effects Research Laboratory and approved for publication. Approval does not signify that the contents reflect the views of the Agency, nor does mention of trade names or commercial products constitute endorsement or recommendation for use.

From: *Methods in Molecular Biology, Vol. 136: Developmental Biology Protocols, Vol. II*
Edited by: R. S. Tuan and C. W. Lo © Humana Press Inc., Totowa, NJ

2. Materials

2.1. Preparation of Medium

1. Custom Modified Biggers BGJ formulation, dry powder (Irvine Scientific, Santa Ana, CA).
2. Sodium bicarbonate (cell culture tested).
3. Monobasic and dibasic sodium phosphate.
4. Sodium hydroxide solution 1 *N*.
5. L-Glutamine (cell culture tested).
6. Bovine serum albumin (BSA) Fraction V powder 98–99% albumin.
7. Penicillin–streptomycin solution (cell culture tested) (10K units penicillin + 10 mg streptomycin/mL).
8. Bottle filter systems for filtration of medium, low-protein-binding filters (nylon or cellulose acetate membranes) 0.2 μm pore size, prefilters are helpful for large-volume applications. Sterile plastic flasks of 500 mL capacity for storage of phosphate-buffered saline (PBS) and medium.

Optional

9. Fetal bovine serum (FBS).
10. Sodium selenite (cell culture tested).
11. apo-Transferrin (cell culture tested).
12. Sodium ascorbate (L-ascorbic acid).
13. All-*trans*-retinoic acid (cell culture tested).

2.2. Prepare Culture Flasks and Gas Medium

1. Dimethylsulfoxide (DMSO).
2. Tissue culture flasks canted neck 25 cm^2 (or 75 cm^2 for rat) plug seal 70 mL capacity. **Note:** Caps must form gas-tight seal, not the style that is ventilated by a filter cap (e.g. Corning Glassworks, Corning, NY #2510025).
3. Medical gas mixture of 50% O_2, 5% CO_2, 45% N_2, analysis to confirm mixture purity.
4. Sterile disposable transfer pipets, individually wrapped, used to gas flasks.
5. Syringe filters, 0.2-μm pore size.

2.3. Culture Setup

1. Dissecting stereomicroscope, fiber-optic illuminator, black rubber stopper cut horizontally to provide ½ in.-high dissection pad or other suitable dark rubber pad (sterilize by storing in 70% ethanol), rocker platform with adjustable tilt rate (side-to-side motion) to fit in incubator, culture incubator to maintain 37°C temperature, vacuum system (pump) to filter medium.
2. Gauze sponges, 70% ethanol, plastic disposable Petri dishes, polypropylene tubes of various sizes, test tube rack.
3. Dissecting tools: For removal of uterus and embryos from uterus: dissecting forceps 4 in. serrated, curved (at least two); curved blunt tip 4½-in. microdissecting scissors, and 4-in. curved sharp-tip microdissecting scissors. For dissection of embryo and removal of midfacial tissue: Dumont forceps #5 stainless steel (at least two pairs, have extra on hand and be aware that these points are easily damaged even in routine use); spring-loaded iridectomy scissors with very fine, small scissor blades such as the Vannas ultramicro scissors 3 in. straight (for mouse) (Roboz Surgical Instrument Co., Rockville, MD, RS-5610, or equivalent) and Vannas extremely delicate 3 in. straight sharp scissors (for rat) (Roboz Surgical RS-5883, or equivalent); disposable #11 scalpels (fine-pointed blade gives good visibility during dissection cuts); microdissecting spatula 5 in. (such as the Roboz RS-6150). All instruments are sterilized by soaking in 70% ethanol just prior

Table 1
Custom Modified Biggers BGJ Formulation (with L-Glutamine 150 mg/mL, without $NaHCO_3$) Irvine Scientific #98337

	mg/L		mg/L
Sodium chloride	6400	p-Aminobenzoic acid	1.5
Potassium chloride	425	Ascorbic Acid	150
Glucose (dextrose) anhydrous	4400	Nicotinic Acid Amide	15
Calcium lactate.5H_2O	600	Phenol red, Na salt	17.04
L-Arginine HCl	70	Thiamine HCl	3
L-Cysteine HCl.H_2O	135	α-Tocopherol Phosphate, 2Na	0.75
L-Glutamine	150	Potassium phosphate, monobasic KH_2PO_4	113
L-Histidine HCl.H_2O	150	Magnesium sulfate, anhydrous $MgSO_4$	161.17
L-Isoleucine	23	Choline chloride	38
L-Leucine	40	Folic acid	0.15
L-Lysine HCl	225	Inositol	0.15
L-Methionine	40	Riboflavin	0.15
L-Phenylalanine	40	Vitamin B-12	0.03
L-Threonine	60	d-Biotin	0.15
L-Tryptophan	30	Pantothenic acid, Ca salt	0.15
L-Tyrosine 2Na.2H_2O	43.25	Pyridoxal HCl	0.15
L-Valine	50		

Source: **ref. *10***.

to use. Small and/or delicate instruments can be wrapped in sterile gauze and soaked with the ethanol.

3. Methods

3.1. Preparation of the Medium and PBS

1. Reconstitute Modified Bigger's BGJ medium, formulation in **Table 1** ***(10)***, with distilled, deionized water as per suppliers instructions to be 1X. (Add 14 g sodium bicarbonate and bring up to 5 L with water, check pH and osmolarity, should be pH 7.0–7.5, osmolarity 265–300 mOs/kg). Store liquid medium at 4°C up to 6 mo.
2. Phosphate-buffered saline: 35.04 g NaCl, 1.38 g $NaH_2PO \cdot H_2O$ (monobasic), 8.04 g $Na_2HPO_4 \cdot 7H_2O$ (Dibasic), add distilled H_2O up to 4 L, adjust pH to 7.4 with 1 *N* NaOH, filter sterilize with 0.2-µm filter, store at 4°C.
3. Supplement the Bigger's BGJ medium prior to use (*see* **Note 1** for preparation of stock solutions):
   0.15 mg/mL L-glutamine
   5 µL/mL penicillin/streptomycin (final concentration = 50 µg streptomycin/mL + 50 units penicillin/mL)
   6 mg/mL bovine serum albumin (add last and mix by gentle inversion)
   Filter sterilize supplemented medium, store at 4°C and discard after 2 wk. (*See* **Note 2** for additional supplements and **Note 3** for cautions regarding supplements.)

3.2. Prepare Culture Flasks and Gas the Medium

1. Add solvent control and compound treatments to aliquots of the supplemented medium (e.g., compounds dissolved in DMSO added to give 1% DMSO final in medium; ethanol

and methanol have teratogenic potential in this culture model *(8)*, and if these solvents are used, controls with solvent in the medium are required to select a no-effect level).

2. Add 9 mL of control or treated medium to each sterile 25-cm^2 culture flask (*see* **Note 4** for larger flasks for rat palate).
3. Direct a stream of medical gas mixture (50% O_2, 5% CO_2, 45% N_2) through a 0.2-µm pore filter (*see* **Note 5**) and flush air from the flask, replacing it with the gas mixture. If the gas flow rate is sufficient to agitate the medium, 10–15 s should be long enough to flush out the air. Immediately cap the flask tightly.

3.3. Culture Setup

1. Place flasks horizontally on a platform (elevating the neck of the flask makes it easier to put tissues inside, a test tube rack works well as a shelf). Loosen cap but leave it resting over flask opening.
2. Dissect tissues as described below, pick up tissue on spatula and insert into flask, touch to medium, and float tissue off the spatula. Up to four tissues per flask (*see* **Note 4**).
3. Gas flasks: After all flasks are loaded with tissues repeat the gassing step and cap tightly.
4. Incubate flasks at 37°C on a rocker platform (side-to-side rock rate approximately 15 tilts/min).

3.4. Culture Maintenance

Replace medium and gas flasks every 24 h. In most cases, the treatments are also included with the fresh medium.

1. Prewarm the complete medium in 37°C water bath.
2. Remove medium from flask by pouring or suction (keep tissues inside flask).
3. Replace with 9 mL fresh prewarmed medium. Gas as before.
4. Replace on rocker in incubator.
5. Cultures can be maintained 4–5 d or until palatal fusion occurs.

3.5. Dissection of the Embryonic Craniofacial Region

1. Timed pregnant female mice are killed on GD12 by CO_2 asphyxiation and/or cervical dislocation (*see* **Note 6**). Embryos are removed from the uterus as quickly as possible and placed in chilled PBS in Petri dishes on ice. If possible, keep the placenta and yolk sac attached to embryo and the circulation intact. This will help delay degeneration of the tissues.
2. Dissect mid-craniofacial region (includes maxillary arch, primary palate, and secondary palatal shelves) (refer to **Notes 7** and **8**):
   a. Place the embryo on a black rubber platform (large rubber stopper that has been cut in half to reduce height to about ½ in.) and remove yolk sac, amnion, placenta with Dumont forceps.
   b. Use Vannas ultramicroscissors and insert one point of the scissors into the oral cavity, and with the other across the outside of the face, cut to separate the mandible and neck from the upper head (line A in **Fig. 1A**). Forceps at the back of the head can be used to brace or hold head in position during the cut. Make a second cut at the level of the eye to remove the upper brain tissues (line B).
   c. Push the other tissues aside and turn the midface over, exposing the palatal shelves (**Fig. 1B**). Use scalpel to remove hindbrain tissue with a cut at the level of the posterior of the palatal shelves (**line C**).
   d. If any mandible, tongue, or hindbrain remains, trim off. Eyes will remain on the final midfacial explant that is now ready to transfer to the culture flask.
3. Transfer explant to flask. Pick up the explant by sliding a dissection spatula under the explant; press down on the rubber mat to assist in pickup. Open cap of flask and insert

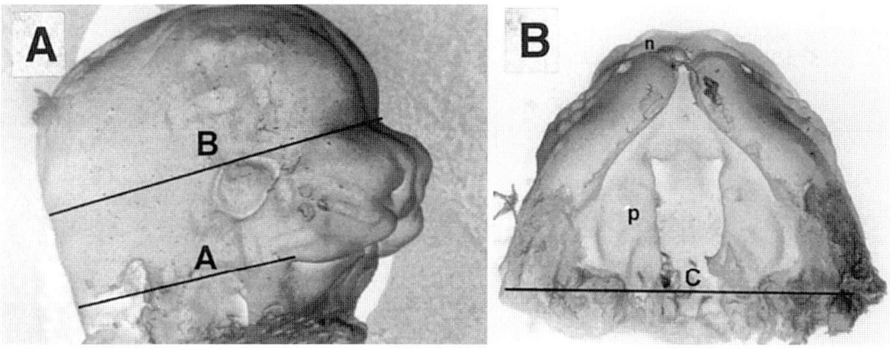

Fig. 1. The gestation day 12 mouse head (plug day = 0) is shown before dissection for palatal organ culture. **(A)** The first incision is made across line A through the oral cavity to separate mandible from the head. The second incision is through line B to remove the upper cranial tissue. **(B)** After turning the midface over, exposing the palatal shelves a final cut is made through line C to remove remaining hindbrain tissues. The resulting midfacial region is placed in culture.

spatula to touch the medium and float tissue off. Embryos from each litter can be distributed across the treatment groups/flasks by placing several flasks on the platform to receive tissues in sequence as the dissections occur. Put up to four tissues per flask.

3.6. Tissue Evaluation

At intervals during culture and at the end of culture, the explants can be examined under a dissecting microscope to track the progress of palatal elevation and fusion. Palates that are adhered cannot be separated even by gentle pulling with Dumont forceps on the primary palatal edges. A palatal seam will remain intact and the adjacent tissue will tear if sufficient force is used.

4. Notes

1. Stock solutions of medium supplements: (Filter sterilize supplements a–d before aliquoting and store as indicated. These stock solutions should be good for 1 yr or more.)
   a. Sodium selenite: 5 mg in 5 mL H_2O; dilute 1:100 to give 10 μg/mL stock and store as 1-mL aliquots at –20°C.
   b. apo-Transferrin: 20 mg in 2 mL H_2O; aliquot 200 μL/tube and store 10-mg/mL stock at –20°C.
   c. Sodium L-ascorbate: 200 mg in 20 mL H_2O; aliquot 1 mL/tube and store 10-mg/mL stock at –20°C.
   d. FBS: dilute 1:1 with medium prepared as above, aliquot 3–5 mL/tube and store 50% serum at –20°C.
   e. Penicillin–streptomycin: Aliquot 1–5 mL/tube of the 10000 U penicillin G/mL + 10 mg streptomycin/mL and store stock at –20°C.
   f. All-*trans*-retinoic acid (RA): Dissolve in DMSO at $1 \times 10^{-2} M$ stock, protect from light, store in amber vials at -80°C. Dilute stock with DMSO in serial dilutions for treatments or as a supplement. Add 1 μL RA/mL medium (DMSO=0.1%).
2. Additional supplements can include:
   a. 5 μL/mL Sodium ascorbate solution (50 μg/mL final)
   b. 1% Fetal bovine serum (50% stock 20 μL/mL medium)
   c. 1 μL/mL selenium solution (final concentration=10 ng/mL)

   d. 1 µL/mL transferrin (final concentration=10 µg/mL)
   e. Glutamine can be increased to 4X normal level
   f. All-*trans*-retinoic acid ($1 \times 10^{-11}M$ final in medium)
3. Palatal shelves show normal development and fuse in the standard control medium without addition of the above-listed supplements. However, depending on the test compound, additional supplements may be required to provide responses to a toxicant that are similar to those observed after an *in utero* exposure. This seems to depend on the mechanism of action of the compound, as the secondary palatal shelves elevate and fuse in the absence of these additives. Caution should be exercised with supplements to assure that normal palatal processes occur in the presence of the additives. Excess fetal bovine serum (>1%) can interfere with palatal fusion in the culture model. Be aware that each new lot of serum, even if it is from the same supplier, needs to be tested in the model to confirm that it supports normal development.
4. For larger species such as rat use 50 mL of medium in a 75-cm^2 flask and place up to 10 tissues/flask. Increasing the number of tissues per flask will decrease the number of specimens with palatal fusion.
5. For gassing the flasks, a syringe filter can be attached to the end of the gas tubing, then cut off the bulb end of a plastic disposable transfer pipet and attach to the outlet of the filter. Inserting the pipet halfway into the flask works well. Replace transfer pipet frequently (e.g., between sets of flasks containing different treatments).
6. Mice can be mated overnight and checked for sperm plug the next morning which is designated GD0. The palatal shelves appear as outgrowths of the maxillary arch between GD11 and GD11.5. The palatal shelves should be clearly visible on GD12 (Theiler stage 20 or Carnegie stage 17) and should be vertical in orientation. A similar developmental stage for rats is found on GD14. Be aware that different strains of mice (and even the same strain in different breeding colonies) may differ by as much as 6–12 h in developmental staging. Confirm the best time for collecting tissues from your animals.
7. The embryos need to be removed from the uterus as quickly as possible and will be easier to dissect if the membranes and placenta remain attached. Any embryos that become separated should be dissected first. The Petri dish is kept on ice to retain integrity of tissues as long as possible. With experience, the average time to dissect a litter can be less than 15 min. In our laboratory, two experienced operators can prepare 80–90 explants for culture in 1 h. The organ culture outcome may depend on expeditious preparation of the explants. It is not recommended to use any embryos that are on ice longer than 40–45 min, waiting to be dissected.
8. Special note for rat palatal organ culture: Rat strains differ in their requirements for successful fusion of the palates in culture. The Fisher 344 rat performs well under the same conditions as the mice (four explants in 9 mL of medium for 4–5 d; culture starts Monday and ends on Thursday or Friday). Sprague-Dawley and CR/Lewis required the larger flasks with 50 mL of medium/10 explants and required media supplemented with $1 \times 10^{-11}M$ all *trans*-retinoic acid. The CR/Lewis rat palates also had a higher percentage of explants fuse if the gas was changed on the fourth day to 95% O_2, 5% CO_2, and the flasks were flushed with the gas in the morning and late afternoon. Tissues were scored the next morning (5) and 80% of the palates were fused.

References

1. Birth defects surveillance data from selected states. (1997) In *Congenital Malformations Surveillance Report, A Report from the National Birth Defects Prevention Network. Teratology* **56,** 1–175.

2. Zimmerman, E. F. (1997) Palate, in *Handbook of Experimental Pharmacology 124/I: Drug Toxicity in Embryonic Development I* (Kavlock, R. J. and Daston, G. P., eds.), Springer-Verlag, New York, pp. 183–208.
3. Abbott, B. D. and Pratt, R. M. (1987) Retinoids and epidermal growth factor alter embryonic mouse palatal epithelial and mesenchymal cell differentiation in organ culture. *J. Craniofac. Gen. Dev. Biol.* **7,** 219–240.
4. Abbott, B. D. and Pratt, R. M. (1987) Human embryonic palatal epithelial differentiation is altered by retinoic acid and Epidermal Growth Factor in organ culture. *J. Craniofac. Gen. Dev. Biol.* **7,** 241–265.
5. Abbott, B. D. and Birnbaum, L. S. (1990) Rat embryonic palatal shelves respond to TCDD in organ culture. *Toxicol. Appl. Pharmacol.* **103,** 441–451.
6. Abbott, B. D. and Buckalew, A. R. (1992) Embryonic palatal responses to teratogens in serum-free organ culture. *Teratology* **45,** 369–382.
7. Abbott, B. D., Lau, C., Buckalew, A. R., Logsdon, T. R., Setzer, W., Zucker, R. M, Elstein, K. H., and Kavlock, R. J. (1993) Effects of 5-fluorouracil on embryonic rat palate in vitro: Fusion in the absence of proliferation. *Teratology* **47,** 541–554.
8. Abbott, B. D., Logsdon, T. R., and Wilke, T. S. (1994) Effects of methanol of embryonic palate in serum-free organ culture. *Teratology* **49,** 122–134.
9. Shuey, D. L., Buckalew, A. R., Wilke, T. S., Rogers, J. M. and Abbott, B. D. (1994) Early events following maternal exposure to 5-fluorouracil lead to dysmorphology in cultured embryonic tissues. *Teratology* **50,** 379–386.
10. Shiota, K., Kosazuma, T., Klug, S., and Neubert, D. (1990) Development of the fetal mouse palate in suspension organ culture. *Acta. Anat.* **137,** 59–64.

22

Palatal Dysmorphogenesis

Quantitative RT-PCR

Gary A. Held and Barbara D. Abbott

1. Introduction

The formation of the secondary palate requires coordinated expression of genes involved in processes such as cell proliferation, differentiation, and production of extracellular matrix proteins. In order to study the mechanisms through which teratogenic agents induce cleft palate, the regulation and expression of these genes needs to be examined and compared between control and treated embryos. The expression of specific mRNAs can be examined by reverse transcription polymerase–chain reaction (RT-PCR) and this chapter presents an application of RT-PCR for quantitating the level of mRNA. The sensitivity of this method allows the investigator to evaluate gene expression in individual embryos using dissected tissues from the specific regions/anatomic structures affected by the teratogen. Even extremely small specimens can usually be used to quantitate mRNAs for several genes. Other methods, such as the Southern blot, generally require pooling of multiple embryos and may not provide information for the isolated, specific target tissue. The quantitative method described in this chapter is derived from the method published by Vanden Heuvel *(1)*. The modifications we introduce improved the performance in our laboratory and may enhance the reliability, specificity, and sensitivity of this approach. This introduction provides an overview of the issues of experimental design that are important to the success of the method, including internal standard design, primer selection, image acquisition and analysis. A basic introduction to the RT-PCR method can be found in one of the many resource and protocol books available on this topic, including a volume edited by McPherson et al. *(2)*.

The method presented in this chapter uses an internal standard RNA sequence (IS) as a control template that is added in known quantities to the total RNA from the embryonic tissue. The IS is specifically designed for each gene under study and the

The information in this document has been funded wholly (or in part) by the U.S. Environmental Protection Agency. It has been reviewed by the National Health and Environmental Effects Research Laboratory and approved for publication. Approval does not signify that the contents reflect the views of the Agency, nor does mention of trade names or commercial products constitute endorsement or recommendation for use.

From: *Methods in Molecular Biology, Vol. 136: Developmental Biology Protocols, Vol. II*
Edited by: R. S. Tuan and C. W. Lo © Humana Press Inc., Totowa, NJ

primers for the IS are the same as those used to amplify the mRNA of interest. Our approach also uses the reverse primer selected to amplify the mRNA as the primer for the reverse transcription step, rather that poly-deoxythymidine. This modification allows us to select primers that amplify a region of mRNA distant from the poly A tail of the molecule. In many cases, we found that attempting to use the poly A primer was not as consistent for reverse transcription of sequences distant from that region. The reverse primer worked well for reverse transcription and provided a good template for the following PCR step. Other quantitative RT-PCR methods may include DNA sequences as standards and these may or may not utilize the same primers as those required to amplify the target mRNA. An advantage of using RNA for an internal standard is that this allows the efficiency of the RT reaction to be included in the evaluation. Some protocols amplify a constitutively expressed gene as an internal standard to compare with the mRNA of the gene of interest. This housekeeping or second gene approach requires that different primer sequences be used for the target mRNA and the standard. Although this allows a naturally existing mRNA to be used as the standard, this approach may lead to inaccurate results if the efficiency of amplification differs for the two primer sets (housekeeping gene and mRNA target). Also, it should be clearly demonstrated that the expression of the "standard" housekeeping gene mRNA does not change depending on the treatment being studied. The major advantage to using the RNA synthetic internal standard with an identical primer sequence as the target mRNA is that this approach eliminates issues related to variations in reverse transcription and PCR efficiency that can occur when different primer pairs are used for the standard and target amplification.

The internal standard used in our method is a synthetically constructed RNA that uses the same primers as the mRNA under study. A serial dilution of the IS is added to a series of reaction tubes containing a constant amount of total RNA (**Fig. 1A**). Thus, both the target mRNA and the internal standard RNA are reverse transcribed and then amplified in the same PCR tube. After the RT-PCR reactions are run, the mRNA and IS products are separated by gel electrophoresis and images are acquired (**Fig. 1B**). The IS is designed to be slightly different in size relative to the amplified mRNA. The intensity of the IS and mRNA bands on the gel are quantitated and a ratio of IS/RNA values is calculated for each lane in the gel. A regression analysis of ratio against known IS concentration provides the point at which the ratio equals 1, and at that point, the amount of mRNA is equal to the amount of IS (a known quantity). Thus, the known IS level provides the level of mRNA as a function of the amount of total RNA in the assay.

The features and design of the IS are illustrated in **Fig. 1C**. A spacer region of pUC19 plasmid sequence is selected to give an IS size slightly different than the target mRNA PCR product. Typically, a difference of about 10–15% works well. The IS is synthesized in a PCR reaction with the pUC plasmid as template and using two composite primers. The forward primer consists of a T7 polymerase promotor, the target mRNA forward primer, and a forward primer for pUC19. The reverse primer consists of a poly dT region, the reverse target mRNA primer, and a reverse pUC19 primer. The pUC19 primers are selected to give a spacer region of the correct length and a GC content similar to the mRNA being measured. We have found that the common cloning vector pU19 is a good source of the spacer region because it is readily available in highly purified form, has both high- and low-GC regions, and we have found it easy to find

PCR primers that give reliable, clean amplifications. Undoubtedly, many other templates would be equally suitable. The RNA internal standard used in the quantitative RT-PCR reactions is synthesized using T7 RNA polymerase in an in vitro transcription reaction. The reverse primer is used for the reverse transcriptions instead of a poly dT primer; the poly dT can be omitted from the composite reverse primer.

One of the critical steps required for successful quantitative RT-PCR is the selection of good primer pairs. There are many programs that assist in the selection of primers that avoid known problems such as extremes in GC content, hairpins, and self-annealing. One program that we have had good success with is Oxford Moleculars' MacVector. We have found the default settings to usually give primers that amplify well with few artifactual bands. After selecting several prospective primer pairs, it is useful to scan Genbank for homologies to known genes. Although two primers rarely match more than the selected gene, any such combinations should be avoided. It is useful to try several primers pairs at this point to determine how well they amplify and how free they are from spurious bands. Spurious PCR bands are a reason to reject the primers at this early stage of protocol development, as optimization frequently does not help much with this problem. It is also very desirable to use primers that do not require a hot start to avoid procedural problems when performing the quantitative RT-PCR reactions.

Attention needs to be given to acquisition of the image of the PCR products on the gel in a manner that prevents compromising the quantitative nature of the data. A method readily available in most laboratories is photography, followed by scanning of the photograph. Although, this can work well, there are many possible pitfalls. A major problem is the short range of a linear response of film to light. In conjunction with this problem, if flat-bed scanners are used for scanning the resulting negatives, as is increasing common, the limited maximum density of the scanners confounds the problem. The photographic/scanner-related problems can be avoided by carefully controlling exposures to ensure that the critical portions of the image are in the linear response range of both the film and scanning equipment. Digital imaging avoids some of the problems of photographic methods due to the linearity of charge-coupled device (CCD) detectors and the speed with which captured images can be evaluated for suitability. The most commonly used digital imaging systems are based on CCD video cameras. Slow-scan CDD cameras are another possible option. Video-based systems have the advantage of ease of focus and framing of the image. Disadvantages of video systems, when compared to slow-scan cameras, are lower sensitivity and less dynamic range. Video cameras that are capable of on-chip integration solve the sensitivity issue, and the dynamic range can be partly compensated by co-adding multiple images. In our laboratory, we have had good results using a Hamamatsu (Bridgewater, NJ) Argus 20 video imaging system, which consists of a video camera capable of on-chip integration and a dedicated controller that allows multiple images to be added together, thus increasing the dynamic range and reducing the noise levels of the image before transferring the images to a computer. We use the gel plotting macro package of NIH Image to determine the relative intensity of the DNA bands.

2. Materials

(See EPA disclaimer regarding sources and products, other suppliers and products are available and may produce similar results.)

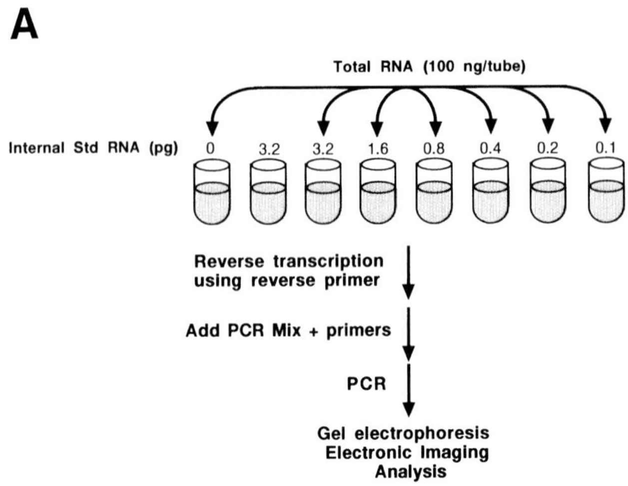

Fig. 1. **(A)** The Internal Standard (IS) is added as a dilution series to reactions containing constant amounts of the total RNA. Both the target mRNA and the IS are reverse transcribed and then amplified in the same PCR tube. The mRNA and IS amplified sequences are separated on a gel. **(B)** The band shown on this example are the amplified IS (upper bands) and/or mRNA (lower bands). The intensity of the IS decreases with the amount (pg) added as the intensity of the mRNA band increases. The quantitation of intensity is shown as peaks in the histogram and the ratio of upper to lower peak is plotted for each level of IS (each lane). Regression analysis provides a formula for estimation of the mRNA level at ratio = 1. In this example $y = -2.069e - 2 + 0.95768x$, with $r^2 = 1.000$, and when $y = 1$ then $x = 1.066$. **(C)** The schematic represents the design of the IS in which pUC DNA is amplified with primers that include a T7 promoter region and that promoter is used to reverse transcribe the cDNA IS to prepare the cRNA IS for use in the RT-PCR reactions.

2.1. Total RNA Preparation

1. TRI® Reagent (Molecular Research Center, Inc., Cincinnati, OH) or similar product to prepare RNA using the Chomczynski *(3)* method.
2. Tissue homogenizer (Brinkmann Polytron® Homogenizer with PT-7 generator for small volumes (Brinkmann Instruments, Inc., Westbury, NY).
3. Eppendorf tubes, chloroform, refrigerated microcentrifuge, isopropanol, autoclaved pipet tips, ethanol, diethylpyrocarbonate (DEPC), DEPC-treated H_2O, 0.3*M* sodium acetate (prepared with DEPC-treated, autoclaved water), low-volume cuvettes, and spectrophotometer to read OD_{260}.

2.2. Primer Oligos

1. Primers for PCR (Genosys Biotechnologies, Inc., The Woodlands, TX). Either synthesize and purify or order purified oligos with sequences determined by computer search programs. Primers <30 bases purified by reversed-phase cartridge and primers for IS (often 60 bases or longer) purified by polyacrylamide gel electrophoresis (PAGE).
2. 1*M* Tris-HCl, pH 8.5 (prepared with DEPC-treated, autoclaved water).
3. 0.5*M* EDTA, pH 8.5 (prepared with DEPC-treated, autoclaved water).

2.3. RT-PCR Reactions

1. AMV Reverse Transcriptase (HC) 23,000 units/mL (Promega, Madison, WI) with 10X buffer mix.
2. Recombinant RNasin Ribonuclease Inhibitor 40 units/µL (Promega).

B

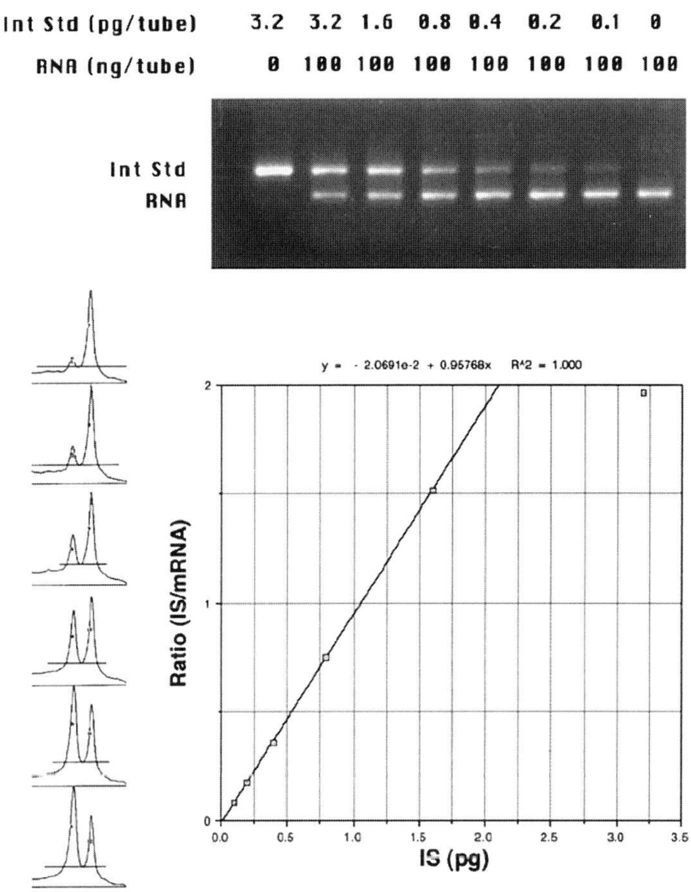
Int Std (pg/tube) 3.2 3.2 1.6 0.8 0.4 0.2 0.1 0
RNA (ng/tube) 0 100 100 100 100 100 100 100
Int Std
RNA
y = - 2.0691e-2 + 0.95768x R^2 = 1.000
Ratio (IS/mRNA)
IS (pg)
0.0 0.5 1.0 1.5 2.0 2.5 3.0 3.5
0 1 2

C

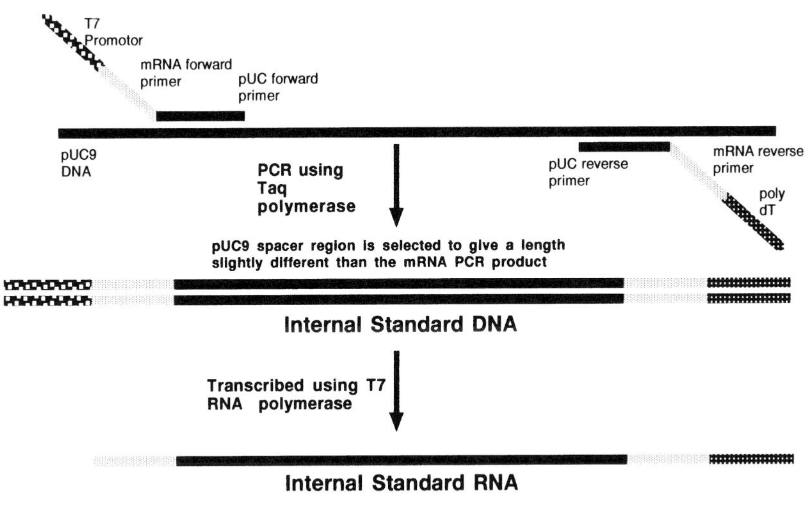
T7 Promotor
mRNA forward primer
pUC forward primer
pUC9 DNA
PCR using Taq polymerase
pUC reverse primer
mRNA reverse primer
poly dT
pUC9 spacer region is selected to give a length slightly different than the mRNA PCR product
Internal Standard DNA
Transcribed using T7 RNA polymerase
Internal Standard RNA

3. dNTPs 10 m*M* each in water, pH 7.0, individual or as a mix.
4. $MgCl_2$, 25 m*M*.
5. 0.2 mL PCR tubes and caps, strips of eight; 1.5-mL sterile autoclaved tubes; sterile filtered pipet tips to accommodate three sizes of pipettors; new pipettors reserved for RT-PCR.
6. Thermocycler with heated lid and block for 96 0.2-mL tubes (MJ Research, Inc, Watertown, MA, USA).
7. *Taq* DNA polymerase (*Thermus aquaticus* strain YT1, 5 units/µL [Promega]) and 10X Thermobuffer.
8. Opti-Prime™ PCR Optimization Kit (Stratagene, Menasha).

2.4. Agarose Gel for High Resolution

1. Metaphor™ Agarose (FMC, Rockland, ME).
2. Electrophoresis apparatus, trays and combs 1.5 mm thick; constant-voltage power source.
3. Tris–borate–EDTA buffer 1X, DNA markers (pGEM® Markers [Promega]); ultraviolet (UV) transilluminator; ethidium bromide (1 mg/mL stock in water); xylene cyanol; bromophenol blue; sucrose.
4. Equipment to acquire digitized images, process, and perform densitometry (Argus-20 SCSI Software for the Hamamatsu Photonic System, NIH Image for Densitometric Analysis.

2.5. Preparation of Internal Standard

1. pUC19 plasmid (Gibco-BRL, Grand Island, NY).
2. *Hin*dIII restriction enzyme and buffer.
3. Columns for purification of PCR products (Wizard™ PCR Preps [Promega]); isopropanol.
4. In vitro transcription kit for large-scale synthesis of short transcript RNAs (Ambion T7-MEGAshortscript™, Ambion, Austin, TX).
5. Phenol:chloroform 1:1 mixture.

3. Methods

3.1. Preparation of Total RNA with TRI Reagent

1. Collect the tissues (optional: weigh tissue) and immediately flash freeze in sterile plastic tube on dry ice. Add 1 mL of TRI Reagent to sample, thaw, and immediately homogenize with Brinkmann Polytron homogenizer low-volume generator. Note that 0.5 mL can be used; just remember to decrease all reagent volumes in half after this step. Replace on dry ice to hold until all samples are prepared. (*See* **Note 1**.)
2. Allow to sit at room temperature for 5 min.
3. Transfer to 1.5-mL tube and add 0.2 mL of chloroform, shake for 15 s, and store at room temperature for 3 min.
4. Centrifuge sample at 4°C for 15 min between 13,000 and 14,000 rpm.
5. Transfer aqueous phase to clean tube and add 0.5 mL isopropanol and store sample at room temperature for 10 min.
6. Centrifuge sample for 10 min as in **step 4**, then remove supernatant and wash pellet with 1 mL 75% ethanol (prepared with DEPC-treated water).
7. Centrifuge at 9000*g* for 5 min at 4°C.
8. At this point, samples can be either dried for immediate resuspension in water or put in long-term storage under 100% ethanol. To store, add 1 mL 100% ethanol and store at –80°C. To recover the RNA, add 100 µL 0.3 M sodium acetate, mix, hold 15 min at -80 C, centrifuge cold at 16,000*g* 15 min, then dry pellet. To dry: Air-dry inverted or under vacuum in clean desiccator, or use SpeedVac.
9. Dissolve pellet in DEPC-treated H_2O. (*See* **Note 2** for volume to use.)

10. Read OD_{260} with a 1:100 or 1:250 dilution of the sample in water (if small-volume cuvettes are not available, there may be problems with low values near the limit of detection). Multiple OD reading by 4 (if dilution is 1:100) or 10 (if dilution is 1:250) to give microgram of total RNA per microliter concentration in the sample.
11. Wrap tube with parafilm and store this stock solution at –80°C.
12. A working dilution needs to be made in DEPC water at a concentration that allows limiting the volume added to the quantitative reaction to 1–5 μL (concentrations of the diluted RNA generally range from 5 to 50 ng/μL).

3.2. Primer Preparation

1. Stock solution: Based on nmol amount of primer in tube, add appropriate volume of DEPC-treated 5 m*M* Tris + 0.1 m*M* EDTA (pH 8.5) to give 100 pmol/μL (*See* **Note 3** about working with RNA and destruction of RNase with DEPC). Store at –20°C.
2. Working solution of primer: Dilute stock 1:10 with DEPC H_2O for use in RT-PCR, label tube with primer gene name and dilution and date. Store working solution (10 pmol/μL) at –20°C.

3.3. Optimize the PCR Reaction Conditions for Each Primer Pair

The optimization can be performed using the buffer supplied with the *Taq* enzyme and a range of Mg levels is tested for each primer pair. It may also be desirable to test using different levels of RNA. Use the following standard RT-PCR protocol for 0.2 mL PCR tubes and adjust the reagent volumes to allow excess reagent (equivalent to two or more reactions), and in the PCR buffer, allow 10 μL total for $MgCl_2$ with or without H_2O. In the following example, six levels of Mg are tested (make buffer for seven tubes). (*See* **Note 4**).

1. The reverse transcription (RT) reaction (1X buffer)

68.6 μL	H_2O
14 μL	10X RT buffer
14 μL	10 m*M* dNTPs
28 μL	25 m*M* $MgCl_2$
3.5 μL	RNasin
2.1 μL	Reverse primer (10 pmol/μL)
2.8 μL	AMV RT HC (20–25 units/μL)

   Per tube add 29 μL 1X buffer (as shown above) and 1 μL total RNA (100 ng or less) (total reaction volume = 20 μL) (*see* **Note 5**).
2. Use PCR apparatus to incubate tubes, add glycerol to wells in PCR block, place tubes in wells, and cap firmly; thermocycler should have heated lid turned on.
3. Incubate at 42°C for 15 min, followed by 95°C for 5 min, and then let sit for a few minutes with apparatus closed.
4. The PCR reaction
   a. To each RT tube, add $MgCl_2$ and water at various levels as follows (10 μL combined volume of $MgCl_2$ + H_2O):

Tube	Water	$MgCl_2$
1	0	10
2	2	8
3	4	6
4	6	4
5	8	2
6	10	0

b. Prepare 1X PCR buffer (for seven tubes):

21 µL	10X *Taq* buffer
2.1 µL	Reverse primer
2.1 µL	Forward primer
113.05 µL	H_2O
1.75 µL	*Taq* enzyme (1.25 U/50 µL reaction)

c. Add 20 µL 1X PCR buffer to each tube (total volume: 20 RT + 30 µL PCR= 50 µL), place in thermocycler; set annealing temperature and times based on information from primer selection program. A general program for a PCR run follows:

Step	°C	Time/notes
1	94	4 m
2	94	15 s dissociation
3	56	1 m primer anneal
4	72	30 s enzyme extension
5	34	Repeat steps 2–4 19 to 34 times for 20–35 total cycles
6	72	5 m
7	4	maintain at 4°C
8		End

The temperature in step 3 will be changed for each primer pair (determined in primer search program as optimal temperature).
The step 5 will be changed for numbers of cycles desired.
The PCR run shown will require 2.5–3 h.

5. Examine PCR products in an electrophoretic gel as detailed below.
6. Repeat with narrowed range of Mg if necessary.
7. Alternatively and in cases where outcomes are not acceptable, use a standard optimization kit (e.g., Opti-Prime PCR Optimization Kit [Stratagene, La Jolla, CA)] to optimize the buffer for pH and potassium and magnesium content. Follow kit directions to test the buffers supplied with the kit.

3.4. Prepare and Run High-Resolution Agarose Gel

1. Depending on the size of the gel apparatus, adjust the agarose volumes as needed and select a comb that allows at least 25 µL of PCR sample + loading buffer (*see* **Note 6**). A 2.8% Metaphor gel will give 10–20 bp resolution of 150–400 bp fragments.
   a. Weigh 2.24 g Metaphor agarose (FMC [Rockland, ME]).
   b. Place 250 mL Erlenmeyer flask with stir bar on platform balance and tare, then add 1X Tris-borate-EDTA (TBE) buffer to bring to 82 g (about 80 mL TBE). Keep scale setting as is, it will be needed later.
   c. Put flask on stir/heating plate, stir vigorously and heat on medium setting.
   d. Bring to boil and boil for 5 min (at too high a temperature, it will boil over very easily), then remove to countertop. Check weight on scale and add more water to bring weight back to 82 g.
   e. While waiting, prepare gel tray by taping across ends (if required by apparatus design you use) and press tape firmly to plastic. If needed for your apparatus, adjust comb to holder so that glass microscope slide moves easily under comb teeth (height of wells).
   f. After weighing dissolved agarose, let sit briefly to allow bubbles to rise to surface, then pour HOT agarose into tray.
   g. Allow to set undisturbed until opaque, usually 20–30 min. Best resolution achieved when gel is refrigerated at least a few hours, preferably overnight. If keeping at 4°C overnight, put small amount buffer on surface.
   h. Place gel in electrophoresis apparatus, use 1X TBE to fill chamber and cover gel.

2. Prepare samples and DNA ladder to run in the gel: use 1.5-mL tubes: label with sample identification.
   a. DNA ladder

0.5 µL	PGEM markers
19.5 µL	Tris/EDTA buffer (10 m*M* Tris/1 m*M* EDTA)
5 µL	tracking dye

   b. PCR sample

20 µL	PCR product (total volume 50 µL), cap and save extra PCR sample
4 µL	bromophenol blue tracking dye

   c. Load all of the sample (24 µL) into the wells of the gel.
3. Run gel with constant-voltage power supply set to 75–100 V (0.09–0.1 mA) (*see* **Note 7**). When dye migrates almost to the end of the gel (1–2 h) or approaches the top of next comb, stop power and remove gel to soak in ethidium bromide.
   a. In a tray, add 50 µL of stock (1 µg/µL) ETBr to 200 mL H_2O for small gel. Place gel in tray and soak for 30 min at room temperature. Remove gel to clean tray filled with water, examine on UV illuminator.
   b. Take gel to densitometer for image analysis, evaluation, and/or quantitation (*see* **Note 8**).
   c. Optional: Photograph each gel or gel section with type 55 p/n Polaroid film. (Gel photography with camera height = 67.5, *f*-stop = 4.5; shutter setting = B; ruler in frame = 1–14 scale in view; filter on camera = orange ethidium bromide filter; exposure time = 2 min, develop time = 30 s; wash negative in sodium sulfite solution (saturated: 101.3 g/500 mL H_2O) briefly. Wash in water 15 min. Dry negative.

3.5. Preparation of Internal Standards (IS)

3.5.1. Linearize Plasmid pUC19 with Restriction Enzyme

1. Digest 5 µg plasmid DNA for 3 h at 37°C:

5 µL	PUC DNA (1 µg/µL)
2 µL	10X buffer for *Hin*dIII
2 µL	*Hin*dIII
11 µL	Water
20 µL	Total

2. Heat digest at 68°C for 15 min to inactivate enzymes. Use 0.8% standard agarose gel to check linearization (should have single band).
3. Dilute the stock linearized PUC 1:10 for use in PCR reactions with IS primer pairs (IS primers) (*see* **Notes 9** and **10**).

3.5.2. Internal Standard cDNA Production from Linearized pUC

1. Prepare PCR buffer (1 tube = 50 µL reaction):

5 µL	10X buffer
6 µL	25 mM $MgCl_2$
2 µL	10 mM dNTPs
35.15 µL	H_2O
48.15 µL	Total

2. Add 1 µL diluted linearized pUC
3. Add 0.3 µL 10 pmol/µL forward IS primer
4. Add 0.3 µL 10 pmol/µL reverse IS primer
   Note: For high yield, prepare four reactions as above.
5. PCR Hot Start: Set PCR program for anneal temperature to 57°C and 30 cycles. Heat PCR tube to 80°C, then add 0.25 µL *Taq* per tube. Start PCR run (refer to PCR protocol).
6. Pool the contents of the four tubes.

3.5.3. Internal Standard cDNA Purification

One method is to use Wizard columns (Promega), as follows:

1. Prepare 80% solution of isopropanol with DEPC water and store at room temperature.
2. Combine the four PCR preps with 100 µL Direct Purification buffer (Wizard Kit) and mix.
3. Add 1 mL of resin from kit and vortex 3X over a 1-min period.
4. Use 3- or 5-mL syringe and 1.5-mL tube. Remove plunger from syringe, attach syringe to minicolumn; pipet resin/DNA into syringe barrel. Put in plunger and slowly and gently push slurry into minicolumn. Remove syringe from column, remove plunger, and reattach syringe to column.
5. Pipet 2 mL of 80% isopropanol into syringe, insert plunger, and gently push isopropanol through column as a wash. Stop pushing when all liquid is through, but do not push air through.
6. Remove syringe and transfer minicolumn onto a 1.5-mL tube. Centrifuge for 20 s at 14,000*g* to dry the column. Transfer minicolumn to new eppendorf tube.
7. Add 50 µL DEPC water by dripping onto column over center of resin bed. Wait 1 minute and spin as before; tube now contains 50 µL IS cDNA in water. Remove and discard column. Estimate concentration by OD_{260}.

3.5.4. In Vitro Transcription of RNA Internal Standard from IS cDNA

1. Perform in vitro transcription using a standard kit (e.g., Ambion T7-MEGAshortscript for (Ambion Inc., Austin, TX) large-scale synthesis of short transcript RNAs).
   All reagents at room temperature except enzyme mix, keep enzyme mix on ice during preparations.

x µL	DEPC water to give final volume = 20 µL
x µL	IS cDNA template = 5 µg
2 µL	10X transcription buffer
2 µL	rATP 75 m*M*
2 µL	rGTP 75 m*M*
2 µL	rCTP 75 m*M*
2 µL	rUTP 75 m*M*
2 µL	T7 MEGAshortscript Enzyme Mix

2. Incubate at 37 C for 2 h.
3. Add 5 µL (10 U/µg template) of RNase-free–DNase (10 U/µL).
4. Incubate 15 min at 37°C
5. Add 50 µL DEPC water and extract with 100 µL buffer-equilibrated phenol:chloroform (DEPC) and save aqueous phase. Add 50 µL DEPC water to phenol phase and re-extract. Combine aqueous phases.
6. Estimate volume with pipettor and add water to aqueous phase to give total volume of 115 µL, then add 15 µL of 5M ammonium acetate + 100 m*M* EDTA (Ambion Kit Stop solution). Add an equal volume of 4°C isopropanol (130 µL) to precipitate RNA. Centrifuge at 14,000*g* for 15 min at 4°C. Pipet or pour off ethanol without disturbing pellet. **When removing EtOH work quickly and watch region where pellet should be lodged.**
7. Add cold 70% ethanol (EtOH) at same volume as used for previous step. Spin 14,000*g* 5 min at 4°C. Pipet or pour off ethanol. Repeat 70% EtOH wash for three washes total. SpeedVac Concentrator (Savant Instruments Inc., Farmingdale, NY) 15–20 min. Resuspend pellet in 20 µL DEPC water.
8. Make 1:250 or 1:100 dilution and read OD_{260}.
9. Calculate concentration as OD × 40 × [dilution correction: (250/1000)or (100/1000)] = _____ µg/µL in original solution. Dilute IS cRNA to provide pg/µL levels for range finding experiments (*see* **Note 11**).

3.6. Quantitative RT-PCR Reaction (See Notes 12–14)

1. RT Buffer for 93 tubes (volume per PCR reaction)

186 µL	10X RT buffer	(2 µL)
372 µL	25 m*M* $MgCl_2$	(4 µL)
186 µL	10 m*M* NTPs	(2 µL)
46.5 µL	Rnasin	(0.5 µL)
818.4 µL	DEPC water	(varies with volume of RNA)
37.2 µL	AMV RT HC	(0.4 µL)
27.9 µL	Reverse Primer	(0.3 µL)
1674 µL	total	(18 µL pool)

2. Add 234 µL to each tube for the seven IS levels.
3. Add 13 µL of IS dilution to each of seven IS tubes (1 µL/PCR reaction).
4. Pipet 19 µL of buffer + IS mix into each tube following chart of each IS position in the strips of tubes. There is enough for 13 tubes, with only 12 tubes at each IS level, this allows 1 extra aliquot.
5. Add 1 µL (50–100 ng/µL) total RNA to each tube; follow chart.
6. Use PCR apparatus to incubate tubes at 42°C for 15 min, then 95°C for 5 min and let sit for a few minutes with apparatus closed.
7. While running RT step, prepare 1X PCR buffer (optimal volume of Mg will need to be determined for each primer set as described previously). This example is for 100 PCR tubes with no additional $MgCl_2$ added to the PCR buffer, anneal temperature of 55°C, and 35 PCR cycles. Each of these parameters needs to be determined for each gene. (Volume of PCR reagents = 30 µL; final volume with RT step is 50 µL.) PCR buffer for 100 tubes (3000 µL total):

10X PCR buffer	300 µL
25 m*M* $MgCl_2$	0 µL
H_2O	2615 µL
Reverse primer (10 pmol/µL)	30 µL
Forward primer (10 pmol/µL)	30 µL
Taq enzyme (1.25 U/each tube)	25 µL

8. After RT step, remove tubes carefully from apparatus. Remove caps from PCR tubes very slowly and by rocking gently and easing off. DO NOT rip off strip end-to-end or in other ways increase the likelihood of aerosol transfer from tube to tube.
9. Add 30 µL PCR reagent mix per tube. Cap firmly. Add glycerol to wells in PCR block if needed. Place tubes in wells. Close PCR lid. Lid heater on; PCR on. Set program with correct times and temperatures and cycle number.
   Example:

Step	°C	min/s
1	94	4 m
2	94	15 s dissociation
3	56	1 m primer anneal
4	72	30 s enzyme extension
5	34	(34 more repeats from step 2)
6	72	5 m
7	4	maintain at 4°C until turned off
8	End	

   The temperature in step 3 will be changed for each primer pair (determined in primer search program as optimal temperature). Step 5 will be changed for numbers of cycles desired.
10. Prepare Metaphor Agarose gel and proceed as described in **Subheading 3.4.**

3.7. Quantitation of the Ethidium-Stained Gels

The specific steps used to acquire the image of the gel and to determine density values for the bands will depend on the equipment and software selected (refer to the introductory comments and to **Note 8**). The basic procedure is as follows:

1. Place stained gel on UV illuminator and acquire an image of the banding patterns, use care to keep the higher-intensity bands within the linear range (i.e., do not allow the highest image intensity, brightest bands, to be at the maximal possible level). The use of a video camera with false color to observe the maximal levels is very helpful and more likely to find the best exposure conditions when compared to the use of photography-based image acquisition.
2. Quantitate the intensity of the bands using a densitometry program (NIH Image; *see* **Note 8**) to give a value for each band.
3. Calculate the ratio of the RNA to the IS values (RNA/IS) for each lane and plot the ratio (Y axis) against IS concentration (X axis). Perform a regression analysis for the best fit (typically, a linear fit to the data with a high correlation coefficient value, r^2) and use the regression formula to obtain the value of X when $Y = 1$ (ratio of IS/RNA=1), where X is the amount of mRNA for the target gene that was present in the total RNA (value is a function of the amount of total RNA used in the RT-PCR reaction).

3.8. Expression of Data as Molecules of mRNA/100 ng Total RNA

The data should be expressed relative to a constant amount of total RNA across all samples. Using molecules of mRNA, rather than a weight-related quantity such as picograms or nanograms, facilitates comparisons between studies. The number of molecules of mRNA represented by a picogram or nanogram amount of IS will depend on the size of the transcript, which, in turn, varies with the position of the primers selected. Thus, the data from different research groups cannot be directly compared with these units. For purposes of setting up the PCR study, we used a concentration of IS (e.g., pg/µL), but the data were all converted to molecules/unit of total RNA for publication. Data expressed as pg mRNA/unit total RNA can be converted to molecules/unit total RNA as follows: Calculate the molecular weight of the amplified IS RNA = # bases (between the forward and reverse primer including primer bases) × 307 (average base molecular weight); divide 1×10^{-12} g (use appropriate number for units used in quantitation [e.g., 1×10^{-15} for femtograms of IS]) by the molecular weight for the amplified region and multiply that by Avagadro's number (6.022045×10^{23} molecules/g molecular weight). This is the correction factor to convert all the data from units of weight to units of molecules.

Example:

The IS is 341 bases long × 307 = 104,687 MW

$1 \times 10^{-12}/104{,}687 = 5{,}752{,}429$ molecules/pg IS

Because the regression analysis determines the concentration of IS that is equal to the level of mRNA in the sample, the amount of mRNA is the sample can now be expressed as molecules of mRNA/100 ng total RNA (or whatever amount of total RNA was loaded in the PCR reactions).

4. Notes

1. Storage of homogenized samples at –80°C is preferred to storing frozen tissues. Homogenate can be stored for several days, but it is best used as soon as possible. For long-term

storage, the prepared total RNA is more stable than either tissue or homogenate and has been used up to 1 yr later with success in our lab.

2. Volume added will depend on the amount of starting material and the desirable concentration needed to perform an optical density analysis to give a precise estimate of total RNA concentration, which may dictate adding as little as 10–20 μL. Embryonic palatal shelves typically yield 0.2% of wet weight as total RNA (e.g., one dissected GD 14 mouse palatal shelf weighs about 1 mg and yields about 2 μg of total RNA). Embryonic liver generally yields at least twice that much total RNA. The following is a formula used in our lab to determine the volume to add to a sample in order to achieve total RNA concentrations of 10–12 μg/μL: volume in μL = tissue wet weight in grams × 166.66. This concentration range is desirable to determine the exact concentration of total RNA in the solution without using all of the sample to perform an OD_{260}.
3. Working with RNA and dealing with RNase enzyme. RNA is vulnerable to destruction by the ubiquitous enzyme RNase. This enzyme may contaminate all laboratory glassware and nonsterile plasticware, as well as being in reagents not certified for molecular biology use as RNase-free. The best way to eliminate the problem is to use only sterile plasticwares and, to autoclave the sterile pipet tips. If glassware is used, it should be autoclaved, and all solutions need to be either treated with diethylpyrocarbonate or made with RNase-free reagents and DEPC-treated water or buffers. Reserve a new set of pipettors exclusively for preparing the RT-PCR reagents and loading the PCR tubes. It is recommended to also reserve a lab bench (some labs even set aside a room exclusively for PCR work) that is used ONLY for this work. Some labs recommend preparing all PCR reactions in a sterile hood that can be cleaned frequently. We have set aside bench space, use dedicated pipettors, and use only filtered pipet tips for setting up RT-PCR. The PCR apparatus is kept in a different room. Once the billions of copies of the amplified target are in the PCR tube at the end of the PCR run, that is the most likely time to contaminate a work area. The product is kept away from the area used to prepare the total RNA and set up PCR reactions. Use different pipettors, pipet tips, and lab space to prepare and load the gels. The filtered tips are not needed at this point anyway. Use extreme caution to avoid contamination of your lab with the amplified PCR products, as cleanup is nearly always impossible to achieve and you will not be able to work with those primers/genes in that lab space again.
4. The temperature for denaturation is usually between 90°C and 95°C, the extension temperature is usually 72°C, and annealing is based on the length and GC content of the primer. The primer selection programs generally provide an optimal temperature and that is usually a good starting point. Adjustments to temperature and the time allowed for annealing can be made based on the outcome of initial screens. For initial tests, run at least 20–35 cycles.
5. Primers are species-specific, in optimization tests, be sure to use total RNA that is appropriate for the primers.
6. In our lab, a 10 × 15-cm gel tray needs 80 mL agarose solution and holds one 15-well, 1.5-mm-thick comb with a 20–25 μL capacity. For gel holding three combs, we use a 20 × 15-cm tray that uses 160 mL agarose solution. Tracking/loading dye used depends on PCR product size. Use a loading dye that suits the size of your amplified DNA. In a 2% Metaphor gel with 1X TBE running buffer, Bromophenol Blue migrates at approximately the rate of a 300-base DNA and xylene cyanol at approximately 4 kb. Prepare either Bromophenol Blue or xylene cyanol as follows: 0.25% bromophenol blue or xylene cyanol = (50 mg per 20 mL H_2O) + 40% sucrose (8 g per 20 mL H_2O).
7. Voltage will depend on gel size and can be up to 150 V for a larger gel. Generally, we run a 10 × 15-cm gel at 90–100 V and a 30 × 14-cm gel at 150 V. Run at the highest voltage that does not heat up the buffer and gel, detectable as excess condensation on the lid over

the apparatus. Run until the loading dye reaches the end of the gel. Because the agarose is fairly costly, we save the gel after use in a ziplock bag at 4°C and then remelt in a microwave oven, boil 5 min, and repour the gel for a second use. Gels are discarded after the second use. The reused agarose performs just as well as the original.

8. Our lab is using the Argus-20 SCSI Software for the Hamamatsu Photonic System with a Macintosh computer to acquire 16-bit digitized images for analysis. The densitometry for the bands is acquired and processed with NIH Image, a public-domain image processing and analysis program for the Macintosh (written by Wayne Rasband at the U.S. National Institutes of Health and available from the Internet by anonymous ftp from zippy.nimh.nih.gov, enter "anonymous" as user name and your e-mail address as the password. The latest version and documentation can be found in the /pub/nih-image directory).
9. Internal standard primers (forward and reverse include the primer sequence of the targeted gene, sequences for PUC plasmid, and a T7 polymerase promoter sequence) are used to amplify a region of the PUC plasmid (refer to **Fig. 1C**). That region is selected to give a cDNA that will be about 50 bases longer or shorter than the gene sequence when amplified using the primers for the target gene.
10. This digest volume may only provide enough DNA for one in vitro transcription reaction. The reaction can be scaled up to improve yield.
11. If cRNA for IS has been stored at –20°C for a long period, it may need to be cleaned up and degraded RNA removed by reprecipitation.
12. The dilutions are set wide for an initial range finding for expression of the target gene in a control total RNA sample. As the range of expression is identified, the IS range is narrowed to give bands for both IS and target mRNA in all reactions. This is desirable, as it means that all reactions contribute to the quantitation, which is based on a ratio of the bands. Lanes with only the IS band or only the mRNA band cannot contribute to the regression analysis. Generally, high-expression genes may be in the pg/µL IS range when 50–100 ng of total RNA is loaded per PCR reaction. The range of IS levels can be adjusted either up or down to provide a suitable PCR outcome. The ideal protocol uses the least amount of total RNA (as low as 20 ng can be used). Small embryonic tissues give limited total RNA and this approach provides the option to examine tissue from individual embryos rather than pooling samples from multiple embryos and may allow multiple genes to be examined for each sample. There are advantages to running the assay with picogram to femtogram levels of IS as well, as the lower ranges use less of the IS cRNA (reducing the need to resynthesize). This assumes that your detection apparatus for imaging the bands on the gel is sufficiently sensitive to operate at low-intensity levels of ethidium bromide staining. In our lab, the large gel holds 3 combs with 15 positions or wells each. For a full RT-PCR run, two large gels can accommodate either 12 samples run with 7 IS levels each or 18 samples with 5 IS levels each. The total PCR reactions for these runs are 84 and 90, respectively, and the 6 combs across 2 gels have 90 wells available. Initially, we ran seven IS levels, but with high-correlation coefficients from the assays, it was considered reasonable to reduce the IS to five levels. This should be determined for each gene studied.
13. The reagents are pooled as much as possible to give high reproducibility across samples and between assays. For example, RT reagent present in all tubes are first pooled (all reagents except the IS and total RNA = 18 µL per reaction tube) and then that pool is divided into aliquots to which each IS added (1 µL per reaction tube). Thus, each sample has an RT buffer and IS level identical to every other sample in that PCR run, as they are from a common stock solution.
14. For quantitative RT-PCR runs, each tube contains 1 µL RNA and 19 µL of pooled reagents. The protocol in the example was set up to run 12 total RNA samples with 7 internal standard levels each (84 reactions). The pooled RT reagent is setup for 93 tubes allowing

excess reagent to avoid running short in the last tube (18 μL per tube = 1674 μL total). That pool is divided across 7 IS levels with excess included (13 reactions per IS tube; 13 × 18 μL each = 234 μL per IS tube). Because 13 reactions are included per IS tube, 13 μL of IS stock is added (1 μL per reaction). Thus, each IS pool now includes enough reagent for 13 reactions (each reaction = 18 μL of pooled reagents + 1 μL IS). The last step is to pipet 19 μL of each IS pool into the seven IS tubes labeled for each sample. Excess reagent from the RT buffer pool can be used to set up controls that have IS but no RNA, or RNA without IS. These indicate that the reaction occurred and that the RNA was in good enough condition to amplify (e.g., no RNase destruction). A control for DNA contamination is recommended and requires a reaction with RNA, without IS and without the RT enzyme. When this tube proceeds into the PCR step, the RNA was not transcribed into DNA; RNA does not amplify under these conditions, so if a PCR product appears in the gel, it would be due to DNA contamination of the total RNA preparation. It is also useful to set up a table or matrix to define the position of each sample and IS level ahead of time for large PCR runs.

References

1. Vanden Heuvel, J. P., Tyson, F. L., and Bell, D. A. (1993) Construction of recombinant RNA templates for use as internal standards in quantitative RT-PCR. *BioTechniques* **14,** 395–398.
2. McPherson, M. J., Quirke, P., and Taylor, G. R. (eds.) (1993) PCR. A Practical Approach. IRL, Oxford University Press, New York.
3. Chomczynski, P. (1993) A reagent for the single-step simultaneous isolation of RNA, DNA and proteins from cell and tissue samples. *BioTechniques* **15,** 532–536.

23

Transspecies Grafting as a Tool to Understand the Basis of Murine Developmental Limb Abnormalities

Sheila M. Bell, Claire M. Schreiner, and William J. Scott

1. Introduction

Elongation along the proximal distal axis of the limb, from either the pectoral or pelvic girdle to the distal tips of the digits, is regulated from the early stages of limb morphogenesis by the formation of a specialized epithelial structure at the limb bud apex, the apical ectodermal ridge (AER). The formation and maintenance of the AER are dependent on signals produced by the underlying mesenchyme ***(1–4)***. In return, the AER maintains the underlying mesenchyme in an undifferentiated, proliferative state. Surgical extirpation of the AER results in truncated limbs; the most distal structures formed can be correlated with the time of AER removal ***(5)***. Thus, in the characterization of abnormal phenotypes in which the limb fails to develop distally complete limbs, a lack of AER formation or regional maintenance may be causative, but the tissue origin of the defect could be either the ectoderm or mesoderm. To determine if the defect is indigenous to the ectoderm or mesoderm, one approach is to perform tissue recombination experiments in which normal ectoderm or mesoderm is isolated and recombined with the complementary tissue isolated from the limb bud of an abnormal embryo. Recombinant limbs are then grafted to the flank of a host chick embryo and allowed to develop. This approach has been used to analyze two chick mutants, *limbless* ***(3)*** and *wingless* ***(6)***, and the mouse mutant *limb deformity (ld)* ***(7)***. In the analysis of the chick mutants, double reciprocal exchanges were performed (i.e., normal ectoderm and mutant mesoderm) as well as mutant ectoderm and normal mesoderm limb recombinants were analyzed. In these studies, the ectoderm isolated from either *limbless* or *wingless* was found to be incapable of promoting normal limb morphogenesis ***(3,6)***. Tissue recombination has been successfully used to examine the ability of rat-limb-derived ectoderm to promote normal morphogenesis of chick mesoderm ***(8)*** and, recently, in the analysis of the mouse mutant *limb deformity (ld)* ***(9)***. The limb buds of *ld*-null animals exhibit an abnormal AER morphology and the mutated gene is expressed in both the limb mesenchyme and AER. Recombination studies in which *ld* limb ectoderm was recombined with chick/quail mesoblasts suggest that the abnormal AER observed in *ld*-null limb buds is attributable to the role of formin proteins in the distal mesenchyme, not in the AER itself ***(7)***. We are currently using this strategy to

From: *Methods in Molecular Biology, Vol. 136: Developmental Biology Protocols, Vol. II*
Edited by: R. S. Tuan and C. W. Lo © Humana Press Inc., Totowa, NJ

determine if the lack of AER formation in the hindlimb buds of the mouse mutant *lgl* is of ectoderm or mesoderm origin ***(10)***. Ideally, one would like to perform this type of analysis as was done for the chick mutants in which recombinant limbs are generated from both mouse mutant limb ectoderm and mesoderm. Unfortunately, mesoderm from rodent limb buds does not become vascularized by the chick host and the recombinant limbs fail to develop.

In designing an experiment of this type, it is desirable to know whether or not the mutant limb bud forms an AER or whether the AER is formed but prematurely regresses, becomes regionally hyperplastic, and so forth. These events can be evaluated by examining the mutant limbs using whole-mount *in situ* hybridization assays ***(11)*** with select genetic markers, Nile Blue staining for localized regions of cell death, and histological evaluation. Pinpointing the gestational stage when proximo-distal patterning first goes awry is beneficial because using ectoderm isolated from limbs at a developmental stage at which the abnormalities are already obvious may not be rescuable.

The following protocol has been adapted to examine mesenchymal–epithelial interactions occurring during the AER inductive stages of murine limb development (**Fig. 1A,B,D,E**). This approach can also be used to examine AER maintenance at later developmental stages. Because of the small size of the mouse limb bud, it is preferential to generate recombinant limbs using quail mesoblasts, which are slightly smaller than that of the chick. This is particularly true if the mouse ectoderm to be recombined is at a developmental stage prior to AER formation [**Fig. 1E** ***(12)***]. The overall objective is to cover as much of the isolated chick or quail mesoblast as possible with mouse-limb-derived ectoderm. Areas not covered by ectoderm will fail to grow, resulting in incomplete limbs usually lacking anterior and or posterior limb structures of the wing or leg.

An additional consideration is whether the donor mesoderm is of wing or leg origin. Recombinant limbs will develop according to the origin of the mesoderm. Mutants like *legless*, in which the hindlimbs are always affected and the forelimbs are variably affected, suggest that different signals may mediate AER induction in the hindlimb versus the forelimb. Thus, for our studies, we only used leg mesoblasts to generate the recombinant limbs. Because the ectoderm of the chimeric limbs is derived from mouse tissue, the developed chimeric limb should lack scales (**Fig. 1F,G**). Similarly, if recombinant limbs are generated with mesoderm isolated from wing buds, the chimeric wing should lack feather buds.

2. Materials

1. Egg incubator: Incubators with rotators are not necessary. The incubator must be able to maintain a constant temperature between 38°C and 41°C and a humidified atmosphere.
2. Dissecting scope: A scope with magnifying ability of at least 25X is recommended.
3. Surgical tools including extrafine watchmaker forceps, Noyes scissors, Vannas microscissors, and tungsten needles (Roboz Surgical Instruments, Washington, DC).
4. Watch glasses 40 mm (Thomas Scientific, Swedesboro, NJ, #9787-B22).
5. Disposable Petri dishes 35 mm and 100 mm.
6. Dremel minimite drill with cutoff wheel No. 409. Dremels are available in most hobby shops and hardware stores.
7. Scotch tape, markers.
8. Horse serum.
9. Penicillin–streptomycin solution (500 U and 500 μg/mL, respectively).

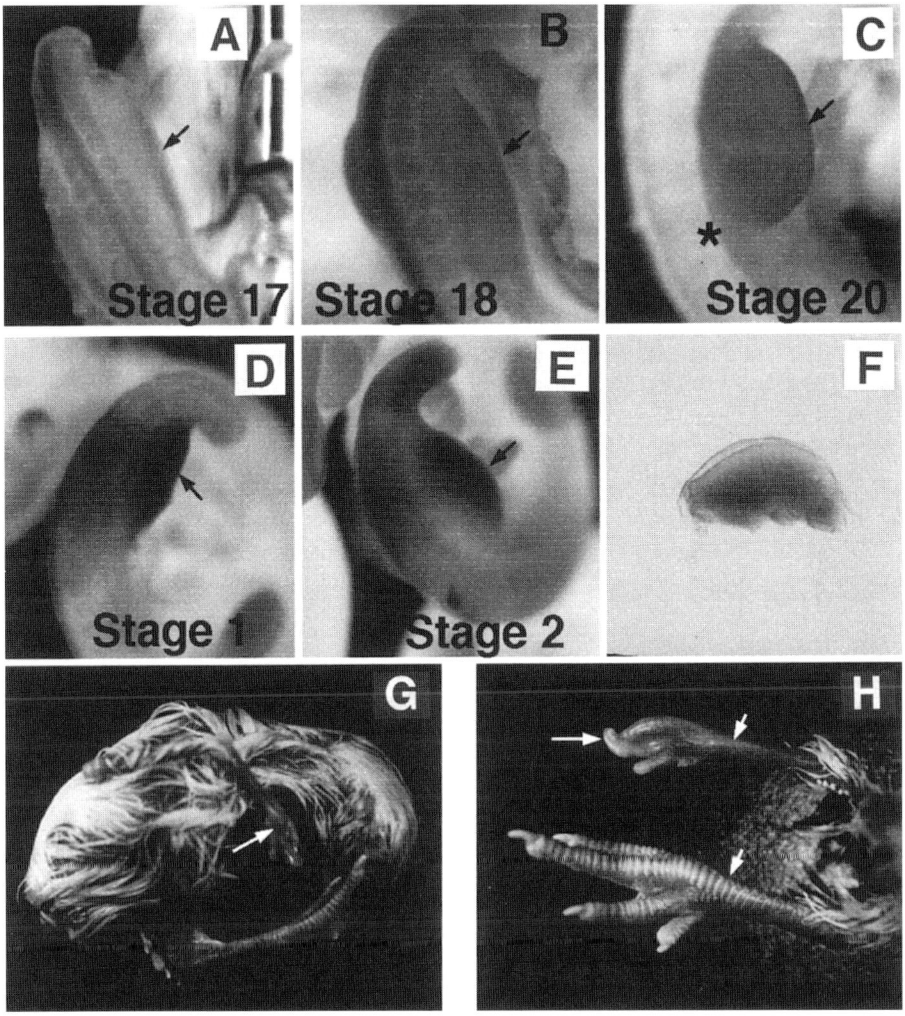

Fig. 1. Stages of chick limb development **(A–C)**, asterisks in (C) denotes site on a host flank to graft the recombinant limb. Stages of mouse limb development **(D,E)**. Appearance of recombinant limb prior to grafting **(F)**. Grafted recombinant limb after 12 d of incubation **(G,H)**. Arrows in (A)–(E) indicate the location of the limb bud. Arrows in (G) and (H) indicate the recombinant limb, short arrows indicate the presence and absence of scales.

10. Buffers: Ca^+- and Mg^+-free phosphate-buffered saline [PBS ($-Ca^{2+}$, $-Mg^{2+}$)], 8 g NaCl, 0.2 g KCl, 0.2 g KH_2PO_4 (anhydrous), 1.14 g Na_2HPO_4 (anhydrous) per liter H_2O, pH with HCl to 7.3, autoclave. For PBS-containing Ca^+ and Mg^+ [PBS ($+Ca^{2+}$, $+Mg^{2+}$)] in addition to the above ingredients, add 0.14 g $MgCl_2 \cdot 6H_2O$, 0.1 g $CaCl_2$, and filter-sterilize.
11. Crude trypsin (Sigma T-7409) (Sigma Chemical, Co., St. Louis, MO) make fresh in PBS ($-Ca^{2+}$, $-Mg^{2+}$) on the day of use.
12. A 5-mL syringe with an 18 gage 1 in. needle.
13. Fertilized white leghorn eggs from SPAFAS farms or other supplier. Fertilized quail eggs can be obtained from Georgia Quail Farms Manufacturing (Savannah, GA).
14. Time-mated pregnant mice.
15. Nile Blue sulfate stock solution that is 1:20,000 w/vol in PBS.

3. Methods

3.1. Incubation

Temperature, humidity, and healthiness of fertilized eggs determine the growth rate of the embryos. Under incubator conditions of 38°C and 60% humidity, the embryos in fertilized white leghorn eggs (SPAFAS) will be at approx stage 17 (**Fig. 1A**) after 72 h of incubation and stage 20 (**Fig. 1B**) after approx 90 h (*see* **Note 1**). Development will occur more rapidly at higher incubation temperatures (**Note 2**).

3.2. Locating the Embryo

On d 2 of incubation, in a darkened room determine the position of the chick/quail embryo in the egg by holding the egg up to a pinpoint light source. In viable eggs, a shadow will be obvious where the embryo is developing. Over the embryo, use a marker to draw a small box on the egg surface. Use the point of a small pair of scissors to make an initial hole into the narrow end of the egg. Insert a syringe with an 18-gage needle gently into the hole and remove approx 2 mL of albumin. This will allow the embryo to drop slightly away from the shell membrane, minimizing damage to the embryo and its surrounding membranes and vasculature when the eggs are windowed (**Note 3**).

3.3. Windowing

On d 3, use a Dremel to cut the drawn box out of the egg shell, remove the pieces using watchmaker forceps (**Note 4**). Any pieces of shell that fall into the hole should be carefully removed. At this time apply 3 drops of a penicillin–streptomycin solution (550 units/mL and 500 μg/mL, respectively) to each egg. Close the window with 3M brand Post-It flags. Secure the colored end of the flag and the other three edges with small pieces of Scotch tape. This prevents the embryos from drying out while in the incubator. *Helpful tip:* The eggs obtained from older flocks of chickens will often have a higher percentage of abnormally developing embryos (i.e., no heads, not turned, etc.). If this occurs, request a flock change from the egg supplier or you will likely have an extremely high mortality rate.

3.4. Staging

Using lighting from above the egg and a dissecting microscope, stage (**Note 5**) the embryo by viewing through the window into the egg. If the embryo has not developed sufficiently, reseal the window with tape and return it to the incubator.

3.5. Isolating Chick Embryos

When the chick or quail embryos to be used as limb mesoderm donors have attained the stage of interest, crack the egg into a 100-mm Petri dish. If careful, the embryo will remain on top of the yolk. Use a fine pair of scissors to cut the embryo free of its attached membranes and vasculature. It is convenient to use a spatula with a spoon end to recover the embryo. Rinse the embryo in PBS and use watchmaker forceps and Vannas scissors to finish removing the surrounding membranes.

3.6. Isolating Donor Mesoblast

To remove stage 16–18 limb buds, make a transverse cut across the embryo at positions anterior and posterior to the limb. This will yield a piece of tissue containing both the left and right limb buds with intervening trunk tissue. Dissect the trunk in half with either watchmaker forceps or 30-gage needles. All subsequent manipulations will be performed in 35-mm Petri dishes set directly onto a bed of ice. Transfer the isolated tissues to a chilled dish containing the 2.0% crude trypsin solution. Incubate for approximately 40 min (**Note 6**). To inactivate the trypsin, transfer the tissue to a fresh dish sitting on ice containing 10% horse serum in PBS with Ca^{2+} and Mg^{2+} and incubate for 30 min. Keeping the limbs cold, use clean watchmaker's forceps to gently strip the chick/quail ectoderm off of the chick/quail mesoderm (**Note 7**). Discard the ectoderm. Use 30-gage needles to cut away extra flank tissue.

3.7. Isolating Donor Mouse Limb Ectoderm

Isolate mouse embryos and dissect free of their surrounding amnion and yolk sac (**Note 8**). Stage the mouse limb buds (**Note 9**). Isolate the limb region as described above for the chick, cutting at positions above and below the limb across the trunk. If using stage 1–2 limb buds, you will need the extra ectoderm from the adjacent flank tissue to help cover the chick mesoblast resulting from the differences in size between the chick/quail and mouse limb buds. To isolate the mouse limb buds along with adjacent flank tissue, cut along the left and right lateral edge of the neural tube using a scissor motion with two 30-gage needles (**Notes 10** and **11**). Discard the neural tissue. In a 35-mm Petri dish sitting on ice, incubate the limb and adjacent trunk tissue with 2.0% crude trypsin in PBS ($-Ca^{2+}$, $-Mg^{2+}$) for 30 min. Transfer the tissue to a fresh dish containing 10% horse serum in PBS ($+Ca^{2+}$,$+Mg^{2+}$) and incubate on ice 30 for min.

3.8. Tissue Recombination

Coordinate the timing of the above trypsinization reactions such that the chick/quail mesoderm and the mouse ectoderm are ready to be recombined at about the same time. Place both the denuded chick mesoderm and the mouse limb bud in a watch glass containing 10% horse serum PBS ($+Ca^{2+}$, $+Mg^{2+}$) sitting on ice. Use clean watchmaker forceps to gently tease the mouse ectoderm away from the mouse mesoderm; beginning at the edges of the somite tissue and gently working the ectoderm free as a sheet of tissue, try to take it off like a glove. This requires practice and patience. Immediately position the mouse ectoderm onto the chick/quail mesoblast, covering as much of the mesoblast as possible. Once recombined, place the watch glass at room temperature. The tissues will shrink and begin to adhere to one another as they warm up, which requires approximately 30 min (**Note 12**). The final recombinant limb should appear as in **Fig. 1F**. If stage 2 or greater mouse limb buds are used, in the center of the tissue a more opaque area should be evident representing either the broad AER precursor, or the AER in older limbs.

3.9. Grafting

Recombinant limbs are grafted onto the flank of stage 20 host chick embryos (**Fig. 1C**, asterisk denotes graft site). At this stage, the host right-wing bud is readily visible. Using watchmaker forceps, make an opening in the overlying membranes just slightly larger than the region over the three to four somites adjacent to the wing bud (**Fig. 1C**) (**Note 13**). Be careful not to damage any major blood vessels in the overlying membranes. Create a wound bed slightly larger than the base of the recombinant limb bud by gently tearing away the ectoderm in the area of interest with watchmaker forceps. Place the base of the recombinant limb onto the wound bed using watchmaker's forceps and a thin blunt tungsten needle as a guide. Once positioned, do not move the egg for 15 min to permit the recombinant tissue to adhere to the host. Tightly seal the window with a 3M flag and scotch tape, then return to a humidified incubator for continued growth. Growth can be monitored for the first 24–36 h by viewing the embryo through the window.

However, viability seems to increase if the eggs are left tightly sealed either with scotch tape or a more adherent tape like electrical tape. Do not attempt to subsequently view an embryo whose window has been sealed by a more adherent tape, as the shell will likely get broken during the tape removal process.

3.10. Analysis

The time postrecombination when the chick hosts should be sacrificed is dependent on the question to be answered. If alterations in mouse AER morphology are of interest, then embryos should be collected within a 48-h time period. Embryos should be dissected free of their associated membranes prior to being fixed for subsequent analysis, such as histological sectioning or *in situ* hybridization assays.

To determine if the mouse ectoderm can promote the formation of distally complete limbs, the hosts should be incubated an additional 7–12 d to permit the formation of bone and cartilage (**Fig. 1G,H**; **Note 14**). Scales normally develop slightly later than feather germs. Thus, chimeric limbs should be allowed to develop for 10–12 d to permit the formation of scales if leg mesoderm are used to generate the recombinant limb. Chimeric limbs derived from wing mesoblasts require only a 7-d incubation period to determine whether feather germs are present or absent. For better visualization of wing feather germs or leg scales, collected specimens can be stained with Nile Blue for 15–30 min. Destain the limbs by subsequently soaking in PBS. Photograph recombinant limbs prior to fixation for the staining of cartilage and bone *(13)* elements in the recombinant limb.

4. Notes

1. Trial incubations should be performed to determine the rate of chick or quail embryo growth in the incubation system setup in the laboratory. This is easily accomplished by placing eggs in the incubator at various times throughout the day and determining the stage achieved at specific times thereafter.
2. The fertilized eggs can be stored at 16°C prior to incubation. Little loss in viability is observed if they are stored less than 1 wk.
3. If the embryo fails to drop away from the shell membrane or if the membranes surrounding the embryo are damaged during the windowing process, discard the egg because it will likely die.
4. When using the Dremel tool, avoid applying excessive pressure or it may saw far enough into the egg that it damages the membranes below or cracks the remainder of the shell. If this happens, the embryos may be used as donors but will not survive extended incubation. Do not use these embryos as the recombinant limb recipients.
5. Most investigators currently stage chick/quail embryos using the system of Hamburger and Hamilton *(14)*, which was recently reprinted *(15)*. The wing and leg buds develop at similar rates, thus the stage of the visible wing is usually also representative of the leg bud.

6. The effectiveness of crude trypsin can vary between lots. For consistency, determine the optimal length of digestion time for a specified lot and then order several vials.
7. After trypsinization, the tissues will be sticky. It is recommended that very clean forceps be used, as traces of tissue from the previous dissections will cause the tissue to stick to the forceps and become unusable.
8. If necessary, either the yolk sac or the remainder of the embryo can be saved for genotyping.
9. We have routinely used the mouse limb staging system defined by Wanek et al. ***(12)***. Some of the early stages (pre-AER formation) of mouse limb development are pictured in **Fig. 1**. Unlike chick limb buds, development of the mouse hindlimb bud lags about 0.5 d behind that of the forelimb bud.
10. To easily grip the needles, attach them to 1-cc disposable syringes.
11. If it is important to know the dorsal and ventral identities of the ectoderm and mesoderm, these can be marked prior to trypsinization with carbon particles.
12. Regions of mesoderm not covered by the overlying ectoderm will not contribute to the developing recombinant limb. If older, stage 20 chick mesoblasts are used, it will likely be necessary to trim the mesoblast down in order for the mouse ectodermal hull to cover the tissue. Significant amounts of trimming will affect the final skeletal pattern and resulting limbs may be missing anterior and/or posterior digits because of the absence of sufficient quantities of mesoderm.
13. The damage done to the membranes should be minimized. However, the membranes should not be allowed to touch the recombinant limb once grafted because as the membranes heal, they can constrict around the limb resulting in it being displaced from the flank.
14. Mortality can be high during this 7–11-d incubation period. It is important to keep the egg tightly sealed to prevent the embryo from becoming dehydrated. Viability is also related to how severely the membranes were damaged during grafting.

References

1. Kieny, M. (1968) Variation de la capcite inductrice de mesoderme et de la competence de l'ectoderme au cours de l'induction primaire du bourgeon de membre, chez l'embryon de Poulet. *Archs Anat. Micr. Morphol. Exp.* **57,** 401–418.
2. Saunders, J. and Reuss, C. (1974) Inductive and axial properties of prospective wing-bud mesoderm in the chick embryo. *Dev. Biol.* **38,** 41–50.
3. Carrington, J. L. and Fallon, J. F. (1988) Initial limb budding is independent of apical ectodermal ridge activity; evidence from a *limbless* mutant. *Development* **104,** 361–367.
4. Zwilling, E. and Hansborough, L. (1956) Interaction between limb bud ectoderm and mesoderm in the chick embryo: III. Experiments with polydactylous limbs. *J. Exp. Zool.* **132,** 219–239.
5. Saunders, J. W. (1948) The proximo-distal sequence of origin of parts of the chick wing and the role of the ectoderm. *J. Exp. Zool.* **108,** 363–404.
6. Carrington, J. L. and Fallon, J. F. (1984) Evidence that the ectoderm is the affected germ layer in the wingless mutant chick embryo. *J. Exp. Zool.* **232,** 297–308.
7. Kuhlman, J. and Niswander, L. (1997) Limb deformity proteins: role in mesodermal induction of the apical ectodermal ridge. *Development* **124,** 133–139.
8. Jorquera, B. and Pugin, E. (1971) Behavior of the mesoderm and ectoderm of the limb bud in the exchanges between chicken and rat (French). C.R. Hebdomad. Seances Academie Sciences—D. *Sci. Natur.* **272,** 1522–1525.
9. Woychik, R. P., Stewart, T. A., Davis, L. G., D'Eustachio, P., and Leder, P. (1985) An inherited limb deformity created by insertional mutagenesis in a transgenic mouse. *Nature* **318,** 36–40.

10. McNeish, J., Scott, W., and Potter, S. (1988) *legless*, a novel mutation found in pHT1-1 transgenic mice. *Science* **241,** 837–839.
11. Wilkinson, D. G. and Nieto, M. A. (1993) Detection of messenger RNA by *in situ* hybridization to tissue sections and whole mounts. *Methods. Enzymol.* **225,** 361–373.
12. Wanek, N., Muneoka, K., Holler-Dinsmore, G., Burton, R., and Bryant, S. (1989) A staging system for mouse limb development. *J. Exp. Zool.* **249,** 41–49.
13. Kuczuk, M. H. and Scott, W. J. (1984) Potentiation of acetazolamide-induced ectrodactyly in SWV and C57BL/6J mice by cadmium sulfate. *Teratology* **29,** 427–435.
14. Hamburger, V. and Hamilton, H. L. (1951) A series of normal stages in the development of the chick embryo. *J. Morphol.* **88,** 49–92.
15. Hamburger, V. and Hamilton, H. L. (1992) A series of normal stages in the development of the chick embryo. 1951 [classical article]. *Dev. Dyn.* **195,** 231–272.

24

Assessment of Laterality Defects in Rodent Embryos

Masahiko Fujinaga

1. Introduction

How the left–right body axis is established during development and how abnormal sidedness of asymmetric body structures i.e., laterality defects or situs inversus (*see* **Note 1**), is induced during development have been questions of great interest in the history of science. A recent discovery of signaling molecules that are asymmetrically expressed before the establishment of morphological asymmetry in the chick embryo by Tabin and colleagues *(1)* has fueled the research of this area. Such investigations have been quickly expanded into mouse and rat embryos, the most commonly used mammalian experimental models (*see* **Note 2**).

When assessing mouse or rat embryos for the existence of laterality defects, it is important to examine embryos at relatively early stages of development, because embryos with laterality defects are associated with a high incidence of other abnormalities and may not survive for long within the uterus or after birth. In the mouse or rat embryo, the first commonly recognized morphological sign of body asymmetry is the looping of the cardiac tube that occurs on gestational day (GD) 8 in the mouse and GD 10 in the rat (*see* **Note 3** for a description of early cardiogenesis and **Note 4** for an asymmetric event that might occur earlier than cardiac looping). The second major morphological sign of body asymmetry is so-called axial rotation (*see* **Note 5** for mechanisms of axial rotation and **Note 6** for a morphological description of axial rotation). After the completion of axial rotation, the establishment of placental circulation is facilitated by rapidly developing umbilical vessels which simultaneously facilitate the process of organogenesis. Therefore, the stage after the completion of axial rotation (i.e., GD 9 in the mouse and GD 11 in the rat) is an ideal time-point to make an assessment of body asymmetry and to obtain a full view of laterality defects before they are affected by other abnormalities and/or maternal factors. This chapter aims to describe the assessment of sidedness of asymmetric body structures at this early developmental stage and to provide background information on those asymmetric developmental events and laterality defects.

From: *Methods in Molecular Biology, Vol. 136: Developmental Biology Protocols, Vol. II*
Edited by: R. S. Tuan and C. W. Lo © Humana Press Inc., Totowa, NJ

2. Materials

1. Timed-pregnant animals: It is important to understand the exact stage of pregnancy of animals to be used (*see* Chapter 8, **Note 7** in Vol. I in this series).
2. Surgical tools for uterine dissection: Large surgical scissors and toothed forceps to cut abdominal skin and muscle, and small surgical scissors and small toothed forceps to dissect the uterus.
3. Disposable Petri dishes e.g., 100 × 15-mm and 60 × 15 × mm size.
4. Phosphate-buffered saline (PBS) solution or other appropriate buffered solution (e.g., Hank's balanced salt solution [HBSS]).
5. Glass dissecting dish with silicon bottom, so-called "silicon Petri dish" (*see* **Note 7**).
6. Surgical tools to dissect embryos from the uterus: fine scissors and fine forceps to dissect deciduae from the uterus, and two pairs of fine forceps to dissect embryos from deciduae (*see* **Note 8**).
7. Dissecting microscopes (*see* **Note 9**).
8. Transfer pipet (*see* **Note 10**).

3. Methods

1. Anesthetize (see Chapter 8, **Notes 10, 11,** and **29,** Vol. I) or kill the animal by appropriate methods.
2. Wet the abdomen with rubbing alcohol and blot with a piece of tissue paper.
3. First, make a transverse abdominal skin incision with scissors; then, make T-shape muscle incisions (*see* Chapter 8, **Fig. 1,** Vol. I).
4. Open the thoracic cavity and incise the heart to ensure the death of the animal.
5. Grab the uterine cervix with toothed forceps from the abdominal side, and transect the uterine cervix with small scissors.
6. While pulling the uterine cervix upward (hold it tight while doing this), cut fat attached to the uterus and resect the uterus, including the ovaries.
7. Rinse the uterus once in PBS in a regular Petri dish to remove blood.
8. Transfer the uterus to a silicon Petri dish containing Hank's balanced salt solution (HBSS), and pin through the uterine cervix and an ovary on one side (*see* Chapter 8, **Fig. 2,** Vol. I).
9. Make incisions in the opposite side of mesometrium in the uterus using a fine pair of scissors, and dissect the deciduae out of the uterus (*see* **Note 11**).
10. Cut off the uterine horn on the dissected side, and repeat **step 9** on the other side.
11. Transfer deciduae into another Petri dish (e.g., 60 × 15 mm) containing HBSS, and dissect the embryos from the decidua using two pairs of fine forceps.
12. Remove Reichert's membrane under a dissecting microscope (*see* **Note 12**).
13. Assess sidedness of the following structures (*see* **Note 13**): (1) chorioallantoic placenta and (2) lower part of body, so-called "tail" (**Fig. 1**).
14. Break the visceral yolk sac and amnion and assess the embryo for sidedness of cardiac looping (**Fig. 2**). Examples of various types of embryos are shown in **Fig. 3**.

4. Notes

1. There is inconsistency in the definition of "situs inversus" among investigators ***(3)***. Whereas some investigators use the term "situs inversus" to describe any type of inversion of asymmetric body structures either complete or partial, others uses it to describe complete inversion only (**Fig. 4**). This chapter follows the former definition.
2. Although many different staging systems are used for mouse and rat embryos, a modified Theiler's staging system ***(4)*** is used in this chapter (**Fig. 5**). In addition, GD is used to

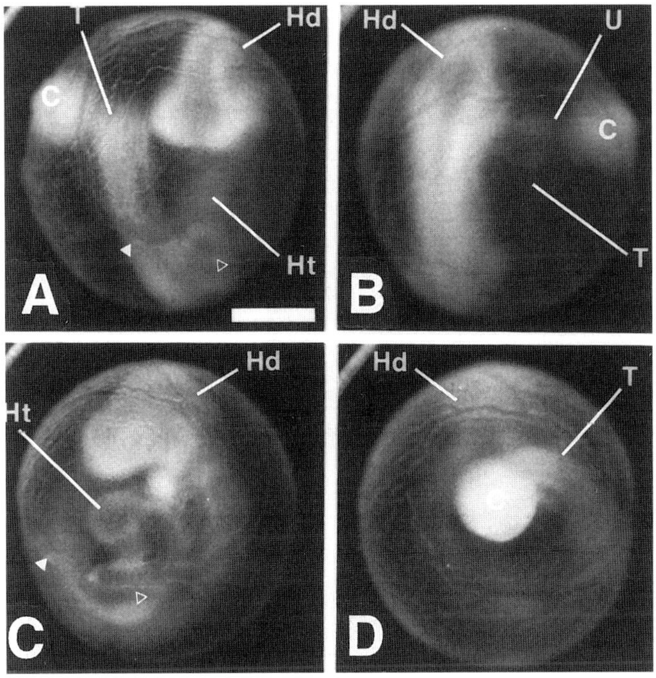

Fig. 1. Normal embryo within the intact visceral yolk sac at stage 15 (approximately 23–24 pairs of somites; GD 9 in mouse and GD 11 in rat). Bar = 1 mm. **(A)** Frontal view. The chorioallantoic placenta and tail are normally located on the right side of the embryo, and the vitelline vessels emerge from the left side of the embryo. The heart (bulboventricular loop) shows a characteristic C-shape dextral loop. **(B)** Back view. The umbilical cord connects to the tail and chorioallantoic placenta on the right side of the embryo. **(C)** Left lateral view. **(D)** Right lateral view. Abbreviations: C, chorioallantoic placenta; Hd, head; Ht, heart; T, tail; U, umbilical vessels. Symbols: Open arrowhead: vitelline vein; closed arrowhead: vitelline artery. (Derived from **ref. *2*.**)

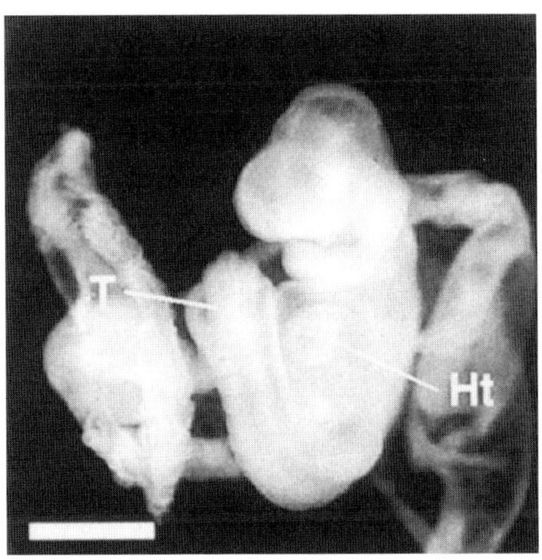

Fig. 2. Frontal view of the normal rat embryo after the visceral yolk sac and amnion have been dissected out at stage 15. Bar = 1 mm. The tail (lower part of body) is flexed to the right side and the bulboventricular loop (heart) shows dextral looping ("C" shape). (Derived from **ref. *2*.**)

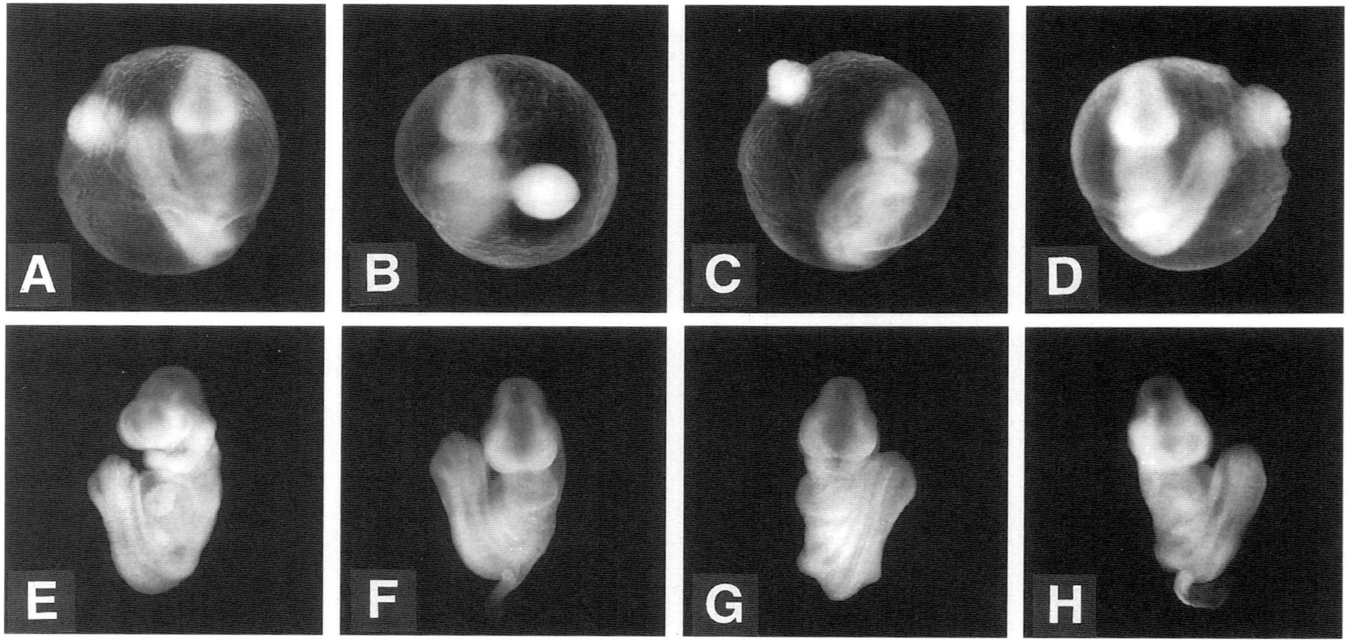

Fig. 3. Rat embryos at stage 15 on GD 11 showing abnormal sidedness of asymmetric structures, the chorioallantoic placenta/vitelline vessels, tail (lower body), and heart, in different combinations. **(A–D)** Frontal view of the embryo within the intact yolk sac. **(E–H)** Frontal view of the embryo after the yolk sac and amnion have been removed. (A) Normal tail and chorioallantoic placenta. (B) Normal tail but inverted chorioallantoic placenta. (C) Normal chorioallantoic placenta but inverted tail. (D) Inverted tail and chorioallantoic placenta. (E) Normal tail and heart. (F) Normal tail but inverted heart. (G) Normal heart but inverted tail. (H) Inverted tail and heart. (Derived from **ref. *3***.)

Terminology used in this manuscript		Classification of normal and abnormal sidedness of asymmetric body structures in human	Terminology used in some other literature
Situs solitus		Normal No complications With complications	Situs solitus
Situs inversus	Partial situs inversus	Partial inversion Bilateral one-sidedness (Isomerism) Bilateral left-sidedness (Levoisomerism) Bilateral right-sidedness (Dextroisomerism) Mixed type	Heterotaxia (Situs ambiguus, Laterallity sequences)
	Total situs inversus	Total inversion With complications No complications	Situs inversus

Fig. 4. Classification and terminology of abnormal sidedness of asymmetric body structures. (Derived from **ref.** ***3***.)

Approx. GD mouse	Approx. GD rat	Modified Theiler's stage	General developmental events	Events related to the LR axis, including the heart, axial rotation and limb buds
5	7	9	Primitive streak appears	
6	8	10a	Amniotic folds become visible	
7	9	10b	Amnion formation finishes	
		$10b_1$	Notochordal process is absent	
		$10b_2$	Notochordal process is present	
		11a	Neural groove becomes visible Allantoic bud begins to develop	(Asymmetry in neural plates*)
		11b	Neural folds become visible	
		11c	Foregut pocket becomes visible	Cardiac primordial cells appear
		12 /s1-2	First pair of somites become visible	
		12 /s3-4		Contraction of cardiac cells begins
8	10	12 /s5-6		Bilateral heart tubes meet in the midline Primitive heart tube begins to beat
		12 /s7-8	Cephalic neural tube begins to close Hematopoiesis begins in the yolk sac	Cardiac looping begins* Twisting of upper body begins*
		13 /s9-10	Allantois makes connection with chorion Yolk sac circulation becomes visible	
		13 /s11-12		Twisting of middle body begins*
		14 /s13-14		
		14 /s15-16	Cephalic neural tube is almost closed Embryo detaches from the yolk sac	Twisting of lower body ("tail") begins*
9	11	14 /s17-18		Axial rotation finishes* Forelimb buds become visible*

Fig. 5. Summary of developmental stages and events based on a modified Theiler's staging system *(4)* when early signs of body asymmetry appear. (Derived from **ref.** ***3***.)

indicate the date of pregnancy, and GD 0 is defined as the day when a copulatory plug was observed.

develops as bilateral cardiac tubes, which gradually approach the midline and fuse to form a single cardiac tube at stage 12/s5-6. The cardiac looping then begins normally to show "dextral looping" around stage 12/s7-8 ***(6,7)***. The detailed description of the looping process is described elsewhere [e.g., by illustrations in the chick ***(8,9)*** and in the mouse ***(10)*** and by scanning electron microscopic pictures in the mouse ***(11,12)***]. Various hypotheses for the mechanism of cardiac looping have been proposed by many investigators and are reviewed elsewhere ***(13–17)***. The mechanisms of cardiac looping are certainly intrinsic to the heart, and cytoskeletal microfilaments appear to play important roles ***(18–22)***.

4. Asymmetry in the embryonic disk at stage 11 in the rat (i.e., the anterior and posterior ends of the neural fold inclining slightly toward the right] was reported by Long and Burlingame ***(6)***. However, Deuchar and Parker ***(23)*** claimed that they did not observe such asymmetry in several hundred embryos that they examined. These investigators suggested that such asymmetry might have been artifact that occurred during fixation. However, they also discussed the possibility that if such asymmetry existed, it might imply some asymmetry in the mechanical properties of the tissues. (No further investigation has been done since then.) In the chick embryo, Hensen's node is known to show asymmetry at the equivalent stage ***(24)***. However, such asymmetry has not been found in the mouse or the rat.
5. In mouse or rat embryos, the embryonic disk develops within a cup-shape "egg cylinder" and consists of ectoderm and endoderm that are inverted compared with most other chordates. Thus, the embryo first develops in a dorsally flexed position and then undergoes rotation to a ventrally flexed position. To date, the underlying mechanisms for asymmetric movement involved in axial rotation are not well understood. Poelmann et al. ***(25)*** reported asymmetric mitotic activity of cells near the rotating part of the neuroectoderm and suggested that asymmetric growth and activity of the neural tube play important roles. Also, asymmetric mitotic activity was demonstrated in the somatopleure by Miller and Runner ***(26)***. Deuchar ***(27)*** suggested that coordinated activity of the somites may play a role. In addition, Poelmann et al. ***(25)*** suggested that looping of the heart is unlikely to be involved in determining the sidedness of axial rotation because the first signs of asymmetric activity within the neuroectoderm are seen before cardiac looping becomes evident. Furthermore, rat embryos subjected to α_1 adrenergic stimulation show various combinations of sidedness of the heart, chorioallantoic placenta, and "tail," ***(28)***. Thus, sidedness of cardiac looping probably occurs by mechanisms independent of those of axial rotation. This emphasizes the need to examine at least three asymmetric structures (i.e., the heart, "tail," and chorioallantoic placenta) to obtain a full picture of sidedness of the embryo at this developmental stage. Understanding the relationships of these structures in determining sidedness requires further investigations.
6. According to Long and Burlingame ***(6)***, axial rotation was first observed in the mouse by Raven in 1894 ***(29)*** and in the rat by Widakowich in 1909 ***(30)***. Since then, many investigators have described this process in the mouse ***(11,12,25,31,32)*** and in the rat ***(23,27,33–35)*** based on the assumption that only one asymmetric movement is involved in this process. However, it has been demonstrated recently that axial rotation of rat embryos actually consists of three movements that start at different stages (**Fig. 6**) ***(2)***. From the top (dorsal) view, they are (1) clockwise twisting of the upper body at stage 12/s7-8, (2) anticlockwise twisting of the middle body at stage 13/s11-12, and (3) anticlockwise twisting of the lower body ("tail") at stage 14/s15-16. Axial rotation is almost completed by stage 14/s17-18, resulting in several asymmetric structures (e.g., chorioallantoic placenta, "tail," and umbilical vessels normally on the right side of the embryo, and vitelline vessels on the left side of the embryo). The vitelline vessels are always located on the opposite side of the chorioallantoic placenta ***(2)***.

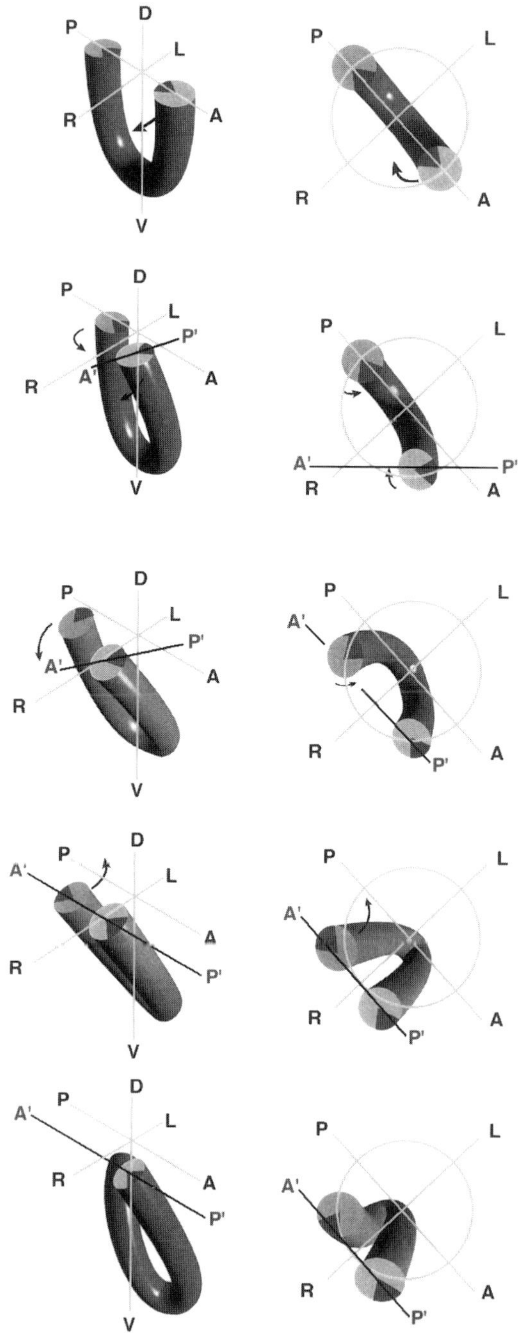

Fig. 6. (*See* color plate 4 appearing after p. 262.) Embryos at different stages during axial rotation from the top view (right column) and three-dimensional right-frontal view (left column); derived from **ref. *2***. This model was created and based on the model presented by Kaufman ***(11,12)***. A dark triangle was added to the illustrations to help identify the dorsal side of the embryo and to be consistent with the illustrations presented by Kaufman. Abbreviations: A, anterior; D, dorsal; L, left; P, posterior; R, right; V, ventral.

Table 1
Incidences of Laterality Defects in Wild-Type Animals

Strain	Incidence	Examination	Investigator
Mouse	3.2% (4/127)	GD 10	Layton ***(36)***[a]
Mouse	5.2% (16/308)	GD 10	Endo and Sakai ***(37)***[a]
Mouse	6.2% (11/177)	GD 11	Endo and Sakai ***(37)***[a]
Rat	0.55% (3/549)	GD 11	Fujinaga et al. ***(38)***[b]
Rat	0.037% (11/30,134)	GD 20/21	Charles River Lab ***(39)***[c]

[a]Only sidedness of the tail (flexion of lower body) was examined.
[b]Only sidedness of the tail and heart was examined.
[c]Indicated as fetuses with situs inversus.

Table 2
Frequency Table of Inverted Asymmetric Structures in Rat Embryos Treated with Phenylephrine, an α_1 Adrenergic Agonist

Inverted CAP	Inverted tail	Inverted heart	No. of embryos	% Affected embryos	Percentage of total
+	+	+	24	31.6	15.8
–	–	+	29	38.2	19.1
+	–	+	0	0.0	0.0
+	+	–	10	13.2	6.6
–	+	–	0	0.0	0.0
+	-	–	8	10.5	5.3
–	+	+	5	6.6	3.3
–	–	–	76	—	50.0
	Total		152	100.0	100.0

Note: +, presence; –, absence.
Embryos were cultured from stage 11a with 4 h or more of 50 μ*M* phenylephrine treatment.
Source: Derived from **ref.** ***28***.

7. A glass dissecting dish (20 × 100 mm) with a silicon base is commercially available (Carolina Biological Supply, Burlington, NC, Catalog No. 62-9016). Alternatively, such a dish can be made by placing silicon sealer on the bottom of a regular glass Petri dish.
8. This author has been satisfactorily using the Genuine Dumont Forceps (Dumont No. 5, BIO INOX, 11 cm long, Cat. No. 11252-20) and fine scissors (Walton, slight curve, strong, narrow blades, very fine points, firm cutting action Moria scissors, 10 cm long, Catalog No. 14077-10, Fine Science Tools [Foster City, CA]).
9. Although any type of dissecting microscope that provides 10–40X magnification is appropriate; those with light coming from the side (usually by using a separate fiberoptic light source) rather than from the bottom of the stage work better.
10. Individually wrapped sterile disposable plastic transfer pipets are commercially available and are very convenient. (Cut the tip depending on the size of embryo.)
11. It is very important not to damage the yolk sacs (Reichert's membrane [i.e., parietal yolk sac] and visceral yolk sac) during this procedure because the embryo is located just beneath the incision site. If the yolk sac is damaged, the embryo will come out of the sac; thus, an accurate assessment of tail and chorioallantoic placenta sidedness will no longer be possible.
12. Again, it is important not to damage the visceral yolk sac when dissecting Reichert's membrane.

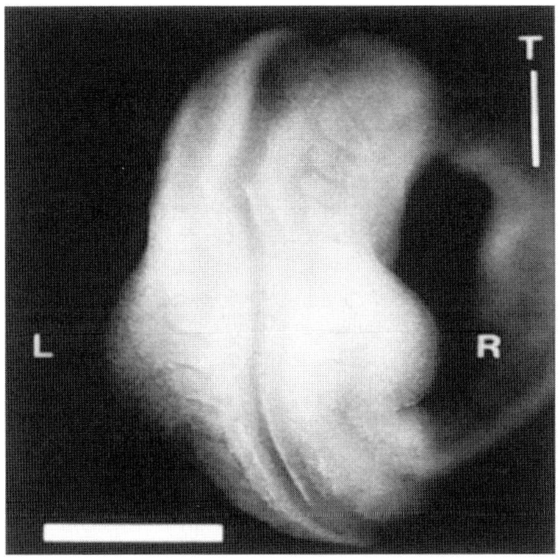

Fig. 7. Photograph of an embryo at stage 15 (GD 11) from the back view. The forelimb bud on the left side is normally larger than that on the right. Derived from **ref.** ***2***. Bar = 0.5 mm. Abbreviations: L, left; R, right; T, tail.

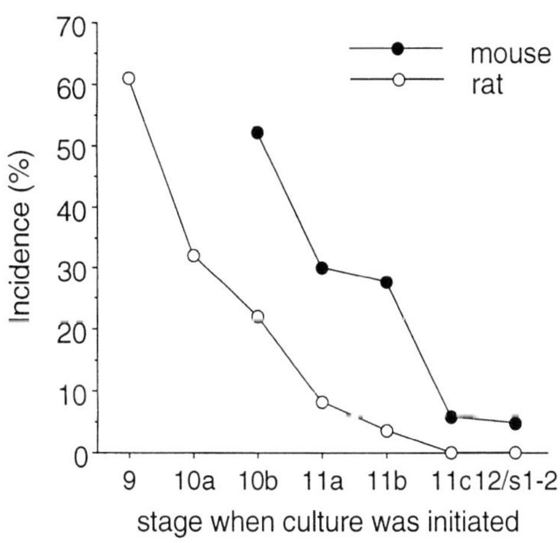

Fig. 8. Incidences of laterality defects/situs inversus in the mouse and rat embryo when cultured in vitro from different stages of development. The figure was created from the data derived from **refs.** ***2,3,28,41***.

13. As discussed earlier (**Note 5**), sidedness of at least these three structures must be assessed to obtain a full picture of the sidedness of the embryo at this developmental stage. The incidences of inversion of these structures in wild-type animals are shown in **Table 1**. It is well known that embryos cultured in vitro beginning before stage 11 show high incidences of inversion of these structures (**Fig. 8**). Also, α_1 adrenergic stimulation at stage 11a is known to cause various types of abnormal sidedness (**Table 2**). There are several other asymmetric structures at this stage. For example, the vitelline vessels normally emerge from left side of the embryo. However, they are always on the opposite side of the chorioallantoic placenta, as discussed earlier (**Note 6**). The forelimb bud is another distinct asymmetric structure (i.e., left-side limb bud is larger than that of the right side [**Fig. 7**]); *see* **ref.** ***41*** for the development of limb buds. However, its sidedness appears to be correlated with sidedness of the chorioallantoic placenta (**Table 3**).

Table 3
Frequency Table of Inverted Asymmetric Structures (Including Sidedness of Larger Limb Bud) in Control and Phenylephrine-Treated Rat Embryos

Chorioallantoic placenta	Tail	Heart	No. of embryos	Sidedness of larger limb bud: Left	Sidedness of larger limb bud: Right
Control (n = 15)					
R	R	N	**15**	**15**	0
Phenylephrine (n = 27)					
R	R	N	**16**	**16**	0
R	R	I	**2**	**2**	0
R	L	N	**1**	**1**	0
R	L	I	**1**	**1**	0
L	R	I	**3**	0	**3**
L	L	I	**3**	0	**3**
L	L	N	**1**	0	**1**

Abbreviations: R, right sided (normal); L, left sided (inverted); N, normal; I, inverted.

Note: Experiments were performed for the purpose of this chapter using the same methods as described in **ref. *28***. Embryos were cultured from stage 11a on GD 9 with or without 100 μ*M* phenylephrine during the whole period of culture and examined at stage 15 on GD 11.

Acknowledgments

I thank Dr. James Scott for editing the manuscript and Dr. Cecilia Lo, who encouraged me to prepare this manuscript.

References

1. Levin, M., Johnson, R. L., Stern, C. D., Kuehn, M., and Tabin, C. (1995) A molecular pathway determining left-right asymmetry in chick embryogenesis. *Cell* **82,** 803–814.
2. Fujinaga, M., Hoffman, B. B., and Baden, J. M. (1995) Axial rotation in rat embryos: morphological analysis and microsurgical study on the role of the allantois. *Teratology* **51,** 94–106.
3. Fujinaga, M. (1997) Development of sidedness of asymmetric body structures in vertebrates. *Int. J. Dev. Biol.* **41,** 153–186.
4. Fujinaga, M., Brown, N. A., and Baden, J. M. (1992) Comparison of staging systems for the gastrulation and early neurulation period in rodents: a proposed new system. *Teratology* **46,** 183–190.
5. Kaufman, M. H. and Navarantam, V. (1981) Early differentiation of the heart in mouse embryo. *J. Anat.* **133,** 235–246.
6. Long, J. A., and Burlingame, P. L. (1938) The development of the external form of the rat, with some observations on the origin of the extraembryonic coelom and foetal membranes. *Univ. Calif. Publ. Zoöl.* **43,** 143–184.
7. Burlingame, P. L. and Long, J. A. (1939) The development of the heart in the rat. *Univ. Calif. Publ. Zoöl.* **43,** 249–320.
8. Patten, B. M. (1922) The formation of the cardiac loop in the chick. *Am. J. Anat.* **30,** 373–397.
9. Stalsberg, H. (1969) The origin of heart asymmetry: right and left contributions to the early chick embryo heart. *Dev. Biol.* **19,** 109–127.
10. Biben, C., and Harvey, R. P. (1997) Homeodomain factor Nkx2-5 controls left/right asymmetric expression of bHLH gene *eHand* during murine heart development. *Genes Dev.* **11,** 1357–1369.

11. Kaufman, M. H. (1992) *The Atlas of Mouse Development.* Academic, San Diego, CA.
12. Kaufman, M. H. (1990) Morphological stages of postimplantation embryonic development, in *Postimplantation Mammalian Embryos. A Practical Approach* (Copp, A. J. and Cockroft, D. L., eds.), IRL, Oxford, pp. 81–91.
13. Stalsberg, H. (1969) The origin of heart asymmetry: right and left contributions to the early chick embryo heart. *Dev. Biol.* **19,** 109–127.
14. Stalsberg, H. (1970) Mechanism of dextral looping of the embryonic heart. *Am. J. Cardiol.* **25,** 265–271.
15. Manasek, F. J. (1976) Heart development: Interactions involved in cardiac morphogenesis, in *The Cell Surface in Animal Embryogenesis and Development* (Poste, G. and Nicolson, G. L., eds.), Elsevier/North-Holland, Amsterdam, pp. 545–598.
16. Manasek, F. J., Kulikowski, R. R., Nakamura, A., Nguyenphuc, Q., and Lacktis, J. W. (1984) Early heart development: a new model of cardiac morphogenesis, in *Growth of the Heart in Health and Disease* (Zak, R., eds.), Raven, New York, pp. 105–130.
17. Taber, L. A., Lin, I.-E., and Clarke, E. B. (1995) Mechanics of cardiac looping. *Del. Dynamics* **203,** 42–50.
18. Manasek, F. J., Kulikowski, R. R., and Fitzpatrick, L. (1978) Cytodifferentiation: a causal antecedent of looping? *Birth Defects: Original Article Series, Vol. 14, No. 7,* pp. 161–178.
19. Itasaki, N., Nakamura, H., Sumida, H., and Yasuda, M. (1991) Actin bundles on the right side in the caudal part of the heart tube play a role in dextro-looping in the embryonic chick heart. *Anat. Embryol.* **183,** 29–39.
20. Itasaki, N., Nakamura, H., and Yasuda, M. (1989) Changes in the arrangement of actin bundles during heart looping in the chick embryo. *Anat. Embryol.* **180,** 413–420.
21. Manning, A. and McLachlan, J. C. (1990) Looping of chick embryo hearts *in vitro. J. Anat.* **168,** 257–263.
22. Price, R. L., Chintanowoges, C., Shiraishi, I., Borg, T. K., and Terracio, L. (1996) Local and regional variations in myofibrillar patterns in looping rat heart. *Anat. Rec.* **245,** 83–93.
23. Deuchar, E. M. and Parker, F. M. (1975) Further observations on axial rotation in rat embryos. *Embryol. Exp.* **1,** 55–68.
24. Wetzel, R. (1929) Untersuchungen am Hühnchen. Die Entwicklung des Keims Während der ersten beiden Bruttage. *Arch. Entwicklungsmech. Organ.* **119,** 188–321.
25. Poelmann, R. E., Mentink, M. M. T., and Van Leeuwen, J. L. (1987) Axial rotation of murine embryos, a study of asymmetric mitotic activity in the neural tube of somite stages. *Anat. Embryol.* **176,** 99–103.
26. Miller, S. A. and Runner, M. N. (1978) Tissue specificity for incorporation of [^{3}H] thymidine by the 10- to 12-somite mouse embryo: alteration by acute exposure to hydroxyurea. *J. Embryol. Exp. Morphol.* **44,** 181–189.
27. Deuchar, E. M. (1975) Reconstitutive ability of axial tissue in early rat embryos after operations and culture in vitro. *J. Embryol. Exp. Morphol.* **33,** 217–226.
28. Fujinaga, M. and Baden, J. M. (1991) Critical period of rat development when sidedness of body asymmetry is determined. *Teratology* **44,** 453–462.
29. Raven, E. (1894) Zur Entwickelung des Nabelstranges der weissen Maus. *Arch. Anat. Entwick.* pp. 293–312. Cited from **ref.** ***6***.
30. Widakowich, V. (1909) Uber die erste Bildung der Körperform bei Entypie des Keimes. *Zeit. Wiss. Zool.* **94,** 240–298.
31. Snell, G. D. (1941) The early embryology of the mouse, in *Biology of the Laboratory Mouse* (Snell, G. D., ed.), Blakiston, Philadelphia, pp. 1–54.
32. Theiler, K. (1972). *The House Mouse. Atlas of Embryonic Development.* Springer-Verlag, New York.

33. Deuchar, E. M. (1969) Effects of transecting early rat embryos on axial movements and differentiation in culture. *Acta Embryol. Exp.*, 157–167.
34. Deuchar, E. M. (1971) The mechanism of axial rotation in the rat embryo: an experimental study in vitro. *J. Embryol. Exp. Morphol.* **25,** 189–201.
35. Deuchar, E. M. (1975) The course of rotation in rat embryos after bisections of the axis. *Acta Embryol. Exp.* **3,** 265–271.
36. Layton, W. M. (1976) Random determination of a developmental process. Reversal of normal visceral asymmetry in the mouse. *J. Heredity* **67,** 336–338.
37. Endo, A., and Sakai, N. (1987) The incidence of tail flexion to the left side in mouse embryos. *Anat. Embryol.* **176,** 133–134.
38. Fujinaga, M., Jackson, E. C., and Baden, J. M. (1990) Interlitter variability and developmental stage of day 11 rat embryos produced by overnight and morning short-period breeding regimens. *Teratology* **42,** 535–540.
39. Anonymous. (1995) Historical control data (1992–1994) for developmental and reproductive toxicity studies using the Crl:CD BR rat. Compiled by MARTA (Middle Atlantic Reproduction and Teratology Association) and MTA (Midwest Teratology Association). Charles River Laboratories, Wilmington, DE.
40. Fujinaga, M. and Baden, J. M. (1991) A new method for explanting early postimplantation rat embryos for culture. *Teratology* **43,** 95–100.
41. Milaire, J. (1985) Histological changes induced in developing limb buds of C57BL mouse embryos submitted in utero to the combined influence of acetazolamide and cadmium sulphate. *Teratology* **32,** 433–451.

25

Cardiac Morphogenesis and Dysmorphogenesis

I. Normal Development

Andy Wessels and Roger Markwald

1. Introduction

In this chapter, we will provide the developmental biologist, interested in cardiac development, a schematic description of heart development. This chapter is not intended to describe in detail every aspect of cardiac morphogenesis in the vertebrate heart or to elaborate extensively on all the species-specific differences that can be found. Such an effort would take up a complete volume in this series. As the mouse (genetic models, i.e., knockouts, transgenes, and trisomies) and the chicken (experimental models, microinjection, and cell-tracing studies) are still the animals of choice for the study of cardiac development, we will concentrate on these two animal models and relate the observations described to relevant data in other species—in particular, human. In order to prevent confusion with the different staging protocols, we have summarized the most frequently used staging systems in **Table 1**. This table should facilitate the comparison of embryonic stages. Hence, chicken embryos are staged according to Hamburger and Hamilton (H/H) *(1)*, the stages of the mouse embryos depicted in this chapter are given in embryonic days (ED) according to Theiler *(2)*, and stages in human development are given in Carnegie Stages (CS) according to O'Rahilly and Mueller *(3)*. We will discuss some of the basic concepts important for the understanding of cardiac development in normal and abnormal situations and briefly touch on the use of molecular markers to trace the fate of subpopulations of cells. In addition, we will describe the development of a few areas in the heart, those most frequently affected in animal model systems for congenital cardiovascular malformations, in a little more detail. If the information concerning normal and abnormal heart development provided in our chapters in this series of Developmental Biology Protocols does not suffice, we strongly encourage developmental biologists to seek assistance by, and collaborate with, cardiovascular embryologists.

A "recipe" for studying the developing heart does not exist and it is beyond the scope of this chapter to describe all the techniques used in the studies in cardiac embryology. Over the years, we have frequently used immunohistochemistry to study normal

From: *Methods in Molecular Biology, Vol. 136: Developmental Biology Protocols, Vol. II*
Edited by: R. S. Tuan and C. W. Lo © Humana Press Inc., Totowa, NJ

Table 1
Comparative Developmental Stages

Carnegie stage	Human (days gestation)	Mouse (embryonic days/ Theiler's stage)	Chicken (Hamburger/Hamilton stage/ days of incubation)
9	20	8–8.5 (12)	7–8 (1.1)
10	22	8.5–9 (13)	10 (1.5)
11	24	9–9.5 (14)	11 (1.8)
12	26	9.5–10.25 (15)	14 (2.2)
13	28	10.25–10.5 (16)	17 (2.6)
14	32	10.5–10.75 (17)	19 (2.9)
15	33	11 (18)	20–21 (3.3)
16	37	11.5 (19)	24 (4)
17	41	12 (20)	26 (4.8)
18	44	12.5 (21)	28 (5.6)
19	47	13 (21)	29–30 (6.4)
20	50	14 (22)	31–32 (7.2)
21	52	14 (22)	34 (8)
22	54	14 (22)	35 (8.7)
23	56	14 (22)	36 (9.6)

A simplified table to compare stages in human, chicken, and mouse embryos (based on **refs.** *1*, *2*, and *3*).

and perturbed cardiac development. In our chapter on congenital malformations (*see* Chapter 15 in this volume) we describe this technique, as used in our laboratory, step by step. For other technical protocols, related to the work described in this chapter (e.g., *in situ* hybridization and scanning electron microscopy), readers are referred to other chapters in this series and to relevant literature published elsewhere.

2. Cardiac Development, from Fields Through Tube in the Four-Chambered Heart

One of the most intriguing aspects of cardiac development is how two bilateral paired fields of splanchnic mesoderm eventually give rise to the developed heart. First, these mesodermal fields, also known as the heart fields, coalesce to form a simple tubular heart. Subsequently, the more or less straight heart tube starts to loop and segments and septa start to take on shape. Eventually, the embryonic heart remodels into a morphologically complex four-chambered heart (**Figs. 1** and **2**). Among the problems that face the cardiac developmental biologist interested in the early stages of development are questions such as the following: What factors are involved in determining the fate of the cells in the heart fields (endocardial, myocardial, epicardial)? What is the nature of the processes determining the organization of the cells of the developing heart? Which conditions determine the way in which the heart tube will eventually gain its specific lateral characteristics?

2.1. The Formation of the Heart Fields

Factors involved in the regulation of the early processes in cardiac development include transcription factors [e.g., GATA-4 *(4)*, Nkx2.5 *(5)*] and growth factors [FGF,

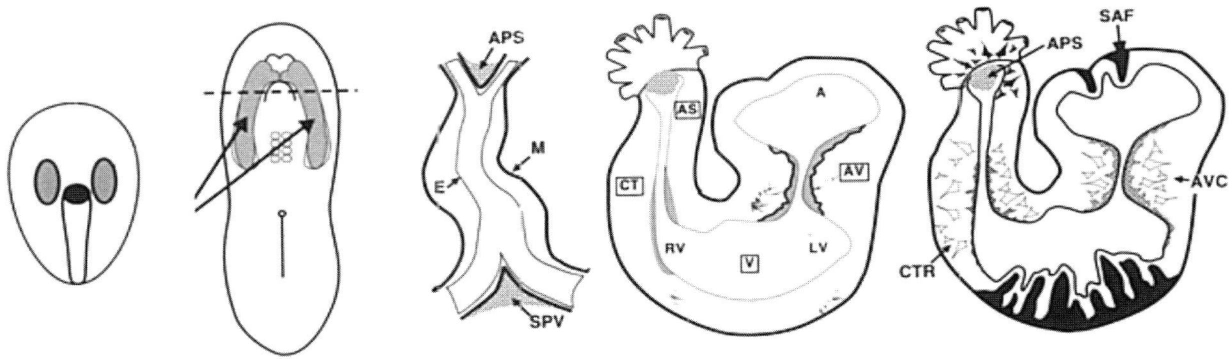

Fig. 1. The progressive formation of the embryonic heart. Two bilateral fields of splanchnic mesoderm (the "heart-forming fields", **A**) coalesce to form the horseshoe-shaped heart primordium **(B)**. The "legs" of the horseshoe (arrows in B) gradually fuse anteriorly to form the straight heart tube **(C)**. In this straight heart tube, the first signs of segment formation can be recognized. During the process of looping, endocardial cushions develop in the atrioventricular (AV) junction and in the outflowtract **(D)**. Looping is followed by a series of developmental events in the heart, which include the formation of the ventricular trabeculae, mesenchymalization of the endocardial cushions, and formation of the cardiac septa **(E)**. AS = aortic sac, AP = aorticopulmonary septum, AV = atrioventricular junction, AVC = atrioventricular cushions, CTR = conotruncal ridges, E = endocardium, LV = left ventricle, M = myocardium, RV = right ventricle, SPV = spina vestibuli.

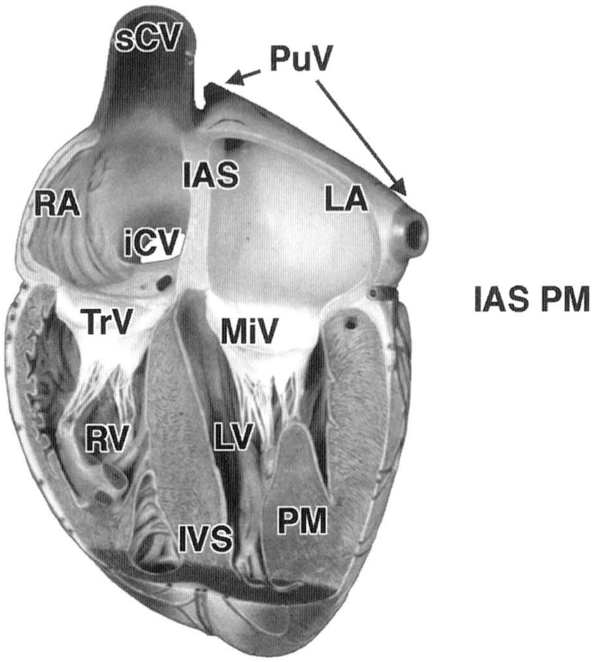

Fig. 2. This model of an adult human heart demonstrates some of the general features of the four-chambered heart as also seen in animal models. The left and right atria are separated by an interatrial septum; the ventricles are separated by a interventricular septum. Valves are located between the atria and ventricles to prevent backflow and are attached to the ventricular myocardium by papillary muscles. Large caval veins drain into the right atrium, and pulmonary veins drain into the left atrium. iCV = inferior caval vein, IAS = interatrial septum, IVS = interventricular septum, LA = left atrium, LV = left ventricle, MiV = mitral valve, PM = papillary muscle, PuV = pulmonary veins, sCV = superior caval vein. TrV = tricuspid valve.

BMP *(6)*]. Several of these factors show a predominant expression in either the left (lefty, nodal, sonic hedgehog) or right (cACT-RIIa) heart field *(7)* but their functions remain largely undetermined. Genes that play a role in the determination of left/right asymmetry [e.g., the genes targeted in the *inv (8)* and *iv (9)* mutations] can alter this asymmetric expression pattern, leading to disturbance in the left/right asymmetry in the body plan. Knowledge of the molecular regulation of early cardiogenesis is accumulating rapidly and any information in this chapter about current knowledge will most likely be obsolete upon publication. In the following subsections, we describe some of he highlights that are of crucial importance to comprehend the most important morphological events that lead to the formation of the four-chambered heart.

2.2. Formation and Looping of the Heart Tube

2.2.1. The Dorsal Mesocardium

The tubular heart is formed by fusion of the left and right splanchnic mesodermal heart-forming fields ventral to the embryonic foregut ***(10)***. After completion of its formation, the tube basically hangs along the dorsal midline freely suspended from the foregut by a mesothelium (**Fig. 3**). This mesothelium is generally referred to as the

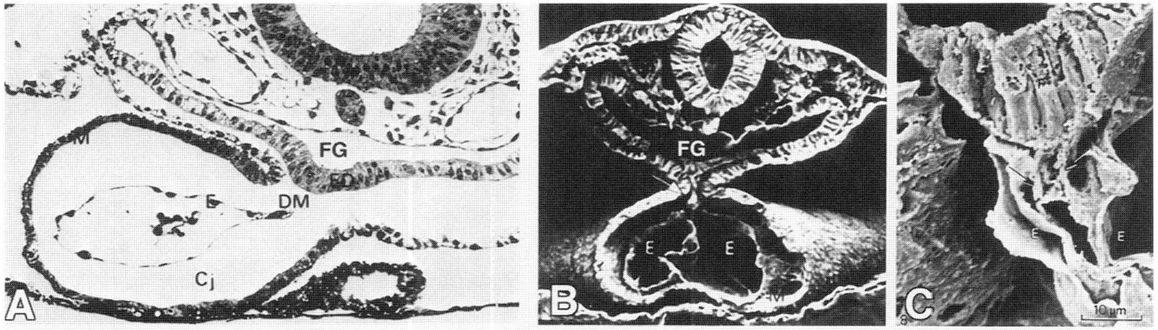

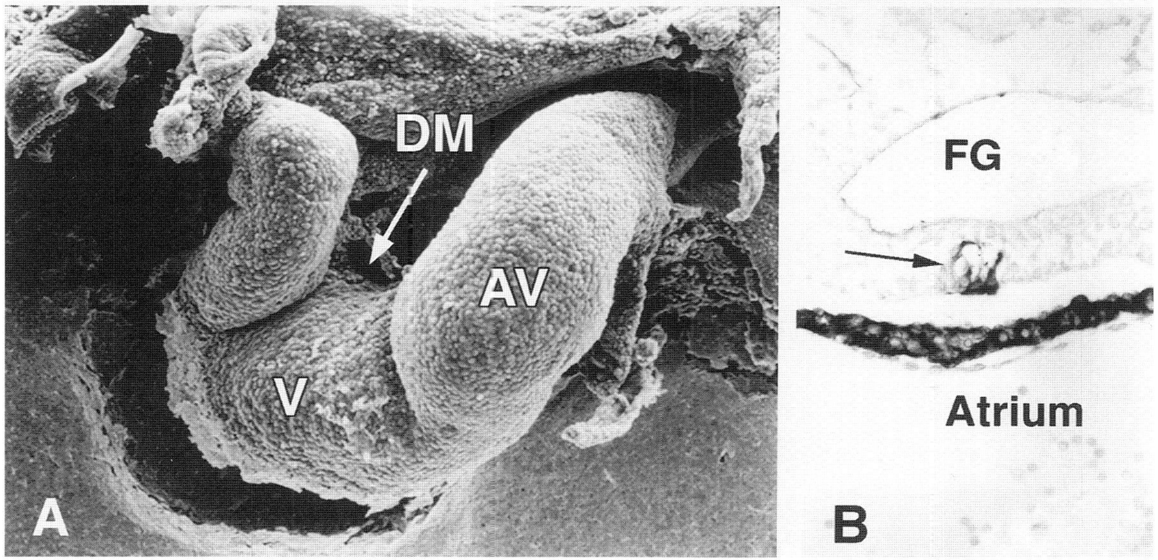

Fig. 3. The disintegration of the primary dorsal mesocardium takes place around stage 12 H/H in the chicken (**A**). The strandlike remnants are distinguishable at the inner curvature of the heart. The site of former attachment of the primary dorsal mesocardium is in the mouse at 9.5 ED characterized by the expression of smooth-muscle actin (**B**). AV = atrioventricular junction, DM = dorsal mesocardium, V = ventricle, FG = foregut, CJ = cardiac jelly, E = endocardium.

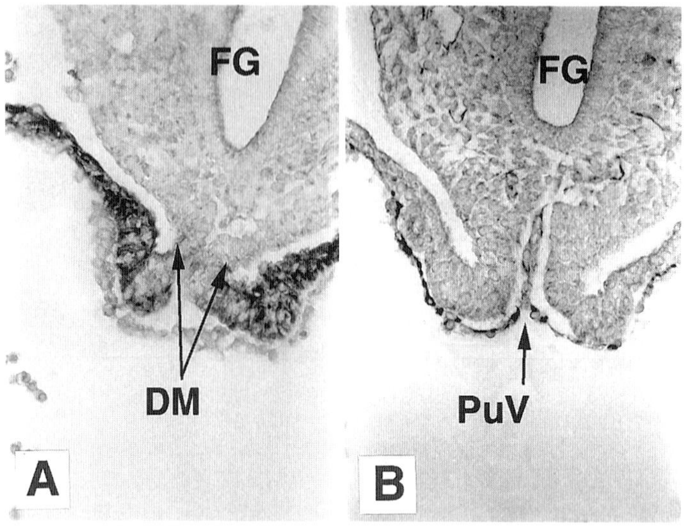

Fig. 4. Serial sections of the atrial region of a 10 ED mouse heart. In this region the remaining dorsal mesocardium forms a gateway for mesenchymal and neural crest cells to migrate into the venous pole of the heart. The pulmonary vein develops from the endothelial strand that persisted after disintegration of the primary dorsal mesocardium **(B)**. This endothelial strand (or primitive pulmonary vein) is in continuity with the endocardium lining the atrial cavity. The sections were stained with an antibody recognizing atrial myosin heavy chain **(A)** and an antibody recognizing endothelial and mesenchymal cells (B). DM = dorsal mesocardium, FG = foregut, PuV = -pulmonary vein.

"primary" dorsal mesocardium *(11)*. Around stage 12 H/H in the chick, the mid-portion of the dorsal mesocardium disintegrates (**Fig. 3**) by processes yet to be elucidated. The persisting "stalk" of dorsal mesocardium connects the venous pole of the atrium and the sinus venosus to the mesenchyme of the developing lung buds and liver. This partial disintegration of the dorsal mesocardium is a crucial event, as it allows the straight heart tube to loop. Both at the dorsal surface of the heart as well as in the tissue located ventral to the foregut, one can find areas indicating the sites of the former attachment. These areas can be identified both using molecular markers (SMA and CK-B; *see* **Fig. 3**) as well as by morphological criteria *(12)*. Although the (persisting) dorsal mesocardium has never gained the attention that other parts (e.g., outflow tract and endocardial cushions) have received over the years, it actually is a very important region of the developing heart. Not only is it the area in which the pulmonary vein develops *(13)* (**Fig. 4**), it also forms a gateway for mesenchymal cell populations [including neural crest cells *(14)*, Poelmann (personal communications), and **Fig. 5**] to migrate into the venous pole of the heart. The cellular and molecular mechanisms involved in the subsequent development of this "intracardiac mesenchymal portion of the dorsal mesocardium (IMPDM)" or spina vestibuli *(15,16)* are very poorly studied. An endothelial to mesenchymal transformation similar to that observed in the endocardial cushions *(17)* has been suggested *(18)*. It is important to note that the dorsal-mesocardium-derived mesenchyme can be immunohistochemically distinguished from developing endocardial cushion tissue *(19)*.

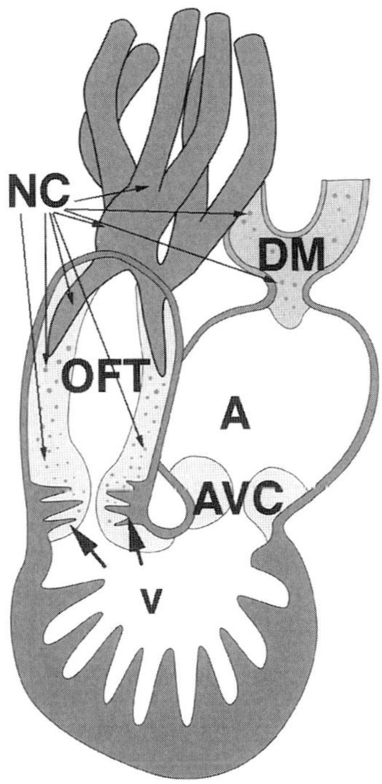

Fig. 5. Neural crest contribution to the developing heart. Neural-crest-derived cells migrate into the heart using two different gateways. In the outflow tract, the neural crest cells reach the heart via the pharyngeal arches, in the venous pole of the heart, the neural crest cells use the dorsal mesocardium as their portal of entry. Isolated neural crest cells descending into the proximal region of the conal ridges of the outflow tract may be involved in the regulation of myocardialization. (some aspects of this figure were adapted from the study *[56]*). A = atrium, AVC = atrioventricular cushions, DM = dorsal mesocardium, NC = neural crest derived cells, OFT = outflow tract, V = ventricle.

2.2.2. Cell Types in the Tubular Heart

At the tubular stage of development, the heart consists basically of two concentric layers of cells (**Fig. 6**). The outer tube is myocardial and is characterized by the expression of myocardial-specific markers such as the cardiac myosin heavy chain [atrial MHC *(20)*]. The inner tube is composed of endothelial (endocardial) cells that express markers specific for the endothelial cell lineage (**Fig. 6**). In the space between the myocardium and endocardium, an acellular space with a high extracellular matrix content, generally referred to as the cardiac jelly or myocardial basement membrane, is found *(21–23)*. Biochemical analysis of the cardiac jelly has demonstrated the presence of some 50 proteins. Among these proteins are collagens, fibronectin, fibrillin, fibulin and growth factors (for a review, *see* **ref.** ***24***).

2.2.3. Early Formation of the Cardiac Segments

As development proceeds, the straight heart tube starts to loop toward the right (d-loop) and the respective compartments become recognizable (**Fig. 7**). After completion of looping, the following serially arranged segments can be described. The most upstream cardiac segment is the sinus venosus. This segment, comprising of the left and right

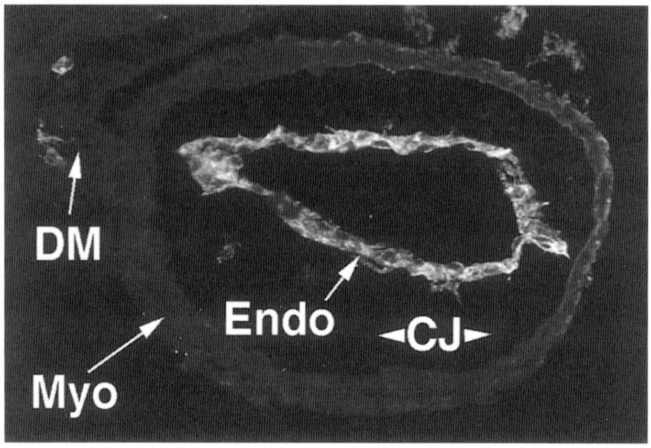

Fig. 6. The tubular heart consists of three layers. The outer epithelial layer consists of myocardial cells and, the inner layer of endocardial cells. In the quail heart, the endothelial layer is characterized by the expression of QH1. Between the myocardial and endothelial cell layers, an acellular cardiac jelly can be found. CJ = cardiac jelly, DM = dorsal mesocardium, Endo = endocardium, Myo = myocardium.

sinus horns, is a prominent structure and collects all the systemic and maternal blood. The sinus venosus empties into the common atrium (the future right and left atrium) through an opening in the dorsal wall known as the sinuatrial foramen. The common atrium is connected to the embryonic left ventricle by a relatively large segment, the atrioventricular canal. The embryonic left ventricle communicates with the embryonic right ventricle through the interventricular foramen. The last segment is the outflow tract, which connects entirely to the right ventricle and consists of a conal (or proximal) and a truncal (or distal) portion. The outflow tract eventually drains the blood into the aortic sac. In the lesser curvature of the heart, all these segments converge and cannot easily be discriminated from each other. Although, for the sake of simplicity, the tubular heart is often illustrated with all segments present, not all segments develop (originate) at the same time. Marking experiments ***(25,26)*** have shown that the future right ventricle is the first segment to develop. The rest of the segments are gradually "added on," with the outflow tract being the last to develop.

2.2.4. Segment-Specific Gene Expression

Over the years, the application of immunohistochemistry and *in situ* hybridization techniques has demonstrated that the respective cardiac segments and their flanking regions are characterized by specific gene expression patterns. Some of these patterns are more or less widespread over the species and seem to be related to the general function (contractility, impulse conduction) of the segments. Other patterns, often species-specific, are much more difficult to understand, as they seem not to relate directly to function. An example of an expression pattern shared among species is the expression of myosin heavy and light chain isoforms within the developing heart. All vertebrate species discussed in this chapter express different myosin heavy chain isoforms in atrial and ventricular myocardia at some time during development [i.e., atrial or alpha, and ventricular or beta MHC ***(20,27,28)***]. However, whereas the expression of

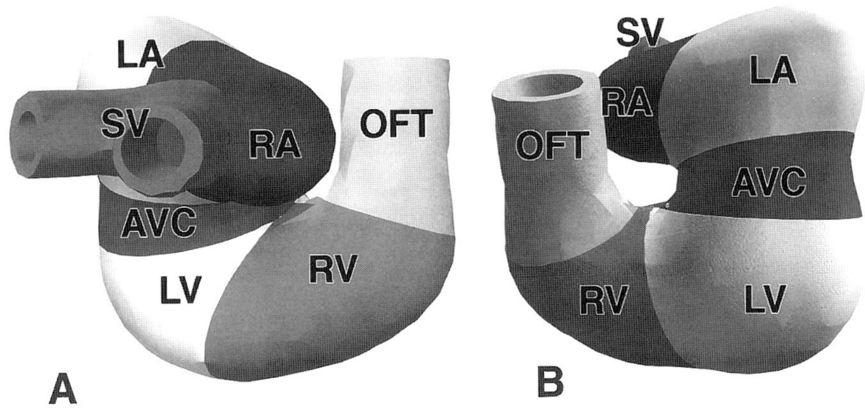

Fig. 7. The embryonic heart in the looped stage of development. All the segments are basically arranged in series. The sinus venosus empties into the right atrium. The right atrium connects to the left atrium. The left atrium drains into the left ventricle through the atrioventricular canal. The left ventricle communicates with the right ventricle and the right ventricle eventually drains into the outflow tract. LA = left atrium, LV = left ventricle, OFT = outflow tract, RA = right atrium, RV = right ventricle, SV = sinus venosus.

ventricular MHC persists in the ventricles postnatally in the larger mammals and in the chick, the expression of ventricular MHC becomes downregulated around birth in the ventricles of mice and other smaller mammals *(29)*. Other examples of general expression patterns include the expression of atrial natriuretic factor (ANF) in late fetal atrial tissue, coexpression of atrial and ventricular myosin heavy chain in the atrioventricular canal *(27,28)*, and specialized connexin isoforms in the atrioventricular conductions system *(30–32)*. Examples of species-specific gene expression include the expression of neurofilament in the developing conduction system of the rabbit heart *(33)*. Recently, new information has emerged about possible mechanisms related to the segmental specific gene expression. These studies show segment-constricted expression of transcription factors and transgenic constructs containing promoter regions of cardiac genes. The work of Srivastava and collaborators has demonstrated left/right differences in gene expression of the heart-specific transcription factors dHAND/eHAND. The construction of knockout mice has shown that these transcription factors are of crucial importance for the development of the segments in which they are expressed, as ablation of their expression leads to maldevelopment of the respective segments *(34,35)*. A completely new and exciting view on segment-specific gene expression is emerging from recent studies using transgenic animals. These studies show that by using truncated promoter constructs reporter-gene activity can be directed to very specific regions (e.g., conduction system, AV junction) of the hearts.

2.3. Cardiac Septation and the Development of Specialized Structures

2.3.1. Tubular Heart Functions Without One-Way Valves and Recognizable Conduction System

In the tubular heart, the myocardium contracts in a peristaltoid fashion (in the human heart, this peristaltic flow start at Carnegie Stage 11). The contraction starts at the inflow segment and makes it way toward the outflow tract, thereby propelling the blood into the forming arterial system. No valves have formed yet in this tubular heart. The cardiac jelly (later the endocardial cushion tissue) basically forms the "stuffing" of the tube. This situation contrasts sharply with the postnatal situation in which the heart can be considered to be a multistep parallel pump with well-developed one-way valves and a morphologically and functionally distinguishable conduction system, now supporting two interdependent, but separate, bloodstreams. This dramatic difference in anatomy and function between the embryonic tubular heart and the developed four-chambered heart confronts us with several basic developmental issues. The first two relate to the formation of the connections between segments which, in the tubular heart, are initially not connected. Thus, our first concern is how the right atrium becomes connected to the right ventricle (i.e., the formation of the right AV junction), and second is how the left ventricle get access to the aortic portion of the outflow tract (i.e., the formation of the left ventricular outflow tract). Other topics that will be discussed in this chapter involve issues related to the processes that lead to the development of subsystems in the heart, specifically the conduction system and the cardiac valvar apparatus.

2.4. Primary Ring and Ventricular Septation; the Formation of the Right AV Junction and the Left Ventricular Outflow Tract

The understanding of the remodeling of the tubular heart into a four-chambered heart was greatly enhanced by the use of immunohistochemical techniques. Using the monoclonal antibodies GlN2 and HNK1/Leu7 *(36,37)*, it was demonstrated how in the looped stages of cardiac development, the transitional zone between the embryonic left and right ventricles, was characterized by the expression of these "neural" markers. This ring of tissue was dubbed the "primary ring" (**Fig. 8**). The studies in the human heart especially provided crucial new insights with respect to this understanding *(37,38)*. By careful three-dimensional computer reconstruction *(39)* of the GlN2 expression in the different stages of development, it was demonstrated that initially the primary ring is located at the junction of the (future) left and right ventricle, thereby separating the upstream portion (i.e., atrium and left ventricle) from the downstream portion (i.e., right ventricle and outflow tract) (**Fig. 9**). At this stage, the embryonic common atrium drains into the left ventricle, and the outflow tract (the future pulmonary trunk *and* aorta) connect entirely to the right ventricle. As development progresses, the atrioventricular junction expands toward the right, thus aligning the developing right atrium with the right ventricle. This expansion "carries" the upper portion of the primary ring into the right atrioventricular junction, thereby forming the right atrioventricular ring. Concomitant with this rightward expansion, the aortic portion of the outflow tract is shifting to the left. This event pushes a portion of the primary ring backward and to the left (this process is also known as "wedging" of the aorta). The part of the primary ring that becomes located in the atrial myocardium behind the aorta has been dubbed the retroaortic root branch. In addition to these structures, from which only remnants can be found in the developed heart *(40)* (**Fig. 10**), part of the primary ring also develops into the atrioventricular conduction system. Interestingly, all of these derivatives of the primary ring can be found in the developed chick heart *(43)*, indicating that the development of the conduction system can be considered to be a phylogenetically preserved developmental event.

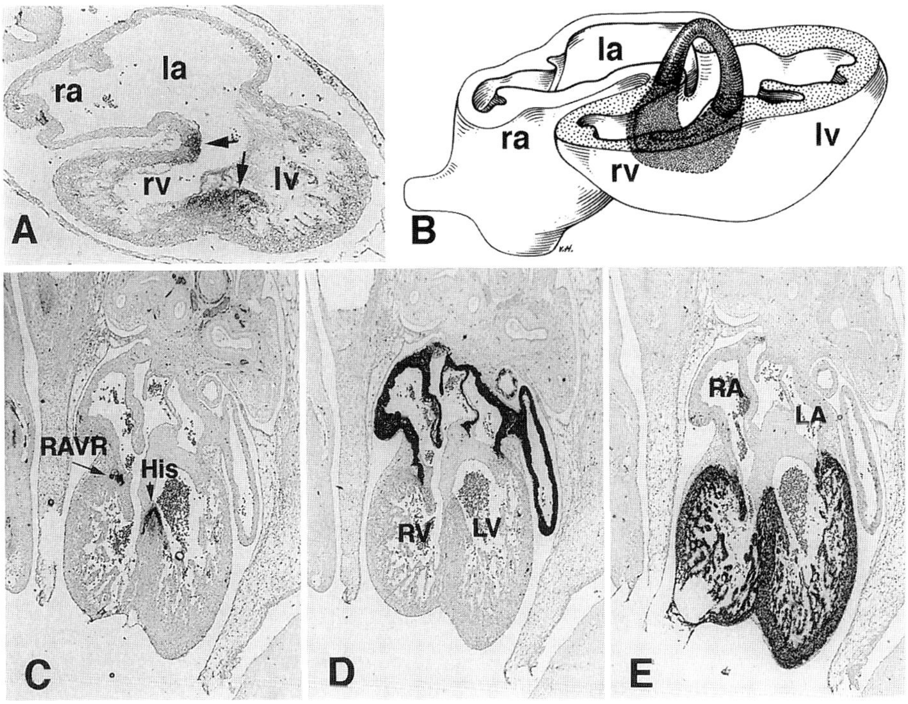

Fig. 8. The primary ring in the developing human heart. Three-dimensional reconstruction **(B)** of GlN 2 expression in the embryonic heart at 5 wk of development **(A)** shows that a ring of tissue surrounding the primary interventricular foramen is characterized by the expression of the GlN2 antigen. At older stages, GlN2 expression is typically found in the right atrioventricular junction and in the developing conduction system **(C)**. (***C***,**D**, and **E**) Serial sections of a heart at 6–7 wk of development incubated with antibodies against GlN2 (C), atrial myosin heavy chain (D), and ventricular myosin heavy chain (E). LA = left atrium, LV = left ventricle, RA = right atrium, RV = right ventricle, RAVR = right atrioventricular ring bundle, His = His bundle.

Although the concept of the primary ring does not explain why the heart develops the way it does, it certainly has helped us to understand how the respective segments of the heart eventually connect to each other. Moreover, it has enabled us also to understand some of the aspects of the ontogenesis of congenital heart malformations. The studies on the development of the primary ring strengthened the notion that many of the congenital malformations can be considered to represent stages of developmental arrest of this remodeling process. Hence, the malformation known as the double-inlet left ventricle (DILV) can be considered to represent persistence of the original embryonic connection of both atria to the left ventricle, whereas persistence of the embryonic connection of both the aorta and the pulmonary trunk to the right ventricle is known as double-outlet right ventricle (DORV). Moreover, situations in which the aorta has only partly shifted to the left ventricle (overriding aorta) can be considered to represent incomplete incorporation (or wedging) of the aorta into the left ventricle. To date, the mechanisms responsible for the normal remodeling processes and factors leading to disturbances of these processes are still unknown. It is to be expected that with increas-

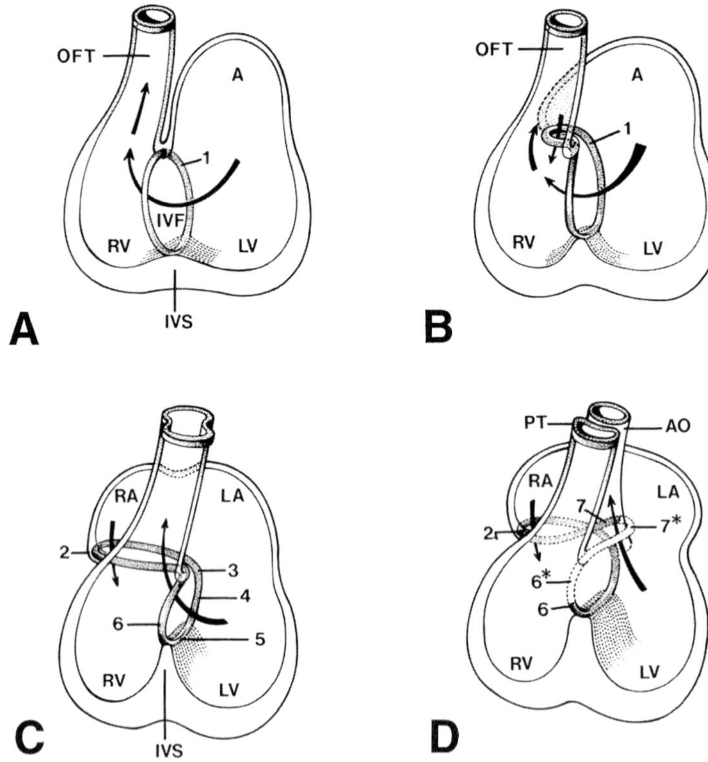

Fig. 9. This series of simplified reconstructions shows the changes of the shape of the primary ring in subsequent stages of human cardiac development. It demonstrates how during development the superior portion of the ring expands to the right (formation of the right AV junction) and how the anterior portion shifts to the left ("wedging" of the aorta). For detailed information on the primary ring and its contribution to septation and segmentation, and the development of the conduction system, *see* **refs.** ***38*** and ***40–42***. AO = aorta, LA = left atrium, LV = left ventricle, OFT = outflow tract, PT = pulmonary trunk, RA = right atrium, RV = right ventricle.

ing numbers of experimental and genetic animal models for congenital heart disease, our knowledge about regulation of these processes will increase.

2.5. Anatomy and Development of the Conduction System

Both the origin and differentiation of the components of the conduction system are still controversial topics, despite all research efforts over the last decades related to this issue. We will try to summarize some of the most recent advancements in this field.

2.5.1. Anatomy of the Conduction System

In the adult mammalian heart (**Fig. 11**), the following components of the conduction system can be distinguished. The sinuatrial node (SAN) is the "pacemaker" of the heart and is located at the junction between the superior caval vein and the right atrium in a structure known as the crista terminalis (or terminal crest). The cardiac impulse, generated in the SAN, spreads over the atrial musculature, initiating contraction of the atria, and eventually converging at the atrioventricular node (AVN). In the mammalian heart,

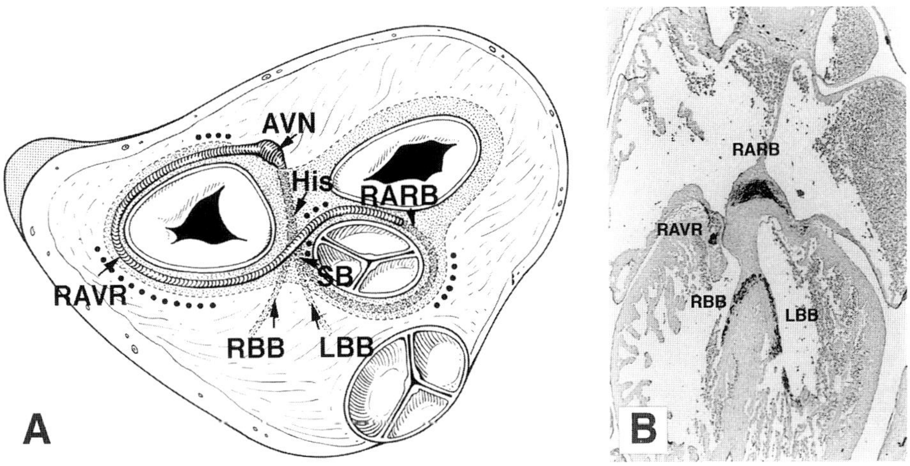

Fig. 10. Remnants of the primary ring can be found in the neonatal human heart (**A**). This schematic shows a heart with the atria "removed." The primary ring is projected on a superior view of the aortic–mitral fibrous unit. The black dots indicate the areas in which remnants of the primary ring were morphologically and immunohistochemically detected. (**B**) shows the expression of neurofilament in components of the primary ring in a rabbit heart at 15 d of development. AVN = atrioventricular node, His = His bundle, LBB = left bundle branch, RARB = retroaortic root branch, RAVR = right atrioventricular ring bundle, RBB = right bundle branch, SB = septal branch (*see* **ref. *40***).

the AVN is the only (persisting) myocardial continuity between the atrial and the ventricular myocardia. In this proximal portion of the atrioventricular conduction system, the impulse conduction is delayed by the specific molecular and/or structural characteristics of the AVN and will travel through the next recognizable structure belonging to the atrioventricular conduction system, the bundle of His (also known as the penetrating bundle). Having reached the ventricular portion of the heart, the cardiac impulse is distributed over the ventricular myocardium by the relatively fast conducting left and right bundle branches and the myocytes of the Purkinje system.

2.5.2. Molecular Delineation of the Conduction System

In recent years, many markers have been described that allow the delineation of the conduction system in a variety of species. Some of these markers do not seem to relate to the conductive properties of the conduction system, whereas others are thought to be extremely important to its function. Many of the conduction system markers seem to be species-specific (usually those not related to function), whereas others show a generally uniform expression among the species. The expression of connexin isoforms, gap junctional proteins involved in the conduction of impulse from cell to cell, forms an important tool for the study of the conduction system. In the mammalian heart, the two major cardiac connexin isoforms, connexin-43 (Cx43) and connexin-40 (Cx40), are abundantly expressed in the fast components of the conduction system (bundle branches and Purkinje system), whereas both are virtually absent from the slow components (i.e., SAN, AVN, and His bundle) ***(30,32,42)***. In addition, Cx43, but not Cx40, is also expressed in the ventricular working myocardium, whereas Cx43 and Cx40 are

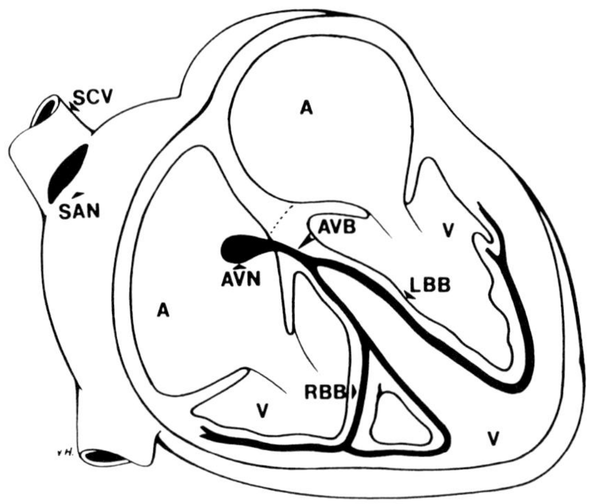

Fig. 11. Schematic of the conduction system in the adult human heart. A = atrium, AVB = atrioventricular (His) bundle, AVN = atrioventricular node, LBB = left bundle branch, RBB = right bundle branch, SAN = sinoatrial node, SCV = superior caval vein, V = ventricle. (From **ref.** ***44***.)

both expressed in atrial myocardium. All the components of the developing and adult conduction system in the rabbit are characterized by the expression of neurofilaments *(46)* (**Fig. 10**), which has not yet been described in the conduction systems of other species. In larger mammals it has been demonstrated that certain phosphorylated isoforms of desmin in cow and pig are specifically expressed in the conduction system myocytes. Other molecules that are related to the ventricular conduction system include Cx45, myosin-binding protein C, creatine kinase ALD 58, PSA–NCAM, EAP-300, and atrial natriuretic factor. For a comprehensive review on the development of the conduction system, we recommend the review by Moorman et al. *(41)*.

2.6. Function and Development of the Cardiac Valves

Well-working one-way valves are crucial for proper function of the postnatal heart. However, despite the fact that the leaflets of the atrioventricular and semilunar valves do not develop until the later stages of embryonic development, the embryonic heart manages very well to establish unidirectional flow. Hence, one of the questions that has captivated cardiovascular biologist for a long time is how the embryonic heart can function without the presence of a well-developed valvar apparatus. The human heart starts to beat in the beginning of the fourth week of development [approximately Carnegie stage 11 *(3)*]. At this stage, the heart can still be considered to represent a tube with cardiac jelly filling the acellular space between the myocardial outer and the endocardial inner mantles of the tube. The contraction form at this stage is peristaltoid, with the contraction starting in the inflow region and stopping in the outflow tract (for a review, *see* **ref.** ***41***). As the heart develops and the cardiac jelly gives rise to the endocardial cushions in the atrioventricular region and the adult contraction pattern in the outflow tract (i.e., after filling with venous blood, the atria contract to fill the ventricles, followed by contraction of the ventricles to propel the blood through the outflow tract into the circulation) becomes recognizable. It is important to realize that although the atrioventricular valves and the valves of the great arteries all develop from

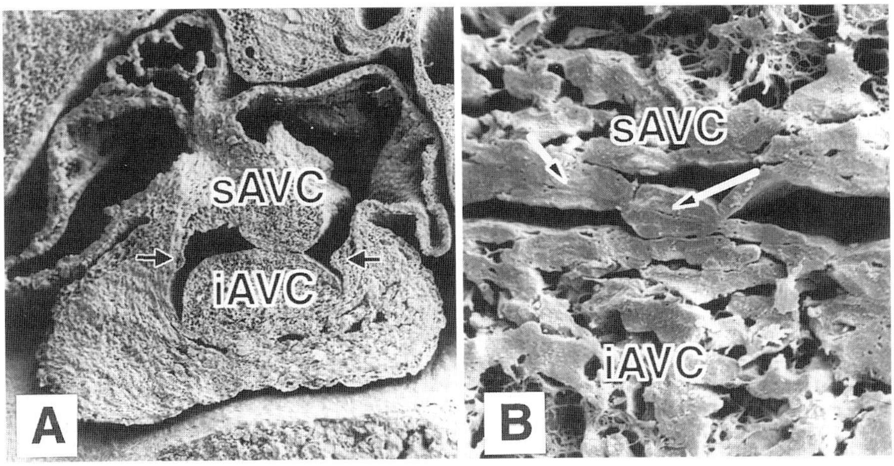

Fig. 12. This scanning electron micrograph of a 13 ED mouse heart "captures" the moment of fusion of the atrioventricular cushions. Note also the relatively small lateral AV cushions (arrows). iAVC = inferior atrioventricular cushion, sAVC = superior atrioventricular cushion.

endocardial cushion material, the way in which they develop and the way they actually function is basically different.

2.6.1. The Atrioventricular Valves

At stage 12–13 H/H, cardiac jelly starts to accumulate *(22)* and forms endocardial cushions at two regions of the tubular heart, the atrioventricular junction, and the outflow tract (**Fig. 1**). In other areas of the heart, most of the cardiac jelly disappears and endocardium becomes plastered against the myocardium. Initially, the atrioventricular cushions and conal ridges are acellular, but beginning at stage 17 H/H (chick), the cushions become populated by mesenchymal cells that derive from the endocardium through an endothelial-to-mesenchymal transdifferentiation *(47)*. At a later stage (around stage 22 H/H), parts of the conal ridges, but not the atrioventricular cushion, also become populated by neural crest cells. The two main atrioventricular cushions (generally referred to as the inferior and superior atrioventricular cushions) fuse to separate the left atrioventricular orifice from the right atrioventricular orifice (**Fig. 12**). This process is very poorly studied. "Fusion" of cushions sounds simple. However, one should realize that this fusion takes place in an environment that is constantly in motion. The first prerequisite for fusion is that the cushions "meet and glue" in a spatio-temporal correct fashion. It is easy to envision how the cushions "meet." At every contraction cycle, they will be squeezed together by the constriction of the myocardial tube, thereby preventing backflow during relaxation. It is more difficult to envision how the process of gluing takes place.

The mechanism underlying this process remains to be elucidated. After fusion, the fused AV cushions form (together with the mesenchymal tissue of the intracardiac mesenchymal portion of the dorsal mesocardium) the central fibrous body and contribute to the formation of the leaflets of the atrioventricular valvar apparatus, in left as well as right AV junctions [*(19)*; *see* **Figs. 12** and **13**]. In addition to the well-known inferior and superior AV cushions, much smaller cushions develop at the lateral aspects

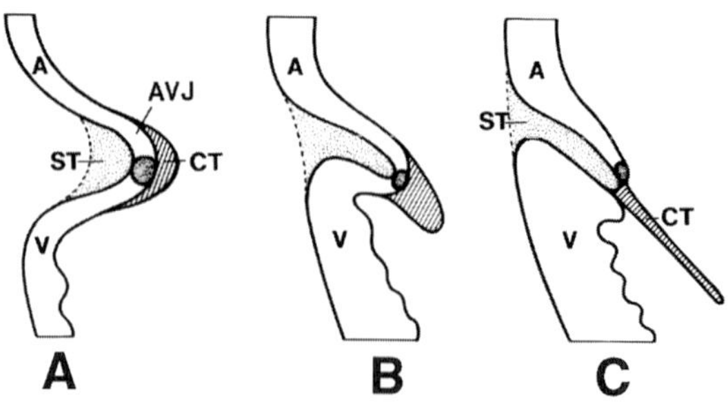

Fig. 13. Schematic showing three important stages in the development of the atrioventricular valves. In the early embryonic stages (e.g., Carnegie stage 14) **(A)**, the atrial myocardium is continuous with the ventricular myocardium through the myocardium of the atrioventricular canal. At the epicardial side, the atrioventricular canal is characterized by the presence of sulcus tissue, whereas at the endocardial side, it is covered by cushion tissue. During the process of atrioventricular remodeling separation between atrium and ventricle becomes established when sulcus- and cushion-derived tissue fuse at the ventricular margin of the atrioventricular canal **(C)**. The processes responsible for this disintegration of the myocardial continuity remain to be elucidated. A = atrium, CT = cushion tissue, ST = sulcus tissue, V = ventricle. (From **ref.** ***19***.)

of the AV canals (**Fig. 12**). These cushions develop at a slightly later stage and are dubbed the lateral atrioventricular cushions ***(19)***. They form the tissue base for the mural leaflet in the left AV junction and for the anterosuperior leaflet in the right AV junction ***(48)***. Although, at the early embryonic stages, cushion formation and the cell transformation are very similar among the different species discussed in this chapter, there are pronounced differences in the formation and maturation of the AV valves as development progresses. In the right AV junction of the chick heart, the initial mesenchymal lateral cushion in the right AV junction becomes muscularized by "myocardialization." This results in the formation of the muscular flap valve of the right AV junction. Interestingly, the left junction in the chick develops almost identically to that in mouse and human. This difference in development between the left and right AV valves is intriguing and the molecular mechanisms underlying this difference (what induces myocardialization of the right lateral cushion) remain to be elucidated. Another interesting difference between chicken and mammals is that in the chick the membranous atrioventricular septum, which initially develops in a way similar to that in the mammals, eventually becomes muscularized. The morphological data available indicate that in addition to myocardialization (ingrowth of existing myocardium into a mesenchymal tissue), this muscularization also occurs by a mesenchymal-to-myocardial transdifferentiation. It is interesting to note that our data strongly suggest that this transdifferentiation also takes place in the mesenchymal intracardiac tissues that are related to the dorsal mesocardium. A critical step in the formation of the AV valves is the delamination of the leaflets from the myocardial walls and interventricular septum ***(48)***. Proper delamination is of crucial importance, as failure of this process to occur can result in inappropriately positioned valves which, in the human, can lead to a congenital heart malformation known as Ebstein's Anomaly. The delamination process gives the heart its "free-moving" leaflets of the (in the human) tricuspid (right) and mitral (left) atrioventricular valves. Their movement is "restricted" by the papillary

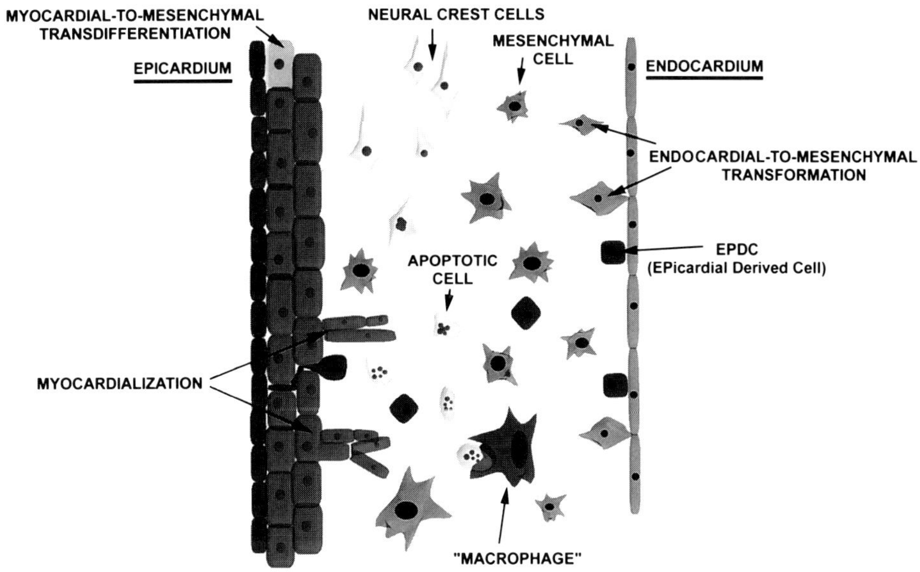

Fig. 14. Cellular contributions to the endocardial cushions of the outflow tract.

muscles that connect the leaflets to the apical region of the ventricles. This attachment prevents the leaflets from "collapsing" into the atrial cavities during ventricular systole.

2.6.2. The Semilunar Valves

Although, historically, the endocardial ridges of the outflow tract and the endocardial cushions in the AV junction were considered to develop similarly, an increasing body of evidence has accumulated over the last couple of years showing that the number of tissues contributing to the endocardial ridges in the outflow tract is probably more extensive than previously thought. The majority of mesenchymal cells in the endocardial ridges derive from the endocardium by an endothelial-to-mesenchymal transformation similar to that in the AV cushions. During the looping stages of development the endocardial ridges become populated by cells migrating from the neural crest (**Figs. 1**, **5**, and **14**), another important source of mesenchymal cells. These cells arrive in the heart via the pharyngeal arches and developing aorticopulmonary septum *(49)*. Recent studies indicate that during normal development, a part of the neural-crest-derived mesenchymal cells undergo apoptosis in the conal region of the outflow tract and are phagocytosed by macrophagelike cells *(50)*. The origin of these macrophagelike cells, however, remains to be elucidated. Other studies have demonstrated that the outflow tract ridges also receive cellular contributions from cells with an epicardial origin (epicardial derived cells [EPDCs]). Another cell type found in these ridges is of a myocardial nature. As the conal ridges fuse to separate the aortic outflow from the pulmonic outflow, myocardial proliferations are found to grow into the proximal portion of the outlet septum. This process is called myocardialization and is responsible for the formation of the muscular portion of the outlet septum *(51)*. Abnormal or incomplete myocardialization has been implied to be one of the mechanisms leading to outflow tract septation malformations, such as the double-outflow right ventricle. Together with the so-called intercalated ridges, they eventually also form the semilunar valves of the aorta and pulmonary trunk.

2.7. Neural Crest in the Developing Heart

The role and the significance of the neural crest in cardiac development remains a controversial topic. Basically, the neural crest has two gateways to migrate into the heart. The first, and best described, is the outflow tract. Here, the cells arrive via the pharyngeal arches into the area of the aorticopulmonary septum and two condensed prongs of condensed mesenchyme which reach deep into the endocardial ridges of the outflow tract ***(49,52,53)***. Neural crest ablation studies have demonstrated the importance of these neural crest cells, as ablation of this cellpopulation results in outflow tract malformations ***(54)*** such as the persistent truncus arteriosus (PTA) and the double-outflow right ventricle (DORV). Mouse mutants in which neural crest contributions to the outflow tract is also impaired (Splotch) show similar malformations ***(55)***. The other cardiac region through which neural crest cells can migrate into the heart is the dorsal mesocardium. Cells traveling through this splanchnic mesodermal corridor contribute to the formation of the innervation of the heart ***(14)*** and are recently reported to be found in the vicinity of the developing atrioventricular conduction system (R. Poelmann, personal communication). Preliminary results obtained by our and other laboratories suggest that neural crest cells may play a role in the spatio-temporal regulation of mycardialization in the outflow septum, possibly through a series of cellular events, including apoptosis, phagocytosis, and growth factor release (*see* **Fig. 14**).

Acknowledgments

The authors thank Dr. C. Drake and Dr. L. Litke for photograhic illustration and A. Phelps, R. Bennington, J. Spruill and Dr. T. Trusk for excellent technical assistance. The authors gratefully acknowledge the financial support of NIH grant HL-52813.

References

1. Hamburger, V. and Hamilton, H. L. (1951) A series of normal stages in the development of the chick embryo. [classical article]. *Dev. Dyn.* **195,** 231–272.
2. Theiler, K. (1989) *The House Mouse: Atlas of Embryonic Development*, 2nd ed. Springer-Verlag, New York, p. 178.
3. O'Rahilly, R. and Muller, F. (1987) *Developmental Stages in Human Embryos.* Carnegie Institute of Washington, Washington, DC.
4. Kelley, C., Blumberg, H., Zon, L. I., and Evans, T. (1993) GATA-4 is a novel transcription factor expressed in endocardium of the developing heart. *Development* **118,** 817–827.
5. Lints, T. J., Parsons, L. M., Hartley, L., Lyons, I., and Harvey, R. P. (1993) Nkx-2.5: a novel murine homeobox gene expressed in early heart progenitor cells and their myogenic descendants [published erratum appears in *Development* **119,** 969], *Development* **119,** 419–431.
6. Lough, J., Barron, M., Brogley, M., Sugi, Y., Bolender, D. L., and Zhu, X. (1996) Combined BMP-2 and FGF-4, but neither factor alone, induces cardiogenesis in non-precardiac embryonic mesoderm. *Dev. Biol.* **178,** 198–202.
7. Levin, M. (1997) Left-right asymmetry in vertebrate embryogenesis. *Bioessays* **19,** 287–296.
8. Yokoyama, T., Copeland, N. G., Jenkins, N. A., Montgomery, C. A., Elder, F. F., and Overbeek, P. A. (1993) Reversal of left-right asymmetry: a situs inversus mutation. *Science* **260,** 679–682.

9. Seo, J. W., Brown, N. A., Ho, S. Y., and Anderson, R. H. (1992) Abnormal laterality and congenital cardiac anomalies. Relations of visceral and cardiac morphologies in the iv/iv mouse. *Circulation* **86,** 642–650.
10. De Ruiter, M. C., Gittenberger-De Groot, A. C., Wenink, A. C., Poelmann, R. E., and Mentink, M. M. (1995) In normal development pulmonary veins are connected to the sinus venosus segment in the left atrium. *Anat. Rec.* **243,** 84–92.
11. Chang, C. (1931) The formation of the interatrial septum in chick embryos. *Anat. Rec.* **50,** 9–22.
12. De Ruiter, M. C., Poelmann, R. E., Mentink, M. M. T., Van Iperen, L., and Gittenberger-De Groot, A. C. (1993) Early formation of the vascular system in quail embryos. *Anat. Rec.* **235,** 261–274.
13. Webb, S., Brown, N. A., Wessels, A., and Anderson, R. H. (1998) Development of the murine pulmonary vein and its relationship to the embryonic venous sinus. *Anat. Rec.*, in press.
14. Filogamo, G. and Peirone, S. M. (1995) Ontogenesis of the structural elements of the heart. *Cardioscience* **6,** 19–23.
15. His, W. (1885) Das Herz, in *Anatomie Menschlicher Embryonen II. Zur Geschichte der Organe* (Vogel, V.v.F.C.W., ed.), 1885: Leipzig.
16. Asami, I. and Koizumi, K. (1995) *Development of the atrial septal complex in the human heart: Contribution of the spina vestibuli*, in *Developmental mechanisms of heart disease* (Clark, E. B. and Takao, A., eds.), Futura, Armonk, NY, pp. 255–260.
17. Markwald, R., Eisenberg, L., Eisenberg, C., and Sugi, Y. (1996) Epithelial mesenchymal transformations in early avian heart development. *Acta Anat.* **156,** 173–186.
18. Gerety, M. and Watanabe, M. (1997) Polysialylated NCAM expression on endocardial cells of the chick primary atrial septum. *Anat. Rec.* **247,** 71–84.
19. Wessels, A., Markman, M. W., Vermeulen, J. L., Anderson, R. H., Moorman, A. F., and Lamers, W. H. (1996) The development of the atrioventricular junction in the human heart. *Circ. Res.* **78,** 110–117.
20. de Jong, F., Geerts, W. J., Lamers, W. H., Los, J. A., and Moorman, A. F. (1987) Isomyosin expression patterns in tubular stages of chicken heart development: a 3-D immunohistochemical analysis. *Anat. Embryol.* **177,** 81–90.
21. Markwald, R. R., Fitzharris, T. P., Bolender, D. L., and Bernanke, D. H. (1979) Structural analysis of cell:matrix association during the morphogenesis of atrioventricular cushion tissue. *Dev. Biol.* **69,** 634–654.
22. Markwald, R. R., Fitzharris, T. P., and Manasek, F. J. (1977) Structural development of endocardial cushions. *Am. J. Anat.* **148,** 85–119.
23. Markwald, R. R., Krook, J. M., Kitten, G. T., and Runyan, R. B. (1981) Endocardial cushion tissue development: structural analyses on the attachment of extracellular matrix to migrating mesenchymal cell surfaces. *Scan. Electron Microsc.* 261–274.
24. Little, C. D. and Rongish, B. J. (1995) The extracellular matrix during heart development. *Experientia* **51,** 873–882.
25. De la Cruz, M. V., Sanchez-Gomez, C., and Palomino, M. A. (1989) The primitive cardiac regions in the straight tube heart (stage 9-) and their anatomical expression in the mature heat: an experimental study in the chick heart. *J. Anat.* **165,** 121–131.
26. de la Cruz, M. V., Sanchez Gomez, C., Arteaga, M. M., and Arguello, C. (1977) Experimental study of the development of the truncus and the conus in the chick embryo. *J. Anat.* **123,** 661–686.
27. Wessels, A., Vermeulen, J. L., Viragh, S., Kalman, F., Lamers, W. H., and Moorman, A. F. (1991) Spatial distribution of "tissue-specific" antigens in the developing human heart and skeletal muscle. II. An immunohistochemical analysis of myosin heavy chain isoform expression patterns in the embryonic heart. *Anat. Rec.* **229,** 355–368.

28. de Groot, I. J., Lamers, W. H., and Moorman, A. F. (1989) Isomyosin expression patterns during rat heart morphogenesis: an immunohistochemical study. *Anat. Rec.* **224,** 365–373.
29. Lompre, A. M., Mercadier, J. J., Wisnewsky, C., Bouveret, P., Pantaloni, C., D'Albis, A., and Schwartz, K. (1981) Species and age dependent changes in the relative amounts of cardiac myosin isoenzymes in mammals. *Dev. Biol.* **84,** 286–290.
30. Gourdie, R. G., Green, C. R., Severs, N. J., Anderson, R. H., and Thompson, R. P. (1993) Evidence for a distinct gap-junctional phenotype in ventricular conduction tissues of the developing and mature avian heart. *Circ. Res.* **72,** 278–289.
31. Van Kempen, M. J., Vermeulen, J. L., Moorman, A. F., Gros, D., Paul, D. L., and Lamers, W. H. (1996) Developmental changes of connexin40 and connexin43 mRNA distribution patterns in the rat heart. *Cardiovasc Res.* **32,** 886–900.
32. van Kempen, M. J. A., ten Velde, I., Wessels, A., Oosthoek, P. W., Gros, D., Jongsma, H. J., Moorman, A. F., and Lamers, W. H. (1995) Differential connexin distribution accommodates cardiac function in different species. *Microsc. Res. Tech.* **31,** 420–436.
33. Gorza, L., Schiaffino, S., and Vitadello, M. (1988) Heart conduction system: a neural crest derivative? *Brain Res.* **457,** 360–366.
34. Srivastava, D., Thomas, T., Lin, Q., Kirby, M. L., Brown, D., and Olson, E. N. (1997) Regulation of cardiac mesodermal and neural crest development by the bHLH transcription factor, dHAND. *Nat. Genet.* **16,** 154–160.
35. Srivastava, D., Cserjesi, P., and Olson, E. N. (1995) A subclass of bHLH proteins required for cardiac morphogenesis. *Science* **270,** 1995–1999.
36. Ikeda, T., Iwasaki, K., Shimokawa, I., Sakai, H., Ito, H., and Matsuo, T. (1990) Leu-7 immunoreactivity in human and rat embryonic hearts, with special reference to the development of the conduction tissue. *Anat. Embryol. (Berl.)* **182,** 553–562.
37. Wessels, A., Vermeulen, J. L., Verbeek, F. J., Viragh, S., Kalman, F., Lamers, W. H., and Moorman, A. F. (1992) Spatial distribution of "tissue-specific" antigens in the developing human heart and skeletal muscle. III. An immunohistochemical analysis of the distribution of the neural tissue antigen G1N2 in the embryonic heart, implications for the development of the atrioventricular conduction system. *Anat. Rec.* **232,** 97–111.
38. Lamers, W. H., Wessels, A., Verbeek, F. J., Moorman, A. F., Viragh, S., Wenink, A. C., Gittenberger-de Groot, A. C., and Anderson, R H. (1992) New findings concerning ventricular septation in the human heart. Implications for maldevelopment. *Circulation* **86,** 1194–1205.
39. Verbeek, F. J., Huijsmans, D. P., Baeten, R. J., Schoutsen, N. J., and Lamers, W. H. (1995) Design and implementation of a database and program for 3D reconstruction from serial sections: a data-driven approach. *Microsc. Res. Tech.* **30,** 496–512.
40. Wessels, A, Mijnders, T. A., de Gier-de Vries, C., Vermeulen, J. L. M., Viragh, S., Lamers, W. H., and Moorman, A. F. M. (1992) Expression of myosin heavy chain in neonatal human hearts. *Cardiol. Young* **2,** 318–334.
41. Moorman, A. F. M., De Jong, F., Denyn, M. M. F. J., and Lamers, W. H. (1998) Development of the cardiac conduction system. *Circ. Res.* **82,** 629–644.
42. Wessels, A., Vermeulen, J. L., Verbeek, F. J., Viragh, S., Kalman, F., Lamers, W. H., and Moorman, A. F. (1992) Spatial distribution of "tissue-specific" antigens in the developing human heart and skeletal muscle. III. An immunohistochemical analysis of the distribution of the neural tissue antigen G1N2 in the embryonic heart; implications for the development of the atrioventricular conduction system. *Anat. Rec.* **232,** 97–111.
43. Vassall-Adams, P. R. (1982) The development of the atrioventricular bundle and its branches in the avian heart. *J. Anat.* **134,** 169–183.
44. Oosthoek, P. W (1994) Microarchitecture of the conduction system and the vasculature in normal and malformed hearts, in *Anatomy and Embryology*. University of Amsterdam/ Rijksuniversiteit Leiden, Amsterdam, p. 147.

45. van Kempen, M. J., Fromaget, C., Gros, D., Moorman, A. F., and Lamers, W. H. (1991) Spatial distribution of connexin43, the major cardiac gap junction protein, in the developing and adult rat heart. *Circ. Res.* **68,** 1638–1651.
46. Gorza, L. and Vitadello, M. (1989) Distribution of conduction system fibers in the developing and adult rabbit heart revealed by an antineurofilament antibody. *Circ. Res.* **65,** 360–369.
47. Bolender, D. L. and Markwald, R. R. (1979) Epithelial-mesenchymal transformation in chick atrioventricular cushion morphogenesis. *Scan Electron Microsc.* 313–321.
48. Lamers, W. H., Viragh, S., Wessels, A., Moorman, A. F., and Anderson, R. H. (1995) Formation of the tricuspid valve in the human heart. *Circulation* **91,** 111–121.
49. Kirby, M. (1993) Cellular and molecular contributions of the cardiac neural crest to cardiovascular development. *Trends Cardiovasc. Med.* **3,** 18–23.
50. Ya, J. (1997) Normal and abnormal development of the outflow tract of the embryonic heart, in *Anatomy & Embryology* University of Amsterdam, Amsterdam, p. 156.
51. van den Hoff, M. J. B., Moorman, A. F. M., Ruijter, J. M., Lamers, W. H., Bennington, R. W., Markwald, R. R., and Wessels, A. (1999) Myocardialization of the cardiac outflow tract. *Dev. Biol.* **212,** 477–490.
52. Thompson, R. P. and Fitzharris, T. P. (1979) Morphogenesis of the truncus arteriosus of the chick embryo heart: the formation and migration of mesenchymal tissue. *Am. J. Anat.* **154,** 545–556.
53. Kirby, M. L., Gale, T. F., and Stewart, D. E. (1983) Neural crest cells contribute to normal aorticopulmonary septation. *Science* **220,** 1059–1061.
54. Kirby, M. L., Turnage, K. L. d., and Hays, B. M. (1985) Characterization of conotruncal malformations following ablation of "cardiac" neural crest. *Anat. Rec.* **213,** 87–93.
55. Conway, S., Henderson, D., and Copp, A. (1997) Pax3 is required for cardiac neural crest migration in the mouse: evidence from the splotch (Sp2H) mutant. *Development* **124,** 505–514.
56. Waldo, K., Miyagawa-Tomita, S., Kumiski, D., and Kirby, M. L. (1998) Cardiac neural crest cells provide new insight into septation of the cardiac outflow tract: aortic sac to ventricular septal closure. *Dev. Biol.* **196,** 129–144.

26

Analysis of Two Aspects of Left–Right Patterning of the Vertebrate Heart

Heart Tube Position and Heart Tube Chirality

Alvin J. Chin

1. Introduction

Although the phenotypes of situs inversus totalis, isolated malposition of the heart (e.g., isolated dextrocardia), and heterotaxy syndrome have been familiar to physicians for decades (**Fig. 1**), the question of how left–right positional information within an embryo is assigned at a molecular level has only become the subject of intense scrutiny in the last decade. In order to fully examine the implications of recently reported experiments, we must first understand which of two aspects of left–right asymmetry they address.

The vertebrate embryo fashions axes first in dorsoventral and anteroposterior directions: thereafter, left and right sides of the embryo can be unambiguously assigned. Manifestations of left right positional information include not only the gross position of organs or structures vis-à-vis the midline, but also the chirality, or handedness, of that organ or structure. In both the heart and the gut, these two aspects of left–right patterning are normally "coupled." For example, knowing that the heart is positioned in the left hemithorax reliably predicts which heart stereoisomer will eventually be formed by the process known as "looping;" normally, the midportion of the heart tube bends to the right, creating a loop (designated D-), which places the ventricle anterior and to the right of the atrium. However, in certain conditions organ chirality appears uncoupled from organ position.

First, isolated dextrocardia in the human (in which only the heart if malpositioned) is frequently accompanied by (normal) D-looping (1). Second, heterotaxy syndrome human patients, in which the laterality of many thoracoabdominal organs is nearly randomized, can display either type of heart looping, D- or L-, regardless of (postlooping [*see* **Note 6**]) heart position *(2)*. Third, in the mouse, homozygosity for *inversus viscerum (iv)*, a mutation that alters *l/r dynein* (an axonemal dynein), results in heterotaxy and in randomization of heart tube looping *(3)*. Fourth, homozygous null activin receptor type IIB mice also appear to model human heterotaxy syndrome *(4)*; e.g., aorta sidedness is independent of (postlooping) heart position in the chest.

From: *Methods in Molecular Biology, Vol. 136: Developmental Biology Protocols, Vol. II*
Edited by: R. S. Tuan and C. W. Lo © Humana Press Inc., Totowa, NJ

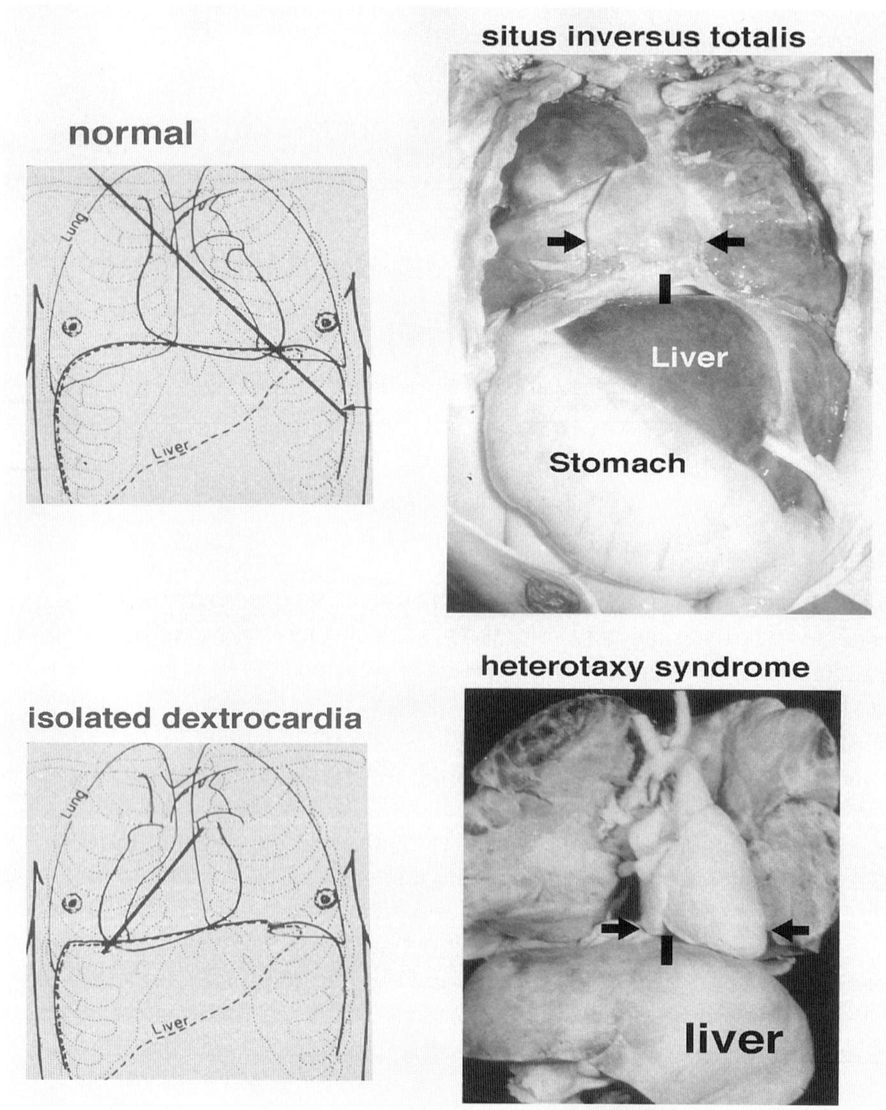

Fig. 1. Varieties of visceral organ lateralization. Top left: normal. The point of the heart (cardiac ventricular apex) lies in the left hemithorax; the heart's base-to-apex direction is denoted by the straight bold line. The liver is right-sided. Modified from **ref. *15***. Top right: situs inversus totalis. Heart (arrows), liver, and stomach sidedness are all mirror image of normal. The vertical bar denotes the midline. Courtesy of P. Weinberg. Bottom left: isolated dextrocardia. The heart's base-to-apex orientation (straight bold line) is to the right; however, all other organs (including the liver and gut) are normally positioned. Bottom right: heterotaxy syndrome. The sidedness of each visceral organ, including the heart (arrows) and the liver, is unpredictable with respect to the midline (vertical bar) and with respect to other organs. Courtesy of P. Weinberg.

The zebrafish system allows the investigator to look at [prelooping (*see* **Note 6**) ***(5)***] heart position and heart tube looping direction separately, as: (1) egg laying can be synchronized, (2) fertilization is external, (3) development is rapid, (4) the embryo is

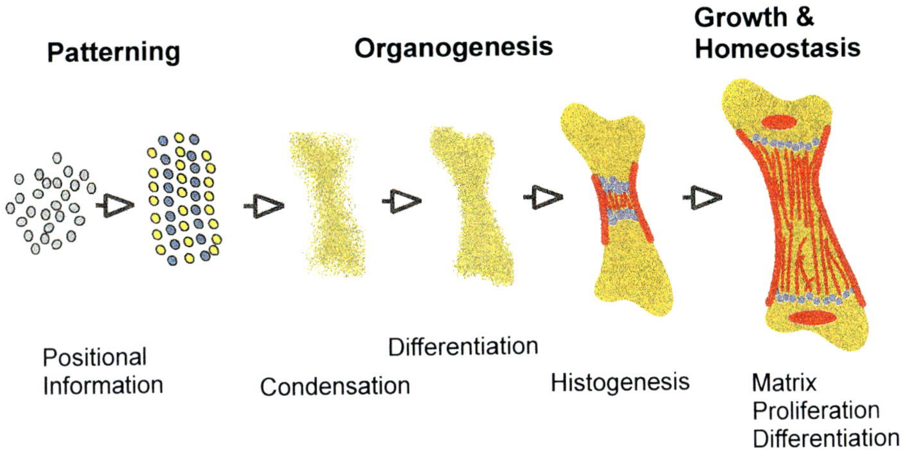

Plate 1, Fig. 1. (*See* discussion in Chapter 8, and full caption on p. 62.) Schematic drawing showing the different steps of skeletal morphogenesis.

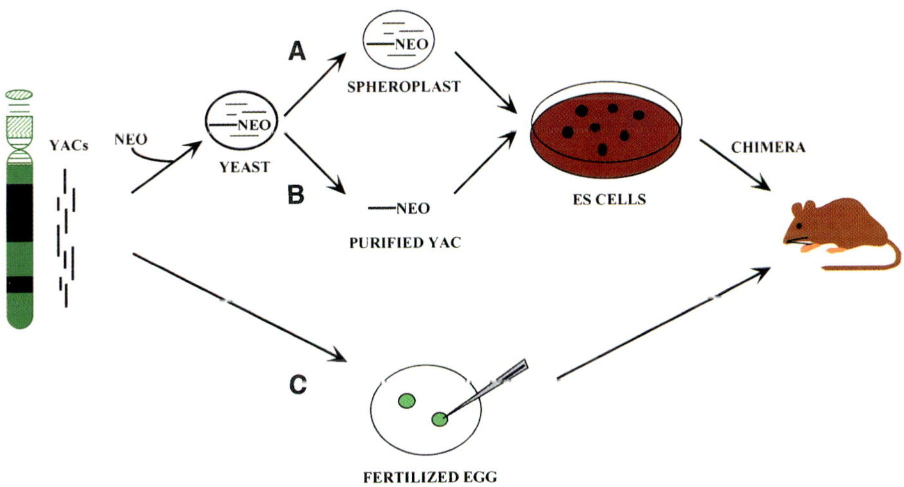

Plate 2, Fig. 1. (*See* discussion in Chapter 42 and full caption on p. 437.) Schematic representation of the three methods for the generation of YAC transgenic mice. **(A)** Spheroplast fusion relies on the introduction of YACs containing a mammalian selectable cassette (NEO) into ES cells by enzymatic digestion of the yeast cell wall and fusion of the resulting spheroplasts with ES cells. **(B)** Lipofection involves the isolation of YACs containing mammalian selectable cassettes away from the other yeast chromosomes. **(C)** Microinjection involves the isolation of YACs away from other yeast chromosomes.

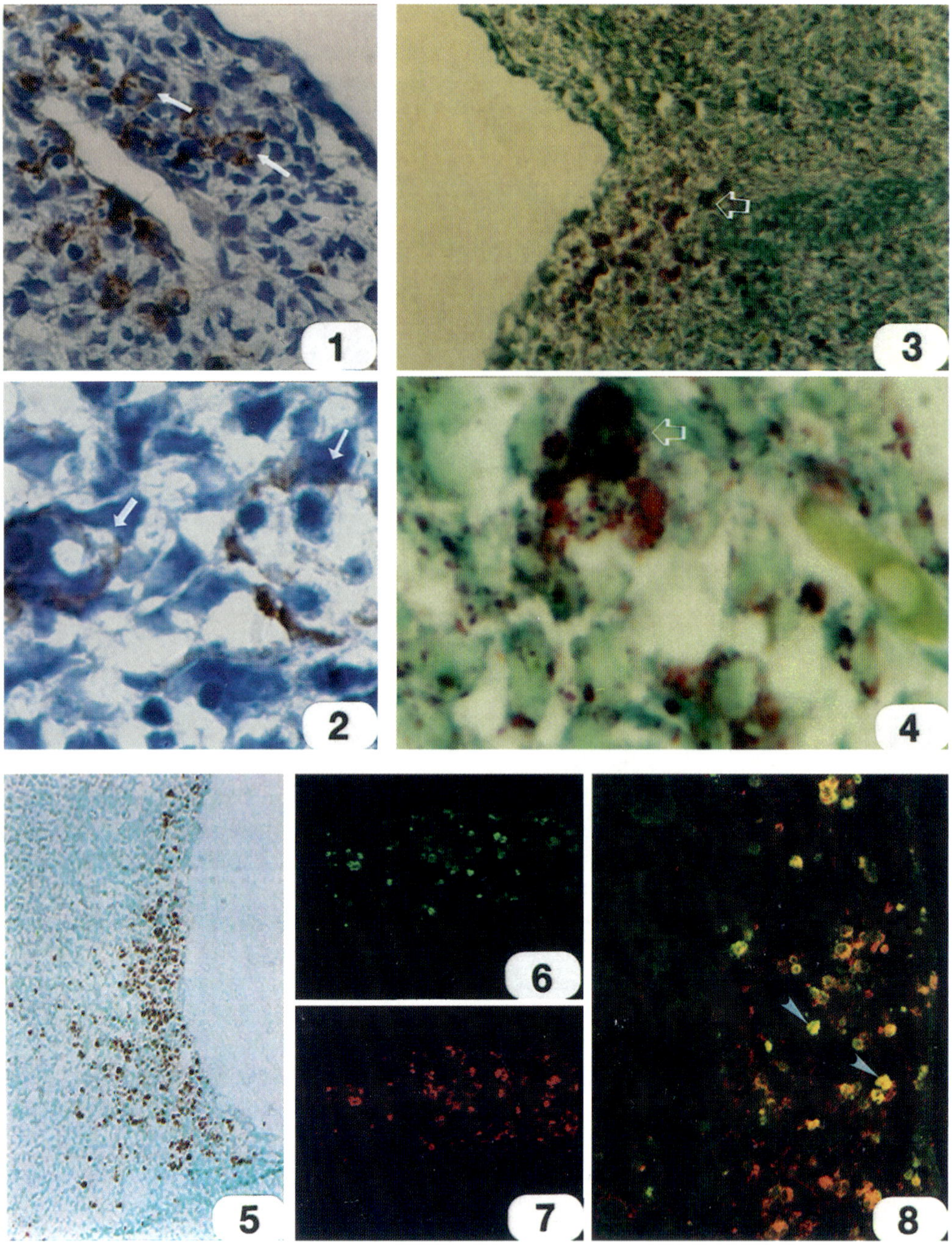

Plate 3, Fig. 1. (*See* discussion in Chapter 13 and full caption on p. 111.) Identification of cell death in d 14.5 mouse limb. (**Panels 1** and **2**) Detection of phagocytic cells by the antibody F4/80. (**Panels 3** and **4**) Lysosomal activity. **(Panel 5)** DNA fragmentation using nonfluoresence end labeling, 40X. **(Panel 6)** DNA fragmentation using fluorescence end labeling. **(Panel 7)** Detection of Cdk5 by Cy3. **(Panel 8)** Labeling and confocal overlay of DNA fragmentation and Cdk5 of Panels 6 and 7 at 40X.

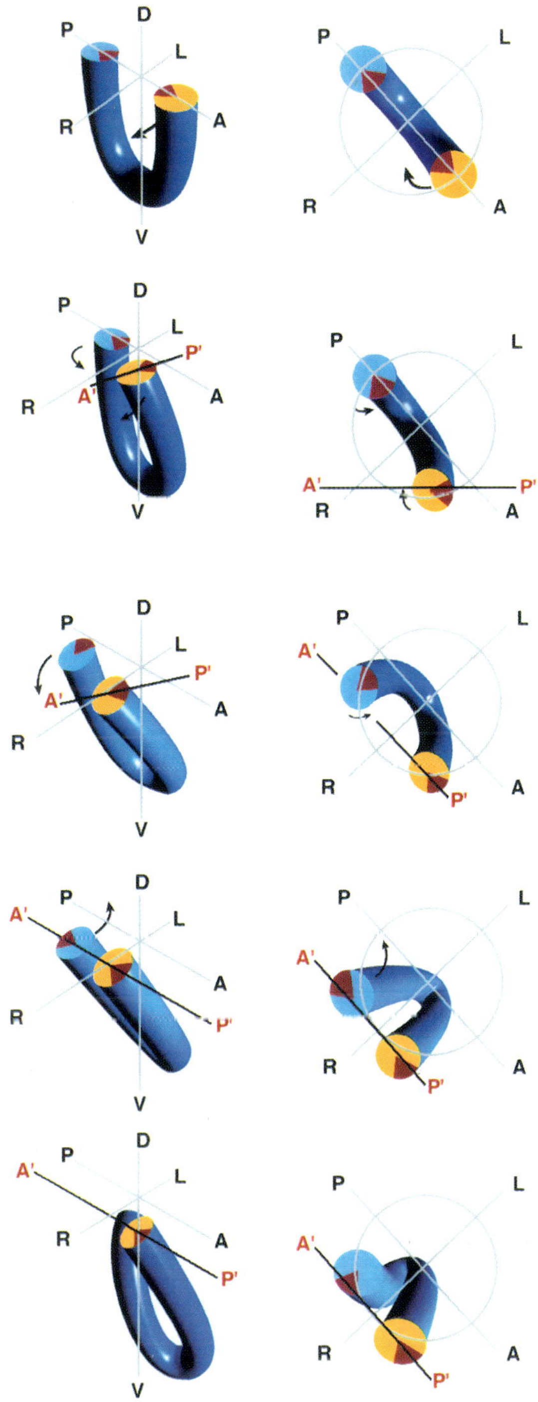

Plate 4, Fig. 6. (*See* discussion in Chapter 24 and full caption on p. 233.) Embryos at different stages during axial rotation from the top view (right column) and three-dimensional right-frontal view (Left column). Abbreviations: A, anterior; D, dorsal; L, left; P, posterior; R, right; V, ventral.

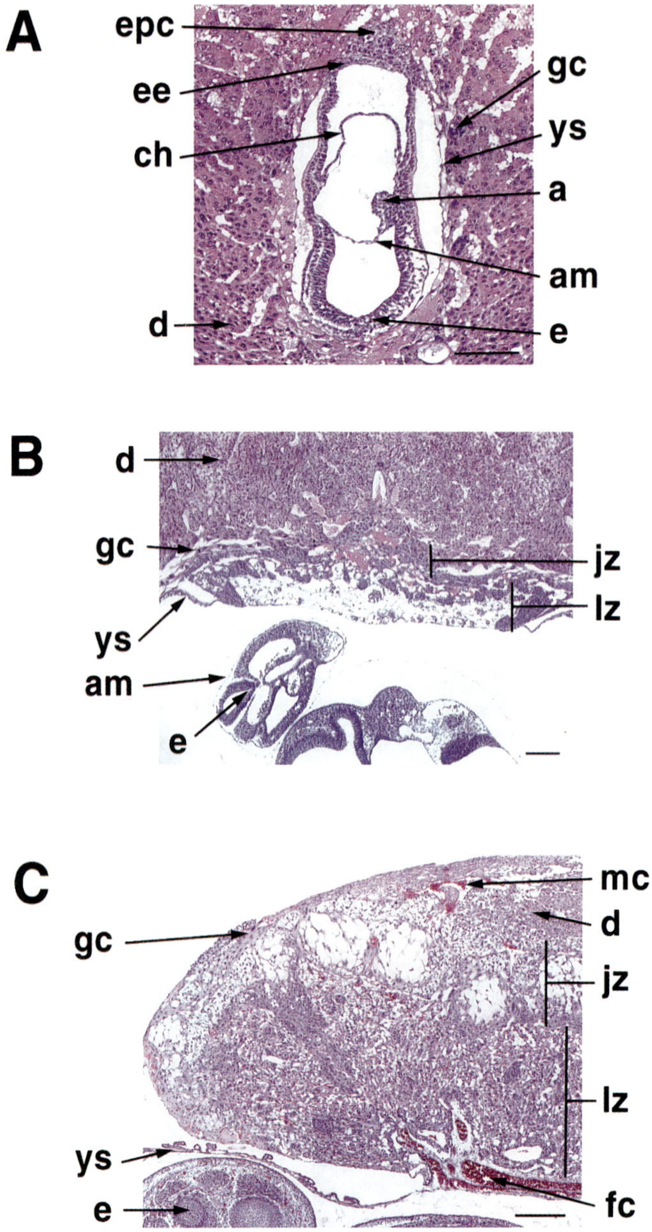

Plate 5, Fig. 2. (*See* discussion in Chapter 20 and full caption on p. 190.) Developing placenta in the mouse. **(A)** A 7.5-dpc gestation site. At this stage, the embryo (e) is undergoing gastrulation. **(B)** A 9.5-dpc gestation site. The labyrinthine (lz) of the mature chorioallantoic placenta is formed. **(C)** A 13.5-dpc placenta.

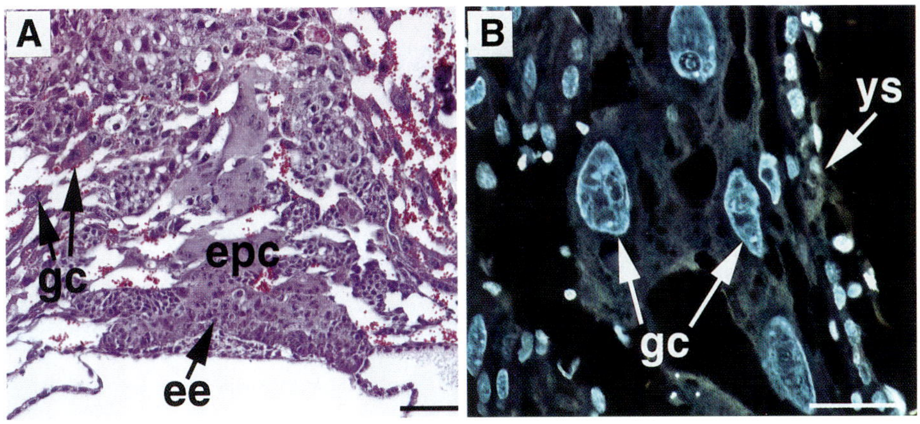

Plate 6, Fig. 3. (*See* discussion in Chapter 20 and full caption on p. 192.) Visualization of abnormal placentation and giant trophoblast cells. **(A)** Hematoxylin and eosin-stained 9.5-dpc gestation site. **(B)** Wild-type 9.5-dpc gestation site stained with Hoechst according to methods. Scale bar = 50 μm.

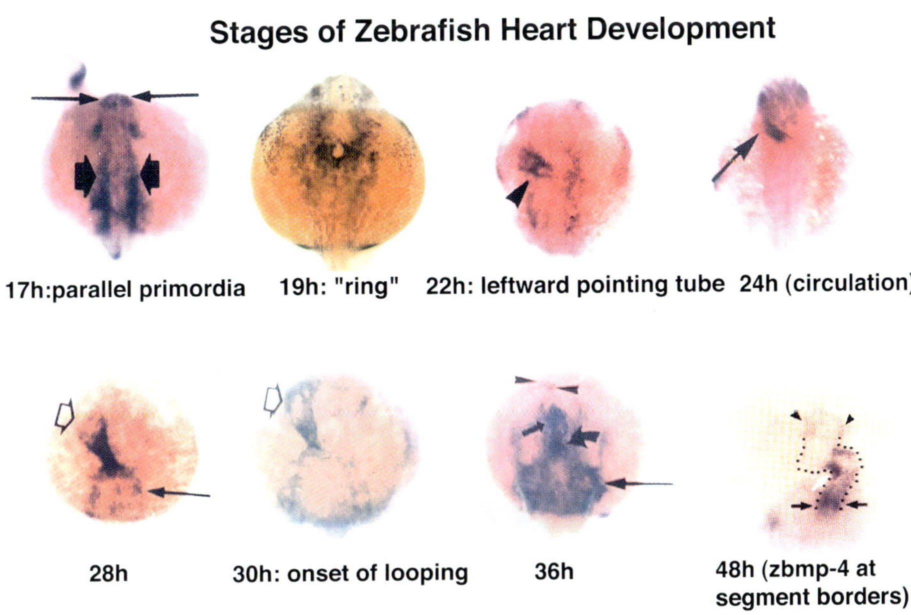

Plate 7, Fig. 2. (*See* discussion in Chapter 26 and full caption on p. 263.) Zebrafish heart development between 17 and 48 h postfertilization. Whole-mount *in situ* hybridization with *zbmp4* riboprobe. All panels are dorsal views with rostral to the top and caudal to the bottom.

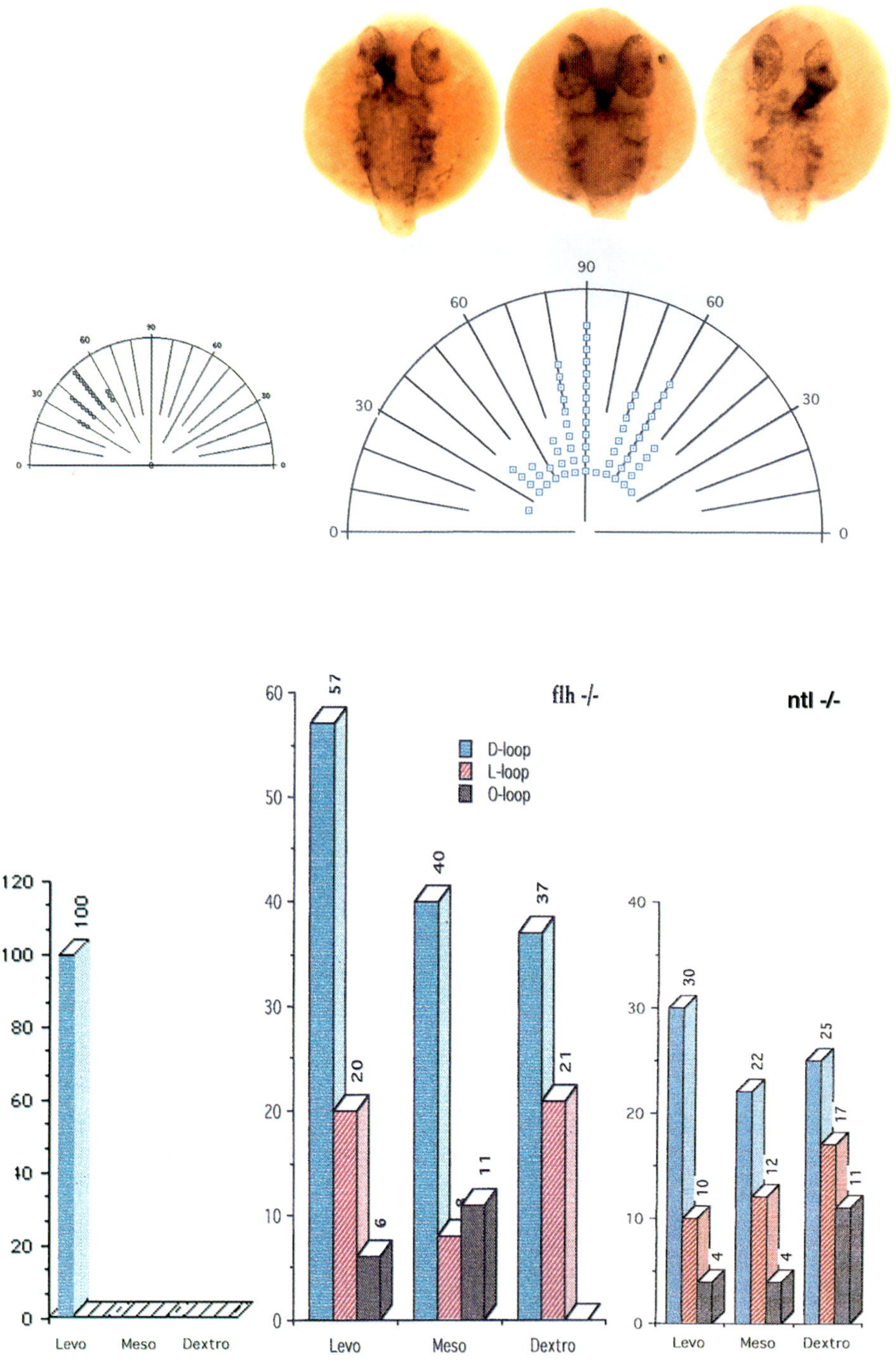

Plate 8, Fig. 3. (*See* discussion in Chapter 26 and full caption on p. 265.) **Top left:** heart tube position in brass wild-type as viewed dorsally (midline is 90°). **Top right:** heart tube position in *floating head (flh)* homozygotes at 23–24 hpf. **Bottom left:** heart tube position and looping direction in Oregon wild-type backgroun. **Bottom center:** distribution of looping directions in *flh* homozygotes with left-sided (levo), midline (meso), and right-sided (dextro) hearts. **Bottom right:** distribution of loping directions in *no tail (ntl)* homozygotes with left-sided (levo), midline (meso), and right-sided (dextro) hearts.

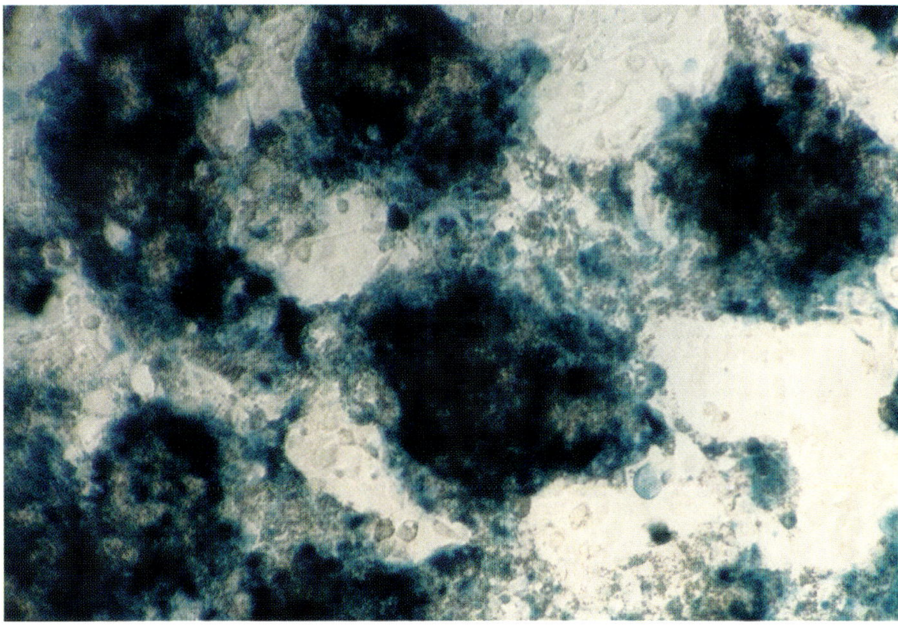

Plate 9, Fig. 1. (*See* discussion in Chapter 38.) CBCs transfected with plasmid pmiwZ by electroporation and stained for exogenous β-galactosidase activity with X-gal after 36 h of culture on a feeder layer of mouse SNL fibroblasts.

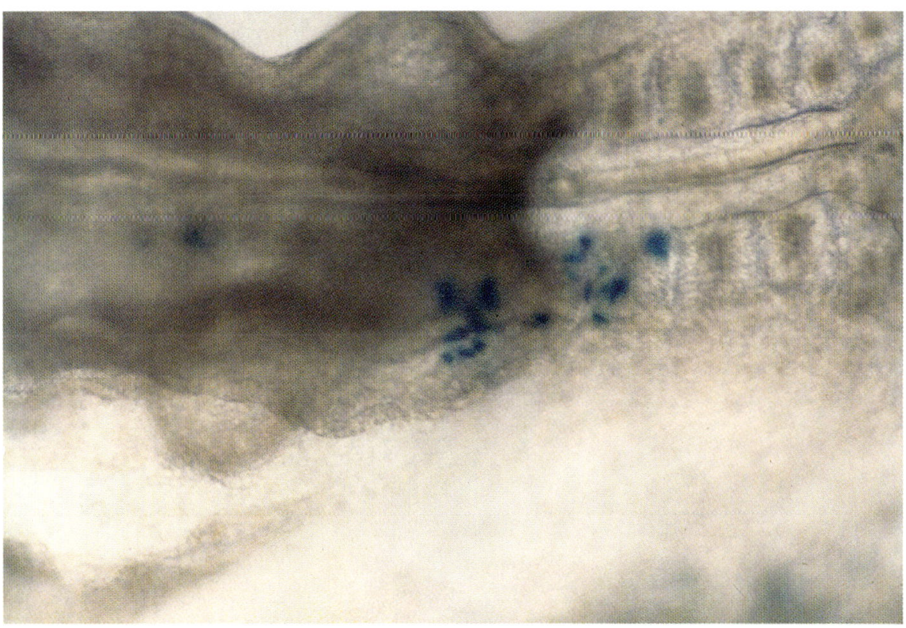

Plate 10, Fig. 2. (*See* discussion in Chapter 38.) Incorporation of CBCs transfected with plasmid pmiwZ by electroporation in a chemic chick embryo. Embryo was stained with X-gal after 72 h of incubation.

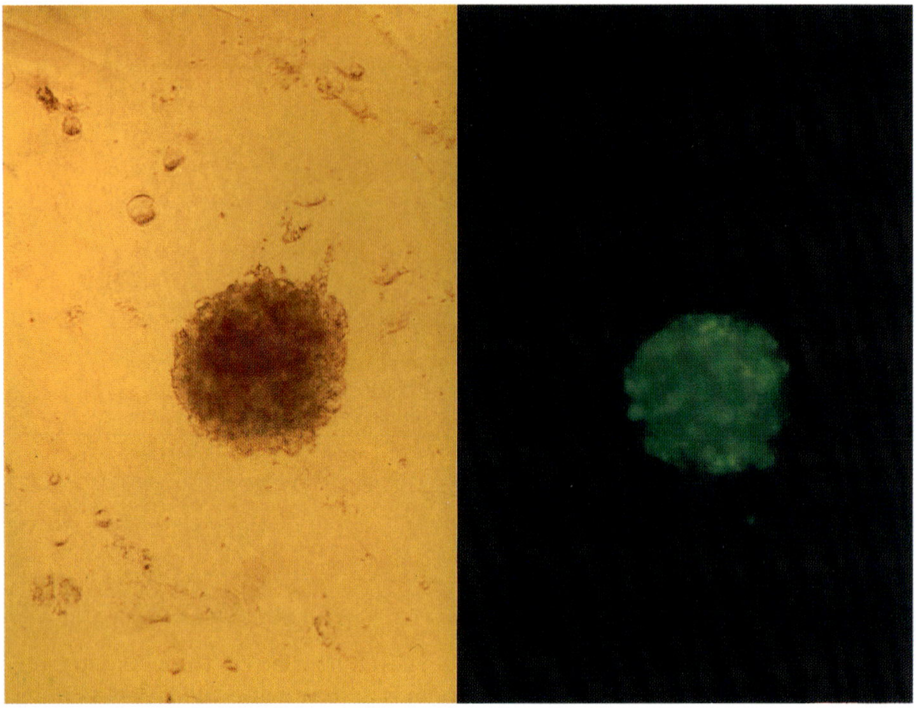

Plate 11, Fig. 1. (*See* discussion in Chapter 39.) **Left panel:** $C_{12}FDG$ staining of a colony of blastodermal cells after 6 d of selection in Zeocin. The cells were transfected with the pZeoV2/*lacZ* plasmid, incubated on SNL feeded cells in the presence of 375 μg Zeocin/mL of medium and stained on the sixth day in culture. **Right panel:** phase contrast of the same colony.

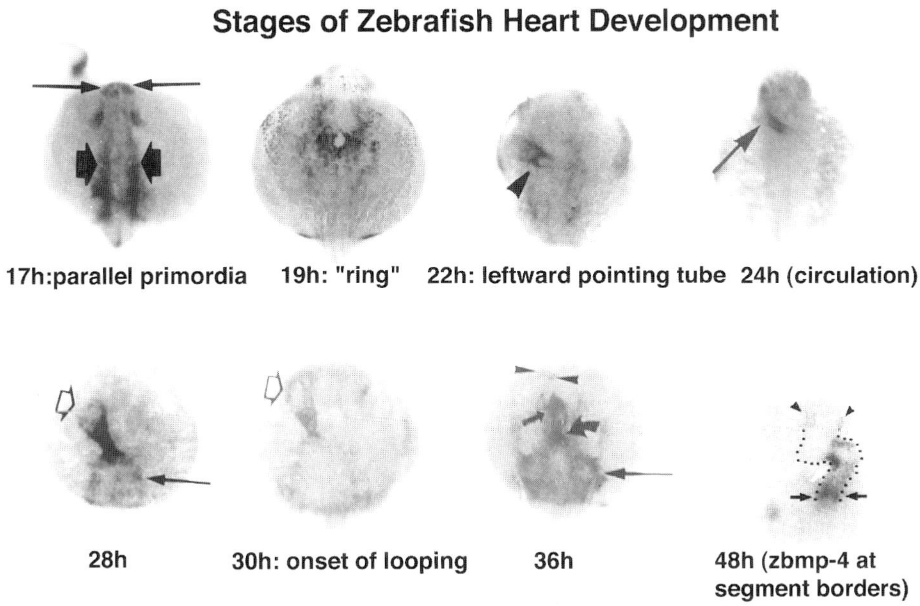

Fig. 2. (*See* color plate 7 appearing after p. 262.) Zebrafish heart development between 17 and 48 h postfertilization. Whole-mount *in situ* hybridization with *zbmp4* riboprobe. All panels are dorsal views with rostral to the top and caudal to the bottom. 17 h: parallel primordia (wide arrows) are seen in the lateral plate mesoderm. 19 h: *zbmp4*-expressing cells collect in the region between the two primordia and form a "ring," which, on closer inspection, is a shallow cylinder with its axis oriented in the dorsoventral direction. 22 h: the heart tube (arrowhead) has elongated and started to point to the left side of the embryo. 24 h and 28 h: the heart tube (arrow) has started to swing back toward the midline. 30 h: the heart tube is nearly midline. Looping commences. 36 h: looping is well underway. *zbmp4* is receding from the heart except at segment boundaries (ventricle-aortic junction, curved arrow; atrioventricular junction, short straight arrow; and sinus venosus-atrial junction, arrowheads). 48 h: *zbmp4* has been extinguished except at segment borders.

transparent, facilitating serial noninvasive inspection, (5) egg batch size is in the hundreds, and (6) these two steps in heart patterning occur roughly 10 h apart (*see* **Fig. 2**). A review of the stages of zebrafish heart development will be followed by an analysis of two mutants that illustrate the promise of the system for the genetic dissection of cardiovascular morphogenesis.

From two parallel primordia in the lateral plate mesoderm, the heart tube assembles at 19 h into a shallow cylinder that appears as a ring when viewed dorsally (**Fig. 2**). Over the next 3 h as the cylinder elongates, it points to the left side of the embryo. By the time the circulation is conspicuous at 24 h, the heart tube is two cell layers thick with myocardium enclosing endocardium. By 30 h, the long axis of the cylinder has migrated back toward a nearly midline position. Looping, the bending of the midportion of the cylinder so that the future ventricle is situated to the right of the future atrium, commences at this time and proceeds until approx 60 h.

Before regular contractions (24 h in our three wild-type strains), the angle of the heart tube vis-à-vis the embryonic midline is best assessed at a particular time point by doing whole-mount *in situ* hybridization (*see* **Subheading 3.**), photographing fixed embryos and using a protractor. After regular contractions are visible, *in situ* hybridization is not necessary; a miniature protractor (inscribed on a transparency) can be placed under each embryo as it viewed under the microscope, allowing serial measurements of live embryos.

Homozygotes for either of the two mutations *floating head (flh)* and *no tail (ntl)*, encoding the transcription factors *Znot* and *Zf-T*, respectively ***(6,7)***, have been reported to display qualitatively abnormal heart tube position ***(8)*** and heart looping ***(9)***. Quantitatively, a plot of the heart tube angle at 23–24 h in *flh* homozygotes [probed with *zbmp4* ***(10)***] reveals that, compared with wild-type embryos, control of heart position is lost (**Fig. 3**). In a second type of experiment, *flh* homozygote and *ntl* homozygote embryos examined by light microscopy at 24 h were segregated into three groups, levocardia, mesocardia, and dextrocardia. Each group was then reexamined at 36–48 h to assay for heart tube looping direction (D-, L-, or O-). For both these mutations, any type of loop can be found in any of the three heart positions, illustrating the fact that normal coupling of heart chirality to heart position has been lost (**Fig. 3**).

2. Materials

1a. Phosphate-buffered saline (PBS), pH 7.4 (composition per liter): 8.0 g NaCl, 0.2 g KCl, 1.44 g Na_2HPO_4, 0.24 g KH_2PO_4.

1b. PBST: PBS containing 0.1% Tween-20 (polyoxyethylene-sorbitan monolaurate, Sigma, St. Louis, MO).

2. Fixative: 4% Paraformaldehyde (Fisher reagent grade, Fisher Scientific, Pittsburgh, PA) in PBS. In order to solubilize, heat to 65°C for approximately 3 h.
3. Digestion solution: Proteinase K (Boehringer-Mannheim, Mannheim, Germany) 10 mg/mL stock solution in PBST. Store at –20°C.
4. Hybridization buffer (Hyb):
   50% formamide (e.g., Boehringer-Mannheim)
   5X standard sodium citrate (SSC) [20X SSC, pH 7.0 (composition per liter): 175 g NaCl, 88.2 g sodium citrate]
   50 µg/mL heparin (sodium salt) Sigma)
   500 µg/mL yeast tRNA (Sigma)
   0.1% Tween-20
   9 m*M* sodium citrate, pH 6.0
   Store at –20°C.
5. Alkaline phosphatase (Alk Phos) Buffer:
   100 m*M* Tris HCl (pH 9.5)
   50 m*M* $MgCl_2$
   100 m*M* NaCl
   0.2% Tween-20
   0.2% Triton X-100 (t-octylphenoxypolyethoxyethanol, Sigma)
   240 mg/200 mL levamisole (Sigma)
   Store at 4°C.
   (Levamisole hydrochloride: inhibitor of mammalian liver, kidney, and bone alkaline phosphatases, it only minimally inhibits the calf intestinal alkaline phosphatase conjugated to the antidigoxigenin antibody, according to the manufacturer.)
6. 5-Bromo-4-chloro-3-indolyl-phosphate (BCIP) (Boehringer-Mannheim):
   Stock solution: 50 mg/mL in dimethylformamide (Fisher). Store at –20°C.
7. 4-Nitroblue tetrazolium chloride (NBT) (Boehringer-Mannheim):
   Stock solution: 75 mg/mL in 70% dimethylformamide. Store at –20°C.

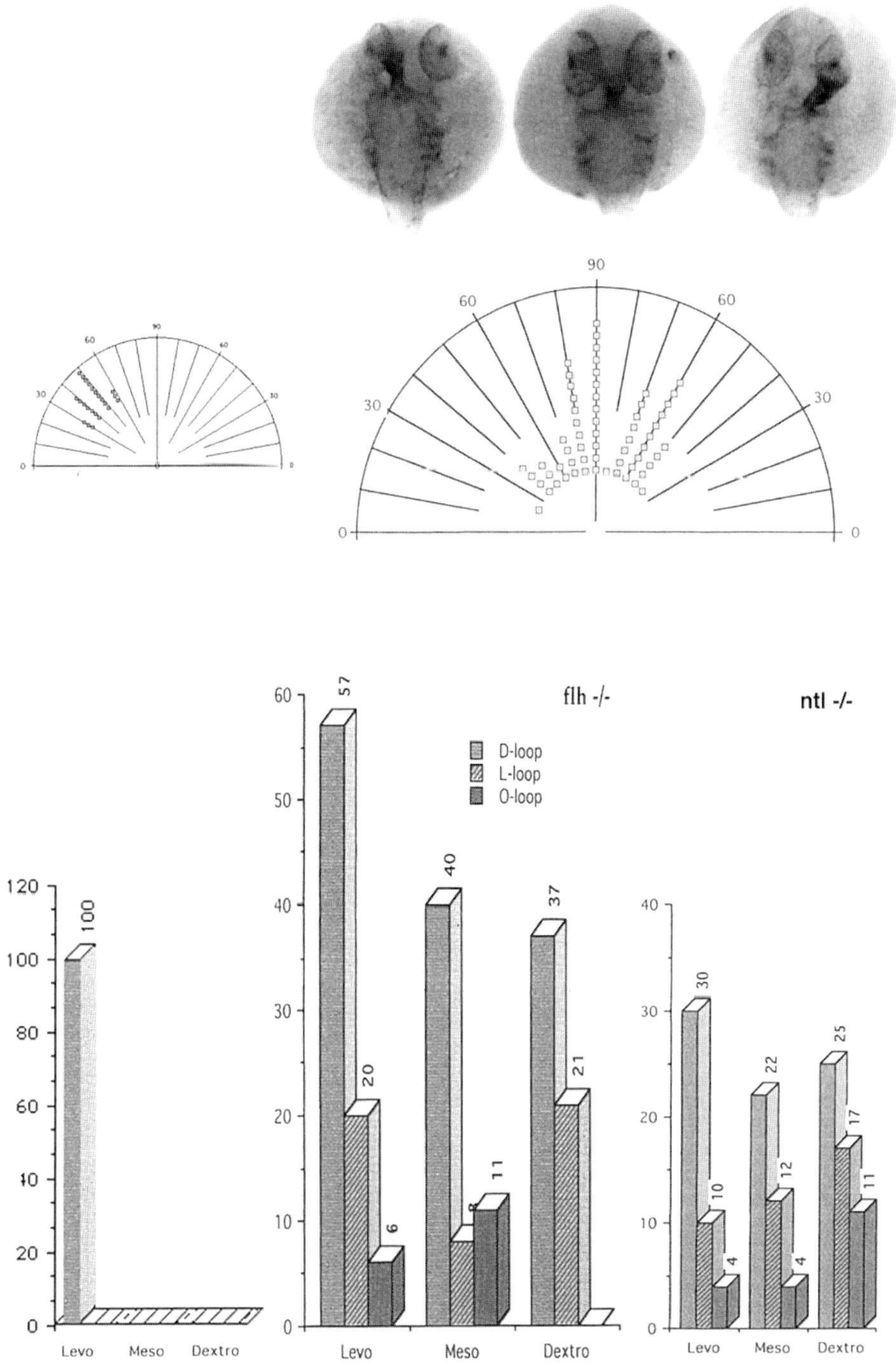

Fig. 3. (*See* color plate 8 appearing after p. 262.) Top left: heart tube position in wild-type zebrafish as viewed dorsally (midline is 90°). Top right: heart tube position in *floating head (flh)* homozygotes at 23–24 hpf, as displayed by *zbmp4 in situ* hybridization. Examples of the three heart tube positions are shown above the plot. Bottom left: heart tube position and looping direction in Oregon wild-type background. Bottom center: Distribution of looping directions in *flh* homozygotes with left-sided hearts (Levo), midline hearts (Meso), and right-sided hearts (Dextro). Bottom right: Distribution of looping directions in *no tail (ntl)* homozygotes with left-sided (Levo), midline (Meso), or right-sided (Dextro) hearts.

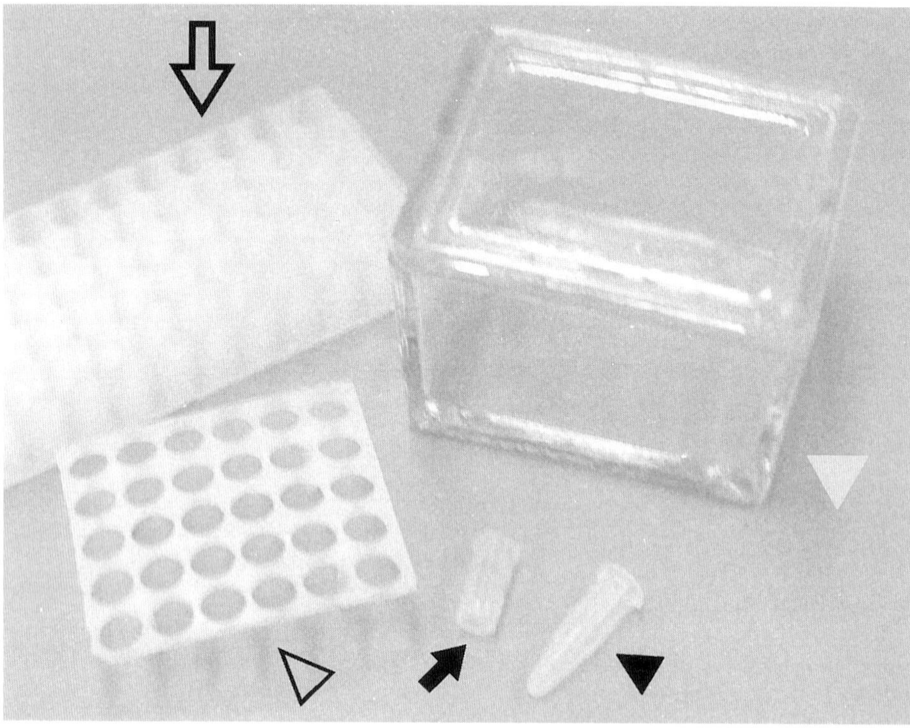

Fig. 4. Accessories for large-scale whole-mount *in situ* hybridization. Racks (open arrowhead) can be made by slicing the top off an ordinary Eppendorf rube rack (open arrow) and attaching small legs (e.g., with Superglue). The mesh-bottom baskets (closed arrow) can be constructed by cutting off the bottom of an Eppendorf tube (closed arrowhead) and gluing on a circle of nylon mesh. A basket-containing rack fits inside a glass staining dish.

8. Digoxigenin-labeled RNA probe for gene of interest, prepared according to standard protocols.
9. Alkaline phosphatase-conjugated anti-digoxigenin antibody (Fab fragments, Boehringer-Mannheim). Store at 4°C.
10. Blocking solution (composition per 100 mL of PBST): 2 mL newborn calf serum (Life Technologies, Inc., Grand Island, NY), 1 mL dimethylsulfoxide (Sigma), 1 g bovine serum albumin (Sigma). Store at –20°C.
11. Glass staining dishes (Fisher).
12. Scintillation vials (FB, 50X16) (Sarstedt, Inc., Newton, NC).
13. Special equipment (for constructing baskets and racks shown in **Fig. 4**):
    Eppendorf tube rack (Fisher).
    Nylon mesh (approximately 0.1 mm pore size).
14. 24-Well, flat-bottom tissue culture plates (Corning, Corning, NY).

3. Methods

Whole-mount *in situ* hybridization for developing zebrafish embryos using digoxigenin-labeled RNA probes [modified from ***(11)***].

1. Fixation (*see* **Notes 1** and **2**): Embryos harvested with chorion intact should be placed in 4% paraformaldehyde-PBS for at least 3 h at room temperature (RT) or 4°C overnight

(ON). Manually dechorionate. It is easier to dechorionate prior to the dehydration, but the protocol will work successfully even if dechorionation is done after dehydration.

2. Dehydration: Methanol (MeOH) 10 min at RT. (Dechorionated embryos may be stored at –20°C for months in glass vials or Eppendorf tubes.)

   Day 1

   Fill glass staining dish with MeOH; lower the rack with baskets into the staining dish; using a wide-bore Pasteur pipet, load no more than 50 embryos into each basket.

   In addition to the baskets (*see* **Note 2**) of embryos to be probed, include one basket with at least 50 other dechorionated embryos (or embryo fragments). The latter embryos will be used to preadsorb the antidigoxigenin antibody.

3. Rehydration:

   5 min 75% MeOH in PBS at RT (agitation is helpful)
   5 min 50% MeOH in PBS at RT
   5 min 25% MeOH in PBS at RT
   4X 5 min PBST at RT

4. Digestion: 30 Min 10 μg/mL proteinase K in PBST at RT (embryos > 48 h) (*see* **Note 3**).
5. Stop reaction: rinse 2X in 2 mg/mL glycine in PBST at RT.
6. Refixation: 30 min in 4% paraformaldehyde in PBS at RT.
7. Rinse: 5X 5 min in PBST at RT. Embryos to be probed should continue to **step 8**. Embryos (or embryo fragments) to be used for preadsorption of the anti-digoxigenin antibody should be removed for adsorption of antibody process (*see* **step 9**).
8. Prehybridization: For each basket, preheat vial containing 450 μL of Hyb buffer. Transfer each basket to individual vial. Incubate 1–2 h in Hyb buffer at 65°C.
9. Hybridization (*see* **Note 4**): Transfer basket to fresh vial with fresh Hyb buffer and approximately 0.5 μg of probe. Incubate at 65°C ON.

   Adsorption of antibody (*see* **Note 5**): Embryos (or embryo fragments) not to be probed should be incubated with 1:400 dilution of antibody in blocking solution (e.g., 12.5 μL of antidigoxigenin antibody and 5 mL of blocking solution) in a glass vial at 4°C ON.

   Day 2

   Transfer each basket back to the rack sitting in glass staining dish with first wash solution (prewarmed to 65°C).

10. Wash:

    10 min 37.5% formamide, 2X SSC at 65°C (prewarmed)
    10 min 25% formamide, 2X SSC at 65°C (prewarmed)
    10 min 12.5% formamide, 2X SSC at 65°C (prewarmed)
    10 min 2X SSC at 65°C (prewarmed)
    2X 30 min 0.2 SSC at 65°C (prewarmed)
    5 min 0.15X SSC, 25% PBST at RT
    5 min 0.10X SSC, 50% PBST at RT
    5 min 0.05X SSC, 75% PBST at RT
    5 min PBST at RT

11. Block: 1 h in blocking solution at RT.
12. Antibody incubation: 3 h with preadsorbed antibody at a 1:4000 dilution in blocking solution at RT (or ON at 4°C).
13. Wash:

    2X rinse with PBST at RT
    3X 15 min PBST at RT
    (Note: can be left in PBST ON at 4°C, if antibody incubation has been completed.)

    Day 3

    Continued washing: 2X 15 min PBST at RT

3X 5 min alkaline phosphatase buffer

Transfer embryos in each basket to individual well of a 24-well microtiter plate using a wide-bore Pasteur pipet.

14. Stain: Incubate embryos in each well with about 500 µL of Alkaline phosphatase buffer that includes substrates BCIP and NBT (3.5 µL BCIP stock and 4.5 µL NBT stock per 1 mL of Alk Phos buffer). Remember to cover the plate with aluminum foil to reduce the light-induced degradation of the BCIP and NBT substrates. Add fresh substrate if you monitor the staining for long periods of time under the microscope.
15. Stop reaction: Rinse 4X in PBST.
16. Refixation: 30 min in 4% paraformaldehyde in PBS.

4. Notes

1. The most critical steps in the *in situ* hybridization protocol are the embryo preparation, permeabilization (digestion), and riboprobe preparation. Aliquots of the paraformaldehyde solution can be stored at –20°C for up to 2 mo but should not be kept at room temperature for more than a few days. Embryos should not be left in paraformaldehyde solution at 4°C for more than 3 d before being dechorionated and transferred to methanol. "Overfixed" embryos are very difficult to permeabilize; this lowers the eventual signal-to-background ration. Dechorionation is easier to do prior to, rather than after, the methanol dehydration.
2. Although *in situ* hybridization can be performed in vials or in multiwell tissue culture plates, the procedure can be greatly facilitated by using staining dishes inside which one can lower a rack capable of holding numerous mesh-bottom baskets (**Fig. 4**). Up to 50 embryos can be placed in each basket. Fluid exchanges are simply done by using a hemostat to lift the entire rack out of a staining dish containing the first solution and quickly insert it into a staining dish with the next solution. This maneuver is less traumatic to the embryos than manually removing solution from individual wells (e.g., with a Pasteur pipet). In addition, solution changes are more complete and uniform.
3. The permeabilization step using proteinase K often needs to customized to the particular tissue, structure, or organ of interest. For example, accessibility of the probe is especially important when investigating a deeply placed organ like the heart and of less importance when the region of interest is superficial, like the caudal fin. To vary the digestion, one can change either the proteinase K concentration or the duration of the digestion.

   20 min 10 µg/mL proteinase K in PBST (embryos 24–48 h)
   10 min 10 µg/mL proteinase K in PBST (embryos 12–24 h)
   5 min 10 µg/mL proteinase K in PBST (embryos 6–12 h)
   5 min 2.5 µg/mL proteinase K in PBST (embryos <6 h)

   With embryos of <6 h, the proteinase K digestion may be omitted if the message of interest is highly abundant.
4. The successful preparation of the RNA probe is most dependent on template quality. Phenol–chloroform extraction of the linearized DNA, followed by ethanol precipitation, is suggested to denature and remove residual RNase and any other extraneous proteins. Short (100–400-nucleotide) probe length tends to result in high background. The conditions of this hybridization protocol are sufficiently stringent that a high signal-to-background ratio is observed with a wide variety of antisense riboprobes. As with all RNA work, it is preferable to wear gloves, at least through the hybridization step in the protocol.
5. The adsorption step reduces nonspecific binding of the antidigoxigenin antibody to the embryos. The dechorionated embryos used for this adsorption need not be intact, but embryos with intact chorions should not be used.
6. Heart position can be assayed pre- or postlooping, and it remains unclear whether these two assessments correlate with each other, as very few species allow both to be made

easily. For example, it is easy to determine postlooping position in the human fetus by ultrasonic means, but it has proved impossible so far to examine prelooping heart position. In the chick, heart position is sufficiently close to the midline both pre- and postlooping that it is difficult to study this aspect of patterning. The mouse heart has been examined both pre- ***(5)*** and postlooping ***(12)***. Prelooping heart position does appear to correlate with the postlooping heart position; both appear to be to the left of the midline. Until additional species have been intensively studied, we do not know whether such correlations are in fact a general rule among vertebrates.

7. Postlooping heart position in the zebrafish is harder to study for two reasons: (a) the heart stays close to the midline from approximately 48 h onward, and (b) pigmentation begins to emerge after 36 h, which makes the heart more difficult to visually inspect. To surmount the latter difficulty, one can utilize pigmentation-defective mutants such as *brass*, *albino*, and *golden* ***(13)***; alternatively, embryos can be raised in 0.2 m*M* 1-phenyl-2-thiourea ***(14)*** starting around 15 h.
8. The criteria for designating (prelooping) heart position in the zebrafish are still controversial. One investigator ***(8)*** has proposed looking at the heart tube (after the commencement of cardiac muscle contractions) from a dorsal view and identifying whether the rostral end of the tube has extended beyond the lateral margins of the neural keel. There are three major limitations to this approach. First, it requires the assumption of a normal sized neural keel; many of the developmental mutants so far identified (e.g., *cyclops*) have abnormal neural keel width, making it preferable to choose a universally applicable criterion (i.e., one that is not as heavily dependent on assumptions). Second, since the time of onset of periodic cardiac contractions (approximately 24 hpf in our wild-type strains) is in the midst of the interval (22–28 hpf) during which movement of the heart tube axis (toward the midline) is most rapid, a short delay in recognition can result in the rostral end of the heart tube already appearing to lie within the lateral margins of the neural keel; it is easy to erroneously construe heart lateralization in this circumstance as "midline," or "meso." Third, it makes the assumption that there is a strict correlation between development time and onset of heart contraction and that this correlation is constant among all mutants.

   An alternative method is to judge heart tube position vis-à-vis the midline; thus, during the interval between 23 and 26 hpt, only contracting heart tubes that are superimposable on the midline when viewed dorsally are deemed meso-; all others are assigned as rightward or leftward.
9. The potential effect of genetic background on heart position or on heart tube looping direction should always be considered. Investigators have already noted the effect of genetic background on postlooping heart position in the mouse ***(4)*** and on the prelooping heart position in the zebrafish ***(8)***.

Acknowledgments

The author thanks David Kozlowski for his comments on the manuscript and Bob Riggleman for suggesting the use of racks of baskets rather than microtiter plates for large-scale, whole-mount *in situ* hybridization experiments. This work was supported by the American Heart Association, Southeastern Pennsylvania Affiliate, and by the national American Heart Association (AHA-9650585N).

References

1. Van Praagh, R., Weinberg, P. M., Smith, S. D., Foran, R. B., and Van Praagh, S. (1989) Malpositions of the heart, in *Moss' Heart Disease in Infants, Children, and Adolescents*, 4th ed. (Adams, F. H., Emmanouilides, G. C., and Riemenschneider, T. A., eds.), Williams & Wilkins, Baltimore, pp. 530–580.

2. Layton, W. M. and Manasek, F. J. (1980) Cardiac looping in early *iv/iv* mouse embryos, in *Etiology and Morphogenesis of Congenital Heart Disease* (Van Praagh, R. and Takao, A., eds.), Futura Publishing, New York, pp. 109–126.
3. Supp, D. M., Witte, D. P., Potter, S. S., and Brueckner, M. (1997) Mutation of an axonemal dynein affects left-right asymmetry in *inversus viscerum* mice. *Nature* **389,** 963–966.
4. Oh, S. P. and Li, E. (1997) The signaling pathway mediated by the type IIB activin receptor controls axial patterning and lateral asymmetry in the mouse. *Genes Dev.* **11,** 1812–1826.
5. Biben, C. and Harvey, R. P. (1997) Homeodomain factor NKx 2-5 controls left-right asymmetric expression of bHLH gene *eHand* during murine heart development. *Genes Dev.* **11,** 1357–1369.
6. Talbot, W. S., Trevarrow, B., Halpern, M. E., Melby, A. E., Farr, G., Postlethwait, J. H., Jowett, T., Kimmel, C. B., and Kimelman, D. (1995) A homeobox gene essential for zebrafish notochord development. *Nature* **378,** 150–157.
7. Schulte-Merker, S., van Eeden, F. J. M., Halpern, M. E., Kimmel, C. B., and Nüsslein-Volhard, C. (1994) *no tail* is the zebrafish homologue of the mouse *T (Brachyury)* gene. *Development* **120,** 1009–1015.
8. Chen, J.-N., van Eeden, F. J. M., Warren, K. S., Chin, A., Nusslein-Volhard, C., Haffter, P., and Fishman, M. C. (1997) Left-right pattern of cardiac *BMP4* may drive asymmetry of the zebrafish heart. *Development* **124,** 4373–4382.
9. Danos, M. C. and Yost, H. J. (1996) Role of notochord in specification of the cardiac left-right axis in zebrafish and *Xenopus*. *Dev. Biol.* **177,** 96–103.
10. Stainier, D. Y. R., Lee, R. K., and Fishman, M. C. (1993) Cardiovascular development in the zebrafish. I. Myocardial fate map and heart tube formation. *Development* **119,** 31–40.
11. Li, Y., Allende, M., Finkelstein, R., and Weinberg, E. S. (1994) Expression of two zebrafish *orthodenticle*-related genes in the embryonic brain. *Mech. Dev.* **48,** 229–244.
12. Icardo, J. M. and Sanchez de Vega, M. J. (1991) Spectrum of heart malformations in mice with situs solitus, situs inversus, and associated visceral heterotaxy. *Circulation* **84,** 2547–2558.
13. Haffter, P., Odenthal, J., Mullins, M. C., Lin, S., Farrell, M. J., and Vogelsang, E., et al. (1996) Mutations affecting pigmentation and shape of adult zebrafish. *Dev. Genes Evol.* **206,** 260–276.
14. Westerfield, M. (1993) *The Zebrafish Book.* University of Oregon Press, Eugene, OR, p. 4.1.
15. St. John Sutton, M. G., Oldershaw, P. J., and Kotler, M. N. (1996) *Textbook of Echocardiography and Doppler in Adults and Children.* Blackwell Science, Cambridge, MA, p. 32.

27

Biologically Based Risk Assessment Models for Developmental Toxicity

Christopher Lau and R. Woodrow Setzer

1. Introduction

Risk assessment is a process conducted by regulatory agencies to safeguard public health from the harmful effects of natural and man-made substances. Thus, a goal of risk assessors is to provide a numerical value of exposure to these substances, below which adverse effects to human health are assumed negligible. This value, termed the reference dose (RfD) or reference concentration (RfC), is commonly derived with the aid of a dose-response model that describes the relationship between the extent of adverse effects and the exposure levels of the toxic agent. For the evaluation of developmental toxicity of an agent, the typical response endpoints include alteration of reproductive outcomes and incidence of malformations *(1)*. In a default situation, where little is known about the agent, the RfD is calculated from

$$RfD = \frac{NOAEL}{UF}$$

where NOAEL is the "no observed adverse effect level" derived from a dose-response study using a laboratory animal model (i.e., the dose level where the response parameters of the treated animals are not statistically different from those of the control cohorts) and UF is the product of multiple uncertainty factors that account for extrapolating from the laboratory animal model to humans, from average humans to a susceptible subpopulation, and from the dose level where the least effect is observed to that where none is observed. Recent introduction of the benchmark dose (BMD) concept *(2)* appears to have gained favor among risk assessors over the NOAEL approach in deriving the reference dose. Instead of utilizing only one data point in the dose-response model for NOAEL, the BMD approach uses data from all experimental doses to create a mathematical expression of the dose-response relationship. The dose level (and its 95% confidence limits) corresponding to a predetermined level of response can be estimated from this mathematical expression. Advantages of the BMD over the NOAEL approach include full utilization of data from a dose-response study, less dependency on the selection of dose levels for evaluation, and consideration of experimental uncer-

From: *Methods in Molecular Biology, Vol. 136: Developmental Biology Protocols, Vol. II*
Edited by: R. S. Tuan and C. W. Lo © Humana Press Inc., Totowa, NJ

tainty by the use of a lower confidence interval. A comparison of the NOAEL and BMD applications to developmental toxicity was recently made (3–6) and readers are referred to those studies for detailed discussion of this topic.

Although advances have been made to improve the estimation of an RfD, the current practice of risk assessment is still hampered by the need to extrapolate experimental values derived from studies using high doses (for observable toxic responses) to realistic human exposures that are typically at very low levels, and to extrapolate the same values across species from laboratory animal models to humans. Indeed, the sensitivity of this approach is limited because the NOAEL or the BMD is derived from empirical observations of some rather extreme toxic outcomes (e.g., perinatal mortality, induction of anatomical defects). How then does one improve the sensitivity of a dose-response model for risk assessment? One concept is to incorporate as much biologically relevant information as possible into the model. Ideally, physiologically based pharmacokinetic (PBPK) data should be applied for the "dose" component of the model to facilitate direct comparisons of exposure levels between tissues and across species, and pertinent biomarkers based on mechanisms of toxicity should be evaluated for the "response" component to enhance the sensitivity of detection for adverse biological outcomes. This approach to integrate pharmacokinetics and pharmacodynamics is termed "biologically based dose-response" (BBDR) modeling for risk assessment *(7)*.

Although the need to include mechanistic information for human health risk assessment has been generally recognized *(8,9)*, the actual practice of such an endeavor proves to be immensely challenging. For developmental toxicity, at least a dozen or so possible mechanisms have been identified to elicit various untoward cellular responses, leading to tissue damages and adverse perinatal outcomes (**Table 1**). This list of toxic mechanisms will undoubtedly be expanded as our understanding of developmental biology is advanced. It would also be naive to assume that a toxic agent exerts its effects through a single mechanism only. From a toxicologist's standpoint, which and how many mechanism(s) should one embrace for modeling? In order to make this exercise tenable, a few assumptions must be made.

1. *A major or critical mechanistic pathway is directly responsible for the adverse developmental outcomes.* Although this assumption is admittedly far too simplistic, it is made out of necessity. Certainly, a toxic agent can act at multiple sites at cellular and molecular levels, and more of these sites will be identified as research tools are refined. However, it is not practical to wait until all information about a specific agent is exhaustively gathered before a risk-assessment process is initiated. In addition, the inclusion of every nuance of an agent's actions and its attendant cellular responses may render the model so complicated that it becomes a mathematical nightmare. Thus, only the "most pertinent" mechanistic steps or events that are likely linked to the toxic outcomes will be captured for modeling.
2. *Perturbation of this mechanistic pathway can be evaluated quantitatively relative to dose and duration of exposure.* Needless to say, the biomarkers for the intervening cellular events associated with the toxic outcomes have to be identified and their changes should bear a relationship to the increasing doses or the concentration × duration ($c \times t$) of exposures.
3. *Homologous mechanisms of toxicity or pathophysiological pathways exist between test animal species and humans.* This assumption is a cornerstone for the use of laboratory animal models for human health risk assessment. Thus, although it is difficult to extrapolate the human risks at high doses (where toxicity is detected in animals), it may be fea-

Table 1
The Anatomy of Developmental Toxicity

Mechanisms	Cellular responses	Tissue effects	Outcomes
Mutation	Cell cycle disruption	Hypoplasia	Impaired growth
Chromosomal breaks	Altered rate of cell replication	Retarded/arrested growth	Deficient physiological and biochemical functions
Mitotic interferences	Reduced macromolecule biosynthesis	Altered differentiation patterns	Anatomical defects
Altered nucleic acid synthesis or function	Excessive cell death	Altered growth patterns due to lack of tissue contact/induction	Death
Lack of precursors, substrates, coenzymes	Failed cell–cell interactions	Altered morphological expression	
Altered energy sources	Impeded morphogenetic movement	Destruction of tissues by pressure, trauma, or vascular stasis	
Enzyme inhibition	Altered pattern formation		
Altered receptor interaction	Mechanical disruption		
Altered gene expression			
Changes in cell membrane integrity			
Fluid–osmolyte imbalance			

Source: Modified from **ref. *10***.

sible to compare perturbations of sensitive cellular events upstream from the overt toxic outcomes between animal and human cells or tissues that are typically evoked at lower doses.

Biologically based dose-response modeling should be an iterative process. Depending on the availability of information, a unique set of assumptions may have to be made for each potential toxicant and for each biological process being modeled. When more details and novel findings come to light, these assumptions can always be revised accordingly.

Although the overall goal of BBDR modeling is to improve low-dose and cross-species extrapolations in risk assessment, the use of these models can be twofold. They can be used as a *descriptive* account for a toxicant's disruption of a specific biological process, and the risk assessed to that particular disruption. In this case, the model will be agent-specific (i.e., one model for each chemical of interest) and the empirical information derived will be restricted to that chemical. For a more ambitious approach, these models can be used in a *predictive* fashion. Thus, if a cascade of biological processes has been identified (even crudely) to account for the genesis of toxicity, the relationship of each perturbed intervening step can be evaluated quantitatively. The extent of perturbation of these biological processes may mark the perimeters of homeostasis at molecular, cellular, and tissue levels, and beyond the threshold of these parameters lies irreversible damage to the organism that would ultimately lead to overt toxic outcomes. The mathematical relationships of these biological processes will then describe the quantifiable tolerance limits of a particular target tissue during development and will reflect the inherent properties of a specific strain and species of organism. Using these established mathematical values, one can predict the toxic potentials of various agents (or classes of chemicals) that affect the same biological pathway. Furthermore, when sufficient information on these biological tolerance limits is gathered for various strains and species, cross-species extrapolation (to humans) would become feasible.

2. Approaches

Because of the diverse mechanisms of developmental toxicity and the lack of a unifying theory even for the etiology of a single birth defect, it is not surprising that advances in a BBDR model are made at a snail's pace. Thus far, several approaches have been attempted to construct such models and these are discussed with examples.

2.1. Retrospective Approach

These models are generated typically based on some simple biological assumptions, and their validity can be tested with the existing database in the literature. Gaylor and Razzaghi ***(11)*** have provided two examples of this exercise. In one case, these investigators postulated a mathematical relationship between the prevalence of cleft palate induced by a chemical and reduction of embryonic and fetal growth. The assumptions made in this model are (1) that embryonic anatomical structures (such as the palate) are under exponential growth during gestation, (2) that structural malformations (cleft palate) are related to reduced or delayed growth, and (3) that a chemical may affect only one stage of development by reducing the growth rate, resulting in palatal shelves too small to make contact, and thus the cleft. The probability of the excess risk (above-background incidence) in producing a cleft palate after exposure to dose *D* of a chemical can be expressed as

$$P(D) - P(0) = (1 - P(0))\,\frac{\exp^{(\beta_0 t' e^{-\alpha D^{\gamma}})} - \exp^{(-\beta_0 t')}}{1 - \beta_0 t'}$$

where $P(D)$ and $P(0)$ are the probabilities of a cleft palate at dose D administered to the pregnant animals and for untreated controls respectively, β_0 is the growth rate constant of palatal cells of control animals, and t' is the time to complete the ith stage of development. α and γ are parameters to be estimated from experimental data. A case study was conducted with data from the National Center for Toxicological Research, where induction of cleft palate in several mouse strains by the herbicide 2,4,5-trichlorophenoxyacetic acid (2,4,5-T) was examined. Although the profile of the dose-response curves differed substantially among the various mouse strains, the values of α and γ calculated from the mathematical model were remarkably consistent, suggesting that the mechanism producing cleft palate was likely to be the same in all strains of mice and that the model could be used to facilitate the extrapolation of experimental results for a specific toxicant across mouse strains. However, the researchers also cautioned that in order to extend the extrapolation of results across species, the values of $P(0)$ and β_0 at t' must be established, a similar mechanism of dysmorphogenesis must be assumed, and pharmacokinetic parameters and intralitter variabilities must also be taken into consideration.

A second example of this modeling approach is provided by Gaylor and Chen ***(12)***. They extended their previous biological assumptions to include that the incidence of structurally normal fetuses is proportional to fetal weight raised to a power. Thus, they postulated the following mathematical relationships:

$$\frac{Q_D}{Q_c} = \left(\frac{W_t}{W_c}\right)^s$$

and

$$W_c = W_0 e^{b_c t}, \qquad W_t = W_0 e^{b_c(1 - aD^g)\, t}$$

where Q_D and Q_c are probabilities of normal structures at dose D and of untreated controls respectively, W_0 is the fetal weight at the time when chemical exposure is initiated, W_t and W_c are weights of treated and control fetuses, respectively, at time t, and b_c, a, and s are constants. The value of g can be obtained from the experimental database. The investigators then put all the constants together so that

$$u_1 = b_c ats, \qquad Q_c = e^{-u_0}$$

and defining $P = 1\text{-}Q$, derive

$$P_D = 1 - e^{-(u_0 + u_1 D^g)}$$

where P_D is the probability of malformation at dose D, u_0 reflects the background incidence of malformation, and u_1 is the potency of the chemical. Using studies published in 42 volumes of the journal *Teratology* as their primary database, the researchers provided the estimated values of u_0, u_1, and g for a variety of chemicals and for exposures across different animal species. Thus, one can compare the values of u_1 and g for a specific chemical between species, for instance, to determine whether there are species differences. However, as noted by the authors of this model, the actual interpretations of the numerical values of these variables remain wide open.

2.2. Theoretical Approach

An example of this approach is provided by Leroux and co-workers ***(13)*** using a cell kinetic model. These modelers theorized that a progenitor cell would either divide, differentiate, or die, with each of these events proceeding at a certain rate. Hence, these biological events can be expressed in mathematical forms, the rate constants of cell division λ, death μ, and differentiation ν can be determined experimentally and the changes induced by exposure to various toxicant doses can be monitored.

$$\lambda_1 = \lambda_{10} e^{\lambda_{11}\text{Dose}}$$

$$\mu_1 = \mu_{10} e^{\mu_{11}\text{Dose}}$$

$$\nu = \nu_0 e^{\nu_1\text{Dose}}$$

They then made the following biological assumptions: (1) that developmental processes involve cell replication, differentiation, and death; (2) that progression of embryogenesis results in an increase of progenitor cells and differentiated cells; (3) that a critical cell mass must be attained (i.e., an insufficient number of these functional cells will result in structural malformations); and (4) that a developmental toxicant may alter the rate parameters of cell kinetics, leading to deficits in a cell population. Thus, with the aid of differential equations, the authors constructed a stochastic model to predict the probability of malformation when the number of differentiated (functional) cells was less than the critical number of cells required at a particular stage of development. The model was then applied to evaluate the neurotoxic effects of methylmercury on brain development in rats, using cell kinetic parameters derived from an embryonic central nervous system (CNS) cell culture system. Although the model produced various dose-response curves according to the input factors, the predicted profile of toxicity for methylmercury, in fact, compared favorably with results derived from an in vivo study. Although the cell kinetic theory may have oversimplified the complex mechanisms of dysmorphogenesis, this example illustrates the process with which one may construct a mathematical/statistical model to approximate the developmental toxicity outcomes.

2.3. Iterative Approach

This is perhaps the most ambitious approach for BBDR modeling; an example is provided by the investigators at the U.S. Environmental Protection Agency ***(14–19)***. In this approach, the modelers assumed, *a priori*, a cascade of biological events that would account for the expression of toxicity for a chemical. The dose-response relationship of these events is then transformed into mathematical terms and fitted to an integrative model. Should a particular part of the model become ill-fitted or illogical, the assumed biological pathway will be reassessed for incorporation of additional or alternative events. Such modeling effort thus requires constant iterations between bench scientists and mathematicians. For their exercise, the Environmental Protection Agency (EPA) researchers selected the chemotherapeutic agent 5-fluorouracil (5-FU) as a prototypic teratogen and examined the pharmacokinetics of the drug as well as the pharmacodynamic responses of the rat fetus at gestational d 14, across a wide dose range. 5-FU appears to be an excellent choice for BBDR modeling for the following reasons: (1) the mechanisms of drug action are known, (2) the intervening cellular events leading to toxicity by and large can be evaluated quantitatively, and (3) similar anatomical defects are seen in animal models and humans after *in utero* drug exposure. **Figure 1** depicts the assumed mechanistic pathways of 5-FU developmental toxicity. 5-FU is converted to its active metabolite 5-fluoro-2'-deoxyuridylic monophosphate (5-FdUMP) and distribution of these compounds can be estimated in both maternal and embryonic compartments. 5-FdUMP inhibits the activity of thymidylate synthetase (TS), the enzyme that catalyzes *de novo* synthesis of thymidylate. This enzyme inhibition thus causes an abrupt imbalance of intracellular nucleotide pools, which, in turn, interrupts DNA synthesis and prolongs the S phase of the cell cycle, leading to inappropriate and untimely cell death. At critical junctures, these cellular deficits are translated into growth retardation and structural malformations. In addition to this pathway, 5-FU has been shown to exert its effects by direct incorporation into nucleic acids, leading to DNA damage and production of nonfunctional proteins. Alternatively, 5-FU may interfere with fetal erythropoiesis, indirectly impeding the developmental processes involving cell replication and differentiation. However, results from experimental studies indicated that the latter two pathways provided only secondary or minor contributions to the overall

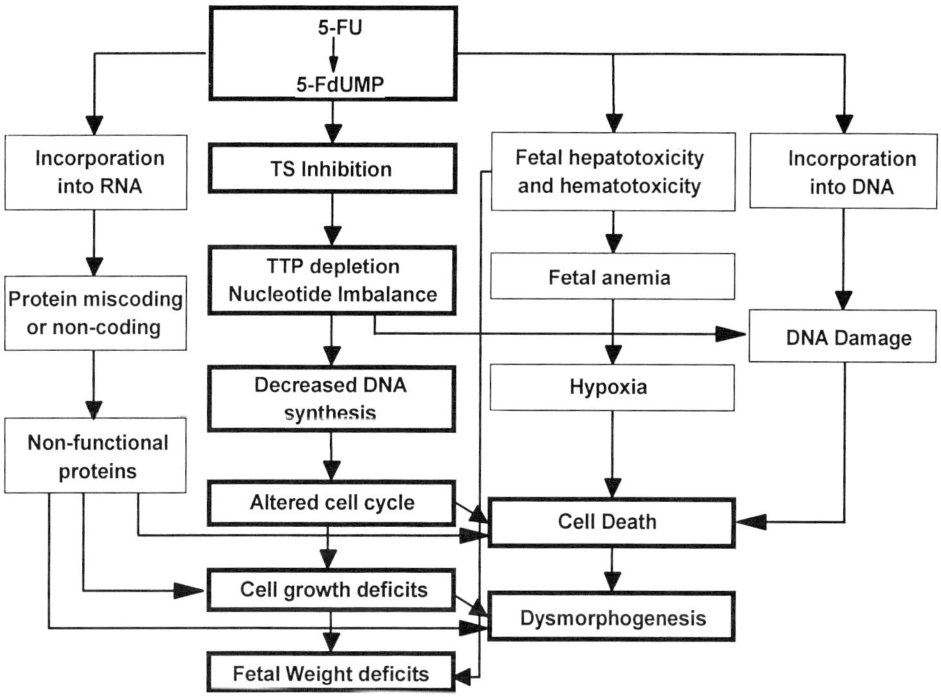

Fig. 1. Proposed mechanistic pathways of 5-FU developmental toxicity *(18)*. Boxes with a heavy border denote parameters selected for BBDR modeling. Abbreviations: 5-FU, 5-fluorouracil; TS, thymidylate synthetase; TTP, thymidylate.

toxicity in the rat fetus; thus, the first assumption of BBDR modeling *(see above)* was invoked and the pathway of enzyme inhibition was chosen for modeling.

In a first attempt at model construction, Shuey et al. *(18)* highlighted one of the structural abnormalities produced by 5-FU, hindlimb defects, and selected the optimal cellular and tissue responses at the intervening steps for examination. Hence, mathematical relationships between pairs of sequential variables were estimated: 5-FU dose and TS inhibition at 1 h after drug exposure; the extent of TS inhibition and percent of cells in the S phase of the cell cycle at 8 h; cells in the S phase and digit morphometry at 24 h; digit morphometry and digital defects at birth. Each of these relationships can be expressed by a Hill equation:

Exposure Dose $\quad C$

↓

Enzyme Inhibition $\quad$ $$TS = 40 + 60\left(\frac{(19.24)^{2.02}}{(19.24)^{2.02} + C^{2.02}}\right)$$

↓

Cell Cycle Alteration $\quad$ $$S = 18.8 + 20\left(\frac{(64.58)^{7.01}}{(64.58)^{7.01} + TS^{7.01}}\right)$$

↓

Growth Reduction

$$D5A = 0.293 + 0.35\left(\frac{(31.07)^{12.21}}{(31.07)^{12.21} + S^{12.21}}\right)$$

↓

Malformation

$$AGEN = 100\left(\frac{(0.392)^{12.04}}{(0.392)^{12.04} + D5A^{12.04}}\right)$$

where *C* denotes the doses of 5-FU, TS denotes enzyme activity in the hindlimb buds, *S* denotes the percentage of cells at the S phase, D5A denotes areas of presumptive fifth digit of the limb, and AGEN denotes the incidence of digital defects at term. Integrating these equations produces dose-response estimates of the 5-FU-induced digital defects; the fact that these estimates approximate the data derived from actual experimental findings lends credence to the assumptions made in this model. However, the current model can only be regarded as empirical or descriptive for one specific chemical, 5-FU. Whether these mathematical relationships truly reflect the biological links responsible for the induction of limb defects remains to be substantiated. This issue perhaps can be addressed by evaluating other agents that may either inhibit TS, alter nucleotide pool balance, interrupt cell replication, or produce excessive cell death. Should these relationships be found consistent among various insults, the model may then be extended from being descriptive to being predictive, and the shapes of these relationships may indeed reflect the thresholds where cellular alterations have exceeded the homeostatic tolerance and are transformed into irreparable damage.

2.4. Considerations of Model Construction

The above examples have illustrated the variety of endpoints that can be used for BBDR modeling of developmental toxicity. These typically include the following (1) mechanistic markers that characterize the initial steps of chemical insults such as alteration of enzyme activity, alteration of receptor binding and activation, interference of cellular signal transduction, alteration of cellular energetics, and alteration of gene expression; (2) biomarkers that reflect normal developmental processes such as cell replication, cell differentiation, cell migration, cell death, and the biochemical regulation of these processes; (3) biological features that indicate adverse developmental outcomes such as anatomical defects, growth deficit, physiological deficiencies, and perinatal mortality. A majority of these endpoints are readily available in the investigative toolbox of developmental toxicologists; although some of these variables are amenable to quantitative assessment, others (such as anatomical defects and death) are more dichotomous in nature and thus more problematic for mathematical modeling. Recent advances in molecular biology have also made significant contributions to bolster the toxicologists' armamentarium. Indeed, a number of changes in gene expression have been linked to the chemically induced aberrant formation of embryonic structures ***(20–22)***. For instance, the *Pax-1* gene has been hypothesized as a molecular target of valproic acid in axial skeletal dysmorphogenesis ***(23)***, the *p53* gene expression is involved in 2-chloro-2'-deoxyadenosine-induced eye defects ***(24)***, and downregulation of the *sonic hedgehog* gene has been implicated in craniofacial defects produced by retinoic acid ***(25)***. Undoubtedly, active research in this area will continue to reveal additional genetic markers that are susceptible to alterations by chemical toxicants and the human homolog of these genes can be similarly evaluated. However, the issues of whether gene expression could be quantified and whether differential gene expression would correspond to cellular deficits, and toxic outcomes must be addressed before these novel biomarkers can be incorporated into BBDR modeling.

In building a model for health effects risk assessment, it is remarkably easy to forget that the ultimate goal of the exercise is to predict the magnitude of a toxic response in humans. To achieve this goal, explicit attention must be given from the start as to how the results from such a model will be applied to the human situation. One of the premises for BBDR modeling is that homologous mechanisms of toxicity exist between laboratory animals and humans. Although it is impossible to verify some of the toxicological endpoints derived from animal models in humans (especially pathological manifestation such as anatomical defects), others are more amenable to validation. These may include pharmacokinetic and metabolic disposition of the compounds, their pharmacodynamic actions (such as alterations of enzyme activities and receptor binding affinity), as well as some of the resultant cellular deviations in replication, differentiation, interaction, and death that can be evaluated in surrogate human cell and tissue cultures or derived from clinical databases. The modelers can then use the values of these corresponding parameters between laboratory animals and human surrogates to provide a basis to scale for those human parameters that are difficult to estimate.

Mathematical models are usually cast in terms of *state variables* and *parameters*. The distinction between the two categories is made largely on the basis of constancy in the time frame of interest: State variables are likely to change appreciably, whereas parameters remain relatively constant. Conceptually, we can think of state variables as quantities such as metabolite concentrations or rates of enzymatic reaction, while parameters might be quantities such as blood flow rates and tissue volumes. This distinction is convenient but may be misleading in specific situations. For example, during development, the volumes of maternal and fetal tissue compartments can vary dramatically from one stage to another. The modeler's task is to select a set of state variables to describe the system and to characterize their interrelationships mathematically so that those state variables of greatest interest are adequately predicted over the range of conditions (represented by values of state variables or parameters) that are likely to occur in real-life exposures. For mechanistic modeling, it is often useful to express relationships in terms of rates of change, so that time plays an integral role in the model. This means that models take the form of differential equations (if changes occur continuously over time) or difference equations (if changes occur at discrete time points). Accordingly, in designing experiments to construct a BBDR model (especially one for developmental toxicity), the time variable, in addition to dose and response, should be explicitly taken into consideration.

The mathematical expression of the model depends completely on the nature and objective of the model. One must bear in mind that a BBDR model for toxicity is designed to capture only the irreversible breakdown of biological mechanisms that ultimately lead to an adverse impact to the organism. Thus, although the mathematical expressions only represent crude snapshots of these biological insults and do not provide detailed information on events that are involved in the maintenance of cellular milieu (e.g., repairs, redundant pathways, and compensatory mechanisms), they may offer rough estimates for the total capacity or tolerance of cells and tissues to handle external perturbations. This point is illustrated in the Shuey et al. example *(18)* where the overall model is expressed as the composition of sequential simple Hill functions for each intervening biological step, and goodness-of-fit is the primary criterion for evaluating the mathematical functions expressing the individual steps. The advantage of this approach is the relative ease of mathematical maneuvers; on the other hand, it completely relies on the selection of appropriate modeling steps that approximate the relevant biological changes.

3. Summary

It is obvious that the task of incorporating mechanistic information into dose-response assessment for developmental toxicity is, by and large, still at its conceptual stage. Our immature understanding of embryogenesis and teratogenesis forces us to make simplifying biological assumptions that may turn out to be erroneous; therefore, these mechanistically based models should be constructed so as to be easily modified as new information becomes available. The data-intensive (and costly) nature of these modeling efforts may also limit their practice to extraordinary situations where, for instance, large segments of the human population are exposed to low levels of a compound and the determination of a safe level of exposure is of utmost importance, or where compounds are of such immense economic or therapeutic value that their use would warrant a concerted effort to minimize the uncertainties inherent in the current methods of extrapolation. Nevertheless, this chapter has presented several attempts to translate the concept of BBDR into practice. When these applications become successful, these models should provide risk assessors with more reliable response indicators at low doses (where human exposures are realistic) and more accurate cross-strain/cross-species comparisons as well as extrapolations across exposure conditions. Although the BBDR models do not necessarily redefine the current practice of risk assessment using either the NOAEL or the benchmark approach, information derived from these efforts should provide a credible scientific basis for the estimates of RfD.

Note

This chapter has been subjected to review by the National Health and Environmental Effects Research Laboratory and approved for publication. Approval does not signify that the contents reflect the views of the Agency, nor does mention of trade names or commercial products constitute endorsement or recommendation for use.

References

1. U. S. Environmental Protection Agency. (1991) Guidelines for developmental toxicity risk assessment. *Fed. Register* **56,** 63,798–63,826.
2. Crump, K. (1984) A new method for determining allowable daily intakes. *Fundam. Appl. Toxicol.* **4,** 854–871.
3. Faustman, E. M., Allen, B. C., Kavlock, R. J., and Kimmel, C. A. (1994) Dose-response assessment for developmental toxicity. I. Characterization of database and determination of the no observed adverse effect levels. *Fundam. Appl. Toxicol.* **23,** 478–486.
4. Allen, B. C., Kavlock, R. J., Kimmel, C. A., and Faustman, E. M. (1994a) Dose-response assessment for developmental toxicity. II. Comparison of generic benchmark dose estimates with no observed adverse effect levels. *Fundam. Appl. Toxicol.* **23,** 487–495.

5. Allen, B. C., Kavlock, R. J., Kimmel, C. A., and Faustman, E. M. (1994b) Dose-response assessment for developmental toxicity. III. Statistical models. Fundam. Appl. Toxicol. 23, 496–509.
6. Kavlock, R. J., Allen, B. C., Faustman, E. M., and Kimmel, C. A. (1995) Dose-response assessment for developmental toxicity. IV. Benchmark doses for fetal weight changes. *Fundam. Appl. Toxicol.* **26,** 211–222.
7. O'Flaherty, E. J. (1997) Pharmacokinetics, pharmacodynamics, and prediction of developmental abnormalities. *Reprod. Toxicol.* **11,** 413–416.
8. Daston, G. P. (1997) Advances in understanding mechanisms of toxicity and implication for risk assessment. *Reprod. Toxicol.* **11,** 389–396.
9. Kavlock, R. J. (1997) Recent advances in mathematical modeling of developmental abnormalities using mechanistic information. *Reprod. Toxicol.* **11,** 423–434.
10. Wilson, J. G. (1973) *Environment and Birth Defects.* Academic, New York.
11. Gaylor, D. W. and Razzaghi, M. (1992) Process of building biologically based dose-response models for developmental defects. *Teratology* **46,** 573–581.
12. Gaylor, D. W. and Chen, J. J. (1993) Dose-response models for developmental malformations. *Teratology* **47,** 291–297.
13. Leroux, B. G., Leisenring, W. M., Moolgavkar, S. H., and Faustman, E. M. (1996) A biologically-based dose-response model for developmental toxicology. *Risk Anal.* **16,** 449– 458.
14. Abbott, B. D., Lau, C., Buckalew, A. R., Logsdon, T. R., Setzer, W., Zucker, R. M., Elstein, K. H., and Kavlock, R. J. (1993) Effects of 5-fluorouracil on embryonic rat palate in vitro: fusion in the absence of proliferation. *Teratology* **47,** 541–554.
15. Elstein, K. H., Zucker, R. M., Andrews, J. E., Ebron-McCoy, M., Shuey, D. L., and Rogers, J. M. (1993) Effects of developmental stage and tissue type on embryo/fetal DNA distributions and 5-fluorouracil-induced cell cycle perturbations. *Teratology* **48,** 355–363.
16. Shuey, D. L., Zucker, R. M., Elstein, K. H., and Rogers, J. M. (1994) Fetal anemia following maternal exposure to 5-fluorouracil in the rat. *Teratology* **49,** 311–319.
17. Shuey, D. L., Buckalew, A. R., Wilke, T. S., Rogers, J. M., and Abbott, B. D. (1994) Early events following maternal exposure to 5-fluororuracil lead to dysmorphology in cultured embryonic tissues. *Teratology* **50,** 379–386.
18. Shuey, D. L., Lau, C., Logsdon, T. R., Zucker, R. M., Elstein, K. H., Narotksy, M. G., Setzer, R. W., Kavlock, R. J., and Rogers, J. M. (1994) Biologically based dose-response models for developmental toxicology: biochemical and cellular sequelea of 5-fluorouracil exposure in the developing rat. *Toxicol. Appl. Pharmacol.* **126,** 129–144.
19. Zucker, R. M., Elstein, K. H., Shuey, D. L., and Rogers, J. M. (1995) Flow cytometric detection of abnormal fetal erythropoiesis: application to 5-fluorouracil-induced anemia. *Teratology* **51,** 37–44.
20. Copp, A. J. (1994) Birth defect: from molecules to mechanisms. *J. R. Coll. Physicians London* **28,** 294–300.
21. Rogers, M. B., Glozak, M. A., and Heller, L. C. (1997) Induction of altered gene expression in early embryos. *Mutat. Res.* **396,** 79–95.
22. Harris, M. J. and Juriloff, D. M. (1997) Genetic landmarks for defects in mouse neural tube closure. *Teratology* **56,** 177–187.
23. Barnes, G. L., Mariani, B. D., and Tuan, R. S. (1996) Valproic acid-induced somite teratogenesis in the chick embryo: relationship with Pax-1 gene expression. *Teratology* **54,** 93–102.
24. Wubah, J. A., Ibrahim, M. M., Gao, X., Nguyen, D., Pisano, M. M., and Knudsen, T. B. (1996) Teratogen-induced eye defects mediated by p53-dependent apoptosis. *Current Biol.* **1,** 60–69.
25. Helms, J. A., Kim, C. H., Hu, D., Minkoff, R., Thaller, C., and Eichele, G. (1997) Sonic hedgehog participates in craniofacial morphogenesis and is down-regulated by teratogenic doses of retinoic acid. *Dev. Biol.* **187,** 25–35.

IV

Screening and Mapping of Novel Genes and Mutations

28

Positional Cloning

Lin Xu and Yefu Li

1. Introduction

A genetic disease gene can be identified by three approaches: (1) Functional cloning in which a disease gene is identified based on biological background of a disease and the gene function without knowledge of chromosomal position of the gene, for example, identification of the globin gene mutations responsible for certain forms of anemia. This cloning strategy relies heavily on detailed information of the disease and the gene. However, such information is not available for the majority of single-gene disorders in human. (2) Positional candidate gene cloning is another approach which can be used to identify a disease gene. This technique combines information of chromosomal location of a disease locus and a candidate gene locus. Once the disease and a candidate gene loci are colocalized on the chromosomal region, the gene will then be cloned and sequenced. Recently, we have mapped two genetic mouse disorders, chondrodysplasia (*cho*) to α1 of type XI collagen (*Col11a1*) locus on chromosome 3 and disproportionate micromelia (*Dmm*) to α1 type II collagen (*Col2a1*) locus on chromosome 15. We have identified the mutation in *cho*, which is a deletion of a cytosine residue in the coding region of *Col11a1* mRNA ***(1)***. In *Dmm*, a deletion of three nucleotides in the coding region of *Col2a1* mRNA was found ***(2)***. (3) Positional cloning is an approach by which a disease gene can be cloned by sole information of its physical location on a chromosome. Because high-density chromosome maps of polymorphic markers of human and mouse have been developed and yeast, bacterial, and phase artificial chromosome libraries of human and mouse are available, positional cloning of a disease gene can be accomplished in a reasonable short time span. In 1986, the first human genetic disease gene, chronic granulomatous disease, was identified by positional cloning ***(3)***. Since then, the number of disease genes cloned by this technique has been dramatically increased. The experimental strategy of the positional cloning includes three basic steps ***(4)***. The first step is to establish a fine genetic linkage map of a disease gene locus by identifying markers that are close to the gene. Currently, 6580 mouse and 5264 human short tandem repeat polymorphic markers and 797 mouse restriction fragment length polymorphic markers have been assigned to chromosomes and are available ***(5,6)***. This makes it possible to map a gene of interest within 1 cM. The second step is to construct a physical map of the gene locus. Yeast artificial chromosome (YAC) ***(7,8)***, bacterial

From: *Methods in Molecular Biology, Vol. 136: Developmental Biology Protocols, Vol. II*
Edited by: R. S. Tuan and C. W. Lo © Humana Press Inc., Totowa, NJ

artificial chromosome (BAC) ***(9,10)***, and phage P1-derived artificial chromosome (PAC) ***(11–13)*** libraries have been utilized by many laboratories to define a genomic region that contains genes of interest. The third step is to search for the gene in the defined genomic region by the techniques of cDNA selection ***(14–16)***, exon trapping ***(17–19)***, or GpC island searching ***(20)***. Although the protocols described in this chapter have been developed for the mouse, they could also be adapted to other species, such as rat, zebrafish, and even humans with the exception of the breeding plan.

2. Materials

2.1. For Isolation of Genomic DNAs for Genetic Linkage Analysis

1. Tissue digestion buffer: 100 m*M* Tris-HCl pH 8.5, 0.2% sodium dodecyl sulfate (SDS), 5 m*M* EDTA pH 8.0, 200 m*M* NaCl, proteinase K
2. Isopropanol
3. 10 m*M* Tris-HCl pH 8.0, 1 m*M* EDTA pH 8.0.

2.2. For Polymerase Chain Reaction Amplification for Genetic Linkage Analysis

1. Primers for polymorphic DNA markers, commercially available from Research Genetics (Huntsville, AL).
2. Polymerase chain reaction (PCR) amplification kit, commercially available from Perkin Elmer (Norwalk, CT).
3. Agarose for molecular biology from Gibco-BRL (Grand Island, NY).

2.3. For isolation of YAC DNA

1. Mouse YAC library, commercially available from Research Genetics.
2. Yeast extract, peptone, dextrose (YPD) medium, 10 g yeast extract, 20 g peptone, 900 mL water, and autoclave 20 min, then add sterilized 100 mL 20% (w/v) dextrose
3. Sorbitol, sodium citrate, EDTA (SCE), 1 *M* sorbitol, 100 m*M* sodium citrate, 60 m*M* EDTA pH 8.0
4. Solution #1, SCE, 14 m*M* β-mercaptoethanol, 80 units/mL of lyticase
5. 50 m*M* Tris-HCl pH 8.0, 20 m*M* EDTA pH 8.0
6. 20% SDS
7. 5 *M* potassium acetate pH 4.8
8. Isopropanol
9. Ethanol
10. 10 m*M* Tris pH 8.0, 1 m*M* EDTA pH 8.0
11. RNaseA
12. 3 *M* Sodium acetate

2.4. Inverse PCR

1. Endonuclease restriction enzymes
2. Phenol and chloroform
3. 3 *M* Sodium acetate
4. Ethanol
5. Agarose molecular biology grade
6. DNA ligase
7. Pairs of inverse PCR primers from YAC left and right arms
8. PCR amplification kit
9. DNA sequencing kit (if necessary)

2.5. cDNA Selection

1. PCR amplification kit
2. A pair of PCR primers flanking the *Eco*RI subcloning site of λgt10
3. Centricon 100 (Amicon [Millipore, Bedford, MA])
4. Agarose for molecular biology
5. DNA purification kit from Bio-Rad (Hercules, CA)
6. 20X SSC (175.3 g of NaCl and 88.2 g of sodium citrate, pH 7.0, in 1 L water)
7. Hybond-nylon$^+$ filter (Amersham, Arlington Heights, IL)
8. 0.5 *M* NaOH, 1.5 *M* NaCl
9. 0.5 *M* Tris-HCl pH7.5, 1.5 *M* NaCl
10. 50X Denhardt's (10 g Ficoll, 10 g poly(vinyl pyrrolidone), 10 g bovine serum albumin in 1 L water)
11. 0.5% SDS
12. 2X SSC/0.1% SDS
13. 1X SSC/0.1% SDS
14. 0.2X SSC/0.1% SDS
15. 0.1X SSC/0.1% SDS
16. PCR amplification kit
17. Agarose for molecular biology
18. A pair of primers inside of the first pair of the primers flanking the *Eco*RI subcloning site of the λgt10 (if necessary)
19. TA cloning kit from Invitrogen (Carlsbad, CA)

2.6. Exon Trapping

See the protocol provided by Gibco-BRL for Exon Trapping Systems.

2.7. Identification of cDNA Fragments of Interest

Zoo Blot from Clontech (Palo Alto, CA).

3. Methods

3.1. Establishing a High-Resolution Genetic Linkage Map of a Locus of Interest

The first step in the positional cloning of a gene of interest is to establish a high-resolution genetic linkage map of the gene locus. A high-resolution linkage map can be accomplished with at least 300 informative meioses in mouse; certainly, the more meioses and polymorphic markers you have, the higher the resolution of the map you will get. In order to have a sufficient number of the polymorphic markers, a mouse breeding plan needs to be established (*see* **Note 1**). For example, if one wants to identify a recessive mutant gene that originally arose from mouse strain C57BL/6j, the best breeding plan is as follows:

F_0 C57BL/6j/C57BL/6j × CAST/Ei/CAST/Ei
F_1 C57BL/6j/CAST/Ei × C57BL/6j/CAST/Ei
F_2 C57BL/6j/C57BL/6j (mutant phenotype)

That is because the polymorphism level of the DNA markers between C57BL/6j and CAST/Ei is about 93%, which means that 93% of DNA markers between these two strains are polymorphic. This breeding plan is called the intersubspecific intercross. The basis for using the intercross breeding plan for identifying a recessive mutant gene is that the mutation is recessive and both heterozygous F1 parents are fertile and homozygous F2 mutant mice can be easily identified shortly after the birth. Another advantage of this breeding is that each F2 mutant animal comes from two informative meiotic events with sex-averaged recombinant frequency. This will give twice as much recombinant information from each animal compared with the backcross breeding. In some cases, the mouse strain CAST/Ei will barely breed with another mouse strain. Alternatively, one can make a breeding plan between C57BL/6j and one of the inbred mouse strains, such as DBA. Considering the polymorphism level of the DNA markers between C57BL/6j and DBA (about 50%), one could use the two different intersubspecific crosses to compensate for a relatively low polymorphism level. Information of the DNA markers among 12 inbred mouse strains are available from Research Genetics.

In mouse breeding, if homozygous parents are fertile, one could also make a backcross plan between one of the homozygous parents and the heterozygous F1 offspring.

F_0 C57BL/6j/C57BL/6j × CAST/Ei/CAST/Ei (mutant phenotype)
F_1 C57BL/6j/CAST/Ei × F_0 C57BL/6j/C57BL/6j (mutant phenotype)
F_2 C57BL/6j / C57BL/6j (mutant phenotype)

There are two advantages in this breeding: (1) 50% of the F2 will be informative homozygotes and (2) because there is a genetic interference during chromosome recombination, the order of the markers on chromosomes should be more reliable.

In this chapter, we use a mouse recessive mutation as an example for describing the experimental strategies. For a dominant mutation in mouse, the experimental procedures are same if the heterozygotes can breed with each other. However, if the heterozygotes cannot breed with each other, the backcross breeding plan or a third inbred mouse strain could be used in the breeding:

F_0 C57BL/6j/C57BL/6j × CAST/Ei/CAST/Ei (mutant phenotype)
F_1 C57BL/6j/CAST/Ei × DBA/DBA (mutant phenotype)
F_2 C57BL/6j/DBA (mutant phenotype)

In this case, each F2 mutant (heterozygote) represents only one informative meiosis.

3.1.1. Isolation of Genomic DNAs for Genetic Linkage Analysis

1. Collect 0.5–1 cm of tail from F2 mutant animals by clipping the tail.
2. Put the tail in a 1.5-mL Eppendorf tube containing 0.5 mL of the solution, 100 m*M* of Tris pH 8.5, 5 m*M* of EDTA pH 8.0, 0.2% of SDS, 200 m*M* of NaCl, and 100 µg proteinase K/mL.
3. Incubate the tube at 55°C 6 h to overnight.
4. Centrifuge the tube at 10,000*g* with the Eppendorf centrifuge at 4°C for 15 min.
5. Transfer the solution to a clean and fresh Eppendorf tube containing 0.5 mL isopropanol.
6. Mix the solution and isopropanol immediately.
7. Leave the tube at room temperature for at least 30 min.
8. Lift out the precipitated DNA with a clean pipet tip.
9. Transfer the DNA to a new tube containing 100 µL TE buffer.
10. Quantitate the DNA samples and make working aliquots for PCR, 50 ng/mL

3.1.2. Polymorphic Markers for Genetic Linkage Analysis

Most of the markers are commercially available from Research Genetics (*see* **Note 2** and **3**). Initially, the number of the markers used for a genome wide search will depend

depend on how many F2 mutant animals (informative meioses) will be screened by PCR. For example, one can start genotyping 20 F2 mutant animals (40 informative meioses) and use about 70 DNA markers, evenly covering the whole mouse genome, corresponding to about 3000 cM (there is one marker in every 40 cM).

The PCR analysis will be carried out in a 20-µL reaction volume of 1X PCR buffer containing 100 ng of genomic DNA, 1 µ*M* of each primer, 0.2 µ*M* of each dNTP, and 0.25 U of *Taq* polymerase (Boehringer Mannheim, Mannheim, Germany). Amplification conditions will be 94°C for 2 min, followed by 35 cycles at 94°C for 1 min, 55°C for 45 s, 72°C for 45 s, and a final 10-min extension at 72°C. The PCR products will be electrophoresed in a 4% agarose gel (Gibco-BRL).

One marker or several markers representing a locus originally coming from the mutant mouse strain should be identified (i.e., carried by most or all of the F2 mutant mice). In other words, only one size of a DNA band representing to the mutant mouse strain should be observed in most or all of the F2 homozygous mutant animals. For analysis of F2 heterozytoes in a dominant disorder, there are two different sizes of DNA markers observed in all of F2 heterozytoes. However, one marker or several markers originally from the mutant strain should be identified. Once the marker or markers are defined, more markers surrounding the locus and more F2 mutant animals will be examined to search for a marker or markers that more closely to the mutation.

3.2. Construction of a Physical Map (YAC Contig) of a Locus

A physical map of a locus on chromosome consists of several overlapping DNA fragments. The size of a locus can be up to a million base pairs. Therefore, a genomic library constructed using a vector that is able to carry a large DNA insert is required. The availability of YAC, BAC, and PAC libraries has made this possible. The average sizes of DNA inserts are different in the libraries: 1 million base pairs in the YAC, 500 thousand base pairs (500 kb) in BAC, and 100 kb in PAC. Each of them has advantages and disadvantages. For example, although the YAC has largest inserts of DNAs, there is a relative high percentage (10–60%) of chimeric DNA clones in YAC libraries. One of the advantages of BAC and PAC is that number of chimeric DNA clones is much lower, because the sizes of DNA inserts are smaller. Therefore, using a combination of the different libraries should be considered for establishing a physical map of a locus (*see* **Note 4**).

Once the closest marker/markers to a disease locus are defined, one should use them to search for the information of YAC contigs from several genome research centers; for example, the Whitehead Institute/MIT Center for Genome Research. If there are no existing YAC contigs, then one should use the most linked markers to screen mouse YAC and/or BAC libraries to construct a physical map of the mutant locus. Mouse YAC and BAC libraries are commercially available from Research Genetics, Inc. YAC or BAC clones corresponding to the markers will be isolated from the libraries by PCR following the detailed protocol provided by Research Genetics, Inc. For constructing a YAC contig (several overlapping YAC clones), it is necessary to identify ends of YAC clones, which should then be used as probes to search for overlapping clones. Finally, a set of YAC clones (YAC contig) will be recovered from the mouse YAC library and sized by pulsed field gel electrophoresis ***(21)*** and characterized by endonuclease restriction enzyme mapping.

3.2.1. Isolation of YAC DNAs

1. Culture the YAC clone in 5 mL of YPD at 30°C, shaking at 250 rpm overnight.
2. Transfer the 5 mL of the culture to a baffled flask containing 50 mL of YPD.
3. Continue to incubate at 30°C, shaking at 250 rpm for 1 or 2 d.
4. Centrifuge the 50 mL culture at 1500*g* with a benchtop centrifuge for 10 min.
5. Resuspend pellet in 20 mL of SCE in a 50-mL tube.
6. Centrifuge the 20 mL of the SCE at 1500*g* with a benchtop centrifuge for 10 min.
7. Resuspend pellet in 4.5 mL of solution 1.
8. Incubate at 37°C for 60 min.
9. Centrifuge the tube at 1500*g* with a benchtop centrifuge for 5 min.
10. Resuspend in 5 mL of 50 m*M* Tris pH 8.0 and 20 m*M* EDTA pH 8.0.
11. Add 250 µL of 20% SDS and mix well to lyse the yeast cells.
12. Incubate at 65°C for 30 min.
13. Add 2 mL of 5 *M* potassium acetate, pH 4.8.
14. Incubate on ice for 60 min.
15. Centrifuge at 1500*g* with a benchtop centrifuge for 10 min.
16. Transfer supernatant (about 7 mL) to a new tube.
17. Centrifuge the supernatant at 1500*g* for 5 min.
18. Transfer about 6 mL of the supernatant to a new tube (make sure not to take any of the precipitated protein).
19. Add 6 mL of room-temperature isopropanol and mix well.
20. Leave the tube at room temperature for 10 min.
21. Centrifuge the tube at 1500*g* for 5 min.
22. Discard the supernatant.
23. Wash pellet with 5 mL of 70% ethanol.
24. Centrifuge the tube at 4000*g* at room temperature for 5 mi
25. Discard the ethanol and dry the pellet in room temperature for 5–10 min.
26. Resuspend the pellet in 1 mL of TE buffer.
27. Treat the YAC DNA with 50 µg RNaseA at 37°C for 60 min to destroy the RNAs.
28. Precipitate the DNA with 1/10 volume of 3*M* sodium acetate and 2.5 volumes of 100% ethanol.
29. Dissolve the DNA in 1 mL of TE buffer.

The YAC DNA is ready for inverse PCR.

3.2.2. Inverse PCR

Inverse PCR is used for isolating ends of YAC clones. There are many experimental protocols of recovering ends of YAC or BAC clones (*see* **Note 5**). It has been reported that inverse PCR has the highest efficiency in recovering the ends ***(22)***. In our laboratory, we have successfully isolated mouse YAC ends by inverse PCR with an efficiency of about 80%. Basically, a pair of inversing oligonucleotide primers from the YAC left or right arm are synthesized. The YAC DNAs are digested with a combination of two or three endonuclease restriction enzymes (*see* **Note 6**) and then ligated under conditions favoring circularization of the digested YAC fragments (20–200 ng of the YAC DNA digested and ligated at room temperature overnight). The restriction enzymes should not recognize any DNA sequence between the primers and original YAC cloning sites. PCR is performed with the circularized DNA fragments as templates and a pair of the primers from each vector arm. The PCR products contain the cloning junction between vector arms and the inserts of mouse genomic DNA (ends of the YAC clones).

1. Digest 200–500 ng of the YAC DNA with a combination of two or three endonuclease restriction enzymes with their appropriate buffers at 37°C overnight.
2. Add distilled water to the tube to make a final volume of 500 mL.
3. Extract the digested DNA with 250 µL phenol and 250 µL chloroform.
4. Transfer upper solution to a new Eppendorf tube.
5. Add 1/10 volume of 3 *M* sodium acetate and 2.5 volumes of ethanol.
6. Leave the tube at –20°C overnight.
7. Centrifuge with Eppendorf centrifuge at 10,000*g* for 20 min.
8. Wash the DNA pellet with 0.5 mL of 70% ethanol.
9. Centrifuge with Eppendorf centrifuge at 10,000*g* for 2 min.
10. Dissolve the DNAs in 30 µL of distilled water.
11. Quantitate concentration of the DNA by gel electrophoresis.
12. Use 50 ng of the digested DNA for re-ligation
    re-ligation condition: 1 µL of ligase (4 U)
    2 µL of ligation buffer
    Water to final volume 20 µL
13. Incubate the tube at room temperature overnight.
14. Add distilled water to the tube for a final volume of 400 µL.
15. Add 1/10 volume of 3 *M* sodium acetate and 2.5 volumes of ethanol.
16. Leave the tube at –20°C overnight.
17. Centrifuge with Eppendorf centrifuge at 10,000*g* for 20 min.
18. Wash the DNA pellet with 0.5 mL of 70% ethanol.
19. Centrifuge with Eppendorf centrifuge at 10,000*g* for 2 min.
20. Dissolve the DNAs in 20 µL of distilled water ready for PCR amplification.
21. Synthesize a pair of oligonucleotide primers from the YAC left or right arm (direction of the two primers is opposite and 20–50 base pairs apart; that is why it is called inverse PCR).
22. PCR analysis: the PCR will be carried out in a 50-µL reaction volume of 1X PCR buffer containing 5 µL of the circularized DNA, 1 µ*M* of each YAC arm primer, 0.2 µ*M* of each dNTP, and 0.25 units of *Taq* polymerase (Boehringer Mannheim). Amplification conditions were 94°C for 2 min, followed by 35 cycles at 94°C for 1 min, 60°C for 1 min, 72°C for 1.5 min, and a final 10-min extension at 72°C. The PCR products will be electrophoresed in a 2% of agarose gel (Gibco-BRL) and analyzed by direct DNA sequence. PCR products contain the cloning junction between vector arms and the inserts of mouse genomic DNA (ends of the YAC clones).

3.3. Cloning a Gene of Interest

When YAC clones are identified, the internal structure of each YAC clone should be verified. Whether a YAC clone is chimeric or not could be examined by PCR. For example, if the YAC clone is recognized by PCR primers from the ends of its overlapping clones, it is likely that the YAC clone is not chimeric. After the YAC clones covering the mutant locus are defined, two molecular cloning methods should be considered to identify a gene of interest in this protocol, cDNA selection, and exon trapping.

3.3.1. cDNA Selection

The cDNA selection method is an approach to identify expressed cDNA sequences from a YAC clone containing genes of interest. Although the underlying principle is the same, there are several different ways to identify the cDNAs from the YAC. Here, we describe the procedure of hybrid selection with a nylon-filter disk. Briefly, double-stranded cDNAs made from mRNAs of normal mouse embryos are subcloned into λgt10. The cDNAs are then amplified by PCR with a pair of primers flanking the *Eco*RI cloning site. The cDNAs are hybridized to the fragments of the defined YAC on a nylon-filter disk. After hybridization, the hybridized cDNAs are eluted and amplified by PCR. Details of this protocol are as follows.

3.3.1.1. Preparation of cDNAs with Known Sequences at Ends

Double-stranded cDNAs (*see* **Note 7** and **8**) originally primed with random primers are subcloned into λgt10 to construct a cDNA library. A procedure to construct the random cDNA library has been well described in **ref. *23***. The subcloning site of λgt10 is the *Eco*RI restriction site. Once the library is made, cDNAs are amplified by PCR. The PCR conditions are as follows: The PCR is carried out in a 0.2-mL PCR tube containing 100 μL reaction volume of 1 X PCR buffer containing 20 ng of the phage DNAs, 1 μ*M* of each primers (a pair of primers flanking EcoRI cloning site, each primer 30 base pairs away from the subcloning site and of a size between 18- and 24-mers and), 0.2 μ*M* of each dNTP, and 0.25 units of *Taq* polymerase (Boehringer Mannheim). Amplification conditions are 94°C for 2 min, followed by 35 cycles at 94°C for 1 min, 60°C for 1 min, 72°C for 1 min, and a final 10-min extension at 72°C. In order to get enough PCR products, one could set up 5–10 PCR reactions. The PCR products are then concentrated by Centricon 100 (Amicon), followed by electrophoresis in a 2% agarose gel (Gibco-BRL) and the collection of DNA fragments between 0.5 and 2 kb using a DNA purification kit (Bio-Rad).

3.3.1.2. Purification of YAC DNA

Because the YAC DNA isolated by the above-described protocol also contains yeast genomic DNAs, the YAC DNA has to be purified before hybridization. This could be accomplished by pulsed-field gel electrophoresis. For a detailed procedure of YAC purification by electrophoresis, one can refer to chapter 5 in the Current Protocols in Human Genetics.

3.3.1.3. Immobilization of the YAC DNA on Nylon Disks

1. Heat the tube containing 5–10 μL of the purified YAC DNA (20 ng/μL) at 95°C for 3 min.
2. Chill on ice.
3. Add 20X SSC pH 7 to make a final solution of 6X SSC.
4. Cut Hybond-nylon$^+$ (Amersham) filter into 5-mm × 5-mm squares.
5. Transfer 1 μL of the DNA solution onto the filter disk (make a marker to distinguish the DNA binding side).
6. Dry the disk at room temperature for 5 min.
7. Denature the DNA by floating the disk on the solution of 0.5 *M* NaOH, 1.5 *M* NaCl for 2 min.
8. Neutralize the disk on the solution of 0.5 *M* Tris, pH 7.5 and 1.5 *M* NaCl for 1 min.
9. Wash the disk by floating on a solution of 5X SSC for 2 min.
10. Immobilize the disk by leaving it in a vacuum at 80°C for 30 min.

The disk is ready for prehybridization.

3.3.1.4. Quenching the Nylon Disk with the Yeast Genomic DNAs by Prehybridization

1. Prepare the prehybridization solution:
   6X SSC
   5X Denhardt's
   0.5% SDS
   Yeast genomic DNA 40 ng/μL (The genomic DNA of a yeast strain used for the construction of the YAC library can be isolated by the procedure described earlier. The genomice DNA should be sheared by using a syringe with a 18-gage needle).
2. Prewarm the solution at 65°C for 30 min.
3. Put the disk in a Eppendorf tube containing 50–100 mL of the solution.
4. Incubate the tube at 65°C for at least 1 h.
5. Wash the disk briefly with 6 X SSC.

The disk is ready for hybridization.

3.3.1.5. Hybridization of the Disks with the cDNA Library

1. Transfer the disk to a fresh tube containing 50 μL of the prehybridization solution.
2. Add the double-stranded cDNAs prepared above to the tube (total 250–500 ng).
3. Hybridize at 65°C for 24 h–48 h with rotation.
4. Wash the disk in.
   2X SSC/0.1% SDS, two times in room temperature
   2X SSC/0.1% SDS, two times at 65°C for 15 min
   1X SSC/0.1% SDS, one time at 65°C for 15 min
   0.2X SSC/0.1% SDS, one time at 65°C for 15 min
   0.1X SSC/0.1% SDS, two times at 65°C for 15 min
5. Transfer the disk to a new 0.2-mL PCR tube containing 50 μL of distilled water.
6. Heat the PCR at 95°C for 5 min.
7. Chill and spin the tube
8. Keep on ice.

The solution is ready for PCR reaction.

3.3.1.6. PCR Amplification Using Nested Primers

The PCR conditions are as follows: The PCR could be carried out in a 0.2-mL PCR tube containing 50 μL of reaction volume of 1X PCR buffer containing 25 μL of the elute solution from above, 1 *M* of each primer flanking *Eco*RI subcloning site, 0.2 μ*M* of each dNTP, and 0.25 units of *Taq* polymerase (Boehringer Mannheim). Amplification conditions include 94°C for 2 min, followed by 35 cycles at 94°C for 1 min, 60°C for 1 min, 72°C for 1 min, and a final 10-min extension at 72°C. The PCR products should then be electrophoresed in 2% of agarose gel (Gibco-BRL).

In some of cases, the PCR product may not be detected by an agarose gel because of the low amount of the DNA product. Therefore, a second round of PCR (nested PCR) will be necessary to generate enough DNA. Because there is a distance at 30 base pairs between each primer and *Eco*RI subcloning site, another pair of primers for the second-round PCR could be synthesized within the first pair of the primers. Then, 5 μL from the first-round PCR could be used for the second-round PCR amplification. The final PCR product could be subcloned into the TA vector for further analysis.

3.3.2. Exon Trapping

It is possible that cDNAs will not be detected by the cDNA selection, especially if the amount of mRNA transcripts of a gene is very low. Alternatively, exon trapping could be another method to identify a gene of interest. Recently, one exon trapping system has been established by Gibco-BRL. Basically, the method is divided into three steps: (1) The YAC genomic DNAs are subcloned as *Bam*HI and *Bgl*II digests into the pSPL3 exon trapping vector. (2) DNAs from pools of either selected recombinant pSPLs clones or entire shotgun minilibraries are transfected into COS-7 cells using the procedure as described in the Exon Trapping System. If the inserted sequence has an exon in the proper orientation, the exon is retained in the mature RNA flanked by known sequence of exons, HIV tat, and β-globin. (3) Total RNAs are isolated 1–3 d after the transfection and RT-PCR of the RNA is performed by using oligonucleotides corresponding to the flanking β-globin sequence.

1. Digest 200 ng of the purified YAC DNA with one of the following endonuclease restriction enzymes: *Eco*RI, *Xho*I, *Not*I, *Xma*III, *Pst*I, and *Bam*HI.
2. Incubate at 37°C for at least 3 h.
3. Extract the DNA with phenol/chloroform once.
4. Wash the DNA pellet with 200 µL of 70% ethanol.
5. Dissolve the DNA in 10 µL of TE buffer.
6. Digest 100 ng of pSPLs vector DNAs with the enzyme used in digestion of the YAC DNA.
7. Treat the digested pSPLs DNA with phosphatase for dephosphorylation.
8. Purify the DNA by agarose gel electrophoresis.
9. Pool 100 ng of the digested YAC and 100 ng of the digested vector DNAs for ligation.
   - 100 ng of the YAC
   - 100 ng of the vector
   - 2 µL of 10 X ligation buffer
   - 1 µL of ligase (4 units)
   - Water up to 20 µL
10. Incubate at 16°C overnight.
11. Transfer the YAC/pSPL3 plasmid to DH10B (Gibco-BRL).
12. Amplify and purify the YAC/pSPL3 plasmid DNAs by miniprep.
13. Transfect 1 µg of the YAC/pSPL3 plasmid DNAs into COS-7 cells using the Lipofectace Reagent as described in the Exon Trapping System.
14. Isolate the total RNA using the Gibco-BRL Trizol reagent after 24 h transfection.
15. Use 2–4 µg of the total RNA for first-strand cDNA synthesis.
16. Perform PCR amplification as described in the Exon Trapping System

3.3.3. Identification of cDNA Fragments of Interest

Each of the PCR products obtained from either the cDNA selection or the exon trapping need to be sequenced and compared with the sequence information in DNA databases using the BLAST network service of the National Center for Biotechnology Information. The cDNAs of putative exons should be labeled as probes to hybridize with mRNAs from a variety of species (zoo blotting). The sequences of the conserved exons should be hybridized with most of the species, including the mouse. In order to identify the mutation, the DNA sequences of the exons will be determined and compared between the normal and the mutant mice.

4. Notes

1. If one plans to identify a mutant gene from a mouse strain other than the 12 inbred mouse strains (see the Research Genetics catalog), one has to determine the polymorphism level of DNA markers among the mouse strains before breeding. One can select several DNA markers randomly distributed on mouse genome and isolate genomic DNAs from several mouse strains out of the 12 inbred strains. Then, the polymorphism level of the DNA markers should be determined by performing PCR with the genomic DNAs and DNA markers. For the breeding, one should use a inbred mouse strain that gives a higher polymorphism level between the inbred strain and the mutant stain.
2. Polymorphisms of DNA markers provided by Genetic Research Inc. are based on the sizes of the DNA fragments (simple sequence length polymorphism [SSLP]). If the size of the markers is less than 4 bps, the difference between the markers has to be detected on polyacrylamide sequence gel.
3. Some of polymorphic markers can only be detected by single-strand conformational polymorphism (SSCP).
4. To obtain as many YAC clones as possible for constructing a physical map, one should search for the clones from several different YAC libraries.
5. Several approaches can be combined for recovering the ends of YAC clones. In our laboratory, we use also Alu PCR to search for the ends. The primers that we used for the PCR are CTTCTGGAGTGTCTGAAGAC, GACTGCTCTTCCGAAGGTCC, CTGGAACTCAC TCTGAAGAC, GTTTACCACTTAGAACACAG, and TTTGCCGCACAAGATTCTGG.
6. In inverse PCR, one should note that the restriction enzymes used for digesting DNAs should not recognize any DNA sequence between the primers and the original YAC cloning sites.
7. For cDNA selection, sources of mRNAs used for making a random priming cDNA library are very important. To identify the gene of interest, the mRNAs of the gene should be relatively abundant in the mRNA population.
8. To construct a random cDNA library for the cDNA selection, one should use total RNAs for cDNA synthesis, because many mRNAs could be lost during mRNA isolation.

Acknowledgments

The authors thank the National Arthritis Foundation, Atlanta, GA; the Charles Hood Medical Foundation, Boston, MA; and Human Growth Foundation, Falls Church, VA; for their generous support of this work.

References

1. Li, Y., Lacerda, D. A., Warman, M. L., Beier, D. R., Yoshioka, H., Ninomiya, Y., Oxford, J. T., Morris, N. P., Andrikopoulos, K., Ramirez, F., Wardell, B. B., Lifferth, G. D., Teuscher, C., Woodward, S. R., Taylor, B. A., Seegmiller, R. E., and Olsen, B. R. (1995) A fibrillar collagen gene, Col11a1, is essential for skeletal morphogenesis [see comments]. *Cell* **80,** 423–430.
2. Pace, J. M., Li, Y. Seegmiller, R. E., Teuscher, C., Taylor, B. A., and Olsen, B. R. (1997) Disproportionate micromelia (Dmm) in mice caused by a mutation in the C-propeptide coding region of Col2a1. *Dev. Dyn.* **208,** 25–33.
3. Royer-Pokora, G., Kunkel, L., Monace, A., Goff, S., Newburger, P., Baehner, R., Cole, F., Curnutte, J., and Orkin, S. (1986) Cloning the gene for an inherited human disorder-chronic granulomatous disease—on the basis of its chromosomal location. Nature **322,** 32–38.
4. Stubbs, L. (1992) Long-range walking techniques in positional cloning strategies. [review]. *Mammal. Genome* **3,** 127–142.
5. Dietrich, W. F., Miller, J., Steen, R., Merchant, M. A., Damron-Boles, D., Husain, Z., Dredge, R., Daly, M. J., Ingalls, K., O'Connor, T., Evans, C., DeAngelis, M., Levinson, D., Kruglyak,

L., Goodman, N., Copeland, N., Jenkins, N., Hawkins, T., Stein, L., Page, D., and Lander, E. (1996). A comprehensive genetic map of the mouse genome. *Nature* **380,** 149–152.

6. Dig, G., Faure, S., Fizames, C., Samson, D., Drouot, N., Vignal, A., Millasseau, P., Marc, S., Hazan, J., Seboun, E., Lathrop. M., Gyapay, G., Morissette, J., and Weissenbach, J. (1996). A comprehensive genetic map of the human genome based on 5,264 microsatellites. *Nature* **380,** 152–154.
7. Larin, Z., Monaco, A. P., and Lehrach, H. (1991) Yeast artificial chromosome libraries containing large inserts from mouse and human DNA. *Proc. Natl. Acad. Sci. USA* **88,** 4123–4127.
8. Schlessinger, D. (1990) Yeast artificial chromosomes: tools for mapping and analysis of complex genomes. *Trends Genet.* **248,** 255–258.
9. Woo, S.-S., Jiang J., Gill, B. S., paterson, A. H., and Wing, R. A. (1994) Construction and characterisation of a bacterial artificial chromosome library of Sorghum Bicolor. *Nuc. Acids Res.* **22,** 4922–4931.
10. Wang, G.-L., Holsten, T. E., Song, W.-Y., Wang, H-P., and Ronald, P. C. (1995) Construction of a rice bacterial artificial chromosome library and identification of clones linked to the Xa-21 disease resistance locus. *Plant J.* **7,** 525–533.
11. Ioannou, P. A., Amemiya, C. T., Garnes, J., Kroisel, P. M., Shizuya, H., Chen, C., Batzer, M. A., and de Jong, P. J. (1994). A new bacteriophage P1-derived vector for the propagation of large human DNA fragments. *Nat. Genet.* **6,** 84–89.
12. Pierce, J. C., Sternberg, N., and Sauer, B. (1992) A mouse genomic library in the bacteriophage P1 cloning system: organization and characterization. Mammal. *Genome* **3,** 550–558.
13. Sternberg, N., Ruether, J., and deRiel, K. (1990) Generation of a 50,000-member human DNA library with an average DNA insert size of 75–100 kbp in a bacteriophage P1 cloning vector. *New Biol.* **2,** 151–162.
14. Morgan, J. G., Dolganov, G. M., Robbins, S. E., Hinton, L. M., and Lovett, M. (1992) The selective isolation of novel cDNAs encoded by the regions surrounding the human interleukin 4 and 5 genes. *Nucl. Acids Res.* **20,** 5173–5179.
15. Lovett, M., Kere, J., and Hinton, L. M. (1991) Direct selection: a method for the isolation of cDNAs encoded by large genomic regions. *Proc. Natl. Acad. Sci. USA* **88,** 9628–9632.
16. Parimoo, S., Patanjali, S. R., Shukla, H., Chaplin, D. D., and Weissman, S. M. (1991) cDNA selection: efficient PCR approach for the selection of cDNAs encoded in large chromosomal DNA fragments. *Proc. Natl. Acad. Sci. USA* **88,** 9623–9627.
17. Auch, D. and Reth, M. (1990) Exon trap cloning: using PCR to rapidly detect and clone exons from genomic DNA fragments. *Nucl. Acids Res.* **18,** 6743,6744.
18. Buciler, A. J., Chang, D. D., Graw, S. L., Brook, J. D., Haber, D. A., Sharp, P. A., and Housman, D. E. (1991) Exon amplification: A strategy to isolate mammalian genes based on RNA splicing. *Proc. Natl. Acad. Sci. USA* **88,** 4005–4009.
19. Duyk, G. M., Kin, S., Myers, R. M., and Cox, D. R. (1990) Exon trapping: a genetic screen to identify candidate transcribed sequences in cloned mammalian genomic DNA. *Proc. Natl. Acad. Sci. USA* **87,** 8995–8999.
20. Lindsay, S. and Bird, A. (1987) Use of restriction enzymes to detect potential gene sequences in mammalian DNA. *Nature* **327,** 336–338.
21 Schwartz, D. C. and Cantor, C. R. (1984) Separation of yeast chromosome-sized DNAs by pulsed field gradient gel electrophoresis. *Cell* **37,** 67–75.
22. Silverman, G. A. (1993) Isolating vector–insert junctions from yeast artificial chromosomes. [review]. *PCR Methods Applic.* **3,** 141–150.
23. Sambrook, J., Fritsch, E. F., and Maniatis, T. (eds.) (1989) *Molecular Cloning*, 2nd ed. Cold Spring Harbor Press, Cold Spring Harbor, NY.

29

Gene Trapping in Embryonic Stem Cells In Vitro to Identify Novel Developmentally Regulated Genes in the Mouse

Gary E. Lyons, Bradley J. Swanson, Melissa A. Haendel, and Joshua Daniels

1. Introduction

The identification of novel, developmentally regulated genes whose products play roles in the differentiation of specific vertebrate tissues and organs can be accomplished using a method called gene trapping *(1–9)*. This technique involves inserting a marker gene, such as β-galactosidase or alkaline phosphatase, into the genome of murine embryonic stem (ES) cells using one of several available methods of transfection. The gene trap construct contains an antibiotic resistance gene such as neomycin to permit the selection of transfected clones. ES cell clones are picked and expanded on multiwell plates before freezing. Screening of the clones for tissue-specific gene expression involves thawing multiwell dishes, expanding them, and allowing them to differentiate in vitro into multiple cell lineages *(10–14)*.

Some cell lineages such as muscle, neurons, and adipocytes can be identified based on morphology when they differentiate in vitro. Others such as cartilage can be easily distinguished with routine histological staining methods. To determine if the gene trap is expressed in a particular differentiated cell type, an antibody to β-galactosidase is used in conjunction with a cell-specific marker. Secondary antibodies are chosen to discriminate between the two primary antibodies such that double-labeled cells can be photographed. RNA is extracted from ES cell clones that show expression in a particular cell type of interest. Using primers in *lacZ*, a partial cDNA can be isolated by 5' rapid amplification of cDNA ends (RACE). This sequence can be compared with DNA databases to determine if it is novel, and then used to screen cDNA libraries to obtain longer and perhaps full-length cDNAs. This cDNA can also be used for *in situ* hybridization on sections of mouse embryos to show that the gene is expressed in the same cell types in vivo as it is in vitro.

This process of selecting for gene trap insertions in cells that can differentiate in vitro *(1)* has a number of advantages over screening for novel genes by making chimeric embryos and looking for interesting patterns in vivo *(2–4,7–9)*. It is more eco-

From: *Methods in Molecular Biology, Vol. 136: Developmental Biology Protocols, Vol. II*
Edited by: R. S. Tuan and C. W. Lo © Humana Press Inc., Totowa, NJ

nomical because animal care costs are eliminated. It can be performed without specialized lab equipment. Several hundred clones can be screened in a few months. Housekeeping genes and previously characterized genes are easily avoided. These aspects allow investigators with limited resources to conduct large-scale screens for novel genes expressed in any embryonic cell type that can differentiate in cultures of ES cells.

2. Materials

2.1. Cells

Murine embryonic stem cells (ES) such as R1, D3, AB1, or other well-characterized lines at as low a passage number as possible, murine STO fibroblasts (American Type Culture Collection [ATCC], Manassas, VA), buffalo rat liver (BRL) cells (ATCC). ES cells are usually obtained from the labs that isolated them (*see* **Note 1**).

2.2. Media Formulations (See Note 2)

1. 10% Calf serum (CS): 50 of mL calf serum, 5 mL of 5000 U/mL penicillin–5000 μg/mL streptomycin (Gibco-BRL [Gaithersburg, MD] or other supplier of tissue culture media and supplements), 5 mL of 200 m*M* L-glutamine, 5 mL of 10 m*M* nonessential amino acids, up to 500 mL with Dulbecco's modified Eagle's medium (DMEM) high glucose; filter sterilize with a 0.22-μm bottle-top filter.
2. 15% Fetal bovine serum (FBS): 75 mL fetal bovine serum, 5 mL penicillin–streptomycin, 5 mL L-glutamine, 5 mL nonessential amino acids, 5 mL βME in 1X phosphate-buffered saline (PBS) (7 μL in 10 mL PBS), up to 500 mL with DMEM high glucose, filter sterilize with a 0.22-μm bottle-top filter. Add 1 μL/mL leukemia inhibitory factor (LIF) [(15); *see* **Note 2**)] when cultures are sparse (i.e., when colonies are small and widely separated).
3. 10% FBS: 50 mL fetal bovine serum, 5 mL penicillin–streptomycin, 5 mL L-glutamine, 5 mL nonessential amino acids, up to 500 mL with DMEM high glucose, filter sterilize with a 0.22-μm bottle-top filter. (*See* **ref. *17***.)
4. BRL-conditioned medium: Seed one vial of BRL cells into two 250-mL tissue culture flasks with 15 mL 10% FBS. Grow to confluence. Collect, sterile filter (to remove cellular debris), and freeze 30 mL of medium every other day. When BRL cells start to peel off from the flask, discard cultures.
5. 15% FBS/BRL-conditioned medium: 30 mL BRL-conditioned medium + 20 mL 22.5% FBS in DMEM with glutamine, Penicillin–streptomycin and nonessential amino acids +1 μL/mL LIF.
6. 0.1% Gelatin: In a beaker, heat 100 mL 1X PBS in the microwave on high for 2 min. Add 0.5 g tissue-culture-grade gelatin and stir to dissolve. Bring up to 500 mL with PBS and sterilize with a 0.22-μm bottle-top filter into a sterile 500-mL bottle. Refrigerate.
7. 10X Phosphate-buffered saline stock: 80 g NaCl, 2 g KCl, 11.5 g Na_2HPO_4, 2 g KH_2PO_4, 800 mL distilled, deionized (dd)H_2O, adjust to pH 7.4 with 10 *N* NaOH to final volume 1 L.
8. Mitomycin C medium: Dissolve 2 mg Mitomycin C (Sigma, St. Louis, MO, USA) in 1 mL of 1X PBS to make the stock solution. Add 0.5 mL stock solution to 99.5 mL 10% CS. The solution can be stored at 4°C and used for 1 wk. The stock can be frozen for 1 mo as aliquots and thawed once.

3. Methods

3.1. Gelatinizing Plates

Add 1 mL 0.1% gelatin to cover entire bottom of 60-mm tissue culture dish. Aspirate excess gelatin. Store at 4°C if not used the same day.

3.2. Thawing Cells

1. Pour 20 mL of media into 50-mL tube in order to make two 100-mm plates.
2. Remove cells from liquid nitrogen and thaw in 37°C water bath a few minutes.
3. As soon as the vial is thawed, transfer all cells into the 50-mL tube containing the media.
4. Gently rock tube back and forth several times to get an even suspension of cells.
5. Transfer 10 mL of cell/media solution into each tissue culture plate.
6. Label, rock back and forth, and place plates into incubator.

3.3. Splitting (Subculturing) Cells (See Note 3)

1. Aspirate medium from tissue culture plates. Wash twice with 5 mL 1X PBS.
2. Add 2 mL 0.05% trypsin/EDTA to each plate. For undifferentiated ES cells, incubate 5 min at 37°C. When making embryoid bodies ([EBs]; *see* **Subheading 3.11.**) incubate ES cells 1 min at room temperature (RT). For STO and BRL cells, incubate at RT.
3. Examine on inverted-phase microscope. Agitate plate by tapping against a solid surface as needed. Continue to do so until all cells are removed from the bottom of the plate.
4. Immediately add 8 mL fresh media to each plate. Repipet to break up clumps of cells.
5. Place cells/media into 15-mL tubes. Mix the cell suspension by rocking. Take an aliquot for counting. Place cells in a clinical centrifuge for 5 min at speed 4.
6. Count cells with hemocytometer. Cells in 1 of 9 squares $\times 10^4$ = cells/mL.
7. Remove media from cell pellet with Pasteur pipet. Resuspend cells in fresh media.
8. Add the cells to a calculated amount of additional media necessary for the number of tissue culture plates. Rock back and forth 10–20 times to get a nice suspension of cells.
9. Pipet cell suspension into plates or flasks, rocking in between. Label and place plates/flasks into incubator.

3.4. Making Feeder Plates for Undifferentiated ES cells

1. Add 5 mL Mitomycin C (Sigma) medium to each confluent 100-mm STO plate.
2. Incubate 2–3 h at 37°C.
3. Wash three times with 5 mL 1X PBS.
4. Split onto gelatinized plates.
   60-mm Plates should be seeded at 1–1.2 million cells per plate.
   24-Well plates should be seeded at 100,000 cells per well.
   STO Cells must completely cover the tissue culture surface before ES cells are added, usually the morning after preparation of the feeder dishes.
5. Seed (1–2) X 10^6 ES cells/60-mm dish.

3.5. Electroporation of ES Cells with Gene Trap Plasmids

1. Aspirate medium from ES cells. Wash twice with 1XPBS.
2. Add 1 mL 0.05% trypsin/EDTA to cells, incubate 5 min at 37°C.
3. Add 3 mL 15% FBS to cells and repipet to break up clumps.
4. Pellet cells in a clinical centrifuge. Resuspend 1×10^7 cells in 1 mL 15% FBS.
5. Add 900 µL of cell suspension to 10 µg of gene trap DNA. Use 100 µL as nontransfected control.
6. Incubate on ice for 15 min. Gelatinize ten 100-mm tissue culture plates.
7. Place ES cell/DNA suspension in sterile 0.4-cm electroporation cuvette.
8. Electroporate at 500 µF, 250 V using a GenePulser (Bio-Rad, Hercules, CA) or another comparable instrument.
9. Add electroporated cells to 100 mL BRL-conditioned medium and seed onto 10 gelatinized plates. Begin selection with 400 µg/mL G418 (Gibco-BRL or other suppliers of tissue culture reagents) in BRL-conditioned medium on the next morning.

10. Selection should be complete and colonies should be visible after 10 d. The untransfected control should have no colonies.

3.6. Infection of ES Cells with Retroviral Gene Traps

1. A suitable retroviral gene trap vector should be used such as ROSAβ-geo or ROSAβ-gal *(2)*. The virus is packaged by a cell line such as GP+E86 into viral particles ***(16)*** and secreted into the surrounding BRL-conditioned medium. This medium is filtered through a 0.2-μm filter prior to use.
2. Viral titer is determined by feeding virus-producing cells with fresh media for 16 h shortly before reaching confluency. ES cells are seeded at a density of 5×10^5 cells/well of a six-well dish and incubated overnight. Serial dilutions of BRL-conditioned medium containing viral particles are used to infect the ES cells that are selected with 0.4 mg/mL G418 24 h after infection. Colonies are counted after 7–10 d of selection after staining with 2% Methylene Blue in water for a few minutes.
3. For infection with the retroviral gene trap, 4 mL of pretitered virus (5×10^3 CFU/mL) in the presence of 4 μg/mL Polybrene (Sigma) is placed over undifferentiated ES cells plated at a density of 10^6 cells per 100-mm tissue culture dish. After 20 h, the medium is replaced with fresh BRL-conditioned medium for 1 d. On the following day, infected cells were selected on BRL-conditioned medium containing the antibiotic G418 (0.4 mg/mL; Gibco-BRL) for 7–10 d. During this period, noninfected cells died and detached from the plate; infected cells generated colonies 7–10 d after infection.

3.7. Picking Clones

1. Using a 1000-μL or a 250-μL micropipet, gently scrape under clone with tip until clone is dislodged from plate. Draw clone into pipet tip.
2. Expel directly into 0.25 mL of trypsin in well of a 24-well plate.
3. Incubate at 37°C for 5 min. Add 1 mL of 15% FBS
4. Repipet a number of times to break up cells, and transfer cells to microcentrifuge tubes.
5. Centrifuge at 1350*g* for 5 min.
6. Aspirate medium and add 1 mL of 15% FBS + G418 (200 μg/mL) + LIF (1 μL/mL).
7. Repipet several times and divide into two multiwell feeder plates.
8. Add 0.5 mL 15% FBS to both.

3.8. Freezing Multiwells

1. Wash with 0.5 mL of 1X PBS.
2. Add 0.25 mL of trypsin/EDTA. Leave on for 1 min at room temperature.
3. Aspirate trypsin.
4. Add 0.1 mL of 1X PBS.
5. Examine under microscope.
6. When ES cells are balled up and just starting to come off, tap plate to release the cells.
7. Add 700 μL of freezing medium (15% FBS + 10% DMSO), and repipet twice.
8. Place two pieces of tape on the multiwell dishes to prevent the tops from separating from the wells.
9. Place in a Styrofoam box, such as a shipping container, in a –80°C freezer. The Styrofoam prevents the cells from freezing too rapidly which could reduce the percent recovery.

3.9. Thawing Multiwells

1. Prepare 24 microcentrifuge tubes with 700 μL of 15% FBS media. Be careful to have them organized or labeled in a manner corresponding to the individual wells of the multiwell.

2. Remove plate from freezer and hold in 37°C water bath until thawed. Outer wells thaw first, so remove from water bath when there are still a few ice crystals in the center wells. Wipe multiwell plate very well with 70% ETOH.
3. Pipet in and out several times with P1000.
4. Transfer contents of wells to corresponding microcentrifuge tubes as quickly as possible. Allowing thawed cells to sit in freezing medium for a long time decreases their viability.
5. Centrifuge at 1350*g* for 5 min.
6. Aspirate media from tubes and add 1 mL of fresh media.
7. Repipet several times and add to new multiwell feeder plate.

3.10. Expanding from Multiwell to 60 mm

1. Wash with 0.5 mL 1X PBS.
2. Add 0.25 mL of trypsin/EDTA.
3. Incubate at 37°C for 5 min.
4. Add 1 mL of 15% FBS.
5. Pipet twice with a 1000-μL micropipet and transfer to Eppendorf tube.
6. Centrifuge for 5 min at 1350*g*.
7. Aspirate media, add fresh 15% FBS, and resuspend.
8. Transfer to 15-mL tube.
9. Add 4 mL of 15% FBS.
10. Resuspend completely and transfer to 60-mm feeder plate. Add LIF to medium while the cells are sparse.

3.11. Making Embryoid Bodies

1. Wash with 2 mL of 1X PBS.
2. Add 1 mL of trypsin and leave at room temperature until cells come off.
3. Add 4 mL of 10% FBS.
4. Using a large-bore pipet (25 mL), gently transfer cells to a 60-mm gelatinized plate.
5. Incubate at 37°C for 15 min (STO cells will sit down, but most ES cells will not).
6. Half of ES cells are removed for freezing.
   Centrifuge at 1350*g* for 5 min.
   Aspirate media and resuspend in 1 mL of freezing medium.
   Transfer cells to a 1.5-mL freezing vial (Nalge Nunc, Rochester, NY, or similar supplier).
   Place in Styrofoam in –80°C freezer.
   Transfer to liquid nitrogen after 24 h.
7. Other half of ES cells are gently transferred with a large-bore pipet to a 100-mm bacterial plate.
8. Add 8 mL of 10% FBS to bacterial plate.
9. After a few days, the EBs will form an outer layer of endoderm. After 4–7 d, some EBs will form an inner ectoderm layer. After 7–10 d, EBs are plated onto gelatin-coated 60-mm tissue culture dishes (*see* **Note 4**).

3.12. Procedure for Staining Embryoid Bodies

If staining floating EBS, use sterile 50-mL tubes. If staining attached differentiated cells, all steps occur in the tissue culture dishes. It is not necessary to perform this procedure in a tissue culture hood.

1. Rinse cells once in 1X PBS.
2. Fix cells for 5 min in 2% formaldehyde and in 1X PBS for 5 min.
3. Wash cells three times for 5–10 min at room temperature in 1X PBS.

4. Make X-gal stain (Ambion Inc., Austin, TX).
   0.5 mL X-Gal (40 mg/mL in dimethylformamide)
   0.105 g Potassium ferrocyanide (Sigma)
   0.105 g Potassium ferricyanide (Sigma)
   0.1 mL of 1 *M* magnesium chloride
   Dilute to 50 mL with 1X PBS (2.5 mL required per 60-mm plate)
5. Stain cells for 4 h to overnight at 37°C.

3.13. Costaining ES Cell Clones for lacZ *and a Cell-Specific Marker (*See Note 5*)*

1. Make fresh 4% paraformaldehyde in 1X PBS. Heat to 60°C with a few drops of 10 *N* NaOH. Bring to volume with H_2O. Filter through a Whatman #1 filter (Whatman, Clifton, NJ) to remove particles.
2. Warm some 1X PBS and the paraformaldehyde in the 37°C bath. Rinse the cells with the PBS, then fix in paraformaldehyde for 15 min at room temperature. Rinse twice in PBS. (After the fix, it is not necessary to warm anything.)
3. Postextract in 0.1% Triton X-100 (Fisher Scientific, Itasca, IL) in PBS for 6 min at room temperature. Rinse in PBS.
4. Make blocking solution: 0.1% Triton X-100, 5% BSA, 2% normal goat serum in PBS. Filter through 0.45-μm filter. Make fresh, can be kept in the refrigerator for 1 wk. Block at room temperature for 2 h or overnight at 4°C.
5. Dilute primary antibodies in blocking solution. We have used the Cappel-Organon Teknika (Durham, NC) rabbit anti-β-galactosidase (1:3000) and Gibco-BRL mouse anti-β-galactosidase (1:500) with success. Spinning before use may eliminate some particulate background. Apply about 250 μL/cover slip on red dental wax. Incubate overnight at 4°C. A number of gene products have been identified in differentiated ES cell cultures ***(18)***
6. Rinse three times for 5 min each in 1X PBS.
7. Apply secondary antibodies in blocking solution (usually 1:200 to 1:1000 is a good dilution for most); place at 37°C for 1 h.
8. Rinse four times for 5 min each in PBS (longer is fine). Mount in mounting solution: 1 mL 1X PBS 1 g 1,4-diazobicyclo-[2.2.2]octane (DABCO), 10 mg *p*-phenylenediamine. Mix until partly dissolved, then add 9 mL glycerol. Store in freezer covered in foil. It is usable until it becomes dark.

3.14. Induction of Certain Differentiated Cell Types in ES Cell Cultures

Although ES cells differentiating in vitro are extremely heterogeneous, culture conditions can be modified to select for certain cell types, such as neurons ***(13,19)***, endothelial cells ***(14)***, hematopoietic lineages ***(11,20)***, and muscle ***(12,21)***. This system can also be used to screen for inducible genes that respond to a hormone such as retinoic acid ***(22)***.

3.15. RNA Extraction

1. Feed cells with fresh medium 2–3 h before extraction.
2. Remove medium from cells. Add 10 mL TRIzol (Gibco-BRL) or another similar product per 100-mm tissue culture dish. Repipet to dissolve cells. This solution can be frozen at –80°C or continue with **step 3**.
3. Follow the protocol given with the particular RNA extraction reagent you decide to use.
4. Always use sterile plasticware. Bake any glassware at least 6 h at 250°C that will contact RNA solutions. Use diethylpyrocarbonate (DEPC)-treated water. 0.1% DEPC in water for 2 h at 37°C, autoclave for 15 min before use. Always wear gloves.

3.16. Protocol for 5' RACE (Rapid Amplification of cDNA Ends; see also ref. 23)

1. This protocol was used to clone regions of gene trapped RNA 5' of the *lacZ* region of the RNA. Use a no-RNA blank as a negative control throughout the entire RACE protocol. Do all of the reactions in the small polymerase chain reaction (PCR) tubes to facilitate the use of the thermal cycler for incubations. Whenever not incubating the reactions, keep them on ice whenever possible.
2. Denaturing the RNA and annealing the first-strand primer: Resuspend your poly A+ RNA in 12λ of RNase free H_2O. Try to start with as much RNA as possible, at least 5 μg. Add 1λ Rnasin (1 U/λ). Add 2λ first-strand primer (20 pmol/λ, or 20 μ*M*; 5' CCG TGC ATC TGC CAG TTT GAG GGG A 3'; this primer binds β-geo 239 base pairs downstream of the β-galactosidase ATG). Incubate at 80°C for 3 min, then at 60°C for 20 min, then chill on ice. Give tubes a quick spin before proceeding.
3. Primer extension: first-strand synthesis

   To the primer-annealed RNA, add the following mixture:
   13λ dNTPs (2 m*M* stock)
   8λ 5X first strand: RT buffer (Gibco-BRL)
   0.5λ RNasin
   1λ M-MuLVReverse transcriptase 200 U/λ (Gibco-BRL)
   2λ 0.1*M* DTT
   1.2λ spermidine (10 m*M*)
   NOTE: Make up a master mix when doing more than one sample, which should be every time because you should have a no-RNA negative control in every RACE experiment.
   Incubate at 37°C for 2 h.
4. Degrading RNA

   After the 2-h incubation at 37°C, add 1 μg of DNAse-free RNase.
   Continue incubating at 37°C for 15 min.
   Give tubes a quick spin and place on ice.
5. Purify cDNA from primers and nucleotides

   Raise volume to 300λ with H_2O
   Spin total volume in Millipore Ultrafree 30,000 NMWL filter (Millipore, Bedford, MA) @ 2000 g for 10 min.
   Remove filtrate from tube.
   Add 300λ ddH_2O to filter and spin at 2000 g for 10 min.
   Collect retentate.
   Wash membrane with 50λ ddH_2O and collect in same tube.
   Lyophilize in a rotary drier until the volume is 20λ.
6. Tailing purified cDNA with dATPs

   To the 20λ purified first strand cDNA, add the following:
   - 10λ 5X terminal transferase buffer (Roche Diagnostics, Indianapolis, IN)
   - 3λ 15 m*M* $CoCl_2$ (Roche Diagnostics)
   - 5λ Terminal deoxyribonucleotide transferase 25 U/λ (Roche Diagnostics)
   - 12.5λ 10 m*M* dATP

   Incubate for 5 min at 37°C and then at 65°C for 5 min to denature the enzyme.
   Dilute to a total volume of 200λ.
7. Second-Strand Synthesis

   To 20λ of the tailed cDNA, add the following:
   - 2λ Anchor oligo (1 pmol/λ, or 1 μ*M*; 5'CGAGGGGGATGGTCGACGGAAG CGACCT$_{18}$ 3')

5λ dNTPs (2 m*M* each)
5λ 10X PCR buffer with Mg
0.5λ *Taq* DNA polymerase (Boehringer or other supplier of *Taq* polymerase)
17.5λ dH_2O

Use the thermal cycler for the following incubation:
94°C 2 min, 37°C 5 min, 5-min ramp up to 72°C, 72°C for 40 min.
You are now ready to perform your first round of PCR. Optimal results were obtained using 3λ of the second-strand synthesis as template along with annealing temperatures 2–4°C below the nearest neighbor T_m.

8. First-Round PCR
   The PCR conditions are as follows:
   2λ each primer (20 m*M*) (5' GTGCCGGAAACCAGGCAA A 3'; 5' GATGGTC GACGGAAGCGA 3')
   5λ dNTPs (2 m*M*)
   4.7λ 10X PCR buffer (Roche Diagnostics) (the template is in 1X PCR buffer)
   3λ template from **step 6**
   0.5λ *Taq* DNA polymerase 5U/λ (Roche or other supplier of *Taq* polymerase)
   32.5λ H_2O (up to a total volume of 50λ)

   Utilize the hot-start technique of not adding the *Taq* polymerase until the reactions have been heated to 94°C for 5 min by adding the enzyme through the mineral oil after heating the tubes. Remember to perform the appropriate positive and negative controls whenever possible. The following is the thermal cycling program used with primers with 66°C T_{ms}: 94°C 5 min → 94°C 2 min → 64°C 2 min → 72°C 2.5 min → 72°C 10 min for 25 cycles
   Nested PCR may be performed if necessary to receive better results in some cases.
   It is also highly advisable to carry-out a positive (and negative) control PCR.
9. Visualize PCR products on ethidium-stained agarose gel. Use a minigel apparatus and a 1.2% agarose gel. Load 8λ of PCR products and use a small molecular weight DNA ladder that will give good resolution from 200–1500 kb. If you see bands that may be your 5' RACE product, you are very fortunate. If you do not give up just yet. In most cases, you should see a smear of PCR products in your experimental lanes (except for the no-RNA control).
10. Screen products with a Southern blot. Regardless of whether you see a band, a smear, or both, it is necessary to verify that the PCR reaction produced the desired 5' RACE product in an amplified quantity, not necessarily a visible band. To do this, you should perform a Southern analysis on the RACE products using the 5' 300 bases of *lacZ* as a probe. Run 4λ of PCR products (including the mock and negative control) or just enough to barely visualize the DNA smear(s) when you stain the gel to photograph it before transferring. Also run 4λ of a 1/50 or 1/100 dilution of the positive control. Use a *lacZ* DNA probe as template for multiprime labeling with ^{32}P to make your probe. Follow the multiprime labeling kit's instructions. The results from the Southern blot should tell you if you produced an amplified 5' RACE product and its size.
11. Prepare a 4X PCR and gel purify the 5' RACE product. Once you have identified the size of your RACE product, precisely repeat the PCR reaction at a 4X volume. Precipitate the PCR products in 1/10 volume 3 *M* sodium acetate pH 5.0 and 2.5 volumes 95% EtOH at –80°C for 30 min. Resuspend the precipitated PCR products and run them on a wide-well agarose minigel. Cut out the PCR products with the size corresponding to your 5' RACE product(s). Purify the DNA from the agarose and resuspend in 50λ. Determine the DNA concentration of the purified DNA on a spectrophotometer and calculate your recovery of PCR product.

12. Cloning of 5' RACE product using a PCR cloning vector. Perform two ligations per 5' RACE product (1:1 and 1:3 molar ratio of vector to insert) and transform bacteria with (1–2) λ of each ligation. Plate the transfected bacteria onto Luria Bertani (LB)/ampicillin plates that have been preplated with 50λ of 20 mg/mL X-Gal in dimethyl formamide for at least 30 min. The next day, pick all white and some pale blue colonies and culture them overnight in 3 mL LB cultures. Make glycerol stocks of all cultures and plasmid miniprep the rest of the culture.
13. Analysis of bacterial clone plasmid DNA. The results from this step should tell you which, if any, of the bacterial clones contain a genuine 5' RACE product. Restriction digest 1/4 of your miniprep DNA with the appropriate enzymes to see if the proper sized insert drops out.

    Measure the concentration of the miniprepped DNA on a spectrophotometer for each of the bacterial clones and prepare 1 nmol/λ solutions. These will be used in the following step as templates for the PCR screening. For each clone, perform three PCR reactions using the following primer pairs if the ES cells the product was derived from is a ROSA β-geo clone; 5'GTGCCGGAAACCAGGCAAA3'/5'GATGGTCGACGGAAGCGA 3', 5'GGG ATCCGCCATGTCACAGA3'/5'GATGGTCGACGGAAGCGA 3', 5'GTGCCGGAAA CCAGGCAAA3'/5'ATGATCTGTGACATGGCGGA3'.

4. Notes

1. Do not use D3 cells from ATCC if you plan to make germline chimeras with your ES cell clones. STO cells work just as well as embryonic fibroblasts and are much easier to maintain. STO cells should not be used after passage 20 because they seem to lose their ability to produce the differentiation inhibiting growth factors with repeated subculturing.
2. Fetal bovine serum (FBS) needs to be prescreened for use with ES cells. Request an aliquot of 50–100 mL of serum from several different suppliers (e.g., Hyclone [Logan, VT], Intergen [Purchase, NY], Sigma, etc.). ES cells should be assayed for plating efficiency, growth, and differentiation in each lot tested. Calf serum should be tested with STO cells. It is desirable to create neomycin-resistant STO cells for maintaining ES cells after cloning. An expression vector containing the neomycin-resistanct cDNA can be transfected into STO and clones selected. LIF can be purchased from a supplier such as Gibco-BRL. Alternatively, a LIF expression vector can be transfected into cells such as COS7 or STO cells and the conditioned media harvested for use with the ES cells. Two antibodies, TROMA-1 (anti-cytokeratin Endo-A) and SSEA-1 (anti-carbohydrate from F9 teratocarcinoma) available from the Developmental Studies Hybridoma Bank, maintained by the Department of Biological Sciences, University of Iowa (Iowa City, IA) are useful for testing batches of FBS and LIF to maintain ES cells in an undifferentiated state.

    Antibody staining of embryonic stem (ES) cells with TROMA1 and SSEA-1 (MC480)

    Grow ES cells on cover slips.

    Wash once in 1X PBS. Fix for 15 min in ice-cold acetone in a glass Petri dish.

    Block with 16% calf serum, 0.1% Tween-20 at RT for 15 min. Drain solution onto a towel with wicking action; do not wash.

    Add antibody in 0.1% serum in PBS. MC480, IgM, supernatant 1:20 or TROMA1, IgG, supernatant 1:20 for 1 h at room temperature or 4°C overnight.

    Wash twice with 1X PBS.

    Add secondary antibody 1:100 in 0.1% calf serum in the PBS in dark for 1 h at RT. For example, horse FITC antimouse IgG (Vector or similar supplier) or goat FITC antimouse IgM (Cappel or similar supplier of immunoreagents).

    Wash twice in 1X PBS.

    Invert in glycerol 50% in PBS and 0.4% *n*-propylgalate or some similar mounting medium with an antibleaching agent.

Examine with FITC filter. MC480 stains undifferentiated cells. TROMA1 stains cells that have started to differentiate into endoderm.

3. The ES cells are not easy to dissociate even after trypsinization and are, therefore, not easy to count accurately using a hemocytometer. STO cells are easily distinguished from ES cells on the hemocytometer because of their much larger size. ES cells should not be grown to greater than 75% confluency, because that will increase the amount of differentiation that occurs. Undifferentiated ES cells grow so rapidly that they need to be subcultured every 2–3 d.
4. High-density cultures, particularly during differentiation, need to be fed every day. Monitor the color of the tissue culture medium on a daily basis.
5. Differentiated ES cells can be immunostained on plastic tissue culture dishes, but the charged plastic tends to give high background with fluorescent antibodies. Better immunostaining results can be achieved by growing differentiated ES cells on glass cover slips. However, the cells do not easily adhere to a glass, noncharged surface. It is necessary to gelatinize the cover slips as described in **Subheading 3.1.** Place the cover slips in 35-mm tissue culture dishes and allow the gelatin to remain on the cover slips for 1 h before removing the excess liquid. Dry the gelatin-coated cover slips for at least 1 h before use. Because differentiated ES cell cultures can vary widely in their complexity and composition of cell types, it may be helpful to add higher concentrations of Triton-X than 0.1% to facilitate permeabilization. Up to 1% Triton-X can be used for the extraction step, and up to 0.3% can be added to the blocking solution.

Acknowledgments

G.E.L. would like to thank Bradley Bodner for his assistance, and Jayne Squirrell for her comments on the manuscript. This work was supported by NIH grants HD29471 to G.E.L., GMO 7507-18 and MH11560 to M.A.H. J.D. was supported by grants from the Muscular Dystrophy Association and the American Heart Association of Wisconsin.

References

1. Baker, R. K., Haendel, M. A., Swanson, B. J., Shambaugh, J. C., Micales, B. K., and Lyons, G. E. (1997) In vitro preselection of gene-trapped embryonic stem cell clones for characterizing novel developmentally regulated genes in the mouse. *Dev. Biol.* **185,** 201–214.
2. Friedrich, G. and Soriano, P. (1991) Promoter traps in embryonic stem cells: a genetic screen to identify and mutate developmental genes in mice. *Genes Dev.* **5,** 1513–1523.
3. Scherer, C. A., Chen, J., Nachabeh, A., Hopkins, N., and Ruley, H. E. (1996) Transcriptional specificity of the pluripotent embryonic stem cell. *Cell Growth Diff.* **7,** 1393–1401.
4. Gossler, A. and Zachgo, J. (1993) Gene and enhancer trap screens in ES cell chimeras, in *Gene Targeting: A Practical Approach* (Joyner, A., ed.), IRL, Oxford, pp. 181–225.
5. Hill, D. P. and Wurst, W. (1993) Gene and enhancer trapping: mutagenic strategies for developmental studies. *Curr. Topics Dev. Biol.* **28,** 181–206.
6. Joyner, A. L. (1991) Gene targeting and gene trap screens using embryonic stem cells: new approaches to mammalian development. *Bioassays* **13,** 649–656.
7. Skarnes, W. C., Auerbach, B. A., and Joyner, A. L. (1992) A gene trap approach in mouse embryonic stem cell: the *lacZ* reporter gene is activated by splicing, reflects endogenous gene expression, and is mutagenic in mice. *Genes Dev.* **6,** 903–918.
8. Takeuchi, T., Yamazaki, Y., Katoh-Fukui, Y., Tsuchiya, R., Kondo, S., Motoyama, J., and Higashinakagawa, T. (1995) Gene trap capture of a novel mouse gene, *jumonji*, required for neural tube formation. *Genes Dev.* **9,** 1211–1222.

9. Wurst, W., Rossant, J., Prideaux, V., Kownacka, M., Joyner, A., Hill, D. P., Guillemot, F., Gasca, S., Cado, D., Auerbach, A., and Ang, S.-L. (1995) A large-scale gene-trap screen for insertional mutations in developmentally regulated genes in mice. *Genetics* **139,** 889–899.
10. Doetschman, T. C., Eistetter, H., Katz, M., Schmidt, W., and Kemler, R. (1985) The *in vitro* development of blastocyst-derived embryonic stem cells lines: formation of visceral yolk sac, blood islands and myocardium. *J. Embryol. Exp. Morphol.* **87,** 27–45.
11. Keller, G. M. (1995) *In vitro* differentiation of embryonic stem cells. *Curr. Opin. Cell Biol.* **7,** 862–869.
12. Baker, R. K. and Lyons, G. E. (1996) Embryonic stem cells and *in vitro* muscle development. *Curr. Topics Dev. Biol.* **33,** 263–279.
13. Strübing, C., Ahnert-Hilger, G., Shan, J., Wiedenmann, B., Hescheler, J., and Wobus, A. M. (1995) Differentiation of plutipotent embryonic stem cells into the neuronal lineage *in vitro* gives rise to mature inhibitory and excitatory neurons. *Mech. Dev.* **53,** 275–287.
14. Wang, R., Clark, R., and Bautch, V. L. (1992) Embryonic stem cell-derived cystic embryoid bodies form vascular channels: an *in vitro* model of blood vessel development. *Development* **114,** 303–316.
15. Williams, R. L., Hilton, D. J., Pease, S., Willson, T., Stewart, C., Gearing, D., Wagner, E., Metcalf, N., and Gough, N. M. (1988) Myeloid leukemia inhibitory factor maintains the developmental potential of embryonic stem cells. *Nature* **336,** 684–687.
16. Soriano, P., Friedrich, G., and Lawinger, P. (1991) Promoter interactions in retrovirus vectors introduced into fibroblasts and embryonic stem cells. *J. Virol.* **6,** 2314–2319.
17. Robertson, E. J. (1987) Embryo-derived stem cell lines, in *Teratocarcinomas and Embryonic Stem Cells: A Practical Approach* (Robertson, E. J., ed.), IRL, Oxford, pp. 71–112.
18. Pederson, R. A. (1994) Studies of *in vitro* differentiation with embryonic stem cells. *Reprod. Fertil. Dev.* **6,** 543–552.
19. Bain, G., Kitchens, D., Yao, M., Huettner, J. E., and Gottlieb, D. I. (1995) Embryonic stem cells express neuronal properties *in vitro*. *Dev. Biol.* **168,** 342–357.
20. Wiles, M. V. (1993) Embryonic stem cell differentiation in vitro, in *Guide to Techniques in Mouse Development*, Methods in Enzymology, vol. 225, Academic, New York, pp. 900–918.
21. Wobus, A. M., Rohwedel, J., Maltsev, V., and Hescheler, J. (1994) *In vitro* differentiation of embryonic stem cells into cardiomyocytes or skeletal muscle cells is specifically modulated by retinoic acid. *Roux's Arch. Dev. Biol.* **204,** 36–45.
22. Forrester, L. M., Nagy, A., Sam, M., Watt, A., Stevenson, L., Bernstein, A., Joyner, A., and Wurst, W. (1996) An induction gene trap screen in embryonic stem cells: identification of genes that respond to retinoic acid in vitro. *Proc. Natl. Acad. Sci. USA* **93,** 1677–1682.
23. Chen, Z. (1996) Simple modifications to increase specificity of the 5' RACE procedure. *Trends Genet.* **12,** 8788.

30

PCR-Based Cloning of Cortically Localized RNAs from *Xenopus* Oocytes

Jian Zhang and Mary Lou King

1. Introduction

RNA localization is an important cellular mechanism for restricting the activities of regulatory proteins to cytoplasmic domains in oocytes and embryos *(1)*. Cloning of these localized RNAs is obviously the first step in determining the functions of the encoded proteins. However, localized RNAs are rare and in low abundance. Their efficient isolation requires a combination of procedures to enrich for these RNAs.

In *Xenopus* oocytes, RNAs have been identified that are localized to both the animal and vegetal hemispheres *(2)*. The isolation procedures detailed below take advantage of two unusual characteristics of vegetally localized RNAs. First, in fully grown *Xenopus* oocytes, these RNAs are insoluble in high salt and detergent and can be recovered in the pellet after a low-speed spin *(3,4)*. A cDNA library constructed from RNA in such a sample was 30- to 50-fold enriched for the desired sequences. Second, vegetally localized RNAs are tightly associated with the cortex *(5)*. Employing a modified and limited reverse transcriptase (RT)–polymerase chain reaction (PCR) method, the RNA population from either animal or vegetal cortices was amplified in a representative fashion and used to screen the enriched cDNA library. The limited RT ensured that most of the cDNAs were 300–800 bp in length, extending from the 3' end of the poly(A)+ RNAs and, therefore, did not discriminate against long transcripts. Using this approach, seven novel localized RNAs were cloned after screening only 7000 phage plaques *(3,6,7)*. Four of these RNAs have been characterized to date. Three are unique to the germline and the fourth is critical to primary germ layer identity *(3,6,7)*. The following procedures were developed for *Xenopus* oocytes, but may also be generally applicable for other cell types found in other species.

2. Materials

2.1. Buffers

1. Extraction buffer A: 1% Triton X-100 (Baker Chemical [Phillipsburg, NJ]), 0.5*M* KCl, 10 m*M* piperazine-N,N'-bis [2-ethanesulfonic acid] (PIPES) (pH 6.8), 5 m*M* magnesium acetate, 1 m*M* ethylene glycol-bis (β-aminoethyl ether)-*N,N,N',N'*-tetraacetic acid (EGTA) (pH 8.0), and 300 mM sucrose (molecular biology grade). Sucrose should be autoclaved separately.

From: *Methods in Molecular Biology, Vol. 136: Developmental Biology Protocols, Vol. II*
Edited by: R. S. Tuan and C. W. Lo © Humana Press Inc., Totowa, NJ

Add 200 μg/mL of heparin to extraction buffer A just before using to inhibit RNase activity. All chemicals should be RNase-free quality.

2. Extraction buffer B: Extraction buffer A without Triton X-100, sucrose, and heparin.
3. Backextraction buffer: 10 m*M* Tris-HCl (pH 7.5), 5 m*M* ethylenediaminetetraacetic acid (EDTA), 0.2-1% sodium dodecyl sulfate (SDS) (molecular biology grade), and 50 m*M* NaCl.
4. Proteinase K buffer (2X) for detergent-soluble fraction (DSF): 100 m*M* Tris-HCl (pH 7.5), 10 m*M* EDTA, 2% SDS (could go as high as 4%), 2*M* urea (could go as high as 8*M*). Urea is added after autoclaving. Add 200 μ*M* Proteinase K (Boehringer Mannheim, Mannheim, Germany) just prior to using.
5. Proteinase K buffer (1X) for detergent-insoluble fraction (DIF): 50 m*M* Tris-HCl (pH 7.5), 5 m*M* EDTA, 1% SDS, 4*M* Urea.
6. P10EM (Cortex isolation buffer): 100 m*M* PIPES (pH 6.9), 10 m*M* EGTA, 1 m*M* magnesium sulfate.
7. OR-2 solution for oocyte isolation: 82.5 m*M* NaCl, 2.5 m*M* KCl, 1 m*M* $CaCl_2$, 1 m*M* $MgCl_2$, 1 m*M* Na_2HPO_4, 5 m*M* *N*-[2-hydroxyethyl]piperazine-*N'*-[2-ethanesulfonic acid (HEPES) (pH 7.8).
8. Reverse-transcription buffer: 50 m*M* Tris-HCl (pH 8.3), 75 m*M* KCl, 3 m*M* $MgCl_2$, 10 m*M* dithiothreitol (DTT), 250 ng/mL primer–adaptor XC1: CTC GAG ACG CTG TCT AGA $(T)_{15}$, 2 units/μL RNasin (Promega, Madison WI), and 2.5 m*M* each of dATP, dGTP, dCTP, and dTTP (Boehringer Mannheim).
9. PCR buffer: 10 m*M* Tris-HCl (pH 9.0), 50 m*M* KCl, 2.5 m*M* $MgCl_2$, 100 mg/mL bovine serum albumin, 0.1% Triton X-100, and 0.5 m*M* of dATP, dGTP, dCTP, and dTTP (Boehringer Mannheim).

2.2. Other Materials and Preparations:

1. Bake the following items and put in the cold room: 9-in. pipets; Dounce homogenizers of proper size, high-speed glass centrifuge tubes (Corex, Corning, Glassworks, Corning, NY)15 mL and 30 mL, polypropylene centrifuge tubes (Falcon [Los Angeles, CA]; 15 mL and 50 mL). Have the following buffers on ice in the cold room: 10X volume extraction buffer A + heparin, PK buffer + Proteinase K (DIF use 1X PK buffer, DSF use 2X PK buffer), extraction buffer B in microcentrifuge tubes (Eppendorf).
2. Fine forceps (Fine Science Tools, [Foster City, CA] Catalog No. 11252-20), No. 10 scalpel blades, Millipore filter (Millipore, Bedford, MA) (pore size 45 mm), small coverslips, Pasteur pipets (pull a few out over a flame to make a small bore), and hair loops (loop of baby's hair held by wax into the tip of a Pasteur pipet), parafilm cut into square centimeters.
3. Phenol, phenol/chloroform/isoamyl alcohol (25:24:1), 3*M* sodium acetate, 10*M* ammonium acetate, diethylpyrocarbonate (DEPC)-treated water, dry ice.

3. Methods

3.1. Isolation of Detergent-Insoluble Fraction (DIF) RNA from Oocytes for cDNA Library Construction (4,8)

1. To obtain oocytes, anesthetize female frogs in 0.5 mg/mL 3-aminaobenzoic acid ethyl ester (Methanesulfonate salt) (Sigma, St. Louis, MO, Catalog No. A-5040) in water until frog is unresponsive (about 3–5 min). Rinse the frogs with water and surgically remove the ovarian fragments through a small cut low in the abdominal region. Keep the oocyte fragments in OR-2 solution at room temperature.
2. To isolate oocytes free of follicles, cut the ovarian fragments into 15–20 oocyte clumps in OR-2 with a pair of small scissors or forceps. Rinse the clumps with fresh OR-2 several times until the solution is free of any yolk or blood. Wash the clumps once with 0.1 *M*

phosphate buffer, pH 7.2. and incubate the oocytes with 0.2% collagenase (Boehringer Mannheim, Mannheim, Germany; Catalog No. 17018-011) in the phosphate buffer for about 30 min at room temperature with constant shaking and stirring (*see* **Note 1**). When most of the oocytes are free from follicles, discard the collagenase and wash the oocytes first with OR-2 plus 0.1% bovine serum albumin (BSA) and then OR-2 solution. Collect all stages of oocytes in 5-mL aliquots. Freeze the oocytes at –80°C if necessary.

3. In a cold room, add 10X volume of extraction buffer A to fresh or frozen oocytes. Break the oocytes by pipetting them up and down with a 9-in. pipet. After 20 min of continuous extraction (*see* **Note 2**), dounce the mixture a few times until no cell debris is observed. After 25 min, pour the extract into the Corex tubes, balance the tubes, and immediately centrifuge the sample at 16,000*g* for 15 min.
4. After centrifugation, the yellowish lipid layer at the top of the tube is discarded by pipeting and any residual lipid was removed with a Kimwipe (Kimberly-Clark, Roswell, GA). Pour the supernatant (DSF) into a Falcon tube prechilled on ice. Gently rinse the pellet (DIF) with 1 mL extraction buffer B. Add 0.5 mL to 1 mL PK buffer (1X) with 200 mg/mL proteinase K to the pellet, resuspend the pellet by pipeting and vortexing, and incubate in a 55°C water bath (*see* **Note 3**). Place 1 mL of the DSF in a tube containing 400 µg/mL Proteinase K in 2X PK buffer of equal volume, and place in the water bath. Vortex the tubes every 5–6 min.
5. Phenol/chloroform extraction at room temperature: Add an equal volume of phenol/chloroform to the DIF and DSF. Vortex and spin at 5000*g* for 15 min. Recover the aqueous phase and repeat the above extraction until the interface is clear of the white participate, indicating that protein has been well extracted into the phenol layer.
6. Add 1/10 volume of 3*M* sodium acetate (pH 5.2) and 2.5 volumes cold 100% ethanol (–20°C) and put at –80°C for at least 1 h or –20°C overnight. Spin 20 min at 10,000*g*. Wash the RNA pellets with 70% cold ethanol. Dry and dissolve the pellet in DEPC-treated water.
7. The DIF RNA should comprise 2% of the total oocyte RNA. About 400 ng of DIF RNA can be recovered from 5 mL of oocytes. Significantly more than 2% will compromise the degree of enrichment obtained for localized RNAs in the cDNA library. Test the quality of the isolated DIF RNA by Northern blot analysis of equivalent amounts of DSF and DIF RNA by hybridizing with [α-^{32}P]-labeled *Vg1*. The *Vg1* RNA should be 30- to 50-fold more concentrated in the DIF than in the DSF sample.
8. Construct a phage cDNA library from 5 µg of DIF RNA by standard laboratory procedures *(8)*.

3.2. PCR Amplification of Cortical cDNA (6)

1. Oocyte isolation *(9)*: Isolate ovarian fragments as described in **Subheading 3.1.** Hand-defolliculate stage VI oocytes from the ovarian fragments in OR-2 using a pair of watchmaker's forceps (**Note 1**).
2. Isolation of cortical RNAs from stage VI oocytes *(5)*: Place oocytes in P10EM before cortical isolation. Cut oocytes in half with a No. 10 scalpel blade. Cut on the vegetal side of the equator for vegetal cortex and on the animal side for animal cortex isolation. Transfer 10–20 halves with a Pasteur pipet onto a cut piece of Millipore filter, yolk side up, in P10EM. Gently open and flatten all the halves with a hair loop.
3. Place a cover-slip piece over the halves and press down with forceps until the pigment of the cortex is visible. Gently lift the cover slip and repeat the above. Pressing several times helps to disperse most of the yolk. Pick up the cover slip with one forcep and hold the filter in the solution with another forcep. Most of the flattened cortices should stick to the filter. Rinse the filter by transferring it through several Petri dishes containing fresh P10EM to help disperse most of the adhering yolky cytoplasm.

4. Gently blow away the rest of the yolky cytoplasm using a steady stream of P10EM from a small-bore Pasteur pipet. Aim the stream of fluid at the center instead of the edges of the cortical halves until the yolky "fuzz" on the cortices is gone. Transfer the filter to fresh P10EM.
5. Free the halves from the filter by gently blowing streams of P10EM at the edges of the cortices. Collect the halves in a small drop of P10EM on parafilm at room temperature. Transfer the parafilm onto dry ice to freeze the drop. Transfer the drop with a prechilled forceps into an Eppendorf tube on dry ice. Cortices can be stored at –80°C.
6. Homogenize the cortices in Eppendorf tubes with a fitted pestle in Proteinase K buffer (1X) containing 250 µg/mL PK. Use 100 µL PK buffer for 20 cortices. Incubate at 55°C for 30 min. Extract once with phenol/chloroform and once with chloroform. Precipitate with 1/10 volume 10 *M* ammonium acetate and 2.5 volumes cold ethanol. Add 20 µg glycogen (Boehringer Mannheim, Catalog No. 901 393) as carrier. Resuspend the RNA pellet in 40 µL of DEPC water (*see* **Note 2**).
7. cDNA synthesis: Denature cortical RNA (equivalent to three cortices) at 70°C for 10 min and quickly chill the tube on ice. Briefly spin the tube at 4°C. Add reverse-transcription buffer with 200 units of SuperScript™ II RNAse H$^-$ reverse transcriptase (Gibco-BRL, Gaithersburg, MD, Catalog No. 18064-014). Incubate at 37°C for 15–20 min, heat inactivate at 70°C for 10 min, and return to ice.
8. For 20 µL cDNA mixture, assemble the dA tailing reaction by adding 8 µL 5X tailing buffer, 2.4 µL $CoCl_2$, 1 µL 10 m*M* dATP, 7.6 µL ddH_2O, and 1 µL terminal transferase (Boehringer Mannheim, Catalog No. 220 582). Incubate at 37°C for 15 min, heat inactivate as above, and continue to PCR or store the tubes at –20°C.
9. For PCR amplification, take one-third of the poly(dA)-tailed cDNA and bring to 100 µL in PCR buffer with 10 units of *Taq* polymerase (Boehringer Mannheim, Catalog No. 18038-018) and 5 ng of the primer–adaptor XC-1. Perform the PCR as follows: 94°C (5 min), 42°C (3 min), and 72°C (6 min) followed by an additional 45 cycles: 94°C (1 min), 53°C (2 min), and 72°C (4 min plus 10 s extension for each cycle).
10. To test the amplification, run equal volumes of vegetal and animal cDNAs on an agarose gel for Southern blot analysis. Hybridize with [α-^{32}P]-labeled *Vg1* to ensure that the cortical cDNAs are amplified in a representative way and fall within the predicted size range of 300–800 bp (**Fig. 1**).
11. Label amplified vegetal and animal cortical cDNAs with [α-^{32}P]-dCTP by random priming. Perform differential screening of the DIF cDNA library. Pick clones that show a strong hybridization signal with the vegetal probe and a weak or no signal with the animal probe.

4. Notes

4.1. Isolation of Detergent Insoluble Fraction (DIF)

1. Different lots or batches of collagenase should be calibrated at each new purchase and the optimal concentration for oocyte defolliculation determined empirically.
2. If the beginning sample is frozen, begin timing the extraction when more than half of the oocytes have thawed. Keep the extraction time as close to 30 min as possible. Too short a time results in incomplete lysis of mitochondria and their recovery in the pellet. Too long a time and the RNA quality is compromised. For early-stage oocytes (I to III), homogenization can be done in 1.5-mL Eppendorf tubes with a fitted pestle in about 25 minutes.
3. After the 30-min extraction step, the remaining steps should be done as quickly as possible. PK buffer can be preincubated at 55°C for a few minutes before being added into the DIF or DSF. PK treatment should proceed for 2 h for stage IV to VI oocytes, and 1–1.5 h for stage I to III oocytes.
4. Construction of DIF cDNA library has been described elsewhere *(8)*.

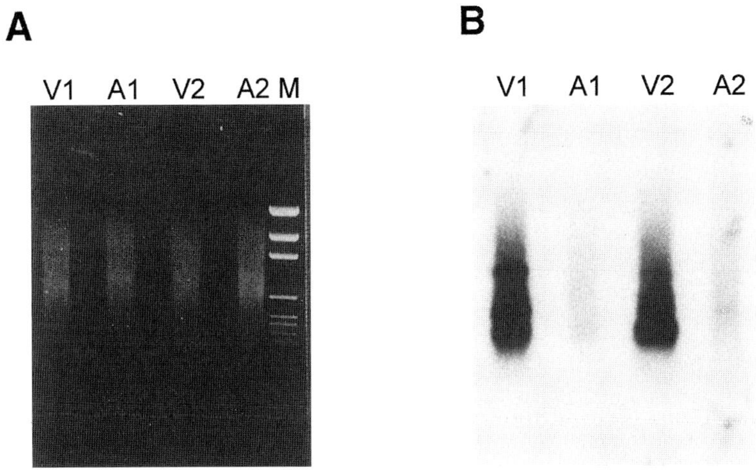

Fig. 1. Representative amplification of 3' ends of cortical cDNAs. **(A)** Amplified cDNAs were fractionated in a 1.2% agarose gel. Five percent (5 µL) of the PCR mixture was loaded onto each lane. Ethidium bromide staining showed equal DNA loading of lanes. A, cDNA from animal cortices; V, cDNA from vegetal cortices. 1 and 2 represent two different PCRs. **(B)** A Southern blot made from the gel shown in A and hybridized with a known localized RNA probe.

4.2. Isolation of Cortices and RT-PCR

1. Oocytes isolated by hand are significantly better for cortical isolation than those obtained by collagenase treatment. The pressure applied to flatten the oocyte halves has to be determined empirically with different oocyte batches.
2. A commercial RNA isolation kit such as Qiagen (Valencia, CA) RNeasy Mini Kit (Catalog No. 74103) can be used for cortical RNA isolation.
3. Any poly(T) primer–adaptor sequence can be used in the RT-PCR reactions as long as the sequence has a suitable T_m for PCR.
4. Some nucleotides are reported to actually inhibit reactions presumably due to contaminants. We use nucleotides from Boehringer Mannheim for RT and PCR.
5. In our lab, labeling the cortical PCR cDNAs by adding α-^{32}P-dCTP before the final five cycles of PCR generated a lot of unwanted products. Therefore, we prefer to label purified PCR DNAs by random-priming methods.

Acknowledgments

The authors are grateful to Dr. Richard Elinson for his intellectual contribution to the isolation of cortices. They thank members of their lab for their help during the development of the procedures described in this contribution. This work was supported by National Institute of Health grant GM 33932 to MLK.

References

1. St Johnston, D. (1995) The intracellular localization of messenger RNAs. *Cell* **81,** 161–170.
2. King, M. L., Zhon, Y., and Bubunenko, M. (1999) Polarizing Genetic Information in the egg: RNA localization in the frog oocyte. *BioEssay* **21,** 546–557.
3. Forristall, C., Pondel, M., Chen, L., and King M. L. (1995) Patterns of localization and cytoskeletal association of two vegetally localized RNAs, *Vg1* and *Xcat-2*. *Development* **121,** 201–208.

4. Pondel, M. D. and King, M. L. (1988) A localized maternal mRNA related to transforming growth factor-β mRNA is concentrated in a cytokeratin-enriched fraction from *Xenopus* oocytes. *Proc. Natl. Acad. Sci. USA* **85,** 7612–7616.
5. Elinson, E. P., King, M. L., and Forristall, C. (1993) Isolated vegetal cortex from *Xenopus* oocytes selectively retains localized mRNAs. *Dev. Biol.* **160,** 554–562.
6. Zhang, J., Houston, D. W., King, M. L., Payne, C., Wylie, C., and Heasman, J. (1998) The role of maternal *VegT* in establishing the primary germ layers in *Xenopus* embryos. *Cell* **94,** 515–524.
7. Houston, D. W., Zhang, J., Maines, J. Z., Wasserman, S. A., and King, M. L. (1998) A *Xenopus DAZ*-like gene encodes an RNA component of germ plasm and is a functional homologue of *Drosophila boule. Development* **125,** 171–180.
8. Mosquera, L., Forristall, C., Zhou, Y., and King M.L. (1993) A mRNA localized to the vegetal cortex of *Xenopus* oocytes encodes a protein with a nanos-like zinc finger domain. *Development* **117,** 377–386.
9. Smith, L. D., Xu. W., and Varnold, R. L. (1991) Oogenesis and oocyte isolation, Xenopus laevis: Methods in Cell Biology, vol. 36: *Practical Uses in Cell and Molecular Biology* (Kay, R. K. and Peng, H. B., eds.), San Diego, CA, Academic, pp. 45–60.

31

Analysis of mRNA Expression During Preimplantation Development

Keith E. Latham, Elena De La Casa, and Richard M. Schultz

1. Introduction

Preimplantation-stage mammalian embryos are very simple morphologically, being composed of mostly round cells that exhibit only a limited range of changes in cell shape, namely those associated with compaction and formation of the trophectoderm cell lineage. This morphological simplicity, however, belies a much greater complexity of events that occur at the molecular level. Throughout preimplantation development, many dramatic and essential gene regulatory events occur that result in fundamental alterations in gene expression. These include, for example, the translational recruitment and/or degradation of maternal mRNAs, transcriptional activation of the embryonic genome at the late one-cell and early two-cell stages, and a second period of gene induction beginning at the late eight-cell stage (*1–5*, and references therein). Inhibiting specific changes in gene expression can completely arrest preimplantation development, indicating that the expression of specific genes at certain times is essential for embryogenesis. Understanding how preimplantation development is controlled and how it is mediated, therefore, requires a detailed knowledge of normal gene expression patterns during this period.

Because preimplantation embryos are so small and because the amount of material available for analysis is accordingly very limited, the analysis of gene expression in preimplantation stage embryos requires the most sensitive methods that can be applied to a small number, or even single embryos. A variety of methods of this type exist, including immunocytochemical methods, which can be combined with confocal microscopy, highly sensitive methods of Western blot analysis using chemiluminescent methods of detection (e.g., **refs.** *6–8*), two-dimensional protein gel analysis of radiolabeled proteins (e.g., **refs.** *1* and *2*), and several different methods employing reverse transcription coupled to polymerase chain reaction (RT-PCR) for the detection of specific mRNAs.

In this chapter, we will describe three RT-PCR methods that have been used widely to perform semiquantitative or quantitative analyses of mRNA expression patterns. These three methods offer distinct advantages over simpler RT-PCR approaches, in

From: *Methods in Molecular Biology, Vol. 136: Developmental Biology Protocols, Vol. II*
Edited by: R. S. Tuan and C. W. Lo © Humana Press Inc., Totowa, NJ

which a specific target sequence is amplified, sometimes exhaustively, without any effort to quantify the target transcript. Such qualitative RT-PCR approaches may reveal the apparent presence of a given transcript, but provide no information concerning temporal changes in transcript abundance or the possible biological relevance to development that the level of expression of a given mRNA detected through such an approach might have. The three methods outlined here are all able to provide information regarding differences in transcript abundance between different stages of development or between different kinds of experimentally manipulated embryo. Two of the methods are also able to provide an indication of the actual transcript abundance within the embryo, which is often critical for assessing biological function (e.g., when attempting to determine the relative importance of one member of a multigene family to development). Each of the three methods has definite advantages in certain situations; likewise, limitations that must be taken into account when evaluating which of the methods would be best suited to a given experimental objective.

The first method that we will describe is a semiquantitative method developed in the laboratory of Dr. Richard Schultz, University of Pennsylvania. That method relies on the addition of a fixed amount of exogenous standard (rabbit globin mRNA) per embryo to each sample before RNA extraction ***(9,10)***. The added globin mRNA then serves as a control for RNA recovery and reverse transcription. Gene-specific PCR is then performed to detect a target cDNA of interest, and PCR to detect the globin cDNA is performed in parallel. The amplification conditions are optimized for each target sequence to ensure that maximum sensitivity is attained, but the number of PCR cycles is limited so that the amount of specific amplicon produced remains linear. After quantifying the amount of target amplicon produced for the specific cDNA of interest and for the globin cDNA, target cDNA values are normalized to correct for differences in the amount of globin amplicon produced in each reaction. The normalized values can then be compared. The advantages of this method are that it is very sensitive and can provide data regarding relative levels of expression between different samples. The disadvantages are that the assay must be optimized for each new target cDNA, only a relatively small number (approx 10–20) of transcripts can be assayed, the number of embryos required for the assay is comparatively large (25–100), no estimates of actual message abundance are obtained, and the possible effects of amplification of nonspecific PCR products, if any, on the amount of specific amplicon produced remain poorly defined.

The second RT-PCR method that we will describe is the Quantitative Amplification and Dot Blot (QADB) method, developed in the laboratory of Dr. Keith Latham, Temple University ***(11)***. This method begins with the quantitative amplification of the 3' termini of an entire mRNA population without altering the relative sequence abundance within the amplified cDNA. The methods for this were first developed by Brady and Iscove for the production of cDNA libraries from single cells ***(12)***. This RT-PCR method is then coupled to a method for quantitative blot hybridization to quantify expression ***(11)***. Aliquots of the amplified cDNA from a few or from many samples are applied to membranes using a dot-blotting procedure. Differences in the amount of DNA applied to each dot are measured and can be accounted for during the quantitation. After blotting, filters are hybridized with specific cDNA probes and the amount of hybridization quantified. Subsequently, relative levels of expression of an mRNA can be determined between samples. Moreover, it is possible to estimate the abundance of each mRNA within the embryo, providing the opportunity to compare abundances between mRNAs. Because the RT step is performed without prior extraction of RNA, the possibility of RNA loss is greatly diminished. Few embryos are required with this method, and, in fact, a single embryo can suffice for the analysis ***(13)***. An essentially unlimited number of transcripts can be assayed (multiple rounds of amplification are possible, many blots can be made, and each blot can be hybridized multiple times), and each blot can contain many samples including embryos subjected to experimental treatments so that a vast amount of information can be obtained. A large number of genes can be assayed on a common set of samples, and once the blots are available and the appropriate cDNA probes are in hand, data for additional transcripts can be obtained quickly. Because the procedure amplifies only the 3' termini of the mRNAs, specificity of hybridization may be enhanced by virtue of the divergent nature of 3' untranslated sequences. The limitations of this method are that, although it is very sensitive (able to detect as little as approx 500 copies per embryo) ***(14)***, it is less sensitive than procedures involving amplification of specific target sequences and some highly rare transcripts are not detected. The hybridization probe must encompass the 3' terminus of the message sequence, and the sequence composition of the 3' terminal portion of the mRNA and any conservation of sequence among gene family members must be taken into account. Hybridization conditions must be carefully controlled, and homologous cDNA probes derived from different species may not be satisfactory because of divergence in the 3' UTR. An accurate determination of probe-specific activity is required for estimating transcript abundance.

The third method that we will describe is the SNuPE (Single Nucleotide Primer Extension) method, which has been developed and used extensively in embryological studies by Dr. Judith Singer-Sam and Dr. Jeff Mann of the Beckman Research Institute of the City of Hope (e.g., **refs. *15–21***). With this method, a target sequence is first amplified and then an additional polymerization reaction is performed with a single radiolabeled nucleotide present in the reaction, that nucleotide being the one that would be incorporated through a one-base extension of a primer annealed to the target sequence. This assay can be used in either of two ways. First, it can be used to compare the relative levels of expression of transcripts derived from two parental alleles that exhibit a sequence polymorphism. The target sequence is amplified and then equal aliquots of the amplified target amplicon are subjected to primer extension using the nucleotide specific for one of the two parental alleles. Quantitation of the amount of nucleotide incorporated for the two parental alleles provides a measure of the relative levels of expression of the two parental alleles. This method has been very valuable for examining expression of imprinted genes ***(20,21)***. The other application of the SNuPE assay is to quantify mRNA abundances. Amplification is performed in the presence of a small amount of exogenously added cDNA identical to the target cDNA except for a single base difference (or a defined amount of exogenous mRNA can be added before RT). The endogenous transcript is then quantified relative to the exogenously added cDNA or mRNA. With appropriate care in constructing a standard curve, it is possible then to calculate mRNA abundance. The advantages of this method are that it is very sensitive and has even been applied to single blastomeres ***(15,20,21)***, and it can provide data about the relative levels of expression of two parental alleles. The disadvantages are that it is somewhat laborious and not well suited to the analysis of a large number of genes and it requires the availability of a sequence polymorphism.

2. Materials

2.1. General

We recommend obtaining ultrapure chemicals for the preparation of all solutions to be used in the lysis of embryos and handling or RNA. Most of our chemicals are purchased from Fluka (Chemie, Buchs, Switzerland). We typically prepare concentrated stock solutions of each component and then combine these to create the working solutions. A reasonable precaution for avoiding degradation of mRNA is to test the working solutions and appropriate dilutions of the stock solutions for RNAse activity. A 7.5-kb tester mRNA can be obtained from Gibco-BRL-Life Technologies (Gaithersburg, MD) (Catalog No. 15621-014) and each solution is tested by combining 0.5 μg of the 7.5-kb tester mRNA with 3.5 μL the test solution for 30 min 37°C. After treatment, the samples are analyzed on RNA gels and the mRNA visualized by ethidium bromide staining to determine whether the mRNA has remained intact. Both untreated tester mRNA and mRNA incubated in water only should be included on the gel. Some reagents interfere with the assay, so that some stock solutions cannot be tested and some reagents (e.g., NP-40) must be omitted from the working solution for the test. We avoid the use of diethylpyrocarbonate for preparing RNAse-free solutions by obtaining water from a Milli-Q water system (Millipore, Bedford, MA). We have used water from both a basic Milli-Q and from a Milli-Q Plus UF system. Both provide water that is effectively free of both DNAse and RNAse activity, provided that the water is filter-sterilized before use. No glassware or glass pipets are used to prepare solutions. Tissue culture plastic and plastic pipets or sterile pipetter tips are used throughout. For adjusting the pH of Tris buffers, the amount of acid needed to bring the pH of a solution to the desired value can be determined empirically. Afterward, that amount of acid is added and a small aliquot of the final buffer is tested for pH. This avoids introducing potentially contaminated pH electrodes into the solution to be used for RNA handling. For RT and PCR reactions, autoclaved tubes and tips are used. To reduce the risk of unwanted DNA contamination, tubes can be exposed to ultraviolet light before use. (Other precautions would include assembling PCR reactions in a location where plasmids or PCR reactions containing the target sequence have not been handled. Likewise, a separate set of pipetters should be used for assembling RT and PCR reactions and for handling the finished PCR reactions.)

2.2. Specific Materials Needed

2.2.1. Semiquantitative RT-PCR Analysis

1. Stock solutions
   1 *M* Tris-HCl, pH 7.4
   1 *M* Tris-HCl, pH 8.3
   1 *M* Tris-HCl, pH 7.9
   1 *M* NaCl
   2 *M* KCl
   1 *M* $MgCl_2$
   (All stock solutions are filter-sterilized before using to make solutions below.)

2. Other solutions
   Guanidine thiocyanate lysis solution (4 *M* guanidine thiocyanate, 0.1 *M* Tris-HCl, pH 7.4, 1 *M* 2-mercaptoethanol, 0.2 µg/µL *Escherichia coli* rRNA [Boehringer Mannheim Biochemicals, Indianapolis, IN], stored in 100-µL aliquots at –70°C)
   Resuspension solution (40 m*M* Tris-HCl, pH 7.9, 10 m*M* NaCl, 6 m*M* $MgCl_2$)
   10X RT buffer (0.25 *M* Tris-HCl, pH 8.3; 0.75*M* KCl, 30 m*M* $MgCl_2$)
   10X PCR buffer (0.1 *M* Tris-HCl, pH 8.3, 0.5 *M* KCl, 14.5 m*M* $MgCl_2$)
   Deoxyribonucleotides (dATP, dGTP, dCTP, dTTP) (100 m*M*, Boehringer Mannheim)
   Mixed deoxyribonucleotide (20 m*M* each)
   100 m*M* Dithiothreitol (DTT) (Gibco-BRL)
   1 *M* Acetic acid
   2 *M* Potassium acetate
   3 *M* Potassium acetate
   RNAse A (20 µg/mL) (Worthington, Freehold, NJ)
   100% Ethanol
   75% Ethanol
3. Other Reagents
   Rabbit globin mRNA (Gibco-BRL)
   RNAsin (Promega, Madison, WI)
   RQ1 DNAse (Promega)
   Phenol (Amresco, Solon, OH)
   oligo(dT_{15}) (Promega) or pd($T_{25–30}$) (Pharmacia Biotech, Piscataway, NJ)
   Gene-specific PCR primers
   Globin PCR primers (α-globin: 5'-GCAGCCACGGTGGCGAGTAT-3' and 5'-GTGGGA CAGGAGCTTGAAAT-3')
   α-[^{32}P]dCTP (3000 Ci/mmol)
   Taq polymerase

2.2.2. Quantitative Amplification and Dot-Blot Analysis

1. Stock solutions
   2 *M* KCl
   1 *M* Tris-HCl, pH 8.3
   1 *M* $MgCl_2$
   10% NP-40
   NP-40 is no longer commercially available, but an equivalent compound (Igepal CA-630) is available from Sigma Chemicals, St. Louis, MO. (All stock solutions are filter sterilized before using to make the following solutions.)
2. Other solutions
   10X RT buffer (0.75 *M* KCl, 0.5 *M* Tris-HCl, pH 8.3, 30 m*M* $MgCl_2$)
   10X PCR buffer (0.1 *M* Tris-HCl, pH 8.3, 0.5 *M* KCl, 25 m*M* $MgCl_2$)
   Embryo lysis solution (1.4 mL of 10X RT buffer, 700 µL 10% NP-40, 11.9 mL water; aliquot and freeze –70°C)
   Deoxyribonucleotides (dATP, dGTP, dCTP, dTTP) (100 m*M*, Boehringer Mannheim)
   Mixed deoxyribonucleotides (25 m*M* each)
   4 m*M* dATP
   5X Terminal transferase buffer (Gibco-BRL)
   1% Triton X-100
3. Other reagents and materials
   pd($T_{25–30}$) (40 µg/mL) (Pharmacia)
   Terminal transferase (Boehringer Mannheim)

Taq polymerase
XT prime (CATGTCGTCCAGGCCGCTCTGGGACAAAATATGAATTCT$_{23}$) 3.3 µg/µL
Nuclease-free bovine serum albumin (BSA, 10 mg/mL) (Promega)
Superscript Reverse Transcriptase (Gibco-BRL)
AMV Reverse Transcriptase (Boehringer Mannheim)
RNAguard (Pharmacia)
Prime RNAse Inhibitor (5'–3', Inc., Boulder, CO)
^{14}C Calibration chips (two chips, 1.11 Ci/g and 0.13 Ci/g attached to plastic strip, available from American Radiolabeled Chemicals, St. Louis, MO)

2.2.3. Single Nucleotide Primer Extension (SNuPE) Analysis

1. Stock solutions
   1*M* Tris-HCl, pH 8.3
   2 *M* KCl
   1 *M* $MgCl_2$
   (All stock solutions are filter-sterilized before using to make solutions below.)
2. Other solutions
   TE (10 m*M* Tris-HCl, 1 m*M* EDTA, pH 8.0)
   5X RT buffer (BRL) (250 m*M* Tris-HCl, pH 8.3, 375 m*M* KCl, 15 m*M* $MgCl_2$)
   10X PCR buffer (Boehringer Mannheim) (100 m*M* Tris-HCl, pH 8.3, 500 m*M* KCl, 15 m*M* $MgCl_2$)
   0.1 *M* DTT
   Deoxyribonucleotides (dATP, dCTP, dGTP, dTTP) (100 m*M*, Boehringer Mannheim)
   10 m*M* dNTPs mix
   15 m*M* $MgCl_2$
   Phenol
   Chloroform
   Ethanol
   75% Ethanol
   Isopropanol
   15% Polyacrylamide–7 *M* urea gel
3. Other reagents
   Ultraspec RNA Isolation System (Biotecx Laboratories, Houston, TX)
   Upstream and downstream primers (20 µ*M*) to amplify a cDNA that includes the allelic difference
   SNuPE primer (20 µ*M*) with the 3' end just 5' to the position of the mismatch
   RNasin, 40 U/µL (Promega)
   SuperScript II Reverse Transcriptase, 200 U/µL (Gibco-BRL)
   Taq polymerase
   α-[^{32}P] dNTPs, 3000 Ci/mmol, corresponding to the mismatched bases and diluted to 2 µCi/µL
   Carrier: mussel glycogen or poly-C (Boehringer Mannheim)

3. Methods

3.1. Semiquantitative RT-PCR Analysis

3.1.1. RNA Extraction

A summary of this method is given in **Fig. 1**. Aliquots (100 µL) of guanidine lysis solution (GLS) are thawed and 0.0125 pg of rabbit globin mRNA are added for each

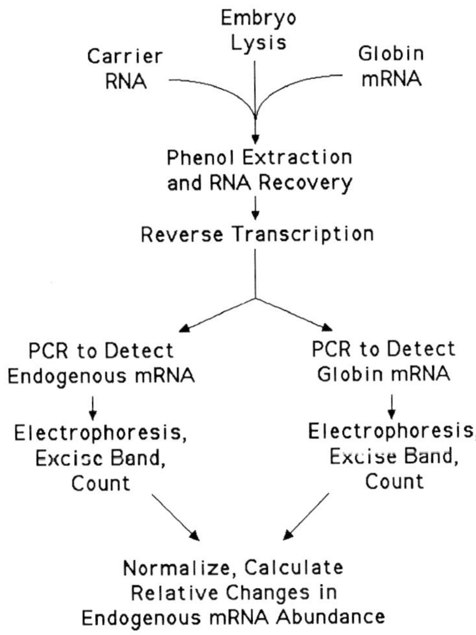

Fig. 1. Diagram summarizing the semiquantitative RT-PCR method.

embryo to by lysed. Embryos are then added in a minimal volume (< 5 µL) and the solution is immediately vortexed vigorously. The RNA is precipitated by the addition of 8 µL of 1 *M* acetic acid and 60 µL of 100% ethanol and incubation at –20°C overnight. After centrifugation at 10,000*g* for 15 min at 4°C, the pellet is washed once with 75% ethanol. The pellet is air-dried briefly, but complete dryness is avoided, as overdrying can render the pellet insoluble. The pellet is resuspended in 20 µL resuspension buffer (RB) and treated with 1 µL (1 U) RQ1 DNAse for 30 min at 37°C. An additional 20 µL of RB is added and the RNA is purified by extraction with 40 µL of RB-saturated phenol. The phases are separated by centrifugation at 10,000*g* and the aqueous phase is transferred to a new 0.5-mL tube. The RNA is precipitated by addition of 5 µL of 3 *M* potassium acetate and 135 µL of 100% ethanol and then incubation at –20°C overnight. After recovery by centrifugation, the RNA pellet is incubated in 75% ethanol for 10 min at room temperature. Centrifugation is performed for 15 min and the pellet is allowed to air-dry briefly, as above. The pellet is dissolved in 5.5 µL of water containing RNAsin (2 U/µL) (*see* **Note 4.1.1**). RNA recovery, which is based on the recovery of the carrier rRNA, is estimated by measuring the A_{260} using 0.5 µL of the sample.

3.1.2. Reverse Transcription

RNA from 50 egg or embryo equivalents is reverse transcribed in 20-µL reactions. Each reaction is assembled by combining 2 µL of 10X RT buffer, 1 µL of the 20 m*M* mixed dNTPs, 2 µL of 100 m*M* DTT, 1 U/µL of RNAsin, 0.4 µL of oligo dT, and water to bring the volume to 19 µL in thin-walled tubes. After an incubation at 37°C for 2 min, 1 µL (200 U) of Superscript is added and reverse transcription performed for 1 h at 42°C. After reverse transcription, the sample is incubated for 5 min at 99°C and then used immediately or stored at –20°C.

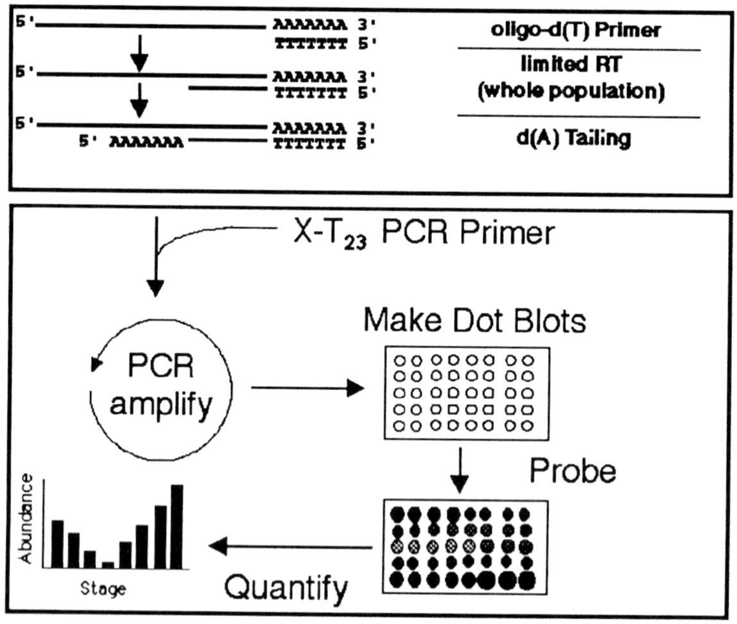

Fig. 2. Diagram summarizing the QADB method.

3.1.3. Polymerase Chain Reaction

Polymerase chain reaction is performed in 100-μL reactions containing 10 μL 10X PCR buffer, 1 μL of mixed 20 m*M* dNTPs, 5 μCi of α-[^{32}P]dCTP, 2.5 U of *Taq* polymerase, 5–10 embryo equivalents of the RT reaction (determined empirically), and 20 pmol of each PCR primer. Amplification is conducted with an initial incubation at 95°C for 1 min, followed by cycles of 95°C for 10 s and 60°C for 15 s, when a Cetus–Perkin Elmer 9600 thermocycler is used (Norwalk, CT). Actual cycle conditions may require modification (e.g., an extended elongation phase) when other instruments are used. The last cycle is concluded with a 6-min incubation at 60°C (*see* **Notes 4.1.2** and **4.1.3**).

3.1.4. Quantitation of PCR Results

The PCR reaction is chilled and centrifuged briefly and then 25 μL are removed and treated with 2 μL of RNAse A solution for 10 min at room temperature. After digestion, 5 μL of 6X gel loading buffer (0.25% Bromophenol Blue and 40% sucrose) are added and the sample is electrophoresed on a 4% agarose gel. Gels are photographed under ultraviolet light, the product bands are excised, and incorporated radiolabel is measured by Cerenkov counting (*see* **Note 4.1.4**).

3.2. Quantitative Amplification and Dot-Blot Analysis

3.2.1. Reverse Transcription

A summary of QADB method is given in **Fig. 2**. Before beginning, the embryo lysis solution (ELS) is augmented with essential reaction components (*see* **Note 4.2.1**). To 96 μL of the above ELS solution are added 1 μL of Prime RNAse Inhibitor, 1 μL of

RNAguard, 1 µL of d(T_{25-30}) (40 µg/mL), and 1 µL of freshly diluted 0.25 m*M* mixed dNTPs (prepared by serial 10-fold dilution of the 25 m*M* stock described earlier). For each sample to be obtained, 7.5 µL of this completed RT mix is placed into a 1.5-mL tube. Embryos or cells are transferred to this lysis buffer in a minimal volume (in just a fraction of 1 µL using a mouth pipet) and all tubes are kept on ice until all the samples are collected. For embryos, zonae pellucidae are removed using acidified Tyrode's solution followed by a wash through bicarbonate-buffered embryo culture medium (usually performed with agarose-coated dishes to prevent embryos from adhering to the dish). After all samples are collected (preferably no more than 20 at a single time), the samples are centrifuged at 10,000*g*, 4°C for 2 min, heated to 65°C for 1 min, incubated at room temperature for 3 min, and then placed on ice for at least 1 min. After a brief centrifugation to recover condensed water vapor, 1 µL of a mixture of 100 U Gibco-BRL Superscript RT and 2 U AMV RT enzymes, diluted in the above RT mix (obtained by combining 2.5 µL of RT mix with 2 µL of Superscript RT plus 0.5 µL avian RT) is added to each sample. Each sample is returned to ice until all reactions are assembled. The reactions are incubated at 37°C for 20 min, heated to 65°C for 10 min (using tube locks to keep tubes closed), chilled on ice for 5 min, and centrifuged for 2 min to collect condensed wader vapor Terminal transferase mix is prepared by combining the following: 4 µL 5X TdT buffer (Gibco-BRL), 1 µL of 4 m*M* dATP, 0.5 µL of TdT enzyme (Boehringer Mannheim), and 4.5 µL of water. To each RT reaction, 8.5 µL of this TdT mix is added. The reaction is then incubated at 37°C for 15 min, heated at 65°C for 10 min, chilled on ice, and the water vapor again recovered by centrifugation. The RT reactions are then used immediately or stored at –70°C (*see* **Notes 4.2.2** and **4.2.3**).

3.2.2. Polymerase Chain Reaction

When samples are to be compared to one another, it is recommended that PCR amplification be performed on all samples at the same time and using a common master PCR mix. This provides for the most accurate normalization between samples later. A PCR mix is prepared by combining for each reaction the following: 5 µL 10X PCR buffer, 2.5 µL of 1% Triton X-100, 1 µL of XT primer, 0.5 µL of nuclease-free bovine serum albumin (10 mg/mL), 2 µL of 25 m*M* mixed dNTPs, 1 µL (5 U) Taq polymerase, 34 µL water, plus 0.4 nCi of α-[^{32}P]dCTP. To 46 µL of PCR mix are added 4 µL of the RT reaction. Mix, cover with oil, and amplify for 25 cycles using 94°C for 1 min, 42°C for 2 min, and 72°C for 6 min, lengthening the elongation phase by 10 s each cycle. Another 1 µL of *Taq* is added to each reaction, and amplification is performed through another 25 cycles. After amplification, 2.5 µL of each reaction are inspected on a 2% agarose gel in order to evaluate the success of the reaction (*see* **Notes 4.2.4** and **4.2.5**).

3.2.3. Blot Preparation

After amplification, 2.5 µL of each PCR reaction are diluted with 300 µL of 0.3*N* NaOH and heated to 65°C for 30 min. After cooling to room temperature, 125 µL of 20X standard sodium citrate (SSC) buffer (3.0 *M* NaCl, 0.3 *M* sodium citrate, pH 7.0) are added and the solution is vortexed and centrifuged briefly. Samples are applied to Nytran (Schleicher and Schuell, Keene, NH) membrane using a Schleicher and Schuell dot blotter. A Nytran membrane is moistened with 10X SSC and placed on top of Whatman 3MM paper (Whatman, Clifton, NJ) moistened with 10X SSC. After assembling the apparatus, each well is washed with 0.5 mL 10X SSC and then the samples are applied. The Nytran filter is then removed and rinsed with 2X SSC, crosslinked with ultraviolet light, and baked at 80°C for 2 h in a vacuum oven. Each blot is then imaged using an overnight exposure and a Fuji BAS 2000 phosphorimager (Fuji, Medical Systems, Stamford, CT) or comparable instrument in order to visualize and quantitate the DNA bound to each dot. The duration of the exposure is accurately recorded. The bound cDNA is then permitted to undergo radioactive decay, typically for several weeks, to minimize the amount of residual signal that will contribute to signals following hybridization. Hybridized blots can be stored for decay to permit additional hybridizations to each blot, with background imaging performed just prior to the next hybridization reaction (*see* **Note 4.2.6**).

3.2.4. cDNA Probe Labeling and Hybridization

cDNA probes should encompass sequences complementary to the 3' terminal portions of mRNAs of interest, as this is the portion of the mRNAs that is reverse transcribed and amplified. If mRNA abundances are to be estimated, it is critical that accurate estimates of probe-specific activities be available. This requires that the concentration of probe DNA being labeled be known with reasonable accuracy. Many techniques exist for isolating specific DNA fragments. Whatever method is employed, it is important that the isolated DNA be free of contaminating substances that might affect absorbance readings, as judged by the $A_{260}:A_{280}$ ratio. We typically gel-purify fragments using ion-exchange paper, such as Schleicher and Schuell NA45 paper. Radiolabeling with α-[^{32}P]dCTP is performed by the random priming method *(22)*, using the Prime-It II kit from Stratagene (La Jolla, CA). We typically label 25 ng of purified probe DNA. After labeling, the amount of incorporated later is ascertained by precipitation with trichloroacetic acid precipitation. Briefly, after labeling, the reaction is diluted to 200 µL with TE buffer (10 m*M* Tris-HCl, pH 7.5, 1 m*M* EDTA). From this, 2 µL are spotted onto glass fiber filters (GF/C Whatman 1822-O24) and allowed to dry at room temperature for about 15 min. The circles are then agitated in 10 mL of a solution containing 10% trichloroacetic acid and 1% sodium pyrophosphate for 10 min at 4°C The buffer is removed and the filter is washed with 10 mL of 95% ethanol for 10 min at 4°C The filter is then dried and acid-precipitable radioactivity quantified by scintillation counting.

Blots are hybridized with 1×10^6 cpm/mL of radiolabeled probe overnight at 65°C in Church hybridization buffer (0.5 sodium phosphate, pH 7.2, 75% sodium dodecyl sulfate [SDS], 1 m*M* EDTA). The blot is then washed twice at 65°C in 40 m*M* sodium phosphate (pH 7.2, 1% SDS) and then at high stringency, typically 65°C in 0.1X SSC and 0.5% SDS. Hybridization and washing conditions may vary with the exact sequence composition of a given probe. The blots are then sealed in plastic and imaged using the Fuji BAS2000 phosphorimager. A ^{14}C calibration chip is included in the exposure with the blot. The duration of exposure is accurately recorded. Care must be taken not to saturate the screen during exposure in order to maintain quantitative validity of the data.

3.2.5. Quantitation

After phosphorimaging of the background and hybridized probe signals, a number of calculations must be performed on the data. Because a number of blots are produced with the method and each can be used repeatedly, we recommend the use of a spreadsheet-type computer program (e.g., Microsoft Excel) in performing these calculations, so that a common template can be used for each probe. The first calculation that is performed is to subtract background signal from the probe hybridization signal for each sample. To do this, both signals are expressed on a per-hour basis and background signal is adjusted for the amount of radioactive decay that occurred prior to hybridization. The probe hybridization signals, now corrected for background, are then converted from phosphorimager units to units of cpm. This is accomplished using data from the ^{14}C calibration chip. Each calibration chip must have previously been exposed alongside dot blots containing a dilution series of known quantities ^{32}P-labeled DNA. In this way, an equivalence can be made between the number of phosphorimager units for the chip and the number of phosphorimager units for a known amount of labeled DNA bound to the dot blot. By exposing the same chip with the blot, one can then convert the individual dot signals to cpm values. Alternatively, pieces of paper with small amounts of ^{32}P can be exposed with the blots, and then subjected to scintillation counting after the exposure. After converting to units of cpm, probe hybridization signals are normalized to adjust for differences in the amount of DNA bound to each dot (determined in the background imaging above). Assuming that no nonuniform loss of DNA occurs from the blots, the probe values for any given blot are always normalized according to the original values for bound DNA obtained for that blot even upon rehybridization.

The above calculations provide quantitative values for expression in units of bound cpm. This provides useful information with regard to changes in mRNA abundances among a group of samples. Because the assay involves hybridization to an entire quantitatively amplified cDNA population, the data, when expressed this way, reflect the relative abundance of a given transcript as a fraction of the total mRNA population. This should be similar to the type of data produced, for example, through Northern blot analysis of poly(A)+ RNA. This presentation of the data, however, does not account for changes in the total amount of poly(A)+ RNA during development, which can be quite substantial during the preimplantation period, nor does it permit comparison of relative abundances between two different mRNAs. For the preimplantation mouse embryo, these needs have been met by calibrating each set of blots using an endogenous mRNA (actin) for which the copy number per embryo is known at a specific stage, and then by taking into account the differences in poly(A)$^+$ RNA between different stages ***(11)***. When assaying the normal temporal patterns of expression of mRNAs during preimplantation development, we typically collect samples from oocytes, eggs, and fertilized embryos at a variety of times postinsemination or postinjection with human chorionic gonadotropin (hCG) for superovulated embryos. The actin mRNA content is estimated at 2.3×10^5 at the early blastocyst stage (97–98 h post-hCG injection for superovulated mice). The total number of poly(A)$^+$ templates is estimated for any given time by plotting the following values: 3.2×10^7 templates per germinal vesicle (GV)-stage oocyte, 1.7×10^7 per egg or newly fertilized (12 h) embryo; 2.4×10^7 per one-cell per (24 h) embryo; 7×10^6 per two-cell (49 h) embryo; 1.3×10^7 per eight-cell (77 h) embryo; 3.4×10^7 per early blastocyst (97 h); and 6.0×10^7 templates per late blastocyst (120 h) ***(11)***. The first in a series of blots is hybridized with the actin probe. This provides a conversion factor *(C)* defined 2.3×10^5 divided by the bound

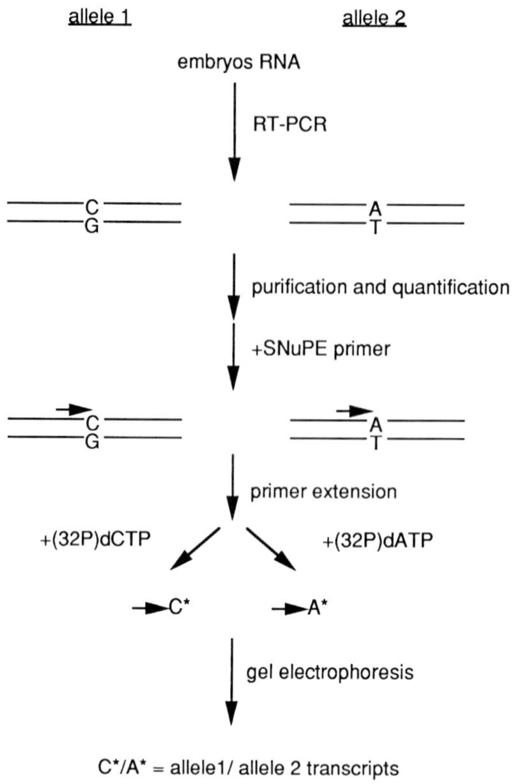

Fig. 3. Diagram summarizing the SNuPE method.

cpm value at approximately 97 h post-hCG. This conversion factor is applied to all of the blots in the series. Each series of blots must be calibrated in this way. Although it may not be necessary, we also recalibrate each series of blots for additional rounds of hybridization. The estimated copy number for a given mRNA is then calculated by the formula: copy number = C $\times$ bound cpm $\times$ (specific activity of actin probe/specific activity of test probe) $\times$ ($N/3.4 \times 10^7$), where N is the number of poly(A)$^+$ templates estimated for a given time. These calculations have provided estimates of mRNA abundances for a number of mRNAs for which estimates had been obtained by Northern blotting (**ref. *11***; Latham, unpublished results). It should be noted that because random primer labeling produces labeled probe molecules of comparatively short length, there does not appear to be any need to account for differences in probe length. It should also be noted that these calculations only provide estimates of copy number, which are intended primarily to aid in the comparison of the abundances of different mRNAs and to obtain an idea of the magnitude of changes in copy number on a per embryo basis rather than as a simple fraction of the total mRNA population present.

3.3. SNuPE Analysis

3.3.1. RNA Isolation

A summary of the SNuPE method is given in **Fig. 3**. Embryos are lysed in 100 µL of a guanidium thiocyanate–phenol solution (e.g., Ultraspec RNA reagent) and the lysate vortexed immediately. After 5 min on ice, 20 µL of chloroform are added and the samples are vortexed again for 30 s and stored 5 min on ice. The homogenate is centrifuged at 12,000*g* (4°C) for 15 min. The aqueous phase is transferred to a new tube and the RNA is precipitated by the addition of an equal volume of isopropanol and carrier (20 µg glycogen or 30 µg poly-C) and stored on ice for 30 min. The RNA is recovered by centrifuging at 12,000*g* (4°C) for 15 min. The RNA pellet is washed twice with 75% ethanol, dried briefly, and resuspended in RNase-free water (*see* **Notes 4.3.1** and **4.3.2**).

3.3.2. Reverse Transcription

Half of each RNA sample is reverse transcribed in a final volume of 20 µL. The RNA and 20 pmol of the downstream primer are combined (total volume 11.5 µL) and heated at 70°C for 10 min, chilled on ice and briefly centrifugated before the addition of 4 µL of 5X RT buffer, 2 µL of 0.1 m*M* DTT, 1 µL of 10 m*M* dNTP mix, and 20 U RNasin (0.5 µL). After incubation at 42°C for 2 min, 200 U of Superscript II are added and mixed by gently pipetting. The reactions are incubated for 50 min at 42°C.

3.3.3. Polymerase Chain Reaction

The RT reactions are inactivated at 99°C for 5 min and equilibrated at 50°C before the addition of 80 µL of PCR mix (8 µL of 10X PCR buffer, 2 µL of 10 m*M* dNTPs mix, 20 pmol of upstream and downstream primers, and *Taq* polymerase). The PCR reaction is performed in a thermal cycler with an initial step of denaturation (95°C, 2 min) and approximately 40 cycles of denaturation–annealing–extension (enough cycles to give a good band on an ethidium bromide-stained gel) (*see* **Note 4.3., item 3**). The PCR products are gel-purified, phenol–chloroform extracted, precipitated with ethanol, and resuspended in TE. The concentration of DNA in the PCR reactions is estimated on an ethidium bromide-stained agarose gel by comparison of band intensity to a known quantity of DNA standard (*see* **Notes 4.3.3** and **4.3.4**).

3.3.4. Quantitative SNuPE Assay

To compare the expression of two alleles to each other, two SNuPE assays are performed for each RNA sample, each one with a different α-[^{32}P]dNTP, in order to compare the signal of the radionucleotide incorporated to each allele product. As controls for background and maximum incorporation, incubate SNuPE reactions with 10–20 ng of the amplified product of each allele. After substraction of background (<1%), the ratio of the two radioactive signals is proportional to the relative amount of each allelic transcript ***(18,23)***. Amplified cDNA (10–20 ng) is mixed with 2 µL of 10X PCR buffer and 2–5 m*M* (final) of $MgCl_2$, 20 pmol of SNuPE primer (heated 90°C, 1 min before use) and *Taq* polymerase to a final volume of 18 µL. Two aliquots of 9 µL are separated and 2 µCi of the α-[^{32}P]dNTP specific of each allele are added (1 µL of dilution 2 µCi/µL). The samples are incubated for one cycle of denaturation, annealing, and synthesis (example: 95°C for 1 min, 42°C for 2 min, 72°C for 1 min) in a thermal cycler and then electrophoresed on a 15% denaturing polyacrylamide gel. The amount of radioactivity of each band can be quantified by phosphorimaging, densitometric scanning of autoradiographs ***(15–19,23,24)***, or by scintillation counting of excised gel fragments following electrophoresis.

Quantitative assays to measure actual mRNA abundance can be performed by addition of an internal standard differing by a base pair from the template. The standard could be a known amount of RNA or DNA, added before the reverse transcription or the PCR reactions, respectively. The internal standard and the template should be amplified at close concentrations, but the assay is quantitative even when the sample represents 0.1–1% of the amount of the internal standard. The sensitivity of the assay varies with different mismatches ***(15,16,19,25)***.

4. Notes

4.1. Semiquantitative RT-PCR Analysis

1. Care must be taken not to overdry the pellets. Overdrying causes the pellets to become practically insoluble.
2. The cycle number for globin (typically 29) and each specific gene of interest must be tested empirically to verify what number of cycles remains within the linear range of amplification (i.e., the linear portion of semilog plots of the amount the reaction product as a function of the number of cycles). This can be done by removing equal aliquots periodically during the amplification reaction and determining the amount of α-[^{32}P]dCTP incorporation into the product band. In addition, because the cycling program provided is a generic one, the cycling conditions may have to be modified for specific primers.
3. During the design of PCR primers, primers must not only be able to function under the conditions described, but they should also produce product bands that contain diagnostic restriction sites that can be used to confirm product identity.
4. It should be noted that because a fixed amount of globin mRNA is added per embryo, the data produced through this method provide comparisons of message abundance on a per-embryo basis.

4.2. Quantitative Amplification and Dot-Blot Analysis

1. One of the strengths of the QADB method is the ability to construct dot-blot arrays that contain a large number of samples. The dot-blotter apparatus accommodates up to 96 samples and one can, in fact, exceed this number by having either two dot blotters or by simply processing two or more blots to cover a series of samples. A single set of PCR reaction products can provide enough material to produce a theoretical yield of 19 blots. Because each blot can be used multiple times, a single set of samples can provide a large amount of information about the pattern of expression of each gene assayed. For this reason, it is worthwhile to design carefully the exact array of samples that will be included in a sample set. As an example of the diverse kinds of information that can be obtained, it may be noted that when we wish to assay the normal temporal pattern of expression of an mRNA during preimplantation development, we employ a set of blots that include samples from GV-stage oocytes, unfertilized eggs, and fertilized eggs at approximately 12, 18, 22, 28, 32, 39, 49, 57, 67, 77, 88, 97, and 120 h post-hCG injection. We also include samples of embryos that were treated with α-amanitin from the mid one-cell stage onward and lysed at 22–49 h post-hCG, and also isolated inner cell mass cells obtained by immunosurgery ***(26)***. For each of these types of samples, between four and six separate samples are obtained in order to minimize the effects of minor variations in reaction yield and to permit the quantitative reliability and variability of the data to be evaluated. By analyzing so many different time-points, we are able to obtain a good understanding of the kinetics of changes in gene expression. By employing α-amanitin-treated samples, we are able to evaluate the degree of contribution of maternally inherited mRNAs to expression observed during the one-cell and two-cell stages and, thus, are able to determine whether a given gene is transcribed from the embryonic genome. For mRNAs that are not expressed as materna mRNAs, the α-amanitin-treated samples also provide an internal control for

detecting any nonspecific hybridization, so that hybridization and washing conditions may be altered if necessary.

2. Another advantage of the QADB method is that it can be applied to the analysis of single embryos. In addition, because the RNA is not extracted through procedures that remove DNA, it is possible to recover enough DNA even from single embryo lysates to permit the genotype of the embryo to be determined. In one description of his method, Brady and Iscove describe the use of a guanidine thiocyanate lysis buffer to permit the simultaneous isolation of both DNA and RNA ***(12)***. Another approach is to centrifuge the sample at high speed after the terminal transferase step, remove an upper portion of the sample for PCR amplification of cDNA, and then use the remaining sample and putative pellet of nuclei for genotype analysis. We have successfully taken such material and amplified the embryonic DNA using the Primer Extension PCR ***(28)*** protocol, and then determined the sex of the embryos by PCR for X-linked and Y-linked gene sequences (Latham, unpublished results).
3. The QADB method can be employed where small numbers of cells or embryos or even single cells or embryos are to be assayed. When this is to be done, the cells or embryos are lysed directly in the RT buffer augmented with NP-40. The QADB method can also be used on RNA purified from tissues or cell cultures, including RNA extracted from very small pieces of tissues, such as fetal organs. For very small tissues, RNA can be extracted using a method like that described for semiquantitative RT-PCR or using one of the commercially available RNA isolation mixes (e.g., Trizol, Gibco-BRL). For very small pieces of tissue, it is not necessary to quantify the amount of RNA obtained.
4. The vast majority of RT-PCR reactions performed with the above procedures produce excellent amounts of amplified material that indeed represent quantitatively the mRNA population of the cells or embryos assayed. A small percentage of reactions, however, may produce material that is not reflective of the mRNA population. This occurs most commonly when the RT reaction conditions are not optimum, and with increasing frequency as the amount of input RNA (or number of cells or embryos) is reduced. A successfully amplified cDNA product should be intensely stained and of heterogeneous size ranging between about 300 and 900 bp on an agarose gel. The product may migrate as two closely spaced bands on the gel, which most likely reflects the presence of some single-stranded product ***(12)***. Unsuccessful reactions frequently produce fainter staining products that migrate as a diffuse smear along the entire gel length, and there is frequently present a band of about 120 bp. Such reactions are also recognizable in that they typically do not produce a significant hybridization signal with any probe applied to the blot.
5. For highly valuable samples, increasing the number of blots obtained would be very beneficial. This can be achieved by amplifying each PCR reaction through an additional 25 cycles ***(27)***. Briefly, 1 µL of the PCR product produced from the first 50 cycles of amplification are amplified in 100 µL reactions containing 10 m*M* of Tris-HCl, pH 8.3, 50 m*M* of KCl, 2.5 m*M* of $MgCl_2$, 100 µg/mL of bovine serum albumin (BSA), 0.2 m*M* of each dNTP, 0.05% of Triton X-100, 0.4 nCi of α-[^{32}P]dCTP, 5 U of Taq polymerase, and 3.3 of XT primer. Excellent preservation of qualitative and quantitative representation of sequences within the reamplified material can be achieved ***(27)***, although there is a slight tendency for diminishment of signals for very rare transcripts.
6. We have investigated the effects of adding less than or more than 2.5 µL of each PCR reaction to the blots. We have found that using a lesser amount may prevent detection or reduce the quantitative reliability of measurements of more rare transcripts. Stronger hybridization signals may be obtained by applying more of the cDNA per dot, but the increase is very disproportionate (e.g., only 30–50% increase with 5 µL of PCR product), and we find it preferable to maximize the number of blots produced. In cases where stron-

ger hybridization signals are required for the assay, however, an increase in the amount of DNA applied to each dot may be a useful option.

4.3. SNuPE Analysis

1. Embryos can be stored in Ultraspec reagent (before choloroform addition) at –80°C.
2. RNA precipitation requires a carrier: Glycogen or poly-C are useful, but other carriers such as *E. coli* ribosomal RNA, may interfere with the cDNA synthesis.
3. The temperature, number of cycles, and $MgCl_2$ concentration should be determined empirically for each primer set to produce a good band within the linear range for product PCR.
4. Upstream and downstream primers should be designed to span introns. Alternatively, DNA digestion with RNAse-free DNase, followed by heat inactivation, can be performed before reverse transcription. RT reactions lacking reverse transcriptase should be PCR amplified as negative controls.

Acknowledgments

This work was supported in part by grants from the National Institutes of Health (HD-22681 and GM-S2332) and the National Science Foundation (#MCB-9630370).

References

1. Latham, K. E., Garrels, J., Chang, C., and Solter, D. (1991) Quantitative analysis of protein synthesis in preimplantation mouse embryos: extensive reprogramming at the one- and two-cell Stages. *Development* **112,** 921–932.
2. Latham, K. E., Garrets, J., Chang, C., and Salter, D. (1992) Analysis of protein synthesis in mouse embryos: construction of a high-resolution, two-dimensional gel protein database. *Appl. Theoret. Electrophoresis* **2,** 163–170.
3. Ram, P. T. and Schultz, R. M. (1993) Reporter gene expression n G2 of the 1-cell mouse embryo. *Dev. Biol.* **156,** 552–556.
4. Thompson, E. M., Legouy, E., Christians, E., and Renard, J.-P. (1995) Progressive maturation of chromatin structure regulates *HSP70.1* gene expression in the preimplantation mouse embryo. *Development* **21,** 3425–3437.
5. Aoki, F., Worrad, D. M., and Schultz, R. M. (1997) Regulation of transcriptional activity during the first and second cell cycles in the preimplantation mouse embryo. *Dev. Biol.* **181,** 296–307.
6. Worrad, D. M., Turner, B. M., and Schultz, R. M. (1995) Temporally restricted spatial localization of acetylated isoforms of histone H4 and RNA polymerase II in the 2-cell mouse embryo. *Development* **121,** 2949–2959.
7. Worrad, D. M., Ram, P. T., and Schultz, R. M. (1994) Regulation of gene expression in the mouse oocyte and early preimplantation embryo: developmental changes in Sp1 and TATA box binding protein, TBP. *Development* **120,** 2347–2357.
8. Worrad, D. M. and Schultz, R. M. (1997) Regulation of gene expression in the preimplantation mouse embryo: temporal and spatial patterns of expression of the transcription factor Sp1. *Mol. Reprod. Dev.* **46,** 268–277.
9. Latham, K. E., Doherty, A. S., Scott, C. D., Schultz, R. M. (1994) *Igf2r* and *Igf2* gene expression in androgenetic, gynogenetic, and parthenogenetic preimplantation mouse embryos: absence of regulation by genomic imprinting. *Genes Dev.* **8,** 290–299.
10. Manajewala, F., Logan, C. Y., and Schultz, R. M. (1991) Regulation of *hsp70* mRNA levels during oocyte maturation and zygotic gene activation in the mouse. *Dev. Biol.* **144,** 301–308.
11. Rambhatla, L., Patel, B., Dhanasekaran, N., and Latham, K. E. (1995). Analysis of G protein α subunit mRNA abundance in preimplantation mouse embryos using a rapid, quantitative RT-PCR approach. *Mol. Reprod. Dev.* **41,** 314–324.

12. Brady, G. and Iscove, N. (1993). Construction of cDNA libraries from single cells. *Methods Enzymol.* **225,** 611–623.
13. Mann, M., Latham, K. E., and Varmuza, S. (1995) Identification of genes showing altered expression in preimplantation and early postimplantation parthenogenetic embryos. *Dev. Genet.* **17,** 223–232.
14. Latham, K. E. and Rambhatla, L. (1995) Expression of X linked genes in androgenetic, gynogenetic, and normal mouse preimplantation embryos. *Dev. Genet.* **17,** 212–222.
15. Singer-Sam, J., Chapman, V., Le Bon, J. M., and Riggs, A. D. (1992) Parental imprinting studied by allele-specific primer extension after PCR: paternal X chromosome-linked genes are transcribed prior to preferential X chromosome inactivation. *Proc. Natl. Acad. Sci. USA* **89,** 10,469–10,473.
16. Singer-Sam, J. Le Bon, J. M., Dai, A., and Riggs, A. D. (1992) A sensitive, quantitative assay for measurement of allele transcripts differing by a single nucleotide. *PCR Methods Appl.* **1,** 160–163.
17. Singer-Sam, J. and Riggs, A. D. (1993) Quantitative analysis of messenger RNA levels: reverse, transcription-polymerase chain reaction single nucleotide primer extension assay. *Methods Enzymol.* **225,** 344–351.
18. Singer-Sam, J. (1994) Quantitation of specific transcripts by RT-PCR SNuPE assay. *PCR Methods Appl.* **3,** S48–S50.
19. Buzin, C. H., Mann, J. R., and Singer-Sam J. (1994) Quantitative RT-PCR assays show Xist RNA levels are low in mouse female adult tissue, embryos and embryoid bodies. *Development* **120,** 3529–3536.
20. Szabó, P. E. and Mann, J. R. (1995) Biallelic expression of imprinted genes in the mouse germ line: Implications for erasure, establishment, and mechanisms of genomic imprinting. *Genes Dev.* **9,** 1857–1868.
21. Szabo, P. E. and Mann, J. R. (1996) Maternal and paternal genomes function independently in mouse ova in establishing expression of the imprinted genes *Snrpn* and *Igf2r*: no evidence for allelic trans-sensing and counting mechanisms. *EMBO J.* **5,** 6018–6025.
22. Feinberg, A. P. and Vogelstein, B. (1983) Addendum: a technique for radiolabeling DNA restriction fragments to high specific activity. *Anal. Biochem.* **137,** 266,267.
23. Lombardo, A. J. and Brown, G. B. (1996). A quantitative and specific method for measuring transcript levels of highly homologous genes. *Nucleic Acids Res.* **24,** 4812–4816.
24. Kuppuswamy, M. N., Hoffmann, J. F., Kasper, C. K., Spitzer, S. G., Groce, S. L., and Bajaj, S. P. (1991) Single nucleotide primer extension to detect genetic diseases: experimental application to hemophilia B (factor IX) and cystic fibrosis genes. *Proc. Natl. Acad. Sci. USA* **88,** 1143–1147.
25. Gilliand, G., Perrin, S., and Bunn, H. F. (1990) Competitive PCR for quantitation of mRNA, *PCR Protocols* (Innis, M. A., Gelfand, D. H., Sninsky, J. J., and White, T. J., eds.), Academic, New York.
26. Solter, D. and Knowles, B. B. (1975) Immunosurgery of mouse blastocyst. *Proc. Natl. Acad. Sci. USA* **72,** 5099–5102.
27. Domachenko, A., Latham, K. E., and Hatton, K. S. (1997) Expression of *myc*-family, *myc*-interacting, and *myc* target genes during preimplantation mouse development. *Mol. Reprod. Dev.* **47,** 57–65.
28. Zhang, L., Cui, X., Schmitt, K., Hubert, R., Navidi, W., and Arnheim, N. (1992) Whole genome amplification from a single cell: implications for genetic analysis. *Proc. Natl. Acad. Sci. USA* **89,** 5847–5841.

32

Differential Screens with Subtracted PCR-Generated cDNA Libraries from Subregions of Single Mouse Embryos

Akihiko Shimono and Richard R. Behringer

1. Introduction

Differential gene expression is an essential mechanism for the development of a single-celled zygote into a patterned multicellular organism with differentiated cell types and tissues. Therefore, the identification of differentially expressed genes provides an important molecular resource for functional studies to define genetic pathways in metazoan development. Differential hybridization or subtraction methods have been used to detect differences in gene expression between distinct cells or tissues. However, these methods are difficult to apply to early stage embryonic tissues because they require relatively large amounts of tissue, cells and mRNA *(1–3)*.

Recent improvements in the polymerase chain reaction (PCR) make it possible to prepare the required amounts of cDNA from small numbers of cells for differential hybridization or subtraction methods *(4,5)*. Indeed, single-cell cDNA libraries have been generated and used for differential hybridization screens to identify genes expressed in rat olfactory neurons and in adult male *Drosophila* neurons *(6,7)*.

Abundant cDNAs that are differentially expressed can be detected very efficiently by differential hybridization screens. However, the isolation of rarer differentially expressed cDNAs requires subtraction techniques prior to differential hybridization screening *(8–10)*. In the method described in this chapter, PCR is used to amplify cDNA from a small cell population and is coupled with a PCR-based subtraction (**Fig. 1**, *11*). This method can be used for comparisons between cDNA pools of small groups of cells (up to 100) from different portions of a single embryo, from the same embryonic tissues at different developmental stages, or from similar tissues of different genotype embryos (*see* **Note 1**).

2. Methods and Materials

2.1. Multicell cDNA Amplification

We have modified the protocol reported by Dulac and Axel *(6)* to amplify cDNAs from a small number of cells. The amplified cDNAs are then used for the subtractions

From: *Methods in Molecular Biology, Vol. 136: Developmental Biology Protocols, Vol. II*
Edited by: R. S. Tuan and C. W. Lo © Humana Press Inc., Totowa, NJ

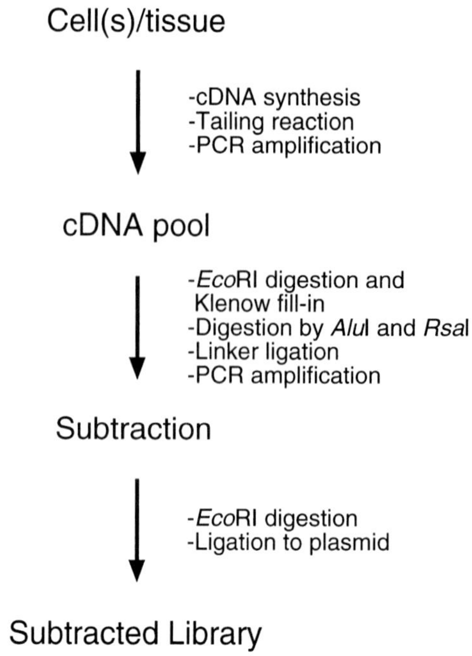

Fig. 1. PCR-based subtraction with cDNA from a small cell population.

(**Fig. 1**). The generation of amplified cDNAs has three basic steps: (1) the first-strand cDNA synthesis, (2) a tailing reaction to add a poly(dA) sequence to the 3' end of the cDNAs, and (3) the PCR amplification of the cDNA with a poly-(dT) primer (**Fig. 2**). One to approximately 100 cells, either as a cell suspension or a tissue region or layer, isolated by enzymatic digestion and/or mechanical dissection, can be used as a source for cDNA synthesis. It is possible that greater than 100 cells will reduce the efficiency of cDNA synthesis and subsequent PCR amplification. A cell suspension or small tissue region is lysed in a buffer containing the poly-(dT) primer. First-strand cDNA synthesis, under limiting conditions, is performed without RNA purification steps by adding reverse transcriptase. Using this method, most cDNA molecules are synthesized to a length of approximately 200–1000 nt, an optimum size for unbiased PCR amplification ***(5)***.

Following heat inactivation of the reverse transcriptase, a poly(dA) sequence is added to the 3' end of the cDNA by terminal transferase. In the final steps, the cDNA molecules are amplified with a poly-(dT) primer that includes a specific sequence that facilitates subsequent reactions. We recommend that oligonucleotides from Oligos Etc., Inc. (Portland, OR) should be used as a source of primers because their high purity and high percentage of full-length oligonucleotides gives a higher amplification efficiency.

In this protocol, the suggested manufacturer's source of various enzymes could be key to the success of the technique. The optimization of the Moloney murine leukemia virus (MMLV) reverse transcriptase is critical and specific lots from Gibco-BRL (Gaithersburg, MD) should be tested. Most of the reagents except for the enzymes should be stored at –80°C.

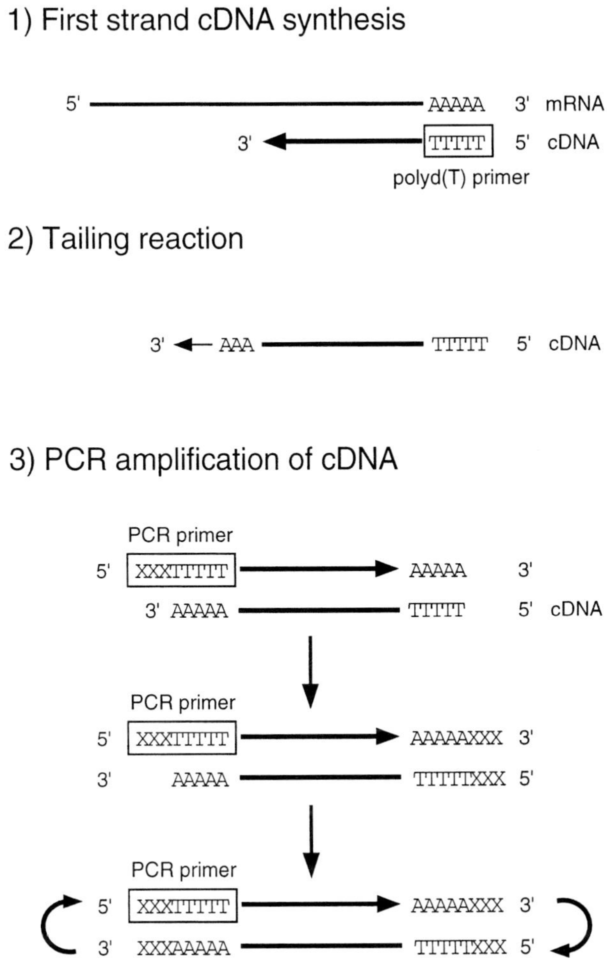

Fig. 2. PCR amplification of cDNA with a poly-(dT) primer.

2.1.1. Protocol 1

The procedure described for Protocol 1 should be completed without stopping.

1. Freshly prepare 100 µL of lysis buffer.
   a. Lysis Buffer

5X MMLV buffer (Gibco-BRL)	20 µL
H_2O (RNase- and DNase-free)	76 µL
NP40	0.5 µL
PrimeRNase inhibitor (3'–5' Inc., Boulder, Co)	1 µL
RNAguard (Pharmacia, Biotech, Piscataway, NJ)	1 µL
1/24 dilution of stock primer mixture	2 µL

   b. Stock Primer Mixture

100 m*M* dATP, dCTP, dGTP, dTTP (Pharmacia)	10 µL each
poly-(dT)24 (1.9 mg/mL)	10 µL
H_2O	30 µL

   c. Enzymes
      Avian myeloblastosis virus (AMV) 25 U/µL (Boehringer Mannheim, Indianapolis, IN)
      MMLV 200 U/µL (Gibco-BRL)

2. Add 4 µL of lysis buffer to thin-walled PCR reaction tubes (Stratagene, La Jolla, CA) and place on ice.
3. Wash cells or tissue with PBS to reduce contamination with blood cells. Note that small tissues in PBS may adhere to plastic.
4. Transfer the sample into the tubes with the 4 µL of lysis buffer by mouth pipetting with a finely pulled glass pipet. The volume of the phosphate-buffered saline (PBS) solution used to transfer the sample should be small (0.2–0.5 µL) because larger volumes will reduce the efficiency of cDNA synthesis. The tubes with the tissue sample in the lysis buffer can be kept on ice for up to 1 h. Also, generate a negative control that does not have the tissue sample added.
5. Lyse the cells by incubating the tubes in a 65°C water bath for 1 min.
6. Remove the tubes from the water bath and place at room temperature for 2 min to allow the poly-d(T) primer to anneal to the mRNA.
7. Spin down the reaction mixture at 4°C for 2 min.
8. Keep the tubes on ice and add 0.5 µL of 1:1 (v:v) mixture of AMV and MMLV reverse transcriptase and then incubate the tubes at 37°C for 15 min (no longer).
9. Place the tubes on ice to stop the reaction. Incubate the tubes at 65°C for 10 min to inactivate the reverse transcriptase.
10. Place the tubes on ice. Spin down the reaction mixture at 4°C for 2 min.
11. Place the tubes on ice and add 4.5 µL of terminal transferase reaction mixture containing 10 U of terminal transferase. Incubate at 37°C for 15 min.
    a. Terminal Transferase Reaction Mixture (Store at –80°C)
       5X terminal transferase buffer (Gibco-BRL) 800 µL
       100 m*M* dATP (Pharmacia) 30 µL
       H_2O (RNase and DNase free) 1.17 mL
    b. Enzyme
       Terminal transferase 25 U/µL (Boehringer Mannheim).
12. Incubate at 65°C for 10 min to inactivate the terminal transferase.
13. Place on ice. Spin down the reaction mixture at 4°C for 2 min and then return the tube on ice.
14. Add 90 µL of freshly prepared ice-cold PCR reaction mixture. Cover surface of the solution with 1 or 2 drops of mineral oil.
    a. PCR Reaction Mixture I

10X PCR buffer II (Perkin-Elmer, Cetus, Norwalk, CT)	10 µL
25 m*M* $MgCl_2$ (Perkin-Elmer). Mg^{2+} concentration should be optimized to generate the maximum amount of amplified cDNA and to reduce nonspecific DNA amplification in a negative control.	10 µL
20 mg/mL bovine serum albumin (BSA) (Boehringer Mannheim)	0.5 µL
100 m*M* dNTP (Pharmacia)	1 µL each
5% Triton-X	1 µL
5 µg/µL Primer. Primer: 5'-ATT**GGATCC**AGGCCGCTCTGGAC AAAATAT**GAATTC**(T)24-3'.	1 µL
H_2O	63.5 µL
Taq polymerase 5 U/µL (Perkin-Elmer)	2 µL

15. Amplify cDNA in a DNA thermal cycler (Perkin-Elmer) using the following program.
    a. First Round of PCR Amplification
       Denature at 94°C for 4 min
       Cycle parameters
          94°C for 1 min
          42°C for 2 min

72°C for 6 min
25 Cycles with 10-s autoextension per cycle
Soak file
Keep at 4°C*.

16. After the first round of PCR cycles has finished, reheat the thermocycler block to 94°C, then add 1 µL of *Taq* polymerase (5 U/µL) to each tube and restart the PCR with the following program:
    a. Second Round of PCR Amplification
       Denature at 94°C for 4 min.
       Cycle parameters
       94°C for 1 min
       42°C for 2 min
       72°C for 6 min
       25 Cycles without extension
       Soak file
       Keep at 4°C.
17. Purify the amplified DNA with phenol–$CHCl_3$ extraction and EtOH precipitation with 0.2 volume of 10 *M* NH_4OAc. Resuspend DNA in 100 µL of H_2O. Store at –80°C.

The 5 µL of the amplified DNA is used to monitor the cDNA quality by agarose gel electrophoresis. An ethidium bromide-stained DNA smear (200 bp to 2.0 kb) with a strong staining intensity should be attained. In many cases, several micrograms of cDNA can be obtained even from a single cell. Amplification of bacterial DNA molecules may be observed in the negative control sample because of DNA contamination in the bacterially derived enzymes that are used in the method.

Southern-blot hybridization of the amplified cDNA is performed to examine the PCR amplification efficiency using ubiquitously expressed and tissue-specific gene probes. Because the cDNA amplification method is biased to amplify the 3' ends of genes, these ubiquitously expressed and tissue-specific gene probes should contain the 3' ends of their cDNAs. The abundance of amplified cDNA recognized by each marker probe is also considered. It is best to perform cDNA amplifications for multiple samples and then choose which cDNA pools are best for the subtraction protocols. If a set of cDNA pools show large differences for the various markers, then they are not appropriate for subtractions.

2.2. Reamplification of cDNA

The subtractions steps described as follows require larger amounts of cDNA (approximately 100 µg) than are amplified by protocol 1. Therefore, the initial amplified cDNAs must be reamplified. The initial amplified cDNA molecules have the same primer sequences on both ends. These sequences must be removed by EcoRI digestion before subtraction because they will interfere with the subtractive hybridization. The *Eco*RI digested ends are filled in by the Klenow fragment to generate blunt ends. Before reamplification, the cDNA is digested by blunt cutting restriction enzymes with 4-bp recognition sequences. This reduces the size of the cDNA in order to increase the efficiency of PCR amplification. A linker is then ligated to both ends of the digested cDNA. In the final step, the cDNA is amplified by PCR with the new linker sequence (**Fig. 3**).

*It is recommended that the second PCR amplification be started as soon as the first PCR has finished.

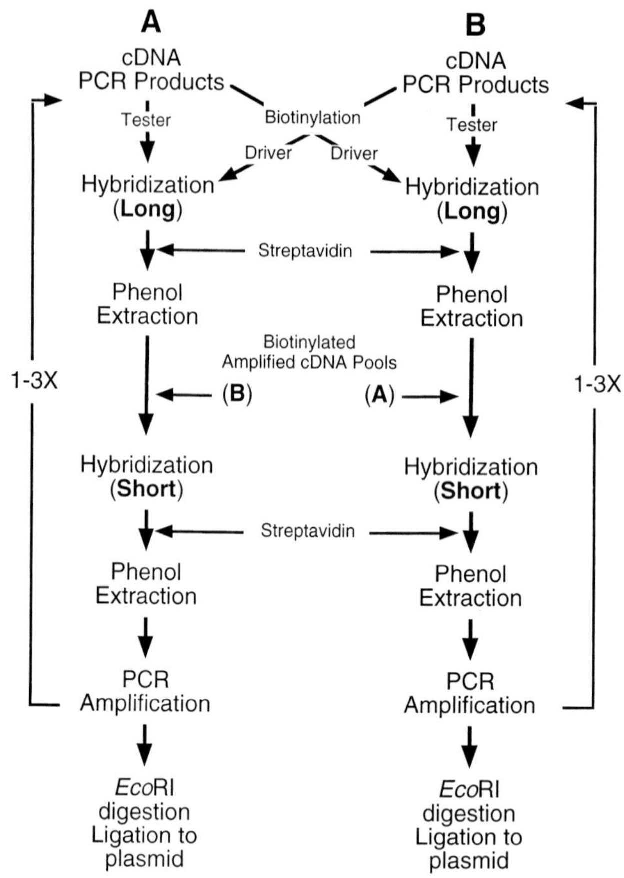

Fig. 3. Subtractive hybridizations.

2.2.1. Protocol 2.1: Primer Sequence Replacement

1. Add 10 µL of cDNA to 90 µL of PCR reaction mixture II to generate double-stranded cDNAs from any remaining single-stranded cDNA from the previous procedure.
   a. PCR Reaction Mixture II

10X PCR buffer II (Perkin-Elmer)	10 µL
25 m*M* $MgCl_2$ (Perkin-Elmer)	10 µL
20 mg/mL BSA (Boehringer Mannheim)	0.5 µL
100 m*M* dNTPs (Pharmacia)	1 µL each
5% Triton-X	1 µL
5 µg/µL Primer. Primer: 5'-ATT**GGATCC**AGGCCGCTCTGGAC AAAATAT**GAATTC**(T)24-3'.	1 µL
H_2O	65.5 µL
Taq polymerase 5 U/µL (Perkin-Elmer)	1 µL

   b. Cycle Parameters
      94°C for 5 min
      42°C for 5 min
      72°C for 30 min
      1 cycle

2. Purify the cDNA by phenol–$CHCl_3$ extraction and EtOH precipitation with 10 *M* NH_4OAc, then redissolve into 25 µL of TE buffer.
3. Digest cDNA with *Eco*RI by incubation at 37°C for 2 h.
   a. *Eco*RI Digestion Reaction

cDNA	25 µL
10X *Eco*RI buffer (Manufacturer's)	5 µL
*Eco*RI (10 U/µL)	5 µL
H_2O	15 µL

4. Purify the cDNA by phenol–$CHCl_3$ extraction and EtOH precipitation with 1/20 volume of 5 *M* NaCl. Then redissolve in 15 µL H_2O. Fill in the ends of the fragments with Klenow fragment by incubating at 17°C for 30 min. After the first fill-in reaction, add 1 µL of the Klenow fragment and incubate at 17°C for another 30 min.
   a. Klenow Fill-in Reaction

cDNA	15 µL
10X Klenow buffer (Manufacturer's)	2 µL
1 m*M* dNTPs	2 µL
5 U Klenow fragment	1 µL

5. Remove the small primer fragments from the cDNA mixture by electrophoresis through 1.5% low-melt agarose. Excise the cDNA between 150 bp and 2.0 kb, add TE buffer up to 500 µL, and incubate the agarose block at 65°C for 10 min. Vigorously mix the cDNA–agarose solution with phenol (prewarmed to 65°C) for 10 min and centrifuge at 4°C for 10 min. This is followed by phenol extraction and, subsequently, phenol–$CHCl_3$ extraction at room temperature. After EtOH precipitation with 5*M* NaCl, redissolve in (10 µL) TE buffer. Use 1 µL of the cDNA to determine its concentration and then dilute with TE accordingly.
6. Make three aliquots of 150 ng of cDNA: (1) without digestion, (2) digested with *Alu*I, and (3) digested with *Alu*I and *Rsa*I at 37°C for 2 h. These digests may be stored at –20°C.
   a. *Alu*I digestion

cDNA	150 ng/17 µL in H_2O
10X Buffer (Manufacturer's)	2 µL
*Alu*I (10U/µL)	1 µL

   b. *Alu*I–*Rsa*I digestion

cDNA	150 ng/16 µL in H_2O
10X Buffer (Manufacturer's)	2 µL
AluI (10 U/µL)	1 µL
RsaI (10 U/µL)	1 µL

2.2.2. Protocol 2.2: Preparation of Phosphorylated Linker

1. Prepare 10 µL of an aqueous solution of 4.6 µg of Oligo 1 and 5.4 µg of Oligo 2.
   Oligo 1; 5'-CTCTTGCTT**GAATTC**GGACTA-3'
   Oligo 2; 5'-TAGTCC**GAATTC**AAGCAAGAGCACA-3'. The 5-bp extension allows for directional ligation of the linker to the cDNA.
2. Add reagents in the following order:

10X Kinase buffer (Manufacturer's)	2 µL
10 m*M* ATP *(not dATP)*	4 µL
500 µg/mL BSA (Promega [Madison, WI] or Boehringer Mannheim)	2 µL
T4 kinase (10 U/µL) (New England Biolabs, Beverly, MA)	2 µL

3. Incubate at 37°C for 1 h.
4. Stop the reaction by adding 1 µL of 0.5 *M* EDTA (pH 8.0) and incubate at 70°C for 5 min.
5. Incubate at 45°C for 10 min. Store at –20°C.

2.3.2. Protocol 2.3: Linker Ligation and DNA Fragment Size Selection

DNA fragment sizes from 200–500 bp are ideal for PCR amplification. Therefore, the cDNA is digested by restriction enzymes with 4-bp recognition sequences. cDNA fragments larger than 1.5 kb should be removed from the cDNA pools because they can reduce PCR amplification efficiency.

1. Mix cDNA solutions (1), (2), and (3) (from **Subheading 2.1., step 6**), each with 1.5 µg of phosphorylated linker. Add 100 µL of H_2O.
2. Extract by phenol, phenol–$CHCl_3$ and precipitate with EtOH using 5 *M* NaCl.
3. Resuspend the cDNA and linker mixture in 13 µL of H_2O.
4. Add the following reagents in the following order:
   10X Ligation buffer 2 µL
   (400 m*M* Tris-HCl, pH 7.5; 100 m*M* $MgCl_2$; 100 m*M* DTT)
   10 m*M* ATP (not dATP) 1 µL
   500 µg/µL BSA 2 µL
   T4 DNA ligase (5 U/µL) 2 µL
5. Incubate at 16°C overnight.
6. Isolate the DNA fragment fraction from 150 bp to 1.5 kb by electrophoresis with 1.5% low-melt agarose as in **Subheading 2.1., step 5**. Resuspend DNA in 100 µL of H_2O. Store at –20°C.
7. Prepare 20–25 tubes of PCR reaction mixture III and amplify using the following conditions:
   a. PCR reaction mixture III
      cDNA solution 1 µL
      10X Buffer (Perkin-Elmer; Type I) 10 µL
      2.5 m*M* dNTPs 8 µL
      Oligo 1 (1 µg/µL)1 µL
      Taq polymerase (Perkin-Elmer) 0.5 µL
      H_2O 79.5 µL
   b. Cycle parameters
      Preheat at 94°C for 4 min
         94°C for 1 min
         50°C for 1 min
         72°C for 2 min
         30 cycles with 25-s autoextension per cycle
      Time-delay file
         72°C for 20 min
      Soak file
         Keep at 4°C
8. Purify cDNA pools by phenol–$CHCl_3$ and precipitate with EtOH using 10 *M* NH_4OAc. Resuspend into 100 µL of H_2O. Generate 1-µg/µL solutions. Store at –20°C.

2.3. Subtraction

A pair of cDNA pools prepared by the above protocols are used for subtraction. Each cDNA pool may be divided into two populations consisting of tester and driver cDNA for reciprocal subtractions. Upon heat denaturation to generate single-stranded cDNA fragments, the tester cDNA from one pool is mixed with an excess amount of photobiotinylated driver cDNA from the other. The single-stranded cDNA fragments that are common between the tester and biotinylated driver hybridize and generate double-stranded cDNAs. Streptavidin is added to the mixture and complexes with the biotinylated molecules (tester–driver hybrids and excess driver). The streptavidin–biotinylated driver complexes are then removed by phenol–$CHCl_3$ extraction. Some tester-specific cDNA fragments are predicted to remain as single-stranded cDNAs or as biotin-negative double-stranded cDNAs. These tester-specific molecules are amplified by PCR for the subsequent subtraction steps or for subcloning (**Fig. 3**).

2.3.1. Protocol 3.1. Photobiotinylation of Driver cDNA

To generate driver cDNA for the subtraction, it is important to incorporate sufficient numbers of biotin molecules into the cDNA molecules. Specific lots of photobiotin reagent should be prescreened for good incorporation into cDNA because this ultimately makes a *significant* impact on the efficiency of the subtraction procedure. It is simple to check the quality of the photobiotin reagent because poor lots do not reconstitute into a homogeneous solution. Moreover, the photoreaction should can repeated until enough biotinylation of the driver cDNA is obtained.

1. Digest the driver cDNA with *Eco*RI to remove the *Eco*RI linkers on the amplified cDNAs. The presence of the *Eco*RI linkers on the driver cDNAs reduces the efficiency of the amplification of driver cDNA by PCR.
   a. *Eco*RI digestion reaction

cDNA	75 μL (μg) [50 μL (μg)]. [ ], for the driver cDNA used for the second and third subtractions.
10X buffer (Manufacturer's)	25 μL
*Eco*RI (40 U/μL)	25 μL
H_2O	125 μL [150 μL]*

2. Incubate at 37°C for 2 h.
3. Add 5 μL of 0.5 *M* EDTA (pH 8.0) to stop the reaction.
4. Purify the cDNA by phenol and phenol–$CHCl_3$ extraction. Precipitate with EtOH using 10 *M* NH_4OAc. Resuspend the cDNA in 75 μL of H_2O.
5. Mix the *Eco*RI-digested cDNA with 75 μL of reconstituted photobiotin on ice.
   a. Photobiotin
      Photoprobe Biotin (Vector Laboratories, Inc., Burlingame, CA) 0.5 mg/vial. Reconstitute the vial with 500 μL of H_2O. Store in the *dark* at –20°C.
6. Expose the mixture on ice to a strong visible light source (350–370 nm) such as that produced by a mercury vapor/sunlamp bulb (Vector Laboratories, Inc.) for 15 min or longer.
7. Add 500 μL of 0.1 *M* Tris-HCl (pH 9.5). To remove free biotin, extract the cDNA solution with *n*-butanol. Repeat four times. Precipitate with EtOH using 5 *M* NaCl. The pellet should be an intense dark orange color (PANTONE® Coated Color Set, 166 CVC [Pantone Inc., Carlstadt, NJ]). Lighter colors of orange color indicate that the degree of biotinylation is insufficient.
8. Resuspend in 75 μL of H_2O.
9. Repeat **steps 5–7** of the photobiotinylation process. If sufficient biotinylation is not obtained, repeat the irradiation or extend the irradiation time.
10. Precipitate with EtOH using 5 *M* NaCl.
11. Resuspend the final cDNA in 12 μL [8 μL] of $H_{10}E_1$ (10 m*M* HEPES pH 7.5, 1 m*M* EDTA pH 8.0).

2.3.2. Protocol 3.2: Subtractive Hybridization

Each subtractive hybridization step consists of a long and a short hybridization substep (**Fig. 3**). For the short subtractive hybridization substep, the original (unsubtracted) biotinylated driver cDNA is used. From the second and subsequent subtraction steps on, this functions to effectively remove abundant cDNA molecules that are common between the tester and driver pools.

1. Mix 2.5 µL (2.5 µg) of tester cDNA with 8 µL (50 µg) of biotinylated driver cDNA. Boil for 3 min, then place on ice for 2 min. Spin down and collect the supernatant.
2. Add 10.5 µL of 2X hybridization buffer (1.5*M* NaCl, 50 m*M* HEPES, pH 7.5, 10 m*M* EDTA, pH 8.0, 0.2% sodium dodecyl sulfate [SDS]). Cover the solution with two drops of mineral oil.
3. Boil for 3 min, then *immediately* transfer to a 68°C water bath. Incubate for 20 h (long hybridization).
4. Add 130 µL of prewarmed (55°C) $H_{10}E_1$ and incubate at 55°C for 5 min.
5. Transfer the aqueous phase to a new tube (room temperature). Add 10 µL of 2 µg/µL Streptavidin. (Reconstitute with 0.15 *M* NaCl/$H_{10}E_1$. Store at 4°C for up to 10 mo.) Stand at room temperature for 20 min.
6. Add an equal volume of phenol–$CHCl_3$ and mix vigorously for 10 min. Spin down for 5 min. After centrifugation, a large amount of particulate material at the solution interface will be observed.
7. Recover the aqueous phase. Repeat **steps 5–6** three times.
8. Add 4 µL (25 µg) of the *original* biotinylated driver cDNA. Precipitate with EtOH using 5 *M* NaCl. Resuspend the pellet in 10 µL of $H_{10}E_1$.
9. Repeat **steps 1–3** and hybridize for 2 h at 68°C. (short hybridization).
10. Repeat **steps 4–7**.
11. Prepare 12–15 tubes. Amplify subtracted cDNA by PCR, using the same cycle conditions as in **Protocol 2.3., step 7**.
    a. PCR reaction mixture IV

Subtracted cDNA solution	5 µL
10X Buffer (Perkin-Elmer; Type I)	10 µL
2.5 m*M* dNTPs	8 µL
Oligo 1 (1 µg/µL)	1 µL
Taq polymerase (Perkin-Elmer)	0.5 µL
H_2O	75.5 µL
(total 100 µL/tube)	

12. Purify cDNA pools by phenol–$CHCl_3$ and precipitate with EtOH using 10 *M* NH_4OAc. Resuspend into 100 µL of H_2O. Generate 1-µg/µL solutions. Store at –20°C.

2.4. Subsequent Subtractions

To examine the efficiency of the subtraction steps, ubiquitously expressed and specifically expressed marker genes should be used as probes for Southern blot hybridization of the subtracted cDNA pools after each subtraction step. The desired result is the reduction in the abundance of the cDNAs of ubiquitously expressed genes after subtraction steps. In addition, control marker genes that should be specific to either the tester or driver should be enriched with the subtraction steps. The subtracted cDNA sample from **Subheading 3.2., step 12** can be used as a tester or driver for subsequent subtraction steps (**Fig. 3**). In many cases, two or three subtraction steps provide a sufficient enrichment. Additional subtraction steps and PCR may increase nonspecific enrichment and decrease specific enrichment. Thus, it is *essential* to perform the suggested controls to monitor the progress of this procedure and the subsequent subtractions.

2.5. Subtracted Library Screening

The subtracted cDNAs are digested with *Eco*RI (present in the linkers), ligated into calf intestinal alkaline phosphatase-treated plasmid vectors, then plated onto X-gal plates to identify bacterial colonies containing cDNA inserts. This subtracted cloned cDNA library should contain cDNAs that are abundantly enriched. These abundantly enriched cDNAs can be identified by colony hybridization, using the initial subtracted cDNAs as a probe. Sometimes these abundantly enriched cDNAs are composed of only a few (two to three) different cDNAs. The subtracted cDNA library also contains cDNAs that are less abundantly enriched and may be contained in the colonies that did not hybridize with the initial subtracted cDNA probe. One way to remove the abundantly enriched cDNAs from the subtracted library and reveal the significance of the less abundantly enriched cDNAs is to add the abundantly enriched cDNAs to the driver for a subsequent subtraction step. To determine if these cDNAs are differentially expressed in the original tissues, they can be used as probes for Southern blots of the original amplified cDNA pools (virtual Northern). Interesting candidate cDNAs can be further evaluated by DNA sequencing, *in situ* hybridization, and Northern blot analysis.

If the final subtracted cDNA pool still has a significant amount of nonspecific background or if only one round of subtraction is used, then it may be useful to use differential hybridization for screening the subtracted libraries.

3. Notes

1. The method described in this chapter is one way to screen for differentially expressed genes from a single cell or a small number of cells using PCR-amplified cDNA coupled with PCR-based subtractive hybridization protocols. The basic screening strategy should be altered according to the expression levels of the genes of interest. When candidate gene expression levels are suspected to be high, such as 0.1% of the cDNA pool, then subtraction may not be necessary and differential hybridization strategies may provide better results. However, if candidate gene expression levels are suspected to be low (e.g., less than 0.01% of the cDNA pool), then subtraction can be useful. One consideration, however, is that the higher sensitivity afforded by subtraction to detect differential expression of rare cDNA molecules may lead to the enrichment of nonspecific cDNA clones. We have also combined suppression subtractive hybridization ***(11)*** with this multicell amplified cDNA method to successfully isolate differentially expressed genes. Thus, different subtraction methods can be used with the initial cDNA amplification procedure using minimal embryonic tissue.

Acknowledgment

We thank Catherine Dulac for the initial single-cell PCR amplification protocol and April Kilburn and Huhui Lee for reading the manuscript. One of the authors (A. S.) was supported in part by a postdoctoral fellowship from the Japan Society for the Promotion of Science for Japanese Junior Scientists. This work was supported in part by a grant from the Human Frontier Science Program.

References

1. Davis, R. L., Weintraub, H., and Lassar, A. B. (1987) Expression of a single transfected cDNA converts fibroblasts to myoblasts. *Cell* **51,** 987–1000.

2. Klar, A., Baldassare, M., and Jessell, T. M. (1992) F-Spondin: a gene expressed at high levels in the floor plate encodes a secreted protein that promotes neural cell adhesion and neurite extension. *Cell* **69,** 95–110.
3. Rothstein, J. L., Johnson, D., DeLoia, J A., Skowronski, J., Solter, D., and Knowles, B. (1992) Gene expression during preimplantation mouse development. *Genes Dev.* **6,** 1190–1201.
4. Brady, G. and Iscove, N. N. (1993) Construction of cDNA libraries from single cells. *Methods Enzymol.* **225,** 611–623.
5. Brady, G., Billia, F., Knox, J., Hoang, T., Kirsch, I. R., Voura, E. B., Hawley, R. G., Cumming, R., Buchwald, M., and Siminovitch, K. (1995) Analysis of gene expression in a complex differentiation hierarchy by global amplification of cDNA from single cells. *Curr. Biol.* **5,** 909–922.
6. Dulac, C. and Axel, R. (1995) A novel family of genes encoding putative pheromone receptors in mammals. *Cell* **83,** 195–206.
7. Amrein, H. and Axel, R. (1997) Genes expressed in neurons of adult male *Drosophila. Cell* **88,** 456–469.
8. Wang, Z. and Brown, D. D. (1991) A gene expression screen. *Proc. Natl. Acad. Sci. USA* **88,** 11,505–11,509.
9. Kanamori, A. and Brown, D. D. (1993) Cultured cells as a model for amphibian metamorphosis. *Proc. Natl. Acad. Sci. USA* **90,** 6013–6017.
10. Shimono, A., Okuda, T., and Kondoh, H. (1999) *N-myc*-dependent repression of *Ndr1*, a gene identified by direct subtraction of whole embryo cDNAs between wild type and *N-myc* mutant mechanisms of development. *Development* **83,** 39–52.
11. Shimono, A. and Behringer, R. R. (1999) Isolation of novel cDNAs by subtractions between the anterior mesoderm of single mouse gastrula embryo. *Dev. Biol.* **209,** 369–380.
12. Diatchenko, L., Lau, Y.F., Campbell, A. P., Chenchik, A., Moqadam, F., Huang, B., Lukyanov, S., Lukyanov, K., Gurskaya, N., Sverdlov, E. D., and Siebert, P. D. (1996) Suppression subtractive hybridization: a method for generating differentially regulated or tissue-specific cDNA probes and libraries. *Proc. Natl. Acad. Sci. USA* **93,** 6025–6030.

33

HPLC-Based mRNA Differential Display

Thomas B. Knudsen

1. Introduction

Analysis of the genetic and molecular mechanisms of embryogenesis requires effective methods to find and isolate genes that are differentially expressed in normal and altered conditions. Comparative genomic screens can lead to the discovery of known or novel transcripts that change in association with critical events. However, the complexity of cellular responses at the genetic level, as well as functional redundancies in cell signaling pathways themselves, make it difficult to predict how a specific genetic or molecular alteration might translate into a particular developmental phenotype.

One approach to this problem is to look for transcripts that are differentially expressed in mutant embryos that lack targeted gene function(s). Differential display refers to the side-by-side comparisons of the mRNAs expressed in different samples based on arbitrary amplification of cDNA sequences from subsets of mRNAs using the polymerase chain reaction (PCR) *(1)*. In conventional differential display, the arbitrary PCR products are labeled with sulfur-35 or phosphorus-33 and displayed on a sequencing gel *(2,3)*. This chapter outlines a modified method whereby native, unlabeled PCR products are displayed by high-performance liquid chromatography (HPLC) *(4)*. It was developed for RNA profiling of early-somite-stage mouse embryos lacking function of the *p53* tumor suppressor gene *(5)*.

2. Materials

1. Reagents
   a. RNeasy miniprep kit (Qiagen Inc., Chatsworth, CA): Contents include lysis buffer RLT with 1% (v/v) 2-mercaptoethanol added just prior to use; wash buffers RW1 and RPE are prepared as instructed by the kit.
   b. DNase I, amplification grade: RNase-free, from Gibco-BRL (Gaithersburg, MD).
   c. Reverse transcriptase: Moloney murine leukemia virus (MMLV) reverse transcriptase (GenHunter Corp., Nashville, TN).
   d. RNAimage kit: Single-base 3'-anchored oligo(dT) primers H-T_{11}M, where M = A, G, or C and a panel of arbitrary primers (i.e., H-AP1 through H-AP8) (GenHunter Corp.).
   e. HPLC mobile phase: Solvent 1 is 25 m*M* Tris-HCl, 1 m*M* EDTA, and 10% acetonitrile (pH 8.0); solvent 2 is the same buffer with 1 *M* NaCl. Both solvents are prepared daily using reagent-grade water and HPLC-grade reagents and filtration at 0.45 μm.
   f. TTE buffer: 0.1 *M* Tris-HCl, 10 m*M* triethylamine, and 1 m*M* EDTA, pH 7.7.

From: *Methods in Molecular Biology, Vol. 136: Developmental Biology Protocols, Vol. II*
Edited by: R. S. Tuan and C. W. Lo © Humana Press Inc., Totowa, NJ

2. HPLC supplies
   a. Main column: Gen-Pak FAX high-performance, anion-exchange column (4.6 × 100 mm) from Waters Associates (Millipore Corporation, Milford MA) protected with an in-line precolumn filter. The main column is flushed with Solvent 1 and stored at 4°C between uses.
   b. Nensorb 20: Nucleic Acid Purification Cartridge (NEN Life Science Products, Boston, MA) preequilibrated with 2 mL HPLC-grade methanol and 2 mL TTE buffer at a flow of 1 drop per 2 s.

3. Methods

3.1. RNA Miniprep

Collect embryonic tissues on ice-cold Hank's balanced saline solution (Sigma Chemical Co., St. Louis, MO) using standard RNase-free conditions (*see* **Note 1**). Lyse embryos in 0.35 mL Lysis Buffer RLT and store them at –70°C. Thaw samples, sonicate at 40 W for 2 s, add an equal volume of 70% ethanol, and apply the suspension to the RNeasy spin column. Centrifuge for 15 s at 8000*g* and wash with 0.7 mL wash buffer RW1 and twice with 0.5 mL wash buffer RPE. Finally, elute the RNA with 0.05 mL diethylpyrocarbonate (DEPC)-treated water and centrifuge for 60 s. If needed, treat RNA with RNase-free DNase I to remove genomic DNA (*see* **Note 2**).

3.2. Reverse Transcription

Reverse transcribe the RNA (0.2 µg) into three distinct display pools using single-base 3'-anchored oligo(dT) primers *(3)*. Add to the reaction buffer 20 µ*M* each of dNTP (dATP, dGTP, dCTP, dTTP) and 0.2 µ*M* of a single 3'-anchored oligo(dT) primer. Reverse transcription is performed in the DNA thermal cycler (65°C, 5 min; 37°C, 60 min; 75°C, 5 min); 100 U of MMLV reverse transcriptase is added to the mixture 10 min into the 37°C step.

3.3. Arbitrary PCR

The 40-µL reactions are set up to contain 4 µL of the cDNA reaction mix, 2 µ*M* each of dNTP, 1.5 m*M* of $MgCl_2$, 50 m*M* of KCl, 0.2 µ*M* of the appropriate oligo(dT) primer, 0.2 µ*M* of arbitrary primer, and 2 U of *Taq* DNA polymerase (Gibco-BRL) in 10 m*M* Tris-HCl, pH 8.3, cycled for 40 rounds (94°C, 30 s; 40°C, 2 min; 72°C, 30 s), followed by final extension at 72°C for 5 min (*see* **Note 3**).

3.4. HPLC Conditions

DNA is eluted from the analytical column at 50°C using a NaCl gradient and a constant flow rate of 0.75 mL/min (*see* **Note 4**). The characteristics of the NaCl gradient must be determined empirically. In our studies, the starting condition was 0.4 *M* NaCl and the NaCl concentration was increased to 0.52 *M* immediately after the sample was injected. The Model 680 Automated Gradient Controller from Waters Associates was used to develop a 0.52 *M*–0.67 *M* NaCl convex gradient (curve no. 4) during a 25-min period. Monitor absorbance at 262 nm, or, if available, use photodiode array detection covering 252–272 nm. Between runs, the column must be flushed with 0.5 mL of 0.1 *N* phosphoric acid and equilibrated to starting conditions for at least 12 min before the next sample is injected.

3.5. Calibration

Daily calibration of the column is required. Use a ladder of restriction fragments generated from digestion of ΦX174 RF DNA with *Hae*III. Typically, 8 of the 11 fragments (72, 118, 194, 234, 603, 872, 1078, and 1353 bp) are separated with baseline resolution and 3 (271, 281, and 310 bp) resolve more or less depending on the age of the column; progressive deterioration of column performance may be noticed after 200 sample injections.

3.6. RNA Profiling

For side-by-side comparison of samples, it is best to alternately inject related PCR reaction mixes for samples A and B that are being compared. An advantage of HPLC over conventional display methods is the ability to move rapidly through different PCR primer combinations (*see* **Fig. 1**). Typically, eight samples can be processed in one afternoon (*see* **Note 5**).

3.7. Peak Analysis

Integrated DNA peaks are directly compared between sample pairs by the formula (A^2+B^2/2AB), where A and B represent the areas of display peaks from the two samples that are being compared. When a difference is found, the samples can be easily repeated to confirm the difference (*see* **Note 6**).

3.8. Peak Isolation

To collect a differential display product, scale up the arbitrary PCR reaction eightfold to 320 µL total; collect the fraction containing the differential peak and reduce it to 0.4 mL in a SpeedVac concentrator (Savant Instruments, Inc., Farmingdale, NY).

3.9. DNA Recovery

Dilute the reduced HPLC fraction to 1 mL with TTE buffer and load it onto a pre-equilibrated Nensorb 20 cartridge. Wash the cartridge with 3 mL of TTE buffer and 3 mL of water; elute the DNA with 0.5 mL of 50% (v/v) methanol (HPLC grade) in water. Evaporate the eluent to dryness and reconstitute the DNA in 15 µL of sterile water.

3.10. Reamplification

A small aliquot of the reconstituted DNA should be subjected to PCR amplification with the appropriate arbitrary primers, using 20 or 200 µ*M* dNTPs (*see* **Note 7**). This PCR product can now be T/A cloned into a sequencing vector such as pCR-TRAP (GenHunter) or pGEM-T Easy (Promega, Madison, WI).

4. Notes

1. RNA species larger than 200 nucleotides are quantitatively isolated by this miniprep. RNeasy (Qiagen) yields high-quality RNA ($A_{260/280}$ = 2.0) from very small amounts of embryonic tissue, sufficient to start with individual mouse embryos on gestational day 8 (GD8) or the equivalent amount of 10–20 µg of protein.
2. Complete removal of genomic DNA is essential. Negative control reactions must be performed where the reverse transcription, either PCR primer, or *Taq* polymerase steps have been omitted. Mitochondrial DNA is prone to contaminate the sample; thus, several hours of DNase digestion may be required to remove it entirely.

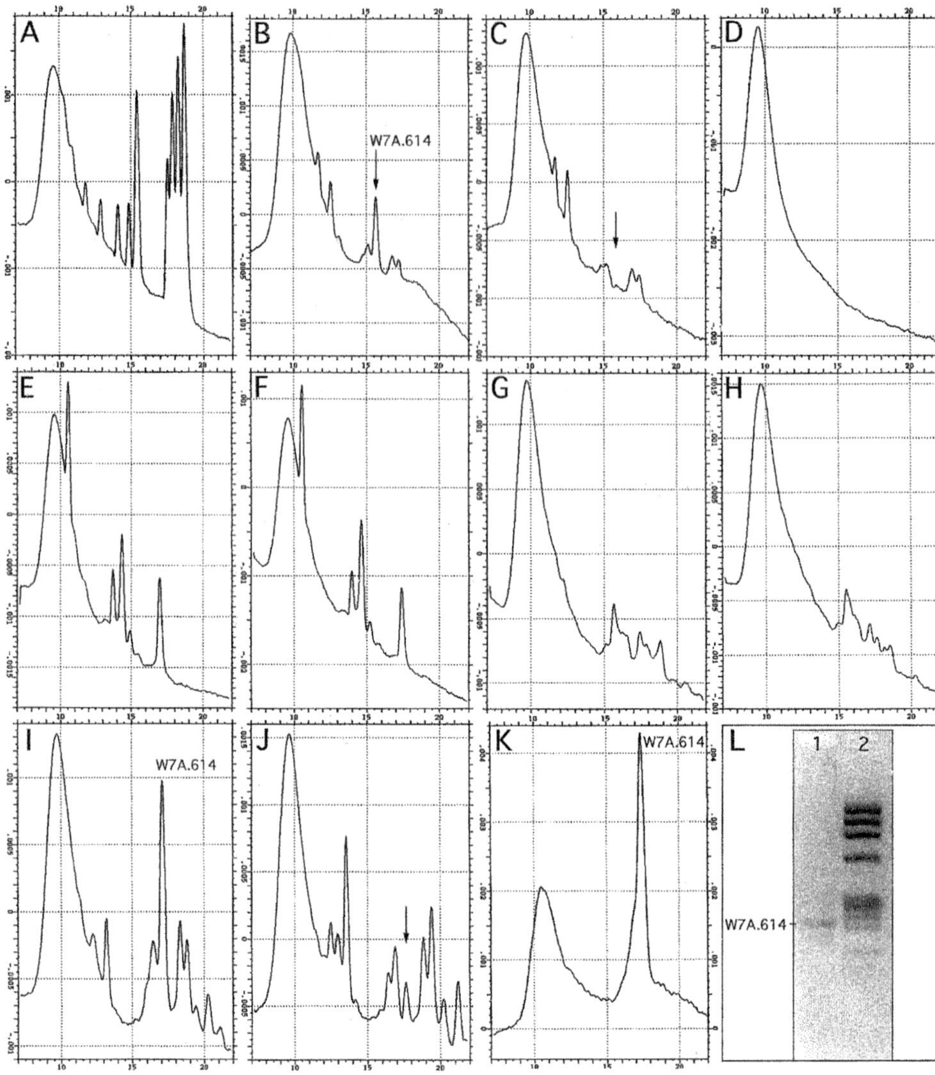

Fig. 1. RNA profiles of p53 (+/+) and p53 (–/–) embryos analyzed by HPLC-based differential display at the 3–7 somite-pair stage of development. Arbitrary PCR products resolved by HPLC showing the region corresponding to the NaCl gradient. **(A)** *Hae*III ΦX174 RF DNA ladder: 72, 118, 194, 234, (271 + 281 + 310), 603, 872, 1078, and 1353 bp. **(B,C)** Display of transcripts in p53 (+/+) and p53 (–/–) embryos, respectively. Underrepresentation of W7A.614 (16S ribosomal RNA) revealed in p53 (–/–) embryos by primer pairs H-AP7 and H-T_{11}A and 2 μ*M* dNTPs. Quantitative PCR and biochemical analyses confirmed a bioenergetic defect in this mutant *(5)*. **(D)** Negative control reaction omitting the reverse-transcription step; baseline fluctuation is due to the NaCl gradient. **(E,F)** Equal display of transcripts in p53 (+/+) and p53 (–/–) embryos, respectively, amplified with primers H-AP1 and H-T_{11}A. **(G,H)** Equal display in the same pools with primer set H-AP7 and H-T_{11}G. **(I,J)** Altered display of W7A.614 with primer pairs H-AP7 and H-T_{11}A and 200 μ*M* dNTP. **(K)** HPLC purification of W7A.614 for cloning and sequencing; the purified DNA was reamplified with primers H-AP7 and H-T_{11}A and 20 μ*M* dNTPs. **(L)** Electrophoresis of purified W7A.614 on 1.5% agarose gel with ethidium bromide staining (lane 1); ΦX174 RF DNA cut with *Hae*III (lane 2). (From **ref. 5.**)

3. The HPLC-based method afforded simple and reproducible patterns of arbitrary PCR products in the 150–600-bp range. Differences between samples could be spotted quickly. Also, the differences were probably kept to a minimum by the low peak complexity (5–10 peaks per primer set). Increasing the dNTP concentration to 20 or 200 μ*M* increases the size and complexity of the HPLC peaks.
4. Running the column at high temperature (i.e., 50–60°C) reduces the effects of secondary structure and lowers the operating pressure ***(4)***. These effects are further promoted by the presence of 10% acetonitrile in the mobile phase, which is optional. The NaCl gradient can be easily adjusted to resolve PCR products in a particular size range, although conditions given above nicely separates PCR products in the 150–600 bp range.
5. Because arbitrary PCR primers hybridize as 8-mers ***(1,3)***, theoretical amplification of cDNA fragments in the 150–600-bp range should sample 1 of every 146 cDNAs (4^8 divided by 450). For the typical mammalian cell expressing 15,000 different mRNA species, an eight arbitrary primer screen should amplify 822 differential display peaks. In practice, 20.3% of the predicted number were actually observed ***(5)***.
6. When single-pass HPLC screening reveals peaks differing by twofold, the display reaction should be repeated twice to verify reproducibility. In our experience, peak differences show up in 16.7% of assays during single-pass differential display. Thus, a general rule to be certain of a difference is to perform triple-pass screening whenever a primer pair reveals a potential difference. This cuts the incidence of false positives below 0.5%.
7. Another stringent test is to repeat the display reaction with higher (20 and 200 μ*M*) dNTP concentrations whenever a reproducible difference is observed in triple-pass screens. This provides PCR reactions under conditions more favorable for the kinetics of Taq polymerase ***(2)***. Finally, candidates that pass these criteria should be tested by quantitative PCR with gene-specific primers or direct analysis by Northern blotting to confirm differences in RNA abundance at the steady state.

Acknowledgment

This work was supported by grant RO1 HD30302 from the National Institute of Child Health and Human Development.

References

1. Liang, P. and Pardee, A. B. (1992) Differential display of eukaryotic messenger RNA by means of the polymerase chain reaction. *Science* **257,** 967–971.
2. Miyashiro, K., Dichter M., and Eberwine, J. (1994) On the nature and differential distribution of mRNAs in hippocampal neurites: implications for neuronal functioning. *Proc. Natl. Acad. Sci. USA* **91,** 10,800–10,804.
3. Liang, P., Zhu, W., Zhang, X., Guo, Z., O'Connell, R.P., Averboukh, L., Wang, F., and Pardee, A.B. (1994). Differential display using one-base anchored oligo-dT primers. *Nucleic Acids Res.* **22,** 5763–5764.
4. Warren, W. and Doniger, J. (1991). HPLC purification of polymerase chain reaction products for direct sequencing. *BioTechniques* **10,** 216–219.
5. Ibrahim, M.M., Razmara, M., Nguyen, D., Donahue, R.J., Wubah, J.A., and Knudsen, T.B. (1998) Altered expression of mitochondrial 16S ribosomal RNA in p53-deficient mouse embryos revealed by differential display. *Biochem. Biophys. Acta* **1403,** 254–264.

V

Transgenesis: *Production and Gene Knockout*

34

Production of Transgenic *Drosophila*

Miki Fujioka, James B. Jaynes, Amy Bejsovec, and Michael Weir

1. Introduction

The generation of transgenic *Drosophila* has become a standard technique for the investigation of a wide variety of biological questions and phenomena. Virtually any cloned gene can be inserted into the genome using the P transposable element system. The gene of interest is first subcloned between P-element ends. The transgene-containing DNA is then injected into the posterior cytoplasm of young embryos in the presence of P-element transposase activity. The DNA enters pole cell nuclei, which give rise to the future germline, where it is integrated into the genome through the action of the P-transposase. The transgene is thereafter stably maintained and inherited. A selectable marker gene (e.g., an eye color marker) is included between the P-element ends to allow identification of transgenic flies and genetic mapping of the transgene's chromosomal location.

Presented here is a basic method for obtaining transgenic flies at high efficiency using a standard transformation vector, pCaSpeR (**ref. *1*** and references therein), which contains the most widely used selectable gene for this technique, the eye color gene, *white*. The basic method described does not, however, depend on the use of this gene, but it can be adapted with minor modifications to use with any of the other available selectable markers, such as *rosy* and *neomycin-resistance* (**Note 1**). We include in the **Subheading 4.** a description of alternative methodologies for preparing DNA and embryos for injection, as well as details that will aid beginners in quickly achieving a reasonably high transformation efficiency.

2. Materials

2.1. Solutions for PEG Preparation of Plasmid DNA

1. Buffer 1: 50 m*M* Tris-HCl (pH 8), 10 m*M* EDTA; store at room temperature; add RNase A to final concentration of 100 µg/mL; store at 4°C for up to 6 mo.
2. Buffer 2: 200 m*M* NaOH, 1% sodium dodecyl sulfate (SDS); make fresh or store at room temperature (RT) in a well-sealed plastic container for up to 3 mo.
3. Buffer 3: 3 *M* potassium acetate (pH 5.5); store at 4°C.
4. PEG solution: 13% polyethylene glycol 8000, 1.6*M* NaOH; filter-sterilize, store at RT.
5. 10 *M* ammonium acetate; store at RT for up to 6 mo, or in aliquots at -20°C.
6. TE: 10 m*M* Tris-HCl (pH 7.2), 1 m*M* EDTA (pH 8); store as 10X stock at RT.

From: *Methods in Molecular Biology, Vol. 136: Developmental Biology Protocols, Vol. II*
Edited by: R. S. Tuan and C. W. Lo © Humana Press Inc., Totowa, NJ

2.2. Collection

1. Grape juice plates: Mix 30 g agar in 1 L of H_2O, melt in microwave; separately mix 360 mL of grape juice, 2 g methyl paraben (*p*-hydroxybenzoic acid, methyl ester [Sigma Chemicals, St. Louis, MO]; a mold inhibitor), 30 g of commercial sugar, and dissolve while warming on a stirring hot plate; add grape juice mix to melted agar, stir well; pour into 100-mm Petri plates.
2. Yeast paste: Mix dry active yeast into water a little at a time, until it becomes the consistency of peanut butter (the ratio of yeast to water varies with the batch of yeast: roughly 30 g in 50 mL of water); add some molasses (the color becomes darker and the consistency softer than peanut butter, but not so thin as to "run" after spreading on a plate (**Note 2**).

2.3. Other Materials and Chemicals

1. Coffee filters, basket type, commercial.
2. Phenol, chloroform.
3. Halocarbon oil, series 700 (*see* **Note 3**) cas#9002-83-9 (Halocarbon Products Corporation, River Edge, NJ).
4. Bleach (commercial).
5. Ultrafree®Probind filter unit (Millipore UFC3-IPH [Millipore, Bedford, MA]), or equivalent; Ultrafree®-MC filter unit, 30,000 NMWL (Millipore UFC3-TTK, polysulfone membrane), or equivalent.
6. Double-stick tape (3M Co., St. Paul, MN), 12 mm wide (catalog no. 021200-01033) (alternative: type 415, [3M Co.] 12 mm width, catalog no. 021200-07148). Note that some other brands of tape can be toxic to embryos.
7. Plastic disposable beakers, 400 mL, Fisherbrand™ Tricornered polypropylene (Fisher Scientific, St. Louis, MO).
8. Glass capillary tubing, Borosil, 1.2 mm outer diameter × 0.6 mm inner diameter Omega dot fiber for rapid fill, 100 mm length, catalog #30-31-1 (FHC, Brunswick, ME) or equivalent.
9. Forceps (Dumont #5 Biologie).

2.4. Required Equipment

1. Compound microscope (inverted, optional), phase contrast, with 10X eyepiece and 40x objective.
2. Zoom stereo microscope with sliding stage (**Fig. 1**).
3. Vibration isolation table (optional).
4. Micropipet puller (Narishige [Tokyo, Japan] model PB-7, or equivalent).
5. CO_2 tank equipped with compressed gas regulator (two-stage, with the output stage set at 15 psi) and hand-held flow controller (optional) ("blow gun," part no. 415-S, from Parker Fluid Connectors [Minneapolis, MN].
6. Micromanipulator (Narishige MO-150, or equivalent).

3. Methods

3.1. Preparation of DNA

DNA for injection is best purified by the polyethylene glycol (PEG) precipitation method (*see* **Note 4**). Thirty milliliters of bacterial culture grown in Terrific Broth (*see*, e.g., **ref. 2**) yields enough DNA for injection. Spin-down bacteria at approx 1000*g*, 15 min., in 50-mL conical bottom tubes and remove as much medium as possible with a micropipet. Suspend the bacteria in 4 mL of buffer 1. Add 4 mL of buffer 2, slowly invert the tube several times (do not shake), and incubate for 5 min at RT. Add 4 mL of ice cold buffer 3 and shake vigorously; leave on ice for 15 min. Spin at approx 2000*g*,

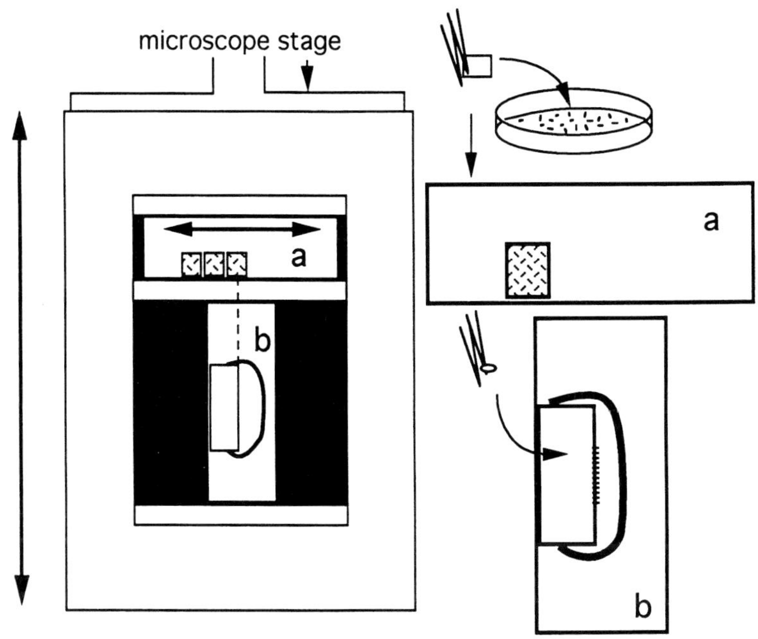

Fig. 1. A sliding stage covering the fixed stage of a stereomicroscope. This stage can move so that slides a and b can be used sequentially while looking through the microscope. The sliding stage is a plexiglass plate with two narrow, parallel "feet," each running the length of an edge. A smooth cardboard support cut to fit inside the feet serves as the runner along which the plexiglass moves (the cardboard support must be thicker than the feet, so that it supports the plexiglass). This support is attached to the microscope stage (tape can be used if the microscope is not dedicated to this task). The center section of the top surface of the plexiglass is covered with black vinyl tape. Three sets of thin, narrow strips of smooth cardboard are attached as in the figure to "frame" slides a and b, as shown. This arrangement allows slide a, which holds embryos prior to dechorionation, to be moved along its length as dechorionation progresses. Each dechorionated embryo is transferred to slide b by moving the stage, which is then moved back to repeat the process. Eggs are transferred to slide a on double-stick tape, as described in the text. It is best to place them near slide b, so that the stage can be moved a minimum distance during each cycle of dechorionation and transfer. Slide b contains a piece of double-stick tape enclosed by a grease pencil mark, as described in the text.

for 15 min. While spinning, prepare another 50-mL conical tube with a funnel containing a coffee filter cut to an appropriate size to fit the funnel (sterilization is not required). Filter solution (gravity flow) and add 12 mL of isopropanol. Mix well (by inverting or vortexing) and spin at 4°C, 2000*g* (or more), for 15 min. Decant the supernatant, and wash with 70% ethanol. Dry the precipitate well. Dissolve in 650 µL of TE containing 20 µg/mL of RNase A and incubate for 30 min at RT. Transfer the DNA solution to a microcentrifuge tube and add an equal volume of PEG solution. Vortex and immediately spin at 8000*g* (or more), 4°C, for 5 min. Discard the supernatant, dissolve the pellet in 400 µL TE, and vortex for 10–20 s. There will be some undissolved pellet. Extract twice with phenol, once with phenol-chloroform, and once with chloroform alone. Add 100 µL of 10 *M* ammonium acetate and 1 mL of ethanol, mix well, and spin-down DNA at 4°C, for 10 min at 8000*g* (or more). Wash pellet with 70% ethanol, respin, and dry the pellet well. Dissolve the DNA in 200 µL TE. (Typical concentration is approx 1 mg/mL).

Measure OD_{260}, and dilute the DNA solution to 100 µg/mL or less (*see* **Note 5**) with H_2O (*see* **Note 6**). After dilution, DNA may be filtered in order to avoid forming bubbles in the needle: transfer diluted DNA (at least 50 µL, without methanol activation of the filter) into the top part of a Probind filter and spin at 8000*g* (or more) for a 2–3 min, or until sufficient solution is filtered (*see* **Note 7**). Add "wings-clipped" (P-transposase-encoding) plasmid DNA to a final concentration of 50 µg/mL. (If a stock of wings-clipped DNA is prepared at a concentration near 1 mg/mL, filtered as described above, adding this will not significantly affect the concentration of the transformation vector DNA; we typically add 0.5 µL of wings-clipped to 10 µL of the transformation construct. Also, *see* **Note 8**.) It may also be useful to add 1/10 volume of McCormick's green food coloring, filter sterilized, to keep track visually of how much DNA is being injected and to help monitor when the needle is beginning to clog. This should be added just before the DNA is used for injection, and the DNA solution should be centrifuged at full speed in a microcentrifuge just before loading into a needle (*see* **Subheading 3.2.**).

3.2. Preparation of Embryos

The recipient fly line is *yw* (*yellow⁻ white⁻*; we generally use y Df(1)w^{67c23}, a white deficiency that cannot spontaneously revert) (*see* **Note 8**). About 200 recently eclosed flies (less than 2–3 d old) are transferred to the collection cage at least 2 d before injection. Cover the open side with a grape juice plate with yeast paste and feed them regularly, at least twice a day (*see* **Note 9**). As collection cages, we use 400-mL disposable beakers (with several small holes made by a 25-gauge needle, to prevent moisture buildup) (*see* **Note 10**). When collecting embryos for injection, change the plate every 30 min. Do not use the first two collections, as they will contain many older embryos (partially developed at the time of laying). Air-dry the plate with collected embryos for about 5 min. Meanwhile, several slides should be prepared for the injection. Put about 3 cm of double-stick tape, cut with scissors to prevent stretching, on the surface of a slide (**Fig. 1, slide b**; if tape is stretched, it will not remain flat). Draw a circle on the slide with a grease pencil (china marker), surrounding most of the tape, to prevent spreading of the oil (*see* **Note 3**). To transfer embryos from the plate, lightly touch eggs on the plate with double-stick tape. Do not touch the surface of the tape with your finger. If the embryos do not stick to the tape, dry the plate more (airflow from microscope light source hastens this process). When the tape picks up many embryos, place it on a slide (**Fig. 1, slide a**) with the egg side up. With narrow-tipped forceps (not too sharp), push lightly on the ventral or lateral side of an embryo, depending on its orientation, so that the pressure is slightly downward and mostly sideways (so as to "roll" the embryo). The chorion should break easily. (Do not move embryos around on the tape, as this will damage them. If you have trouble removing embryos from the chorion, wet the forceps with a damp Kimwipe [Kimberly-Clark, Roswell, GA].) Pick up the dechorionated embryo by touching it lightly with one tip of a forceps. While the embryo is on the forceps, adjust its direction (see below) by touching it to other embryos, chorions, or micropyles ("antennae") on the slide, so that you do not have to move it after you put it down on the tape. (*See* **Note 11** for an alternative dechorionation and embryo-transfer procedure. *See* **Note 12** for an alternative injection procedure without dechorionation.)

Embryos should be lined up on the slide previously prepared with double-stick tape, with a spacing of about 10 embryos/cm (*see* **Fig. 1**). The posterior of each embryo should be on the tape, whereas the anterior part (one-third to one-half) should hang over the edge of the tape, for better survival rate (possibly due to better gas exchange through the oil). The lateral side of the embryo should be down on the tape so that you can identify the dorsal and ventral side during injection. While learning, line up the embryos on one slide for a maximum of 2–3 min, then proceed to desiccation and injection, because if more time is taken, the dehydration of the embryos will vary too much. A skilled worker can line up 10–15 embryos/slide with consistent dryness.

After dechorionation and placement on the slide, the embryos are desiccated and then covered with halocarbon oil. For desiccation, prepare several small plastic boxes with lids (e.g., a small slide case) half filled with Drierite (W. A. Hammond Drierite Co., Xenia, OH) ahead of time. (Drierite will last much longer in such containers if a light coating of petroleum jelly or silicon grease is applied to seal the edges of the box.) Place the slide with embryos in a desiccation box. Dryness is probably the most critical variable affecting the survival rate. At the beginning of each set of injections, several slides should be tested to determine how long embryos should be dessicated on that day (optimal time can vary depending on ambient temperature and humidity). This is done by trying a series of desiccation times (e.g., 3, 4, and 5 min) and checking by actual injection. If the vitelline membrane is pushed by an injection needle and does not come back to its original shape immediately, the embryo is too dry. On the other hand, if the needle can pierce the embryo smoothly, but cytoplasm leaks out immediately after the needle is pulled out, it is too wet. The quickest way to find out the appropriate desiccation time is to record the temperature and humidity each time you inject, along with the corresponding desiccation time. In our laboratory, it seems that the desiccation timing correlates better with temperature than with humidity. Typical desiccation times in our laboratory are 4–5 min at 25°C and 5–6 min at 24°C.

Immediately following desiccation, pipet oil onto the slide, covering the line of embryos from one end to the other. If the oil is too thick, the embryos will die from lack of gas exchange (*see* **Note 3**), whereas if it does not cover them completely, they will die as a result of drying. If several slides are prepared with embryos prior to injection (we typically prepare five or six at a time), the covered embryos should be placed into a humidified chamber before (as well as after) injection. If an 18°C incubator is conveniently available, it can be used to slow the development of the embryos waiting to be injected.

3.3. Injection

Because the injection must be done before pole cells form, it is better to use eggs within 1 h of collection. You can use eggs up to 1.5 h after collection, however, there will usually be a significant number of older embryos, which should be killed during the injection process by piercing with the needle. Alternatively, to avoid clogging the needle with debris from the embryos, record the position of overaged embryos and remove them after injection using forceps.

Needles should be prepared ahead of time. Precise adjustment of the micropipet puller is required. The pulled capillary must be thin enough so that the thin, rounded end is not visible to the naked eye. The stretched part of the needle (each "half" after pulling) should be about 1 cm long. Once the puller is adjusted to give good needles that work well after breaking (*see* later in this paragraph), which may take some experience to determine, readjustment is usually not necessary. Thus, it is useful to have a puller dedicated to this task. Otherwise, it is a good idea to "pull" a large number of needles at once, so that when a batch turns out to be optimal, it will last for some time. While dechorionating the embryos, one or more needles are filled with DNA. Pipet 0.3 μL of DNA onto the large open end of the needle. A simple way to do this is to tape the needle on the edge of a bench or table with the point hanging downward at about a 30° angle to vertical. The DNA can be pipetted onto the upper end as a "bead," which will slowly disappear as the liquid moves into the needle by capillary action. Alternatively, DNA can be transferred into the large end of the needle using a drawn out Pasteur pipet. Make sure there are no bubbles left in the solution in the needle, particularly near the fine end (typically, bubbles will initially form, but will move away from the tip if the DNA solution is not too viscous). If there are bubbles, discard this needle and use another. Attach the needle to the compressed gas line using a small-gauge plastic tubing, with an appropriate control device in between (**Note 13**), then mount it on the micromanipulator. Mount a slide with the lined-up dechorionated embryos on the stage, with their posterior ends facing the needle. Under the microscope, break the fine end of the needle: Focus on the tip, then move the stage so that the edge of the glass slide just comes into contact with the needle tip (the needle and the edge of the slide are perpendicular to each other) . Move the slide away and see if the tip is broken by placing it into the oil and releasing pressure (liquid should appear as a "bubble" in the oil). The broken tip should be as sharp as possible, while allowing DNA solution to flow out when pressure is applied. If the needle tip is too wide, the success rate will be low as a result of the leakage of cytoplasm after injection, causing both a low hatching rate and sterility. (With our pressure delivery system, this usually manifests itself as the DNA solution flowing "back" along the outside surface of the needle, a result of the high flow rate, rather than forming a bubble in the oil surrounding the tip. The tip should be less than 3 μm in diameter, and preferably less than 1 μm.) During injection, the needle should be inserted smoothly. On most setups, it works best to move the stage rather than to use the micromanipulator controls. In order to line up the needle with each embryo in the vertical plane, however, the "up and down" micromanipulator control is used. The needle should be inserted into the dorsal half of the posterior end, where pole cells will form. Touch the embryo to the needle (the end of the needle should be in focus) to determine where the needle makes contact, then adjust the height of the needle with the micromanipulator, if necessary. When the needle makes contact at the desired point, insert the needle by moving the stage fairly quickly and firmly, so that the needle enters smoothly and quickly. Pull the embryo back so that the needle tip is as close to the vitelline membrane as possible, then release pressure to inject the DNA. Inject enough to see easily, as "movement" in the cytoplasm, but not so much as to cause leakage of cytoplasm when you pull the needle out, which is done, again, by moving the microscope stage.

Sometimes, the needle becomes clogged by debris from the injected embryos. You can often rebreak the needle to restore flow, by, again, lightly touching the tip of the needle with the edge of the slide, testing for a restored opening by applying pressure and observing the emergence of solution. Of course, the second break may result in too broad a tip for successful injection, requiring use of a new needle. In anticipation of this problem, we usually prepare at least two needles with the DNA solution at the outset.

After injection, the embryos should be kept at 25°C for more than 22 h, or at 18°C for more than 36 h, under moist conditions (**Note 14**). A plastic container with a tightly fitting lid, containing a wet sponge, is suitable for this. After embryos hatch, pick them up using a dissection needle or the same type of forceps used to dechorionate and place them gently onto the surface of a food vial. When agar-based food is used, the surface should be softened and moistened by scratching the surface with a dissection needle and adding a drop of water. Also, sprinkle a small amount of yeast onto the surface to provide easy-to-eat food for the often weaker than normal larvae.

If you notice larvae dying as a result of bacterial infection, which usually manifests itself as a brown spot spreading from the posterior, you can add antibiotic to the yeast paste used to feed the larvae (an antibiotic not routinely used for other purposes in the lab is recommended, such as kanamycin in our case, to reduce the chance of selecting antibiotic-resistant bacteria).

3.4. Crosses

Separate adult flies within 1 d of eclosion to prevent mating among themselves. To establish independent transformant lines, a very efficient procedure is to cross the "injected" adults individually with a double balancer stock (e.g., *w/w*; *CyO/Sco*; *TM3/D*) containing four marked chromosomes. Examine adult flies in the next generation for w^+ transformants (**Note 15**). Despite the presence of recessive lethal chromosomes, this cross does not apparently decrease the yield of fertile transgenic flies relative to a wild-type cross. Individual transformant flies are crossed with a single balancer stock (e.g., *yw/yw* ; *CyO/Sco* ; +/+, or (and) *yw/yw*; +/+ ; *TM3/D*) to identify which chromosome the transgene is on, based on the segregation pattern of the eye color relative to the dominant markers. This cross also facilitates the establishment of an independent line. (For example, if the transgene proves to be on the second chromosome, P[*w*+]/*CyO* transformants resulting from the cross to the *CyO* stock are subsequently mated, and non-*Cy* progeny, homozygous for the transgene, are then mated to establish a stock.) For this cross, male transgenic flies are preferable, as they usually yield more progeny and do not have to be separated from their siblings while still virgin. Therefore, we usually use only males from vials that yield multiple transformants (all from a single original injected fly). If many transformants result from a vial, it is possible to establish more than one independent line from the single injected parent. However, in this case, there is an increased probability of multiple insertions within one individual, which is usually undesirable. Therefore, we keep transformants from vials with only a few transformed siblings in cases where enough transformants are obtained from such vials. Three transformed individuals from a single injected parent, mated individually to the single balancer stock, are usually enough to ensure a viable line, particularly if they are males. Thus, we generally stop screening from a vial once we have taken three transformants from it.

3.5. Efficiency

The efficiency of obtaining transgenic flies may vary depending on the size of the injected plasmid. In our hands, a similar transformation efficiency is obtained with plasmids that range in size from 12 to 22 kb. The average hatching rate is 40–70%, and the eclosion rate (from hatched embryos) is 60–80%, whereas the overall efficiency of generating transgenic flies (from injected embryos) is 5–20%. We usually inject 40–50 embryos for an individual construct in order to get 5 or more independent lines. In different lines, the eye color phenotype from the *mini-white* gene in pCaSpeR varies depending on the chromosomal insertion site of the transgene: Eye colors can range from light yellow to orange or (rarely) red. These different levels of *mini-white* expression often correlate with the levels of expression of the adjacent transgene.

Sometimes transformation frequencies are much lower than expected. In fact, with some constructs, no transformants are obtained despite injection of large numbers of embryos (and using different DNA preparations of the same construct). This can be the result of dominant (poisoning) effects of the injected transgene that kill the would-be transformants or cause sterility in both sexes. Sometimes, simply reducing the concentration of the injected plasmid solves the problem, possibly because expression of the transgene before insertion is the major source of killing or sterility. If the problem persists despite using very low concentrations (approx 20 μg/mL), under conditions where other constructs injected in parallel do give transformants at a reasonable frequency, more drastic solutions may be considered. Strategies to overcome the problem include using a GAL4 system (**ref.** ***3***), in which the transgene is placed under the control of GAL4 UAS sequences, such that the initially established transgenic line does not express the gene. The transgene is subsequently induced by crossing in a second transgene that expresses the GAL4 activator. An alternative strategy involves placing an FRT cassette containing poly-adenylation signals (inserted between two FRT sequences) between the promoter and the open reading frame of the transgene (**refs.** ***4*** and ***5***). After obtaining transformants, the FRT cassette can be removed by introducing a transgene expressing FLPase (which catalyzes recombination between the two FRT sequences, thereby removing the inhibitory poly-A signal). When using either of these strategies to express transgenes with poisoning effects, it is best to maintain the transgenic stock in a form that does not allow expression of the transgene (i.e., without the GAL4 expression transgene, or with the FRT cassette intact) in order to minimize the accumulation of modifying loci that neutralize the function (poisoning effect) of the transgene.

If transformation frequencies are low with a transgene whose protein product is expected to be functionally neutral (e.g., a *LacZ* reporter gene), this can be the result of regulatory sequences of the transgene causing repression of the nearby *mini-white* gene, so that transformants have undetectable *white*$^+$ eye color. If transformants are recovered, they generally have very faint eye color and the integration sites are probably biased toward those that augment expression of both *mini-white* and the transgene. We have found that this problem can be overcome by including Glass binding sites (**ref.** ***6***) upstream of the promoter of the *mini-white* selectable marker gene. The Glass sites increase the expression of *white*, because of induction by the Glass protein in the eye–antennal disk (and other photoreceptor-containing tissues, as well as some cells in the brain; *see* **ref.** ***6***), thereby allowing detection of transformant flies despite the repressive effect of the transgene. (pCaSpeR-GL, a transformation vector with introduced Glass sites, is available upon request from M.F. and J.B.J.) However, this approach may also increase the probability of recovering transgene insertions into heterochromatic or other repressive DNA, which may result in repression of transgene expression.

4. Notes

1. Some details of the method describe the use of a pCaSpeR-based vector carrying the *mini-white* gene as a selectable marker. To use *rosy* as a selection marker, the injection stock is typically *rosy*^{506}, and the vector is Carnegie 20 (**ref.** *7*). To use neomycin resistance, virtually any stock can be used. Examples of vectors and typical selection conditions are given in **ref.** *8*.
2. There are several alternative recipes for embryo collection plates. Another common recipe uses 2% agar, glacial acetic acid (2 drops per 100 mL) and ethanol (2 drops per 100 mL)—the acetic acid and ethanol are both added when the agar solution has cooled after boiling. Just before use, each plate is seeded with some grains of dried baker's yeast, and glacial acetic acid which is blotted almost dry with Kimwipes.
3. While the injected embryos are incubating, the oil will tend to spread, causing the embryos to be uncovered and to dehydrate. Drawing a circle around the double-stick tape with a china marker (grease pencil) will restrict the spread. However, a thick layer of oil will also kill embryos, so use just enough oil to keep the embryos slightly covered until they hatch. Using series 400 halocarbon oil (same Catalog number), which is lighter weight, eliminates the possibility of applying too thick a layer, but it makes it more difficult to keep embryos covered as the oil spreads.
4. Plasmid DNA purified by a CsCl gradient or commercial column may also be used for injection. However, these methods sometimes yield a viscous DNA solution that causes bubbles to remain in the needle, resulting in a blockage of flow during injection. Highly purified DNA (e.g., twice CsCl gradient purified) is not necessary.
5. A DNA concentration greater than 500 µg/mL may dramatically reduce the surviving rate of the larvae. Even with DNA concentrations as low as 20 µg/mL, transgenic efficiency can be high. DNA prepared by this PEG method has a tendency to have a lower concentration than its A_{260} reading would indicate.
6. Phosphate-buffered injection solution (e.g., 5 m*M* KCl, 0.1 m*M* PO_4^{3+}, pH 7.8) is often used to dilute DNA for injection. However, 1X TE or less, which is generated by dilution of the PEG-prep DNA with sterile, deionized water to a final concentration of around 100 µg/mL, does not apparently affect survival rates of larvae or the efficiency of obtaining transformants.
7. DNA prepared by CsCl gradient or by commercial column (e.g., Qiagen, Chatsworth, CA) may have some difficulty in passing through the filter, and a significant portion may be lost. Also, as warned by the manufacturer, when a plasmid is more than 10 kb, you may lose one-tenth to one-half of it on the column. After filtration, run a sample of the DNA on an agarose gel to check the final concentration. You will not usually need to readjust the concentration within this range of loss. However, if desired, DNA may be concentrated by using a 30,000 NMWL filter (Millipore).
8. Instead of the *yw* recipient line, DNA can be injected into *w; SbΔ2–3/TM6* embryos, which contain a chromosomally supplied transposase activity. In this case, it is not necessary to add P-transposase-encoding plasmid DNA. Using chromosomally supplied transposase results in very high transformation frequencies, often with multiple insertions. In addition, when such a stock is used, half of the injected flies will be homozygous for recessive lethal mutations, reducing the survival rate accordingly.
9. Do not allow grape juice plates to dry too much after pouring, as the flies must obtain moisture from them, and they will not lay as readily on them if they are too dry. Yeast paste should be freshly spread in a thin layer over roughly one-fourth of the surface of the plate, to provide food for the flies during collection. The cage should be changed at least every 4 or 5 d, because in a dirty cage, the flies will begin to lay on the side of the cage rather than on the collection plate. Flies older than about 12 d should not be used for collection.

10. Alternatively, embryo collections can be made in inverted 1/2-pint glass (or plastic) milk bottles containing no food, but taped closed with a collection plate (agar/yeast in lids from 35-mm Petri dishes—*see* **Note 2**). Then, 100–200 adults 2–10 d old should be transferred to new food 2 d before collections are made.
11. Dechorionation using Clorox is also common. Pipet water onto the plate. Gently brush the surface with a paint brush (brush size #3, camel hair) and pour the eggs into a small collection basket. Rinse with water and transfer to 50% Clorox for 2.5 min, with intermittent agitation. Rinse well with water and transfer embryos using a brush to a molasses agar block (2% agar, 25% molasses; made by cutting a section out of a poured plate with a razor blade). Using forceps, line up dechorionated embryos at the edge of the molasses agar block with micropyles pointing off the end of the block. Then, pick up the embryos with the "glue-strip" end of a cover slip. (These cover slips are made as follows: Place double-sided 3M tape, Scotch brand, in heptane overnight. Pipet an elongated drop [20 µL] of glue onto each cover slip and allow to air-dry. Store the cover slips in a dust-free chamber for use the same day.) The cover slip is then taped to a microscope slide and placed in the dehydration chamber before injection, as described in the text.
12. For a description of how to inject without dechorionation, *see* **refs. *9*** and ***10***.
13. A 60-mL syringe can be used instead of a compressed gas system, with some loss of control of the amount injected.
14. In previous injection protocols, incubation at 18°C was recommended following injection. However, in our hands, incubation at 25°C does not cause a noticeable decrease in either hatching rate or overall efficiency.
15. Do not expect the injected flies themselves to exhibit the w^+ phenotype, nor all the progeny of an injected fly (typically 1–20% of progeny in a single "positive" vial will be transformed). Also, transformant flies often eclose later in the next generation, after the initial "burst" of adults in a vial. It may still be worthwhile to continue screening for up to 10 d after the first adults emerge.

Acknowledgments

The authors thank the late Tadaatsu Goto for advise and encouragement, and Laurel Appel and Joel D'Angelo for useful comments. This work was supported by NIH grant GM42752 and ACS grant JFRA-430 to M.W., Basil O'Connor Starter Scholar Research Award FY95-1107 from the March of Dimes Birth Defects Foundation and NSF Grant 96-00539 to A.B., and NIH grant GM50231 and NSF grant IBN9808931 to J.B.J.

References

1. Thummel, C. S., Boulet, A. M., and Lipshitz, H. D. (1988) Vectors for Drosophila P-element-mediated transformation and tissue culture transfection. *Gene* **74,** 445–456.
2. Sambrook, J., Fritsch, E. F., and Maniatis, T. (1989) *Molecular Cloning: A Laboratory Manual, 2nd ed.* Cold Spring Harbor Laboratory Press, Cold Spring Harbor, New York.
3. Brand, A. H. and Perrimon, N. (1993) Targeted gene expression as a means of altering cell fates and generating dominant phenotypes. *Development* **118,** 401–415.
4. Basler, K. and Struhl, G. (1994) Compartment boundaries and the control of Drosophila limb pattern by hedgehog protein. *Nature* **368,** 208–214.
5. Buenzow, D. E. and Holmgren, R. (1995) Expression of the Drosophila gooseberry locus defines a subset of neuroblast lineages in the central nervous system. *Devel. Biol.* **170,** 338–349.
6. Ellis, M. C., O'Neill, E. M., and Rubin, G. M. (1993) Expression of Drosophila *glass* protein and evidence for negative regulation of its activity in non-neuronal cells by another DNA-binding protein. *Development* **119,** 855–865.

7. Spradling, A. C. (1986) P element-mediated transformation, in *Drosophila, A Practical Approach* (Roberts, D. B., ed.), IRL, Oxford, England, pp. 175–197.
8. Xu, T. and Rubin, G. M. (1993) Analysis of genetic mosaics in developing and adult Drosophila tissues. *Development* **117,** 1223–1237.
9. Bartoszewski, S. and Gibson, J. B. (1994) Technical notes: Injecting un-dechorionated eggs of Drosophila melanogaster under ethanol. *Droso. Info. Newsletter* 14. Available online at URL http://flybase.bio.indiana.edu:82/docs/news/DIN/dinvol14.txt
10. Robertson, H. M., Preston, C. R., Phillis, R. W., Johnson-Schlitz, D. M., Benz, W. K., and Engels, W. R. (1988) A stable genomic source of P element transposase in Drosophila melanogaster. *Genetics* **118,** 461–470.

35

Transgenic Manipulation of the Sea Urchin Embryo

Jonathan P. Rast

1. Introduction

The *cis*-regulatory regions of developmentally active genes offer an accessible inroad from which to approach an understanding of the transcriptional networks that coordinate embryogenesis *(1)*. In a developmental context, *cis*-regulatory regions control not only transcriptional amplitude but also spatial and temporal aspects of gene activation. The binding sites that impart this multidimensional control are often numerous and the proteins that recognize them may participate in complex interactions among themselves (e.g., *see* **ref. 2**). In order to unravel the processes that control the transcription of any particular gene, transgenic experiments placing a variety of reporter constructs into authentic developmental contexts are needed. The sea urchin larva has proven to be a highly tractable model system for this type of research. Thousands of transgenic embryos can routinely be produced by an investigator during a single day's work, resulting in the acquisition of data from 10 or more reporter constructs. Gametes are available in immense numbers enabling the isolation of rare transcription factors purely by their affinity to their DNA binding sites (*3,4*; *see also* Coffman and Leahy in Developmental Biology Protocols, Vol. I). Thus it is possible to efficiently climb up developmental regulatory networks once the function of the *cis*-regulatory region of a gene of interest has been defined. In *Strongylocentrotus purpuratus*, the purple sea urchin, the development of a feeding larva takes place over a period of about 3 d. The initial development of this larva is relatively simple, resulting in only about 800 cells and requiring the expression of only approximately 100–200 transcription factors *(5)* and roughly 8500 genes in total *(6)* during gastrulation. Thus, a global understanding of the gene regulatory networks that control larval development represents a relatively approachable problem.

The methods described here are taken from work that is routinely done in our laboratory with *S. purpuratus*. However, other sea urchin species have been used in transgenic experiments (e.g., *Lytechinus variegatus, L. pictus,* and *Paracentrotus lividus*) and these techniques are more or less widely applicable. Although alternative methods have been used to successfully introduce foreign DNA into egg or zygote cytoplasm (e.g., particle gun *[7]*), microinjection of linear DNA has proven to be the most universally useful technique and is described here. Upon injection into egg or zygote cytoplasm,

From: *Methods in Molecular Biology, Vol. 136: Developmental Biology Protocols, Vol. II*
Edited by: R. S. Tuan and C. W. Lo © Humana Press Inc., Totowa, NJ

linear DNA is quickly ligated into a high-molecular-weight concatemer ***(8)***. Later this concatemer appears to be incorporated into the genome ***(9)***, usually in a single nucleus during the second, third, or fourth cleavage division resulting in the formation of a mosaic embryo ***(10)***. Injection directly into the zygotic nucleus ***(11)*** or multiple injection into the cytoplasm ***(12)*** results in somewhat more widespread mosaic incorporation, but it is technically more difficult than a single cytoplasmic injection and is useful only in special cases. Whereas transgenes have been shown to persist through metamorphosis ***(9,11)*** and to be present in the gametes of mature transgenic animals ***(9)***, the large majority of sea urchin transgenic investigations have been focused on the embryonic development of injected specimens. An array of complementary techniques and tools has been developed that enhance the usefulness of this transgenic model system. These include (1) well characterized embryonic promoters that are available to target expression to all or parts of the developing larva ***(13)***, (2) well-developed *in situ* hybridization ***(14,15)*** and green fluorescent protein (GFP) ***(13)*** technologies for direct visualization of spatial expression patterns, and (3) the ability to quantify expression levels from small numbers of transgenic animals using enzymatic assays ***(2)***. Additionally, because DNA is concatenated upon injection, coinjected constructs are incorporated simultaneously, enabling targeted expression of multiple transgenes during embryogenesis. GFP constructs are particularly useful in this regard as visualization of GFP spatial expression can be employed to select mosaic animals that have incorporated a less visible, coinjected construct in a desired region of the embryo ***(13)*** and, along with *in situ* hybridization define where transgenes are present but are not expressed.

2. Materials

The following is a list of materials needed for gamete preparation and egg injection. Materials and techniques required for maintenance of adults or for larval rearing beyond the initiation of feeding are not included here, but they are available elsewhere ***(16,17)***.

1. Filtered seawater (FSW) is made by passing natural or artificial seawater through a 0.45-μm Millipore filter (Millipore, Bedford, MA).
2. Glassware that is suitable for embryological work (i.e., free of detergents and contaminating chemicals). Best if soaked overnight in 0.1 *M* EDTA followed by a thorough rinsing with deionized H_2O.
3. 0.5 *M* KCl for injection of adults to induce spawning.
4. PABA-FSW: 150 mg of *p*-aminobenzoic acid (PABA; Sigma Chemicals, St. Louis, MO, cat. no. A-6928) is dissolved in 500 ml of FSW to make a 2 m*M* solution. PABA is used to inhibit crosslinking of the fertilization envelope to allow easier needle penetration.
5. Penicillin/streptomycin: Antibiotics are diluted into PABA–FSW (or FSW for embryo culture) to give a final concentration of 20 U/mLpenicillin and 50 μg/mL streptomycin. A 1000X stock solution can be made in FSW and stored in aliquots at –20°C.
6. Coated Petri dishes for injection are made from the lids of 60 mm × 15mm polystyrene Petri dishes (Falcon, Los Angeles, CA, 1007). It is important to use only the lids, as they are sufficiently shallow to allow clearance of the microinjection needle during injection. These are treated with 10 ml of a 1% protamine sulfate (Sigma, cat. no. P-4380) in water solution for precisely 1 min, followed by extensive washing with deionized H_2O at room temperature and air-drying (*see* **Note 1**). Coated Petri dishes are stored dry in a dust-free closed container.

7. Mouth pipeting apparatus: An adapter apparatus for mouth pipeting is constructed by attaching a length of small latex tubing (approx 60 cm; 5mm outer diameter [O.D.], 3 mm inner diameter [I.D.]) to a short section of larger tubing (approx 2.0 cm; 12 mm O.D., 6 mm I.D.) of sufficient diameter to hold the large end of a Pasteur pipet. A section of the conical end of a 1000-µL micropipet tip serves well as an adapter between the two tubing sizes. The opposite end of the tubing is fitted with a mouthpiece. It may be helpful to place a small section (approx 2 cm) of glass capillary (or a comparable small-diameter restriction) in the tubing to dampen pressure changes and allow for smoother pressure control.
8. Pasteur pipets: both short (14.6 cm) and long form (22.9 cm).
9. Diamond or tungsten carbide stylus for cleanly breaking pipets.
10. Needles glass: Needles are drawn from Omega DOT borosilicate 1.0-mm O.D./0.75-mm I.D. glass capillaries (e.g., Frederick Haer and Co. [Bowdoinham, ME, cat. no. 03-30-0]). Prior to pulling, the glass capillaries are washed for 1 h in hot 35% nitric acid then extensively washed in distilled water. A final rinse is made in 0.2 µm filtered dH_2O and the glass tubes are placed in a screw-cap glass vial and oven dried at 100°C. Needles are made with a Flaming-Brown needle puller such that, upon breaking, their tip diameter is approx 0.4–0.9 µm.
11. Loading pipet tip: microloading pipet tips (such as Eppendorf microloaders cat. no. 5242956.003) or similarly sized, drawn-out Pasteur pipets.
12. A cold plate for chilling of eggs in Petri dishes can be made by placing a piece of plexiglass over ice in a Styrofoam cooler or by using a refrigerated flowthough plate system. Egg temperature should not exceed 16°C.
13. Pulsed pressure (Picospritzer) or continuous-flow injection apparatus.
14. Micromanipulation apparatus: We use a Leitz three-dimensional micromanipulation device (Leica Microsystems, Deerfield, IL) with a Narishige (Narishige, Tokyo, Japan) axial drive. This is used with an inverted microscope fitted with a chilled stage. The entire setup is mounted on a vibration-damping platform. An upright microscope can also be used as long as the objective lens working distance allows sufficient space for the injection needle.
15. A 16°C incubator for sea urchin embryo culture.

3. Methods

3.1. Preparation of the Injection Solution

Injection solution stock components (1 *M* KCl, sterile glycerol, and sterile dH_2O) are each filtered through 0.2-µm Millipore syringe filters. Carrier DNA (**Note 2**) is prepared by complete restriction digestion of sea urchin genomic DNA followed by two phenol–chloroform–isoamyl alcohol (25:24:1) extractions, two chloroform–isoamyl alcohol (24:1) extractions, and an ethanol–sodium acetate precipitation ***(18)***. Stock solutions should be kept at –20°C. Plasmid construct DNA is prepared by standard alkaline lysis procedures followed by purification on a CsCl gradient or Qiagen Maxiprep columns (Qiagen, Chatsworth, CA). Constructs are linearized by restriction digest, followed by phenol–chloroform extraction and precipitation as described above. DNA is then quantified by spectrophotometry and analyzed by agarose gel electrophoresis to verify complete digestion. The injection solution is combined as in **Table 1**.

Mixed injection solutions can be stored for some time at –20°C but are not routinely kept for more than about 1 mo. Prior to needle filling it may be helpful to spin down the injection solution in a microcentrifuge for 5 min and to take liquid from the surface. This procedure eliminates particulates that may clog the needle.

Table 1
Preparation for 10 µL of Injection Solution

	Amount added	Final concentration
1 *M* KCl	1.2 µL	120 m*M*
Sterile glycerol	2.0 µL	20%
Construct DNA	~50 ng/10 µL injection solution[a]	2500 copies/2 pL[b]
Carrier DNA	250 ng	Approx fivefold mass excess to construct
dH_2O	to 10 µL	

[a]Mass depends on construct size and desired copy number.
[b]Two picoliters is the standard injection volume. Copy number can be varied (*see* **Note 3**).

3.2. Acquiring and Maintaining Gametes

Sex determination in most sea urchin species can only be made by observations of gonads or gametes. Gametes are shed from gravid sea urchins in one of three ways:

Method 1: While wrapped in a paper towel, the animal is vigorously shaken for 5–10 s. The animal is the placed upright (i.e., oral side down) and kept moist. Spawning will occur within a few minutes if the animal is sufficiently gravid.
Method 2: If Method 1 fails then electrical stimulation is attempted. The animal is placed upright and electrodes carrying 12 V (AC) are applied to the test in the area near the gonopores. Five-second duration stimulations are repeated in three or four places. The animal may then be shaken as described in Method 1. Gamete shedding should follow within a few minutes. Methods 1 and 2 have the advantage of a high survival rate for the spawning adults. They are the methods of choice for transgenic experiments, as individual sea urchins may be used to supply gametes for repeated injection sessions, reducing egg batch variability among experiments.
Method 3: If Methods 1 and 2 fail to produce gametes or a great number of gametes are needed (this is usually not the case for transgenic experiments), then KCl injection can be used. Injection of isotonic KCl causes gonadal muscle contraction and near-complete shedding of gametes. For this procedure, a total of 0.75–1.0 mL of 0.5 *M* KCl is injected in two to three spots through the peristomeal membrane into the perivisceral cavity. Do not exceed 1 mL total injection. The animal is then gently shaken, placed upright, and monitored for gametes. Survival rates after KCl injection are typically low and, in any event, all gametes are usually shed in the process.

Once gametes begin to appear at the gonopores, the sex of the specimen can be determined. Eggs are yellow to orange in color, and sperm appears whitish. When a female is identified, the animal is inverted onto a beaker (75–150 mL depending on sea urchin size) that has been filled to the top with FSW and placed in a tray surrounded by ice. The animals gonopores should be submerged in the FSW. Eggs will begin to "rain" down through the FSW and, upon close examination, appear as individual particles. If a male was mistakenly used, sperm will appear milky. Gamete shedding usually proceeds for 10–30 min. FSW should periodically be poured over the spawning animal to keep it moist. Eggs should be kept at 4°C and will remain fertilizable for up to 4 h.

Sperm is collected by pipeting directly off the top of spawning males. For long-term storage, it is important that the sperm is not activated by contact with seawater. Undiluted sperm (referred to as "dry" sperm) is stored on ice or refrigerated at 4°C. Kept in this manner, sperm is adequate for fertilization of injected eggs for approx 7 d.

3.3. Rowing Eggs for Injection

The ability to fix eggs to a protamine-sulfate-coated Petri dish with a mouth pipet (rowing) in an accurate and uniformly distributed manner is the most important factor contributing to the number of eggs that can be injected in a given period of time. Upon spawning, sea urchin eggs are enveloped in a jelly coat that hydrates and expands upon contact with seawater. This layer can interfere with the adherence of eggs to the coated Petri dish if it is not removed (*see* **Note 4**). A short wash in acidic seawater (pH 5.0) will remove most of the jelly coat. For this procedure, eggs are placed in 10 mL of FSW to which is added 27 µL of 0.3 *M* citric acid. After swirling for approx 1 min, the eggs are transferred to a second Petri dish containing 10 mL of PABA–FSW, then to a third dish coated with 2% Noble agar (Difco Laboratories, Detroit, MI) (*see* **Note 5**) also containing 10 mL of PABA–FSW. Care should be taken not to overexpose the eggs to acid seawater treatment, as this will reduce viability.

A rowing pipet is made by quickly drawing out a short-form Pasteur pipet over a small Bunsen burner flame and breaking the resulting fiber at the desired diameter. The inside diameter of the drawn region should be equal to or slightly smaller than the eggs to be rowed and terminate in a region of this diameter that is approx 0.75 cm in length. This will require some trial-and-error attempts, but the time taken to produce a suitable rowing pipet will be more than made up for by that saved later in the rowing and injection process. Often a drawn pipet will narrow toward the orifice and the opening diameter can be chosen by progressive breaking. Once a rowing pipet is made, eggs are drawn from above the Petri dish bottom into the small open end of the pipet by negative pressure using the mouth pipeting apparatus described. Eggs can be separated from some debris by gentle swirling of the Petri dish until they gather in the center. Drawing a volume of clean seawater into the pipet prior to egg pickup often allows for smoother egg deposition. Once a sufficient number of eggs are in the pipet, positive pressure is applied until a steady flow of eggs is obtained. It may be helpful to hold the pipet in a vertical orientation with the opening pointed down for 30 s or so to allow the eggs to concentrate near the drawn end. When a continuous, single-file flow of eggs is obtained, they are deposited onto a protamine-treated dish containing 10 mL of 2 m*M* PABA–FSW (with penicillin and streptomycin). This is accomplished by slowly applying positive pressure while moving the pipet over the surface of the dish and observing under a stereomicroscope at approx 20X magnification. The hand holding the rowing pipet should be steadied by resting on a platform at the level of the microscope stage. Eggs should be spaced so that they are approximately half an egg diameter apart (to allow room for fertilization envelope elevation) and in as straight a line as is possible. An overhang in the end of the broken pipet can be helpful for gently seating the eggs in place while rowing. A sharp edge of the pipet can be used to steady the opening along the rowing line by lightly scoring the plastic Petri dish. When possible it may be easier to slowly move the Petri dish while holding the pipet tip stationary. Eggs are rowed in a line that intersects the center of the dish and is approximately two-thirds of the dish diameter in length. Typically, a total of about 200 eggs, but up to 400 eggs (in 2 parallel rows), may be rowed per dish, depending on the speed at which the operator is able to inject. Following rowing, a 3-mm score is made on the dish bottom near the top of the row with the broken end of a glass pipet. This will be used to break open the end of the injection needle.

3.4. Fertilization

Eggs are fertilized prior to injection. Although this limits the time available for injection (10–20 min) before the cortical layer becomes impenetrable, it reduces needle-clogging problems and eliminates injection-induced egg activation. After fertilization, sufficient time remains for the injection of 200–400 eggs per dish, depending on batch variation and the skill of the operator. Care should be taken to keep the eggs cool during and after fertilization, as temperature reduction delays the toughening of the cortical layer. Following rowing, eggs are fertilized, by the controlled addition of a sperm suspension. "Dry" sperm is diluted at 1:200 in FSW. This suspension will remain active for up to 1 h if kept on ice. About 15 μL of sperm suspension are added with a micropipet in a line that is about 3 mm to the side of the rowed eggs. The sperm is allowed to actively diffuse toward the rowed eggs. This process decreases the probability of polyspermy and allows for the addition of a minimal amount of sperm that during culture, can become substrate for bacterial growth. Fertilization is monitored by elevation of the fertilization membrane and is usually complete after about 1 min. If incomplete fertilization is observed (<90%), a second addition of sperm from a fresh dilution is advisable. As much of the first sperm suspension as possible should be removed and fresh PABA–SW with antibiotics should be added if necessary.

3.5. Injection

The method that follows describes the use of a Picospritzer pulsed pressure injection device. Continuous-flow devices are also routinely used in our laboratory with excellent results (*see* **Note 6**). Injection needles are filled from their large end with 1 or 2 μL of DNA solution using a micropipet fitted with a microloading tip. Alternatively, a drawn-out Pasteur pipet may be used for this purpose. The solution should automatically fill the needle tip by capillary action and displace any air bubbles.

As quickly as is possible after fertilization is complete, the Petri dish is moved to the chilled microscope stage for injection. Typically, a 10X objective lens is used that allows 15–20 eggs to be visualized at a time. Once the Petri dish is in place, the focal plane is adjusted as near as possible to the liquid surface. The needle at an angle to the horizontal of approx 20° is centered by eye (this can most easily be done by watching for the glint of the needle in the microscope illumination) and lowered until it just touches the liquid surface. After visually finding the needle tip in the focal plane, it is lowered while observing under the scope until it is near the dish bottom. Care must be taken to assure that the dish lip does not contact and, subsequently, break the needle shaft. With the Picospritzer set in the middle of its useful pulse-time range (approx 15 ms), the needle is broken by running its tip against the score on the Petri dish bottom. This is best done by moving the needle along its axis with the axial drive control. Simultaneous with breakage, the foot pedal pulse control is activated and flow from the needle is monitored. Once breakage has occurred, it is important that the needle be pulsed about once every 5 s to counteract diffusive dilution at the needle tip where the DNA sample is in direct contact with seawater. The eggs are brought into focus at their widest section and the needle tip is brought to this plane. Injection is easiest slightly above this level. The needle tip can be brought to this point using the fine vertical control. A 2-pL injection corresponds approximately to a cleared volume with a diameter of one-fifth to one-fourth of that of the egg. The injection of an excessive volume may result in immediate rejection of the DNA solution, egg lysis, or reduced viability (**Note 3**). Injection of too low a volume may result in a lowered frequency of integration. Injection volume can be adjusted by varying pulse-time length (typically between 8 and 30 ms). It is often helpful, especially as the cortical layer begins to toughen, to pulse the needle after pushing it slightly against the egg surface in order to break through the membrane. If the needle clogs, then it may be possible to rescue it by rebreaking on the scratch in the Petri dish. Rarely, an egg may dislodge from the Petri dish and become impaled on the end of the needle. This problem can often be solved by raising the needle above the liquid surface, at which point the egg is often retained at the meniscus, freeing the needle.

Following injection, uninjected zygotes are removed by aspiration with a drawn-out Pasteur pipet. For this purpose and for the subsequent manipulation of developing embryos, a wider-bore pipet with an opening of approximately 0.2 mm is useful. A diamond stylus can be used to score this size pipet so that a clean break is formed.

3.6. Embryo Culture and Observation

In order to prevent evaporation, up to three injected Petri dishes are placed on top of moistened tissue paper in a 150-mm Petri dish. *S. purpuratus* embryos are cultured at 16°C. Cleavage is often delayed in injected embryos. Embryos should be transferred to clean penicillin–streptomycin FSW (without PABA) after 24–48 h to avoid bacterial contamination. Twenty-four-well microtiter plates are often convenient for culture at this point. Feeding begins at 3–5 d after fertilization. Methods for long-term culture through to mature adult have been developed ***(16,17)***. Typically, >60% and often >90% of injected embryos will develop normally and express the injected transgenes.

Embryos can be observed most easily under cover slips supported by small clay feet. The feet are made by scraping the corners of a glass cover slip (18 mm × 18 mm, No. 1) against a smooth block of modeling clay. The cover slip is mounted on a microscope slide and depressed to leave a space that is slightly greater than the diameter of the embryos. Embryos are mouth-pipeted under the cover slip from the side where they are drawn in along with seawater by capillary action. If the cover slip is placed at a slight tilt, the embryos will become wedged and held stationary for viewing. The orientation of the embryo can be manipulated by carefully moving the cover slip and rolling the specimen.

4. Notes

1. The protamine sulfate solution can be stored at 4°C and reused. Overexposure of the dish to protamine sulfate solution may result in restricted fertilization membrane elevation and lead to abnormal cleavage. Insufficient washing may cause precipitation of negatively charged molecules at the needle opening during injection.
2. Carrier DNA is useful both to avoid nonspecific losses and to isolate linearized transcription constructs from one another upon concatemerization. An increase in efficiency for the latter usage may be gained if the carrier has cohesive ends that are compatible with those of the construct. Thus, when possible, the carrier should be digested with the same enzyme as is used to linearize the experimental construct. It should be noted, however, that this is by no means necessary, as a number of very successful transgenic studies have been made using different enzymes to cut carrier and construct DNA.
3. Above approx 300 ng/μL (about 10-fold higher concentration than is usually injected), DNA injection becomes toxic and abnormal development is evident in most embryos

injected *(8)*. Thus, care should be taken to limit the amount of DNA injected both by lowering copy concentration and the injection volume. It may be helpful to titrate toxicity by experimenting with different DNA concentrations.

4. For *S. purpuratus*, the mechanical action of egg manipulation and drawing into the rowing pipet typically dejellies the eggs to an extent adequate for rowing, and acid treatment may be omitted. If the eggs will not stick to the Petri dish, more extensive dejellying may be necessary.
5. To make agar-coated plates, 60-mm Petri dishes are treated with a 2% solution of Noble agar (dissolved in filtered dH_2O) by filling and immediately emptying. Once the adherent agar has gelled, 10 mL of FSW is added for approximately 10 min to equilibrate the agar, then replaced with fresh FSW. Coated dishes are made just prior to use.
6. For continuous-flow injection, a DNA solution flow rate of approximately 4 pL/s should be maintained. Care must be taken with this setup type to maintain a consistent 2-pL injection from egg to egg.

Acknowledgments

The author is grateful to R. A. Cameron and C.-H. Yuh for helpful comments and suggestions regarding the manuscript, C. V. Kirchhamer for initial training in sea urchin injection, and E. H. Davidson under whose guidance the technique has been developed. J. P. R. is supported by a NRSA fellowship GM-18478.

References

1. Arnone, M. I. and Davidson E. H. (1997) The hardwiring of development: organization and function of genomic regulatory systems. *Development* **124**, 1851–1864.
2. Yuh, C. -H. and Davidson, E. H. (1996) Modular *cis*-regulatory organization of *Endo-16*, a gut-specific gene of the sea urchin embryo. *Development* **122,** 1069–1082.
3. Calzone, F. J., Höög, C., Teplow, D. B., Cutting, A. E., Zeller, R. W., Britten, R. J., and Davidson, E. H. (1991) Gene regulatory factors of the sea urchin embryo. I. Purification by affinity chromatography and cloning of P3A2, a novel DNA binding protein. *Development* **112,** 335–350.
4. Coffman, J. A., Moore, J. G., Calzone, F. J., Britten, R. J., Hood, L. E., and Davidson, E. H. (1992) Automated sequential affinity chromatography of sea urchin embryo DNA binding proteins. *Mol. Marine Biol. Biotechnol.* **1,** 136–146.
5. Harrington M. G., Coffman, J. A., Calzone, F. J., Hood, L. E., Britten, R. J., and Davidson, E. H. (1992) Complexity of sea urchin embryo nuclear proteins that contain basic domains. *Proc. Natl. Acad. Sci. USA* **89**, 6252–6256.
6. Davidson, E. H. (1986) *Gene Activity in Early Development*, 3rd ed., Academic Press, Orlando, FL.
7. Akasaka, K, Nishimura, A., Hijikata, K., Iuchi, Y., Junji, M., Takahashi, M., Morikawa, H., and Shimada, H. (1995) Introduction of DNA into sea urchin eggs by particle gun. *Mol. Marine Biol. Biotechnol.* **4,** 255–261.
8. McMahon, A. P., Flytzanis, C. N., Hough-Evans, B. R., Katula, K. S., Britten, R. J., and Davidson, E. H. (1985) Introduction of cloned DNA into sea urchin egg cytoplasm: replication and persistence during embryogenesis. *Develop. Biol.* **108,** 420–430.
9. Flytzanis, C. N., McMahon, A. P., Hough-Evans, B. R., Katula, K. S., Britten, R. J., and Davidson, E. H. (1985) Persistence and integration of cloned DNA in postembryonic sea urchins. *Dev. Biol.* **108,** 431–442.
10. Hough-Evans, B. R., Britten, R. J., and Davidson, E. H. (1988) Mosaic incorporation and regulated expression of an exogenous gene in the sea urchin embryo. *Dev. Biol.* **129,** 198–208.

11. Franks, R. R., Hough-Evans, B. R., Britten, R. J., and Davidson, E. H. (1988) Direct introduction of cloned DNA into the sea urchin zygote nucleus, and fate of injected DNA. *Development* **102,** 287–299.
12. Livant, D. L., Hough-Evans, B. R., Moore, J. G., Britten, R. J., and Davidson, E. H. (1991) Differential stability of expression of similarly specified endogenous and exogenous genes in the sea urchin embryo. *Development* **113,** 385–398.
13. Arnone, M. I., Bogarad, L. D., Collazo, A., Kirchhamer, C. V., Cameron, R. A., Rast, J. P., Gregorians, A., and Davidson, E. H. (1997) Green fluorescent protein in the sea urchin: new experimental approaches to transcriptional regulatory analysis in embryos and larvae. *Development* **124,** 4649–4659.
14. Ransick, A., Ernst, S., Britten, R. J., and Davidson, E. H. (1993) Whole mount *in situ* hybridization shows *Endo*16 to be a marker for the vegetal plate territory in sea urchin embryos. *Mech. Dev.* **42,** 117–124.
15. Kirchhamer, C. V., and Davidson, E. H. (1996) Spatial and temporal information processing in the sea urchin embryo: modular and intramodular organization of the *CyIIIa* gene *cis*-regulatory system. *Development* **112,** 333–348.
16. Cameron, R. A. and Hinegardner, R. T. (1978) Early events of metamorphosis in sea urchins, description and analysis. *J. Morphol.* **157,** 21–32.
17. Leahy, P. S. (1987) Laboratory culture of *Strongylocentrotus purpuratus* adults, embryos and larvae. *Methods Cell Biol.* **27,** 1–13.
18. Sambrook, J., Fritsch, E. F., and Maniatis, T. (1989) *Molecular Cloning: A Laboratory Manual.* Cold Spring Harbor Press, Cold Spring Harbor, New York.

36

Transgenic Zebrafish

Shuo Lin

1. Introduction

The zebrafish has become a popular model system for genetic studies of vertebrate development. Zebrafish are amenable to such studies because the generation time of zebrafish is short, adults are small and easy to maintain, and females typically lay up to several hundred eggs per mating. Early development occurs rapidly outside the mother and the embryos are transparent, making the observation and manipulation of the embryos quite simple *(1)*. Recently, large-scale chemical mutagenesis screens in zebrafish have identified over 1000 mutations, including hundreds of genes essential for embryogenesis *(2,3)*. However, to maximally benefit from genetic analysis of embryogenesis using zebrafish, a number of additional challenges must be overcome. First, the chemically induced mutant genes must be identified. Second, the mutations must be analyzed in vivo at the cellular and molecular levels. Third, those functional genes that have been missed by the two screens must be identified. The generation of transgenic zebrafish can provide useful tools for overcoming each of these challenges.

Transgenic technology provides proof that a mutant gene has been identified by rescuing the mutant phenotype through reintroduction of a functional candidate gene. To this end, a transgenic technology that allows the correct temporal and spatial expression of genes of interest must be established. The ectopic expression of transgenes in whole animals allows one to study gain-of-function phenotypes, an approach that is often used to address the biological role of a gene at the cellular and molecular levels. Disruption of endogenous genes by random insertion of a transgene allows one to rapidly identify mutant genes and study loss-of-function phenotypes. Finally, transgenic techniques using reporter genes linked to specific control sequences make it possible to identify the regulatory elements responsible for spatial and temporal gene expression patterns. This approach often leads to the identification of transcription factors that function in genetic cascades underlying important biological processes.

Since the generation of the first transgenic zebrafish in 1988 by microinjecting naked DNA *(4)*, numerous successes of producing transgenic zebrafish have been reported *(5–11)*. Although approaches using retroviral infection *(12)*, microprojectiles *(13)*, and electroporation *(14,15)* have been developed to introduce transgenes into the zebrafish

From: *Methods in Molecular Biology, Vol. 136: Developmental Biology Protocols, Vol. II*
Edited by: R. S. Tuan and C. W. Lo © Humana Press Inc., Totowa, NJ

genome, microinjection of naked DNA is the only method that is able to produce transgenic zebrafish that express transgenes in multiple generations *(5–9)*. For this reason, only protocols of DNA microinjection are summarized in this chapter.

2. Materials

2.1. Equipment

1. Flaming/Brown Micropipette Puller: Model P-91, Sutter Instrument Co., Novato, CA.
2. Microinjection capillaries (1 mm), cat. no. GC100-10, Clark Electromedical Instruments, Reading, UK.
3. Micromanipulator: model MN-151, Narishige Scientific Instrument Lab, Tokyo, Japan; magnetic stand: model GJ-1; CT-1 Teflon tubing; HI-6-1 stainless-steel injection holder (for 1-mm glass capillary),
4. MICRO-MATE interchangeable syringe (10cc), Popper and Sons, Inc., New Hyde Park, NY.
5. Plastic two-way stopper for 1-mm Teflon tubing.
6. Stereomicroscopes and fluorescent microscopes.
7. Mating cages: Breeding traps and dividers, cat. no. CD-203, Aquaculture Supply, FL; polycarbonate mouse cage (24 × 14 × 13 cm), cat. no. 6601177, Nalge Nunc International, Rochester, NY.
8. Microinjection plates *(10)*: Pour 12.5 mL of warm 1.2% agarose (prepared in double-distilled water, ddH_2O) into a lid or bottom dish of a 60-mm Petri plate. Place a glass microscope slide (5 cm × 7.5 cm) onto the agarose at a 10–20° angle (**Fig. 1A**). After the agarose has solidified at room temperature, remove the glass slide to make a groove. Multiple plates can be made at one time and should be stored in ddH_20 at 4°C. Injection plates can be used many times if they are washed with 70% ethanol followed by ddH_2O after each use. Used plates are covered with plastic wrap and stored at 4°C.
9. Glass Pasteur pipets with rubber bulbs.
10. Glass microscope slides (5 × 7.5 cm).
11. Glass cover slips.
12. 100 mm × 20-mm and 60 mm × 15-mm Petri dishes and 24-well tissue culture dishes.

2.2. Reagents, Buffers, and Kits

1. Holtfreter's solution: 7.0 g NaCl, 0.4 g $NaHCO_3$ (sodium bicarbonate), 0.2 g $CaCl_2$ (anhydrous) or 0.24 g $CaCl_2 \cdot 2H_2O$, 0.1 g KCl. Dissolve in 2 L ddH_2O, add 30 µL concentrated HCl (pH should be 6.5–7.0). Make fresh Holtfreter's solution daily.
2. Pronase: Cat. no. P 5147, Sigma Chemicals, St. Louis, MO. Make 30 mg/mL stock solution in ddH_2O, pH 7.0. Self-digest the stock solution at 37°C for 1 h and store 10-mL aliquots at –20°C.
3. GENECLEAN II Kit, cat. no. 1001-400, BIO 101 Inc., Vista, CA.
4. KCl: 1*M* in ddH_2O, autoclave and store as 1-mL aliquots at –20°C.
5. Tetramethyl-rhodamine dextran, cat. no. D-1817, Molecular Probes (Eugene, OR). Stock solution is 12.5% in ddH_2O. Store at 4°C.
6. Qiagen (Chatsworth, CA) brand or other plasmid DNA preparation kits.
7. Light mineral oil, cat. no. 0121-1, Fisher Scientific, Pittsburgh, PA.
8. DNA extraction buffer: 10 m*M* Tris pH 8.2, 10 m*M* EDTA, 200 m*M* NaCl, 0.5% sodium dodecyl sulfate (SDS), and 200 µg/mL pronase.
9. Tricane: 0.16% in ddH_2O (adjust to pH 7.0 using 1.0 *M* Tris base).

2.3. Stock Fish for Microinjection

The postinjection embryo survival rate is often influenced by the quality of zebrafish eggs obtained. To ensure that there is a sufficient number of healthy eggs for microin-

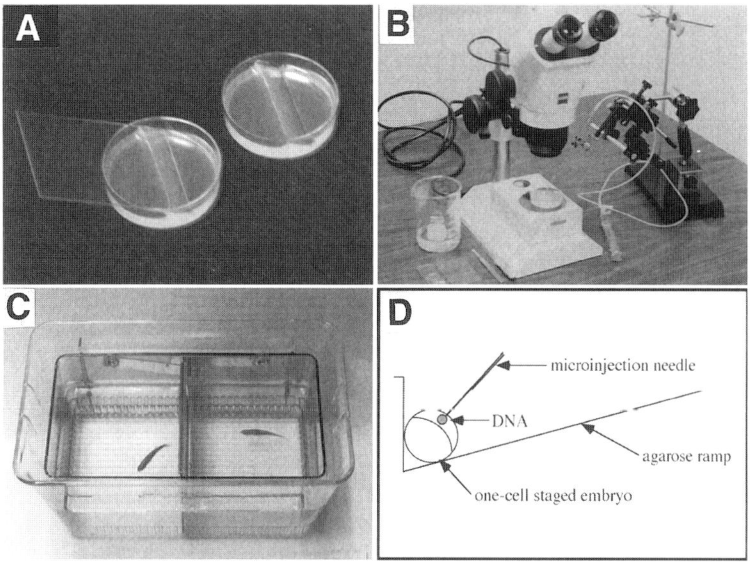

Fig. 1. **(A)** A microinjection plate is made by inserting a glass microscope slide into warm agarose at a 10–20° angle and letting the agarose solidify. The slide is removed, leaving a groove in the agarose. During microinjection, embryos are positioned along this groove so that they are supported when penetrated by the microinjection needle. **(B)** This is a typical zebrafish microinjection system. A micromanipulator is used to control needle movements during the microinjection process. The DNA is pushed through the needle by a syringe containing mineral oil. Embryos, the needle, and the microinjected DNA are all visible under a stereomicroscope, allowing for precise control of the microinjection process. The micromanipulator is controlled by the right hand and the microinjection plate is moved on the X and Y axes using the left hand. **(C)** The day prior to microinjection, male and female zebrafish are placed in a breeding trap inside of a mouse cage containing fish water. They are separated by a divider and covered with a lid (not shown). The next morning, after the fish are placed in a new mouse cage filled half-way with fresh fish water, the divider is removed. After eggs are laid, they are immediately collected and dechorionated. **(D)** The embryos are microinjected by first penetrating the cell wall with the microinjection needle and then pushing gently on the syringe plunger until the microinjected DNA occupies approximately one-fifth of the cytoplasm volume. The needle is immediately pulled out of the cytoplasm and the next egg is microinjected.

jection, dedicated stocks of healthy and highly productive breeding fish should be established. When young fish reach sexual maturity, a preselection can be made based on a pairwise mating test to determine which fish will provide the highest quality eggs. The selected males and females are maintained separately and fed four to five small meals daily, as opposed to three larger meals, because fish fed smaller and more frequent meals tend to produce more eggs. The diet should consist of both brine shrimp and flake food. Stock fish for microinjection should be used for mating only once a week. However, periodical mating is necessary to sustain the productivity of breeding stocks because fish that have not mated for several weeks often produce no eggs or bad eggs. Thus, fish should be routinely mated even during weeks when no eggs are needed for microinjection.

3. Methods

3.1. Preparation of DNA Fragments for Microinjection

Plasmid DNA is prepared using a Qiagen kit or other plasmid DNA preparation procedure following the manufacturer's instructions. DNA for microinjection must be linearized with the appropriate restriction enzyme(s) and, if possible, insert DNA fragments should be purified free of vector sequences. This can be done by excision of the desired DNA fragment from an agarose gel containing 0.5 μg/mL ethidium bromide followed by purification using a GENECLEAN II Kit. At the final step, DNA is dissolved in nuclease-free 1× TE (pH 8.0) at a concentration 5–10 times higher than the final concentration to be used for microinjection. The DNA solution is spun for 30 s using a microcentrifuge and two-thirds of the volume is transferred to a new 1.5-mL Eppendorf microcentrifuge tube. Effort should be made to avoid taking any precipitate that can later clog the microinjection needle during DNA loading. Two microliters of the DNA should be checked by electrophoresis and ethidium bromide staining to make sure that the DNA is pure and structurally intact. The purity of the DNA to be microinjected is a critical factor in determining the number of embryos that survive the microinjection procedure.

For microinjection, the DNA should be diluted to 50–100 μg/mL in 0.1 *M* KCl containing tetramethyl-rhodamine dextran (0.125%). This dye serves as an indicator that allows one to discriminate uninjected embryos under a fluorescent microscope. Our experiences indicate that higher concentrations of DNA can increase gene expression levels to a certain degree but often result in lower postinjection embryo survival rates. The prepared DNA solution can be stored at 4°C for up to 2 wk or at –20°C for a longer time.

3.2. Microinjection Setup

The microinjection system consists of a stereomicroscope, a micromanipulator, a magnetic stand, a syringe, a plastic two-way stopper, teflon tubing, a needle holder, microinjection capillaries, and a microinjection plate (**Fig. 1B**). The micromanipulator is attached to the magnetic stand, which is securely fastened to a metal surface. The syringe is filled with 3–4 mL of mineral oil and is directly connected to the two-way stopper. This stopper is connected to approx 3 ft of 1-mm Teflon tubing that, at the other end, is connected to the microinjection needle holder. It takes considerable effort to attach the tubing to the needle holder. However, a tight fit is required to ensure that the microinjection system remains intact when under the pressure caused by pulling and pushing the syringe plunger. The needle holder is attached to the micromanipulator.

3.3. Preparation of the Microinjection Needle

Microinjection needles are made with a Flaming/Brown Micropipette Puller using fine glass microinjection capillaries. The tip of the microinjection needle should be approx 0.05–0.15 mm. Prior to microinjection, a needle is opened by breaking its tip under a stereomicroscope using a clean, sharp forcep. It will take practice to consistently break the needle at a position optimal for microinjection. A needle broken too short will be too large to puncture the embryos effectively and will likely destroy the embryos. Conversely, if a needle is broken too long, it will be difficult to load with DNA and will be more prone to clogging. The needle is attached to the needle holder and filled with mineral oil by pushing oil from the syringe, through the Teflon tubing, and into the needle. Effort should be made to avoid trapping air bubbles in the microinjection tube and needle.

The DNA prepared for microinjection is spun on high speed in a microcentrifuge for 2 min to sediment any contaminating particulates. Depending on how many eggs one plans to inject, 2–10 μL of DNA is transferred onto a clean plastic cover slip and is overlaid with a few drops of mineral oil to prevent evaporation. The DNA is loaded by placing the needle tip into the DNA solution using the micromanipulator and by gently pulling the 10cc syringe plunger. If loading occurs too slowly or the needle has too small of an opening, the needle can become clogged with debris. If this happens, the needle can be broken again using forceps as described earlier and loading can be resumed. However, in most cases it is better to start with a new needle. Because loading is a slow process, a clamp may be used to hold the syringe barrel during loading.

3.4. Collection and Preparation of Eggs

In the late afternoon of the day before microinjection, and at least 30 min after their last feeding, a male and one or two females are transferred from their fish tanks to a mouse cage containing a breeding trap. The male and female are separated using a divider (**Fig. 1C**) and covered to prevent them from leaping out of the cage. One pair of fish can normally produce one to several hundred eggs, but sometimes no eggs or bad eggs are produced. Consequently, 10–15 pairs should be set up to ensure that a sufficient number of eggs will be obtained for microinjection. On the day of microinjection, after the lights have come on in the morning, the breeding trap (with the fish and the divider still inside) is moved to a new mouse cage filled halfway with fresh fish water. The fish are joined by removing the divider. Females will usually begin laying eggs after only a few minutes. Eggs are collected in a 50-mL beaker approx 10 min after being laid. The fish water in the beaker is replaced with 15–20 mL of a diluted pronase solution (approx 1 mg/mL in Holtfreter's solution) prewarmed at 28–30°C and the beaker is placed in a 28–30°C incubator for several minutes. The beaker can be gently swirled to help embryos fall out of their chorions. When approx 5–10 chorions drop off of the embryos (chorions can be seen floating in the solution), the eggs are washed 8–10 times with Holtfreter's solution to remove all traces of chorions and pronase. One should avoid treating the embryos with pronase too long because this will decrease the embryo survival rate. How many mating pairs of fish to release at one time is determined by the speed of the microinjection process. For example, someone who microinjects quickly may need to release three pairs at once, whereas a slower worker may only need to release one pair each time.

3.5. Microinjection

A microinjection plate is filled with Holtfreter's solution. The dechorionated embryos are gently transferred into the middle groove of the microinjection plate using a glass Pasteur pipet. Approximately 200 embryos can be lined up in the groove of one plate. The microinjection needle should be positioned with the help of the micromanipulator. The needle goes into the cytoplasm of single-cell embryos from the top of the cell (**Fig. 1D**) and the DNA solution is microinjected by gently pushing the syringe plunger. The microinjected DNA normally occupies approximately one-fifth the volume of the single-cell mass, which can be seen under the stereomicroscope (**Fig. 1D**). Because many of the embryos will not be in the correct orientation for microinjection, their position in the microinjection plate can be adjusted using the microinjection needle.

Microinjected embryos are transferred into several 100-mm × 20-mm Petri dishes containing fresh Holtfreter's solution and incubated at 28–30°C. Four to five hours after microinjection, unfertilized and deformed eggs can be easily recognized and should be removed. Developed embryos (approx 50% epiboly) are transferred into a 24-well tissue culture plate in 1/2X Holtfreter's solution (five embryos per well) and grown overnight at 28–30°C. The next day, dead and deformed embryos are removed and healthy embryos are transferred into 1/4X Holtfreter's solution. After 48 h of development, microinjected embryos are placed into regular fish water. The care and maintenance of microinjected embryos is a critical part of the microinjection process and will directly influence the number of surviving embryos (*see* **Note 1**).

3.6. Identification of Germline Transgenic Founder Fish by Polymerase Chain Reaction

Although it is possible to screen germline transgenic zebrafish by examining expression of a reporter gene (*see* **Notes 2** and **3**), such as green fluorescent protein (GFP) or *lacZ*, polymerse chain reaction (PCR) still represents the most commonly used method to detect germline transmission of transgenes. To reduce the labor involved in this process, eggs from multiple pairs of fish are pooled, as described later. Potential transgenic founder fish (10–12 mo old) are mated to each other or to a nontransgenic wild-type fish. Because transmission of a transgene to the F1 generation is often mosaic (ranging from 2% to 90%), it is necessary to collect approx 100 eggs from each founder fish. F1 embryos are grown 24 h at 28–30°C and treated with pronase to remove their chorions as described above. Approximately 100 embryos are transferred to an Eppendorf tube containing 1 mL of DNA extraction buffer, vortexed briefly, and incubated with rotation at 55°C for 3–4 h. The tubes are vortexed briefly, once per hour during this incubation. The samples are then centrifuged at 14,000 rpm for 10 min at room temperature using a microcentrifuge. Ten samples are pooled by combining 20 μL of extraction solution from each sample in a new Eppendorf tube. The remaining DNA extraction solution from each sample should be saved at –20°C until further use. The combined DNA mixtures are precipitated with the addition of 300 μL of 100% ethanol followed by microcentrifugation. Preciptates are washed with 70% ethanol, vacuum dried, and dissolved in 100 μL of 1× TE (pH 8.0).

Using the isolated genomic DNA, standard PCR can be performed to identify pools of DNA samples that contain a transgene. A number of controls should be included in the PCR reactions. First, an internal control that detects an endogenous zebrafish gene should be included in each reaction to assess if the PCR reaction worked. Second, a positive control that contains one transgenic embryo per 500 embryos should be included to assure that the PCR conditions used are sensitive enough to detect such a low transgene concentration. This can be achieved by mixing a previously identified transgenic embryo with nontransgenic embryos or mixing transgene DNA with DNA isolated from nontransgenic fish at appropriate ratios. Finally, a negative control should be included to address possible contamination.

Once a positive pool is identified, PCR can be performed using the saved DNA extraction solutions from each individual sample to identify the transgenic founder fish. Using the pooling method, it is possible to screen 200 transgenic families per week. After a founder fish harboring a transgene is identified by PCR, other methods, such as Southern blotting and examination of transgene expression, should be performed on the progeny to confirm the identity of the founder fish.

3.7. Generation of Homozygous Transgenic Fish

The founder transgenic (F0) fish can be bred to generate homozygous transgenic populations that can be maintained for generations without loss of the integrated transgenes. F0 fish are crossed with wild-type fish and their progeny (F1) grown to sexual maturity. The presence of a transgene can be detected by analyzing reporter gene expression in F1 fish. Alternately, live transgenic F1 fish can be identified by performing PCR on genomic DNA isolated from caudal fin clips. F1 fish are anesthetized using a 0.16% tricaine stock solution diluted further in fish water by adding 4.2 mL stock solution per 100 mL fish water. The amount of tricaine to use varies with water volume and number of fish, but should be used sparingly so as not to kill the fish. With a razor blade, one-half to two-thirds of the tails are cut off and placed in separate Eppendorf tubes, each containing 100 μL of DNA extraction buffer. The fish are placed in separate mouse cages numbered to correspond to the Eppendorf tube containing the tail cut. Each tube is vortexed thoroughly and then incubated with rotation at 55°C for 3–4 h. The tubes are vortexed briefly once per hour during this incubation and then are centrifuged at 14,000 rpm for 20 min at room temperature in a microcentrifuge. DNA samples are diluted 1:10 for PCR. If desired, multiple DNA samples can be pooled to minimize the number of PCR reactions.

Positively identified F1 brothers and sisters are mated. An F1 pair should normally produce 25% transgenic homozygote positives, 50% heterozygotes, and 25% wild-type homozygotes among the F2 progeny. F2 embryos are grown to sexual maturity and individual F2 fish are mated to wild-type fish. The F3 progeny are analyzed for the transgene either by assaying reporter gene expression or by analyzing individual F3 eggs by PCR, as described earlier. A homozygous F2 fish should pass the transgene to 100% of its offspring when mated with a wild-type fish. Identified homozygous F2 fish can be mated to each other to produce a large homozygous population.

4. Notes

1. Postinjection care is a critical factor that determines the production rate of transgenic zebrafish. During the first 2 wk after injection, the baby fish are first fed with live paramecia and then with a mixture of paramecia and brine shrimp. Water should be changed every other day to minimize infection. After 2 wk, injected fish are transferred into tanks with circulating water and fed with brine shrimp. A fine screen is placed on the tank outlet to prevent baby fish from being sucked into the circulating system. After 1 mo, the fish can be fed with mixed flake food and brine shrimp.
2. Although it is possible to obtain expression of transgenes after germline transmission using heterogeneous promoters, it appears to be more effective to use zebrafish promoters to recapitulate the expression pattern of a gene of interest *(9)*.
3. By microinjecting DNA constructs containing GFP ligated to zebrafish promoters/enhancer sequences, transient gene expression assays can be performed in living zebrafish embryos to identify regulatory elements responsible for tissue specific expression pat-

terns *(16)*. Using the injection method described above, two persons can generate more than 1000 microinjected embryos that transiently express GFP in a single day. This makes it possible to obtain statistically significant data for analyzing promoter/enhancer activity for each construct.

Acknowledgments

The author thanks members of his laboratory for experimental contributions and comments. Some of the protocols were developed while he was working as a postdoctoral fellow in Nancy Hopkins' laboratory at MIT and he wishes to thank her for advice and support.

References

1. Kimmel, C. B. (1989) Genetics and early development of zebrafish. *Trends Genet.* **8,** 283–288.
2. Driever, W., Solnica-Krezel, L., Schier, A. F., Neuhauss, S. C., Malicki, J., Stemple, D. L., Stainier, D. Y., Zwartkruis, F., Abdelilah, S., Rangini, Z., Belak, J., and Boggs, C. (1996) A genetic screen for mutations affecting embryogenesis in zebrafish. *Development* **123,** 37–46.
3. Haffter, P., Granato, M., Brand, M., Mullins, M. C., Hammerschmidt, M., Kane, D. A., Odenthal, J., van Eeden, F. J., Jiang, Y. J., Heisenberg, C. P., Kelsh, R. N., Furutani-Seiki, M., Vogelsang, E., Beuchle, D., Schach, U., Fabian, C., and Nusslein-Volhard, C. (1996) The identification of genes with unique and essential functions in the development of the zebrafish, Danio rerio. *Development* **123,** 1–36.
4. Stuart, G. W., McMurray, J. V., and Westerfield, M. (1988) Replication, integration and stable germ-line transmission of foreign sequences injected into early zebrafish embryos. *Development* **103,** 403–12.
5. Stuart, G. W., Vielkind, J. R., McMurray, J. V., and Westerfield, M. (1990) Stable lines of transgenic zebrafish exhibit reproducible patterns of transgene expression. *Development* **109,** 577–584.
6. Amsterdam, A., Lin, S., and Hopkins, N. (1995) The Aequorea victoria green fluorescent protein can be used as a reporter in live zebrafish embryos. *Dev. Biol.* **171,** 123–129.
7. Bayer, T. A., and Campos-Ortega, J. A. (1992) A transgene containing lacZ is expressed in primary sensory neurons in zebrafish. *Development* **115,** 421–426.
8. Lin, S., Yang, S., and Hopkins, N. (1994b) lacZ expression in germline transgenic zebrafish can be detected in living embryos. *Dev. Biol.* **161,** 77–83.
9. Long, Q., Meng, A., Wang, H., Jessen, J. R., Farrell, M. J., and Lin, S. (1997) GATA-1 expression pattern can be recapitulated in living transgenic zebrafish using GFP reporter gene. *Development* **124,** 4105–4111.
10. Culp, P., Nusslein-Volhard, C., and Hopkins, N. (1991) High-frequency germ-line transmission of plasmid DNA sequences injected into fertilized zebrafish eggs. *Proc. Natl. Acad. Sci. USA* **88,** 7953–7957.
11. Gibbs, P. D., Gray, A., and Thorgaard, G. (1994) Inheritance of P element and reporter gene sequences in zebrafish. *Mol. Marine Biol. Biotechnol.* **3,** 317–326.
12. Lin, S., Gaiano, N., Culp, P., Burns, J. C., Friedmann, T., Yee, J. K., and Hopkins, N. (1994a) Integration and germ-line transmission of a pseudotyped retroviral vector in zebrafish. *Science* **265,** 666–669.
13. Zelenin, A. V., Alimov, A. A., Barmintzev, V. A., Beniumov, A. O., Zelenina, I. A., Krasnov, A. M., and Kolesnikov, V. A. (1991) The delivery of foreign genes into fertilized fish eggs using high-velocity microprojectiles. *FEBS Lett.* **287,** 118–120.

14. Muller, F., Lele, Z., Varadi, L., Menczel, L., and Orban, L. (1993) Efficient transient expression system based on square pulse electroporation and in vivo luciferase assay of fertilized fish eggs. *FEBS Lett.* **324,** 27–32.
15. Powers, D. A., Hereford, L., Cole, T., Chen, T. T., Lin, C. M., Kight, K., Creech, K., and Dunham, R. (1992) Electroporation: a method for transferring genes into the gametes of zebrafish (Brachydanio rerio), channel catfish (Ictalurus punctatus), and common carp (Cyprinus carpio) *Mol. Marine Biol. Biotechnol.* **1,** 301–308.
16. Meng, A., Tang, H., Ong, B. A., Farrell, M. J., and Lin, S. (1997) Promoter analysis in living zebrafish embryos identifies a cis-acting motif required for neuronal expression of GATA-2. *Proc. Natl. Acad. Sci. USA* **94,** 6267–6272.

37

Production of Avian Chimeras and Germline Transmission

Kristin L. Woods, Scott Schau, Mary Ellen Clark, Jacqueline A. Bonselaar, and Robert J. Etches

1. Introduction

Fertilization in chickens occurs in the infundibulum of the reproductive tract within 15 min after ovulation of the yolk. The first cleavage division occurs 5 h later as the egg enters the shell gland. As the shell is calcified during the next 20–22 h, the embryo develops into a radially symmetric dome of cells (the blastoderm) containing approx 60,000 cells. Development within the oviduct and in the first few hours after the egg is laid has been described by Eyal-Giladi and colleagues *(1,2)*; these stages of development are given in roman numerals followed by E G&K. Later stages of development, which occur after embryos are placed into an incubator at 37.5 C and 50% relative humidity, are described by Hamburger and Hamilton *(3)*; these stages of development are given in arabic numbers followed by H&H.

Chimeric chickens can be produced by injecting dispersed blastodermal cells from stage X (E–G&K) embryos into the subgerminal cavity of recipient embryos at the same stage of development *(4)*. The resulting chimeras contain donor-derived cells in both somatic and germline lineages. Compromising recipient embryos with irradiation reduces the contribution of the recipient to the chimera and, therefore, the relative contribution of donor cells to the somatic tissues and the germline significantly increases *(5)*. Alternatively, recipient embryos can be compromised by physically removing the central region of the blastoderm in which the precursors of the germline are believed to reside *(6,7)*. The rate of germline transmission by chimeras constructed from physically compromised recipients is similar to the rate of germline transmission by chimeras constructed from irradiated recipients.

Blastodermal cells can be grown in culture, cryopreserved and thawed, transfected in vitro, and specifically propagated using selective media *(8–12)* before they are used as donor cells in the production of chimeras. Hence it may be possible to use chimeras as intermediates in the production of transgenic chickens. If the transfected donor cells contribute to the germline of a chimera, their offspring could serve as founders of a line of transgenic birds. In order for this approach to be practical, an efficient method of producing germline chimeric chickens is required.

From: *Methods in Molecular Biology, Vol. 136: Developmental Biology Protocols, Vol. II*
Edited by: R. S. Tuan and C. W. Lo © Humana Press Inc., Totowa, NJ

2. Materials

1. Buffers:
   a. PBS-G: Phosphate-buffered saline (PBS) containing 5.6 m*M* glucose, Gibco-BRL, Gaithersburgh, MD, cat. no. 11500) pH 7.4.
   b. CMF-PBS: calcium and magnesium free PBS, (Gibco-BRL cat. no. 21600) containing 2% chicken serum (CS) (Gibco-BRL cat. no. 16110-082) pH 7.4.
   c. CMF-T/E: CMF-PBS containing 0.05% Trypsin and .53 n*M* EDTA.
2. Medium: OptiMEM complete: OptiMEM (Gibco-BRL cat. no. 51985-034) containing 10% FBS (Gibco-BRL, ES Cell Qualified Fetal Bovine Serum cat. no. 16141-079), 100 U penicillin, and 100 U streptomycin.
3. Filter paper rings: Whatman #1 filter paper cut into rings with the inside portion slightly larger than the embryo (*see* **Note 1**).

3. Methods

1. Embryo isolation: Crack open freshly laid fertilized eggs (*see* **Note 2**) and remove excess albumen from the yolk by sifting between the two shell halves. With the vitelline membrane intact, place the yolk into a 100-mm Petri dish for embryo isolation. After placing a filter paper ring around the embryo, the vitelline membrane is cut with a pair of dissection scissors (Fine Science Tools, Foster City, CA, cat. no. 15011-12) using the outside of the filter paper ring as a guide. The embryo in the filter paper ring is then removed with a pair of forceps (Canadawide Scientific, cat. no. 38370-14 and placed into a 100-mm petri dish containing PBS-G. The adhering yolk is gently washed from the embryo by PBS-G from a Pasteur pipet and by gentle swirling (*see* **Note 3**). The central area of the embryos may now be loosened and removed from the vitelline membrane by isolating the central disk with a hair loop (*see* **Note 4**). Aspirate the central disk of the embryo into a Pasteur pipet and transfer into a sterile 15-mL tube containing 3 mL PBS-G with 2% CS on ice (*see* **Note 5**).
2. Embryo dispersion: After all embryos have been isolated and chilled on ice for at least 10 min, replace PBS-G with an equal volume of cold CMF–PBS (*see* **Notes 6** and **7**). Incubate on ice for 10 min. Replace CMF–PBS with cold CMF–T/E. Incubate on ice for 10 min.
3. Add an equal volume of Dulbecco's modified Eagle's medium (DMEM) containing 10% FBS to the tube to inhibit the action of trypsin. Cells may be dispersed into a single-cell suspension using gentle pipeting.
4. Centrifuge for 4 min (300g), remove supernatant, and resuspend in the desired amount of OptiMEM complete or DMEM. Cell concentration can be determined using a hemocytometer.

3.1. Isolation of Blastoderms

This section outlines the protocols developed for the isolation and dispersion of blastodermal cells (CBCs) from chick embryos. The protocol has been optimized for the isolation of embryos from stage X (E-G&K) blastoderms ***(4,9)***, but may be useful for isolation of embryos from stages 2 to 32 (H&H) with slight modifications.

3.2. Production of Chimeras with Irradiation

1. *Preparation of recipient eggs:* Compromise recipient embryos by exposing eggs to approx 550–600 rads from a ^{60}Co source within 1 h after oviposition and within 2 h before injection. After exposure to irradiation, leave eggs on their long axis for 1 h to allow the embryo to float to the top of the egg. Wipe the recipient eggs with 70% ethanol. Using a small drill and stone grinder, remove a 1-cm^2 section of shell at the widest point along the long axes of the egg, leaving the membrane intact.
2. Using a sharp scalpel, cut a 6-mm x 6mm square out of the shell membrane.

3. *Injection of donor cells into recipients:* Using a mouth pipet, inject approx 500 cells in 2–5 μL of medium by means of a microcapillary pipet drawn to 30–50 μm. Insert the pipet and inject the cells into the subgerminal cavity of the recipient blastoderm. Injections can be done under a dissecting microscope, although experience has shown that equal success can be achieved without visual aids. Successful injections are indicated by the spread of dye in the medium throughout the fluid in the subgerminal cavity.
4. *Closing the recipient egg for incubation:* Cover the hole in the shell with two pieces of donor membrane and let dry (*see* **Note 8**). Use contact cement to seal the edges of the membrane to the shell and let dry. Cover the area with surgical membrane and incubate at 37°C and a relative humidity of 50% with an hourly rotation through 90° for 84 h (3.5 d).
5. *Transfer to a surrogate shell after 84 h of incubation:* Prepare surrogate eggshells from an egg that is 25 g larger than the egg containing the recipient embryo by removing the blunt end of the egg (*see* **Note 9**). Crack the egg containing the recipient embryo into a bowl lined with cling wrap, being careful not to break the yolk. Form a pouch with the cling wrap and transfer the embryo, yolk, and albumen from the bowl into the surrogate shell. Cut off extra cling wrap along outer edge of hole in the surrogate eggshell. Gently remove the remaining cling wrap leaving the yolk and albumen in the surrogate shell. Cover the hole at the top of the surrogate eggshell using a square of cling wrap and seal the film to the shell with albumen.
6. Incubate at 37°C and 50% relative humidity with an hourly rotation through 60° from d 3.5 to d 20. Move the egg on the d 20 to a hatcher and maintain 37°C and 85% relative humidity until the chick emerges from the egg. Poke holes through the Clingwrap when the chick begins to breathe. Loosen the wrap from the eggshell as the chick starts to emerge.

3.3. Production of Chimeras by Physically Compromising Recipient Embryos

1. Wipe the recipient eggs with 70% ethanol. Using a small drill and stone grinder remove a 1-cm^2 section of shell at the widest point along the long axes of the egg, leaving the membrane intact.
2. Using a sharp scalpel, cut a 6-mm × 6-mm square out of the shell membrane.
3. *Physically compromising the recipient blastoderms:* Using a finely tapered glass pipet, remove about 500 cells from the center of the recipient blastoderm. Discard pipet.
4. *Injection of donor cells into the recipient blastoderms, closing recipient eggs, and transfer to a surrogate shell: see* **Subheading 3.2., steps 2–6**. The donor cells are injected into the hole made by physically compromising the recipient embryo.

4. Notes

1. The filter paper rings are made by punching a 6-mm hole in the center of an 11-mm circle of filter paper. They can be made by cutting around a hole from a paper punch. Alternatively, a custom-made punch that cuts concentric holes may be constructed for this purpose.
2. Embryos between stages IX (E–G&K) and XI (E–G&K) can be obtained from freshly laid eggs. The protocol in **Subheading 3.1.** is also useful for embryos up to 4 d of incubation. In order to obtain embryos from earlier stages, hens may be injected with arginine vasotocin or prostaglandin *(13)* to induce premature oviposition, or hens may be sacrificed and embryos harvested from the oviduct. For later stages, the eggs should be incubated under standard conditions.
3. Donor cells isolated from the central disk produce germline chimeras more efficiently than cells from the surrounding area opaca.

4. If the embryo is placed in PBS-G at a 45° angle, most of the yolk will float off the embryo. The remaining yolk can be removed under a dissection microscope using gentle pipeting and/or a hair loop.
5. Up to 10 embryos may be placed in one 15-mL tube containing 3 mL PBS-G with 2% CS. The tube and PBS should be prechilled on ice.
6. From this point on, all work should be conducted in a sterile environment.
7. The embryos will settle to the bottom of the tube. Care should be taken not to disturb the embryos when removing solution from the tube. When adding solution, the embryos are resuspended and rinsed.
8. *Preparation of membrane squares:* Using forceps remove the juxtaposed inner and outer membrane from an eggshell and cut it into squares that are just larger than the hole drilled into the side of the recipient eggs (approx 1.3 cm^2). Place membrane squares in PBS to keep them from drying out.
9. *Preparation of surrogate shell* ***(14)****:* Using a template, mark a ring around the blunt end of the egg. Mark the weight of the egg on the sharp end of the egg. Using a small drill with a fine circular blade, cut the blunt end of the egg off, discarding contents. To ensure that not too much of the egg is removed, use a circular template that has a 40-mm circumference for eggs that weigh 65–75 g. For larger eggs use a template with a circumference of 42 mm.

Acknowledgments

The financial support of the Natural Sciences and Engineering Research Council of Canada and the Ontario Ministry of Agriculture and Food are gratefully acknowledged.

References

1. Eyal-Giladi, H., and Kochav, S. (1976) From cleavage to primitive streak formation: a complementary and normal table and a new look at the first stage of the development of the chick I. *Gen. Morphol. Dev. Biol.* **49,** 321–337.
2. Kochav, S. M., Ginsburg, M., and Eyal-Giladi (1980) From cleavage to primitive streak formation: a complementary and normal table and a new look at the first stage of the development of the chick II. Microscopic anatomy and cell population dynamics. *Dev. Biol.* **79,** 296–308.
3. Hamburger, V. and Hamilton, H. L. (1951) A series of normal stages of development of the chick embryo. *J. Morph.* **88,** 49–92.
4. Petitte, N. J., Clark, M. E., Liu, G., Verrinder Gibbins, A. M., and Etches, R. J. (1990) Production of somatic and germline chimeras in the chicken by transfer of early blastodermal cells. *Development* **108,** 185–189.
5. Carsience, R. S., Clark, M. E., Verrinder Gibbins, A. M., and Etches, R. J. (1993) Germline chimeric chickens from dispersed donor blastodermal cells and compromised recipient embryos. *Development* **117,** 669–675.
6. Schau, S.N., Bonselaar, J., Etches, R. J. (1997) Production of germline chimeric chickens from physically compromised recipient embryos and dispersed donor blastodermal cells. *Poultry Sci.* 76(Supplement 1,10) (Abstr.)
7. Kagami, H., Tagami, T., Matsubara, Y., Harumi, T., Hanada, H., Maruyama, K., Sakurai, M., Kuwana, T., and Naito, M., (1997) The developmental origin of primordial germ cells and the transmission of the donor-derived gametes in mixed-sex germline chimeras to the offspring in the chicken. *Mol. Reprod. Dev.* **48,** 1–10.
8. Brazolot, C. L., Petitte, J. N., Etches, R. J., and Verrinder Gibbins, A. M. (1991) Efficient transfection of chicken cells by lipofection, and introduction of transfected blastodermal cells into the embryo. *Mol. Reprod. Develop.* **30,** 304–312.

9. Fraser, R. A., Carsience, R. S., Clark, M. E., Etches, R. J., and Verrinder Gibbins, A. M. (1993) Efficient incorporation of transfected blastodermal cells into chimeric chicken embryos. *Int. J. Dev. Biol.* **37,** 381–385.
10. Etches, R. J., Clark, M. E., Toner, A., Liu, G., and Verrinder Gibbins, A. M. (1996) Contributions to somatic and germline lineages of chicken blastodermal cells maintained in culture. *Mol. Rep. Devel.* **45,** 291–298.
11. Pain, B., Clark, M.E., Nakazawa, H., Sakurai, M. Samarut, J. and Etches, R. J. (1996) Long term culture and characterization of avian embryonic stem cells with multiple morphogenetic capabilities. *Development* 122, 2339-2348.
12. Kino, K., Pain, B., Leibo, S. P., Cochran, M., Clark, M. E., and Etches, R. J. (1997) Production of chicken chimeras from injection of frozen-thawed blastodermal cells. *Poultry Sci.* **76,** 753–760.
13. Watt, J. M., Pettite, J. N., and Etches, R. J. (1993) Early development of the chick embryo. *J. Morphol.* **215,** 165–182.
14. Etches, R. J., Clark, M. E., Verrinder Gibbins, A. M., and Cochran, M. B. (1996) Production of chimeric chickens as intermediates for gene transfer, in *Transgenic Animals: Generation and Use* (Houdebine, L. M., ed.), Harwood Academic, Amsterdam, pp. 75–81.

38

Incorporation of Genetically Modified Cells in Chicken Chimeras

Laura D. Zajchowski, S. Mansoor Mohammed, Qingxia Wei, and Robert J. Etches

1. Introduction

The production and study of transgenic mice has proven invaluable in dissecting the complex genetic regulatory events that control mammalian development. Although the chick embryo is a classic model system in developmental biology, its scope is limited by the inability to routinely and efficiently produce transgenic birds. One proposed strategy toward accessing the avian genome is the use of transgenic chimeric intermediates. As described in Chapter 37, dispersed chicken blastodermal cells (CBCs) from stage X (E–G&K) *(1)* embryos produce both somatic and germline chimeras when injected into the subgerminal cavity of recipient stage X (E–G&K) embryos. Transfection of the donor cells in vitro could lead to the production of chimeras capable of transmitting the transgene to their offspring. Avian transgenesis through chimeric intermediates is theoretically sound because CBCs can be efficiently transfected using both lipofection and electroporation and both lipofected and electroporated CBCs are capable of contributing to somatic and germline lineages *(2–5)*. Transfection using liposome reagents has the disadvantages of requiring extended incubations and optimization of DNA/lipid ratios for each construct used. Although electroporation requires access to relatively expensive equipment, it is a rapid, physical method that may have fewer adverse effects on the pluripotentiality of transfected CBCs. Therefore, both transfection protocols are described in this chapter.

Although the production of germline transgenic chimeras is the ultimate goal of producing chimeras with genetically modified CBCs, valuable information can be derived from chimeras produced by the injection of stably transfected somatic cell lines into adult chickens. Such chimeras do not transmit the genetic modification to their offspring, but they do provide a useful simulation model of somatic transgenesis. DT40 cells are a virally transformed pre-B-cell line derived from SC Hy-Line™ International Dallas Center, IA hens. The rapidity and ease with which these cells can be cultured make them ideal candidates for generating stably transfected cell lines. DT40 cells also possess the valuable characteristic of unusually high rates of homologous recombination *(6)*. When DT40 cells are injected intravenously into adult SC Hy-Line

From: *Methods in Molecular Biology, Vol. 136: Developmental Biology Protocols, Vol. II*
Edited by: R. S. Tuan and C. W. Lo © Humana Press Inc., Totowa, NJ

birds, they colonize the bird in the form of a leukemia, producing a chimera. DT40/SC Hy-Line chimeras are particularly useful for studying the in vivo trafficking of recombinant proteins of interest (e.g., those proteins that are deposited in egg yolk or albumen), and they provide a useful tool in determining which protein-encoding DNA constructs would be suitable candidates for the production of transgenic birds. Specifically, DT40 cell lines secreting recombinant human IgG3 and recombinant human IgA have been used to produce DT40/SC Hy-Line chimeras that deposit the recombinant human immunoglobulins into egg yolk *(7)* (and albumen in the case of human IgA). This chapter also outlines protocols for producing DT40/SC Hy-Line chimeras and the preparation of immunoglobulin extracts from egg yolk and albumen.

2. Materials

2.1. Reagents for Cell Culture

1. Dulbecco's modified Eagle's medium (DMEM) supplemented with 8% heat-inactivated and light-treated fetal bovine serum (FBS), 100 U/mL penicillin, and 100 μg/mL streptomycin.
2. Phosphate-buffered saline (PBS), pH 7.4.

2.2. Reagents for Transfection of CBCs

1. 2X Cytomix: Add in the following order: 27 mL sterile water, 5 mL 2.4 *M* KCl, 0.5 mL of 30 m*M* $CaCl_2$, 5 mL 200 m*M* K_2HPO_4/KH_2PO_4, pH 7.6, 5 mL of 0.5 *M* HEPES, pH 7.6, 2.5 mL of 80 m*M* EGTA, pH 7.6, and 5 mL 100 m*M* $MgCl_2$. Sterilize each of the above stock solutions by autoclaving. A 2X cytomix is stable at room temperature for several weeks.
2. 1X Cytomix: For 1 mL of 1X cytomix, add 500 μL of 2X cytomix, 80 μL of 50 m*M* ATP, pH 7.6, and 1.6 mg of reduced glutathione powder. Make up to 1 mL with water. Adjust the pH of the ATP stock solution with 5 *M* KOH, filter-sterilize, and store at –20°C. 1X cytomix containing ATP and glutathione should be freshly prepared for each electroporation.
3. 1-mm Gap electroporation cuvettes: Cuvets from BTX (Bio-Rad, Hercules, CA) and Invitrogen (Carlsbad, CA) have been used with equal success.
4. OptiMEM (Gibco-BRL, Gaithersburgh, MD): Do not add serum or antibacterial agents when using this medium for lipofection.
5. Lipofectin reagent (Gibco-BRL).

2.3. Reagents for Extraction of Immunoglobulins from Yolk and Albumen

1. Acidified distilled water, pH 2.5: Adjust pH with concentrated HCl.
2. Saturated ammonium sulfate: Add 720 g of ammonium sulfate to 1 L of distilled water. Stir over low heat until dissolved. Store at 4°C.

3. Methods

3.1. Transfection of CBCs by Electroporation

1. Collect blastoderms from freshly laid, unincubated eggs and dissociate as described in Chapter 37.
2. Pellet cells by centrifuging at 170*g* for 4 min. Resuspend in 10 mL of room temperature 1X cytomix without ATP or glutathione (*see* **Note 1**). Centrifuge as above.
3. Wash cells again as in **step 2**.
4. Resuspend cell pellet at 2–5 × 10^6 cells/mL of 1X cytomix with ATP and glutathione Add 10 μg of plasmid DNA per 70 μL (*see* **Note 2**) and incubate at room temperature (*see* **Note 3**) for 10 min.

5. Transfer 70 μL of the cell/DNA suspension into a 1-mm gap electroporation cuvette. Electroporate (*see* **Note 4**) at room temperature using three 99-μs square wave pulses at 200 V with 1-s intervals.
6. Immediately following electroporation, add 1 mL of DMEM + 8% FBS to each electroporation cuvette, mix gently with the pipeter, and incubate cells in the cuvette at room temperature for 10 min.
7. Transfer the cells from the electroporation cuvette into a 15-mL centrifuge tube and pellet as above.
8. Resuspend cells at approximately 4–6 × 10^6 cells/mL in DMEM + 8% FBS (usually in 200–250 μL if approx 50 blastoderms were originally harvested).
9. Inject electroporated cells as described in Chapter 37 to produce chimeras (*see* **Note 5**).

3.2. Transfection of CBCs by Lipofection (see Note 6)

1. Collect blastoderms from freshly laid, unincubated eggs and dissociate as described in Chapter 37.
2. Wash the cells once in serum-free OptiMEM.
3. Collect cells by centrifuging at 170*g* for 4 min.
4. In a four-well culture plate, seed cells at a density of 1.5 × 10^5 cells/well in 0.8 mL of serum-free OptiMEM.
5. Prepare the following solutions in 15-mL polystyrene (*see* **Note 7**) tubes:

   Solution A: For each well to be transfected, dilute 1–5 μg of DNA (*see* **Note 8**) into 100 μL of serum free OptiMEM.

   Solution B: For each well to be transfected, dilute 2–25 μg of Lipofectin™ reagent into 100 μL of serum-free OptiMEM and allow to stand at room temperature for 30–45 min.
6. Combine solutions A and B in a 15-mL polystyrene tube (*see* **Note 8**) and mix gently. Incubate at room temperature for 10–15 min.
7. Add 200 μL of the Lipofectin/DNA mixture to each well of cells, mixing gently to ensure uniform distribution. Incubate 4 h (*see* **Note 9**) at 37°C and 5.5% CO_2 in a humidified incubator.
8. Pellet cells by centrifugation as above and completely remove the supernatant. Wash cells once in DMEM + 8% FBS to remove the Lipofectin reagent.
9. Resuspend cells at approximately 4–6 × 10^6 cells/mL in DMEM + 8% FBS (usually in 200–250 μL if approx 50 blastoderms were originally harvested) and inject as described in Chapter 37 to produce chimeras (*see* **Note 10**).

3.3. Production of Somatic Chimeras with DT40 Cell Lines

1. Maintain DT40–IgG3and DT40–IgA cells in suspension in 25 mL culture flasks at 5 × 10^6 cells/mL in DMEM + 8% FBS (*see* **Note 11**).
2. Collect cells by centrifugation at 170*g* for 5 min. Discard the supernatant.
3. Wash the cells once in PBS and pellet as in **step 2**.
4. Resuspend the cells at 1 × 10^7 cells/mL in PBS.
5. Inject 1 mL of either DT40–IgG3 or DT40–IgA cells intravenously into the wing vein of an SC Hy-Line hen (*see* **Note 12**).
6. Collect eggs and store at 4°C until analyzed.

3.4. Extraction of Immunoglobulins from Egg Yolk

1. Crack the egg into a 100-mm Petri dish. Gently pick up the intact yolk and pierce the yolk membrane, collecting the yolk in a 50-mL centrifuge tube. Be careful not to include the yolk membrane. Save the albumen in a 50-mL centrifuge tube.

2. Dilute the yolk with four volumes of acidified distilled water, pH 2.5. Place at –20°C overnight (*see* **Note 13**).
3. Thaw the frozen yolk samples at room temperature and centrifuge at 25,000*g* for 30 min at 4°C.
4. Carefully decant the supernatant and discard the lipid pellet.
5. Precipitate egg yolk immunoglobulins by adding an equal volume of saturated ammonium sulfate dropwise to the supernatant and place the samples at –20°C overnight (*see* **Note 14**).
6. Thaw and centrifuge samples as in **step 3**. Discard the supernatant.
7. Resuspend the pellet in 5–10 mL of PBS.

3.5. Extraction of Immunoglobulins from Albumen

1. Add an equal volume of saturated ammonium sulfate dropwise to the collected albumen fraction. Leave at 4°C overnight.
2. Centrifuge at 25,000*g* for 30 min at 4°C. Discard the supernatant.
3. Resuspend the pellet in 5–10 mL of PBS.
4. The yolk and albumen extracts can be used without further purification (*see* **Note 15**) in conventional enzyme-linked immunosorbent assays to detect the presence of the specific immunoglobulin of interest.

4. Notes

1. Electroporation is performed in cytomix buffer ***(9)*** to obtain maximum cell viability. This buffer was originally developed to sustain chemically permeabilized cells in metabolic studies and its composition closely resembles the cytoplasmic concentration of most important ions. The initial washing steps remove contaminating trypsin, culture medium, and fetal bovine serum.
2. Plasmid DNA can be either supercoiled or linearized. Generally transfection with supercoiled DNA gives better transient expression, which may be useful for short-term cell tracking studies. Transfection with linearized DNA has previously been shown to increase the frequency of stable integration in other cell types ***(10,11)***, although this has not been specifically demonstrated in the case of CBCs.
3. All steps of the electroporation procedure are carried out at room temperature. If CBCs are electroporated on ice, there is a large decrease in cell viability (from 80–90% viability at room temperature to 40–50% viability on ice).
4. This protocol was developed using the BTX Electrosquareporator T820. Other square-wave pulse generators may also be applicable.
5. When cultured in vitro for 36 h, electroporated CBCs have transfection efficiencies of appro 70% and cell viability is 80–90% (*see* **Fig. 1**). CBCs electroporated with a *lacZ* reporter (plasmid pmiwZ; H. Kondoh, Nagoya University, Nagoya, Japan) were injected into recipient stage X blastoderms. After 3 d of incubation surviving embryos were harvested and stained with X-gal to detect transfected donor cells. Of 59 embryos stained, 21 showed evidence of intraembryonic staining (e.g., *see* **Fig. 2**) and most (49) contained transfected donor cells in extraembryonic tissues.
6. The lipofection protocol using Lipofectin reagent is largely based on the recommendations of the manufacturer, Gibco-BRL (Gaithersburg, MD). The described protocol was optimized specifically for CBCs.
7. Polypropylene tubes should not be used, as the DNA–lipid complexes bind to the polypropylene.
8. The relative amounts of DNA and Lipofectin reagent that are combined must be carefully optimized for constructs of different sizes in order to ensure proper formation of the DNA–lipid complexes and optimal transfection efficiencies.

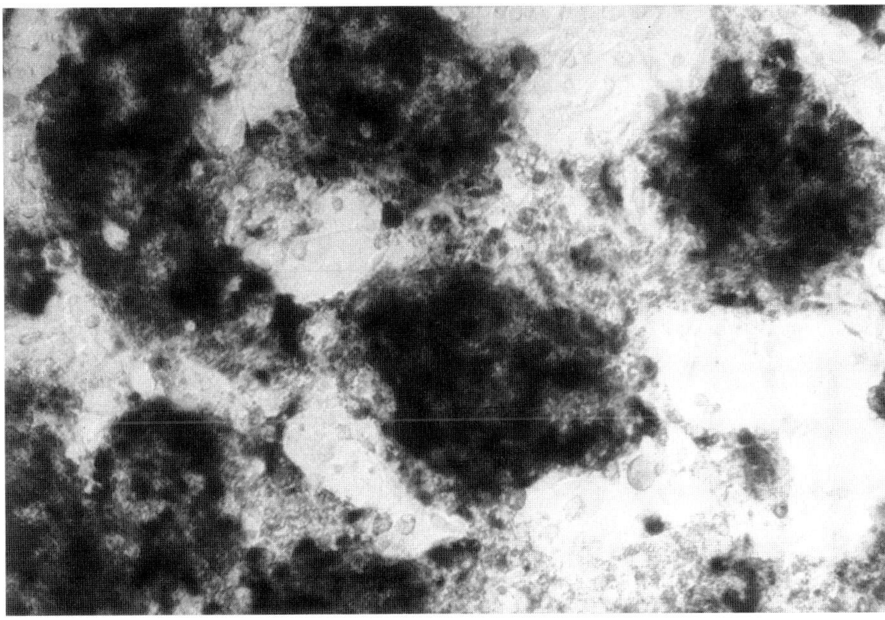

Fig. 1. (*See* color plate 9 appearing after p. 262.) CBCs transfected with plasmid pmiwZ by electroporation and stained for exogenous β-galactosidase activity with X-gal after 36 h of culture on a feeder layer of mouse SNL fibroblasts.

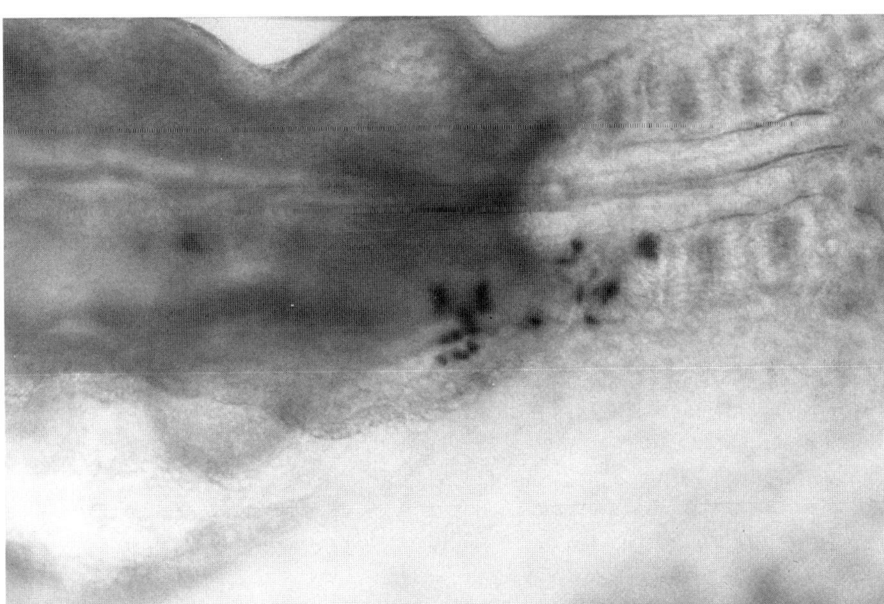

Fig. 2. (*See* color plate 10 appearing after p. 262.) Incorporation of CBCs transfected with plasmid pmiwZ by electroporation in a chimeric chick embryo. This embryo was stained with X-gal after 72 h of incubation.

9. Incubations longer than 4 h result in decreased cell viability.
10. A transfection efficiency of approx 40% and cell viability of approx 80% is obtained after lipofection of CBCs. CBCs transfected with the *lacZ* reporter pmiwz were injected into

recipient blastoderms and 45 embryos were collected and stained with X-gal *(3)*. All 45 embryos possessed extraembryonic cells expressing bacterial β-galactosidase and 15 embryos displayed intraembryonic staining.

11. The DT40 cell lines are usually split 1:5 into fresh culture medium every 3–4 d in order to maintain the desired cell density.
12. If subcutaneous tumors develop at the site of injection, this generally indicates that at least some of the cells were delivered subcutaneously. Deposition of human immunoglobulin in the egg is usually low or undetectable in these cases.
13. Freezing and thawing of the yolk samples aids in the separation of the lipid and aqueous layers. Yolk samples can be stored frozen indefinitely.
14. Be sure to add the ammonium sulfate in a dropwise fashion. If it is added too quickly, the recovery of immunoglobulins decreases significantly.
15. If an immunoglobulin preparation of increased purity is desired, additional steps such as alcohol precipitation can be used. For further details *see* **ref.** ***12***.

Acknowledgments

We thank Dr. Y. Mine (University of Guelph) for his assistance in preparing protocols for purification of immunoglobulins, Dr. S. Morrison, L. Wims and K. R. Trinh for preparing the DT40–IgG and DT40–IgA cell lines, and Mr. B. Austin (BTX) for advice regarding electroporation protocols. The financial support of the Natural Sciences and Engineering Research Council, the Ontario Ministry of Agriculture and Food and the National Institutes of Health are gratefully acknowledged.

References

1. Eyal-Giladi, H. and Kochav, S. (1976) From cleavage to primitive streak formation: a complementary normal table and a new look at the first stages of the development of the chick I. General morphology. *Dev. Biol.* **49,** 321–337.
2. Brazolot, C. L., Petitte, J. N., Etches, R. J., and Verrinder Gibbins, A. M. (1991) Efficient transfection of chicken cells by lipofection, and introduction of transfected blastodermal cells into the embryo. *Mol. Reprod. Dev.* **30,** 304–312.
3. Fraser, R. A., Carsience, R. S., Clark, M. E., Etches, R. J., and Verrinder Gibbins, A. M. (1993) Efficient incorporation of transfected blastodermal cells into chimeric chicken embryos. *Int. J. Dev. Biol.* **37,** 381–385.
4. Etches, R. J., Clark, M. E., Zajchowski, L., Speksnijder, G., Verrinder Gibbins, A. M., Kino, K., Pain, B., and Samarut, J. (1997) Manipulation of blastodermal cells. *Poultry Sci.* **76,** 1075–1083.
5. Zajchowski, L. D. and Etches, R. J. (1997) Efficient transfection of chicken blastodermal cells by electroporation . Proceedings of *Transgenic Animals in Agriculture*, August 23–27, Granlibakken, California (Abstact).
6. Buerstedde, J.-M. and Takeda, S. (1991) Increased ratio of targeted to random integration after transfection of chicken B cell lines. *Cell* **67,** 179–188.
7. Mohammed, S. M., Bonselaar, J., Etches, R. J., Trinh, R., Wims, L., and Morrison, S. (1997) Uptake of human immunoglobulins into the egg yolk of laying hens following intravenous injection of a tansfected chicken pre-B cell line. Proceedings of *Transgenic Animals in Agriculture*, August 23–27, Granlibakken, California (Abstract).
8. Petitte, J. N., Brazolot, C. L., Clark, M. E., Liu, G., Verrinder Gibbins, A. M., and Etches, R. J. (1993). Accessing the chicken genome using germline chimeras, in *Manipulation of the Avian Genome.* (Etches, R. J. and Gibbins, A. M., eds.), CRC, Boca Raton, FL.

9. Van Den Hoff, M. J. B., Christoffels, V. M., Labruyere, W. T., Moorman, A. F. M., and Lamers, W. H. (1995) Electrotransfection with "intracellular" buffer, in *Methods in Molecular Biology* (Nickoloff, J. A., ed.), Humana, Totowa, NJ, pp. 185–197.
10. Toneguzzo, F., Hayday, A. C., and Keating, A. (1986) Electric field-mediated DNA transfer: Transient and stable gene expression in human and mouse lymphoid cells. *Mol. Cell Biol.* **6,** 703–706.
11. Davis, A. C., Wims, M., and Bradley, A. (1992) Investigation of coelectroporation as a method for introducing small mutations into embryonic stem cells. *Mol. Cell. Biol.* **12,** 2769–2776.
12. Akita, E. M. and Nakai, S. (1992) Immunoglobulins from egg yolk: Isolation and purification. *J. Food Sci.* **57,** 629–634.

39

Long-Term Culture of Chicken Blastodermal Cells (CBCs) and Selection of Transfected CBCs Using Antibiotic Resistance

Qingxia Wei, Kristin L. Woods, and Robert J. Etches

1. Introduction

Chicken blastodermal cell (CBC) culture is a useful tool for the production of germline and somatic chimeras. The long-term culture of CBCs may be used for studying the in vitro effects of various cytokines and growth factors at different stages of embryonic development *(1)*. In addition, cell culture may prove to be a useful tool for the introduction of a genetic modification into the genome through transfection *(2)* and the selection of the cells carrying the modified gene conferring antibiotic resistance and/or the ability to grow in selective media before the cells are introduced into a recipient embryo in which they would colonize the germ line *(3)*.

Currently, the use of a feeder layer is necessary to obtain maximum proliferation of CBCs. When the culture of feeder cells is prepared, their proliferation is inhibited by exposure to γ-irradiation or mitomycin C. The feeder layer used by our laboratory is a STO-derived line designated SNL (kindly provided by Dr. Alan Bradley, Baylor College of Medicine) that has been transfected with leukemia inhibitory factor (LIF). CBCs cultured on this or a similar feeder layer for up to 7 d are able to contribute to both somatic tissues and germline of chimeras *(1)*.

This chapter is divided into two sections. The first section outlines the protocols for long-term culture of CBCs developed for studying the in vitro effects of various cytokines and growth factors. The second section outlines the protocols for selection of genetically modified CBCs using antibiotic resistance.

2. Long-Term Culture of CBCs

2.1. Materials

2.1.1. Buffers

1. *CMF–PBS + 2% CS:* Calcium- and magnesium-free PBS (CMF–PBS, Gibco-BRL, [Gaithersburgh, MD] cat. no. 21600) containing 2% (v/v) chicken serum (CS, Gibco-BRL cat. no. 16110-082), pH 7.4.
2. *T/E + 2% CS:* CMF–PBS containing 0.2% (w/v) trypsin (Sigma Chemical Co., St. Louis, MO), 0.04% EDTA, and 2% (v/v) CS.

From: *Methods in Molecular Biology, Vol. 136: Developmental Biology Protocols, Vol. II*
Edited by: R. S. Tuan and C. W. Lo © Humana Press Inc., Totowa, NJ

2.1.2. Medium

1. *OptiMEM complete:* OptiMEM (Gibco-BRL, cat. no. 51985-034) containing 4% (v/v) fetal bovine serum (FBS, Gibco-BRL, cat. no. 16141-079), 2% (v/v) CS, and 100 U/mL penicillin–streptomycin (1X P/S).
2. *DMEM complete:* Dulbecco's modified Eagle's medium (DMEM, Gibco-BRL, cat. no. 12100-046) supplemented with 10% (v/v) FBS, 2 m*M* L-glutamine, and 1X P/S.

2.2. Methods

2.2.1. Preparation of Feeder Cells

1. *Initial SNL mouse fibroblast culture:* From liquid-nitrogen storage, thaw about 1×10^6 SNL cells at 37°C. Dilute the freezing medium by slowly adding five times the volume of DMEM complete. Collect the cells by centrifugation at 300*g* for 10 min and resuspend them in DMEM complete. Plate cells at 1×10^4 cells/cm^2 on a standard cell culture dish (i.e., 60-mm dish, *see* **Note 1**).
2. *Passage:* At confluence, discard a half volume of medium, add 0.5 mg/mL of Mitomycin C stock solution to the final concentration of 10 µg/mL to stop cell division, swirl to cover cell surface and incubate at 37°C, 6% CO_2 for 2 h (*see* **Note 2**). Remove all of the medium. Rinse cells twice with cold CMF + 2% CS (5–10 mL for a 60-mm dish). Discard the CMF + 2% CS, and add another 5–10 mL CMF +2% CS and incubate for 10 min at room temperature (RT). Aspirate the CMF + 2% CS, and add 2 mL cold T/E + 2% CS, swirl to cover the cells, and incubate for 5 min at RT. Add 5–10 mL DMEM complete to each dish to inactivate the trypsin. Collect the cells by centrifugation at 300*g* for 10 min. Remove supernatant and resuspend cells in DMEM complete. Determine the cell concentration by counting cells on hemocytometer. SNL mouse fibroblast are then ready to be plated (5×10^4 cells/mL) (*see* **Note 3**).

2.2.2. Plating of CBCs

Remove medium from SNL feeder cells prepared as described in **Subheading 2.2.1., step 1**. Plate CBCs on top of feeders at the desired cell concentration. A concentration of 4×10^4 cells/cm^2 is recommended for maximum proliferation of CBCs.

2.2.3. Passage of CBCs

Remove CBC medium, rinse with cold CMF–PBS + 2 % CS, and add cold T/E + 2% CS. Incubate for 8 min at RT. Remove cells from the plate using gentle pipeting and collect in centrifuge tube. Add an equal volume of OptiMEM complete to inactivate trypsin. Centrifuge at 300*g* for 4 min. Remove supernatant and resuspend cells in OptiMEM complete (*see* **Note 4**).

3. Selection of Transfected CBCs Using Antibiotic Resistance

3.1. Materials

1. Plasmid: pZeoSV2/*lacZ*.
2. Lipofectin regent (Gibco-BRL, cat. no. 18292-011).
3. Zeocin™ (Invitrogen, cat. no. 45-0430).
4. X-gal (Gibco-BRL, cat. no. 15520-018).
5. ImaGene Green C_{12}FDG*lacZ* Gene Expression Kit (Molecular Probes, Inc., Eugene, OR, cat. no. I-2904).

6. Buffers:
   a. PBS: Phosphate-buffered saline (Gibco-BRL, cat. no. 21300-025).
   b. CMF-PBS: Calcium- and magnesium-free PBS (Gibco-BRL, cat. no. 21600).
   c. X-gal staining buffer: 5 m*M* ferrocyanide, 5 m*M* ferricyanide, 2 m*M* $MgCl_2$, dilute in CMF–PBS, pH 7.4.
7. Medium: OptiMEM complete medium and DMEM complete medium.

3.2. Methods

3.2.1. Preparation of CBCs and SNL Feeder Cells

1. *Preparation of CBCs:* Isolate area pellucidae of stage X (E–G & K,1976) embryos from freshly laid Barred Plymouth Rock (BR) eggs and dissociate with 0.25% trypsin/0.04% EDTA (w/v) (*see* Chapter 37).
2. *Preparation of SNL feeder cells:* The procedures for initial SNL mouse fibroblast culture and passage are the same as described in **Subheading 2.2.1.** Plate about $(5–6) \times 10^5$ mitomycin C-treated SNL feeder cells in each 35-mm culture dish containing OptiMEM complete medium and incubate for 24 h before the CBCs are added.

3.2.2. Transfection and Cell Culture

1. Incubate the dissociated CBCs (1.5×10^5) in 500 µL OptiMEM containing 11 µg Lipofectin regent in the presence or absence of 3 µg linearized pZeoSV2/*lacZ* for 4 h at 37°C in 6% CO_2 (*see* Chapter 38). pZeoSV2/*lacZ* can be cut by *Rca*I and encodes β-galactosidase and expresses the stable protein, which confers resistance to Zeocin™.
2. Collect transfected and nontransfected CBCs by centrifugation at 300*g* for 4 min.
3. Resuspend cells in OptiMEM complete medium, plate them on SNL feeder cells, and coculture with feeder cells for 48 h at 37°C in 6% CO_2.

3.2.3. Cell Selection (see **Note 5**)

1. After culturing transfected CBCs on SNL feeder cells for 2 d, aspirate the old medium, add fresh OptiMEM complete containing Zeocin at a concentration of 375 µg/mL.
2. Change the medium every day or at 2-d intervals according to the color of the medium (*see* **Note 6**).
3. After selecting for 4 d (at sixth day of culture), count the number of surviving CBC colonies, incubate them in fresh OptiMEM complete for further detection of expression of bacterial *lacZ* gene.

3.2.4. Detection of Expression of Bacterial lacZ Gene

The expression of the bacterial *lacZ* gene can be detected by X-gal staining or fluorescent staining with the Imagene Green C_{12}FDG *lacZ* Gene Expression Kit.

1. *X-gal staining:* Discard the culture medium. Wash the cells twice with PBS. Add 1 mL X-gal staining solution containing 25 µL X-gal stock solution (20 mg/mL) and 975 µL X-gal staining buffer into each 35-mm dish, incubate for 3 h at 37°C in 6% CO_2 (*see* **Note 7**).
2. *Fluorescent staining:* The ImaGene Green C_{12}FDG *lacZ* Gene Expression Kit allows detection of *lacZ* expression in single CBCs through the enzymatic hydrolysis of 5-dodecanoylaminoflurescein di-β-D-galactopyranoside (C_{12}FDG) to the highly fluorescent lipophilic fluorescein derivative 5-(*N*-dodecanoyl)aminofluorescein. The procedure for staining is as follows. After selecting the cells for 4 d (at sixth day of culture), remove the culture medium, replace with prewarmed (37°C), 200 m*M* chloroquine diphosphate containing DMEM medium supplemented with 10% FBS and incubate the cells at 37°C in

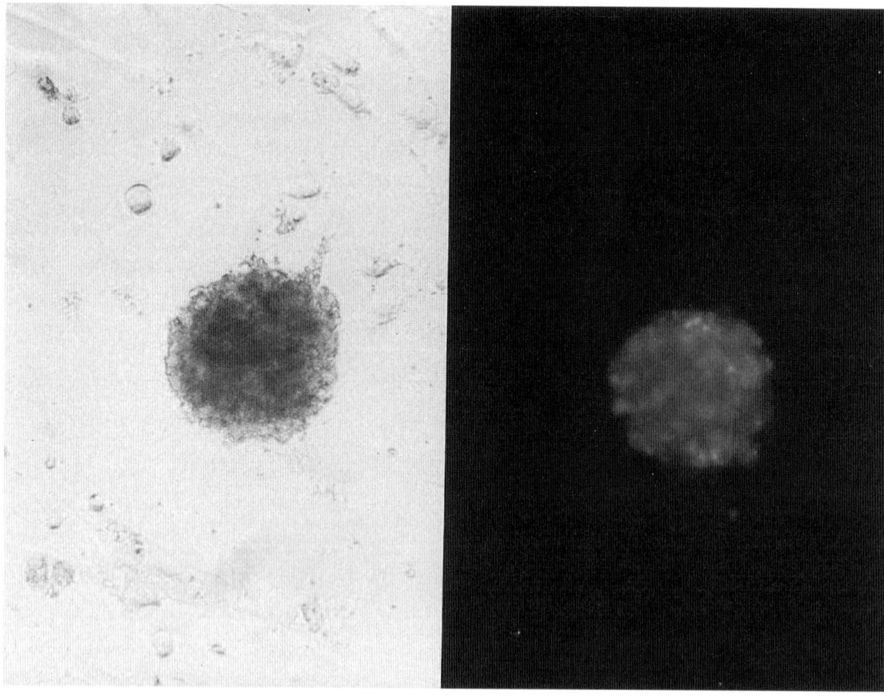

Fig. 1. (*See* color plate 11 appearing after p. 262.) **Left panel:** C_{12}FDG staining of a colony of blastodermal cells after 6 d of selection in Zeocin. The cells were transfected with the pZeoSV2/*lacZ* plasmid, incubated on SNL feeder cells in the presence of 375 µg Zeocin/mL of medium and stained on the sixth day in culture. **Right panel:** Phase contrast of the same colony.

6% CO_2 for 1 h to inhibit endogenous, lysozomal β-galactosidase activity. Then replace the medium with prewarmed, filtered (0.2 µm Acrodisc; HT Tuffryn polysulfone membrane; the substrate will bind to some other types of filters) DMEM + 10% FBS medium containing 60 µ*M* C_{12}FDG and incubate the cells for 1–1.5 h (*see* **Note 8**). Replace the medium with OptiMEM complete medium containing 1 m*M* phenylethyl-β-D-thigalactopyranoside (PETG) to inhibit any further β-galactosidase activity. Observe fluorescence of the CBCs under a fluorescence microscope equipped with a high-pressure mercury lamp, 485-nm bandpass excitation and 525-nm long-pass detection filters. The C_{12}FDG staining of a surviving colony after 4 d of selection (at sixth day of culture) is shown in **Fig. 1**.

4. Notes

1. Depending on the type of culture dish used, it may be helpful to coat the plate with a substratum such as gelatin or poly-L-lysine. Cell culture dishes are usually coated with 0.1% gelatin at least 2 h before use.
2. Mytomycin C can be inactivated by adding 50 mL of 50% NaOH for every 0.5 mg mitomycin C and incubating the mixture overnight. The proliferation of SNL feeder cells can also be stopped by exposing them to 2800 rads of γ-radiation from a ^{60}Co source.
3. Feeder cells will be attached to the plate and ready for the addition of CBCs 12 h after plating. Both CBCs and SNL cells are incubated at 37°C, 6% CO_2.
4. Chicken blastodermal cells should be passaged every 3–5 d according to the age of the culture and the rate of proliferation. For each passage, CBCs should be plated on SNL feeders as described.

5. Previous work in our laboratory showed that CBCs cultured for 7 d could contribute to both somatic tissue and germline. Therefore, systems have been designed, therefore to accommodate selection within 6 d of culture.
6. The medium must be changed when the medium turns yellow.
7. X-gal stock solution can be stored at –20°C. Dilute to 0.5 mg/mL in staining buffer just before use. Staining buffer is light sensitive but can be stored at RT for up to 2 mo in the dark.
8. After 1 h fluorescence can be detected in most cells expressing the *lacZ* gene product. Cells that have very low levels of expression may require longer incubation.

Acknowledgments

The financial support of the Natural Sciences and Engineering Research Council of Canada and the Ontario Ministry of Agriculture and Food are gratefully acknowledged

References

1. Pain, B, Clark, M. E., Shen, M., Nakazawa, H., Sakurai, M., Samarut, J., and Etches, R. J. (1996) Long-term in vitro culture and characterisation of avian embryonic stem cells with multiple morphogenetic potentialities. *Development* **122,** 2339–2348.
2. Brazolot, C. L., Petitte, J. N., Etches,, R. J., and Verrinder Gibbins, A. M. (1991) Efficient transfection of chicken cells by lipofection, and introduction of transfected blastodermal cells into the embryo. *Mol. Reprod. Dev.* **30,** 304–312.
3. Etches, R. J. and Verrinder Gibbins, A. M. (1997) Strategies for the production of transgenic chickens, in *Methods in Molecular Biology Vol. 62, Recombinant Gene Expression Protocols* (Tuan, R., ed.), Humana, Totowa, NJ, pp. 433–449.

40

Nuclear Transplantation and Cloning in Mammals

Keith E. Latham and Mark E. Westhusin

1. Introduction

Nuclear transplantation has been used for many years as an investigative tool with which to study the interactions between the nucleus and cytoplasm in early embryos and with which to characterize the developmental potency of embryonic and adult cell nuclei. As reviewed in a recent book on the subject by Professor Marie DiBerardino *(1)*, nuclear transplantation has been accomplished in a variety of organisms, both unicellular and multicellular, and both vertebrate and invertebrate. Nuclear transplantations were first accomplished in mammals during the early 1980s, using techniques developed originally by James McGrath and Davor Solter in studies using laboratory mice *(2)*. That methodology has since been used to improve our understanding of many interesting aspects of mammalian biology, including genomic imprinting, postfertilization epigenetic modifications of the parental chromosomes, embryonic genome activation, and nuclear reprogramming. The majority of these studies have involved nuclear transplantations into enucleated fertilized one-cell-stage recipients. Cloning in mice has not been possible using one-cell-stage embryos as recipients, but some studies have employed oocytes or two-cell-stage embryos as recipients to produce live offspring (e.g., *3–6*). With oocyte recipients, the greatest success has been achieved using serial transplantation protocols.

Even though cloning by nuclear transplantation in mice has been difficult to achieve, most likely because of the comparatively precocious nature of preimplantation development among rodents, similar difficulties seem to have been largely avoided in domestic livestock species, in which cloning by nuclear transplantation has been quite successful using slightly modified forms of the original McGrath and Solter method *(7–12)*. The past several years have witnessed ever-increasing success with cloning in domestic species, with regard to both the efficiency of production of viable offspring in some experiments and the ability to obtain offspring using nuclei from progressively later stages, most recently using adult somatic cell nuclei *(8)*. These studies have helped to support the idea of genetic totipotence of somatic cells in mammals and have also revealed to us the power of the oocyte cytoplasm to reprogram nuclei, which may offer a valuable system for investigating the molecular mechanisms that regulate cellular commitment and differentiation.

From: *Methods in Molecular Biology, Vol. 136: Developmental Biology Protocols, Vol. II*
Edited by: R. S. Tuan and C. W. Lo © Humana Press Inc., Totowa, NJ

In this chapter, we describe the techniques for performing nuclear transplantation in mice, which continues to be valuable for investigating early nuclear–cytoplasmic interactions and genome potency. In addition, we describe the methods used for cloning in cattle, as a representative of domestic livestock species that illustrates well the potential value of nuclear transplantation in these species for both basic and applied studies.

2. Materials

2.1. Nuclear Transfers and Experiments in Mice

2.1.1. Embryo Culture Media

The success of any nuclear transplantation experiment is critically dependent on the culture medium being used. Both the formulation of the medium and the accuracy and care employed in preparing the medium are key factors in supporting the survival of the embryos. One important element is the source of water that is used to make the medium. There is considerable variation in the source of water used in different laboratories. We find the simplest, most reliable, and most suitable source of laboratory water to be that obtained from a Millipore water purification system, such as the MilliQ or MilliQ plus UF system used in combination with either reverse osmosis or deionization as a pretreatment. The 18 MΩ resistance water produced is very low in pyrogens and free of DNAses and RNAses, making it suitable for molecular biological and biochemical methods as well as for embryo culture.

Culture medium formulations that may be used are grouped into either of two types. One category includes bicarbonate-buffered media that are used for long-term culture of the embryos. The other includes HEPES-buffered media that are suitable for manipulating the embryos at the microscope. For culturing embryos from the one-cell-stage onward, we recommend either culture for 2 d in CZB medium ***(13)*** until the eight-cell-stage, when embryos are switched to a modified Whitten's medium ***(14)***, or culture for the entire period in the newly developed KSOM + amino acids medium (KSOM/aa; *see* **ref. *15***). As a HEPES-buffered medium, we prefer the M2 formulation ***(16)*** or HEPES-CZB medium, which is especially good for micromanipulation ***(17)***. A separate set of high-quality chemicals is kept for use in preparing embryo culture media. Tissue culture plastic and plastic pipets, sterile pipetter tips, and sterile, disposable plastic spoons are used throughout medium preparations. No glassware should be used. For adjusting the pH of HEPES-buffered media, the amount of NaOH needed to bring the pH of a solution to the desired value can be determined empirically. Afterward, that amount of NaOH is added and a small aliquot of the final medium is tested for pH. This avoids introducing potentially contaminated pH electrodes into the medium to be used for embryo culture. When bringing media to volume, it is useful to prepare the media in preweighed flasks, so that water can be added by weight. This permits more consistent media preparations than relying on volumetric indicators on the flask (*see* **Note 1**).

2.1.2. Other Materials

1. Hyaluronidase: 200 U/mL in M2 medium (Sigma, St. Louis, MO, cat. no. H3884; approx 800 U/mg). Note: Other sources or preparations may be detrimental to embryos subjected to nuclear transplantation.
2. Cytochalasin B (Sigma) dissolved in ethanol at 5 mg/mL, stored at –70°C in 20-µL aliquots.

3. Demecolcine (Sigma) dissolved in water at 0.2 mg/mL, stored at –70°C in 20-µL aliquots.
4. Inactivated Sendai virus ***(18)***.
5. 35-mm and 60-mm tissue culture dishes.
6. Light mineral oil (Fisher Scientific Co., Pittsburgh, PA).
7. 50-centistokes (cs) and 100-cs silicone oil (Sigma).
8. Prosil.
9. Hydrofluoric acid.
10. NP-40 (no longer commercially available, but an equivalent compound, Igepal CA-630, is available from Sigma).

2.1.3. Equipment

1. Stereo microscope: A good stereo microscope for the isolation and sorting of embryos during the isolation procedure is essential. The microscope base should be equipped with a light source and mirror configuration that permits the pronuclei to be visualized. An example of such a microscope is the Olympus SZH10 Stereo Zoom microscope with a planapo objective and 7–70X zoom and 15X eyepieces (Olympus America, Melville, NY) mounted on a base with a Gimbal-type mirror mount. With this base, the mirror slides from front to back, and tilts from side to side as well as from front to back, enabling an oblique illumination to be achieved with which the pronuclei are visible.
2. Compound microscope: Nuclear transplantations can be performed with either a standard compound microscope with an upright configuration or using an inverted microscope. An upright microscope should be equipped with 10 or 15X eyepieces, a 10X objective and a 40X long working distance objective as well as Hoffman modulation contrast optics with a long-working-distance 40X objective. Inverted microscopes may be similarly equipped.
3. Manipulation chamber: The choice of microscope will influence a number of aspects of the procedure, but perhaps most importantly, the design of the chamber for the manipulations. The design of the chamber is intended to permit the embryos to remain separated according to what specific procedure has been performed. When using an upright microscope, embryos are placed in droplets suspended from cover slips supported on two strips of glass at front and back, leaving the sides open for insertion of pipets. Such a chamber can be constructed from quartz glass, with a base cut to 4 cm × 4.5 cm × 3-mm thick, with two 3 mm × 3 mm × 3-cm strips of glass attached parallel to the long dimension with a separation of 2.5 cm (**Fig. 1**). The strips of glass can be attached to the base using glass glue, which can be cured using ultraviolet light for a permanent bond (e.g., Norland optical adhesive). The cover slips are treated briefly with Prosil, rinsed with water, air-dried, and stored in a dust-free container before use. Small droplets are placed on the cover slip, the cover slip inverted and adhered to the glass strips using silicone vacuum grease, and the space between the cover slip and the base filled with 100-cs-viscosity silicone oil. For inverted microscopes, a glass frame may be attached to a 5 × 7.5-cm Prosil-treated glass microscope slide, droplets placed inside the frame, and the droplets then covered with oil.
4. Pipets: Two kinds of pipets are used to manipulate the mouse embryos. For both, borosilicate capillaries (R-6, 150 mm × 0.65 mm inner diameter [id], 1.0 mm outer diameter [od]) are used. For holding pipets, a capillary is heated and drawn to produce a long section that is 100–150 µm in diameter (od). This can be done by heating the pipet on a vertical pipet puller with the solenoid turned off, and then allowing the pipet to be drawn out by gravity. The pipet is then broken to a convenient length using a deFonbrune-type microforge (e.g., Alcatel, San Jose, CA) with a thick filament (0.12–0.3 mm). This is done by allowing the side of the pipet to fuse with a small glass bead on the end of the filament, turning off the heat, and then rotating the pipet upward. With practice, the pipet can be made to break

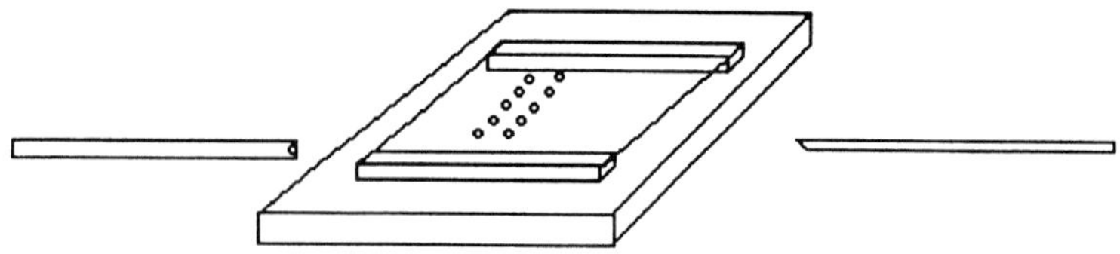

Fig. 1. Design of chamber for nuclear transplantation. A cover slip is supported by two strips of glass at front and back. Droplets of manipulation medium are suspended from the cover slip and the space between the cover slip, and glass slide is filled with oil, as described in the text. Holding pipet (left) and enucleation pipet (right) are inserted into the space for manipulation.

with a comparatively flat surface, which can then be polished by heating to leave an opening of 20–30 μm. Enucleation pipets are produced by pulling the same type of capillary tubing on a Flaming-Brown Type Pipet Puller (Sutter Instruments, Carlsbad, CA, P-87). The tip is broken at a diameter of approx 25–30 μm on the microforge using a thin filament (0.1 mm), and then beveled briefly at an angle of approx 30° from vertical (BV-10 beveler; Sutter Instruments) using a fine-grade ("D") abrasive disk. It is helpful to lower the tip onto the beveling wheel using a micromanipulator, while visualizing contact with the wheel by watching the convergence of two shadows from the tip (produced by two cables of a fiber-optic light source) as the tip approaches the wheel. The pipet is lowered until the tip contacts the wheel and bends slightly. After beveling, the tip is washed three times quickly with a 10-fold dilution of hydrofluoric acid and six times with water. The thickness of the tip is then thinned by immersion in the acid for approx 20–30 s while expelling air from the tip. This is followed by six more washes with water and six washes with ethanol. A small spike is then pulled at the tip of the bevel by briefly touching the tip lightly to the filament (not the glass bead) and then pulling away. Care must be taken to control the heat of the filament so that only a small spike is pulled and the tip is neither deformed nor melted closed. The spike is broken off against the holding pipet immediately prior to use, leaving a sharp edge to facilitate penetration of the zona pellucida (**Fig. 2**). After pulling the spike, the tip of the pipet is treated by washing six times thoroughly with 100% NP-40 followed by 24 washes with water. Because of the viscosity of the NP-40, adequate time must be allowed for it to enter and coat an adequate length of the tip. During the washes, the tip of the pipet should be agitated vigorously by rapid dipping in a large beaker of water to remove NP-40 that adheres to the outside of the capillary. The outside of the pipet is then wiped gently with an absorbent tissue to remove excess water and any remaining NP-40. When using an inverted microscope, the tips of the pipets may be angled slightly to permit a horizontal approach to the embryos (*see* **Note 1**).

5. Incubators: Embryos develop in vitro best when cultured under an atmosphere of reduced oxygen. Typically, a gas mixture of 5% CO_2, 5% O_2, and 90% nitrogen is used. The embryos can be cultured in small drops (approx 50 μL) of medium on plastic tissue culture dishes under light mineral oil (Fisher). To minimize the amount of gas mixture used, the dishes are placed inside a plastic modular incubator (Billups-Rothenberg, Del Mar, CA) that can be flushed with the gas mixture and then kept in a standard tissue-culture incubator. The modular incubator is humidified by placing a small dish of water in the bottom. Keeping the embryos in this chamber also protects them from any fumes that might be in the laboratory. The outer tissue-culture incubator can be maintained at 5% CO_2 to provide a convenient place where dishes of embryos can be kept while manipulations are being performed.
6. Micromanipulators and base: Leitz (Ernst Leitz, Wetzlar, Germany) micromanipulators and base are ideal for these procedures, especially when an upright microscope is used.

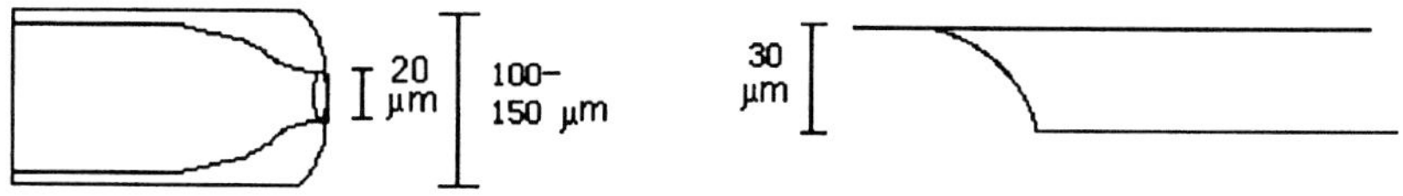

Fig. 2. Design of pipets for nuclear transplantation in mice. At left is shown the tip of a holding pipet, and at right is shown the tip of the enucleation pipet with spike.

Other manipulators and bases may be suitable, but it is strongly recommended that the manipulators be equipped with a joystick type of handle for fine manipulations.

7. Syringes: The holding pipet is attached to a 60-cc syringe with plastic tubing using, for example, a Leitz instrument holder, with no oil. The enucleation pipet should be filled with 50-cs-viscosity silicone oil and inserted into a Leitz instrument holder that is attached by plastic tubing to a microinjector, for example, a Narishige IM-6 (Narishige USA Inc., Sea Cliff, NY) equipped with a 3-mL glass syringe. The syringe, tubing, and instrument holder are filled with 100-cs silicone oil. A narrow-gauge spinal tap needle is useful for filling the enucleation pipet with oil. Care must be taken to avoid any air bubbles between the syringe and pipet tip.

2.2. Nuclear Transfers and Cloning in Cattle

2.2.1. Culture Media and Working Solutions

1. BOMC medium *(18)* modified by omitting glucose and supplementing with 6 mg/mL bovine serum albumin (BSA, Fraction V, Sigma), 0.1 m*M* minimal essential medium nonessential amino acids (MEM NEAA), and 1% v/v penicillin–streptomycin solution (10,000 U/mL penicillin G sodium/10,000 μg/mL streptomycin; Gibco-BRL, Grand Island, NY), 10 m*M* glycine, 1 m*M* alanine (MBMOC; *see* **ref. *19***).
2. Tyrode's Lactate HEPES (TL HEPES, *see* **ref. *20***) supplemented with 4 mg/mL BSA (Fraction V; Sigma) and 1% v/v penicillin–streptomycin stock solution (Gibco-BRL).
3. Zimmerman's Cell Fusion Medium: 0.28*M* sucrose, 0.5 m*M* $Mg(CH_3CO_2)_2 \cdot 4H_2O$, 0.1 m*M* $Ca(CH_3CO_2)_2$, 1 m*M* K_2HPO_4, 0.1 m*M* glutathione, and 0.01 mg/mL BSA *(21)*.
4. Pronase E: 0.05% (Protease type XXV; Sigma) in TL HEPES.
5. Cytochalasin B: 5 μg/mL (Sigma) in TL HEPES.
6. Hoechst 33342, 5 μg/mL (Bisbenzimide, Sigma) in TL HEPES.
7. TL HEPES supplemented with 20% fetal bovine serum (FBS; Gibco-BRL).

2.2.2. Supplies and Equipment

1. Supplies: Supplies needed for bovine nuclear transfers are essentially those normally used when working with mammalian embryos in the laboratory. Transfer of oocytes and embryos between solutions can be accomplished using a variety of instruments, ranging from capillary tubes or Pasteur pipets connected to tubing and a mouthpiece to a more sophisticated-5 μL Drummond pipetter (Drummond Scientific, Broomall, PA). Whatever the system used for movement of embryos, it is important that a sterility be maintained. It is a common practice in our laboratory to keep a beaker of boiling distilled water available for periodically rinsing pipets.
2. Microscopes: Microscope requirements for nuclear transfer/cloning in cattle are similar to those described above for mice. Procedures for cloning cattle normally do not require visualization of pronuclei, because metaphase II oocytes are most commonly used as recipient cytoplasts. However, it is important to confirm that enucleation of unfertilized oocytes has been successful, and for this procedure, a compound microscope fitted with

the necessary equipment to perform fluorescence microscopy is important. The most common fluorescent dye used to confirm oocyte enucleation is Hoechst 33342 (Bisbenzimide), which excites at 355 nm and emits at 465 nm.

3. Micromanipulators: As with nuclear transfer in mice, a set of Leitz micromanipulators (one each of right- and left-handed models) is an excellent choice for micromanipulation of bovine embryos and oocytes. However, nuclear transfer in cattle can be performed using either a stereoscope or a compound microscope, and the best choice of manipulators may depend on the microscope that is used. Narishige micromanipulators (Narishige) equipped with a joystick-type handle work very well and can be mounted directly onto a compound microscope: Leitz manipulators are preferred when working with a stereoscope but can also be fitted to work with a compound microscope.
4. Aspiration syringes: Syringes used to control aspiration in embryo-holding pipets and enucleation pipets vary considerably from simple plastic syringes to finely controlled micrometer syringes such as those manufactured by Hamilton (Hamilton, Inc., Reno, NV). The type of syringe selected for use is unimportant as long as the technician can maintain precise control. The system most commonly used in our laboratory for holding embryos is a 30-cc plastic syringe (Air-Tite Products, Inc., Virginia Beach, VA) connected to approx 3 ft of silastic tubing (Dow Corning Corp., Midland, MI, cat. no. 508-002, 0.51 mm id, 0.94 mm od); the tubing is passed through a Leitz instrument holder and connected directly to the holding pipet for use. Aspiration during oocyte enucleation requires finer control, so a Beaudouin syringe (Alcatel, Micro Instruments Ltd.; Oxford, UK) is used for this procedure. Aspiration controls for holding and enucleation pipets are best located opposite the micromanipulator on which the pipets are mounted so that one hand can be used to control movement while the other controls aspiration. Whereas air- and oil-filled systems are equally effective for use with holding and enucleation pipets, air-filled systems save time and are easier to maintain.
5. Manipulation chamber: Although the manipulation chamber described above can also be used for nuclear transfer in cattle, it is much more common to perform embryo manipulations in either a Petri dish filled with the appropriate culture medium or in drops of medium contained in a Petri dish and covered with sterilized mineral oil. Procedures for nuclear transfer are generally carried out at room temperature, but there may be some advantage to maintaining temperature at 37–39°C, which can be accomplished easily by equipping the microscope with a heated stage.
6. Electrofusion system: Even though a number of electroporation/electrofusion machines are on the market and will likely suffice for fusing recipient oocytes with donor karyoplasts, the most common systems used are those manufactured by BTX, Inc. (San Diego, CA) *(see below)*. These systems provide a variable AC current, which aids with alignment of the cells during electrofusion, and a DC current, which is used to induce cell membrane fusion. Most are equipped with both manual and automatic controls. An optional piece of equipment that comes with the BTX system is the "Optimizer," which allows one to visualize and record the electrical activity and/or confirm that the parameters selected are those that were delivered. The extra expense of this option may be warranted in some situations, but in the majority of the cases, it is unnecessary. If performance problems are suspected, an oscilloscope can be used to measure the output. The requirements for an adequate electrofusion system are quite simple and can likely be manufactured in most departmental maintenance/electrical shops. Electrical output required from the machine will vary depending on the electrofusion chamber used and is primarily dependent on the distance between electrodes. Fusion chambers commonly vary in electrode distance from 200 μm to several millimeters. The system purchased or manufactured should provide a variable-voltage DC pulse, and the length and number of pulses should

also be adjustable. We have used two different sizes of electrofusion chambers and two different AC settings. For a 200-μm chamber, we use 6 V at 600 kHz, and for a 3.2-mm chamber, we use 25–30 V. A DC current pulse of approx 1–2 kV/cm is adequate. The DC pulse length is normally in the range of 10–100 μs.

7. Pipets and micromanipulation tools: Probably the most important requirement for efficient nuclear transfers is the quality of the microtools. Clean tools of the proper size and shape are critical. To ensure that microtools are clean and sterile before use, large quantities may be manufactured, sterilized, stored, and discarded after each nuclear transfer procedure. However, with proper care, microtools can be reused several times. This requires sterilization with boiling distilled water before and after each procedure.

The pipets needed for nuclear transfer in cattle and methods for constructing these are similar to those described above for mice; however, the microtools are normally a bit larger. The od of the holding pipet should be 150–200 μm. A holding pipet with a diameter approximating that of the embryo/oocyte seems to work best. Two types of enucleation pipets can be used, depending on the particular methods employed. A sharpened-beveled pipet with an od of 30–40 μm can be used to penetrate the zona pellucida and enucleate oocytes by removing the first polar body and metaphase chromosomes. Alternatively, a glass knife can be used to cut a slit in the zona pellucida of approx two-thirds the circumference of the oocyte, followed by enucleation using a blunt-ended pipette with an od of approx 50 μm. The glass knife is produced by two steps. First, a pipet is pulled using a vertical pipette puller as described in **Subheading 2.1.3., step 4**. This pipet is then secured to the microforge and oriented perpendicular to the microfilament. The microfilament is turned on and the pipet lowered so that its sharp end touches the melted glass bead. As the pipet begins to melt and fuse with the glass bead, it is quickly drawn upward to produce a solid glass shaft approx 400–600 μm in length. The blunt-ended enucleation pipet is produced using similar techniques for producing a holding pipet with the exception of breaking the pipette off so that the od is smaller (50 μm vs 200 μm), and it is normally not fire polished. Pipets to transfer donor cells/karyoplasts into the perivitelline space are manufactured using the same method as used to produce a blunt-ended enucleation pipet. The od will vary but should be slightly larger than the cell to be transferred. If a sharpened/beveled pipet is used for enucleation, this tool can also be used in most cases for injection of nuclear donor cells into the perivitelline space.

Tools for alignment and agar chipping (*see* **Subheading 3.**) are made from Pasteur pipets. The alignment tool is a blunt-ended instrument used for manual alignment of oocytes and donor cells after placement into the fusion chamber. It can be pulled by hand over a flame. The end should be small enough to fit inside the fusion chamber, between the electrodes, and before use is fire polished to seal the end. Agar-chipping pipets are used to embed nuclear transfer embryos in agar prior to transfer into an intermediate host (most commonly the sheep oviduct) for embryo culture. These pipets are also pulled by hand, and then cut bluntly using a microforge so that the od is approx 250 μm.

3. Methods

3.1. Nuclear Transplantation in Mice

1. Selection of embryos: After isolation, embryos are examined under a stereomicroscope that permits visualization of pronuclei. Select from among the embryos those for which

pronuclei can be seen. For experiments in which parental genomes must be identified, preference is given to eggs that exhibit clearly asymmetric location of the two pronuclei with respect to the polar body. The maternal pronucleus is usually closest to the polar body. When viewed at higher magnification under Hoffman optics (or Nomarski), the maternal pronucleus is usually smaller in size than the paternal pronucleus. Using these two criteria together permits unambiguous identification of the two parental genomes (*see* **Note 1**).

2. Enucleation: After selection, the embryos to be used are incubated for 30 min at 37°C in CZB or Whitten's medium containing 5 µg/mL cytochalasin B and 0.1–0.2 µg/mL demecolcine, added fresh at the time of transfer. Embryos are then put singly into separate droplets of M2 medium, supplemented with 5 µg/mL cytochalasin B and 0.1–0.2 µg/mL demecolcine, in the manipulation chamber as previously described above. When an upright microscope is used together with a chamber that has hanging drops and horizontal access from the two sides, the embryo transfer pipet can be bent at approx 90° angle to facilitate insertion of the pipet into the space between the cover slip and the base. For enucleation, the embryo is held on the holding pipet by gentle suction, with the polar body adjacent to the holding pipet and the two pronuclei in the same, or nearly the same, plane of focus (**Fig. 3**). This orientation serves two purposes: First, it permits the embryo to be picked up later at the same spot so that the original opening in the zona pellucida made by the enucleation pipet is accessible and known to lie directly opposite the holding pipet; second, this orientation facilitates the identification of the two pronuclei by clarifying their positions relative to the polar body. The microscope is focused on the edge of the embryo opposite the polar body. The tip of the enucleation pipet is brought into the same plane of focus and slowly inserted through the zona pellucida without penetrating the plasma membrane. Applying firm suction with the holding pipet to place tension on the zona pellucida may help zona penetration. As the enucleation pipet is inserted, care should be taken to keep the tip directly opposite the holding pipet, as judged by the curvature of the indentation in the egg surface. There is usually a slight recoil of the zona pellucida as it tears and the pipet tip enters, which is a useful indicator of zona penetration. Care must be taken not to contact the plasma membrane on the opposite side of the egg, as this usually leads to disruption of the plasma membrane and lysis of the egg. Once the tip of the enucleation pipet is inside the perivitelline space, it is maneuvered until it is adjacent to the nucleus that is to be removed. This is accomplished without penetrating the plasma membrane. The plasma membrane will deform to envelop the tip of the enucleation pipette, so that it can be placed where desired. Verification that the pipet tip is located correctly can be obtained by gently pushing with the pipet and making sure that the tip does not pass in front of or behind the nucleus, but rather that the nucleus is displaced slightly by the pushing. Gentle suction is used to draw up a portion of the plasma membrane, a small amount of intervening cytoplasm, and the nucleus. The amount of cytoplasm withdrawn should be minimized. The pipet is then slowly withdrawn, and the membrane-bound karyoplast containing the nucleus will remain within the pipet, and then separate from the egg, as the membranes heal.
3. Transplantation: After aspirating the nucleus to be transplanted, the tip of the pipet is relocated to a drop containing the inactivated Sendai virus (diluted in M2, e.g., 1:8, but the dilution factor must be determined empirically). A small amount of virus in M2 medium is aspirated. The pipet tip is then inserted through the opening in the zona pellucida of a previously enucleated recipient. The virus and karyoplast are then gently expelled into the perivitelline space. Care must be taken that the tip of the pipette is truly within the perivitelline space before expelling the karyoplast. Two good indications that this is the case will be the slight recoil of the zona pellucida as the tip re-enters the perivitelline space, and the movement of the egg within the zona as fluid begins to be expelled. After expel-

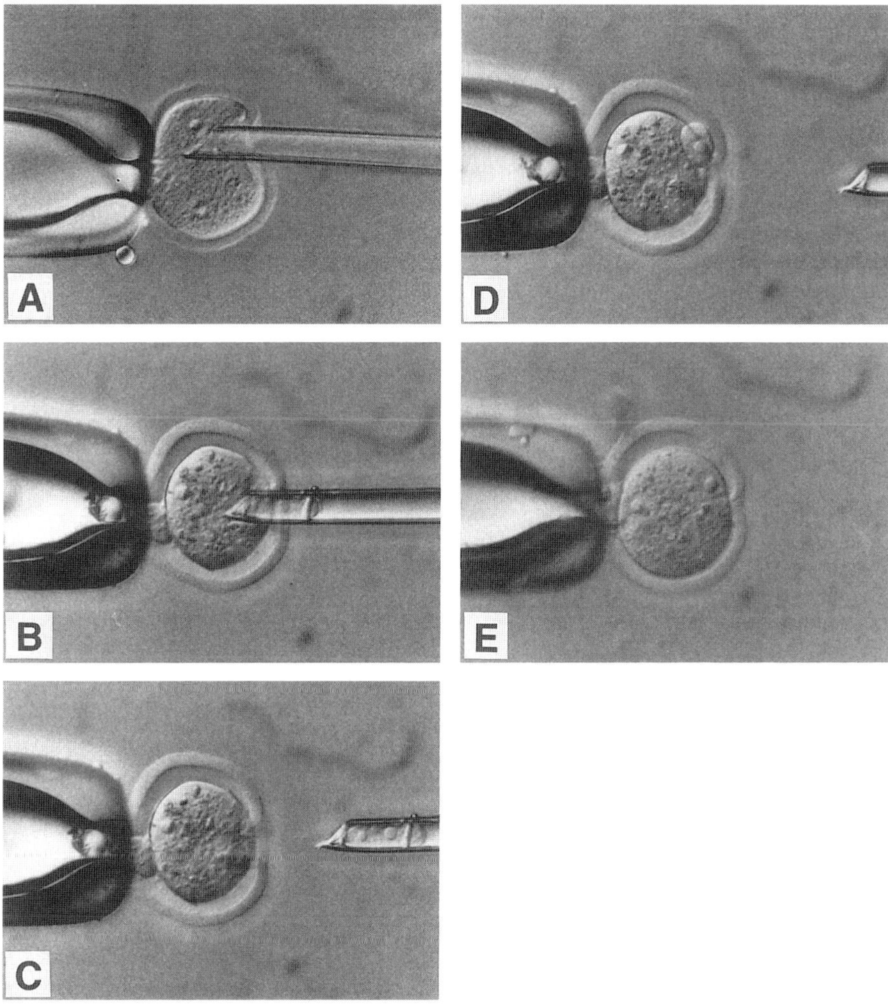

Fig. 3. Nuclear transplantation in mice. **(A)** Enucleation pipet positioned adjacent to one of the two pronuclei. **(B)** Aspiration of a portion of the cell membrane, some intervening cytoplasm, and a pronucleus. **(C)** withdrawal of pipet with karyoplast inside. Note thin cytoplasmic strand that will separate as pipet is withdrawn, leaving the cell membrane and karyoplast membrane intact. **(D)** Karyoplast transferred to the perivitelline space of the recipient embryo. **(E)** Completed transfer after fusion of karyoplast with the recipient embryo.

ling the karyoplast, the enucleation pipet is removed. If an excess amount of cytoplasm was aspirated with the nucleus, then a small anucleate portion of the karyoplast may remain within the pipet by the time the nucleus is within the perivitelline space. Withdrawal of the pipette at that point will simply cause this excess cytoplasm to come away with the pipet, without harm to the karyoplast, reducing further the amount of cytoplasm transferred with the nucleus (*see* **Note 1**).

4. Recovery: After transfer, embryos are carefully removed from the manipulation chamber and incubated in culture medium at 37°C. Care must be taken that no compression occurs when pipetting to remove the embryos from the chamber. The embryo transfer pipet should be approx twice the diameter of the embryos and should not become fouled with oil inside

it. Fusion of the karyoplast with the recipient cytoplasm usually occurs within 30 min. On occasion, cytoplasmic restructuring as the embryo recovers from treatment with the cytoskeletal inhibitors may cause the karyoplast to be forced from the perivitelline space before fusion occurs. This can be avoided by simply returning the embryos to drops of culture medium containing cytochalasin B and colcemid until fusion has occurred. After fusion occurs, the embryos are washed five to eight times through drops of culture medium to eliminate the drugs. Embryos are then cultured further in modular incubators as previously described.

5. Alternative methods: Rather than using inactivated Sendai virus, electrofusion can be used to unite karyoplast and recipient cytoplasm. Techniques for this have been described ***(22)***. We find it useful to remove the polar bodies during the enucleation step to avoid the possibility of their fusing with the recipient cytoplasm. We find that with a 1-mm chamber a single pulse of 90 V for 10 μs works well without damaging the embryo. Additional pulses can be given if required. A short 1–2 s AC pulse of 8 V can be used to help alignment of embryos in the electric field. A comparatively new instrument (BTX ECM 2000 embryo manipulation system) has recently been marketed to assist with electrofusion following nuclear transfers, and permits the AC pulse to begiven before the DC pulse. Simpler instruments can also be used. Sendai virus promotes fusion very efficiently, and the overall success of fusion and embryo development may be less with electrofusion, but some investigators may prefer not to have to prepare the inactivated virus and test it for efficacy. Another variation on the original procedure outlined above is to cut the zona pellucida with a glass needle, which is inserted into the perivitelline space and rubbed against the holding pipet to create the tear, and then aspirate the nucleus with a nonbeveled enucleation pipet ***(24)***. This eliminates the need to bevel the pipet, but adds an extra step to the procedure.

3.2. Nuclear Transfers and Cloning in Cattle

3.2.1. Preparation and Enucleation of Recipient Oocytes

Two sources of unfertilized oocytes can be used as recipients for nuclear transfer in cattle: in vivo matured and in vitro matured. In-vivo-matured oocytes can be obtained by using standard superovulation procedures for cattle followed by surgical removal and flushing of oviducts with culture medium to collect the ovulated ova in a Petri dish or other vessel. This procedure involves the administration of twice daily injections of follicle stimulating hormone (e.g., FSH-P, Schering Corp., Kenilworth, NJ) over a period of 4 d. Total FSH dose will vary with the age and breed of the donor cows but will normally range from 25 to 50 mg total dose. Prostaglandin F2 alpha (Lutalyse; UpJohn Co., Kalamazoo, MI; 5 mg) should be administered in conjunction with the fifth and sixth injection of FSH-P to induce regression of the corpus luteum and promote ovulation; 4000 IU of human chorionic gonadotropin (hCG) (Sigma) should also be administered approx 12 h after the final FSH-P injection to induce synchronous ovulation of unfertilized ova. Surgical removal of oviducts and collection of oocytes should be performed 24–36 h after hCG injection.

Even though in-vivo-matured oocytes work well as recipients, acquisition of these involves extensive labor and extreme expense. Therefore, in cattle, it is more common to use in-vitro-matured oocytes. These can be obtained by collecting oocytes from ovaries obtained at the abattoir. Oocytes are collected by follicular aspiration with a needle and syringe or a needle connected via tygon tubing to a vacuum pump or by slicing and mincing of ovaries with a scalpel followed by thorough rinsing with culture medium. Follicular aspirates should be collected in Petri dishes or some other type of sterile vessel and examined under a stereo microscope. Only oocytes exhibiting three or more layers of compact cumulus cells should be used as recipient oocytes. The acceptable oocytes should be rinsed several times through TL HEPES medium and transferred to 1-mL tissue-culture wells containing 250 µL of maturation medium (TCM 199 supplemented with FBS and gonadotrophins as previously described). The oocytes are cultured in vitro for 24 h in an atmosphere of 5% CO_2 and air. Normally, 50 oocytes are matured in each well. Alternatively, oocytes can be matured in 50-µL drops of maturation medium, under oil, with 10 oocytes per drop.

Approximately 24 h after the initiation of in vitro maturation, oocytes are removed from culture and the cumulus cells removed. This can be accomplished by repeated movement of the oocytes into and out of a glass pipette with an inner diameter just large enough to accommodate an oocyte without cumulus cells. These pipets are normally constructed by pulling Pasteur pipets, and then connected to a mouthpiece via rubber tubing. The addition of hyaluronidase to the medium used for cumulus removal may aid somewhat with this procedure, but hyaluronidase treatment is not very effective for removal of cumulus cells from bovine oocytes. An alternative method for removal of cumulus cells is vortexing. Several hundred oocytes are vortexed for 3 min in a 15-mL plastic conical test tube containing 750 µL of TL HEPES. This technique works very well for removal of the majority of cumulus cells, but a few cells may remain that can be removed easily by pipetting. Even after vortexing and pipetting, bovine oocytes will likely still contain a few tightly attached cumulus cells and their zona pellucida will be sticky, making the ova difficult to handle during micromanipulation. The problem with sticky zonae and removal of tightly attached cumulus cells can be remedied by a brief exposure to 0.05% pronase dissolved in TL HEPES. However, care must be taken when pipetting the oocytes in a solution of pronase, because overdigestion will cause the zonae to dissolve. Normally 1–2 min is adequate, but the best approach is to remove the oocytes from pronase as soon as the zonae appear to be thinning (i.e., when the perivitelline space becomes larger) and transfer them to TL HEPES containing 20% FBS.

The age of in-vitro-matured oocytes used for nuclear transfer will vary somewhat depending on the technique that is used. Most recent work involves the use of oocytes matured for approx 24 h, so that once the cumulus cells are removed, the oocytes are ready to be enucleated. Oocytes that are matured for 36–42 h also work well as recipients. These are normally obtained by removing the oocytes from maturation medium at 24 h, removing the cumulus cells, and then culturing the oocytes in TCM 199 supplemented with 10% FBS in an atmosphere of 5% CO_2 and air for 12–16 h. Oocytes cultured for 36–42 h are commonly referred to as "aged" oocytes, whereas those matured for only 24 h are often referred to as "young" oocytes. Oocytes that have been aged undergo oocyte activation readily, in many cases simply by exposing them to room temperature during micromanipulation and certainly when fused with a donor cell by electrofusion. Young oocytes are more difficult to activate, and most nuclear transfer protocols involving the use of young oocytes also include the application of some type of artificial oocyte activation procedure. Whether young or aged oocytes are used will depend on the particular application for which nuclear transfer is being used.

For example, young oocytes may be preferred when nuclear transfer is being used to gather scientific data on the effects of cell cycle synchrony on nuclear reprogramming and precise control over the time of oocyte activation is needed. If genetically identical cattle are to be produced by cloning using preimplantation-stage nuclei, then it may be easier to work with aged oocytes.

Two different procedures can be used for enucleation of bovine oocytes to prepare recipient cytoplasts. The first procedure, which is also the most common, involves a technique similar to that described for mice. While a holding pipet is used to secure the oocyte, a sharpened pipet with an od of approx 30 μm is introduced through the zona pellucidae and the first polar body and adjacent cytoplasm are aspirated into the enucleation pipet and removed. It is important to remove the smallest amount of cytoplasm possible. Removal of a large fraction of the oocyte cytoplasm can result in nuclear transfer embryos with reduced cell numbers ***(25)***. The second method for enucleating bovine oocytes is to cut a rent in the zona pellucida spanning approx two-thirds the circumference and then to use a blunt pipette to aspirate and remove a portion of the cytoplasm containing the metaphase chromosomes (**Fig. 4**). Although this procedure works well, it is more time consuming and generally results in the removal of more cytoplasm. The advantage of this method is that transfer of nuclear donor cells into the perivitelline space following enucleation may be more easily accomplished because of the large rent in the zona. In oocytes that are enucleated with a sharpened pipet, transfer of donor cells must occur through the same puncture as made by the enucleation pipet, and this small tear in the zona pellucida is sometimes difficult to visualize and align with the nuclear transfer pipette. With a bit of practice, this is normally not a problem. Also, donor cells can be put in the manipulation chamber with the oocytes during enucleation, so that one can pick up and transfer the donor cell into the perivitelline space with the same pipet used for enucleation, without moving the oocyte or altering its alignment.

It is important to confirm that oocyte enucleation has been successful in order to avoid the production of triploid embryos. This is normally accomplished by using a fluorescent dye, such as Hoechst 33342. After enucleation, oocytes can be incubated in culture medium containing 5 μg/mL Hoechst 33342 for 15–30 min, observed under a fluorescent microscope, and then those oocytes that have been successfully enucleated are selected. Metaphase chromosomes are easily visualized using this approach and those ova in which the chromosomes are still present should be discarded. Another approach that works well and seems to have no detrimental effect is to incubate the oocytes in Hoechst stain before enucleation and then to perform the enucleation using a fluorescent microscope to confirm enucleation immediately after the first polar body and cytoplasm are removed. This approach can also be used to locate the metaphase chromosomes and align the oocyte properly before enucleation, especially if the first polar body is absent, which is often the case when aged oocytes are used. Whichever approach is used for confirming oocyte enucleation, it is important that exposure to Hoechst staining and ultraviolet (UV) light be minimized, because overexposure can damage the recipient oocyte and inhibit development following nuclear transfer.

3.2.2. Nuclear Donors

To date, the cells most commonly used as nuclear donors to produce genetically identical cattle (cloning) by nuclear transplantation have been those of preimplanta-

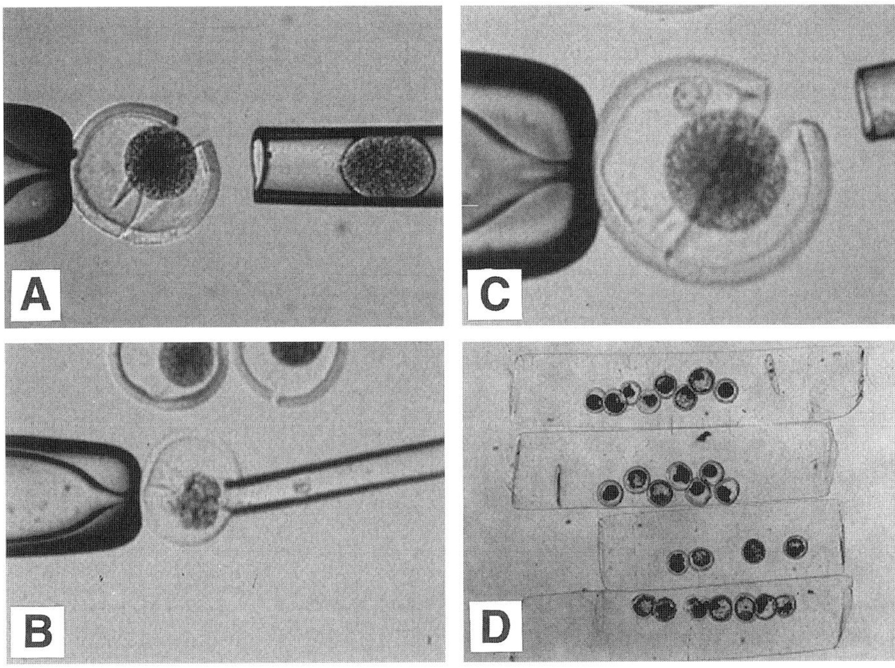

Fig. 4. Nuclear transplantation in cattle. **(A)** enucleation of oocyte. **(B)** Aspiration of donor blastomere. **(C)** Donor blastomere injected into perivitelline space. **(D)** Agar chips containing blastocysts derived from nuclear transplant bovine embryos.

tion-stage embryos. Embryos ranging from the eight-cell-stage to blastocyst (inner cell mass cells) have been used successfully to produce genetically identical calves. Embryos can be obtained using either in vivo or in vitro procedures. Protocols for the production of in-vivo-produced embryos are similar to those described for the collection of in-vivo-produced oocytes with the exception that, after superovulation, the cows are artificially inseminated 12 and 24 h after standing estrus is observed, and embryos are collected nonsurgically by flushing the uterine horns with culture medium on d 4–7, with the day of observed estrus designated as d 0. In-vitro-produced embryos can be obtained using standard protocols for in vitro maturation, fertilization, and embryo culture. Embryos that are themselves a product of nuclear transfer can also be used as nuclear donors (serial nuclear transfer). This procedure is used when the goal is to obtain large numbers of genetically identical embryos/offspring. Recently, other cell types (e.g., gonial cells, fetal cells, fibroblasts) have be used as nucleus donors in an attempt to bypass the need for serial cloning procedures to produce large numbers of identical animals and to use such cells in combination with genetic engineering to produce transgenic cattle. In most cases, the cells are placed in culture for an extended period of time and passaged several times before use. In other cases, they are used shortly after collection. Cell lines derived following extended culture of inner cell masses have also been used *(9)*.

3.2.3. Nuclear Transplantation

Whatever the cell type selected for use as a nuclear donor, transfer of the cell(s) into the perivitelline space is accomplished using a system similar to that used for enucleation of oocytes, except that the transfer pipet is smaller, with an id approximating the size of the cell to be transferred. It is not essential to have a pipet specifically made for the injection of donor cells into the perivitelline space; as mentioned above, the same sharpened pipet used for enucleation can be used. Once the cell is transferred into the perivitelline space (**Fig. 4**), it is important to be sure that the membranes of the recipient oocyte and the donor cell are in close contact. If, after transferring the nucleus donor cell into the perivitelline space, it is evident that the membranes of the oocyte and cell are not in contact, the transfer pipet can be used to push on the outside of the zona pellucida to bring the membranes together.

3.2.4. Electrofusion

Once the donor cell is positioned inside the perivitelline space of the recipient oocyte and the membranes of the ova and donor cell are in contact, electrofusion is used to fuse the recipient and donor cell membranes and thus complete the transfer. Before electrofusion, the oocyte–donor cell pairs are transferred to a medium specifically formulated for electrofusion. Normally, this medium contains very few if any electrolytes, and the main ingredient is a sugar such as sucrose or mannitol. Zimmerman's cell fusion medium works quite well (*see* **Subheading 2.2.1.**). Oocyte–donor cell pairs are equilibrated in cell fusion medium and then transferred to an electrofusion chamber filled with fusion medium. Proper alignment of the oocyte and donor cell membranes so that they are parallel at their point of contact to the electrodes is essential. This is accomplished manually with an alignment tool and by applying an AC current through the fusion machine, which also helps to pull the oocyte and donor cell into close contact. Once alignment is accomplished, fusion is induced by applying a DC current. Parameters selected for inducing cell fusion will vary. In general, a single DC pulse of 1 kV/cm for 25–30 µs is sufficient for fusing embryonic cells to recipient oocytes. Smaller cells require a higher voltage to induce fusion, but a higher voltage also causes more oocytes to lyse. Multiple pulses may also be applied to help increase fusion efficiency.

After applying the fusion pulse, oocyte–donor cell pairs should be transferred to culture medium and maintained at 37°C. It is common to culture the embryos after fusion in medium containing 20% FBS. After 1 h of culture, the oocyte–donor cell pairs should be examined to determine fusion efficiency. With larger donor cells, such as those obtained from an 8- to 16-cell embryo, fusion efficiencies should be approx 100%, whereas with smaller cells collected from the inner cell mass or fetal cells, fusion rates may be as low as 50%. It is very difficult, if not impossible, to obtain high efficiencies of fusion when using small cells, because the higher voltage needed to induce fusion of small cells with the oocyte increases oocyte lysis.

3.2.5. Oocyte Activation

Normal embryogenesis following nuclear transfer requires the activation of recipient oocytes in order to initiate the program for early development. The timing of oocyte activation and the method employed to induce activation will vary depending on oocyte age and the donor cell used. In general, however, the goal is to mimic normal oocyte activation and the associated events that occur during fertilization and to synchronize the cell cycles of the donor cell and recipient ova. When cloning embryos by fusing blastomeres to recipient ova, it is best to activate the recipient ova 4 h before fusing with donor cells.

This is because most donor blastomeres will be in the S phase of their cell cycle, and activation of recipient ova 4 h before cell fusion should provide good synchronization of donor cells with recipient ova. In addition, by 4 h after oocyte activation, MPF activity in the oocyte will have declined and the oocyte will begin DNA synthesis. Another approach is to synchronize donor and recipient cell cycles at metaphase. This approach requires manipulating the donor cells to induce their entry into G_1 or, as more recently reported, G_0. Incubation of donor cells in compounds such as nocodazole or serum starvation can be used to synchronize cells in G_1/G_0. This practice is normally not used when preimplantation-stage embryos are used as donor cells, but it is common, and perhaps the method of choice, when using established cell lines as donor cells for nuclear transfer. If donor cells in G_1 or G_0 are used as nuclear donors, it is probably best that oocyte activation be induced at the same time as, or perhaps even after the donor cell has been fused with the recipient ovum, but oocytes activated before fusion of donor cells in G_1/G_0 with recipient ova can also be used. The optimum time for oocyte activation when using donor cells in G_1 or G_0 has not been determined, but it is most common to induce activation at the time of cell fusion or shortly thereafter.

Methods for inducing oocyte activation vary and depend somewhat on the protocol and the age of the oocyte being used for nuclear transfer. Electropulsing will induce oocyte activation, and when the protocol involves oocyte activation at the time of cell fusion, the electric pulse used for electrofusion is normally adequate to induce oocyte activation. Electropulsing can also be used to activate oocytes before cell fusion. When aged oocytes are used, exposure of oocytes to room temperature will normally cause oocyte activation. Therefore, it is important to realize that if aged oocytes are used for nuclear transfer and micromanipulation is performed at room temperature, then it is likely that the oocytes will be activated during the enucleation process, well before fusion of donor cells with enucleated ova. If the desire is to use aged oocytes but to avoid oocyte activation before cell fusion, it is necessary to perform all manipulations at approx 37°C. Using "young" oocytes will also avoid activation during enucleation. Electrofusion will activate these "young" oocytes, but it is not as efficient as some other methods. Exposure for a few minutes to calcium ionophore A23187, ethanol, or inhibitors of protein synthesis are alternative means of activating bovine oocytes ***(26–28)***. A popular and reliable method for activating bovine oocytes for nuclear transfer involves a 5-min exposure to 5 µ*M* ionomycin (Calbiochem-Novabiochem, La Jolla, CA) followed by a 3-h incubation in 1.9 m*M* 6-dimethylaminopurine (DMAP; Sigma) ***(28)***.

Overall, although several different protocols can be used for oocyte activation in nuclear transplant embryos, the most common and perhaps most reliable approach involves electropulsing or room-temperature exposure of "aged" oocytes 4 h prior to electrofusion or exposure to ionomycin followed by DMAP when "young" oocytes are used. Oocyte activation should occur at least 4 h before electrofusion, when the goal is to synchronize cells in the S phase of the cell cycle. When donor cells are in G_1 or G_0, oocyte activation can be induced before electrofusion, but it is probably most efficient to induce activation at or near the time of electrofusion.

3.2.6. Embryo Culture

Following electrofusion, nuclear transfer embryos are normally cultured for 6–7 d until reaching the compact morula or blastocyst stage and then either transferred into surrogate mothers for development to term or cryopreserved and then thawed and transferred at a later date. Two different methods can be employed for culture of nuclear transfer embryos: (a) in vitro embryo culture or (b) culture in an intermediate host such as the sheep oviduct. For culture in the sheep oviduct, the embryos must be embedded in agar before transfer (**Fig. 4**). Embedding nuclear transfer embryos in chips of agar minimizes the potential immune response from the host and ensures that blastomeres are not extruded through the tear in the zona, and thus lost from the embryo during the cleavage stages before compaction. Agar is dissolved in boiling 0.9% NaCl to produce two solutions, a 1% agar solution and a 1.2% solution. The 1% solution is poured into a small Petri dish and the temperature monitored. Once the temperature reaches 37°C, nuclear transfer embryos are transferred into the liquid agar using a glass pipet constructed specifically for making agar chips (*see* **Subheading 2.1.3., step 4**). While the embryos are in the agar, the tip of the pipet is quickly filled with culture medium supplemented with 20% FBS followed by a small air bubble. The pipet is then reintroduced into the agar and the embryos drawn into the pipet tip so that they are closely aligned. The pipet containing the embryos within the agar is then removed from the Petri dish and allowed to cool at room temperature. Once the agar has hardened within the pipet, the agar chip containing the embryos is expelled into another Petri dish filled with culture medium. It is important to limit the size of the agar chips, as they may break when transferred and/or collected from the sheep oviduct. Normally 10–12 embryos is the maximum number that should be placed into a single chip. Once the embryos are embedded in agar, the second solution (1.2% agar) is decanted into a Petri dish. When this solution reaches 37°C, the agar chip containing the embryos is transferred into the Petri dish using a larger pipet (5-µL Drummond pipettor works well for this step). After a few seconds and a bit of mixing, the agar chip is drawn back into the pipet and allowed to cool at room temperature before expelling it into a Petri dish filled with culture medium. The end result is a double-layered agar chip containing the nuclear transfer embryos. The ends of the outer layer should be trimmed with an 18-gage needle or a scalpel blade in order to minimize the size of the chip.

Agar chips should be stored in culture medium such as TL HEPES at room temperature until they can be transferred into an intermediate host. Even though rabbits can be used as intermediate hosts, it is more common to use mature, nonpregnant ewes. Transfer of the agar chips containing the nuclear transplant embryos is performed surgically while the animal is under general anesthesia. The reproductive tract should be exposed through a ventral midline incision and the oviducts ligated at the uterotubal junction with #5 braided silk. Care should be taken that the ligature is not too tight and does not constrict any supporting vessels. Agar chips are aspirated in a minimum amount of medium using a chip transfer pipet such as a 25-µL Drummond pipetter (set to deposit 5 µL). With the aid of a pair of small forceps, the pipet should be advanced into the oviduct as far as possible without causing trauma to the oviduct, and the plunger depressed slowly to deposit the chips in the oviduct as the pipet is being retracted. The transfer pipet is then removed from the oviduct and examined under a stereoscope to ensure that the agar chip(s) have been delivered. Several chips can be transferred into a single oviduct, but the transfer of more than two to three chips increases the chance of damage or loss of the chip and embryos. It is common to insert a progesterone implant (Synchromate B; CEVA Corp., Overland Park, KS) into the ewe to maintain an elevated progesterone level for the duration of in vivo incubation.

Recovery of the agar chips is accomplished by a second mid-ventral laparotomy after 6–7 d of culture. This is accomplished by reentry into the incision site and exteriorization of the reproductive tract. A flushing tube, composed of silastic tubing connected to a plastic straw with a flanged end, is inserted 2–3 cm into the oviduct via the infundibulum and secured with an elastic band ligature around the ampulla. The oviduct is flushed with culture medium, retrograde, using a 30-cc syringe and blunted 18-gage needle, directed into the tip of the uterine horn. Medium containing the agar chips is collected into a sterile bowl or Petri dish. The chips are recovered with the aid of a microscope. Once the chips are recovered, nuclear transplant embryos can be carefully dissected from the chips using two 20-gage needles connected to tuberculin syringes.

For culture of nuclear transfer embryos in vitro, a number of options are available, which are basically those used for in vitro culture of embryos produced by in vitro maturation and in vitro fertilization. It is important to realize, however, that embryos produced by nuclear transfer may be more sensitive to the in vitro culture environment than embryos produced by in vitro fertilization. Evidence of this comes from a series of experiments conducted in our laboratory. Nuclear transfer embryos cocultured with BRL cells in TCM 199 developed to the blastocyst stage much less efficiently than those cultured in the ligated sheep oviduct or those cocultured with oviductal cells in MBMOC (*see* **Subheading 2.2.1., step 1**), even though there was no difference in the efficiencies of development among the culture systems when embryos produced by in vitro fertilization were used (unpublished observations). Other laboratories have reported using CR1 *(29)* to culture bovine nuclear transfer embryos successfully to the blastocyst stage. In our laboratory, the method of choice has been that reported by Moore and Bondioli involving coculture with bovine oviductal cells.

Oviductal epithelial tissue can be collected from oviducts obtained at the abattoir using a number of different methods. Oviducts should be obtained from cows early in pregnancy, preferably those exhibiting a recent ovulation. Some cells can be collected by simply flushing the oviducts with culture medium into sterilized bowls, then removing the cells from the washings using some type of glass pipet. This method works quite well, but the number of epithelial cells that can be collected is limited. The procedure used in our laboratory allows for the collection of large numbers of cells from only a few oviducts and involves using a pair of sterilized forceps to hold one end of the oviduct while another pair of forceps is used to squeeze the oviduct and "strip" the cells out into a sterile test tube. Whatever the method employed, it is important that the oviduct be dissected free of all connective tissue before performing the cell collection procedure. Once the cells have been collected, they should be washed three times in TL HEPES, resuspended 1:50 in TCM 199 supplemented with 10% FBS and 1% v/v penicillin–streptomycin stock solution, and incubated in 50-mL culture flasks (Nalgene Nunc Int., Rochester, NY) at 39°C in a humidified atmosphere of 5% CO_2 and air for 24 h prior to use. For culture of embryos, 20-µL drops of MBMOC (*see* **Subheading 2.2.1., step 1**) are prepared in 60-mm culture plates (Costar, Cambridge, MA) 24 h prior to the addition of embryos. Drops should be overlaid with sterilized mineral oil or an alternative, such as Dow Corning 200 Fluid (Specialized Products, Houston, TX) and incubated at 39°C in a humidified atmosphere of 5% CO_2 and air. For coculture of embryos, oviductal cell aggregates exhibiting ciliary movement and/or vesicle formation should be selected and washed in MBMOC, then approx 1 µL of packed cells transferred into each culture drop. Nuclear transfer embryos should be washed through MBMOC and then transferred into the culture drops containing the oviductal cells with not more than 15 embryos per drop. Culture dishes should then be transferred to modular incubators (Billups-Rothenberg) and sealed in a humidified environment of 5% CO_2 and air at 39°C for 7 d. Following in vitro culture, the embryos should be evaluated for development to the blastocyst stage and those that appear to be normal transferred to synchronized recipient cows using standard procedures for nonsurgical embryo transfer or cryopreserved (*see* **Note 2**).

4. Notes

1. Nuclear transplantation in mice.
   a. When isolating the embryos, extreme care must be taken to minimize insults to the embryo. For example, the concentration and duration of treatment of Hyaluronidase must be kept minimal. We typically release cumulus oocyte complexes (COCs) into a large pool of room temperature M2 medium (approx four to six drops) and then add one to two drops of the Hyaluronidase solution. The COCs are agitated by gentle pipetting to facilitate removal of the cumulus cells and shorten the incubation time. Subsequently, the embryos should be carefully washed through at least four drops of M2 medium and five drops of CZB or KSOM medium, with minimal carryover to remove hyaluronidase and HEPES.
   b. Extreme care must be exercised in preparing the culture media. Media that permit efficient development of normal fertilized embryos may not be of suitable quality to permit development, for example, of androgenetic or parthenogenetic embryos. Because of this, each batch of culture medium should be tested for its ability to support development of the types of embryos being produced in the procedure. A stringent indicator of medium quality used in our laboratory is the development of androgenetic embryos (e.g., C57BL/6 derived).
   c. There are many "tricks" involved in performing nuclear transplantations that are very difficult to describe. These tricks relate to how to position the embryo, how to penetrate stubborn zonae pellucidae, and how to accomplish the transfers with minimal cytoplasm accompanying the nucleus. With practice, persons performing nuclear transplantations should acquire the knowledge of these tricks over a period of several months. If possible, it is recommended that students view videotapes of transplantations performed by a skilled individual so that they can observe some of the more subtle aspects of the procedures.
   d. The preparation of the enucleation pipets is critical to the success of the method. The pipets must be the correct diameter, and the correct shape, and have just the right sized spike pulled at their tips. Sufficient washing with acid is necessary to clean the pipet, but excess acid treatment may reduce its suitability. Similarly, the NP-40 treatment is also very important. We have experimented with a number of siliconizing agents to reduce adherence between the karyoplast and the pipet, but we find that NP-40 treatment is superior. We also find that for best results, the enucleation pipets should be prepared within a few days of use.
2. Nuclear transfers and cloning in cattle.

   As with nuclear transfer in mice, preparation and use of high-quality microtools is essential for performing nuclear transfer in cattle. It will normally take a technician several weeks to months of practice to become proficient at performing the necessary embryo/oocyte manipulations, and it is critical that this learning phase be carried out using high-quality tools. Do not attempt to learn the procedures using inferior tools, as it will just lead to frustration.

Very few experiments have sought to optimize the methods for in vitro culture of bovine embryos produced by nuclear transfer. The method described involving coculture with oviductal cells is the method of choice in our laboratory, but recent improvements in culture methods for bovine embryos warrant their trial for nuclear transfer embryos *(30,31)*. If cryopreservation of nuclear transfer embryos is planned, it is important to realize that methods used for in vitro embryo culture can have a dramatic effect on the survival of cryopreserved embryos. Although in our laboratory coculture with BRL cells *(32)* has not proven as effective for promoting blastocyst development as coculturing with oviductal cells, embryos cultured with BRL cells survive cryopreservation much better than those cultured with oviductal cells *(33)*. It may be advantageous to culture nuclear transfer embryos in BRL cells for the final day of culture if cryopreservation of the embryos is anticipated.

A typical outcome following transfer of bovine embryos produced by embryo cloning is a pregnancy efficiency of approx 40% at 35 d of gestation, but the incidence of abortion is higher than normal, resulting in a calving efficiency of only approx 25%. A large number of calves obtained after nuclear transfer exhibit large birth weights *(34)*. Therefore, it is extremely important that extra care be taken in the management of recipients. Careful records must be maintained in order to predict the expected day of calving. Also, veterinary services must be made available in anticipation of the need for assisted delivery and/or cesarean section.

Acknowledgments

K.L. was supported by NSF grant no. MCB-9630370. M.W. was supported by the NIH/NICHD as part of the National Cooperative Program on Nonhuman In Vitro Fertilization and Preimplantation Development (HD21970) and the Texas Coordinating Board of Higher Education, Advanced Research Advanced Technology Program.

References

1. DiBerardino, M. A. (1997) *Genomic Potential of Differentiated Cells.* Columbia University Press, New York. pp. 180–213.
2. McGrath, J., and Solter, D. (1983) Nuclear transplantation in the mouse embryo by microsurgery and cell fusion. *Science* **220,** 1300–1302.
3. Kono, T., Tsunoda, Y., Watanabe, T., Nakahara, T. (1989) Development of chimaeric two-cell mouse embryos produced by allogenic exchange of single nucleus from two- and eight-cell embryos. *Gamete Res.* **24,** 375–384.
4. Tsunoda, Y., and Kato, Y. (1997) Full-term development after transfer of nuclei from 4-cell and compacted morula stage embryos to enucleated oocytes in the mouse *J. Exp. Zool.* **278,** 250–254.
5. Kwon, O. Y., and Kono, T. (1996) Production of identical sextuplet mice by transferring metaphase nuclei from four-cell embryos. *Proc. Nati. Acad. Sci. USA* **93,** 13,010–13,013.
6. Cheong, H.-T., Takahashi, Y., and Kanagawa, H. (1993) Birth of mice after transplantation of early cell-cycle-stage embryonic nuclei into enucleated oocytes. *Biol. Reprod.* **48,** 958–963.
7. Campbell, K. H. S., McWhir, J., Ritchie, W. A., and Wilmut, I. (1996) Sheep cloned by nuclear transfer from a cultured cell line. *Nature* **380,** 64–66.
8. Wilmut, I., Schnieke, A. E., McWhir, J., Kind, A. J., and Campbell, K. H. (1997) Viable offspring derived from fetal and adult mammalian cells. *Nature* **385,** 810–813.
9. Stice, S. L., Strelchenko, N. S., Keefer, C. L., and Matthews, L. (1996) Pluripotent bovine embryonic cell lines direct embryonic development following nuclear transfer. *Biol. Repr.* **54,** 100–110.

10. First, N. L., Sims, M. M., Park, S. P., Kent-First, M. J. (1994) Systems for production of calves from cultured bovine embryonic cells. *Repr. Fert. Dev.* **6,** 553–562.
11. Bondioli, K. R. (1993) Nuclear transfer in cattle. *Mol. Repr. Dev.* **36,** 274,275
12. Westhusin, M. E., Levanduski, M. J., Scarborough, R., Looney, C. R., and Bondioli, K. R. (1992) Viable embryos and normal calves following nuclear transfer into Hoechst stained enucleated bovine demi-oocytes. *J. Reprod. Fert.* **95,** 475–480.
13. Chatot, C. L., Ziomek, C. A., Bavister, B. D., Lewis, J. L., and Torres, I. (1989) An improved culture medium supports development of random-bred 1-cell mouse embryos in vitro. *J. Repr. Fert.* **86,** 679–688.
14. Abramczuk, J., Solter, D., and Koprowski, H. (1977) The beneficial effect of EDTA on development of mouse one-cell embryos in chemically defined medium. *Dev. Biol.* **61,** 378–383.
15. Ho, Y., Wigglesworth, K., Eppig, J. J., and Schultz, R. M. (1995) Preimplantation development of mouse embryos in KSOM: augmentation by amino acids and analysis of gene expression. *Mol. Reprod. Devel.* **41,** 232–238.
16. Hogan, B. F., Costantini, F., and Lacy, E. (1986) *Manipulating the Mouse Embryo.* Cold Spring Harbor Laboratory Press, Cold Spring Harbor, NY. pp. 107,108.
17. Kuretake, S., Kimura, Y., Hoshi, K., and Yanagimachi, R. (1996) Fertilization of mouse oocytes injected with isolated sperm heads. *Biol. Reprod.* **55,** 789–795.
18. Giles, R. E., and Ruddle, F. H. (1973) Production of Sendai virus for cell fusion. *In Vitro* **9,** 103–710.
19. Brinster, R. L. (1972) Cultivation of the mammalian egg, in *Growth, Nutrition and Metabolism of Cells in Culture Vol. II,* (Rothblat, G. and Cristofalo, V., eds.), Academic Press, New York, pp. 251–286.
20. Moore, K. and Bondioli, K. R. (1993) Glycine and alanine supplementation of culture medium enhances development of in vitro matured and fertilized cattle embryos. *Biol. Reprod.* **48,** 833–840.
21. Bavister, B. D., Leibfried, M. L., and Lieberman, G. (1983) Development of preimplantation embryos of golden hamster in a defined culture medium. *Biol. Reprod.* **28,** 235–247.
22. Wolfe, B. A. and Kraemer, D. C. (1992) Methods in bovine nuclear transfer. *Theriogenology* **37,** 5–15.
23. Tsunoda, Y., Kato, Y., and Shioda, Y. (1987) Electrofusion for the pronuclear transplantation of mouse eggs. *Gamete Res.* **17,** 15–20.
24. Tsunoda, Y., Yasui, T., Nakamura, K., Uchida, T., and Sugie, T. (1986) Effect of cutting the zona pellucida on the pronuclear transplantation in the mouse. *J. Exp. Zool.* **240,** 119–125.
25. Westhusin, M. E., Collas, P., Marek, D., Sullivann, E., Stepp, P., Pryor, J., and Barnes, F. (1996) Reducing the amount of cytoplasm available for early embryonic development decreases the quality but not the quantity of embryos produced by in vitro fertilization and nuclear transplantation. *Theriogenology* **46,** 243–325.
26. Minamihashi, A., Watson, A. J., Wastson, P. H., Church, R. B., and Schultz, G. A. (1993) Bovine parthenogenetic blastocysts following in vitro maturation and oocyte activation. *Theriogenology* **40,** 63–76.
27. Chian, R. C. and Sirard, M. A. (1996) Protein synthesis is not required for male pronuclear formation in bovine zygotes. *Zygote* **4,** 41–48.
28. Susko-Parrish, J. L., Leibfried-Rutledge, M. L., Northey, D. L., Schutzkus, V., and First, N. L. (1994) Inhibition of protein kinases after an induced calcium transient causes transition of bovine oocytes to embryonic cycles without meiotic completion. *Dev. Biol.* **166,** 729–739.
29. Rosenkrans, R. C. and First, N. L. (1991) Culture of bovine zygotes to the blastocyst stage: effects of amino acids and vitamins. *Theriogenology* **35,** 266.

30. Biggers, J. D., Summers, M. C., and Mcginnis, L. K. (1997) Polyvinyl alcohol and amino acids as substitutes for bovine serum albumin in culture media for mouse preimplantation embryos. *Human Reprod. Update* **3,** 125–135.
31. Keskintepe, L. and Brackett, B. G. (1996) In vitro developmental competence of in vitro-matured bovine oocytes fertilized and cultured in completely defined media. *Biol. Reprod.* **55,** 333–339.
32. Voelkel, S. A. and Hu, Y. X. (1992) Effect of gas atmosphere on the development of one-cell bovine embryos in two culture systems. *Theriogenology* **37,** 1117–1131.
33. Voelkel, S. A., Hu., Y. X., Moore, K., and Bondioli, K. R. (1992) Freeze survival of bovine embryos produced by in vitro maturation, fertilization, and culture of oocytes. *Theriogenology* **37,** 317.
34. Wilson, J. M., Williams, J. D., Bondioli, K. R., Looney, C. R., Westhusin, M. E., and McCalla, D. F. (1995) Comparison of birth weight and growth characteristics of bovine calves produced by nuclear transfer (cloning), embryo transfer, and natural mating. *Animal Reprod. Sci.* **38,** 73–83.

41

Production of Transgenic Mice

Jean Richa

1. Introduction

The term "transgenic" has been used since the beginning of the 1980s to define animals that had incorporated foreign DNA into their genome *(1,2)*. Currently, transgenic animals include various species such as fish, poultry, rats, rabbits, pigs, sheep, goats, cattle, and mice *(3,4)*. Mice remain the preferred system for exploring new trends in transgenic research mainly because they are easy to maintain, have short generations (6–8 wk on average), and produce large numbers of progeny (eight on average). Although gene transfer into mice can be achieved by several methods, such as retroviral infection, embryonic stem (ES) cell transfer or cell/embryo aggregation, the direct microinjection of purified DNA into the male pronucleus of fertilized mouse oocytes has been the most widely used technique for generating transgenic mice, with a rate of 5–30% of the pups born harboring the transgene *(5)*. Several books and review articles have discussed many aspects of this technology, and they provide good references for further readings *(3–11)*.

Our method, although requiring selected equipment, offers a very efficient system of transgenic mouse production. In our hands, this system has allowed us to provide fast turnover for injection of DNA constructs (3–4 wk) while maintaining an average rate of 20% in transgenic efficiency (Transgenic & Chimeric Mouse Facility, at the following website address: http://www.med.upenn.edu/tcmf/).

2. Materials

1. Culture media: We use M2 and M16 as our HEPES- and bicarbonate-buffered embryo culture media, respectively. The recipes and procedures for preparing the stocks as well as making the M2 and M16 media can be found in **ref.** *7*.
2. Hyaluronidase: Use Sigma product H-3884 (Sigma Chemicals, St. Louis, MO). Add 500 U/mL of M2 solution (3–5 mL are often enough for several days). Mix well and store at 4°C.
3. Avertin: This anesthetic is made by mixing 1 g of 2,2 tribromoethanol (Sigma-Aldrich, cat. no. T4,840-2) with 1 mL of tert-amyl alcohol (Sigma-Aldrich, cat. no. 24,048-6). Use a 15-mL polypropylene tube wrapped in foil. Mix well until all crystals are dissolved (it takes several hours). Aliquot and freeze at –20°C. Once the aliquot is thawed, it should be used in the following dilution: Add 100 µL of the aliquot to 4 mL of phosphate-buffered saline (PBS) solution (Life Technologies, cat. no. 14190-136). The dose per animal is

From: *Methods in Molecular Biology, Vol. 136: Developmental Biology Protocols, Vol. II*
Edited by: R. S. Tuan and C. W. Lo © Humana Press Inc., Totowa, NJ

adjusted to the body weight; a mouse of 25 g needs 0.4 mL of the diluted anesthetic. However, every new batch should be tested carefully because each preparation may vary in potency; any overdose will result in the immediate or delayed (by a day or two) death of the animal.

4. Hormones: Pregnant mare serum gonadotropins (PMSG) and human chorionic gonadotropins (HCG) are purchased from Sigma; cat. no. C-1063 for 1000 units (IU) of PMSG and G-4408 for 2500 IU of HCG. The powder is dissolved in PBS in its original vial to the final dilution of 1,000 IU/mL and then diluted, also with PBS, to 50 IU/mL, aliquoted and frozen at –20°C.

3. Methods

3.1. Preinjection

1. Mice
   a. Superovulation: In order to obtain large number of oocytes, it is best to use 3-wk-old female mice. However, the yield may vary between strains; the inbred being less productive than F1 hybrids, and some hybrid strains being better than others (check **ref.** ***7*** for details). Also, testing various doses of hormones can enhance the yield of superovulated oocytes. The mice are routinely injected intraperitoneally with 5 IU of PMSG between 12:30 and 1:30 PM followed by 5 IU of HCG 46–48 h later. After the injection of HCG, the females are placed in the presence of a fertile male at the ratio of one female/male. The animal room should have an automatic setting of light/dark cycle having 12 h of light and 12 h of darkness (or at least 10 h of darkness and 14 h of light). On the day following the HCG injection, the females are examined for the presence of vaginal plugs, a coagulation of the seminal fluid, indicating that mating has occurred. Females with plugs are removed and sacrificed, the oviducts are excised and the fertilized oocytes are retrieved from the ampullae.
   b. Stud males: A set of 20-30 fertile males (7–8 wk old) are housed one per cage and used only to obtain fertlized oocytes from superovulated females. The strain usually matches that of the females. Such males are replaced once a year to maintain efficient productivity.
   c. Vasectomized males: A set of 20–25 males can be purchased already vasectomized from various suppliers at the age of 8 wk, or can be acquired at 5 wk of age if vasectomy is to be done in house (chech **ref.** ***7*** for details of the surgery). Such males are housed individually and are replaced once a year, again to maintain efficient productivity.
   d. Pseudopregnant females: There is no unique strain of choice, the ultimate factor being good nursing and care of the litter. Our choice is CD-1 females (Charles River) acquired at the age of 6 wk. They are placed in the cage of the vasectomized male (three females to one male) and checked every day thereafter for the presence of vaginal plug. They are rotated within the colony until they reach 30–32 g. At that level of body weight, the females become slightly obese and inefficient in pregnancy, so they would be discarded.
2. Pipets
   a. Injection pipet: Such pipets are prepared from glass capillaries 1 mm outer diameter containing a glass filament (World Pecision Instruments Inc., Sarasota, FL, cat. no. TW100F-4). Settings on the pipet puller (Sutter Instruments, Carlsbad, CA, Model P-80) can be adjusted to produce a very fine tip with a long and slightly uniform shank (as in **Fig. 1**). The tip is treated according to **ref.** ***12*** by dipping in hydrofluoric acid solution (VWR Scientific, West Chester, PA, cat. no. JT9572-1) for a few seconds and then dipping into methanol to check for the presence of air bubbles, indicating the opening of the pipet tip. This is followed by dipping in glass treat (Alltech Associates, Deerfield, IL, cat. no. 9700) and then

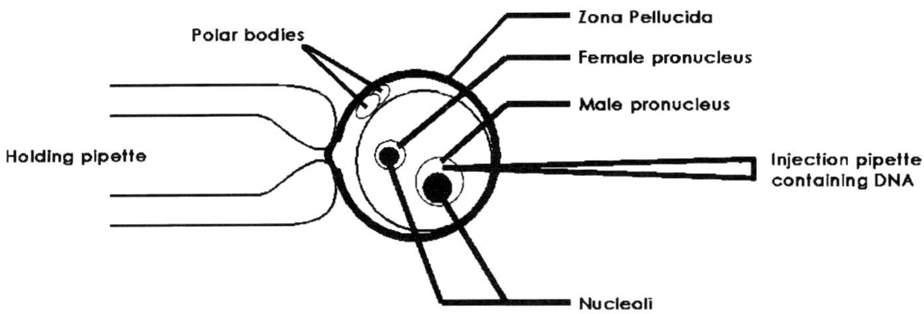

Fig. 1. Oocyte injection. The holding pipet immobilizes the fertilized oocyte while the injection pipet penetrates the zona pellucida, the plasma membrane, and the nuclear membrane before the DNA solution is released into the male pronucleus.

rinsed in methanol and air-dried. Such treatment reduces the stickiness of the tip during injection and reduces the frequency of replacing the injection pipet.

b. Holding pipet: These pipets are prepared from capillaries of 1 mm outer diameter lacking an inside filament (World Precision Instruments, cat. no. TW100-4). The glass is heated over a pilot flame of a Bunsen burner and pulled to obtain an outer diameter of 90–100 µm. (It may be necessary to pull many of these and then select the appropriate size later.) Twisting the pulled glass and touching opposite ends of the pull will produce a blunt-ended break of the glass. With the help of the heated filament of a microforge (Narishige, Tokyo, Japan, Model MF-9), the blunt tip is polished and the tip opening is reduced to 20 µm. Such dimensions allow for good grip of the oocytes and reduce the visual obstruction that a pipet with larger outer diameter may cause.
c. Handling/transfer pipet: These pipets are prepared from 50-µL capillaries (Fisher Scientific, Pittsburgh, PA, cat. no. 21-164-2G). The glass is heated over a Bunsen burner and pulled. The pulled glass is then twisted to touch opposite ends of the pull, which would result in a blunt-end break. Such pipets are then connected to a mouthpiece (supplied with the capillaries) fitted with a safety fiter, and used to handle oocytes and embryos during injetion, and later during embryo transfer.

3. Egg harvest: Excised oviduct is transferred to a droplet of hyaluronidase solution and immobilized with a fine forceps, and the ampulla is broken using another fine forceps; the mass of oocytes will ooze out into the medium. Ten minutes later (by the time all oviducts are processed), the oocytes are collected and washed three times by serial transfer through droplets of M2 medium, using a transfer/handling pipet filled with the culture medium. At this time, the fertilized eggs are sorted out, transferred to a culture dish containing droplets of M16 medium under oil, and placed in the incubator, awaiting DNA injection.
4. Injection setup: The DNA solution is prepared according to either protocol described in the Appendix. The injection pipet is backfilled by dipping the unpulled end in the DNA solution and then fitted into the pipet holder connected to a microinjector (Eppendorf Scientific, Westburg, NY, cat. no. 5242). The assembly is then secured onto the microscope unit of a micromanipulator (Eppendorf, Model 5170). The holding pipet is fitted onto a left-handed manual micromanipulator (Narishige, Model MN-151), and both manipulators are fixed on an inverted Zeiss microscope (Axiovert 10, Zeiss, Jena, Germany) equippped with Nomarski optics. Prepare two 35-mm culture dishes with several 50-µL droplets of M16 under mineral oil, and label one "noninjected" and the other "injected" to help sort the oocytes later.

3.2. Injection

1. Use a glass depresion slide to deposit a drop of M2 medium and cover it with mineral oil (Sigma, cat. no. M-5904).
2. Fit slide onto the microscope stage holder and adjust the positions of the injection and holding pipets, so that they both are at a 45° angle to the stage and are immersed in M2 medium.
3. Set the presure on the microinjector to "purge" and push the amber button to clear the tip of the injection pipet.
4. Set the pressure to continuous and adjust the level to assure a 50% increase in the volume of the pronucleus (this is achieved by test injecting a few oocytes).
5. Deposit several oocytes (20–30) in the well at one end of the field of view (north), the injection and holding pipettes are at east and west, respectively.
6. Bring the holding pipet to the same focal plane of the oocytes and aspirate to hold one in place. Rotate the oocyte by repeatedly aspirating and expelling, until the male pronucleus is at the opposite end of the holding pipet.
7. Bring the pronuclear membrane in focus using the fine focusing knob of the microscope.
8. Bring the tip of the injection pipette to the same focal plane using the the slow-controlled joystick of the Eppendorf micromanipulator (*see* **Notes 1** and **2**).
9. Under the control of the joystick, proceed slowly by pushing the tip of the injection pipet through the zona pellucida, plasma membrane, and the pronuclear membrane (*see* **Fig. 1**), while the oocyte is held firmly by the holding pipet. As the male pronucleus swells, retract the injection pipet and release the oocyte from the grip of the holding pipet in the south end of the field (*see* **Note 3**). Repeat the process for as many oocytes as you have in the well.
10. Once the batch of oocytes is injected, use the handling/transfer pipet to collect them from the well and deposit them in a droplet of M16 medium under mineral oil and place them in a tissue-culture incubator (Fisher Scientific, Model 116885H), at 37°C and in a 5% CO_2 atmosphere and 98% humidity.
11. At the end of the injection session, examine the droplets in the culture dish and remove the dead oocytes (*see* **Note 4**); the remainder are kept in the incubator overnight. Maintain records of the survival rate; this would be helpful in assessing the injection efficiency (*see* **Note 5**).

3.3. Postinjection

1. On the following day, the cultures are checked for developmental progress and the number of two-cell embryos is recorded. This number will detremine how many pseudopregnant females are needed for embryo transfer.
2. Sort the two-cell embryos by batches of 20–22 and place them in separate drops of media.
3. Prepare and sterilize all surgical tools needed, using a dry sterilizer (Inotech, Biosystems, Rockville, MD, Steri 250):

   2 fine forceps (cat. no. RS-5010)
   2 serrated forceps, (cat. no. RS-5135)
   2 serafine (cat. no. RS-5471)
   1 curved scissors (cat. no. RS-5883)
   1 hemaostatic forceps (cat. no. RS-7331)

   All from Roboz Surgical Instruments, and a pack of sterile 3.0 suture attached to a curved needle (Ethicon, Rockville, MD, cat. no. 632H).
4. Weigh the mouse and inject (ip) the appropriate amount of Avertin (*see* **Note 6**).
5. Once the plane of anesthesia is reached, usually within a few minutes (test by pinching hind limb), the mouse is placed on its side (with its left side up), and the hair is clipped over a 1-cm^2 area where the top of the hind leg meets the bottom of the rib cage.

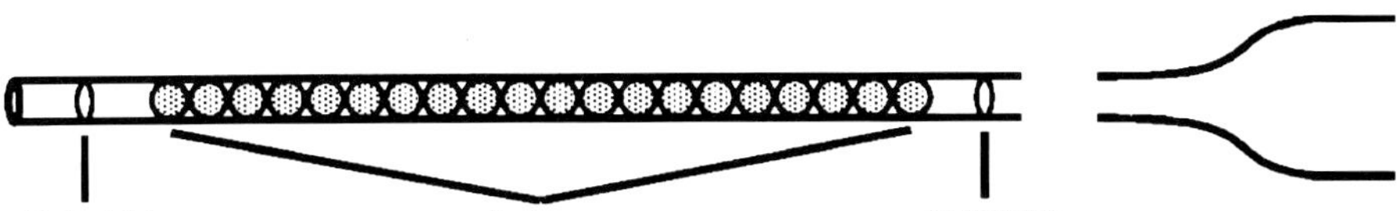

Fig. 2. Transfer pipet. The transfer pipet, connected to a mouthpiece, is used to collect the injected eggs/embryos (about 20 of them). Air bubbles are aspirated before and after the eggs/embryos to serve as a marker during the transfer.

6. Disinfect the clipped area with 70% ethanol followed by chlorhexidine diacetate (Nolvasan; Aveco, cat. no. A0626F).
7. Use the tip of the scissors to cut a small incision in the skin and the abdominal muscle. Drape the area with a sterile piece of gauze that has an opening to match the size of the incision.
8. Hold the peritoneum (abdominal muscle) with one serrated forceps and, with the other forceps, reach into the abdominal cavity and pull the white fat pad underneath. This will expose the ovary, oviduct, and part of the uterine horn.
9. Lay the organs on the gauze and clasp the fat pad with one serafine and the mesentary of the uterus with the other serafine.
10. Orient the mouse, laying on its side, under a dissecting microscope (Zeiss, Model Stemi 2000C) with the head pointing West. Turn on the fiber-optic light (Fostec, Auburn, NY, Model FO 150) and adjust light intensity.
11. Focus on the oviduct/bursa (the membrane that envelops the ovary and oviduct) complex. With the fine forceps, break the bursa to expose the infundibulum (the opening of the oviduct). Sometimes, a gentle probing of the ovary and the oviduct is necessary to expose the infundibulum if it lies underneath (*see* **Notes 7** and **8**).
12. Collect one batch of embryos (20–22) in the transfer pipet as follows: Aspirate M2 media into the pipet, followed by air, followed by the embryos, then some air and then by media. The layout of the transfer pipet should now appear as in **Fig. 2**.
13. Insert the transfer pipet into the infundibulum and hold the latter, containing the pipet with the fine forceps.
14. Expel the embryos into the oviduct with a gentle blow; you should be able to see air bubbles weave their way into the lumen of the oviduct and rest in the ampulla (as in **Fig. 3**).
15. Withdraw the transfer pipet and replace the organs into the abdominal cavity.
16. With the help of the hemostatic forceps, suture the peritoneum and then the skin with evenly spaced discontinuous stitches.
17. Place the mice in a clean cage under a heat lamp, at an intensity level low enough to keep them warm. When they awake and start moving around, the mice can be transferred to the colony. Make sure you check on them daily, during the first 3 d postsurgery, to make sure the wound is healing properly and that there is no sign of infection.
18. Around d 17 after transfer, place each pregnant female into a separate labeled cage, and include a 1-in. square of packed cotton (Ancare Corp., Belmore, NY, cat. no. NES 3600) to serve as nesting material. Place a sign on the cage prohibiting anyone from changing the cage (*see* **Note 9**).
19. Perform a daily check on the mice to note when pup delivery takes place, how many are born, and whether there are any signs of cannibalism or neglect on the part of the mother (*see* **Note 10**).
20. Once the pups reach the age of 1 wk, the sign "No Disturbance" can be removed and cages are changed.
21. At d 10 after birth, a small piece of the tail (0.5–1 cm) can be snipped off with a sterile sharp scissors and pups are numbered by toe clipping, using the map provided in **ref. *6***.

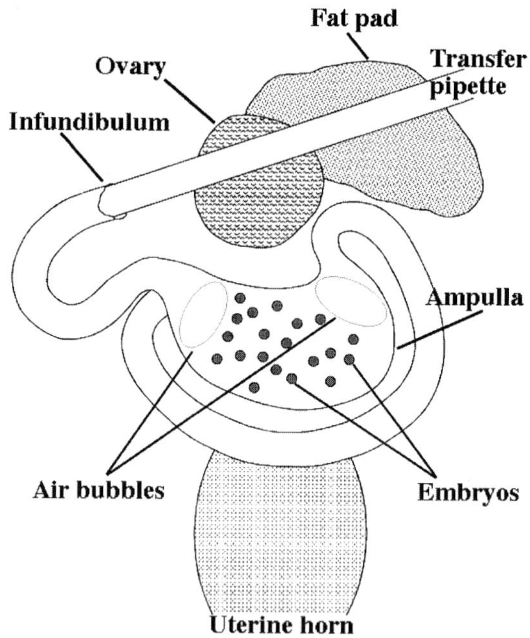

Fig. 3. Embryo transfer. The transfer pipet is inserted into the infundibulum where the eggs/embryos are released gently into the oviduct. The presence of air bubbles in the ampulla indicate a successful transfer (the embryos cannot be visualized but are included here for representation).

(Although toe clipping for animal identification is discouraged by the US Food and Drug Administration, evaluation and final approval for such procedure rests with the Institutional Animal Care and Use Committee [IACUC] overseeing the animal facility).

22. Tail clips are then processed to extract DNA using the procedure published in **ref. *6***, and the DNA is screened for the presence of the transgene using the appropriate sequences for polymerase chain reaction (PCR) or Southern/slot blots.
23. Once founder mice are identified, the remainder of the animals are sacrificed to reduce colony maintenance and cage cost (*see* **Note 11**).
24. At the age of sexual maturity (6 wk for females and 8 wk for males), founders are set up to breed so that transgene transmission through the germline can be assessed, using the same screening procedure as that for the founders.

4. Notes

1. The choice of equipment (microscope and micromanipulators) can affect the efficiency of DNA injection, from the ability to clearly visualize the pronuclei, to the ease of controlling the micromanipulators, although the latter would depend mainly on dexterity and experience.
2. Vibration, transmitted from the floor to the pipet tip through the micromanipulator, can present an uncontrollable hazard to the eggs and cause their lysis after injection. This may be eliminated by acquiring a nitrogen-gas-cushioned table (Vibraplane, Model 1201).
3. Swelling of the pronucleus should be rapid but not sudden. High pressure settings invariably lead to a sudden burst of DNA solution into the pronucleus, and lysis of the plasma membrane as the pipet is being retracted. This will be followed by the leakage of the egg cytoplasm into the perivitelline space.

4. A high incidence of oocyte death immediately following DNA injection, is invariably a sign of inappropriate injection technique; either the pipet tip is too large or the pipet is getting sticky and needs to be replaced.
5. Overnight culture of injected oocytes can be used to reveal toxicity either from the DNA sample or from the culture media. A sample of buffer-injected oocytes cultured side by side with DNA-injected sample can serve as control for DNA toxicity. In our hands, a test injection of two different DNA samples followed by an overnight culture of the injected oocytes revealed 80% development to the two-cell-stage of one batch and only 45% in the other—a clear indication of DNA toxicity.
6. The dose of avertin should be tested for every new batch. The effect of anesthesia should take place within 1 min of injection and the plane of anesthesia should be maintained for 30–45 min. Diluted samples of avertin can be stored at 4°C for 24 h wrapped in foil and still remain effective. Older samples should not be used, as they may become contaminated and cause peritonitis in the mouse.
7. Before proceeding in releasing the embryos into the oviduct, look for proper signs of early pregnancy, such as the presence of ampulla in the oviduct or mature follicles (white patches) on the ovary.
8. As you attempt to break the bursa, avoid the main blood vessel that runs across it; this will reduce the amount of bleeding that can obscure the location of the infundibulum.
9. After separating pregnant females, in preparation for delivery, be careful to rotate the position of the water bottle. If the nesting female places the nest just underneath the spout, a touch of cotton fibers will wick the water into the cage (flood); this will result in the death of the pups.
10. During the first 12 h after delivery, it is very important to check on the nursing female and its litter. Absence of white stomach (milk) is an indication of inadequate nursing; this may reflect a neglectful mother or of a female incapable of nursing. In either case, a foster mother (a lactating female) should be substituted to avoid the loss of the litter.
11. Every litter must be weaned by the age of 4 wk to avoid overcrowding of the housing cage.
12. When purifying DNA for microinjection, it is strongly recommended that powder-free gloves be used. Powdered gloves are most often associated with the presence of particulate matter that gets lodged at the tip of the injection pipet and renders it useless.

Appendix

1. Purification of DNA for Microinjection (*see* **Note 12**)
   a. Recombinant plasmid should be purified by CsCl gradient (for details, see **ref. *13***). The insert is then released with the appropriate restriction enzymes. It is critical to remove all vector sequences, as they can significantly alter the expression of transgenes.
   b. Separate the insert of interest from the vector on an agarose gel run in Tris/acetate/EDTA (not Tris/borate/EDTA) buffer.
   c. Excise the gel slice containing the transgene fragment and electroelute the DNA.
   d. Recover the DNA fragment by ethanol precipitation and then pass it through an ion exchange column (e.g., Elutip, Schleicher & Shuell, Keene, NH, cat. no. 27360b) according to the manufacturer's instructions.
   e. Ethanol-precipitate the DNA by adding 1/10 vol of 3 *M* NaAc and 2–2.5 volume ethanol.
   f. Resuspend the DNA in injection buffer (10 m*M* Tris/0.1 m*M* EDTA, pH 7.5). This buffer must be prepared with mQ (Millipore, Marlboro, MA; Model Milli-Q UF plus) or double-distilled water to maintain high injection effficiency.
   g. Estimate the DNA concentration by comparing the ethidium bromide staining of a sample run on an agarose gel, next to a standard of known concentration such as λ DNA cut with *Hin*dIII.
   h. Adjust the DNA concentration to 1–5 ng/μL with injection buffer.

2. Alternate Procedure for DNA Purification (*see* **Note 2**)
   a. After separating the fragment on agarose gel run in Tris/acetate/EDTA, excise the fragment of interest from the gel.
   b. Process the gel slice through Geneclean II kit (BIO 101, Vista, CA, cat. no. 1001-400) or Qiaex II gel extraction kit (Qiagen, Valencia, CA, cat. no. 20021) according to the manufacturer's instructions.
   c. Ethanol-precipitate the recovered DNA and resuspend in Elutip buffer.
   d. Pass through Elutip column, according to manufacturer's instructions.
   e. Ethanol-precipitate the recovered DNA and resuspend it in injection buffer (as in **Notes 5** and **6**).
   f. Estimate the DNA concentration and adjust it to 1–5 ng/µL, using injection buffer.

Acknowledgments

I am grateful to Nancy E. Cooke and Stephen A. Liebhaber for their helpful comments, and their guidance in the operation of the facility. Also, I would like to thank Andre W. Harris for preparing the figures, and Peifu He for his technical assistance in all the stages of the transgenic service.

References

1. Gordon, J. W., et al. (1980) Genetic transformation of mouse embryos by microinjection of purified DNA. *Proc. Natl. Acad. Sci. USA* **77,** 7380–7384.
2. Wagner, T. E., et al. (1981) Micro-injection of a rabbit b-globin gene into zygotes and its subsequent expression in adult mice and their offspring. *Proc. Natl. Acad. Sci. USA* **78,** 6376–6380.
3. Pinkert, C. A. (1994) *Transgenic Animal Technology: A Laboratory Handbook.* Academic Press, Cold Spring Harbor, NY.
4. Houdebine, M. L. (ed.) (1997) Genertaion and use, in *Transgenic Animals.* Harwood Academic Publishers.
5. Palmiter, R. D. and Brinster, R. L. (1986) Germ-line transformation in mice. *Ann. Rev. Genet.* **20,** 465–499.
6. Hogan, B., Costantini, F., and Lacy, E. (1986) *Manipulating the Mouse Embryo, A Laboratory Manual.* Cold Spring Harbor Laboratory, Cold Spring Harbor, NY.
7. Hogan, B., Beddington, R., Costantini, F., and Lacy, E. (1994) *Manipulating the Mouse Embryo, A Laboratory Manual.* Cold Spring Harbor Laboratory Press, Cold Spring Harbor, NY.
8. Wasserman, P. M. and DePamphilis, M. L. (1993) Guide to techniques in mouse development. *Meth. Enzymol.* **225.**
9. Monastersky, G. M. and Robl, J. M. (eds.) (1995) *Strategies in Transgenic Animal Science.* American Society for Microbiology Press, Washington, D.C.
10. Camper, S. A. (1987) Research applications of transgenic mice. *Biotechniques* **5,** 638–650.
11. Brinster, R. L., Chen. H. Y., Trumbauer, M. E., Yagle, M. K., and Palmiter, R. D. (1985) Factors affecting the efficiency of introducing foreign DNA into mice by microinjecting eggs. *Proc. Natl. Acad. Sci. USA* **82,** 4438–4442.
12. Fox. N. and Solter, D. (1988) Expression and regulation of the pituitary- and placenta-specific human glycoprotein hormone alpha-subunit gene is restricted to the pituitary in transgenic mice. *Mol. Cell. Biol.* **8,** 5470–5476.
13. Sambrook, J., Fritsch, E. F., and Maniatis, T. (1989) *Molecular Cloning: A Laboratory Manual.* Cold Spring Harbor Laboratory Press, Cold Spring Harbor, NY.

42

Production of Transgenic Mice with Yeast Artificial Chromosomes

Aya Jakobovits, Bruce T. Lamb, and Kenneth R. Peterson

1. Introduction

The stable introduction of recombinant DNA into the germline of mammals is one of the major technological advances in biology and has greatly facilitated the study of gene function in development and disease, as well as the development of biotherapeutics. Landmark experiments conducted in the early 1980s established the feasibility and reproducibility of stably introducing DNA by microinjection into the male pronucleus of single-cell mouse embryos *(1–5)*. The utilization of transgenic mice now represents a major strategy for the examination of genetic and biologic questions in vivo, both in basic research and industry.

It is well accepted that transgenes containing genomic DNA with introns and critical regulatory sequences are more properly expressed in vivo than cDNA-based constructs *(6–9)*. Before 1993, a substantial shortcoming of standard transgenic technology had been the inability to reproducibly introduce preparations of DNA fragments larger than 40–50 kilobases (kb). This problem seriously limited efforts to study large genes, gene complexes, regulatory sequences, and higher-order genomic structure encompassed on DNA fragments larger than could be cloned into plasmids and/or cosmids. However, advances in yeast artificial chromosome (YAC) and gene-transfer technologies have overcome this obstacle.

Yeast artificial chromosomes (YACs) are cloning vectors with a DNA capacity of over 2 megabases (Mb) that are stably maintained and propagated in yeast *(10)*. YACs are readily tractable to genetic alterations by homologous recombination in yeast, permitting the efficient introduction of mutations, including deletions, insertions, and nucleotide substitutions *(11,12)*. Homologous recombination also allows the removal of undesired DNA sequences contained on YACs introduced during the creation of YAC libraries (chimeric YACs), as well as the production of larger YACs from multiple smaller YACs *(13)*. YACs have proven invaluable in mapping the genomes of numerous organisms and in the identification and cloning of disease genes.

Over the past several years, numerous publications have reported the introduction of YACs (as well as other cloning vectors with a large DNA capacity such as P1s and bacterial artificial chromosomes [BACs]) into the mouse germline (for a review, *see*

From: *Methods in Molecular Biology, Vol. 136: Developmental Biology Protocols, Vol. II*
Edited by: R. S. Tuan and C. W. Lo © Humana Press Inc., Totowa, NJ

refs. ***14–16***). This chapter outlines the three different protocols developed for the introduction of YACs into the mouse germline (*see* **Fig. 1**) including spheroplast fusion of yeast containing a YAC with totipotent mouse embryonic stem (ES) cells, lipofection of purified YAC DNA into ES cells, and microinjection of purified YAC DNA into single-cell mouse embryos. In addition, the unique attributes of each protocol in terms of the distinct advantages and disadvantages of a particular method will be briefly discussed.

2. Spheroplast Fusion

2.1. Introduction

Spheroplast fusion entails the fusion of yeast spheroplasts containing a YAC with ES cells (*see* **Fig. 1A**) ***(13,17–21)***. A mammalian selectable cassette (*HPRT*, *neomycin resistance* [*neo*^{r}], etc.) is first targeted onto the YAC vector arm by homologous recombination in yeast (protocols detailing homologous recombination in yeast as well as general yeast methods are described elsewhere; *see* **refs.** ***[12,22,23]***). The cell wall of the yeast is removed by enzymatic digestion followed by a polyethylene glycol (PEG)-mediated fusion with the ES cells. Selection for the integrated YAC results in approx 30–50% of the ES colonies retaining a majority of the DNA insert as well as both YAC vector arms, primarily in one to two copies. Of these ES clones, 30–80% (depending on the intrinsic stability of the YAC) contain intact YACs with no deletions or rearrangements. Not surprisingly, a subset of the fused ES colonies also contain varying amounts of yeast chromosomal DNA. YACs up to 1020 kb in length have been introduced intact into transgenic mice via spheroplast fusion ***(13)***.

The technique of spheroplast fusion has a number of advantages: No YAC DNA isolation is necessary, a relatively high percentage of YACs are transferred to ES cells intact, the integrity and/or expression of genes on the YAC can be determined in ES cells prior to introduction into mice and large YACs (>1 Mb) have been introduced intact. One of the potential disadvantages of shperoplast fusion is the transfer of yeast genomic DNA in a subset of fused ES cells and the effect this may have on the ES cells and the YAC transgenic mice. However, thus far, cointegrated yeast genomic sequences have not affected the ability of ES cells to differentiate properly or to transmit the YAC through the mouse germline and have also not affected apparent gene function ***(17–20)***. Another disadvantage of spheroplast fusion is the relatively time-consuming transmission of genetically altered ES cells through the mouse germline.

2.2. Materials

1. Materials for yeast growth and phenotype testing are described elsewhere ***(22)***.
2. Materials for preparation of yeast spheroplasts:
   a. STC: 1 *M* sorbitol, 10 m*M* Tris-HCl pH 7.5, 10 m*M* $CaCl_2$; composition per 100 mL, 50 mL of 2 *M* sorbitol, 1 mL of 1 *M* Tris-HCl, pH 7.5, 1 mL of 1 *M* $CaCl_2$. Bring to final volume with distilled water and autoclave.
   b. SPEM: 1 *M* sorbitol, 10 m*M* sodium phosphate, pH 7.5, 10 m*M* EDTA, pH 8.0, 30 m*M* β-mercaptoethanol (BME); composition per 10 mL, 5 mL of 2 *M* sorbitol, 0.1 mL of 1 *M* sodium phosphate, pH 7.5 (make by mixing sterile 1 *M* dibasic and monobasic sodium phosphate to the appropriate pH), 0.2 mL of 0.5 *M* EDTA, pH 8.0, 21 μL of 14 *M* BME (Sigma, St. Louis, MO, cat. no. M7522). The sorbitol, sodium phosphate, and EDTA

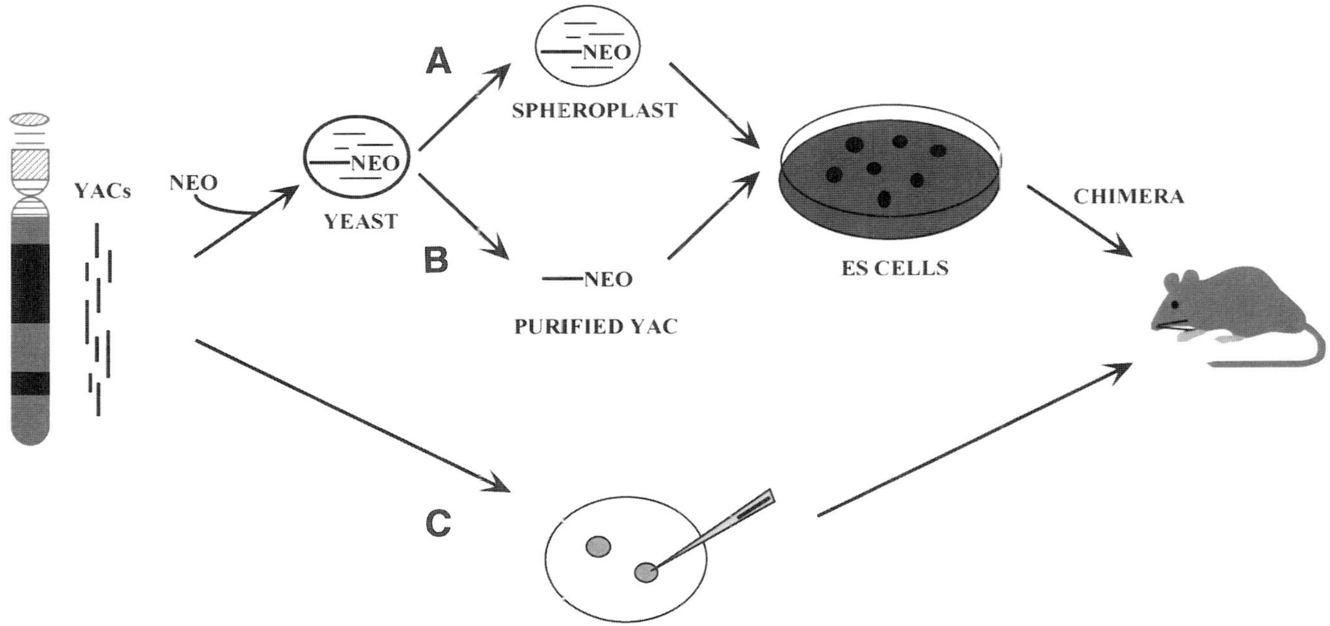

Fig. 1. (*See* color plate 2 appearing after p. 262.) Schematic representation of the three methods for the generation of YAC transgenic mice. **(A)** Spheroplast fusion relies on the introduction of YACs containing a mammalian selectable cassette (NEO) into ES cells by enzymatic digestion of the yeast cell wall and fusion of the resulting spheroplasts with ES cells. Please note that spheroplast fusion has typically utilized HPRT for positive selection in ES cells, but *neo*^{r} (NEO) can also be utilized and has been included in the figure for the sake of clarity. **(B)** Lipofection involves the isolation of YACs containing mammalian selectable cassettes away from the other yeast chromosomes and subsequent introduction of purified YAC DNA into mouse ES cells via lipid-mediated transfection. ES cells containing YACs are subsequently introduced into host blastocyst-staged embryos and germline transmission of the YAC through the resulting chimeras generates YAC transgenic mice. **(C)** Microinjection involves the isolation of YACs away from the other yeast chromosomes, purification, and concentration of the YAC DNA with modifications to limit the shearing of high-molecular-weight DNA in solution and direct transfer of YAC DNA into the male pronucleus of mouse eggs.

(SPE) are mixed together and brought to 100 mL with distilled water and autoclaved. The BME is added just prior to use.

c. Zymolyase 20T: 100 mg/mL in 10 m*M* sodium phosphate, pH 7.5; composition per 1 mL, 100 mg zymolyase 20T (ICN, Costa Mesa, CA, cat. no. 320921) dissolved in 1 mL of 10 m*M* sodium phosphate, pH 7.5 (see above recipe for 1 *M* sodium phosphate, pH 7.5). Mix well (solution always stays turbid) and aliquot. Solution is stable for several months at –20°C.

3. Materials for ES tissue culture:
   a. ES media: Dulbecco's modified Eagle's medium (DMEM), 100 U/mL penicillin, 100 μg/mL streptomycin, 2 m*M* L-glutamine, 15% heat inactivated ES-qualified fetal calf serum (FCS), 100 μ*M* BME, 10^3 U/mL murine leukemia inhibitory factor (LIF); composition per 500 ml, 415 mL DMEM, high glucose (4500 mg/L) with sodium pyruvate (110 mg/L; JRH Biosciences, cat. no. 51444-78P), 5 mL 100 X penicillin-streptomycin solution (Gibco-BRL, Gaithersburgh, MD, cat. no. 15140-122), 5 mL 200 m*M* L-glutamine (JRH Biosciences, cat. no. 59202), 75 mL ES-qualified FCS (heat inactivated at 56°C for 30 min; *see* **Note 1**), 4 μL of 14 m*M* BME, 50 μL of 10^7 U/mL LIF (ESGRO™; Gibco-BRL, cat. no. 13275-029). ES media should be stored at 4°C and can be kept for several weeks. Serum-free DMEM media (SFM) is simply DMEM lacking the serum and other additions.
   b. Gelatin: 0.1% gelatin; composition per 500 mL, 0.5 g gelatin (Sigma, cat. no. G1890) in 500 mL picopure water. Gelatin will go into solution upon autoclaving and is stable for several months at room temperature. For the preparation of gelatin-coated 10-cm tissue-culture plates, 10 mL of 0.1% gelatin is incubated on the plate for 15–30 min at room temperature and then aspirated just prior to plating cells.
   c. PEG Solution: 50% polyethylene glycol 1500 (Boehringer Mannheim, Indianapolis, IN, cat. no. 783-641), 10 m*M* $CaCl_2$; composition per 5 mL, one 5-mL PEG 1500 bottle, 50 μL of 1 *M* $CaCl_2$. Warm solution to 37°C prior to use.
   d. HAT-containing ES media: 100 μ*M* hypoxanthine, 400 n*M* aminopterin, 16 μ*M* thymidine; add 10 mL 50 X HAT reconstituted supplement (Sigma, cat. no. H0262) to 500 mL ES media. HAT selection will only work on HPRT-deficient ES cell lines such as E14TG2a or E14.TG3B1 ***(24)*** when the *HPRT* gene is present on the YAC ***(13,17,19)***.
   e. G418-containing ES media: 225 μg/mL G418 (a neomycin analog; *see* **Note 2**); add 0.56 mL 200 mg/mL G418 (Geneticin; Gibco-BRL, cat. no. 11811-031; made by dissolving to 200 mg/mL active concentration G418 in 20 m*M* HEPES, pH 7.4 followed by filter sterilization, aliquoting and storage at –20°C) to 500 mL ES media. G418 selection will only work when the neo^r gene is present on the YAC.

2.3. Methods

1. Growth of yeast strains containing YACs: Inoculate a fresh colony of a YAC-containing yeast strain from selective plates into 10 mL selective media. Grow O/N at 30°C at 250 rpm on an orbital shaker (*see* **Note 3**). Transfer the growing culture to 500 ml of selective media and grow for an additional 24 h (*see* **Note 4**). Count cells with a hemocytometer to determine density (there should be approx 10^7 cells/mL or an OD_{600} of 1; *see* **Note 5**).
2. Preparation of yeast spheroplasts: Pour desired amount of culture (2.5×10^8 yeast spheroplasts are needed for each fusion) into 50-mL sterile conical falcon tubes and pellet cells in a Sorvall table-top centrifuge at 3000 rpm for 5 min. Wash cells once with 20 mL sterile water and resuspend pellet by vortexing or by triturating with a pipet. Pellet cells as described earlier and resuspend in 20 mL of 1 *M* sorbitol. Pellet cells once again and resuspend yeast cells in SPEM at 5×10^8 cells/mL. Take 10 mL of the sample into 90 mL

of 5% SDS and another 10 mL of the sample into 1 *M* sorbitol. Count cells from each sample—they should be the same density (*see* **Note 6**). Warm yeast suspension at 30°C. Add 1.5 μL zymolyase 20T/mL of yeast cells and gently mix (*see* **Note 7**). Incubate cells at 30°C taking 10 μL samples into 90 μL 5% sodium dodecyl sulfate (SDS) every 5 min as described previously. Determine the drop in cell number and continue until approx 90% of the cells are lysed in 5% SDS. Spheroplasts can take 5–20 min for completion (*see* **Note 7**).

Immediately pellet the spheroplasted cells at 1000 rpm for 5 min. Pour off supernatant carefully—the pellet should be very loose and some loss of cells will occur. Resuspend pellet by gentle trituration with a pipet (do not vortex). Wash spheroplasts twice with 20 mL STC, gently resuspending the pellet each time. Finally, resuspend the spheroplasts in STC at 2.5×10^8 cells/mL. Keep spheroplasts at room temperature until ready to fuse to ES cells.

3. Preparation of ES cells and fusion with yeast spheroplasts: A detailed description of the steps involved in ES cell manipulations and generation of transgenic mice utilizing ES cells are beyond the scope of this chapter (for a detailed description, *see* **refs. *25–27***). Plate ES cells at 6×10^6 cells/10-cm plate, coated with a mouse primary fibroblast feeder layer. After approx 48 h, trypsinize the cultures and plate the cells onto gelatin-coated plates at 10^7 cells/10-cm plate for 16–24 h. Feed ES cultures 4 h before fusion. Trypsinize ES cells, wash three times with SFM (pellet at 1500 rpm for 5 min in tissue culture centrifuge) and bring to a final concentration of 5×10^6 cells/ml in SFM.

   Pellet 1 mL (2.5×10^8) of yeast spheroplasts in a 15-mL conical tube at 1500 rpm for 5 min. Aspirate all STC buffer and apply gently (horizontal as possible) 1 mL of ES cells (5×10^6 cells), without disturbing the yeast pellet (*see* **Note 8**). Pellet ES cells at 1200 rpm for 3 min and aspirate all media. The copellet can be loosened by gently tapping on tube. Add 0.5 mL of PEG solution drop-wise to the copellet while gently mixing with the 1-mL pipet tip and then pipet once gently. Incubate at room temperature for 90 s and then slowly add 5 mL SFM from the bottom of the tube and incubate for 30 min at room temperature. Spin cells at 1200 rpm for 3 min and resuspend in 10 mL complete ES media and plate 5×10^6 cells (one tube) onto one 10-cm feeder-coated plate (*neo*^{r} feeders, if G418 selection is planned). Change to fresh media the following morning and start drug (HAT, G418) selection 48 h postfusion. Change media every 2–3 d and pick ES colonies at 10–15 d postfusion onto feeder-coated wells for expansion and for DNA (*see* **Note 9**) and/or gene expression analysis. ES cells containing integrated, intact YACs are utilized to produce chimeric mice by standard protocols ***(26)***, and transmission of the YAC through the germline results in the production of YAC transgenic mice.

3. Lipofection

3.1. Introduction

Lipofection requires the isolation of YACs from yeast cells, their purification away from the other yeast chromosomes by preparative pulsed-field gel electrophoresis (PFGE), the complexing of the YAC DNA with various lipid reagents and transfection into ES cells (*see* **Fig. 1B**) ***(23,28–33)***. The selection for the presence of the YACs in ES cells can be accomplished through the previous introduction of selectable cassettes (*HPRT*, *neo*^{r}) into the YAC vector arm via homologous recombination in yeast (protocols detailing homologous recombination in yeast as well as general yeast methods are described elsewhere; *see* **refs. *12,22,23***). Lipofection with YACs containing a selectable cassette results in approx 10–30% of the drug resistant ES colonies retaining an integrated, intact, and unrearranged copy (or multiple copies) of the YAC. In addition, selected ES clones frequently contain fragmented YACs. The overall efficiency of lipofection varies considerably depending on the YACs, the DNA preparation and the ES cell lines utilized, but generally 1–20 drug-resistant ES colonies are obtained per microgram of input YAC DNA [*(29)*; K. A. Bardel, L. A. Shaprio, and B. T. Lamb, unpublished observations]. YACs up to 1000 kb in length have been introduced intact into transgenic mice via lipofection (71; K. A. Bardel, L. A. Shapiro, and B. T. Lamb, unpublished observations).

The lipofection of purified YACs into ES cells has its advantages; for example, no (or little) yeast DNA is transferred into ES cells, the integrity and expression of genes on the YAC can be determined in ES cells prior to introduction into mice, and large YACs (>1 Mb) have been introduced intact. The disadvantages include the often tedious isolation of YAC DNA by PFGE, the fragmentation of YAC DNA during isolation and transfection and the relatively long time for transmission of modified ES cells through the mouse germline.

3.2. Materials

1. Materials for yeast growth and phenotype testing are described elsewhere ***(22)***.
   a. YPD media: 1% yeast extract, 2% peptone, 2% dextrose; composition per liter: 10 g yeast extract (Difco), 20 g peptone (Difco), 20 g dextrose. Dissolve components in 800 mL distilled water, bring volume up to 1 L, and autoclave 20–40 min, 121°C, 15 psi. For solid media, add 20 g of bactoagar (Difco) prior to autoclaving.
2. Materials for agarose plug preparation:
   a. Solution A: 50 m*M* EDTA; composition per liter: 100 mL of 0.5 *M* EDTA, pH 8.0 (disodium salt), 900 mL distilled water.
   b. Solution B: 1 *M* Sorbitol, 20 m*M* EDTA, pH 8.0, 10 m*M* Tris-HCl, pH 7.5, 14 m*M* BME, 1 mg/mL zymolyase 20T: composition per 50 mL, 25 mL of 2 *M* sorbitol, 2 mL of 0.5 *M* EDTA, pH 8.0, 0.5 mL of 1 *M* Tris-HCl, pH 7.5, 50 µL of 14 *M* BME (Sigma, cat. no. M7522), 50 mg zymolyase 20T (ICN, cat. no. 320921). Bring up to final volume with sterile water. This solution should be made fresh just prior to use.
   c. Solution C: 1 *M* sorbitol, 20 m*M* EDTA, pH 8.0, 10 m*M* Tris-HCl, pH 7.5, 14 m*M* BME, 1% low melting point (LMP) agarose; composition per 50 mL: 25 mL of 2 *M* sorbitol, 2 mL of 0.5 *M* EDTA, pH 8.0, 0.5 mL of 1 *M* Tris-HCl, pH 7.5, 50 µL of 14 *M* BME, 0.5 g SeaPlaque GTG LMP agarose (FMC, Rockland, ME, cat. no. 50110). The sorbitol, EDTA, Tris-HCl, and LMP agarose are brought up to volume with sterile water and heated to a boil in a microwave until the agarose is completely dissolved. The BME is then added and the solution is equilibrated at 37°C.
   d. Solution D: 1% lithium dodecyl sulfate (LDS), 100 m*M* EDTA, pH 8.0, 10 m*M* Tris-HCl, pH 8.0; composition per liter: 10 g dodecyl lithium sulfate (Sigma, cat. no. L4632), 200 mL of 0.5 *M* EDTA, pH 8.0, 10 mL of 1 *M* Tris-HCl, pH 8.0. Bring up to volume with sterile water. This solution can be stored at room temperature for several months.
   e. Solution E: 50 m*M* EDTA, pH 8.0, 10 m*M* Tris-HCl, pH 8.0; composition per liter: 100 mL of 0.5 *M* EDTA, pH 8.0, 10 mL of 1 *M* Tris-HCl, pH 8.0. Bring up to volume with distilled water. This solution can be stored at room temperature for several months.
3. Materials for preparative PFGE and isolation of YAC DNA:
   a. 5X TBE: 45 m*M* Tris, 45 m*M* boric acid, 1 m*M* EDTA, pH 8.0; composition per liter: 54 g Tris base, 27.5 g boric acid, 20 mL of 0.5 *M* EDTA, pH 8.0. Bring up to volume with distilled water.
   b. Agarose: SeaPlaque GTG and Nusieve GTG agarose (FMC, cat. no. 50080).
   c. YAC stabilization buffer: 20 m*M* Tris-HCl, pH 7.5, 1 m*M* EDTA, pH 8.0, 0.5 m*M* spermine; composition per 100 mL: 2 mL of 1 *M* Tris-HCl, pH 7.5, 200 µL of 0.5 *M*

EDTA, pH 8.0, 50 μL of 1 *M* spermine tetrahydrochloride (Sigma, cat. no. S2876; dissolved in water and filter sterilized). Bring up to volume with sterile water.

   d. β-agarase I (1 U/μL; New England Biolabs, Beverly, MA, cat. no. 392L).

4. Materials for ES tissue culture:
   a. ES media: DMEM, 100 U/mL penicillin, 100 μg/mL streptomycin, 0.1 m*M* non-essential amino acid solution, 1 m*M* sodium pyruvate, 15% heat inactivated ES-qualified fetal calf serum, 100 μM BME, 10^3 U/mL murine LIF; composition per 500 mL, 405 mL DMEM, high glucose with L-glutamine (Gibco-BRL, cat. no. 11965-092), 5 mL of 100X penicillin-streptomycin solution (Gibco-BRL, cat. no. 15140-122), 5 mL of 10 m*M* MEM nonessential amino acids solution (Gibco-BRL, cat. no. 11140-076), 5 mL of 100 m*M* MEM sodium pyruvate solution (Gibco-BRL, cat. no. 11360-070), 75 ml of ES-qualified fetal calf serum (Gibco-BRL, cat. no. 16141-079, heat inactivated at 56°C for 30 min; *see* **Note 1**), 5 mL of 0.1 m*M* BME (7 μL of 14 *M* BME stock mixed in 10 mL of picopure water, filter sterilized), 50 μL of 10^7 U/mL LIF (ESGRO™; Gibco-BRL, cat. no. 13275-029). ES media should be stored at 4°C and can be kept for several weeks.
   b. Reduced serum media: OptiMEM-I media (Gibco-BRL, cat. no. 31985-070).
   c. Lipid transfection reagent: Lipofectin (Gibco-BRL, cat. no. 18292-037; 1 μg/μL).
   d. G418-containing ES media; *see* **Subheading 2.2., item 3**.

3.3. Methods

1. Growth of yeast strains containing YACs: Inoculate a fresh colony of a YAC-containing yeast strain from a selective plate into 8 mL of selective media. Grow O/N at 30°C in a roller-drum or on an orbital shaker (225 rpm). Transfer the growing culture (the OD_{600} should be approx 1; *see* **Note 5**) to 200 mL of culture of YPD and grow for an additional 24 h. Count cells with a hemocytometer to determine density (there should be approx 10^8 cells/mL or an OD_{600} of 10; *see* **Note 5**).
2. Preparation of high-molecular-weight DNA plugs for preparative PFGE: Pellet cells in 250-mL bottles in a Sorvall GSA rotor at 2500 rpm for 15 min. Resuspend cells in 100 mL of solution A by pipeting and/or vortexing and pellet again. Resuspend cells in 3–4 mL of solution B warmed to 37°C (*see* **Note 10**). Add an equal volume of solution C cooled to 37°C and mix by gentle trituration with a 10-mL pipet and immediately aliquot onto strips of bent parafilm (V-shaped) sitting on ice (*see* **Note 11**).

   Let plugs solidify for 5–10 min and cut into 3–5-mm slices with a clean razor. Slide plugs into a 50-mL conical tube, add sufficient solution B to cover the plugs (6–8 mL), and incubate at 37°C for 2 h (*see* **Note 12**). Carefully remove buffer from plugs, add 25 mL buffer D and incubate at 37°C for 1 h. Replace with 25 mL of fresh buffer D and incubate at 37°C O/N. Wash four times with 25 mL of solution E at room temperature for 1 h on a slow-speed orbital shaker (*see* **Note 13**). Plugs may be stored at 4°C for up to 1 yr. To determine the quality of the plug DNA for YAC isolation, a standard PFGE is run (CHEF-DR III [Bio-Rad, Hercules, CA, cat. no. 170-3695], 1% Seakem LE agarose [FMC], 0.5X TBE, 12°C, 6 V/cm, 120° angle, 60-s switch for 14 h, 90-s switch for 10 h).
3. Preparative PFGE and YAC DNA isolation: Assemble preparative gel by placing a 30-well comb (Bio-Rad, cat. no. 170-3628) with all of the teeth taped together in a large casting stand (21 × 14 cm; Bio-Rad, cat. no. 170-3704). Position comb so that it is in complete contact with the bottom of the casting stand. It is imperative that the casting stand is clean and level. Prepare 225 mL of 1% SeaPlaque GTG agarose in 0.5X TBE by completely melting in microwave oven and pour into casting stand after cooling slightly. After the gel has solidified at room temperature, place at 4°C for 15–20 min to completely harden. Place gel in clean, dry, level chamber, carefully remove the comb, and gently pull apart the top portion of the gel so that the well is much wider in the middle and yet is still

connected at both ends of the gel. Remove any remaining buffer in the well by gently blotting with paper towels.

Load preparative gel by completely melting 3–3.5 mL of plugs at 68–70°C for 15–20 min in a 17 × 100-mm polypropylene round-bottom tube (Falcon, Lincoln Park, NJ, cat. no. 2059) and slowly add to the center of the well using a 5-mL serological pipet with a few millimeter of the tip removed with a hot razor (*see* **Note 14**). Using a wall of the casting stand as a guide, carefully lower the top edge of the well so that it is even with the sides of the gel. Let the plugs solidify in the well for 10–15 min, and then add 2 L of chilled 0.5X TBE (*see* **Note 15**). Run preparative PFGE (CHEF-DR III, 1% Seaplaque GTG agarose, 0.5X TBE, 12°C, 6 V/cm, 120° angle; conditions will vary dependent on the size of the YAC; *see* **Note 16**). Remove gel from apparatus, cut 15-mm strips from the both sides of the gel and store remainder covered with saran wrap at 4°C. Stain the gel strips in 0.5X TBE with 1 μg/mL ethidium bromide for 30 min on a rotating platform. Destain in water for 30 min at room temperature (*see* **Note 17**). Locate the YAC band under ultraviolet (UV) illumination and mark the position by slightly cutting the gel on either side of the band with a razor. Reassemble the gel and carefully cut out the region of the unstained preparative gel corresponding to the YAC band. In addition, cut out a band corresponding to a neighboring yeast chromosome to serve as a marker lane for the second-dimension gel electrophoresis. Cut excised bands into four equal-sized pieces (approx 4.5 cm) and store in solution E (*see* **Subheading 3.2., item 2**). Excised bands can be stored for several weeks at 4°C. Stain and destain the remaining gel to ensure that the YAC band was properly excised.

To concentrate the YAC DNA *(34)* (*see* **Note 18**), place the four gel slices together in an 12 × 14 cm (Owl, Woburn, MA, model B2) gel casting stand at a 90° angle to the PFGE, along with a single gel slice of the marker yeast chromosome on either side of the YAC-containing gel slices. Prepare 150 mL of 4% Nusieve GTG agarose in 0.5X TBE, melt in the microwave oven, and cool slightly. Pour into casting stand until the molten gel is level with the top of the gel slices and let solidify. Electrophorese in 800 mL of 0.5X TBE at 6 V/cm (84 V for a 14 cm gel) for 12–14 h at 4°C. Excise the marker lanes and stain *(see above)* to localize and mark the position of the DNA within the Nusieve gel. Remove the corresponding location in the YAC-containing samples (0.5–1 cm) and place in solution E (*see* **Subheading 3.2., item 2**). The excised, concentrated YAC band can be stored for several weeks at 4°C.

To release the YAC DNA from the agarose, first determine the weight of the agarose plug. Place the plug in a 17 × 100 mm polypropylene round-bottom tube (Falcon) and wash twice for 30 min at 4°C in 5 mL of YAC stabilization buffer (*see* **Note 19**), followed by an O/N wash at 4°C. Transfer the agarose plug to a new 15-mL tube, being careful to remove all of the buffer and melt at 68°C for 15–20 min (*see* **Note 20**). Equilibrate sample at 40°C for 15 min and add 3 U β-agarase I/100 mg of gel by gently swirling in with pipet. Prewarm the appropriate volume of β-agarase I at 40°C prior to addition. Incubate for 1 h at 40°C, gently swirl the mixture with a pipet, and continue incubation for another 1–2 h. To determine whether the agarose is completely digested, place the tube on ice for 10 min (*see* **Note 21**). To determine the concentration and integrity of the DNA, run 5–10 μL on a 0.8% standard agarose gel using λ/*Hin*dIII makers as a known concentration standard, and 30–40 μL analyzed by PFGE, respectively. YAC DNA is agarased just prior to transfection and is stored at 4°C for no more than a few days.

4. Lipid-mediated transfection of ES cells: A detailed description of the steps involved in ES cell manipulations and generation of transgenic mice utilizing ES cells are beyond the scope of this chapter (for a detailed description, *see* **refs. *25–27***). Plate ES cells at 3×10^6 cells/10-cm plate, coated with a mouse primary fibroblast feeder layer. After 48 h,

trypsinize ES cells, resuspend in complete ES media (without penicillin-streptomycin; *see* **Note 22**) and gently triturate to form a single-cell suspension. Count cells with a hemocytometer. Spin down ES cells at 1500 rpm for 5 min in a tissue-culture centrifuge and resuspend at 2.5 × 10^6 cells/mL in Optimem I media. Add 1 mL of purified YAC DNA (approx 1–2 μg) to a 17 × 100-mm polystyrene round-bottomed tube (Falcon, cat. no. 2057) using a wide-bore pipet. Mix in 50 μL of lipofectin by gently swirling the YAC DNA solution with the pipet (*see* **Note 23**). Incubate mixture at room temperature for 45 min to form lipid-DNA complexes. Aspirate media from a 10-cm tissue-culture plate coated with a primary fibroblast feeder layer and add 8 mL Optimem I media. Add 1 mL ES cell suspension (2.5 × 10^6 cells) to the plate and then slowly add the lipid-DNA complexes while gently swirling the plate. Place plates in 37°C incubator for 4 h. Add 10 mL complete ES media (without penicillin-streptomycin), bringing the final volume up to 20 mL. Place plates back into incubator for 14–16 h and then replace the media and lipid-DNA complexes with 10 mL of complete ES media (with penicillin-streptomycin). At 24–36 h posttransfection trypsinize thecells and place onto new neo^r feeder-coated 10-cm plates and add complete ES media (with penicillin streptomycin) with 225 μg/mL G418. Feed plates every 2–3 d with G418-containing ES media and pick ES colonies at 10–15 d posttransfection onto feeder-coated wells for expansion and for DNA (*see* **Note 9**) and/or gene expression analysis. ES cells containing integrated, intact YACs are utilized to produce chimeric mice by standard protocols ***(26)***, and transmission of the YAC through the germline results in the production of YAC transgenic mice.

4. Microinjection

4.1. Introduction

YACs (as well as P1s and BACs) can also be introduced into mice by direct microinjection into single-cell mouse embryos (*see* **Fig. 1C**) ***(34–61)***. The preparation of YAC DNA suitable for microinjection has seen several major advances: purification by preparative PFGE, enzymatic digestion of agarose in the presence of high salt and/or polyamines to protect against the shearing of high molecular-weight DNA in solution and concentration (up to 1–5 ng/mL) by low-speed ultrafiltration ***(23,35,62)***, dialysis with sucrose ***(36)***, or a second-dimension electrophoresis ***(34,63)***. The concentrated YAC DNA is subsequently injected into the male pronucleus and offspring are scored for the presence of various portions of the YAC. The efficiency of introducing YACs into mice via microinjection has varied considerably between studies, but the most complete published data ***(34,61)*** suggests that although the number of transgenics per liveborn pups is similar to what is observed for standard transgenesis (approx 10–25%), a much larger number of injected oocytes (five- to 10-fold) are required to generate the same number of liveborn pups. These observations suggest that there is a certain toxicity/lethality associated with the microinjection of purified YAC DNA into mouse embryos. YAC transgenic mice produced by microinjection frequently contain fragmented YACs with single or multiple copies of DNAs integrated at either one or multiple sites in the genome. YACs up to 670 kb in length have been introduced intact into transgenic mice via microinjection ***(56)***.

The microinjection of YACs into single-cell mouse embryos has a number of advantages: the production of transgenic mice via microinjection is relatively rapid, no (or little) yeast DNA is transferred into mice, and any resulting mice can be tested immediately for the complementation of existing mouse mutants ***(37,41,44,47)*** or other phenotypic effects. The disadvantages include the tedious and technically difficult isolation and concentration of intact YAC DNA, the shearing and fragmentation of YAC DNA during microinjection, and the possibility that larger YACs (>1 Mb) may be extremely difficult to microinject because of shearing and the physical constraints of molarity of extremely large DNA molecules.

4.2. Materials (see *Note 24*)

1. Materials for yeast growth:
   a. YPD media: *see* **Subheading 3.2., item 1**.
2. Reagents for agarose plug preparation:
   a. Plug molds (Bio-Rad).
   b. 50 m*M* EDTA, pH 8.0: Dilute 0.5 *M* EDTA, pH 8.0 stock 1:10.
   c. Solution 1: 1 *M* sorbitol, 20 m*M* EDTA, pH 8.0, 14 m*M* β-mercaptoethanol (BME), 2 mg/mL zymolyase-20T; composition per 10 mL: 10 mL of 1 *M* sorbitol, 400 μL of 0.5 *M* EDTA, pH 8.0, 10 μL of 14 *M* BME (Sigma, cat. no. M7522), 20 mg zymolyase-20T (ICN, cat. no. 320921). This solution should be made fresh just prior to use.
   d. Solution 2: 1 *M* sorbitol, 20 m*M* EDTA, pH 8.0, 2% SeaPlaque GTG low-melting point agarose (LMPA; FMC, cat. no. 50110), 14 m*M* BME; composition per 10 mL: 10 mL of 1 *M* sorbitol, 400 μL of 0.5 *M* EDTA, pH 8.0, 0.2 g agarose, 10 μL of 14 *M* BME. The sorbitol, EDTA, and LMPA are heated to a boil and equilibrated to 50°C, the BME is added and solution is held at 50°C until needed.
   e. Solution 3: 1 *M* sorbitol, 20 m*M* EDTA, pH 8.0, 10 m*M* Tris-HCl, pH 7.5, 14 m*M* BME, 2 mg/mL zymolyase-20T; composition per 100 mL: 100 mL of 1 *M* sorbitol, 4 mL of 0.5 *M* EDTA, 1 mL of 1 *M* Tris-HCl pH 7.5, 100 μL of 14 *M* BME, 200 mg zymolyase-20T. Make fresh day of use.
   f. LDS: 1% Dodecyl lithium sulfate, 100 m*M* EDTA, pH 8.0, 10 m*M* Tris-HCl, pH 8.0; composition per 500 mL: 5 g dodecyl lithium sulfate (Sigma, cat. no. L4632), 100 mL of 0.5 *M* EDTA, 5 mL of 1 *M* Tris-HCl, pH 8.0, and water to 500 mL. Filter-sterilize; can store at room temperature for several months.
   g. 1X NDS: 0.5 *M* EDTA, 10 m*M* Tris, 1% *N*-laurylsarcosine, pH 9.0; composition per 500 mL: 93 g EDTA (di-sodium salt), 0.6 g Tris base, 5 g *N*-laurylsarcosine. Mix EDTA and Tris base in 350 mL of water, adjust pH to greater than 8.0 with 100–200 NaOH pellets. Add N-laurylsarcosine (predissolved in 50 mL water) and pH to 9.0 with 5 *M* NaOH solution. Bring volume up to 500 ml with water and filter-sterilize. Store at 4°C for up to 1 yr. Working stock is 0.2X NDS; dilute to this concentration using sterile water. This solution may be stored as for 1X NDS.
   h. TE, pH 8.0: 10 m*M* Tris-HCl, pH 8.0, 1 m*M* EDTA, pH 8.0; composition per liter: 10 mL of 1 *M* Tris-HCl, pH 8.0, 2 mL of 0.5 *M* EDTA, pH 8.0.
3. Materials for preparative PFGE and isolation of YAC DNA:
   a. 0.5X TBE: *See* **Subheading 3.2., item 3**, for recipe for 5X TBE and dilute accordingly. Autoclave before use.
   b. 2% Absolve solution (NEN Dupont, Boston, MA, cat. no. NEF971). Prepare according to manufacturer's instructions.
   c. Agarose MP (Boehringer-Mannheim, Indianapolis, IN, cat. no. 1-388-983).
   d. Seaplaque GTG, Nusieve GTG, and Seakem GTG agarose (FMC, cat. no. 50110, 50080, and 50070, respectively).
   e. Injection buffer: 10 m*M* Tris-HCl, pH 7.5, 250 μ*M* EDTA, pH 8.0, 100 m*M* NaCl; composition per liter: 10 mL of 1 *M* Tris-HCl, pH 7.5, 0.5 mL of 0.5 *M* EDTA, pH 8.0, 20 mL of 5 *M* NaCl. Autoclave solution before use.

f. β-agarase I (1 U/μL; New England Biolabs, cat. no. 392L).
g. 0.2 μ*M* Acrodisk syringe filters (Gelman, Ann Arbor, MI, cat. no. 4192).

4.3. Methods

1. Growth of yeast strains containing YACs: Inoculate 200 mL YPD with 1 mL of an O/N culture of the yeast strain containing the YAC. Incubate O/N at 30°C with shaking (225–250 rpm). Chill appropriate number of plug molds on ice (*see* **Note 25**). Count cells with a hemocytometer. The culture should be saturated at approx 1×10^8 cells/mL.
2. Preparation of high-molecular-weight DNA plugs for preparative PFGE: Spin down cells at 600*g* (*see* **Note 26**) for 5 min in a 250-mL polypropylene centrifuge bottle at room temperature. Pour off media and resuspend pellet in 80 mL of 50 m*M* EDTA, pH 8.0. Spin-down cells again, pour off supernatant, and resuspend pellet in 20 mL of 50 m*M* EDTA, pH 8.0. Transfer suspension into a preweighed 50-mL Falcon tube. Spin-down cells again (*see* **Note 27**) and decant off supernatant. Weigh the yeast pellet and tube, subtract the weight of the tube, and assume a density of 1 (weight = volume) for the yeast pellet. Add solution 1 to give 8×10^9 cells /mL. The yeast will comprise 75–90% of the volume. Add an equal volume of solution 2. The yeast concentration in the plug will be 4×10^9 cells/mL. Mix rapidly, without introducing bubbles and pipet aliquots into chilled plug molds with a blue pipet tip. Leave on ice for 10 min to allow agarose to set.

   Transfer plugs to 50 mL Falcon tube containing 40 mL of solution 3 (*see* **Note 28**). Incubate at 37°C for 2 h with occasional gentle rocking. Remove supernatant with 25-mL pipet, add 40 mL of LDS (*see* **Note 28**) and incubate at 37°C for 1 h with occasional gentle rocking. Replace LDS with fresh solution and incubate at 37°C O/N without agitation. Remove LDS and add 40 mL of 0.2X NDS (*see* **Note 28**). Incubate at room temperature for 2 h with agitation. Replace 0.2X NDS with fresh solution and incubate as just described. Wash plugs twice with 40 mL of TE, pH 8.0 (*see* **Note 28**) for 30 min at room temperature with agitation. Remove second TE wash and add 20 mL of fresh TE, pH 8.0. Plugs may be stored at 4°C for at least 1 yr.
3. Preparative PFGE and YAC isolation:

   Day 1: Prepare CHEF Gel apparatus for PFGE. Autoclave 4 L of 0.5X TBE the day before the PFGE run. Cool to room temperature or 4°C, because the temperature will be maintained at 12°C during the electrophoresis run. Set up gel system (CHEF-DR II; Bio-Rad cat. no. 170-3612). Make sure that the electrophoresis chamber is level using a bubble level. Circulate 1 L of sterile water through the electrophoresis chamber and refrigeration unit for 30 min to 1 h to clean the unit prior to adding sterile 0.5X TBE. Drain water from PFGE apparatus and fill with 2 L sterile 0.5X TBE. Equilibrate to 12°C.

   Cast the gel for PFGE. Soak gel casting stand parts and comb in 2% Absolve for 1 h. Rinse well with sterile water. Assemble gel casting stand wearing gloves. Tape enough teeth of the comb to obtain a preparative lane of approx 7 cm (five to six teeth of Bio-Rad 15-tooth comb; *see* **Note 29**) or use a preparative comb (Bio-Rad, cat. no. 170-3623). Casting stand should be level. Prepare 200 mL of 0.5% agarose MP solution: 1 g agarose in 200 mL of sterile 0.5X TBE. Melt by boiling on hot plate or in microwave oven. Cool to 58°C and pour. After the gel has solidified and reached room temperature, refrigerate for 15–30 min. This will prevent the wells from collapsing in this low percentage gel when the comb is removed.

   Load preparative yeast plugs cut to 7-mm height next to one another in the preparative well. In the standard size wells flanking the preparative well load a small amount of the preparative plug, yeast chromosome markers (New England Biolabs or Bio-Rad), and λ midrange marker I (New England Biolabs). Seal all wells with sterile 0.8% Seaplaque LMPA in 0.5X TBE. Run the PFGE using the following conditions:

CHEF-DR II (Bio-Rad), 12°C, 200 V, 60-s switch for 20 h (*see* **Note 30**). We have used these conditions to successfully isolate 155-kb, 248-kb, 450-kb, and 650-kb YACs. *Day 2:* Locate the YAC band. When the gel run is complete, cut off the marker lanes flanking the preparative lane and stain them on a rotating platform in 0.5X TBE containing 0.5 µg/mL ethidium bromide for 30 min. Destain in 0.5X TBE for 30 min. Locate the YAC band under UV illumination (long or short wave) and mark the position by cutting out a notch encompassing the YAC band with a sterile scalpel, razor blade, or glass cover slip. Reassemble the gel and cut out the unstained region of the preparative lane containing the YAC DNA. Cut out a second region of the preparative lane encompassing a yeast chromosome of approximately the same size as the YAC to serve as a marker lane for the second gel run. Stain and destain the remaining gel as described above to ensure that the YAC band was properly excised.

Concentration of YAC DNA by second-dimension agarose gel electrophoresis (*see* **Note 31**). Presoak a mini-gel apparatus and gel casting stand in 2% Absolve for at least 1 h or O/N on d 1. Prepare 50 mL of 4% Nusieve GTG agarose solution: 2 g agarose in 50 mL of sterile 0.5X TBE. Melt by boiling on hot plate or in microwave oven. Cool to 68–72°C before pouring. Place YAC slice and marker slice in the mini-gel casting stand at a 90° angle to the PFGE run so that the DNA will migrate the length of the PFGE gel slice into the Nusieve gel. Pour the gel around them. Electrophorese in sterile 0.5X TBE at 2.4 V/cm (48 V for 20 cm mini-gel apparatus) for 16 h (*see* **Note 32**). *Day 3:* Excise YAC plug. Excise the marker lane and stain as described above to localize the DNA within the Nusieve gel. Locate the corresponding position in the YAC-containing lane and cut out a Nusieve gel plug containing the YAC (usually 0.5 cm; *see* **Note 33**). Stain the remaining YAC lane to demonstrate excision of the YAC plug. Place YAC slice in a preweighed 50-mL Falcon tube and weigh to determine the weight of the plug. Add 40–50 mL injection buffer and equilibrate for 1–2 h at room temperature with occasional rocking (*see* **Note 34**).

Agarase treatment: Transfer the gel slice to a sterile, DNase-free 1.5-mL microfuge tube with sterile forceps. Melt the agarose at 68°C for 10 min, rapidly transfer to 42.5°C and equilibrate for 5 min at this temperature. Add 2 U of β-agarase I/100-mg gel slice using a wide-bore yellow pipet tip. Prewarm the appropriate volume of agarase to room temperature in the yellow tip prior to addition. Mix by gently releasing air bubbles into the solution three times from a wide-bore yellow pipet tip. Incubate for 3 h (or up to O/N) at 42.5°C. Place on ice for 10 min to check for undigested agarose. Only if many lumps are present should the agarase treatment be repeated. Small amounts of undigested agarose are removed by filtration just prior to microinjection. To determine DNA concentration, check 5 µL on a 0.8% agarose gel using λ/*Hin*dIII markers of known concentration as a standard. Check the integrity of the YAC DNA by running 20–25 µL of purified DNA on a PFGE gel followed by Southern-blot analysis. The agarose and PFGE conditions should be chosen to allow optimum resolution of the YAC to be microinjected.

4. Microinjection into fertilized mouse oocytes: A detailed description of the steps involved in murine transgenesis are beyond the scope of this chapter (for a detailed description, *see* **ref.** *7*), but we will highlight treatment of the YAC DNA leading up to microinjection. The purified DNAs should be used as soon as possible, but can be stored for approx 2–4 wk at 4°C without a detectable increase in degradation. Our YAC DNA preparations generally have concentrations of 5–10 ng/µL. Injections at this concentration reduce embryonic survival and produce a greater frequency of transgenics with deleted YAC copies ***(64)*** (K. R. Peterson, unpublished observations). The latter problem may be due to shearing of the YAC DNA as this higher-viscosity DNA solution passes through the needle. We dilute our

DNAs to 1–2 ng/µL in injection buffer and filter the solution through a 0.2 µ*M* Acrodisk. The viscosity of the DNA solution is reduced at this concentration, thus facilitating flow through the injection needle. Filtration removes any undigested agarose, preventing blockage in the needle and does not seem to have a detrimental effect on YAC integrity. Transgenic animals are identified by hybridization of slot blots of tail-derived DNA. Our efficiency of transgenesis is 10–14% for 155- and 248-kb YACs. Up to 25% of these contain only intact YAC copies and up to 60% contain both intact and deleted copies ***(65)*** (K. R. Peterson, unpublished data; *see* **Note 9**).

5. Notes

1. Batches of ES-qualified serum are generally tested by individual laboratories for the ability to promote high plating efficiency and undifferentiated growth of ES cell lines ***(27)***.
2. The optimal concentration of G418 should be determined for each particular ES cell line.
3. Yeast should be grown using a media-to-flask ratio of 1:5 to 1:10 to ensure proper aeration and growth.
4. Inoculating fresh yeast cultures allows for most optimal growth of the yeast and preparation of spheroplasts. Alternatively, yeast cells frozen at 10^8 cells/mL can be inoculated into selective media and grown for 14 h at 30°C.
5. The exact correlation between cell number and OD_{600} varies slightly between yeast strains.
6. As yeast spheroplasts lyse in 5% SDS, this method will determine what percentage of cells are spheroplasted. Alternatively, the extent of spheroplasting can be determined by measuring the OD_{600} of a 1/100 dilution in distilled water. When the OD_{600} is 1/10 of the starting value, spheroplasting is approx 90% complete.
7. The exact amount of zymolyase and optimal incubation times for the preparation of spheroplasts needs to be determined for each particular yeast strain, cell density, cell volume and enzyme batch. Aliquots of a particular tested batch of zymolyase can be stored at –20°C.
8. The yeast:ES cell ratio can vary from 50:1 to 25:1.
9. A major problem of YAC transgenics is the determination of structural integrity of the YAC transgene copies in selected ES cells and in YAC-containing mice ***(65)***. In order to unambiguously demonstrate YAC integrity, a variety of experimental approaches should be undertaken. A combination of PCR, PFGE and conventional Southern-blot hybridization using probes (including species-specific repetitive elements) spanning the YAC is informative, but does not definitively prove the continuity of individual YAC copies, unless restriction sites are chosen that flank the transgene or insert DNA cloned in the pYAC vector. Such sites may be utilized if a restriction map is available or, alternately, unique restriction sites may be introduced at the insert–vector junctions of the YAC prior its use in producing transgenics ***(66)***. RecA-assisted restriction endonuclease (RARE) analysis may also be used to visualize entire inserts, but this approach is technically demanding and expensive for routine analysis of large numbers of transgenic lines ***(67)***. Finally, modifications of fluorescence *in situ* hybridization (FISH) may be utilized to examine the orgnization of YAC transgene insertion sites ***(68)***, although this approach requires expertise with FISH, as well as multiple FISH probes spanning the YAC.
10. The exact amount of buffer B to be utilized should be empirically determined as the conditions vary considerably for different yeast strains and/or YACs. If the plugs are too concentrated, the YACs will not be separable by PFGE, whereas if the plugs are too dilute, the final purified YAC DNA preparation will not be concentrated enough for efficient lipofection.
11. Alternatively, high-molecular-weight DNA plugs can also be formed in plug molds (*see* **Subheading 4.2., item 2**).

12. Plugs can be left in solution B O/N, although we have generally obtained the best results when incubated for only 2 h.
13. High-molecular-weight DNA plugs look best by PFGE if allowed to remain in fresh solution E O/N at 4°C.
14. The exact amount of molten plugs to add to the well is determined empirically, but is generally within 3–3.5 mL. It is imperative that the plugs are completely melted before loading the preparative gel, as incompletely melted plugs will result in streaks and unevenness in the gel. The preparative PFGE gel can also be loaded by carefully pouring the molten plugs into the well. The sample will not leak from the large, exapanded well unless the gel is torn.
15. It is convenient to maintain a carboy of distilled water in a 4°C cold room to prepare the 0.5X TBE.
16. Although the exact parameters for YAC DNA isolation are determined empirically, approximate guidelines are as follows: for a 250-kb YAC, a 25-s switch for 29 h; for a 450-kb YAC, a 45-s switch for 37 h; for a 650-kb YAC, a 60 sec switch for 48 h; and for a 1000-kb YAC, a 85-s switch for 49 h.
17. Resolution of PFGE gels will improve with a destaining O/N at 4°C.
18. Alternate methods of YAC DNA concentration include low-speed ultrafiltration ***(23,35,62)*** and dialysis with sucrose ***(36)*** following agarase digestion of the gel slice.
19. Alternate buffers for maintaining the integrity of high-molecular-weight DNA contain other positively charged molecules, including high salt ***(62)***, poly-L-lysine ***(28–30)***, and hexamine cobalt chloride ***(69)***.
20. It is imperative to completely melt the agarose or the subsequent agarase and lipid-mediated transfection steps will not proceed efficiently.
21. If large amounts of agarose remain undigested at this step, the agarase treatment can be performed a second time.
22. Gibco-BRL recommends performing lipid-mediated transfections in the absence of antibiotics (penicillin–streptomycin). ES media containing penicillin-streptomycin is added the morning after the transfection to protect against possible contamination of the DNA with bacteria.
23. Numerous lipid-based transfection reagents are available on the market. Lipofectin (Gibco-BRL) has provided the most consistent results in delivering YAC DNA prepared as described above into ES cells (K. A. Bardel, L. A. Shaprio, and B. T. Lamb, unpublished observations), although not all of the commercially available lipids have been tested.
24. Kilobase- to megabase-size YAC DNA is susceptible to mechanical shear and enzymatic degradation. To ensure recovery of intact YAC DNA for microinjection, care must be taken to maintain sterility and minimize photodamage and shearing forces. High-quality water (Milli-Q-filtered, Millipore, Bedford, MA) should be used in the preparation of all solutions, buffers, and gels. When possible, solutions should be sterilized by autoclaving or filtration, or made from sterile stock solutions.
25. Bio-Rad plug molds shape 10 plugs of approx 0.25 mL each. Thus, each mold can accommodate 2.5 mL of a yeast cell-agarose suspension.
26. 1900 rpm in a Sorvall GSA rotor.
27. 1770 rpm in a Sorvall HS-4 rotor.
28. Adjust volume of solution, if necessary, so that a ratio of 8 mL solution/mL plug is maintained.
29. The preparative well should be placed as close to the center of the gel as possible to avoid anomalous migration. If DNA migration is uneven, incomplete YAC DNA excision will result, reducing DNA concentration for microinjection. Two YACs of the same or different size may be purified on one gel using separate preparative wells and the PFGE conditions described here.

30. The PFGE conditions given in this protocol are generally applicable for purification of most YACs. The 0.5% agarose MP will fractionate DNAs in the 150-kb to 1.6-Mb size range ***(63)***. Alternate agaroses, gel concentrations, and electrophoresis conditions may be utilized to optimize separation and resolution of YACs from yeast chromosomes. These conditions may be determined empirically or Bio-Rad technical services can suggest conditions tailored to maximally fractionate a given size range for isolation of a given size YAC. For example, we routinely use 1% Seakem GTG agarose, 0.5X TBE, 14°C, 200 V, 14-s switch for 24 h to purify 155- and 248-kb YACs.
31. Alternate methods of YAC DNA concentration include low-speed ultrafiltration and dialysis with sucrose following agarase digestion of the PFGE gel slice.
32. If other PFGE conditions are utilized, it may take longer for the YAC DNA to completely migrate from the PFGE gel slice into the Nusieve gel. Use of multiple marker lanes is recommended in this case to monitor the run and determine when it is complete.
33. Do not include any of the PFGE gel slice. Agarase will not digest standard (non-low-melting point) agarose.
34. It is imperative that high ionic strength be maintained to prevent breakage of YAC DNA at the high temperatures required for agarase treatment and during passage of the DNA through the microinjection needle ***(35,62)***. YAC DNA prepared in the presence of 100 m*M* NaCl is sufficiently resistant to shear ***(62)***, although polyamines, such as spermine and spermidine, often are included as well ***(54,70)***. Not enough data exist to suggest that the addition of polyamines affords any further degree of resistance to shearing than with high salt alone.

Acknowledgments

The authors wish to thank the following for their participation in the development of these protocols: S. Klapholz (A. J.); K. A. Bardel, L. M. Call, S. R. Chideya, and L. S. Kulamc (B. T. L.); and A. Gnirke, C. Huxley, C. H. Clegg, and H. S. Haugen (K. R. P.). The development of these protocols was supported in part by the grants from the National Institutes of Health, including AG08012 and AG14451 (B. T. L.), and DK45365 and HL53750 (K. R. P.).

References

1. Wagner, T. E., Hoppe, P. C., Jollick, J. D., Scholl, D. R., Hodinka, R. L., and Gault, J. B. (1981) Microinjection of a rabbit β–globin gene into zygotes and its subsequent expression in adult mice and their offspring. *Proc. Natl. Acad. Sci. USA* **78,** 6376–6380.
2. Wagner, E. F., Stewart, T. A., and Mintz, B. (1981) The human β–globin gene and a functional thymidine kinase in developing mice. *Proc. Natl. Acad. Sci. USA* **78,** 5016–5020.
3. Harbers, K., Jahner, D., and Jaenisch, R. (1981) Microinjection of cloned retroviral genomes into mouse zygotes: integration and expression in the animal. *Nature* **293,** 540–542.
4. Gordon, J. W., Scangos, G. A., Plotkin, D. J., Barbosa, J. A., and Ruddle, F. H. (1980) Genetic transformation of mouse embryos by microinjection of purified DNA. *Proc. Natl. Acad. Sci. USA* **77,** 7380–7384.
5. Brinster, R. L., Chen, Y., Trumbauer, E., Senera, A. W., Warren, R., and Palmiter, R. D. (1981) Somatic expression of herpes thymidine kinase in mice following injection of a fusion gene into eggs. *Cell* **27,** 223–231.
6. Jaenisch, R. (1988) Transgenic animals. *Science* **240,** 1468–1474.
7. Gordon, J. W. (1993) Production of transgenic mice, in *Guide to Techniques in Mouse Development Vol. 225,* (Wassarman, P. M. and DePamphilis, M. L., eds.), Academic Press, San Diego, CA, pp. 747–771.

8. Brinster, R. L., Allen, J. M., Behringer, R. R., Gelinas, R. E., and Palmiter, R. D. (1988) Introns increase transcriptional efficiency in transgenic mice. *Proc. Natl. Acad. Sci. USA* **85,** 836–840.
9. Choi, T., Huang, C., Gorman, C., and Jaenisch, R. (1991) A generic intron increases gene expression in transgenic mice. *Mol. Cell. Biol.* **11,** 3070–3074.
10. Burke, D., Carle, G., and Olson, M. (1987) Cloning of large segments of exogenous DNA into yeast artificial chromosomes. *Science* **236,** 806–812.
11. Spencer, F., Ketner, G., Connelley, C., and Hieter, P. (1993) Targeted recombination–based cloning and manipulation of large DNA segments in yeast. *Methods* **5,** 161–175.
12. Hieter, P., Connelly, C., Shero, J., McCormick, M. K., Antonarakis, S., Pavan, W., and Reeves, R. (1990) Yeast artificial chromosomes: promises kept and pending, in *Genome Analysis Volume I: Genetic and Physical Mapping* (Davies, K. E., and Tilghman, S., eds.), Cold Spring Harbor Laboratory Press, Cold Spring Harbor, NY, pp. 83–120.
13. Mendez, M. J., Green, L. L., Corvalan, J. R. F., Jia, X.–C., Maynard–Currie, C. E., Yang, X.–D., et al. (1997) Functional transplant of megabase human immunoglobulin loci recapitulates human antibody response in mice. *Nature Genet.* **15,** 146–156.
14. Lamb, B. T. and Gearhart, J. D. (1995) YAC transgenics and the study of genetics and human disease. *Curr. Opin. Genet. Dev.* **5,** 342–348.
15. Peterson, K. R., Clegg, C. H., Li, Q., and Stamatoyannopoulos, G. (1997) Production of transgenic mice with yeast artificial chromosomes. *Trends Genet.* **13,** 61–66.
16. Jakobovits, A. (1994) Humanizing the mouse genome. *Curr. Biol.* **4,** 761–763.
17. Jakobovits, A., Moore, A. L., Green, L. L., Vergara, G. J., Maynard–Currie, C. E., Austin, H. A., and Klapholz, S. (1993) Germ–line transmission and expression of a human–derived yeast artificial chromosome. *Nature* **362,** 255–258.
18. Mendez, M. J., Abderrahim, H., Noguchi, M., David, N. E., Hardy, M. C., Green, L. L., et al. (1995) Analysis of the structural integrity of YACs comprimising human immunoglobulin genes in yeast and in embryonic stem cells. *Genomics* **26,** 294–307.
19. Green, L. L., Hardy, M. C., Maynard–Currie, C. E., Tsuda, H., Louie, D. M., Mendez, M. J., et al. (1994) Antigen–specific human monoclonal antibodies from mice engineered with human Ig heavy and light chain YACs. *Nature Genet.* **7,** 13–21.
20. Davies, N. P., Rosewell, I. R., Richardson, J. C., Cook, G. P., Neuberger, M. S., Brownstein, B. H., Norris, M. L., and Brüggemann, M. (1993) Creation of mice expressing human antibody light chains by introduction of a yeast artificial chromosome containing the core region of the human immunoglobulin k locus. *Biotechnology* **11,** 911–914.
21. Zweigerdt, R., Braun, T., and Arnold, H.–H. (1997) Faithful expression of the *Myf–5* gene during mouse myogenesis requires distant control regions: A transgene approach using yeast artificial chromosomes. *Devlop. Biol.* **192,** 172–180.
22. Guthrie, C., and Fink, G. R. (1991) *Guide to Yeast Genetics and Molecular Biology.* Academic, San Diego, CA.
23. Lamb, B. T., Sisodia, S. S., Lawler, A. M., Slunt, H. H., Kitt, C. A., Kearns, W. G., Pearson, P. L., Price, D. L., and Gearhart, J. D. (1993) Introduction and expression of the 400 kilobase *amyloid precursor protein* gene in transgenic mice. *Nature Genet.* **5,** 22–30.
24. Tsuda, H., Maynard–Currie, C. E., Reid, L. H., Yoshida, T., Edamura, K., Maeda, N., Smithies, O., and Jakobovits, A. (1997) Inactivation of the mouse *HPRT* locus by a 203–bp retroposon insertion and a 55–kb gene–targeted deletion: Establishment of new HPRT–deficient mouse embryonic stem cell lines. *Genomics* **42,** 413–421.
25. Ramirez–Solis, R., Cavis, A. C., and Bradley, A. (1993) Gene targeting in embryonic stem cells. *Methods Enzymol.* **225,** 855–878.
26. Stewart, C. L. (1993) Production of chimeras between embryonic stem cells and embryos, in *Guides to Techniques in Mouse Development Vol. 225,* (Wassarman, P. M. and DePamphilis, M. L., eds.), Academic, San Diego, CA, pp. 823–855.

27. Abbondanzo, S. J., Gadi, I., and Stewart, C. L. (1993) Derivation of embryonic stem cell lines, in *Guides to Techniques in Mouse Development Vol. 225,* (Wassarman, P. M. and DePamphilis, M. L., eds.), Academic, San Diego, CA, pp. 803–823.
28. Lee, J. T., Strauss, W. M., Dausman, J. A., and Jaenisch, R. (1996) A 450 kb transgene displays properties of the mammalian X–inactivation center. *Cell* **86,** 83–94.
29. Lee, J. T., and Jaenisch, R. (1996) A method for high efficiency YAC lipofection into murine embryonic stem cells. *Nucleic Acids Res.* **24,** 5054–5055.
30. Strauss, W. M., Dausman, J., Beard, C., Johnson, C., Lawrence, J. B., and Jaenisch, R. (1993) Germ–line transmission of a yeast artificial chromosome spanning the murine *Col1A1* (*a1(I)Collagen*) locus. *Science* **259,** 1904–1907.
31. Lamb, B. T., Call, L. M., Slunt, H. H., Bardel, K. A., Lawler, A. M., Eckman, C. B., Younkin, S. G., Holtz, G., Wagner, S. L., Price, D. L., Sisodia, S. S., and Gearhart, J. D. (1997) Altered metabolism of familial Alzheimer's disease–linked amyloid precursor protein variants in yeast artificial chromosome transgenic mice. *Hum. Mol. Genet.* **6,** 1535–1541.
32. Pearson, B. E., and Choi, T. K. (1993) Expression of the human b–amyloid precursor protein gene from a yeast artificial chromosome in transgenic mice. *Proc. Natl. Acad. Sci. USA* **90,** 10578–10582.
33. Choi, T. K., Hollenbach, P. W., Pearson, B. E., Ueda, R. M., Weddel, G. N., Kurahara, C. G., Woodhouse, C. S., Kay, R. M., and Loring, J. F. (1993) Transgenic mice containing a human heavy chain immunoglobulin gene fragment cloned in a yeast artificial chromosome. *Nature Genet.* **4,** 117–123.
34. Schedl, A., Larin, Z., Montoliu, L., Thies, E., Kelsey, G., Lehrach, H., and Schütz, G. (1993) A method for the generation of YAC transgenic mice by pronuclear microinjection. *Nucleic Acids Res.* **21,** 4783–4787.
35. Peterson, K. R., Clegg, C. H., Huxley, C., Josephson, B. M., Haugen, H. S., Furukawa, T., and Stamatoyannopoulos, G. (1993) Transgenic mice containing a 248–kb yeast artificial chromosome carrying the human β–globin locus display proper developmental control of human globin genes. *Proc. Natl. Acad. Sci. USA* **90,** 7593–7597.
36. Gaensler, K. M. L., Kitamura, M., and Kan, Y. W. (1993) Germ–line transmission and developmental regulation of a 150–kb yeast artificial chromosome containing the human β–globin locus in transgenic mice. *Proc. Natl. Acad. Sci. USA* **90,** 11381–11385.
37. Antoch, M. P., Song, E.–J., Chang, A.–M., Vitaterna, M. H., Zhao, Y., Wilsbacher, L. D., Sangoram, A. M., King, D. P., Pinto, L. H., and Takahashi, J. S. (1997) Functional identification of the mouse circadian *Clock* gene by transgenic BAC rescue. *Cell* **89,** 655–667.
38. Bungert, J., Dave, U., Lim, K.–C., Lieuw, K. H., Shavit, J. A., Liu, Q., and Engel, J. D. (1995) Synergistic regulation of human β–globin gene switching by locus control region elements HS3 and HS4. *Genes Dev.* **9,** 3083–3096.
39. Frazer, K. A., Narla, G., Zhang, J. L., and Rubin, E. M. (1995) The apolipoprotein(a) gene is regulated by sex hormones and acute–phase inducers in YAC transgenic mice. *Nature Genet.* **9,** 424–431.
40. Frazer, K. A., Ueda, Y., Zhu, Y., Gifford, V. R., Garofalo, M. R., Mohandas, N., et al. (1997) Computational and biological analysis of 680 kb of DNA from the human 5q31 cytokine gene cluster region. *Genome Res.* **7,** 495–512.
41. Hamilton, B. A., Smith, D. J., Mueller, K. L., Kerrebrock, A. W., Bronson, R. T., Berkel, V. V., et al. (1997) The *vibrator* mutation causes neurodegeneration via reduced expression of PITPa: positional complementation cloning and extragenic suppression. *Neuron* **18,** 711–722.
42. Heard, E., Kress, C., Mongelard, F., Courtier, B., Rougeulle, C., Ashworth, A., et al. (1996) Transgenic mice carrying an *Xist*–containing YAC. *Hum. Mol. Genet.* **5,** 441–450.

43. Hiemisch, H., Schutz, G., and Kaestner, K. H. (1997) Transcriptional regulation in endoderm development: characterization of an enhancer controlling *Hnf3g* expression by transgenesis and targeted mutagenesis. *EMBO J.* **16,** 3995–4006.
44. Hodgson, J. G., Smith, D. J., McCutcheon, K., Koide, H. B., Nishiyama, K., Dinulos, et al. (1996) Human huntingtin derived from YAC transgenes compensates for loss of murine huntingtin by rescue of the embryonic lethal phenotype. *Hum. Mol. Genet.* **5,** 1875–1885.
45. Huxley, C., Passage, E., Manson, A., Putzu, G., Figarella–Branger, D., Pellissier, J. F., and Fontes, M. (1996) Construction of a mouse model of Charcot–Marie–Tooth disease type 1A by pronuclear injection of human YAC DNA. *Hum. Mol. Genet.* **5,** 563–569.
46. Liu, Q., Bungert, J., and Engel, J. D. (1997) Mutation of gene–proximal regulatory elements disrupts human ε–, γ–, and β–globin expression in yeast artificial chromsome transgenic mice. *Proc. Natl. Acad. Sci. USA* **94,** 169–174.
47. Manson, A. L., Trezise, A. E., MacVinish, L. J., Kasschau, K. D., Birchall, N., Episkopou, V., Vassaux, G., Evans, M. J., Colledge, W. H., Cuthbert, A. W., and Huxley, C. (1997) Complementation of null CF mice with a human CFTR YAC transgene. *EMBO J.* **16,** 4238–4249.
48. Matsuura, S., Episkopou, V., Hamvas, R., and Brown, S. D. M. (1996) *Xist* expression from an *Xist* YAC transgene carried on the mouse Y chromosome. *Hum. Molec. Genet.* **5,** 451–459.
49. McCormick, S. P. A., Ng, J. K., Taylor, S., Flynn, L. M., Hammer, R. E., and Young, S. G. (1995) Mutagenesis of the human apolipoprotein B gene in a yeast artificial chromosome reveals the site of attachment for apolipoprotein(a) *Proc. Natl. Acad. Sci. USA* **92,** 10,147–10,151.
50. McCormick, S. P., Ng, J. K., Cham, C. M., Taylor, S., Marcovina, S. M., Segrest, J. P., Hammer, R. E., and Young, S. G. (1997) Transgenic mice expressing human ApoB95 and ApoB97. Evidence that sequences within the carboxyl–terminal portion of human apoB100 are important for the assembly of lipoprotein. *J. Biol. Chem.* **272,** 23616–23622.
51. Montoliu, L., Schedl, A., Kelsey, G., Zentgraf, H., Lichter, P., and Schutz, G. (1994) Germ line transmission of yeast artificial chromosomes in transgenic mice. *Reprod. Fertil. Dev.* **6,** 577–584.
52. Peterson, K. R., Li, Q. L., Clegg, C. H., Furukawa, T., Navas, P. A., Norton, E. J., Kimbrough, T. G., and Stamatoyannopoulos. (1995) Use of yeast artificial chromosomes (YACs) in studies of mammalian development: Production of β–globin locus YAC mice carrying human globin developmental mutants. *Proc. Natl. Acad. Sci. USA* **92,** 5655–5659.
53. Schedl, A., Beermann, F., Thies, E., Montoliu, L., Kelsey, G., and Schütz, G. (1992) Transgenic mice generated by pronuclear injection of a yeast artificial chromosome. *Nucleic Acids Res.* **20,** 3073–3077.
54. Schedl, A., Montoliu, L., Kelsey, G., and Schütz, G. (1993) A yeast artificial chromosome covering the tyrosinase gene confers copy number–dependent expression in transgenic mice. *Nature* **362,** 258–261.
55. Schedl, A., Ross, A., Lee, M., Engelkamp, D., Rashbass, P., Heyningen, V. V., and Hastie, N. D. (1996) Influence of *PAX6* gene dosage on development: overexpression causes severe eye abnormalities. *Cell* **86,** 71–82.
56. Smith, D. J., Zhu, Y., Zhang, J., Cheng, J.–F., and Rubin, E. M. (1995) Construction of a panel of transgenic mice containing a contiguous 2–Mb set of YAC/P1 clones from human chromosome 21q22.2. *Genomics* **27,** 425–434.
57. Ainscough, J. F.–X., Koide, T., Tada, M., Barton, S., and Surani, M. A. (1997) Imprinting of *Igf2* and *H19* from a 130 kb YAC transgene. *Development* **124,** 3621–3632.
58. Yang, W. Y., Model, P., and Heintz, N. (1997) Homologous recombination based modification in *Esherichia coli* and germline transmission in transgenic mice of a bacterial artificial chromosome. *Nature Biotech.* **15,** 859–865.
59. Mullins, L. J., Kotelevtseva, N., Boyd, A. C., and Mullins, J. J. (1997) Efficient Cre–lox linearisation of BACs: applications to physical mapping and generation of transgenic animals. *Nucleic Acids Res.* **25,** 2539–2540.

60. Nielsen, L. B., McCormick, S. P. A., Pierotti, V., Tam, C., Gunn, M. D., Shizuya, H., and Young, S. G. (1997) Human apolipoprotein B transgenic mice generated with 207– and 145–kilobase pair bacterial artificial chromosomes. *J. Biol. Chem.* **272,** 29752–29758.
61. Wutz, A., Smrzka, O. W., Schweifer, N., Schellander, K., Wagner, E. F., and Barlow, D. P. (1997) Imprinted expression of the *Igf2r* gene depends on an intronic CpG island. *Nature* **389,** 745–749.
62. Gnirke, A., Huxley, C., Peterson, K., and Olson, M. V. (1993) Microinjection of intact 200– to 500–kb fragments of YAC DNA into mammalian cells. *Genomics* **15,** 659–667.
63. Maule, J. C., Porteous, D. J., and Brookes, A. J. (1994) An improved method for recovering intact pulsed field gel purified DNA of at least 1.6 megabases. *Nucleic Acids Res.* **22,** 3245–3246.
64. Brinster, R. L., Chen, H. Y., Trumbauer, M. E., Yagle, M. K., and Palmiter, R. D. (1985) Factors affecting the efficiency of introducing foreign DNA into mice by microinjecting eggs. *Proc. Natl. Acad. Sci. USA* **82,** 4438–4442.
65. Peterson, K. R. (1997) Production and analysis of transgenic mice containing yeast artificial chromosomes, in *Genetic Engineering* (Setlow, J. K., ed.), Plenum Press, New York, pp. 235–255.
66. Fairhead, C., Heard, E., Arnaud, D., Avner, P., and Dujon, B. (1995) Insertion of unique sites into YAC arms for rapid physical analysis following YAC transfer into mammalian cells. *Nucleic Acids Res.* **23,** 4011–4012.
67. Ferrin, L. J. and Camerini–Otero, R. D. (1991) Selective cleavage of human DNA: RecA–assisted restriction endonuclease (RARE) cleavage. *Science* **254,** 1494–1497.
68. Rosenberg, C., Voltz, A. K., Lawler, A. L., Lamb, B. T., Stetten, G., and Gearhart, J. D. (1996) Alterations of yeast artificial chromosome transgenic sequences in stretched embyronic stem–cell chromatin visualized by fluorescence in situ hybridization. *Cyotogenet. Cell. Genet.* **75,** 67–70.
69. Kovacic, R. T., Comai, L., and Bendich, A. J. (1995) Protection of megabase DNA from shearing. *Nucl. Acids Res.* **23,** 3999–4000.
70. Couto, L. B., Spangler, E. A., and Rubin, E. M. (1989) A method for the preparative isolation and concentration of intact yeast artificial chromosomes. *Nucleic Acids Res.* **17,** 8010.
71. Lamb, B. T., Bardel, K. A., Kulame, L. S., Anderson, J. J., Holtz, G., Wagner, S. L., Sisodia, S. S., and Hoeger, E. J. (1999) Amyloid production and deposition in mutant *amyloid precursor protein* and *presenilin-1* yeast artificial chromosome transgenic mice. *Nature Neurosci.* **2,** 695–697.

43

Gene-Targeting Strategies

Siew-Sim Cheah and Richard R. Behringer

1. Introduction

Gene targeting in mouse embryonic stem (ES) cells has become a routine methodology to study gene function in vivo *(1)*. Indeed, gene-targeting core facilities have been established at numerous institutions to facilitate the generation of targeted ES cell lines and the production of mouse chimeras. Although these facilities have centralized many of the unique skills and reagents that are required to generate "knockout" mice, the investigator must still generate a thoughtful strategy to mutate the gene of interest and to subsequently characterize the resulting mutant mice.

For the novice, the details that are required to design the optimal strategy to generate a targeted mutation are usually not considered because there are so many targeting strategies that are possible *(2)*. The novice often becomes dazzled by these options and chooses strategies that are highly complex and therefore likely to fail. In addition, beginners often become so excited about generating their first knockout mutation that they rush to generate their gene-targeting vector without carefully considering all of the consequences. Inevitably, these shortcuts come back to haunt them, resulting in unnecessary and costly delays, and many times, starting over to retarget the locus using a different strategy. By investing the time in a thoughtful design of a gene targeting strategy, one actually saves valuable time and money at subsequent stages of the experiment. In this chapter, we provide a fundamental strategy to generate a null allele for a protein-coding gene by gene targeting in ES cells.

2. General Gene-Targeting Strategy

Typically, the first desired mutation to generate in a protein-coding gene is a null allele. There are generally two situations to consider. In the first situation, the protein-coding exons of your gene span a relatively small genomic distance (20 kb or less). In the second situation, the protein-coding exons of your gene span a relatively large distance (greater than 20 kb). For a small gene, the gene-targeting strategy is simple, delete all of the protein coding exons *(3–5)*. Using this strategy, you guarantee the generation of a null allele because it is impossible to generate a protein product from the targeted locus (**Fig. 1**). In this case, once you have verified that the DNA coding sequences are deleted, it is not necessary to perform mRNA or protein analyses for the

From: *Methods in Molecular Biology, Vol. 136: Developmental Biology Protocols, Vol. II*
Edited by: R. S. Tuan and C. W. Lo © Humana Press Inc., Totowa, NJ

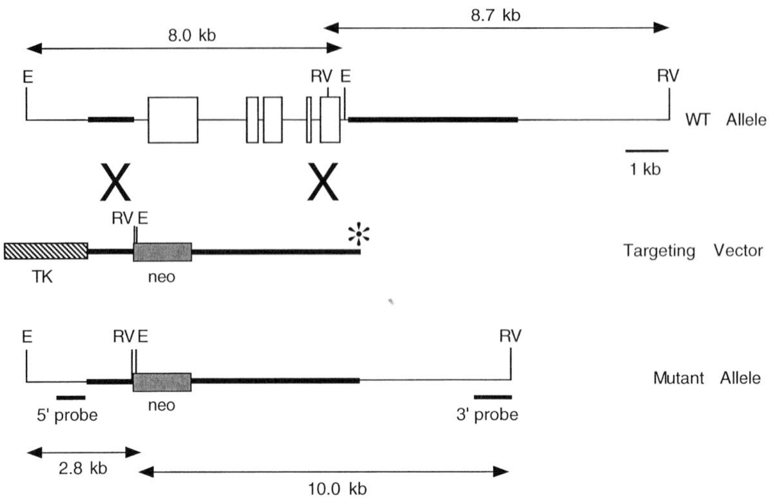

Fig. 1. Deletion of all protein coding exons for a small gene. The example shown is for the generation of a null allele for *Lim1* ***(5)***. All of the protein coding exons (open boxes) were replaced by a neomycin (neo) cassette to ensure the generation of a null allele. In this case, the total amount of homology (5.5 kb) was divided as a 1.2-kb 5' arm and a 4.3-kb 3' arm (thick lines). A thymidine kinase (tk) cassette was placed adjacent to the 5' arm of homology. The site of vector linearization is indicated (∗). In this situation, 5' and 3' external probes were used to identify targeting events. The sizes of the wild-type (WT) and mutant bands for Southern analysis are indicated. E, *Eco*RI; RV, *Eco*RV.

deleted gene. Thus, your time can be allocated for phenotypic characterization studies. If coding sequences remain, mRNA and protein studies of the mutated gene can sometimes be difficult and confusing ***(6,7)***. In many situations, biochemical or molecular biological studies have identified domains that are essential for the activity of a specific protein in vitro. This leads some to only delete specific domains because they believe that by doing so, any partial protein that is generated would have no function. This is a fundamental error in logic because partial protein products may have unknown functions. For a larger protein-coding gene, this logic does have some validity because of the constraints imposed by targeting a large gene. However, for small genes, we strongly urge that all protein-coding exons be deleted.

Mutating a larger gene requires more thought because simple gene targeting strategies are less efficient for generating very large deletions (>20 kb) ***(8–10)*** and may require sophisticated gene-targeting skills that the novice has yet to develop ***(11,12)***. In addition, large deletions may remove other genes or regulatory sequences of neighboring genes that reside within large genes. The general strategy we suggest is to generate up to a 20-kb deletion to remove as many of the protein-coding exons as possible, including the exon containing the translation initiation codon. We also suggest that the transcription initiation site, if known, also be included in the targeted deletion (**Fig. 2**). Although this does not guarantee the generation of a null allele, it increases its likelihood ***(13,14)***.

The general strategy we describe uses a replacement gene-targeting vector ***(15)***. The standard features of a replacement vector are a plasmid backbone containing a positive

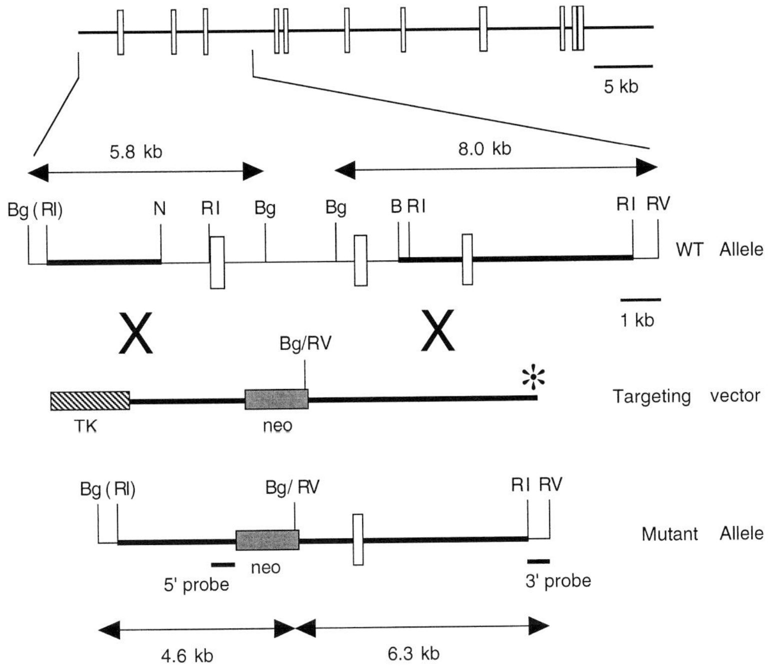

Fig. 2. Deletion of the initial protein coding exons for a large gene. The example shown is for the *Bmpr* locus ***(13)***. The coding exons (open boxes) of this gene span 38 kb (top). Therefore, the first two coding exons were deleted. This region (6.3 kb) was replaced by a neo cassette. A total of 9.0 kb of homology was divided as a 3.0-kb 5' arm and a 6.0-kb 3' arm. A tk cassette was placed on the 5' arm of homology. The site of vector linearization is indicated (*). A 3' external probe was used for the initial identification of homologous recombinants. Once these clones were identified they were expanded for subsequent analysis with an internal probe to characterize the structure of the 5' recombination event. The sizes of the wild-type (WT) and mutant bands for Southern analysis are indicated. Bg, *Bgl*I; N, *Nco*I; (RI), *Eco*RI from phage multicloning site; RV, *Eco*RV.

selection cassette placed between two regions of chromosomal homology and a negative selection cassette adjacent to one of the homologous arms (**Fig. 3**). A positive/negative selection scheme is employed to enrich for homologous recombinants ***(16)***. For a replacement strategy, the gene-targeting vector is linearized outside of the regions of homology before introduction into mouse ES cells by electroporation. Drug-resistant ES cell colonies are placed into 96-well tissue-culture plates for expansion and analysis by Southern blot ***(17)***. Once the correctly targeted ES cell clones have been identified, they are used for the generation of mouse chimeras ***(18)*** or for in vitro differentiation ***(19)*** or the generation of teratomas ***(20)***.

3. Method

3.1. Isolate Multiple Clones for the Gene from a Mouse Genomic Library

In constructing your targeting vector, it is important to use genomic DNA that is isogenic to the mouse ES cell line that will be used for the gene-targeting experiments to facilitate the maximum frequency of homologous recombination events ***(21)***. For

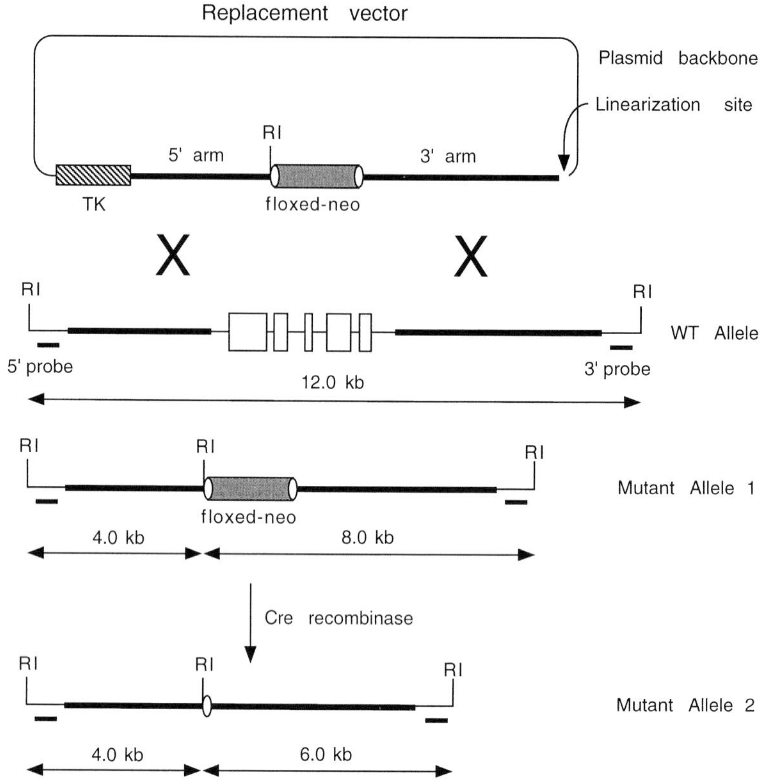

Fig. 3. General design of a replacement gene-targeting vector. The floxed neo and tk cassettes can be placed in either forward or reverse orientation relative to the orientation of transcription for the gene to be targeting. Mutant allele 1 represents the targeted locus with neo and Mutant allele 2 represents the targeted locus that has had the neo, cassette removed by cre recombinase. The sizes of the wild-type (WT) and mutant bands for Southern analysis are indicated.

historical reasons, the majority of mouse ES cell lines are derived from 129 inbred mouse strains *(22)*. It is essential to note that there are many substrains of 129 that can be quite genetically diverse, especially the 129/SvJ substrain *(23)*. Therefore, it is important to determine the precise strain and substrain of mouse used to generate the ES cell line that you will use in your gene-targeting experiments and screen a genomic library from that same strain/substrain. Currently, one very popular mouse ES cell line called R1 is derived from a (129/Sv × 129/SvJ)F_1 embryo *(24)*. The 129 strain mouse genomic libraries are commercially available from numerous sources (Stratagene, La Jolla, CA; Genome Systems, St. Louis, MO; Research Genetics, Huntsville, AL).

It is important to isolate multiple genomic clones for your gene, especially if it is a large gene. This will provide more genomic sequences and, thus, more options in designing a targeting vector. In addition, be aware that genomic libraries may contain single clones that contain DNA fragments from different regions of the genome. The coligated genomic inserts in these clones can eliminate all chances of generating or recognizing a targeted mutation. Generate a detailed restriction map for each genomic clone and verify that this map truly matches the chromosomal locus by Southern-blot hybridization of isogenic genomic DNA derived from cells or tissues. There have been many instances when correct gene-targeting had been obtained, but results of the Southern analysis were not as predicted because of an erroneous restriction map.

3.2. Replacement Vector Design

Once you have obtained and characterized genomic clones for the gene of interest, you must design your gene-targeting strategy "on paper." We urge beginners to develop a gene-targeting strategy that employs a Southern-blot analysis to identify targeting events. The 96-well tissue-culture plate method for the rapid screening of ES cell clones by Southern analysis is exceedingly easy for beginners to master very quickly *(17)*. Do not use a polymerase chain reaction (PCR) genotyping strategy because it imposes constraints on your gene-targeting vector design. In addition, we have found that PCR genotyping is more problematic for beginners than Southern-blot analysis. Furthermore, PCR results must always be confirmed by Southern analysis. In your design, identify genomic regions that can serve as Southern blot probes to recognize targeting events. Do not generate your targeting vector until your potential probes have been verified on Southern blots of isogenic genomic DNA. Once the probes have been verified, then construct your gene-targeting vector as shown in **Fig. 3**. The details of each aspect of the replacement vector are discussed in the following.

3.2.1. Choice of Homologous Regions for the Targeting Vector

The total amount of chromosomal homology should be between 5 and 8 kb *(15)*. Smaller amounts of homology may reduce the targeting frequency and significantly larger amounts of homology become unwieldy for targeting vector construction and homologous recombinant identification. The amount of homologous sequences for each arm of homology in the vector should be about half of the total amount of homology. For example, a vector with a total of 6 kb of homology may have two arms of homologous sequence that are approx 3 kb each.

Each arm of homology in the targeting vector should flank the protein-coding exons and other sequences that you wish to delete. Remember that neighboring genes may lie very close to your gene of interest *(25)*. Therefore, try not to delete beyond the known sequences of your gene.

3.2.2. Positive Selection Cassette

We suggest that you use a neomycin phosphotransferase (neo) gene expression cassette for selection with G418 that has the mouse phosphoglycerate kinase (PGK) promoter and the polyadenylation signal from the bovine growth hormone (bpA) gene (PGKneobpA) because it is more efficient than other neo expression cassettes that are typically used in gene-targeting experiments *(26)*. Although there are other positive selectable markers that can be used, (e.g., hygromycin, puromycin, etc.), we suggest that the beginner use neomycin because it is routinely used by most laboratories and most readily available feeder cell lines, on which ES cells grow, are G418 resistant. The neo cassette can be placed in either forward or reverse orientation relative to the direction of transcription of your gene.

The presence of a selectable marker cassette with its exogenous promoter has been documented to alter gene expression at targeted loci *(27)*. Therefore, the neo cassette in the strategy we outline should be flanked by loxP sequences (floxed) to provide the option of subsequently removing the neo cassette with cre recombinase *(28)*. You lose nothing by cloning a floxed–neo expression cassette into your gene-targeting vector and gain many potential benefits. The option to remove the neo cassette using simple methods can potentially save effort and the costs of regenerating another mutant should any question arise about the altered transcription of your gene or neighboring genes *(29)*. The floxed–neo cassette strategy also provides the opportunity to remove the selectable marker cassette from the targeted locus in vitro for subsequent retargeting of the remaining wild-type allele with the original targeting vector and drug selection to generate homozygous mutant ES cell lines, a process called "marker recycling" *(30)*. The floxed–neo cassette strategy also allows one to easily generate two different alleles for your gene of interest, one with neo and one without, that can be distinguished by Southern blot or PCR. The availability of two different alleles can be exploited for potential chimera experiments *(31,32)*. The floxed–neo cassette can be removed in ES cells by transient cre expression in vitro *(30)* by the injection of a cre expression vector into the pronuclei of eggs containing the floxed gene *(33)* or by simply crossing mice carrying the floxed gene with cre expressing transgenic mice *(34)*.

3.2.3. Negative Selection Cassette

We suggest that you use a herpes simplex virus thymidine kinase gene expression cassette for negative selection with gancyclovir or FIAU (1-2-deoxy-2-fluroro-1-β-D-arabinofuranosyl-5-iodouracil). The MC1TKpA vector works very well with FIAU, providing a five- to 10-fold enrichment for targeting events *(35)*. An alternate and convenient negative selection cassette is one that expresses the diphtheria toxin A-chain (DT-A) gene *(36)*. The advantage of the DT-A strategy is that it is not necessary to add any drug selection to the culture medium because expression of DT-A in cells is toxic.

The beginner should note that even after positive/negative drug selection, many of the resulting ES cell colonies are nontargeted (random integration) events. This occurs because the negative selection cassette used in the gene-targeting vector is not expressed, possibly because of damage or integration into chromosomal regions that inhibit its expression. The negative selection cassette can be placed in either forward or reverse orientation relative to the direction of transcription of your gene.

3.2.4. Site of Targeting Vector Linerization

Gene-targeting vectors are routinely introduced into ES cells by electroporation in a linearized form. Thus, in the design of your targeting vector, a unique restriction enzyme site must be present as a site for linearization. For a replacement gene-targeting vector, the linearization site must be outside of the regions of homology. Typically, the linearization site is located at the junction of one of the homologous arms and the plasmid backbone. We suggest that the linearization site should not be at the junction between the negative selection cassette and the plasmid backbone (**Fig. 3**) to reduce the chances of the negative selection cassette from being degraded by nucleases after introduction into cells, causing an increase in the background of nontargeted clones.

3.2.5. Restriction Enzyme Sites for Southern Analysis

To identify homologous recombinants, Southern analysis using a diagnostic restriction enzyme digest with a probe that is external to the regions of homology (external probe) included in the targeting vector must be performed. Therefore, it is important to consider restriction enzyme sites that can be used to differentiate between targeted and nontargeted events during the construction of the targeting vector. A unique restriction enzyme site should be introduced by the positive selection cassette. Remember that this restriction enzyme site should be located outside of the floxed–neo expression cassette so that after cre recombinase expression, the diagnostic restriction enzyme site will still remain. In this way, a diagnostic digest will yield a smaller DNA fragment for the mutant allele in comparison to the wild-type allele that can be recognized by the external probe. This is essential because partial restriction enzyme digestions can occur in the cruder DNA samples that are prepared using the 96-well tissue-culture plate method for the rapid screening of recombinant ES cells. Thus, if the mutant allele yields a DNA fragment that is larger than the wild-type allele, a partial restriction enzyme digest can cause difficulties during the Southern analysis. We suggest choosing restriction enzymes that work efficiently on these cruder DNA preparations for the most consistent results for genomic Southerns. The restriction enzymes that have worked well for us include *Bam*HI, *Bgl*I, *Bgl*II, *Eco*RI, *Eco*RV, *Pst*I, and *Sst*I, whereas *Kpn*I, *Xba*I, *Xho*I, and *Xmn*I have proven to be problematic.

3.2.6. 5' and 3' External Probes for the Identification of Targeted ES Cell Clones

It is important to use both 5' and 3' external probes that can recognize the structure of the gene targeting events on the 5' and 3' arms of homology. It is not uncommon for correct targeting to be obtained on one arm of homology but not the other. If both 5' and 3' external probes cannot be found, the initial targeting event can be identified using one external probe that confirms the structure of the targeting event on one side of homology. Once this subset of targeted ES cell clones is identified, a probe within the region of targeting vector homology (internal probe, for example neo) can be used to verify the structure of the targeting event on the other side of homology. Do not use the PGK promoter sequences of the neo expression cassette as an internal probe because it is derived from mouse and will recognize the endogenous mouse PGK gene.

4. Summary

Gene targeting in mouse ES cells is a powerful method for studying gene function in vivo. For the novice, this combination of molecular biology, specialized tissue-culture cell lines, and mouse reproductive biology can be daunting. We present a straightforward, one might say constrained, guide for novices of gene targeting to generate a null allele in large or small protein-coding genes. The method we outline has evolved from years of experience of training and advising beginners on this powerful technology. We believe that a good design for a gene-targeting strategy ultimately saves time, money, and research animal lives. Once you feel comfortable with a fundamental knockout, we suggest you consider the new and exciting gene-targeting variations that can be used to address important biological questions *(2)*.

Acknowledgments

The authors thank Yuji Mishina and Maki Wakamiya for helpful comments on the manuscript.

References

1. Torres, M. (1998) The use of embryonic stem cells for the genetic manipulation of the mouse. *Curr. Top. Dev. Biol.* **36,** 99–114.
2. Bradley, A. and Liu, P. (1996) Target practice in transgenics. *Nature Gen.* **14,** 121–123.
3. Chen, Z.-F. and Behringer, R. R. (1995) *twist* is required in head mesenchyme for cranial neural tube morphogenesis. *Genes Dev* **9,** 686–699.
4. Rivera-Pérez, J. A., Mallo, M., Gendron-Maguire, M., Gridley, T., and Behringer, R. R. (1995) *Goosecoid* is not an essential component of the mouse gastrula organizer but is required for craniofacial and rib development. *Development* **121,** 3005–3012.
5. Shawlot, W. and Behringer, R. R. (1995) Requirement for *Lim1* in head-organizer function. *Nature* **374,** 425–430.
6. Moens, C. B., Auerbach, A. B., Conlon, R. A., Joyner, A. L., and Rossant, J. A. (1992) Targeted mutation reveals a role for *N-myc* in branching morphogenesis in the embryonic mouse lung. *Genes Dev.* **6,** 691–704.
7. Hasty, P., Bradley, A., Morris, J. H., Edmondson, D. G., Venuti, J. M., Olson, E. N., and Klein, W. H. (1993) Muscle deficiency and neonatal death in mice with a targeted mutation in the myogenin gene. *Nature* **364,** 501–506.
8. Mombaerts, P., Clarke, A. R., Hooper, M. L., annd Tonegawa, S. (1991) Creation of a large genomic deletion at the T-cell antigen receptor beta-subunit locus in mouse embryonic stem cells by gene targeting. *Proc. Natl. Acad. Sci. USA* **88,** 3084–3087.
9. Zhang, H., Hasty, P., and Bradley, A. (1994) Targeting frequency for deletion vectors in embryonic stem cells. *Mol. Cell. Biol.* **14,** 2404–2410.
10. Tsuda, H., Maynard-Currie, C. E., Reid, L. H., Yoshida, T., Edamura, K., Maeda, N., Smithies, O., and Jakobovits, A. (1997) Inactivation of the mouse HPRT locus by a 203-bp retroposon insertion and a 55-kb gene-targeted deletion: establishment of new HPRT-deficient mouse embryonic stem cell lines. *Genomics* **15,** 413–421.
11. Ramírez-Solis, R., Liu, P., and Bradley, A. (1995) Chromosome engineering in mice. *Nature* **378,** 720–724.
12. Li, Z. W., Stark, G., Gotz, J., Rulicke, T., Gschwind, M., Huber, G., Muller, U., and Weissmann, C. (1996) Generation of mice with a 200-kb amyloid precursor protein gene deletion by Cre recombinase-mediated site-specific recombination in embryonic stem cells. *Proc. Natl. Acad. Sci. USA* **93,** 6158–6162.
13. Mishina, Y., Suzuki, A., Ueno, N., and Behringer, R. R. (1995) *Bmpr* encodes a type I bone morphogenetic protein receptor that is essential for gastrulation during mouse embryogenesis. *Genes Dev.* **9,** 3027–3037.
14. Mishina, Y., Rey, R., Finegold, M. J., Matzuk, M. M., Josso, N., Cate, R. L., and Behringer, R. R. (1996) Genetic analysis of the Mullerian-inhibiting substance signal transduction pathway in mammalian sexual differentiation. *Genes Dev.* **10,** 2577–2587.
15. Hasty, P. and Bradley, A. (1993) Gene targeting vectors for mammalian cells, in *Gene Targeting: A Practical Approach* (Joyner, A. L., ed.), IRL, Oxford, pp. 1–31.
16. Mansour, S. L., Thomas, K. R., and Capecchi, M. R. (1988) Disruption of the proto-oncogene *int-2* in mouse embryo-derived stem cells: a general strategy for targeting mutations to non-selectable genes. *Nature* **336,** 348–352.
17. Ramírez-Solis, R., Rivera-Pérez, J., Wallace, J. D., Wims, M., Zheng, H., and Bradley, A. (1992) Genomic DNA microextraction: a method to screen numerous samples. *Analytical Biochem* **201,** 331–335.
18. Wood, S. A., Allen, N. D., Rossant, J., Auerbach, A., and Nagy, A. (1993) Non-injection methods for the production of embryonic stem cell-embryo chimaeras. *Nature* **365,** 87–89.
19. Robertson, E. J. (ed.) (1987) Embryo-derived stem cell lines, in *Teratocarcinomas and Embryonic Stem Cells: A Practical Approach*, IRL, Oxford, pp. 71–112.

20. Shaw-White, J. R., Denko, N., Albers, L., Doetschman, T. C., and Stringer, J. R. (1993) Expression of the *lacZ* gene targeted to the *HPRT* locus in embryonic stem cells and their derivatives. *Transgenic Res.* **2,** 1–13.
21. te Riele, H., Maandag, E. R., and Berns, A. (1992) Highly efficient gene targeting in embryonic stem cells through homologous recombination with isogenic DNA constructs. *Proc. Natl. Acad. Sci. USA* **89,** 5128–5132.
22. Papaioannou, V. and Johnson, R. (1993) Production of chimeras and genetically defined offspring from targeted ES cells, in *Gene Targeting: A Practical Approach* (Joyner, A. L., ed.), IRL, Oxford, pp. 1–31.
23. Threadgill, D. W., Yee, D., Matin, A., Nadeau, J. H., and Magnuson, T. (1997) Genealogy of the 129 inbred strains: 129/SvJ is a contaminated inbred strain. *Mamm. Genome* **8,** 390–393.
24. Nagy, A., Rossant, J., Nagy, R., Abramow-Newerly, W., and Roder, J. C. (1993) Derivation of completely cell culture-derived mice from early-passage embryonic stem cells. *Proc. Natl. Acad. Sci. USA* **90,** 8424–8428.
25. Dresser, D. W., Hacker, A., Lovell-Badge, R., Guerrier, D. (1995) The genes for a spliceosome protein (SAP62) and the anti-Mullerian hormone (AMH) are contiguous. *Hum. Mol. Genet.* **4,** 1613–1618.
26. Soriano, P., Montgomery, C., Geske, R., and Bradley, A. (1991) Targeted disruption of the *c-src* proto-oncogene leads to osteopetrosis in mice. *Cell* **64,** 693–702.
27. Fiering, S., Epner, E., Robinson, K., Zhuang, Y., Teiling, A., Hu, M., Martin, D. I., Enver, T., Ley, T. J., and Groudine, M. (1995) Targeted deletion of 5'HS2 of the murine beta-globin LCR reveals that it is not essential for proper regulation of the beta-globin locus *Genes Dev.* **9,** 2203–2213.
28. Sauer, B. and Henderson, N. (1988) Site-specific DNA recombination in mammalian cells by the CRE recombinase in bacteriophage P1. *Proc. Natl. Acad. Sci. USA* **85,** 5166–5170.
29. Olson, E. N., Arnold, H. H., Rigby, P. W., and Wold, B. J. (1996) Know your neighbors: three phenotypes in null mutants of the myogenic bHLH gene *MRF4*. *Cell* **85,** 1–4.
30. Abuin, A. and Bradley, A. (1996) Recycling selectable markers in mouse embryonic stem cells. *Mol. Cell. Biol.* **16,** 1851–1856.
31. Quinn, J. C., West, J. D., and Hill, R. E. (1996) Multiple functions for *Pax6* in mouse eye and nasal development. *Genes Dev.* **10,** 435–446.
32. Rivera-Pérez, J. A., Wakamiya, M., and Behringer, R. R. (1999) *Goosecoid* acts cell autonomously for the maintenance of mesenchymal tissues during craniofacial development. *Development* **126,** 3811–3821.
33. Sunaga, S., Maki, K., Komagata, Y., Ikuta, K., and Miyazaki, J. I. (1997) Efficient removal of loxP-flanked DNA sequences in a gene-targeted locus by transient expression of Cre recombinase in fertilized eggs. *Mol. Reprod. Dev.* **46,** 109–113.
34. Lasko, M., Sauer, B., Mosinger, B. Jr, Lee, E. J., Manning, R. W., Yu, S. H., Mulder, K. L., and Westphal, H. (1992) Targeted oncogene activation by site-specific recombination in transgenic mice. *Proc. Natl. Acad. Sci. USA* **89,** 6232–6236.
35. McMahon, A. P. and Bradley, A. (1990) The *Wnt-1 (int-1)* proto-oncogene is required for development of a large region of the mouse brain. *Cell* **62,** 1073–1085.
36. McCarrick, J. W., Parnes, J. R., Seong, R. H., Solter, D., and Knowles, BB. (1993) Positive-negative selection gene targeting with the diphtheria toxin A-chain gene in mouse embryonic stem cells. *Transgenic Res.* **2,** 183–190.

44

Chimeric Animals and Germline Transmission

Jaspal S. Khillan

1. Introduction

Gene targeting by homologous recombination in pleuripotent embryonic stem (ES) cells provides a powerful tool to introduce specific mutations in the genome of intact animal *(1–5)*. It, therefore, allows to unravel the function of genes that control development and differentiation in the organism. Generally, a gene targeting cassette that contains positive–negative selection markers is prepared in which an exon of the target gene is interrupted by the gene for neomycin resistance, which serves as a positive selection marker *(6–8)*. A herpes simplex virus thymidine kinase (HSV TK) gene is fused at either one or both ends of the genomic sequences for negative selection. The targeting cassette is introduced into ES cells, and the cells in which the endogenous gene is disrupted are selected with G418 and gancyclovir or FIAU *(6,8)*. The cells in which the DNA is inserted randomly will die as a result of the incorporation of gancyclovir or FIAU, the thymidine analogs, by the thymidine kinase gene, which blocks the DNA synthesis. The selected ES cell clones are then used to prepare chimeric animals for germline transmission.

The chimeric animals are prepared either by direct microinjection of ES cells into the cavity of a blastocyst *(1)* or by aggregation of ES cells with morula followed by transfer into pseudopregnant females. Various aggregation methods to prepare chimeric animals are described in literature *(9–11)* (*see* **Note 1**). Here I will describe a one-step coculture method recently established in my own laboratory *(12)*.

2. Materials

2.1. Animals

1. Four to five week old C57BL6 or FVB/N female mice as embryo donors and 8- to 10-wk-old stud males of the same strains.
2. Five- to six-wk-old female mice of CD1 or NIH-GP strains as pseudopregnant recipient mothers.

2.2. Hormones

1. Pregnant mare serum gonadotropin (PMSG).
2. Human chorionic gonadotropin (HCG).

From: *Methods in Molecular Biology, Vol. 136: Developmental Biology Protocols, Vol. II*
Edited by: R. S. Tuan and C. W. Lo © Humana Press Inc., Totowa, NJ

A solution of 50 IU/mL for each hormone is prepared in 0.9% NaCl. One-milliliter aliquots are stored frozen, protected from light up to 4–6 wk (*see* **Note 2**).

2.3. Anesthesia

Avertin is the most commonly used anesthesia in mice. A stock solution of 100% avertin is prepared by mixing 10 g of tribromo ethanol with 25 mL tertiary amyl alcohol. The stock solution is diluted to 2.5% in water before use and stored in the dark at room temperature for prolonged periods. About 0.015–0.017 mL/g body weight is administered to anesthetize adult mice of 20–25 g weight.

2.4. Embryo Culture Medium

A modified Whitten's medium, M16 medium *(13)*, is used to culture preimplantation embryos. The composition of the medium is shown in **Table 1**. The medium can be stored up to 2 wk at 4°C. The medium, can also be purchased from Gibco-BRL (Grand Island, NY) (Brinster's BMOC-3 medium).

2.5. Tissue Culture Medium

1. Blastocyst microinjection medium: Dulbecco's modified Eagle medium (DMEM) w/o sodium pyruvate, w/o sodium bicarbonate, 10% fetal bovine serum (FBS) and 25 m*M* HEPES pH 7.2. The osmolarity of the medium should be around 290–300 mosmol.
2. Primary fibroblast (PF) culture medium: This medium contains 10% FBS and 1% penicillin/streptomycin (pen/strep) in DMEM.
3. Embryonic stem cell (ES) medium: The medium is prepared by mixing 90 mL of FBS, 6 mL of 10 m*M* nonessential amino acids, 3 mL of 100X pen/strep, 1 mL of β-mercaptoethanol, and 60 μL lypmhocyte inhibitory factor (LIF; available from Gibco-BRL as ES-GRO cat. no. 3275SB), which provides 1000 IU/mL, with 500 mL of DMEM. The medium should be used within 2–3 wk (*see* **Note 3**).
4. ES cell freezing medium: The 2X freezing medium contains 20% dimethyl sulfoxide (DMSO), 20% FBS, and 60% ES medium.
5. For morula aggregation, the embryo culture medium is supplemented with bovine serum albumin (BSA) to a final concentration of 10 mg/mL.

2.6. Microtools (See Note 4)

1. Isolation and transfer pipet: Glass capillaries (World Precision Instruments, Inc., Sarasota, FL, 1 mm internal diameter [id]) are prepulled on a gas burner. The prepulled capillary is cut with a diamond pencil so that the opening of the needle is about 120–140 μm id. The tip of the capillary is polished on a microforge with a 100- to 120-μm opening. The same pipet is used for transfer of embryo into pseudopregnant females.
2. Preparation of coculture dish: A 35-mm petridish with 8–10 microwells, 0.3–0.5 mm outer diameter (od) at the top and 0.1–0.2 mm at the bottom, are constructed in a circle of 0.5 cm (**Fig. 1**). The microwells are constructed by pressing the blunt end of a fire polished glass Pasteur pipet against the bottom of a 35-mm Petri dish. The Petri dishes can also be purchased from Genome System Inc. (St. Louis, MO) (*see* **Note 5**).

2.7. Equipment

1. Microforge.
2. CO_2 incubator.
3. Stereomicroscope.
4. Surgical instruments.

Table 1
Embryo Culture Medium

Compound	m*M*	g/L
NaCl	94.66	5.533
KCl	4.78	0.356
$CaCl_2 \cdot 2H_2O$	1.71	0.252
KH_2PO_4	1.19	0.162
$MgSO_4 \cdot 7H_2O$	1.19	0.293
$NaHCO_3$	25.0	2.101
Sodium lactate	23.28	2.610
Sodium pyruvate	0.33	0.036
Glucose	5.56	1.000
Bovine serum albumin		4.000
Penicillin G (100 U/mL)		0.060
streptomycin sulfate (50 µg/mL)		0.050
Phenol red		0.010

Note: The aggregation medium is supplemented with BSA to the final concentration of 10 mg/mL.

Fig. 1. A coculture Petri dish. Microwells are constructed on the surface of a 35-mm petridish in a circle of about 0.5 cm diameter. Each Petri dish contains 10 wells with a diameter of about 0.2 mm at the bottom and about 0.5 mm at the top.

2.8. Radiolabeled Compounds

1. A ^{32}P-labeled deoxyribonucleotide triphosphate to prepare gene-specific probes and testing the recombination event in ES cells and in embryos by Southern-blot analysis (*see* **Note 6**).

3. Methods

Chimeric animals can be prepared either by microinjection of ES cells into blastocyst *(1)* or by aggregation of ES cells with morula *(9–12)*. Following are the steps involved in generating chimeric animals by the one-step aggregation method *(12)*.

3.1. Preparation of Primary Fibroblasts

1. To establish primary fibroblast cells (*see* **Note 3**), normal or transgenic females with a neomycin-resistance gene are crossed with respective males. Sacrifice mothers at 14–16 d of pregnancy and remove embryos from the uterine horns.
2. Wash embryos once in HEPES saline and dissect individual embryo to remove head and soft tissues such as liver heart and other viscera.
3. Wash carcass twice in HEPES saline buffer followed by mincing into small pieces and transfer to a clean 150-mL flask.
4. Trypsinize tissue with 30 mL trypsin/EDTA for 30 min with constant stirring. Add additional 30 mL of trypsin/EDTA and stir again for 30 min.
5. Add 60 mL of medium with 10% FBS and dissociate tissue by vigorous pipeting and transfer into 50-mL falcon tubes to allow larger tissue pieces to settle.
6. Transfer supernatant to 100-mm Petri dishes (about one dish for each embryo). After 3–4 d, the confluent cells are harvested. Most of the cells are frozen for future use. Propagate one Petri dish to prepare feeder layer cells. The cells are generally used up to fifth passage because their efficiency to support ES cells declines after the fifth passage.

3.2. Preparation of Feeder Plates

1. To use as feeder layers, PF cells are inactivated either by treating with mitomycin C or by irradiation *(5)* (*see* **Note 7**). The PF cells are irradiated at 4000–6000 rads. The cells are stored in individual vials, 1×10^6 cells/vial, in liquid nitrogen.
2. Treat 60-mm or 100-mm plates with 0.2% gelatin for 2 h.
3. Transfer one vial of frozen PF cells into a 60-mm plate and two vials of cells into a 100-mm plate.
4. Allow cells to settle overnight and change the medium the next day. The cells form a monolayer on the bottom of the Petri dish.
5. Switch over to ES cell medium about 2 h before the transfer of ES cells. The plates can be used up to 10 d after preparation.

3.3. Preparation of Gene Targeting Vector

1. A gene-targeting cassette is constructed in which an exon of the gene is interrupted by neomycin-resistance gene that also serves as a positive selection marker *(6–8)* (**Fig. 2**).
2. Typically, about 5–10 kb of the genomic DNA is cloned into a targeting vector in such a way that the neomycin gene is flanked by the genomic sequences.
3. On one or both ends of the genomic sequences, a gene for herpes simplex virus thymidine kinase (HSV TK) is ligated as a negative selection marker. Both of the selector genes are directed by a constitutive promoter that express in ES cells (e.g., phosphoglycero kinase [PGK]).
4. The targeting cassette is then cut out from the vector sequences and purified on agarose gel (**Fig. 2**).

3.4. Electroporation of DNA into ES Cells

1. Thaw one vial of ES cells on a 60-mm plate with a feeder layer and culture till confluency (*see* **Note 8**) (**Fig. 3**). Trypsinize cells and wash one time with fresh ES cell medium. Resuspend $1–2 \times 10^7$ cells in 700 µL of phosphate buffer saline (PBS).

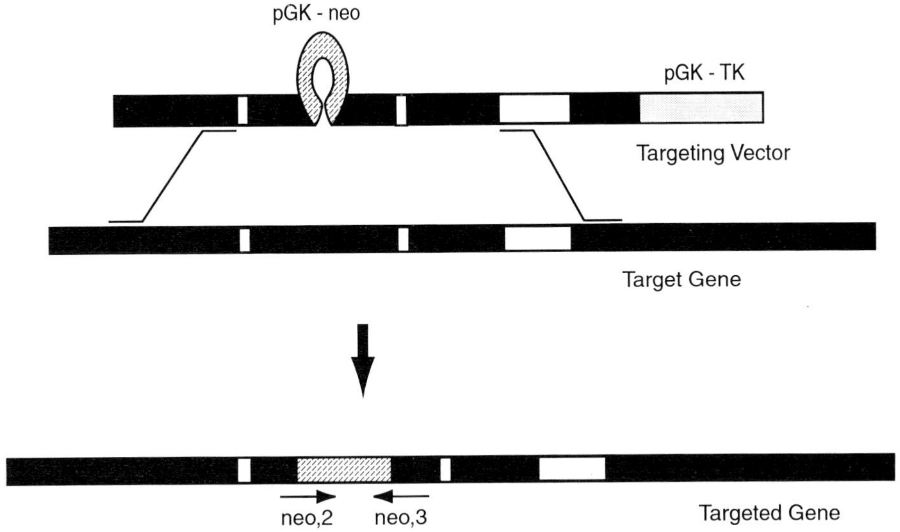

Fig. 2. The gene targeting cassette (**top panel**) with gene for neomycin resistance (neo) and herpes simplex thymidine kinase (HSV-TK). The neomycin gene is introduced into one of the exons (black box), whereas the HSV TK gene is ligated outside the region of the homology. The middle panel represents the chromosomal DNA. The bottom panel shows that after homologous recombination, the HSV-TK gene is excluded.

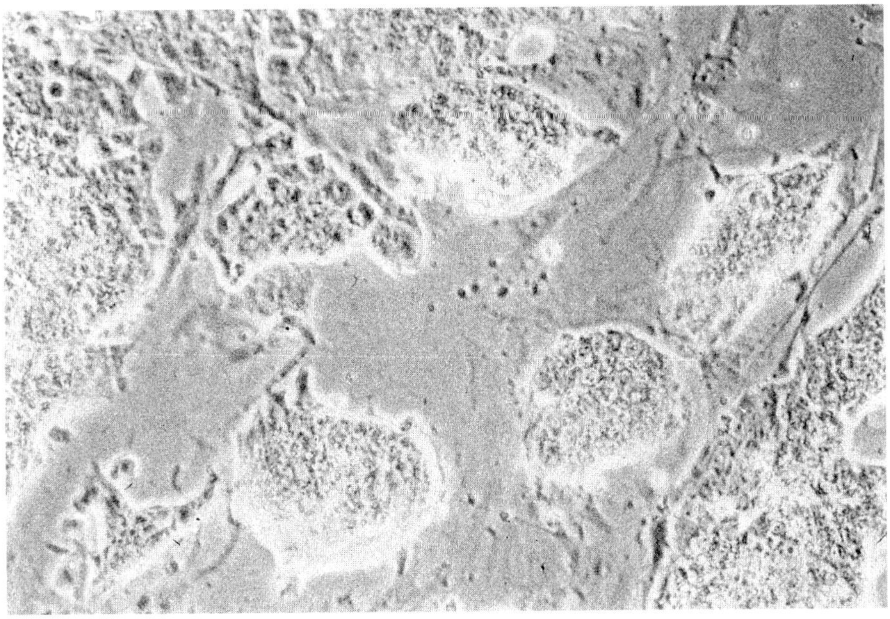

Fig. 3. The culture of ES cells after 48 h of culture. The cells were cultured over the feeder layer. The round morphology of the colonies indicates undifferentiated cells.

2. Transfer cells to a 800 µL electroporation vial and mix 25 µg linearized DNA dissolved in 100 µL of PBS. Electroporate DNA at 280 V and 500 µF capacitance (Bio-Rad Pulsar Electroporator, Bio-Rad, Hercules, CA).

3. Allow cells to stand at room temperature for 10 min followed by dilution in 5 mL of ES medium. Transfer cells to six 100-mm Petri dishes with feeder layer.
4. After 48 h, the cells are selected by treating with ES cell medium containing 350 μg/mL of G418 and 2 μm of gancylovir or 0.2 μ*M* FIAU.
5. Change medium every day. After about 8–10 d the individual colonies can be seen growing in the petridish *(5)*.
6. The individual colonies are harvested with a yellow tip and trypsinized in 50 μL of trypsin. Trypsinized cells are transferred to a well of 24-well plate and the cells are allowed to grow for 2–3 d.
7. Confluent cells from each well are harvested by trypsinization. One half of the cells are transferred to a new 24-well plate with 2X ES cell freezing medium and the other half is transferred back to the same petridish for further culture. The cells are cultured for 3–5 d to isolate DNA.
8. The plate with freezing medium are stored at –70°C till the screening is complete.

3.5. Screening of ES Clones

1. After the cells in the 24-well plate are confluent, add 100 μL of lysis buffer (10 mM Tris pH 7.5, 10 m*M* EDTA, 10 m*M* NaCl, 0.5% sarcosyl, and 1 mg/mL proteinase K) and transfer lysate into a 1.5-mL eppendorf tube and incubate overnight at 55°C.
2. Add equal volume of 1:1 phenol chloroform mix and shake well to mix the contents. Centrifuge and transfer supernatant to another tube and precipitate DNA with an equal volume of isopropanol.
3. Pellet DNA at 13,000*g* in a microfuge centrifuge. Wash pellet with 70% ethanol and dissolve in 10 m*M* Tris and 1 m*M* EDTA buffer.
4. The DNA is analyzed either by PCR or by Southern-blot analysis.

3.6. Preparation of ES Cells for Aggregation with Morula

1. Thaw ES cells in which the gene is knocked out from the frozen 24-well plates and transfer onto 60-mm plates with feeder layer.
2. Change medium daily and harvest cells after 48 h or as soon as the ES cells are confluent by adding 1–2 mL of trypsin/EDTA followed by incubation for 5–8 min at 37°C. Add 10 mL of ES medium .
3. Pipet cells up and down several times to break the lumps. The cells should be dispersed as individual cells.
4. Transfer cells to a 50-mL tube followed by centrifugation at 1000*g* for 5 min.
5. Resuspend in 10 mL of ES medium and allow to stand for 10 min on ice for feeder cells to settle down.
6. Collect top 5 mL of the cells for the preparation of chimeras.
7. The remaining cells are cultured again on new PF plates for use on following day and a part of the cells are frozen in liquid nitrogen for future use.

3.7. Preparation of Vasectomized Males

1. Anaesthetize adult 6- to 8-wk-old male mice with avertin. Make a horizontal incision in the abdominal region at the level of hind legs. Pull out testis on each side along with the vas deference.
2. Cut about half cm of the vas deference and push organs back into the peritoneal cavity, followed by closing the skin with the wound clips. The testis descend into the scrotal sac automatically.
3. Allow 5–7 d for mice to recover and then mix with the females.

4. Check females daily for the vaginal plug which is formed from the coagulation of proteins secreted in the semen. The plugs generally fall off after 12–14 h of mating. About 30 vasectomized males are sufficient to provide three to four pseudopregnant females every day.

3.8. Preparation of Pseudopregnant Females

1. Mix two to three CD1 female mice with vasectomized males.
2. Check plugs the next day.
3. Use 2.5-d pseudopregnant females for blastocyst transfer (*see* **Note 9**).

3.9. Collection of Morulas

1. Superovulate 8–10 FVB/N or C57BL6 female mice by injecting 5 IU PMSG intraperitoneally. After 48 h, inject 5 IU HCG (*see* **Note 10**).
2. Mix female mice with the stud males overnight and check females for vaginal plugs the next morning. Use only the plugged females for embryo isolation. Unplugged females can be recycled after 3–4 wk.
3. The next morning separate plugged females from the vasectomized males as foster mothers (*see* **Note 11**).
4. Prepare the following 60-mm Petri dishes with embryo culture medium and incubate at 37°C for about 1 h.
   First Petri dish—approx 5 mL medium.
   Second Petri dish—approx 8–10 mL of medium.
5. Sacrifice pregnant females by cervical dislocation and flush abdominal arca with 70% ethanol. The females are sacrificed after 2.5 d of mating for morula isolation and after 3.5 d for blastocyst isolation.
6. Open peritoneal cavity, cut out each uterine horn separately, and transfer into the first Petri dish.
7. Using a 3-mL syringe filled with medium, flush uterine horns individually in a fresh 60-mm Petri dish with about 1 mL medium per uterine horn.
8. Collect morulas or blastocysts with the help of an isolation pipet and wash in fresh medium.
9. Transfer embryos to a 35-mm Petri dish with fresh medium and culture in a CO_2 incubator at 37°C until further use.

3.10. Preparation of Plates for Aggregation of ES Cells and Morula

1. Transfer ES cells (**Subheading 3.4.**) to a 1.5-mL microcentrifuge tube.
2. Centrifuge at 1000 rpm in a microcentrifuge for 2 min at room temperature followed by washing two times in 1 mL of morula aggregation medium (*see* **Subheading 2.5., item 5**).
3. Resuspend cells in the same medium at a concentration of $1–2 \times 10^4$ cells/mL.
4. Transfer 100 µL of ES cell suspension over the microwells and cover with paraffin oil.
5. Allow cells to settle for 10 min at room temperature. About 15–20 cells settle down in each microwell.

3.11. Preparation of Zona-Free Morula and Aggregation with ES Cells:

1. While the cells are settling down in microwell plates, collect 30–50 healthy morulas in a minimum volume of medium.
2. Transfer embryos into a drop of 100 mL acid Tyrode solution (137 m*M* NaCl; 2.7 m*M* KCl; 0.5 m*M* $MgCl_2 \cdot 6H_2O$; 5.6 m*M* glucose; 1.6 m*M* $CaCl_2 \cdot 2H_2O$, and 0.4% polyvinylpyrrolidine pH 2.5) ***(14)*** in a 35-mm Petri dish. Up to 100 embryos can be processed in the same volume.
3. As soon as the zona pellucida is dissolved, add 1 mL of coculture medium to neutralize the acidic pH.
4. Wash cells in 5–7 mL of fresh aggregation medium (*see* **Note 12**).

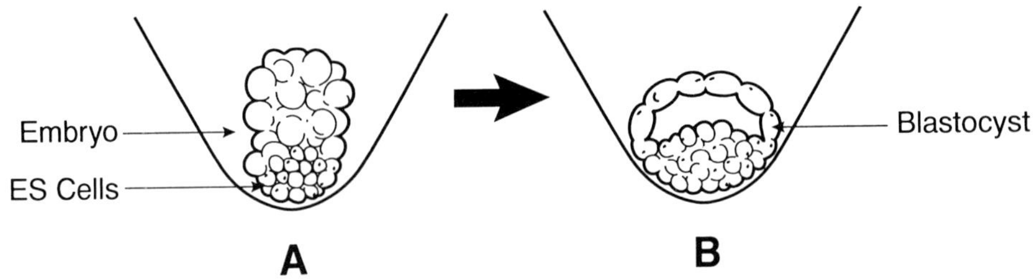

Fig. 4. Model for the integration of ES cells into a morula. The bottom of the microwell provides an area for tight cell–cell contact and simulates the formation of ES cell aggregates (**A**). Transfer of embryos over these cells establishes cell–cell–embryo contact and further growth of the embryo therefore results in integration of ES cells into the embryo (**B**).

5. Transfer one embryo each to individual microwells over ES cells (**Fig. 4**) and incubate overnight in the CO_2 incubator.
6. Collect blastocyts after overnight culture and wash in 5 mL of aggregation medium.
7. Store embryos in incubator until transfer into pseudopregnant mothers.
8. The animals with more than 90% chimerism can be generated reproducibly by this method and most of the animals display germline transmission.

3.12. Transfer of Embryos into Pseudopregnant Mothers

1. Anesthetize a 2.5-d pseudopregnant female by injecting 0.4–0.5 mL of 2.5% avertin and wipe the back with 70% ethanol.
2. Shave fur on the back and make an incision of about 1 cm in the mid region. Uterine horns on both sides can be reached from the same incision.
3. Place mouse under a dissecting microscope.
4. Cut off connective tissue under the skin very carefully and make an incision in the peritoneal membrane on one side and locate the pink-colored ovary under the membrane.
5. Pull out reproductive organs by pulling the periovarian fat that surrounds the ovary.
6. Mount uterine horn carefully onto a cotton applicator to provide support to the organs.
7. Place Petri dish with embryos under another dissecting microscope and collect 10 embryos in an embryo transfer needle using a minimum volume of medium. A few air bubbles preceding the embryos help to control the flow (*see* **Note 13**).
8. With a 25-gage needle, make a hole near the tip of the uterus. Keeping the hole in view, insert the transfer needle into the uterus and, slowly deliver the embryos inside.
9. Remove cotton applicator and with the help of a new applicator, push organs back into the peritoneal cavity.
10. Close the opening with a surgical suture followed by clamping the skin with wound clips. If the cut in the membrane is not very large, the suturing may not be necessary.
11. Place animal under a lamp until recovery. Pups are born after 16–17 d. After 7–10 d the degree of chimerism can be judged by the coat color (**Fig. 5**).

3.13. Germline Transmission

1. Mate chimeric animals with wild-type females at the age of 7–8 wk (*see* **Note 14**). The germline transmission is obvious from the coat color of the newborns.
2. Check for coat color after 7–10 d. Typically, the crossing of a chimeric male with a C57BL6 female will produce a progeny that has an agouti coat color, which represents the contribution of ES cells to the germline.
3. At the age of 3 wk a piece of tail is excised to isolate DNA.

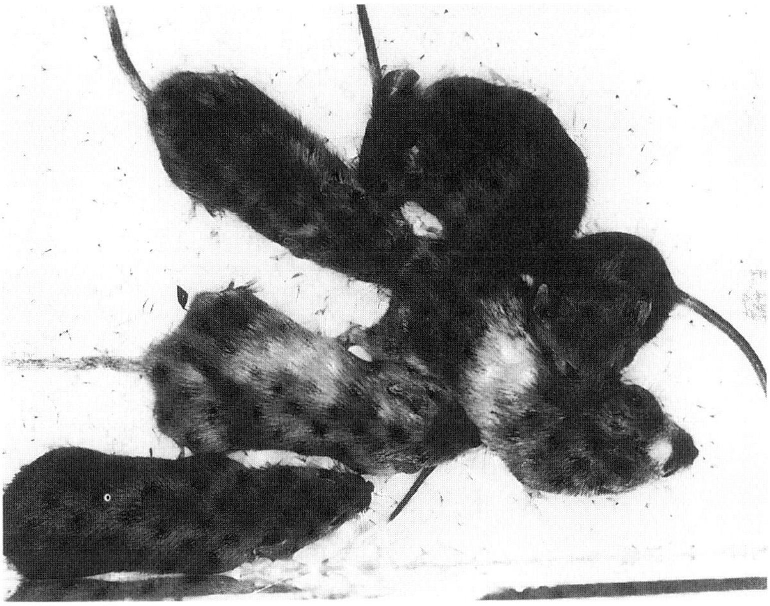

Fig. 5. Chimeric animals prepared from aggregation of FVB/N embryos (albino) with R1 cells derived from 129SV (agouti) strain of mice. Three chimeras displayed almost complete ES-cell-derived phenotype, whereas the other three were between 70% and 90% chimeric.

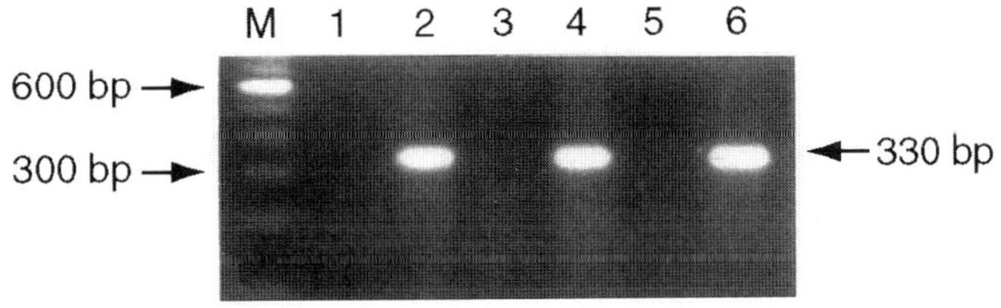

Fig. 6. PCR on DNA isolated from the progeny of chimeric animal crossed with a wild-type FVB/N female. The DNA was amplified using primers neo 2 and neo 3, specific for the neomycin gene shown in **Fig. 2**. The DNA from three pups amplified a band of 330 bp, indicating the germline transmission of the mutant allele.

4. Confirm germline transmission either by polymerase chain reaction PCR or by Southern analysis.
5. Progeny obtained in the first cross are heterozygous for the mutant allele, therefore breeding of two heterozygotes will generate homozygous progeny in which both the alleles are disrupted. (*see* **Note 15**).

3.14. Analysis of Animals

1. Cut a small piece of tail of a 2- to 3-wk-old animals and isolate total DNA ***(13)***.
2. Mince tail tissue and digest in lysis buffer (50 m*M* Tris-HCl pH 8.0, 100 m*M* EDTA, 100 m*M* NaCl, 1.0% sodium dodecyl sulfate (SDS), and 35 mg/mL proteinase K) for 8–10 h.
3. Extract DNA with phenol:chloroform followed by precipitation with ethanol.
4. Dissolve DNA in TE buffer and analyze either by Southern blot ***(13)*** or by PCR analysis. (**Figs. 5** and **6**, respectively).

4. Notes

1. Although microinjection of ES cells into blastocysts is a quite efficient method, it has many limitations, such as it requires specialized training and skills and also needs many expensive and sophisticated instruments. The morula aggregation provides an alternate method to prepare the chimeric animals, which does not require any major equipment.
2. The function of PMSG is equivalent to the follicle-stimulating hormone, whereas HCG serves the function of a leutinizing hormone.
3. The ES cells are available as established cell lines from several different laboratories (e.g., R1 cells from Dr. A. Nagy from Canada and J1 from Dr. Rudolph Jaenisch, MIT, MA). The cells are also commercially available (RW4 cells from Genome Systems Inc., St. Louis, MO). The lines are established from the inner cell mass (ICM) of a blastocyst. These cells are pleuripotent and can be maintained in the undifferentiated state indefinitely. When introduced into a host blastocyst, ES cells can contribute to the formation of all organs of the developing embryo, including testis and ovary. The microinjected cells integrate into the ICM of the host blastocyst and contribute to the formation of different tissues. The animal born from such embryo is chimeric (i.e., contains tissue formed from microinjected cells as well as cells from the host blastocyst). The transmission of the gene to the next generation occurs only if the microinjected cells contribute to the germline.

   The ES cells cultures are carried out on a layer of mitotically inactive fibroblast feeder cells in ES cell culture medium. The pleuripotency of cells can be maintained by adding lymphocyte inhibitory factor (LIF) in the medium which prevents differentiation of ES cells. The fibroblasts feeder cells secrete LIF. In general, the ES cells require about 2000 IU of LIF/mL. The feeder cells provide about 1000 IU and, therefore, the cells are supplemented by exogenous LIF which is commercially available from Gibco-BRL.
4. For the micromanipulation of embryos for aggregation procedure, only one type of micropipets is required, the embryo isolation and transfer needle. The internal diameter of the micropipet is about 100–120 μ*M*. The Glass capillaries for the micropipets are washed overnight in 100% ethanol and rinsed with dH_2O followed by heating at 180°C.
5. The shape of the microwell is such that about 15–20 cells settle at the bottom to from an aggregate. The denuded embryo therefore is placed over the cells.
6. Radiolabeled compounds are health hazards; therefore, they must be handled behind a thick protective plastic shield.
7. The feeder layer cells are mitotically inactivated before preparing feeder plates. The cells are treated with mitomycin C, 10 μg/mL, for 2 h. The cells are washed extensively in PF culture medium and can be used directly to prepare feeder plates. Alternatively, the cells are irradiated at 4000–6000 rads and aliquoted as 1 × 106 cells and stored in liquid nitrogen.
8. For the ES cell culture, the medium must be changed every day and the cells must be splitted after 48 h otherwise the cells tend to differentiate.
9. To ensure the availability of 4–5 foster mothers, generally 30 males are sufficient. The day before the collection of foster mothers, approx three CD1 females 5–6 wk of age are mixed with males. The foster mothers are separated everyday by screening the females for vaginal plugs. The females can be allowed to stay with the males during the period the foster mothers are required. However it is recommended to separate females from males over the weekends and mix again when the experiment is to be resumed. Discard females that are more than 12-wk-old and are fat. The efficiency of pregnancy decreases with age. The unused females can be recycled.
10. Female mice are superovulated to obtain synchronized embryos. Usually, 8–10 female mice are sufficient to obtain 30–50 good compact morulas.
11. For the transfer of blastocyts, the foster mothers are generally collected a day later so that the blastocysts are transferred to 2.5-d pseudopregnant females, which is required for the embryo to adjust to the mother's environment.

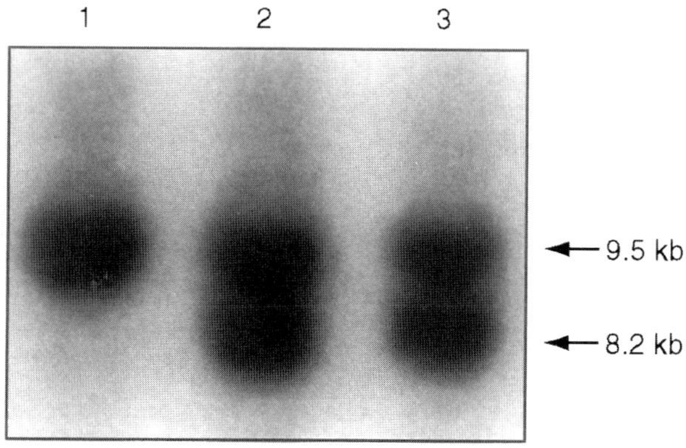

Fig. 7. Total DNA isolated from three independent ES cell clones was analyzed by Southern-blot analysis. The clones were targeted to introduce a null mutation in the gene for collagen type II (COL2A1). The normal allele generates a 9.5 kb band with the COL2A1 specific probe after digestion with *Bam*HI restriction enzyme. A second band of 8.2 kb is produced from the targeted allele. The equal intensity of both the bands represent that one allele is disrupted. Similar pattern was observed when the DNA from the progeny was analyzed.

12. For ES cells to integrate with the embryo, the zone a pellucida must be removed from the embryo. The tyrode solution is highly acidic, therefore it is extremely important that the embryo be exposed for minimum amount of time. The embryos are constantly watched under a steromicroscope during the treatment. It is more helpful to have a pipet full of medium so that as soon as the zona pellucida is dissolved, the tyrode solution is diluted with the fresh medium. After the removal of zona pellucida, the embryos tend to stick to each other, a high concentration of BSA (10 mg/mL) in the aggregation medium prevents adhesion of denuded embryos to each other.
13. Generally, 10 embryos are transferred to each female in 1 uterine horn only. On the average, the efficiency of implantation is about 50%; therefore, five pups in the one uterine horn is an appropriate number. However, the extra embryos can be transferred to the other uterine horn, making a similar incision in the other side of the peritoneal cavity.
14. Almost all of the established ES cell lines are of male origin, therefore the chimeric animals are mostly males. This provides an advantage because males can be mated with several females to obtain positive progeny more efficiently and quickly.
15. If the targeted gene is essential for the development, the homozygous progeny will show either developmental abnormalities or lethal phenotype after birth.

References

1. Bradley, A., Evans, M., Kaufman, M. H. and Robertson, E. (1984) Formation of germ-line chimeras from embryo-derived teratocarcinoma cell lines. *Nature* **309,** 255,256.
2. Capecchi, M. R. (1989) The new mouse genetics: altering the genome by gene targeting. *Trends Genet.* **5,** 70–76.
3. Evans, M. J. and Kaufman, M. H. (1981) Establishment in culture of pleuripotential cells from mouse embryos. *Nature* **292,** 154–56.
4. Martin, G. R. (1981) Isolation of a pleuripotent cell line from early mouse embryos cultured in medium conditioned by teratocarcinoma stem cells. *Proc. Natl. Acad. Sci. USA* **78,** 7634–7638.
5. Robertson, E. J. (1987) in *Teratocarcinomas and Embryonic Stem Cells: A Practical Approach* (Robertson, E. J., ed.), IRL, Oxford, UK, pp. 71–112.

6. Hasty, P., Crist, M., Grompe, M., and Bradley, A. (1994) Efficiency of insertion versus replacement vector targeting varies at different chromosomal loci. *Mol. Cell. Biol.* **14,** 8385–8390.
7. Mansour, S. L., Thomas, K. R., and Capecchi, M. R. (1988) Disruption of the proto-oncogene int-2 in mouse embryo-derived stem cells: a general strategy for targeting mutations to non-selectable genes. *Nature* **336,** 348–352.
8. Mortensen, R. (1993) In *Current Protocols in Molecular Biology, Vol. 1.* (Susubel, F. M., et al., eds.), Wiley, New York, pp. 9. 15. 1–9. 17. 3.
9. Wood, S. A., Pascoe, W. S., Schmidt, C., Kemler, R., Evans, M. J., and Allen, N. D. (1993) Simple and efficient production of embryonic stem cell-embryo chimeras by coculture. *Proc. Natl. Acad. Sci. USA* **90,** 4582–4585.
10. Nagy, A., Gocza, E., Diaz ,E. M., Prideaux, V. R., Ivanyi, E., Markkula, M., and Rossant, J. (1990) Embryonic stem cells alone are able to support fetal development in the mouse. *Development* **110,** 815–821.
11. Nagy, A., Rossant, J., Nagy, R., Abramow-Newerly, W., and Roder, J. C. (1993) Derivation of completely cell culture-derived mice from early-passage embryonic stem cells. *Proc. Natl. Acad. Sci. USA* **90,** 8424–8428.
12. Khillan J. S. and Bao, Y. (1997) Preparation of animals with a high degree of chimerism by one-step coculture of embryonic stem cells and preimplantation embryos. *BioTech* **22,** 544–549.
13. Hogan, B., Beddington, R., Costantini, F., and Lacy, E. (1994) *Manipulating the Mouse Embryo.* Cold Spring Harbor Labs., Plainview, NY.

45

Conditional Gene Knockout Using Cre Recombinase

Yunzheng Le and Brian Sauer

1. Introduction

The directed introduction of null mutations into defined genes has proven invaluable in elucidating gene function in a variety of experimental organisms. In the last decade or so this approach has been extended to mice *(1)* by the combined use of homologous recombination in murine embryonic stem (ES) cells to precisely target a mutation to a desired gene and subsequent derivation of mice carrying the targeted gene alteration from the genetically manipulated ES cells (e.g., by injection of gene-modified ES cells into blastocysts with subsequent germline transmission). In most instances null, or knockout (KO), mutations have been generated in mice by either simple insertion of a *neo* selectable marker in the target gene or *neo* insertion coupled with deletion of a critical region of the target gene. Targeted null mutations in a gene of interest, however, can lead to embryonic lethality in mice, thus obscuring the particular role of that gene in a target tissue or in the adult.

Site-specific recombination strategies allow circumvention of the problem of embryonic lethality by directing gene ablation in a spatially and temporally controlled manner. Because the Cre recombinase of phage P1 catalyzes efficient excisive recombination in mammalian cells, it has been become a useful tool for generating a conditional KO *(2,3)*. In addition, Cre-mediated excision has been useful both for targeted activation of genes in transgenic mice and for elimination of the selectable drug marker that is necessarily left in the genome after homologous gene targeting in ES cells *(3,4)* (*see* **Note 1**).

The 38 kDa Cre recombinase catalyzes DNA recombination between specific 34-bp sequences called *loxP* *(5)*. This site exhibits two inverted 13-bp repeats and a central asymmetric 8-bp core region that confers an overall directionality to the site (**Fig. 1**, *see* **Note 2**). Cre-mediated recombination between two directly repeated *loxP* sites on a DNA molecule results in excision of DNA between the two *loxP* sites. Mutant sites, such as *loxC2*, having base changes in the outer 4 bp of one of the inverted repeats (positions 1–4 or 10–13) are also recognized by Cre recombinase *(6)*. The *loxC2* site is a useful alternative to *loxP* when the site will be placed into transcribed sequences because it decreases stability of the 13-bp inverted repeat-derived hairpin in RNA. Such hairpins in the 5' untranslated RNA region can diminish translation.

From: *Methods in Molecular Biology, Vol. 136: Developmental Biology Protocols, Vol. II*
Edited by: R. S. Tuan and C. W. Lo © Humana Press Inc., Totowa, NJ

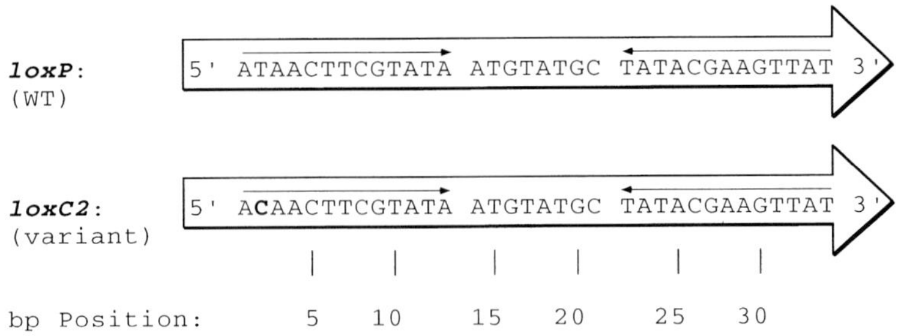

Fig. 1. Recombination sites for Cre recombinase. The wildtype *loxP* and variant *loxC2* sites are shown, with base pair (bp) positions indicated below. The 13-bp inverted repeats are indicated by the thin arrows above the sequence. The *loxC2* site differs from the wt site by replacing the T at position 2 with a C, and is named accordingly. By convention, the sense of the overall directionality of the *lox* site is as indicated by the large arrow. Note that *loxG33* would be functionally equivalent to *loxC2*, because of symmetry.

1.1. Strategies for Genomic Manipulation

Conditional gene deletion allows assessment of a gene's function in a target tissue without disturbing expression of that gene in nontarget tissues. Two components are required: a target mouse carrying a gene-modified allele of the gene to be ablated, and a *cre* transgenic mouse that expresses Cre under the control of a promoter with the desired spatial and temporal pattern of expression. Exact placement of *lox* sites in the target gene will depend both on the type of deletion event desired and on constraints imposed by the structure of the target gene. An example of a target-gene modification strategy in ES cells for the generation of a conditional knockout is shown in **Fig. 2A** (*see* **Note 3**). The *neo* selectable marker, flanked by two directly repeated *lox* sites (a *lox*^{2} *neo* cassette), is placed at one of the deletion endpoints (shown here in the first intron) and a third *lox* site is placed at the other deletion endpoint (shown here in the 5' leader before the first ATG in exon I) (*see* **Note 4**). Standard homologous recombination in ES cells is used to modify the target-gene locus (shown here by a double crossover event). Because *neo* could interfere with the correct expression of the target gene (depending on its placement in a particular strategy), it may be prudent to remove it by a limited Cre-mediated recombination event that removes the *neo* interval but leaves intact the genomic interval that is to be later targeted for deletion. Removal of *neo* can be effected in ES cells by transient transfection with a Cre-expressing construct.

As shown in **Fig. 2B**, conditional gene KO is accomplished by the mating of two animals, one having the *lox*-modified target locus (most conveniently a homozygote) and the other carrying the *cre* transgene and, in addition, one copy of either a null allele in the target locus or a copy of the *lox*-modified target locus. Cre-mediated recombination then excises the target gene from the genome in a manner that reflects the (tissue-specific) pattern of expression of the *cre* gene. Penetrance of expression of the *cre* transgene will govern the ratio of deletion vs nondeletion on a per cell basis in the target tissue (*see* **Note 3**).

A similar mating strategy is used for targeted activation of a transgene or endogenous gene in a tissue-specific, developmental or inducible fashion (**Fig. 3**). A "STOP"

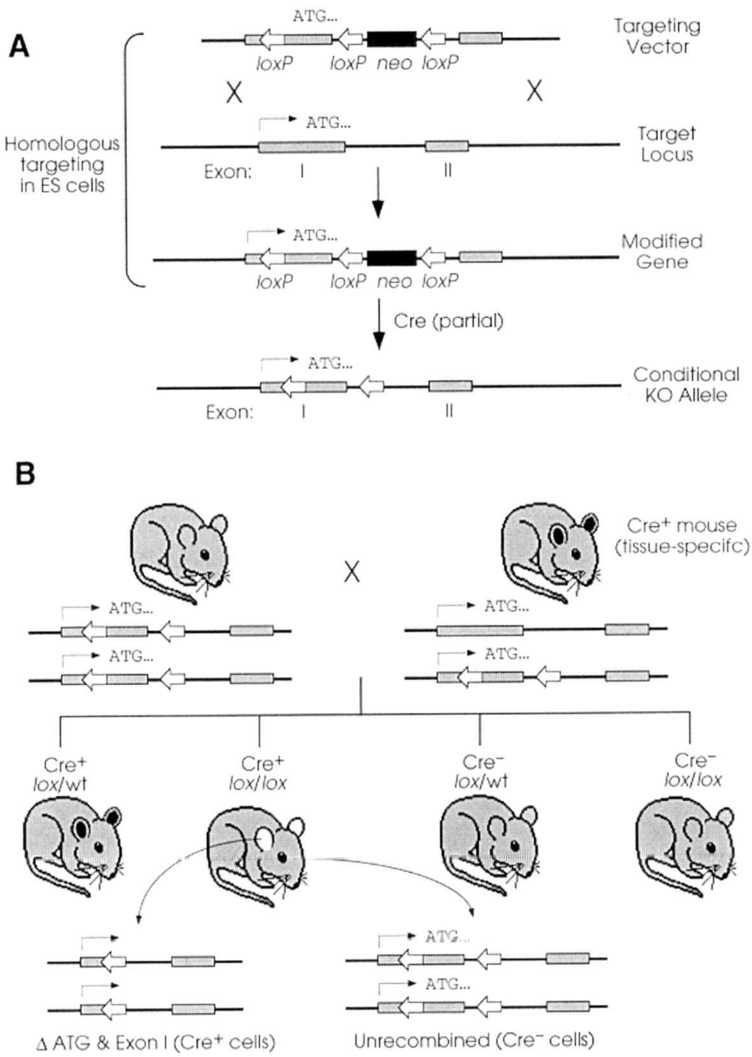

Fig. 2. A conditional KO strategy. **(A)** Targeting of the genomic locus by homologous recombination generates a modified gene that carries the correctly placed *lox* sites (indicated by white arrows) and the *neo* selectable marker (black rectangle). A second transfection step with a *cre* expression plasmid can be used to give partial excision. ES cells with the desired genomic structure are injected into blastocysts to give chimeric founders. **(B)** Mating strategy for a conditional KO. The desired double "transgenic" inherits a conditional allele from one parent and the *cre* transgene and the second copy of the conditional allele from the other. Note that the *cre* transgenic parent used in this cross is the product of a previous cross of a *cre* transgenic with the conditional knockout mouse so that it is both heterozygous at the target knockout locus and positive for the *cre* transgene. Tissue-specific Cre expression is shown here by the black ears. In the Cre^+ mouse homozygous for the conditional allele, productive recombination in the ear is represented by the white ears.

cassette flanked by two directly repeated *loxP* sites (a lox^2 STOP cassette) is placed between the promoter and the gene to be activated ***(4)***. STOP is designed to block gene

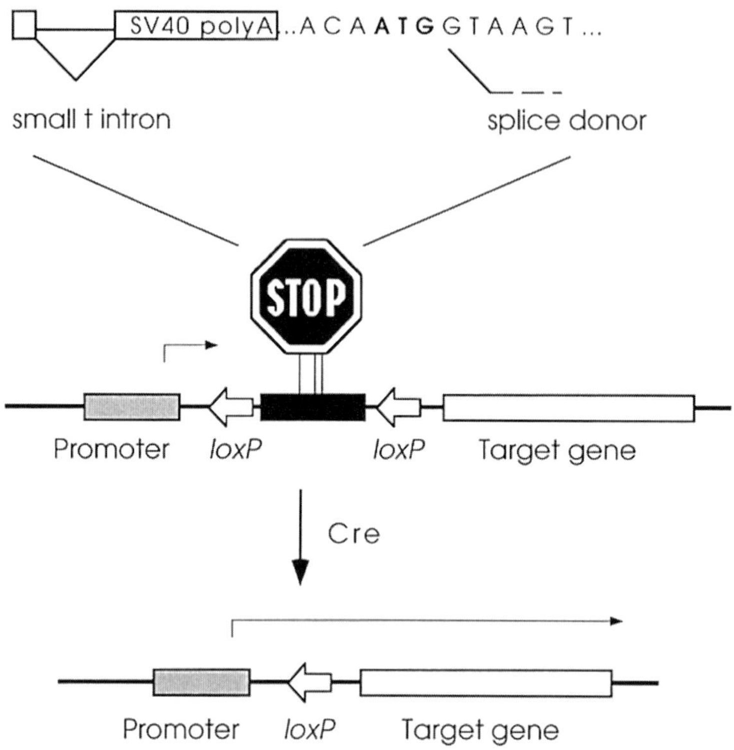

Fig. 3. Recombinational activation of gene expression. A synthetic STOP sequence (GenBank Accession No. U51223), is flanked by two directly repeated *lox* sites (white arrows) and is placed between a promoter and the target gene to be activated. Cre expression removes the STOP signal to allow target gene expression, leaving behind a single 34-bp *lox* site.

expression and consists of a stuffer region (from the yeast *HIS3* gene), the SV40 polyadenylation region, and errant optimized ATG translational start and splice donor signals. Cre-mediated excision removes STOP thus permitting target-gene expression under the control of the adjacent promoter. Two animals are required: one expressing Cre with the desired spatial and temporal pattern and the other either carrying a lox^2 STOP-equipped transgene or, alternatively, an animal modified by homologous gene-targeting in ES cells to carry the lox^2 STOP cassette in the desired endogenous gene. Mating of these two animals results in activation of the target gene in those cells that both express (or have expressed in a progenitor cell) Cre recombinase and are capable of expressing the target gene, as determined by the specificity of the promoter used in the *lox* target strain.

To target recombination to a particular tissue, Cre should be specifically expressed in the target tissue. This can be achieved by making a transgenic mouse with *cre* under the control of a promoter with the desired expression characteristics. Alternatively, the *cre* gene can be targeted to be under the control of an endogenous promoter in the genome by using homologous recombination in ES cells (a "knock-in"). The pattern of Cre mouse expression should be checked using a Cre-specific antibody *(2,7)*, rtPCR, or *in situ* RNA hybridization. This is important because some promoters are susceptible to position effects, and resulting founder lines may express the transgene in a mosaic fashion in the target tissue or misexpress the transgene in an unwanted tissue. Gene ablation (or gene activation) will be difficult to achieve in the majority of cells in a target tissue if Cre is expressed in a variable and mosaic fashion. *cre* knock-ins may thus be preferred unless the promoter used in a transgenic strategy has already been established to behave with the desired expression pattern in a transgenic (e.g., by monitoring expression of a *lacZ* reporter gene).

2. Materials

2.1. Detection of the cre Transgene by PCR

1. DNA oligonucleotide primers are stored at –20°C at a stock concentration of 3 μ*M*.
   Sense: 5' AGG TGT AGA GAA GGC ACT TAG C 3'
   Antisense: 5' CTA ATC GCC ATC TTC CAG CAG G 3'.
2. 10X PCR buffer: 500 m*M* KCl, 100 m*M* Tris-HCl pH 8.3, 20 m*M* $MgCl_2$, stored at –20°C.
3. dNTP's: 12.5X stock solution, 2.5 m*M* of each in H_2O, stored at –20°C.
4. AmpliTaq polymersase (Perkin-Elmer, Norwalk, CT) at 250 U/50 μL, stored at –20°C.
5. Perkin Elmer 9600 thermocycler, or equivalent, and 0.2 mL PCR reaction tubes.

2.2. Antibody Detection of Cre Protein in Cells

1. Anti-Cre mAb 7.23 (Berkeley Antibody Co., Richmond, CA), 1 mg/mL.
2. Formaldehyde (Electron Microscopy Sciences, Ft. Washington, PA), 16% solution.
3. Normal goat serum (Life Technologies, Inc., Gaitherburg, MD).
4. FITC-conjugated goat anti-mouse IgG1 polyclonal serum (Southern Biotech., Birmingham, AL).
5. Phosphate buffered saline (PBS); PBS/B (PBS + 0.5% BSA + 0.01% NaN_3); PBS/B + S (PBS/B + 0.5% saponin). Filter with 0.2-micron filter.
6. Epifluorescence microscope for FITC detection.

2.3. Gene Popout in ES Cells by GFPcre

All cell culture reagents are from Life Technologies, Inc. unless otherwise indicated.

1. Culture dishes: Falcon 3002 (6-cm).
2. Phosphate buffered saline without calcium and magnesium.
3. 0.2% Gelatin (Sigma) in PBS, autoclaved.
4. Irradiated mouse embryonic fibroblasts (EFB, made in the laboratory).
5. ES medium: DMEM suplemented with 20% fetal calf serum (FCS, Hyclone, Logan, UT), 1000 U/mL ESGRO, 2 m*M* L-glutamine, 0.1 m*M* of nonessential amino acids and 0.1 m*M* of β-mecaptoethanol, with 50 U/mL of penicillin, and 50 μg/mL of streptomycin.
6. 0.1% trypsin in PBS with 0.05 m*M* EDTA.
7. Plasmid pBS500, EF1a-GFP*cre* ***(8)***.
8. 2.5 *M* $CaCl_2$.
9. 50 m*M* BES solution: 50 m*M*, *N,N*-bis[2-hydroxyethyl]-2-aminoethanesulfonic acid, 280 m*M* NaCl, and 1.5 m*M* Na_2HPO_4, adjust pH to 6.96 with HCl.
10. Epifluorescence microscope for FITC (GFP) detection.

3. Methods

3.1. Detection of the cre Transgene by PCR

After identification of *cre* transgenic animals by Southern blotting, it is convenient to monitor *cre* transgene transmission by PCR, using standard genomic "tail DNA" ***(9)***.

PCR amplification yields a 411 bp product diagnostic for *cre* (GenBank Accession No. X03453).

1. In a 0.2-mL reaction tube add: 50–500 ng of genomic DNA, 5 μL of 10X PCR buffer, 4 μL of 10X dNTPs, 5 μL of each sense, and antisense primers, and H_2O to 50 μL total volume.
2. Denature 5 min at 94°C.
3. Add 0.5 μL of AmpliTaq polymerase.
4. Perform 25 or 30 cycles of amplification: 30 s at 94°C, 30 s at 63°C, 1 min at 72°C.

3.2. Antibody Detection of Cre Protein in Cells

Indirect immunofluorescence with anti-Cre specific antibody is used to assess tissue specific expression and also the level of mosaicism of expression in a population of cells. Detection by antibody quickly identifies Cre-expressing animals that may be suitable for further analysis *(7)*.

1. Cre-expressing cells are washed once with PBS and fixed with PBS + 2% formaldehyde for 20 min at RT. All steps are at RT unless otherwise indicated.
2. Wash twice with PBS and once with PBS/B.
3. Permeabilize with PBS/B + S for 6 min.
4. Block with PBS/B + 5% normal goat serum for 30 min.
5. Incubate with anti-Cre mAb 7.23 (diluted 1/100 in PBS/B + S) for 15 min. Incubate longer if necessary.
6. Wash three times with PBS/B + S.
7. Incubate with FITC-conjugated goat antimouse IgG1 polyclonal serum (10 μg/mL in PBS/B + S containing 7% normal goat serum) for 20 min.
8. Wash three times with PBS/B + S.
9. Examine by epifluorescent microscopy.

3.3. Gene Popout in ES cells by GFPcre

After targeted modification of a locus in ES cells by homologous recombination, removal of the selectable marker (for example *neo*) may be desired in order to preclude interference by the marker gene on correct gene expression at the target locus. Incorporation of a lox^2 *neo* cassette into the homologous targeting vector allows subsequent Cre-mediated removal of the *neo* gene from the targeted locus. This step is facilitated by use of a functional fusion between Cre and an enhanced fluorescent derivative of the green fluorescent protein (GFP) of *Aequorea victoria* ***(10)***. Transient transfection with the GFP*cre* fusion construct pBS500 *(8)* results in cells that are simultaneously GFP^+ and Cre^+ (and, hence, committed to excision of *neo*). Because the transfection efficiency of ES cells is often low, it is convenient to enrich for productively transfected ES cells by using fluorescence (**Fig. 4**). We have found that gene transfer by calcium phosphate co-precipitation with DNA ***(11)*** to give somewhat more routinely efficient transient transfection than electroporation. Alternatively, standard electroporation of ES cells ***(12)*** with pBS500 (EF1a-GFP*cre*) can be used.

1. Plate 2×10^6 ES cells on a gelatinized 6-cm dish seeded with 2×10^6 irradiated EFB two days before the transfection. Change medium next day and also on the second day, 3 h before transfection.
2. Combine 9 μg pBS500 DNA with H_2O (total volume for DNA and H_2O is 165 μL), add 18.3 μL 2.5 *M* $CaCl_2$ and mix. Add the DNA solution dropwise into 183 μl BES solution, mixing gently and continuously. Incubate 20 min at RT.

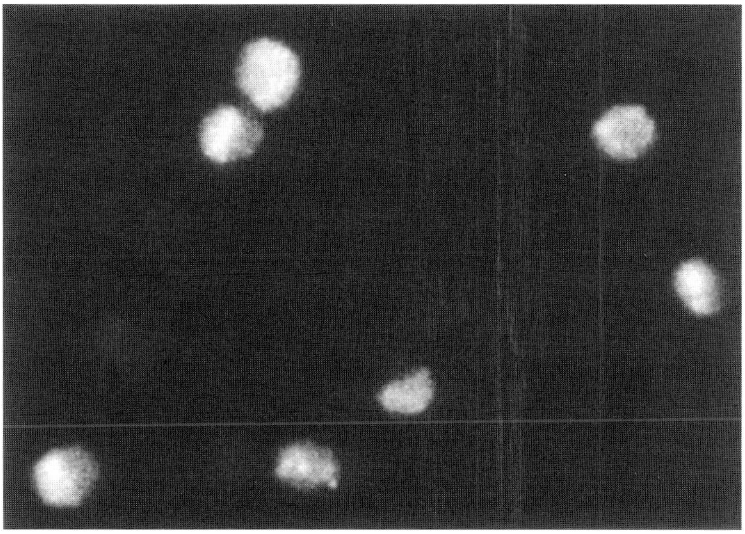

Fig. 4. GFPCre expression in ES cells. ES cells were transiently transfected with pBS500 (EF1α-GFP*cre*) by $CaPO_4$ coprecipitation. Two days later cells were trypsinized and examined for fluorescence.

3. Wash the dish of ES cells with 5 mL PBS, add 2 mL 0.1% trypsin solution and incubate 2 min at RT. Remove the trypsin solution and incubate the dish at 37°C for 5 min. Add 4 mL medium to inactivate the trypsin. Count the cells and dilute the ES cells in the medium to 3×10^5 cell/mL. This ES cell suspension will be mixed with the DNA/$CaPO_4$ coprecipitate.
4. Mix the DNA/$CaPO_4$ coprecipitate solution with 5 mL of the ES cell suspension (1.5×10^6 ES cells total). Plate the mixture on a 6-cm gelatinized dish and incubate overnight at 37°C.
5. Wash the transfected ES cells with DMEM and replace with fresh medium 16 h after transfection.
6. Fluorescence should be observed in about 2–7% of the GFP*cre*-transfected cells 2 d after transfection. After trypsinization, fluorescent (Cre^+) ES cells can be either sorted manually at this time with a micromanipulator, or by FACS.

4. Notes

1. Genomic manipulation strategies with the distantly-related DNA recombinase FLP (from yeast 2-μ circle) are conceptually similar to those for Cre recombinase. Somewhat greater variability in efficiency of recombination in mammalian cells has been reported for FLP ***(13,14)***.
2. There are two "ATG's" in the conventional orientation of *loxP* as shown in **Fig. 1**. However, in the reverse orientation there are <u>no</u> ATGs, hence the reversed orientation is preferred when inserting the *lox* sequence into the 5' RNA leader region of a gene.
3. In the conditional KO strategy, the *cre* transgenic parent in the "knockout activation" cross is heterozygous at the target KO locus for the conditional allele. Alternatively, the *cre* parent can be heterozyous at the target locus for a wt and a complete null. In this case, it should be verified that animals heterozygous for the null allele at this locus are phenotypically wt.
4. To evaluate the recombination potential of plasmids containing two *loxP* sites, it is convenient to transform the plasmid into the Cre-expressing *E. coli* strain BS591 [F^- *recA1 endA1 hsdR17* δ*lac*(*lacZYA-argF*)*U169 supE44 thi-1 gyrA96* (λ *imm434 nin5* X1-*cre*)].

Plasmid DNA isolated from this strain will have undergone Cre-mediated recombination and can be checked by a simple restriction digest ***(15)***. Alternatively, and more rigorously, *lox* regions should be sequenced in the final plasmid construction.

5. PBS + 0.1% Triton X-100 for 6 min also gives permeabilization of cells for indirect immunofluorescent detection of Cre.
6. Cre protein can be detected in target tissues by standard Western blotting or immunoprecipitation by using the anti-Cre mAb 7.23 at dilutions of 1/1000 and 1/150, respectively.
7. The pattern of Cre expression from a *cre* transgenic or knock-in animal may also be determined using the GFP*cre* gene. This potentially would allow direct detection of Cre but most likely would only be useful in situations in which the promoter used is known to give relatively strong expression (to allow direct detection of fluorescence) in the target tissue. Since only a single catalytic event is required for productive recombination in a cell, the actual amount of Cre required per cell is not high, and antibody detection is potentially more sensitive than reliance on fluorescence from GFPCre.
8. As an alternative to excision of the *loxP*-flanked *neo* gene in ES cells by transient transfection with a *cre* plasmid, ES cells can be injected directly into blastocysts to give a chimeric founder that is then mated with the EIIa-*cre* mouse ***(16)***. The EIIa promoter directs Cre expression only in fertilized zygotes and early embryos, and is not expressed post-implantation, thus allowing *neo* removal from the germline. In multi-*lox* transgene arrays, both partial and complete deletion events can be obtained that are then transmitted through the germline.
9. GFPCre localizes to the cell nucleus ***(8)*** since Cre carries an endogenous nuclear localization signal (unpublished).
10. Following sorting of ES cells transiently transfected with GFP*cre*, resulting colonies should be confirmed for excisive recombination by Southern or a suitable PCR assay. Most (≥80%) fluorescent cells will give rise to colonies carrying the desired *lox*-delimited deletion. DNA analysis will also confirm that the colony is not mosaic for both recombinant and nonrecombinant cells.

References

1. Gordon, J. W., Harold, G., and Leila, Y. (1993) Transgenic animal methodologies and their applications. *Hum. Cell* **6,** 161–169.
2. Sauer, B., and Henderson, N. (1988) Site-specific DNA recombination in mammalian cells by the Cre recombinase of bacteriophage P1. *Proc. Natl. Acad. Sci. USA* **85,** 5166–5170.
3. Gu, H., Marth, J. D., Orban, P. C., Mossmann, H., and Rajewsky, K. (1994) Deletion of a polymerase beta gene segment in T cells using cell type-specific gene targeting. *Science* **265,** 103–106.
4. Lakso, M., Sauer, B., Mosinger, J., B., Lee, E. J., Manning, R. W., Yu, S.-H., Mulder, K. L., and Westphal, H. (1992) Targeted oncogene activation by site-specific recombination in transgenic mice. *Proc. Natl. Acad. Sci. USA* **89,** 6232–6236.
5. Hoess, R. H., and Abremski, K. (1990) The Cre-*lox* recombination system, in, *Nucleic Acids and Molecular Biology, vol. 4.* (Eckstein, F. and Lilley, D. M. J., eds.), Springer-Verlag, Berlin, pp. 99–109.
6. Sauer, B., Whealy, M., Robbins, A., and Enquist, L. (1987) Site-specific insertion of DNA into a pseudorabies virus vector. *Proc. Natl. Acad. Sci. USA* **84,** 9108–9112.
7. Schwenk, F., Sauer, B., Kukoc, N., Hoess, R., Müller, W., Kocks, C., Kühn, R., and Rajewsky, K. (1997) Generation of Cre recombinase-specific monoclonal antibodies to characterize the pattern of Cre expression in *cre*-transgenic mouse strains. *J. Immunol. Meth.* **207,** 203–212.

8. Gagneten, S., Le, Y., Miller, J., and Sauer, B. (1997) Brief expression of a GFP*cre* fusion gene in embryonic stem cells allows rapid retrieval of site-specific genomic deletions. *Nucleic Acids Res.* **25,** 3326–3331.
9. Gendron-Maguire, M., and Gridley, T. (1993) Identification of transgenic mice. *Meth. Enzymol.* **225,** 794–799.
10. Heim, R., Cubitt, A. B., and Tsien, R. Y. (1995) Improved green flourescence. *Nature* **373,** 663–664.
11. Chen, C., and Okayama, H. (1987) High-efficiency transformation of mammalian cells by plasmid DNA. *Mol. Cell. Biol.* **7,** 2745–2752.
12. Ramírez-Solis, R., Davis, A. C., and Bradley, A. (1993) Gene targeting in embryonic stem cells. *Meth. Enzymol.* **225,** 855–878.
13. Feiring, S., Kim, C. G., Epner, E. M., and Groudine, M. (1993) An "in-out" strategy using gene targeting and FLP recombinase for the functional dissection of complex DNA regulatory elements: analysis of the β-globin locus control region. *Proc. Natl. Acad. Sci. USA* **90,** 8469–8473.
14. O'Gorman, S., Fox, D. T., and Wahl, G. M. (1991) Recombinase-mediated gene activation and site-specific integration in mammalian cells. *Science* **251,** 1351–1355.
15. Sauer, B., and Henderson, N. (1988) The cyclization of linear DNA in *Escherichia coli* by site-specific recombination. *Gene* **70,** 331–341.
16. Lakso, M., Pichel, J. G., Gorman, J. R., Sauer, B., Okamoto, Y., Lee, E., Alt, F. W., and Westphal, H. (1996) Efficient *in vivo* manipulation of mouse genomic sequences at the zygote stage. *Proc. Natl Acad. Sci. USA* **93,** 5860–5865.

46

Application of *Cre/loxP* in *Drosophila*

Site-Specific Recombination and Transgene Coplacement

Mark L. Siegal and Daniel L. Hartl

1. Introduction

The use of site-specific recombinases has revolutionized the genetic analysis of development and has made possible the precise engineering of genomes *(1,2)*. In *Drosophila*, the FLP/*FRT* system, introduced by Golic and Lindquist *(3)*, has been used (1) to generate genetic mosaics by mitotic recombination as well as by "flip-outs" *(3–5)*, and (2) to generate defined chromosomal rearrangements *(6,7)*. In yeast and mammalian cells, site-specific recombination has also been used to mediate targeting of exogenous DNA to genomic docking sites *(8)*. Although such targeted integration is by nature an inefficient process—as a result of the favoring of intramolecular over intermolecular recombination—this limitation has been overcome in these systems by the ability to introduce DNA into a large number of cells simultaneously and to select for rare integration events by chemical means. In *Drosophila*, such an approach is not currently available, although an approximation of targeted integration of exogenous DNA has been used, with varying efficiency, to mobilize *FRT*-flanked DNA already in the genome to a specific *FRT* target site elsewhere *(9)*.

An efficient, reliable targeted-integration system in *Drosophila* would reap enormous practical benefits. Because transgenes are subject to genomic position effects, independent integrations of the same transformation construct may differ quantitatively and qualitatively in their expression *(10–13)*. Thus, the ability to compare transgenes at the same position in the genome would increase the power of an experiment to detect subtle expression differences between the transgenes. Furthermore, some experiments absolutely require that transgenes be present at the same position, such as those investigating pairing or location-dependent phenomena—including transvection *(14)*, meiotic recombination *(15)*, and position-effect variegation *(16)*—and those seeking to trace allele frequencies over many generations in experimental populations so as to make inferences concerning the relative effects of the transgenes on reproductive fitness *(17)*.

In this chapter, we outline the use of a system we have developed to place pairs of transgenes at the same position in the *Drosophila* genome *(18)*. This system of

From: *Methods in Molecular Biology, Vol. 136: Developmental Biology Protocols, Vol. II*
Edited by: R. S. Tuan and C. W. Lo © Humana Press Inc., Totowa, NJ

transgene coplacement relies only on the efficient procedures of *P*-element-mediated germline transformation and the excision reactions of the FLP and Cre site-specific recombinases. The pair of transgenes is introduced on a single, specially-designed *P* element so that the action of FLP eliminates one transgene, whereas the action of Cre eliminates the other transgene. Thus, one may derive, from each insertion of the original *P* element, pairs of strains containing one or the other of the transgenes, in the same genomic position and orientation, and flanked by the same sequences. In addition to an outline of the coplacement method, we present, here, detailed information on the Cre constructs we have introduced into *Drosophila*. Because it has the same mode of action as FLP, Cre may be used in all applications for which FLP is appropriate. More importantly, as this chapter demonstrates, the use of Cre and FLP in concert can perform functions that neither one alone can.

2. Materials

2.1. Plasmids

1. The plasmid pP[wFl], shown in Fig. 1A, is a P-element transformation vector that contains two unique cloning regions in which to place each of the transgenes to be compared. In addition, the vector contains the mini-white selectable marker (19), positioned such that it is excised along with the desired transgene upon FLP or Cre recombination. Thus, each insertion of the original transgene-containing P[wFl] construct is identified by the white$^+$ phenotype it confers to otherwise w$^-$ flies, whereas desired excisions are identified by reversion of the phenotype to white$^-$.
2. The plasmid pP[SFl], shown in Fig. 1B, is a derivative of pP[wFl] in which the mini-white marker has been removed, leaving a unique SphI restriction site in its place. This vector is useful in cases where a selectable marker other than mini-white is desired, or where the transgenes themselves confer a selectable phenotype.

2.2. Fly Stocks Expressing Cre

1. The Cre-expressing transgene used for transgene coplacement and other applications in Drosophila is a chimeric construct, described in ref. 18, in which the Cre coding sequence is flanked by sequences derived from the active mariner transposable element Mos1 (20), which provides a promoter sequence as well as a translation initiation consensus sequence (21) and a polyadenylation signal sequence. Additionally, an hsp70 promoter lies upstream of the Mos1 promoter. The chimeric Cre construct was cloned into both the mini-white-containing P-element vector, pCaSpeR4 (19), and the mini-yellow-containing P-element vector, pCar-y (22), to produce P[w$^+$, cre] and P[y$^+$, cre], respectively. Despite carrying the same Cre construct, these Cre sources have different properties (see Note 1).
2. Because the Cre sources described above are highly expressed without heat shock (see Table 1 and Note 2; ref. 18), we mutagenized flies carrying a second-chromosome insertion of P[y$^+$, cre] to generate Cre sources with inducible expression. The resulting strains (see below and Note 2) are designated P[y$^+$, cre*].
3. Available stocks include the following (genotypes not specifically described here are in ref. 23; P[ry$^+$, hsFLP] is described in ref. 3):
   - i. insertions of *P[w$^+$, cre]* on each major autosome
   - ii. insertions of *P[y$^+$, cre]* on each major chromosome
   - iii. *y w, P[y$^+$, cre]; CyO/Sco*
   - iv. *y w, P[y$^+$, cre]; TM3 Sb/D*
   - v. *y w; CyO, P[w$^+$, cre]/Sco*

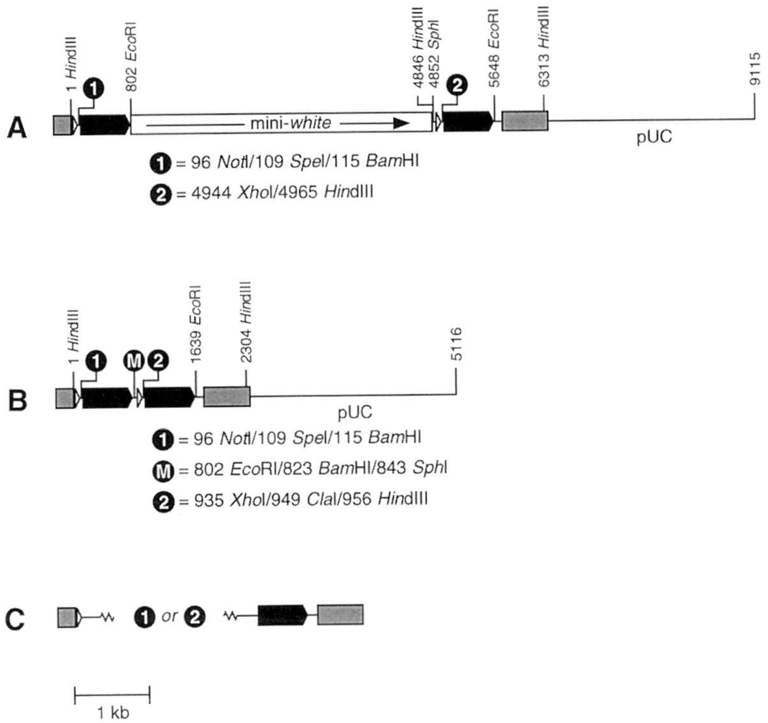

Fig. 1. Transgene coplacement plasmid vectors: **(A)** The 9.3-kb vector, p*P*[*wFl*], showing *P*-element ends (shaded boxes), *loxP* sites (open arrowheads), *FRT* sites (solid arrows), key restriction sites, and the two unique cloning regions (note that the *Hin*dIII site in region 2 is not unique). Numbering starts with the *Hin*dIII site upstream of mini-*white*, as it does for the pCaSpeR4 vector from which it is derived (GenBank accession number for pCaSpeR4 is X81645). **(B)** The 5.3-kb vector, p*P*[*SFl*], showing the same features as for p*P*[*wFl*]. In addition, a third unique cloning region, "M", is indicated, into which may be placed a marker of choice (note that the *Bam*HI site in region 1 is no longer unique). **(C)** Footprint of recombination of either *P*[*wFl*] or *P*[*SFl*] by either Cre or FLP. Recombination by Cre leaves the transgene cloned into region 2, flanked on one side by a *loxP* site and on the other side by an *FRT* site; recombination by FLP leaves the transgene cloned into region 1, flanked by the same sequences. Details of construction of p*P*[*wFl*] and p*P*[*SFl*] are available on our World Wide Web site at http://www.oeb.harvard.edu/hartl/lab/, or upon request from the authors.

vi. *y w*; *P*[y^+, cre^*]

vii. *y w*; *CyO*, *P*[y^+, cre^*]/*Sco* (remobilization of *P*[y^+, cre^*])

viii. *y w*; *TM6B*, *P*[w^+, *cre*]/*MKRS*, *P*[ry^+, *hsFLP*]

2.3. All molecular techniques, including cloning, Southern hybridization, and fluorescent in situ hybridization to polytene chromosomes are performed using standard reagents and protocols (24,25).

2.4. Flies are raised on standard cornmeal-molasses medium and at 25°C unless otherwise noted.

3. Methods

1. Cloning of transgene coplacement constructs: Introduce into the unique cloning sites of either p*P*[*wFl*] or p*P*[*SFl*] the pair of transgenes to be coplaced (*see* **Fig. 1** and **Notes 3** and **4**). The transgene introduced into cloning site 1 will be excised by Cre, and the transgene introduced into cloning site 2 will be excised by FLP.

Table 1
Germline Excision Frequencies of Various Cre and *loxP* Combinations

	loxP target			
	P[^*w*$^{+}$^]		*P*[^*a*>*w*$^{+}$^*m*>]	
Cre source	no h.s.	h.s.	no h.s.	h.s.
y w, *P*[*y*$^{+}$, *cre*]; *Cyo*/*Sco*	160/160 (100.0)	98/98 (100.0)	76/106 (71.7)	95/104 (91.3)
y w; *CyO*, *P*[*w*$^{+}$, *cre*]/*Sco*	345/345 (100.0)	†	88/117 (75.2)	127/130 (97.7)
yw; *P*[*y*$^{+}$, *cre*] [a]	298/298 (100.0)	258/258 (100.0)	171/233 (73.4)	149/157 (94.9)
y w; *CyO*, *P*[*y*$^{+}$, *cre**]/*Sco*	96/277 (34.7)	344/388 (88.7)	0/83 (0.0)	15/270 (5.6)
y w; *TM6B*, *P*[*w*$^{+}$, *cre*]/ *MKRS*, *P*[*ry*$^{+}$, *hsFLP*]	175/175 (100.0)	†	156/315 (49.5)	151/152 (99.3)

Note: Two *loxP* target transgenes, described in ***(18)***, were used to assay Cre activity. *P*[^*w*$^{+}$^] contains a mini-*white* gene flanked by *loxP* sites, whereas *P*[^*a*>*w*$^{+}$^*m*>] is a derivative of *P*[*wFl*] in which the *Adh* genes of *D. affinidisjuncta* (*"a"*) and *D. melanogaster* (*"m"*) are cloned into cloning sites 1 and 2 (for an explanation of notation, *see* **Note 8**). Each Cre stock was crossed separately with both *P*[^*w*$^{+}$^] and *P*[^*a*>*w*$^{+}$^*m*>] stocks, and these crosses were brooded once. The first brood was heat shocked (h.s.) as first-instar larvae for 1 h at 37°C, and the second brood was not heat shocked (no h.s.). Appropriate progeny genotypes were backcrossed to *y w* flies to assay germline excision frequencies, reported as the fraction of progeny in which the *loxP* target was excised (percentages are in parentheses). Those crosses exhibiting lethality upon heat shock (*see* **Note 2**) are marked by †.

*a*This is the second-chromosome insertion of *P*[*y*$^{+}$, *cre*] that was mutagenized to produce *P*[*y*$^{+}$, *cre**].

2. Transformation of constructs: Use standard means ***(26,27)*** to generate *P*-element-mediated germline transformants of the transgene coplacement constructs. By genetic segregation and molecular hybridization analyses, identify single-insertion lines that carry the construct on known chromosomes.
3. Selective elimination of transgenes: Pass the original transformed constructs through flies that express either FLP or Cre. If p*P*[*wFl*] is used, loss of white^{+} phenotype in the progeny of such flies will indicate excision of the desired sequences. Induction of FLP expression requires heat shock; generally it is sufficient to heat shock the first-instar larvae in glass vials for 1 h in a circulating water bath at 37°C ***(3)***. Depending on transgene size, heat shock for Cre expression may or may not be necessary (*see* **Table 1** and **Note 2**).
4. Creation of fly stocks to assay: Generate homozygous or balanced lines that carry the FLP- or Cre-mediated excision derivatives. For many applications it will be necessary to create lines that are also homozygous for mutations in the gene of interest. An example of such a crossing scheme involving coplaced *Alcohol dehydrogenase (Adh)* genes is shown in **Fig. 2**.
5. Confirmation of site-specific recombination: By molecular means, confirm that the desired recombination events have taken place. *See* Fig. 8 in Siegal and Hartl ***(18)*** for an example of fluorescent *in situ* hybridization analysis of coplaced *Adh* genes. As shown in **Fig. 3**, Southern hybridization may be used to confirm the identity of the unexcised transgene and to confirm that flanking genomic sequences are the same for each pair of excision

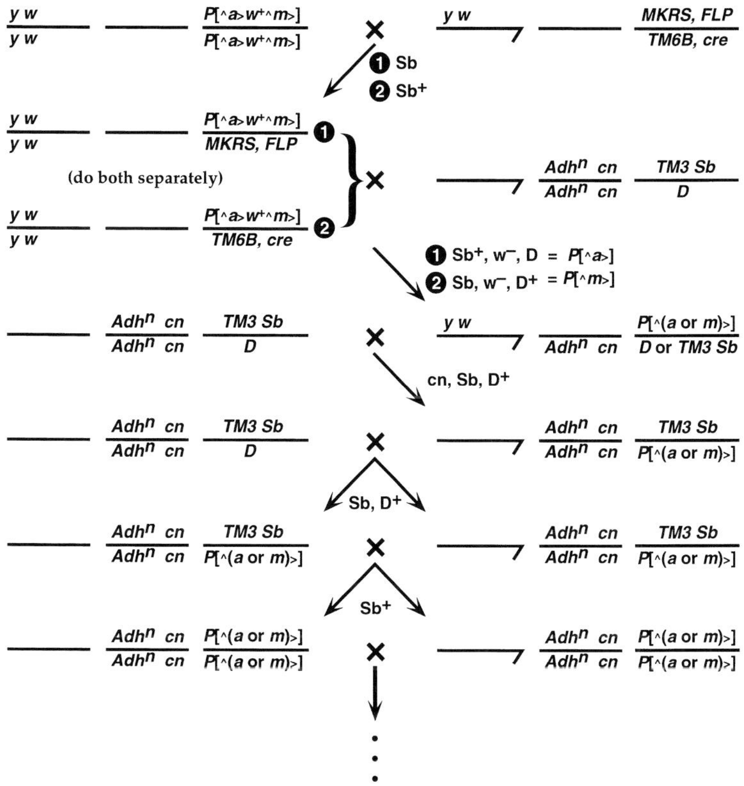

Fig. 2. Crossing scheme for coplaced *Adh* genes of *D. affinidisjuncta* ("*a*") and *D. melanogaster* ("*m*"): A third-chromosome insertion of the *Adh*-containing p*P*[*wFl*] derivative, *P*[^*a*>*w*$^+$^*m*>], is subjected to Cre or to FLP by crossing with males of the strain *y w*; *MKRS*, *P*[*ry*$^+$, *hsFLP*]/*TM6B*, *P*[*w*$^+$, *cre*] (here the *hsFLP* and *cre* constructs are abbreviated "*FLP*" and "*cre*"; for an explanation of the ">" and "^" notation, *see* **Note 8**). Males are used so as to avoid maternal-effect Cre expression in *FLP* progeny (*see* **Note 1**). Excision derivatives are identified in the next generation by loss of *w*$^+$. These derivatives are made homozygous on the third chromosome while the second chromosome is made homozygous for a null allele of *Adh*. Note that the final stocks have a standardized genetic background, as all major chromosomes are from known sources: X, Y, and second chromosomes from *Adhn cn*; *TM3 Sb*/*D* and third chromosomes from the *y w* stock originally used for transformation of *P*[^*a*>*w*$^+$^*m*>].

derivatives. PCR analysis using primers specific to each transgene may also be a useful means of characterizing excision derivatives.

6. Gene-expression assays: Assay gene expression in the pairs of lines, by appropriate means. Choice of sample size (number of pairs of lines and number of assayed individuals per line) depends on the variability associated with the measurement across genomic locations and on the correlation in expression between transgenes located at the same genomic position (*see* **Note 6**). Our results for ADH specific activity suggest that, in general, this correlation is indeed very high (data not shown).

4. Notes

1. There is a maternal effect associated with the Cre-expressing transgenes, *P*[*w*$^+$, *cre*] and *P*[*y*$^+$, *cre*], in that *cre*$^-$ progeny of *cre*-heterozygous mothers exhibit recombination at *loxP*

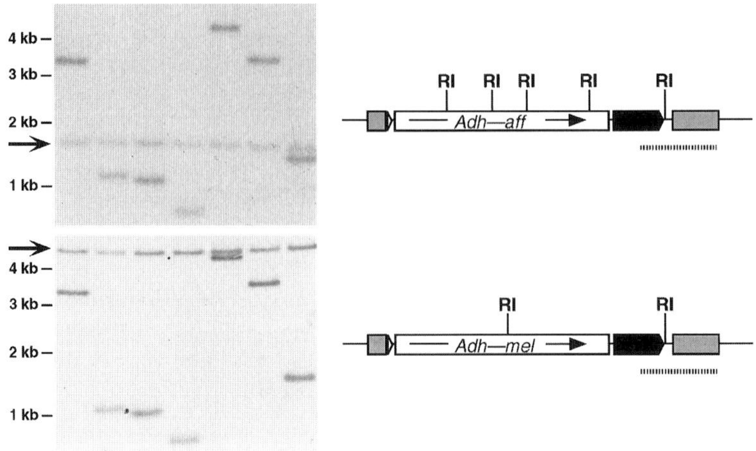

Fig. 3. Confirmation of site-specific recombination: The FLP-mediated excision derivative of *P*[^*a*>*w*+^*m*>], carrying the *Adh* gene of *D. affinidisjuncta*, is shown schematically on the right, above the corresponding Cre-mediated excision derivative, carrying the *D. melanogaster Adh* gene. Symbols are as in **Fig. 1**, and *Eco*RI sites (RI) are indicated. The extent of the probe sequence used for the Southern hybridizations shown at left is indicated by the striped bars. Genomic DNA was isolated from seven pairs of strains representing FLP or Cre derivatives of third-chromosome insertions of *P*[^*a*>*w*+^*m*>]. DNA was digested with *Eco*RI, separated on an agarose gel, blotted onto a nylon filter, and probed with the indicated sequence. Those strains expected to be left with the *D. affinidisjuncta Adh* are shown in the top autoradiograph, and those expected to be left with the *D. melanogaster Adh* are shown in the bottom autoradiograph; strains in both blots are ordered the same, according to the original *P*[^*a*>*w*+^*m*>] insertion strain from which they derive. The Southern blots confirm the identity of each *Adh* gene (constant bands across all lanes, noted by arrows), as well as the presence of the same 3'-flanking genomic sequences for each pair of strains (corresponding variable bands for each lane in top and bottom autoradiographs).

targets ***(18)***. The maternal effect is much stronger for *P*[*w*+, *cre*] than for *P*[*y*+, *cre*]. While this difference should not affect choice of Cre source when the *cre* transgene is present in the same genotype as the *loxP* target, there may be certain crossing schemes in which the maternal effect may be useful; in such cases, *P*[*w*+, *cre*] should be used.

2. As can be seen in Table 1, the apparent degree of heat-shock inducibility of the Cre sources is dependent on the amount of DNA intervening between *loxP* sites in the target. For small targets (e.g., *P*[^*w*+^]), the excision reactions of *P*[*w*+, *cre*] and *P*[*y*+, *cre*] are practically 100% efficient without heat shock. In fact, for unknown reasons, heat shock is lethal when *P*[^*w*+^] is present in the same genotype as *P*[*w*+, *cre*]. This may suggest that the Drosophila genome contains cryptic *loxP* sites, which may participate in illegitimate recombination events when Cre concentration is extremely high. Excision efficiency without heat shock drops off for larger targets (e.g., *P*[^*a*>*w*+^*m*>]), but heat shock restores efficiency to high levels, generally without a significant effect on viability. The *P*[*y*+, *cre**] strain has lower baseline Cre activity in the germline (34.7% excision of *P*[^*w*+^]), but high activity upon heat shock (88.7% excision of *P*[^*w*+^]). This difference is even more pronounced in the soma; heat-shocked flies carrying both *P*[*y*+, *cre**] and *P*[^*w*+^] have predominantly white eyes with small patches of color, whereas in the absence of heat shock such flies

have predominantly white$^+$ eyes with small patches lacking color (not shown). Therefore, this mutant recombinase may be particularly useful for application of Cre/*loxP* to mosaic analysis.

3. Depending on the size of the genes to be coplaced, the resulting plasmid construct may be rather large, and care should be taken both in its construction and in large-scale isolation of pure DNA for embryo injections. In our hands, the construction of large plasmids is often facilitated by using a three-piece ligation strategy ***(28)***, which obviates the need to handle large restriction fragments and also yields a low background of undesired clones. Additionally, some *P*-element-containing plasmids are prone to rearrangements when grown in large bacterial cultures. In our hands, pooling small cultures (2 mL each) after overnight growth is a means of avoiding this problem. Despite these caveats, cloning and transformation of large constructs is routine in most laboratories. For our 20-kb *Adh*-containing *P* element (23-kb plasmid), 133 fertile G_0 adults yielded 17 (12.8%) independently transformed G_1 progeny.
4. The structure of *FRT*- and *loxP*-containing plasmids can be tested by recombination in bacteria, before one proceeds with Drosophila embryo injection ***(29,30)***.
5. The arrangement of *loxP* and *FRT* sites in *P*[*wFl*] and *P*[*SFl*] suggests other potentially useful configurations. For example, one may reduce the size of a transgene-coplacement construct if upstream sequences are shared between the two genes to be compared, as when two cDNA sequences are driven by the same promoter. In such a case, the promoter may be placed upstream of the first *loxP* site and the two cDNAs placed in cloning sites 1 and 2. Because the *loxP* site is only 34-bp in length, it is minimally disruptive. Plasmids are now available ***(31)*** containing *loxP* and *FRT* sites in various combinations, thus facilitating the cloning of custom-made coplacement constructs.
6. Quantitative analysis of expression differences between coplaced transgenes can make use of the paired-comparisons *t* test (or the equivalent two-way analysis of variance), which confers greater power than the standard *t* test (or one-way ANOVA). The power of the statistical test increases as the coefficient of variation of measurements across genomic positions decreases, and increases as the correlation in expression between transgenes at the same position in the genome increases.
7. Qualitative analysis of transgene expression should precede quantitative analysis, because some genomic insertion sites may cause misexpression of the transgenes, and should therefore be excluded from the quantitative analysis.
8. The ">" notation for *FRT* sites was introduced by Golic and Lindquist ***(3)***, so that the orientation of repeated *FRT* sites can readily be designated; ">X>" means that X is flanked by directly-repeated *FRT* sites, whereas ">X<" means that X is flanked by inverted *FRT* sites. The "^" notation for *loxP* sites is modelled after this notation, and allows the analogous designations "^X^" and "^X$^{\vee}$."

Acknowledgments

The authors thank the members of the Hartl lab for helpful discussions and technical advice. M. L. S. is a Howard Hughes Medical Institute Predoctoral Fellow. This work was supported by National Institutes of Health grants GM-33741 and HG-01250 to D. L. H.

References

1. Kilby, N. J., Snaith, M. R., and Murray, J. A. H. (1993) Site-specific recombinases: tools for genome engineering. *Trends Genet.* **9,** 413–421.
2. Simpson, P. (1993) Flipping fruit-flies: a powerful new technique for generating *Drosophila* mosaics. *Trends Genet.* **9,** 227,228.

3. Golic, K. G. and Lindquist, S. (1989) The FLP recombinase of yeast catalyzes site-specific recombination in the Drosophila genome. *Cell* **59,** 499–509.
4. Golic, K. G. (1991) Site-specific recombination between homologous chromosomes in *Drosophila. Science* **252,** 958–961.
5. Xu, T. and Rubin, G. M. (1993) Analysis of genetic mosaics in developing adult *Drosophila* tissues. *Development* **117,** 1223–1237.
6. Golic, K. G. (1994) Local transposition of *P* elements in *Drosophila melanogaster* and recombination between duplicated elements using a site-specific recombinase. *Genetics* **137,** 551–563.
7. Golic, K. G. and Golic, M. M. (1996) Engineering the Drosophila genome: chromosome rearrangements by design. *Genetics* **144,** 1693–1711.
8. Sauer, B. and Henderson, N. (1990) Targeted insertion of exogenous DNA into the eukaryotic genome by the Cre recombinase. *New Biol.* **2,** 441–449.
9. Golic, M. M., Rong, Y. S., Peterson, R. B., Lindquist, S. L., and Golic, K. G. (1997) FLP-mediated DNA mobilization to specific target sites in *Drosophila* chromosomes. *Nucleic Acids Res.* **25,** 3665–3671.
10. Spradling, A. C. and G. M. Rubin. (1983) The effect of chromosomal position on the expression of the Drosophila xanthine dehydrogenase gene. *Cell* **34,** 47–57.
11. Wilson, C., Bellen, H. J., and Gehring, W. J. (1990) Position effects on eukaryotic gene expression. *Annu. Rev. Cell Biol.* **6,** 679–714.
12. Laurie-Ahlberg, C. C. and Stam, L. F. (1987) Use of *P*-element-mediated transformation to identify the molecular basis of naturally occurring variants affecting *Adh* expression in *Drosophila melanogaster. Genetics* **115,** 129–140.
13. Brennan, M. D. and Dickinson, W. J. (1988) Complex developmental regulation of the *Drosophila affinidisjuncta* alcohol dehydrogenase gene in *Drosophila melanogaster. Dev. Biol.* **125,** 64–74.
14. Wu, C.-T. (1993) Transvection, nuclear structure, and chromatin proteins. *J. Cell Biol.* **120,** 587–590.
15. McKim, K. S. and Hawley, R. S. (1995) Chromosomal control of meiotic cell division. *Science* **270,** 1595–1601.
16. Henikoff, S. (1994) A reconsideration of the mechanism of position effect. *Genetics* **138,** 1–5.
17. Árnason, E. (1991) Perturbation-reperturbation test of selection *vs.* hitchhiking of the two major alleles of *Esterase-5* in *Drosophila pseudoobscura. Genetics* **129,** 145–168.
18. Siegal, M. L. and Hartl, D. L. (1996) Transgene coplacement and high efficiency site-specific recombination with the Cre/*loxP* system in Drosophila. *Genetics* **144,** 715–726.
19. Pirrotta, V. (1988) Vectors for P-mediated transformation of Drosophila in "Vectors: A Survey of Molecular Cloning Vectors and Their Uses" (Ed. R. L. Rodriguez and D. T. Renhardt) Pp. 437–456. Butterworths, Boston.
20. Lohe, A. R., Lidholm, D.-A., and Hartl, D. L. (1995) Genotypic effects, maternal effects and grand-maternal effects of immobilized derivatives of the transposable element *mariner. Genetics* **140,** 183–192.
21. Cavener, D. R. (1987) Comparison of the consensus sequence flanking translational start sites in *Drosophila* and vertebrates. *Nucleic Acids Res.* **15,** 1353–1361.
22. Patton, J. S., Gomes, X. V., and Geyer, P. K. (1992) Position-independent germline transformation in *Drosophila* using a cuticle pigmentation gene as a selectable marker. *Nucleic Acids Res.* **20,** 5859–5860.
23. Lindsley, D. L. and G. G. Zimm. (1992) *The Genome of Drosophila Melanogaster.* Academic, San Diego.
24. Ausubel, F. M., Brent, R., Kingston, R. E., Moore, D. D., Smith, J. A., Seidman, J. G., and Struhl, K. (Eds.) (1993) *Current Protocols in Molecular Biology, Vol. 1.* Greene Publishing Associates and Wiley-Interscience, New York.

25. Vieira, J., Vieira, C. P., Hartl, D. L., and Lozovskaya, E. R. (1997) A framework physical map of *Drosophila virilis* based on P1 clones: applications in genome evolution. *Chromosoma* **106,** 99–107.
26. Spradling, A. C. and Rubin, G. M. (1982) Transposition of cloned P elements into *Drosophila* germ line chromosomes. *Science* **218,** 341–347.
27. Rubin, G. M. and Spradling, A. C. (1982) Genetic transformation of *Drosophila* with transposable element vectors. *Science* **218,** 348–353.
28. Siegal, M. L., Petrov, D. A., and De Aguiar, D. (1996) Triple-ligation strategy with advantages over directional cloning. *BioTechniques* **21,** 614–619.
29. Snaith, M. R., Kilby, N. J., and Murray, J. A. H. (1996) An *Escherichia coli* system for assay of Flp site-specific recombination on substrate plasmids. *Gene* **180,** 225–227.
30. Buchholz, F., Angrand, P.-O., and Stewart, A. F. (1996) A simple assay to determine the functionality of Cre or FLP recombination targets in genomic manipulation constructs. *Nucleic Acids Res.* **24,** 3118,3119.
31. Snaith, M. R., Murray, J. A. H., and Boulter, C. A. (1995) Multiple cloning sites carrying *loxP* and *FRT* recognition sites for the Cre and Flp site-specific recombinases. *Gene* **166,** 173,174.

Index

B

C